The Supercontinuum Laser Source

Robert R. Alfano
Editor

The Supercontinuum Laser Source

With 250 Illustrations

Springer-Verlag
New York Berlin Heidelberg
London Paris Tokyo

Robert R. Alfano
Distinguished Professor of Science and Engineering
Institute for Ultrafast Spectroscopy and Lasers
Physics and Electrical Engineering Departments
The City College of New York
New York, NY 10031, USA

Library of Congress Cataloging-in-Publication Data
The Supercontinuum laser source / Robert R. Alfano, editor.
p. cm.
Includes bibliographies and index.
Contents: Theory of self-phase modulation and spectral broadening / Y.R. Shen, Guo-Zhen Yang—Supercontinuum generation in condensed matter / Q.Z. Wang, P.P. Ho, R.R. Alfano—Ultrashort pulse propagation in nonlinear dispersive fibers / Govind P. Agrawal—Cross-phase modulation: a new technique to control the spectral, temporal, and spatial properties of ultrashort pulses / P.L. Baldeck, P.P. Ho, R.R. Alfano—Simple models of self-phase and induced-phase modulation / Jamal T. Manassah—Self steepening of optical pulses / B.R. Suydam—Self-focusing and continuum generation in gases / Paul B. Corkum and Claude Rolland—The utilization of UV and IR supercontinua in gas-phase subpicosecond kinetic spectroscopy / J.H. Glownia, J. Misewich, P.P. Sorokin—Applications of supercontinuum / R. Dorsinville, P.P. Ho, J. Manassah, R.R. Alfano—Pulse compression and femtosecond continuum / C. Shank and A. Johnson.
1. Laser pulses, Ultrashort. 2. Nonlinear optics. I. Alfano. R.R.
QC689.5.L37S87 1989
621.36′6—dc19 88-8481

Printed on acid-free paper

Cover shows supercontinuum generation of intensity vs. wavelength for 1 mm of carbon tetrachloride liquid excited by a 120-fs, 625-nm laser pulse. Photo by R. Alfano, A. Katz, and P. Ho.

Typeset by Asco Trade Typesetting Ltd., Hong Kong.
Printed and bound by Edwards Brothers, Inc., Ann Arbor, Michigan.
Printed in the United States of America.

9 8 7 6 5 4 3 2 1

ISBN 0-387-96946-2 Springer-Verlag New York Berlin Heidelberg
ISBN 3-540-96946-2 Springer-Verlag Berlin Heidelberg New York

To my father, Alfonso L. Alfano
and my father-in-law, Samuel J. Resnick
whose advice I deeply miss.

Preface

This book deals with both ultrafast laser and nonlinear optics technologies. Over the past two decades, we have seen dramatic advances in the generation of ultrafast laser pulses and their applications to the study of phenomena in a variety of fields. It is now commonplace to produce picosecond (10^{-12} s) pulses. New developments have extended this technology into the femtosecond (10^{-15} s) time region. Soon pulses consisting of just a single cycle will be produced (i.e., 2 fs at 600 nm). These ultrafast pulses permit novel investigations to study phenomena in many disciplines. Sophisticated techniques based on these laser pulses have given rise to instruments with extremely high temporal resolution. Ultrafast laser technology offers the possibility of studying and discovering key processes unresolved in the past. A new era of time-resolved spectroscopy has emerged, with pulses so fast that one can now study the nonequilibrium states of matter, test quantum and light models, and explore new frontiers in science and technology. Ultrashort light pulses are a potential signal source in future high-bit-rate optical fiber communication systems. The shorter the pulses, the more can be packed into a given time interval and the higher is the data transmission rate for the tremendous bandwidth capacity of optical fiber transmission.

Nonlinear optics is an important field of science and engineering because it can generate, transmit, and control the spectrum of laser pulses in solids, liquids, gases, and fibers. One of the most important ultrafast nonlinear optical processes is the supercontinuum generation—the production of intense ultrafast broadband "white light" pulses—that is the subject of this book.

The first study on the mechanism and generation of ultrafast supercontinuum dates back over 19 years to 1969, when Alfano and Shapiro observed the first "white" picosecond pulse continuum in liquids and solids. Spectra extended over $\sim 6000 \text{ cm}^{-1}$ in the visible and infrared wavelength region. They attributed the large spectral broadening of ultrafast pulses to self-phase modulation (SPM) arising from an electronic mechanism and laid down the formulation of the supercontinuum generation model. Over the years, the improvement of mode-locked lasers led to the production of wider supercontinua in the visible, ultraviolet, and infrared wavelength regions using various materials.

The supercontinuum arises from the propagation of intense picosecond or shorter laser pulses through condensed or gaseous media. Various processes are responsible for continuum generation. These are called self-, induced-, and cross-phase modulations and four-photon parametric generation. Whenever an intense laser pulse propagates through a medium, it changes the refractive index, which in turn changes the phase, amplitude, and frequency of the pulse. However, when two laser pulses of different wavelengths propagate simultaneously in a condensed medium, coupled interactions (cross-phase modulation and gain) occur through the nonlinear susceptibility coefficients. These coupled interactions of two different wavelengths can introduce phase modulation, amplitude modulation, and spectral broadening in each pulse due to the other pulse using *cross-effects*.

An alternative coherent light source to the free electron laser, the supercontinuum laser source, can be wavelength selected and coded simultaneously over wide spectral ranges (up to 10,000 cm^{-1}) in the ultraviolet, visible, and infrared regions at high repetition rates, gigawatt output peak powers, and femtosecond pulse durations.

Ultrafast supercontinuum pulses have been used for time-resolved absorption spectroscopy and material characterization. Supercontinuum generation is a key step for the pulse compression technique, which is used to produce the shortest optical pulses. Future applications include signal processing, three-dimensional imaging, ranging, atmospheric remote sensing, and medical diagnosis.

Thus far, a great deal of information on supercontinuum technology has been obtained and has enhanced our understanding of how intense optical pulses propagate in materials. These developments are most often found in original research contributions and in review articles scattered in journals. Textbooks do not cover these subjects in great detail. There is a need for a book that covers the various aspects of ultrafast supercontinuum phenomena and technology.

This book reviews present and past progress on the experimental and theoretical understanding of ultrafast nonlinear processes responsible for supercontinuum generation and related effects such as pulse compression and ultrashort pulse generation on a picosecond and femtosecond time scale. The content of the chapters in the book is a mixture of both theoretical and experimental material. Overviews of the important breakthroughs and developments in the understanding of supercontinuum during the past 20 years are presented. The book is organized into 10 chapters.

Summarizing the highlights of the 10 chapters of the book:

In Chapter 1, Shen and Yang focus on the theoretical models and mechanisms behind supercontinuum generation arising mainly from self-phase modulation.

In Chapter 2, Wang, Ho, and Alfano review the experiments leading to the supercontinuum generation in condensed matter over the past 20 years.

In Chapter 3, Agrawal discusses the effects of dispersion on ultrafast light pulse propagation and supercontinuum generation in fibers.

In Chapter 4, Baldeck, Ho, and Alfano cover the latest experimental observations and applications of the cross-interactions in the frequency, time, and space domains of strong pulses on weak pulses.

In Chapter 5, Manassah reviews the theoretical models giving rise to many phenomena from self-phase and induced modulations.

In Chapter 6, Suydam highlights the effect of self-steepening of pulse profile on continuum generation.

In Chapter 7, Corkum and Rolland review the work on supercontinuum and self-focusing in gaseous media.

In Chapter 8, Glownia, Misewich, and Sorokin utilize the supercontinuum produced in gases for ultrafast spectroscopy in chemistry.

In Chapter 9, Dorsinville, Ho, Manassah, and Alfano cover the present and speculate on the possible future applications of the supercontinuum in various fields.

In Chapter 10, Johnson and Shank discuss pulse compression from the picosecond to femtosecond time domain using the continuum and optical dispersive effects of gratings, prisms, and materials.

The reader will find that these chapters review the basic principles, contain surveys of research results, and present the current thinking of experts in the supercontinuum field. The volume should be a useful source book and give young and seasoned scientists, engineers, and graduate students an opportunity to find the most necessary and relevant material on supercontinuum technology in one location.

I hope these efforts will stimulate future research on understanding the physics behind supercontinuum technology and exploring new applications.

I wish to thank all the expert contributors for their cooperation in this endeavor. Most thought it would not be completed. Special thanks goes to Mrs. Megan Gibbs for her administrative and secretarial assistance. I gratefully acknowledge T. Hiruma for his continued support. I pay particular tribute to my friend Stan Shapiro, who missed seeing the outgrowth of our first work in this field 20 years ago.

R.R. ALFANO

Contents

Contributors

GOVIND P. AGRAWAL, Institute of Optics, University of Rochester, Rochester, New York 14627, USA

R.R. ALFANO, Departments of Physics and Electrical Engineering, City College of the City University of New York, New York 10031, USA

P.L. BALDECK, Department of Electrical Engineering, City College of the City University of New York, New York 10031, USA

PAUL B. CORKUM, Division of Physics, National Research Council of Canada, Ottawa, Ontario, K1A OR6 Canada

R. DORSINVILLE, Department of Electrical Engineering, City College of the City University of New York, New York 10031, USA

J.H. GLOWNIA, IBM Research Division, Thomas J. Watson Research Center, Yorktown Heights, New York 10598, USA

P.P. HO, Department of Electrical Engineering, City College of the City University of New York, New York 10031, USA

A.M. JOHNSON, AT&T Bell Laboratories, Holmdel, New Jersey 07733, USA

JAMAL T. MANASSAH, Department of Electrical Engineering, City College of the City University of New York, New York 10031, USA

J. MISEWICH, IBM Research Division, Thomas J. Watson Research Center, Yorktown Heights, New York 10598, USA

CLAUDE ROLLAND, Division of Physics, National Research Council of Canada, Ottawa, Ontario, K1A OR6 Canada

C.V. Shank, AT&T Bell Laboratories, Holmdel, New Jersey 07733, USA

Y.R. Shen, Department of Physics, University of California, Berkeley, California 94720, USA

P.P. Sorokin, IBM Research Division, Thomas J. Watson Research Center, Yorktown Heights, New York 10598, USA

B.R. Suydam, Los Alamos National Laboratory, Los Alamos, New Mexico 87545, USA

Q.Z. Wang, Department of Physics, City College of the City University of New York, New York 10031, USA

Guo-Zhen Yang, Institute of Physics, Academy of Sciences, Beijing, China

1
Theory of Self-Phase Modulation and Spectral Broadening

Y.R. SHEN and GUO-ZHEN YANG

1. Introduction

Self-phase modulation refers to the phenomenon in which a laser beam propagating in a medium interacts with the medium and imposes a phase modulation on itself. It is one of those very fascinating effects discovered in the early days of nonlinear optics (Bloembergen and Lallemand, 1966; Brewer, 1967; Cheung et al., 1968; Lallemand, 1966; Jones and Stoicheff, 1964; Shimizu, 1967; Stoicheff, 1963). The physical origin of the phenomenon lies in the fact that the strong field of a laser beam is capable of inducing an appreciable intensity-dependent refractive index change in the medium. The medium then reacts back and inflicts a phase change on the incoming wave, resulting in self-phase modulation (SPM). Since a laser beam has a finite cross section, and hence a transverse intensity profile, SPM on the beam should have a transverse spatial dependence, equivalent to a distortion of the wave front. Consequently, the beam will appear to have self-diffracted. Such a self-diffraction action, resulting from SPM in space, is responsible for the well-known nonlinear optical phenomena of self-focusing and self-defocusing (Marburger, 1975; Shen, 1975). It can give rise to a multiple ring structure in the diffracted beam if the SPM is sufficiently strong (Durbin et al., 1981; Santamato and Shen, 1984). In the case of a pulsed laser input, the temporal variation of the laser intensity leads to an SPM in time. Since the time derivative of the phase of a wave is simply the angular frequency of the wave, SPM also appears as a frequency modulation. Thus, the output beam appears with a self-induced spectral broadening (Cheung et al., 1968; Gustafson et al., 1969; Shimizu, 1967).

In this chapter we are concerned mainly with SPM that leads to spectral broadening (Bloembergen and Lallemand, 1966; Brewer, 1967; Cheung et al., 1968; Lallemand, 1966; Jones and Stoicheff, 1964; Shimizu, 1967; Stoicheff, 1963). For large spectral broadening, we need a strong SPM in time (i.e., a large time derivative in the phase change). This obviously favors the use of short laser pulses. Consider, for example, a phase change of 6π occurring in 10^{-12} s. Such a phase modulation would yield a spectral broadening of

$\sim 100\ \mathrm{cm}^{-1}$. In practice, with sufficiently intense femtosecond laser pulses, a spectral broadening of $20{,}000\ \mathrm{cm}^{-1}$ is readily achievable by SPM in a condensed medium, which is essentially a white continuum (Alfano and Shapiro, 1970). The pulse duration of any frequency component (uncertainty limited) in the continuum is not very different from that of the input pulse (Topp and Rentzepis, 1971). This spectrally superbroadened output from SPM therefore provides a much needed light source in ultrafast spectroscopic studies—tunable femtosecond light pulses (Busch et al., 1973; Alfano and Shapiro, 1971). If the SPM and hence the frequency sweep in time on a laser pulse are known, then it is possible to send the pulse through a properly designed dispersive delay system to compensate the phase modulation and generate a compressed pulse with little phase modulation (Treacy, 1968, 1969). Such a scheme has been employed to produce the shortest light pulses ever known (Fork et al., 1987; Ippen and Shank, 1975; Nakatsuka and Grischkowsky, 1981; Nakatsuka et al., 1981; Nikolaus and Grischkowsky, 1983a, 1983b).

Self-phase modulation was first proposed by Shimizu (1967) to explain the observed spectrally broadened output from self-focusing of a Q-switched laser pulse in liquids with large optical Kerr constants (Bloembergen and Lallemand, 1966; Brewer, 1967; Cheung et al., 1968; Jones and Stoicheff, 1964; Lallemand, 1966; Shimizu, 1967; Stoicheff, 1963). In this case, the spectral broadening is generally of the order of a hundred reciprocal centimeters. Alfano and Shapiro (1970) showed that with picosecond laser pulses, it is possible to generate by SPM a spectrally broadened output extending over $10{,}000\ \mathrm{cm}^{-1}$ in almost any transparent condensed medium. Self-focusing is believed to have played an important role in the SPM process in the latter case. In order to study the pure SPM process, one would like to keep the beam cross section constant over the entire propagation distance in the medium. This can be achieved in an optical fiber since the beam cross section of a guided wave should be constant and the self-focusing effect is often negligible. Stolin and Lin (1978) found that indeed the observed spectral broadening of a laser pulse propagating through a long fiber can be well explained by the simple SPM theory. Utilizing a well-defined SPM from an optical fiber, Grischkowsky and co-workers were then able to design a pulse compression system that could compress a laser pulse to a few hundredths of its original width (Nakatsuka and Grischkowsky, 1981; Nakatsuka et al., 1981; Nikolaus and Grischkowsky, 1983a, 1983b). With femtosecond laser pulses, a strong SPM on the pulses could be generated by simply passing the pulses through a thin film. In this case, the beam cross section is practically unchanged throughout the film, and one could again expect a pure SPM process. Fork et al. (1983) observed the generation of a white continuum by focusing an 80-fs pulse to an intensity of $\sim 10^{14}\ \mathrm{W/cm^2}$ on a 500-μm ethylene glycol film. Their results can be understood by SPM along with the self-steepening effect (Manassah et al., 1985, 1986; Yang and Shen, 1984).

Among other experiments, Corkum et al. (1985) demonstrated that SPM and spectral broadening can also occur in a medium with infrared laser pulses. More recently, Corkum et al. (1986) and Glownia et al. (1986) have independently shown that with femtosecond pulses it is even possible to generate a white continuum in gas media.

The phase modulation induced by one laser pulse can also be transferred to another pulse at a different wavelength via the induced refractive index change in a medium. A number of such experiments have been carried out by Alfano and co-workers (1986, 1987). Quantitative experiments on spectral superbroadening are generally difficult. Self-focusing often complicates the observation. Even without self-focusing, quantitative measurements of a spectrum that is generated via a nonlinear effect by a high-power laser pulse and extends from infrared to ultraviolet are not easy. Laser fluctuations could lead to large variations in the output.

The simple theory of SPM considering only the lower-order effect is quite straightforward (Gustafson et al., 1969; Shimizu, 1967). Even the more rigorous theory including the higher-order contribution is not difficult to grasp as long as the dispersive effect can be neglected (Manassah et al., 1985, 1986; Yang and Shen, 1984). Dispersion in the material response, however, could be important in SPM, and resonances in the medium would introduce pronounced resonant structure in the broadened spectrum. The SPM theory with dispersion is generally very complex; one often needs to resort to a numerical solution (Fischen and Bischel, 1975; Fisher et al., 1983). It is possible to describe the spectral broadening phenomenon as resulting from a parametric wave mixing process (in the pump depletion limit) (Bloembergen and Lallemand, 1966; Lallemand, 1966; Penzkofer, 1974; Penzkofer et al., 1973, 1975). In fact, in the studies of spectral broadening with femtosecond pulses, four-wave parametric generation of new frequency components in the phase-matched directions away from the main beam can be observed together with the spectrally broadened main beam. Unfortunately, a quantitative estimate of spectral broadening due to the parametric process is not easy. In the presence of self-focusing, more complication arises. Intermixing of SPM in space and SPM in time makes even numerical solution very difficult to manage, especially since a complete quantitative description of self-focusing is not yet available. No such attempt has ever been reported. Therefore, at present, we can only be satisfied with a qualitative, or at most a semi-quantitative, description of the phenomenon (Marburger, 1975; Shen, 1975).

This chapter reviews the theory of SPM and associated spectral broadening. In the following section, we first discuss briefly the various physical mechanisms that can give rise to laser-induced refractive index changes responsible for SPM. Then in Section 3 we present the simple physical picture and theory of SPM and the associated spectral broadening. SPM in space is considered only briefly. Section 4 deals with a more rigorous theory of SPM that takes into account the higher-order effects of the induced refractive index change. Finally, in Section 5, a qualitative picture of how self-focusing can influence and enhance SPM and spectral broadening is presented. Some semiquantitative estimates of the spectral broadening are given and compared with experiments, including the recent observations of supercontinuum generation in gases.

2. Optical-Field-Induced Refractive Indices

The material response to an applied laser field is often nonlinear. An explicit expression for the response is not readily available in general. Unless specified otherwise, we consider here only the case where the perturbative expansion in terms of the applied field is valid and the nonlocal response can be neglected. We can then express the induced polarization in a medium as (Shen, 1984)

$$\begin{aligned}
\mathbf{P}(t) &= \mathbf{P}^{(1)}(t) + \mathbf{P}^{(2)}(t) + \mathbf{P}^{(3)}(t) + \cdots, \\
\mathbf{P}^{(1)}(t) &= \int \chi^{(1)}(t - t')\cdot \mathbf{E}(t')\,dt' \\
&= \int \chi^{(1)}(\omega)\cdot \mathbf{E}(\omega)\,d\omega, \\
\mathbf{P}^{(n)}(t) &= \int \chi^{(n)}(t - t_1, \ldots, t - t_n) : \mathbf{E}(t_1)\ldots\mathbf{E}(t_n)\,dt_1\ldots dt_n \\
&= \int \chi^{(n)}(\omega = \omega_1 + \omega_2 + \cdots + \omega_n) : \mathbf{E}(\omega_1)\ldots\mathbf{E}(\omega_n)\,d\omega_1\ldots d\omega_n,
\end{aligned} \tag{1}$$

where the applied field is

$$\mathbf{E}(t) = \int \mathbf{E}(\omega)\,d\omega \quad \text{with } \mathbf{E}(\omega) \propto \exp(-i\omega t) \tag{2}$$

and the nth-order susceptibility is

$$\begin{aligned}
\chi^{(n)}(t - t_1, \ldots, t - t_n) = \int \chi^{(n)}(\omega = \omega_1 + \cdots + \omega_n)\exp[i\omega_1(t - t_1) \\
+ \cdots + i\omega_n(t - t_n)]\,d\omega_1\ldots d\omega_n.
\end{aligned} \tag{3}$$

We note that, strictly speaking, only for a set of monochromatic applied fields can we write

$$\mathbf{P}^{(n)} = \chi^{(n)}(\omega = \omega_1 + \cdots + \omega_n) : \mathbf{E}(\omega_1)\ldots E(\omega_n). \tag{4}$$

In the case of instantaneous response (corresponding to a dispersionless medium), we have

$$\mathbf{P}^{(n)} = \chi^{(n)} : [E(t)]^n. \tag{5}$$

Here, we are interested in the third-order nonlinearity that gives rise to the induced refractive index change. We consider only the self-induced refractive index change; extension to the cross-field-induced change should be straightforward. Thus we assume a pulsed quasi-monochromatic field $\mathbf{E}_\omega(t) = \mathscr{E}(t)\exp(-i\omega t)$. The third-order nonlinear polarization in a medium, in general, takes the form

$$\mathbf{P}_\omega^{(3)}(t) = \int \Delta\boldsymbol{\chi}(t - t') \cdot \mathbf{E}_\omega(t')\, dt' \tag{6}$$

with $\Delta\boldsymbol{\chi}(t - t') = \int \boldsymbol{\chi}^{(3)}(t - t', t - t'', t - t''') : \mathbf{E}_\omega(t'')\mathbf{E}_\omega^*(t''')\, dt''\, dt'''$. If the optical field is sufficiently far from resonances that the transverse excitations are all virtual and can be considered as instantaneous, we can write

$$\mathbf{P}_\omega^{(3)}(t) = \Delta\boldsymbol{\chi}(t) \cdot \mathbf{E}_\omega(t),$$

$$\Delta\boldsymbol{\chi}(t) = \int \boldsymbol{\chi}^{(3)}(t - t') : |\mathbf{E}_\omega(t')|^2\, dt'. \tag{7}$$

In the dispersionless limit, the latter becomes

$$\Delta\boldsymbol{\chi}(t) = \boldsymbol{\chi}^{(3)} : |E_\omega(t)|^2. \tag{8}$$

Equation (8) is a good approximation when the dispersion of $\Delta\boldsymbol{\chi}$ is negligible within the bandwidth of the field. The optical-field-induced refractive index can be defined as

$$\Delta\mathbf{n} = (2\pi/n_0)\Delta\boldsymbol{\chi} \tag{9}$$

where n_0 is the average linear refractive index of the medium. With $\Delta n \equiv n_2|E_\omega|^2$, we have $n_2 = (2\pi/n_0)\chi^{(3)}$.

A number of physical mechanisms can give rise to $\Delta\boldsymbol{\chi}$ or $\Delta\mathbf{n}$ (Shen, 1966). They have very different response times and different degrees of importance in different media. We discuss them separately in the following.

2.1 Electronic Mechanism

Classically, one can imagine that an applied optical field can distort the electronic distribution in a medium and hence induce a refractive index change. Quantum mechanically, the field can mix the electronic wave functions, shift the energy levels, and redistribute the population; all of these can contribute to the induced refractive index change. For a typical transparent liquid or solid, n_2 falls in the range between 10^{-13} and 10^{-15} esu. For gases at 1 atm pressure, $n_2 \sim 10^{-16}$ to 10^{-18} esu far away from resonances. The response time is of the order of the inverse bandwidth of the major absorption band ($\sim 10^{-14}$ to 10^{-15} s in condensed media) except for the population redistribution part. As the optical frequency approaches an absorption band, n_2 is resonantly enhanced. In particular, when the population redistribution due to resonant excitation is significant, the enhancement of n_2 can be very large, but the time response will then be dominated by the relaxation of the population redistribution. In a strong laser field, saturation in population redistribution and multiphoton resonant excitations can become important. The perturbative expansion in Eq. (1) may then cease to be valid. For our discussion of SPM in this chapter, we shall assume that the laser beam is deep in the transparent region and therefore all these electronic resonance effects on the induced refractive index are negligible.

2.2 Vibrational Contribution

The optical field can also mix the vibrational wave functions, shift the vibrational levels, and redistribute the populations in the vibrational levels. The corresponding induced refractive index change Δn is, however, many orders (~ 5) of magnitude smaller than that from the electronic contribution because of the much weaker vibrational transitions. Therefore, the vibrational contribution to Δn is important only for infrared laser beams close to vibrational resonances. For our discussion of SPM, we shall not consider such cases.

If the laser pulse is very short (10 fs corresponding to a bandwidth of 500 cm^{-1}), the vibrational contribution to Δn can also come in via Raman excitations of modes in the few hundred cm^{-1} range. The Raman transitions are also much weaker than the two-photon electronic transitions, so their contributions to the self-induced Δn are usually not important for the discussion of SPM unless femtosecond pulses are used.

2.3 Rotation, Libration, and Reorientation of Molecules

Raman excitations of molecular rotations can, however, contribute effectively to Δn. This is because the rotational frequencies of molecules are usually in the few cm^{-1} region except for the smaller molecules. Thus, even with a monochromatic field, one can visualize a Raman process (in which absorption and emission are at the same frequency ω) that is nearly resonant. (The difference frequency of absorption and emission is zero, but it is only a few cm^{-1} away from the rotational frequencies.) In condensed media, the rotational motion of molecules is, however, strongly impeded by the presence of neighboring molecules. Instead of simple rotations, the molecules may now librate in a potential well set up by the neighboring molecules. The librational frequencies determined by the potential well are often in the range of a few tens of cm^{-1}. The modes are usually heavily damped. Like the rotational modes, they can also contribute effectively to Δn via the Raman process.

Molecules can also be reoriented by an optical field against rotational diffusion. This can be treated as an overdamped librational motion driven by the optical field. More explicitly, molecular reorientation arises because the field induces a dipole on each molecule and the molecules must then reorient themselves to minimize the energy of the system in the new environment.

All the above mechanisms involving rotations of molecules can contribute appreciably to Δn if the molecules are highly anisotropic. Typically, in liquids,

n_2 from such mechanisms falls in the range between 10^{-13} and 10^{-11} esu, with a response time around 10^{-11} s for molecular reorientation and $\sim 10^{-13}$ s for libration. In liquid crystals, because of the correlated molecular motion, n_2 can be much larger, approaching 0.1 to 1, but the response time is much longer, of the order of 1 s. The rotational motion is usually frozen in solids, and therefore its contribution to Δn in solids can be neglected.

2.4 Electrostriction, Molecular Redistribution, and Molecular Collisions

It is well known that the application of a dc or optical field to a local region in a medium will increase the density of the medium in that region. This is because the molecules in the medium must squeeze closer together to minimize the free energy of the system in the new environment. The effect is known as electrostriction. The induced density variation $\Delta\rho$ obeys the driven acoustic wave equation, and from $\Delta n = (\partial n/\partial\rho)\Delta\rho$ the induced refractive index change can be deduced. For liquids, we normally have $n_2 \sim 10^{-11}$ esu with a response time of the order of 100 ns across a transverse beam dimension of ~ 1 mm.

Molecules will also locally rearrange themselves in a field to minimize the energy of induced dipole–induced dipole interaction between molecules in the system. Whereas electrostriction yields an isotropic Δn, this molecular redistribution mechanism will lead to an anisotropic Δn. Molecular correlation and collisions could also affect molecular redistribution. A rigorous theory of molecular redistribution is therefore extremely difficult (Hellwarth, 1970). Experimentally, molecular redistribution is responsible for the anisotropic Δn observed in liquids composed of nearly spherical molecules or atoms in cases where the electronic, electrostrictive, and rotational contributions should all be negligible. It yields an n_2 of the order of 10^{-13} esu with a response time in the subpicosecond range. In solids, the molecular motion is more or less frozen, so the contribution of molecular redistribution to Δn is not significant.

2.5 Other Mechanisms

A number of other possible mechanisms can contribute to Δn. We have, for instance, laser heating, which increases the temperature of a medium and hence its refractive index; photorefraction, which comes from excitation and redistribution of charged carriers in a medium; and induced concentration variation in a mixture.

We conclude this section by noting that there is an intimate connection between third-order nonlinearities and light scattering (Hellwarth, 1977): each physical mechanism that contributes to Δn (except the electronic mechanism) is also responsible for a certain type of light scattering. The third-order susceptibility from a given mechanism is directly proportional to the scattering cross section related to the same mechanism, and the response time is

TABLE 1.1

Physical mechanism	Magnitude of third-order nonlinearity n_2 (esu)	Response time τ (s)
Electronic contribution	10^{-15}–10^{-13}	10^{-14}–10^{-15}
Molecular reorientation	10^{-13}–10^{-11}	$\sim 10^{-11}$
Molecular libration and redistribution	$\sim 10^{-13}$	$\sim 10^{-13}$
Electrostriction	$\sim 10^{-11}$	$\sim 10^{-6}$*

* For a beam radius of ~1 mm.

inversely proportional to the linewidth of the scattering mode. Thus from the low-frequency light scattering spectrum, one can predict the value of n_2 for the induced refractive index. For example, in most liquids, light scattering shows a Rayleigh wing spectrum with a broad background extending to a few tens of cm^{-1}. This broad background is believed to arise from molecular libration, redistribution, and collisions (Fabellinski, 1967), but the details have not yet been resolved. For our semiquantitative prediction of n_2 and the response time, however, we do not really need to know the details if the Rayleigh wing spectrum of the medium is available. A broad and strong Rayleigh wing spectrum is expected to yield a large n_2 with a fast response.

In Table 1.1 we summarize the results of our discussion of the various physical mechanisms contributing to Δn. It is seen that in nonabsorbing liquid, where all the mechanisms could operate, electrostriction and molecular reorientation may dominate if the laser pulses are longer than 100 ns; molecular reorientation, redistribution, and libration may dominate for pulses shorter than 100 ns and longer than 1 ps; molecular redistribution and libration and electronic contribution may dominate for femtosecond pulses. In transparent solids, usually only electrostriction and electronic contribution are important. Then for short pulses the latter is the only mechanism contributing to Δn.

3. Simple Theory of Self-Phase Modulation and Spectral Broadening

For our discussion of SPM of light, let us first consider the case where the propagation of a laser pulse in an isotropic medium can be described by the wave equation of a plane wave:

$$\left(\frac{\partial^2}{\partial z^2} - \frac{n_0^2}{c^2}\frac{\partial^2}{\partial t^2}\right)E = \frac{4\pi}{c^2}\frac{\partial^2}{\partial t^2}P^{(3)}, \tag{10}$$

where

$$E = \mathscr{E}(z,t)\exp(ik_0 z - i\omega_0 t),$$
$$P^{(3)} = \chi^{(3)}|E|^2 E,$$

and n_0 is the linear refractive index of the medium. In the simple theory of SPM (Cheung et al., 1968; Gustafson et al., 1969; Shimizu, 1967), we use the usual slowly varying amplitude approximation by neglecting the $\partial^2\mathscr{E}/\partial t^2$ term on the left and keeping only the $(4\pi/c^2)\chi^{(3)}|\mathscr{E}|^2\mathscr{E}$ term on the right of Eq. (10), which then becomes

$$\left(\frac{\partial}{\partial z} + \frac{n_0}{c}\frac{\partial}{\partial t}\right)\mathscr{E} = -\frac{4\pi\omega_0^2}{i2k_0c^2}\chi^{(3)}|\mathscr{E}|^2\mathscr{E}. \tag{11}$$

The approximation here also assumes an instantaneous response of $\chi^{(3)}$. Letting $z' \equiv z + ct/n_0$ and $\mathscr{E} = |\mathscr{E}|\exp(i\phi)$, we obtain from the above equation

$$\frac{\partial|\mathscr{E}|}{\partial z'} = 0,$$

$$\frac{\partial\phi}{\partial z'} = \frac{2\pi\omega_0^2}{c^2k_0}\chi^{(3)}|\mathscr{E}|^2. \tag{12}$$

They yield immediately the solution

$$|\mathscr{E}| = |\mathscr{E}(t)|, \tag{13a}$$

$$\phi(z,t) = \phi_0 + \frac{2\pi\omega_0^2}{c^2k_0}\chi^{(3)}|\mathscr{E}(t)|^2 z. \tag{13b}$$

Equation (13a) implies that the laser pulse propagates in the medium without any distortion of the pulse shape, while Eq. (13b) shows that the induced phase change $\Delta\phi(t) = \phi(z,t) - \phi_0$ is simply the additional phase shift experienced by the wave in its propagation from 0 to z due to the presence of the induced refractive index $\Delta n = (2\pi/n_0)\chi^{(3)}|\mathscr{E}|^2$, namely $\Delta\phi = (\omega/c)\int_0^z \Delta n\, dz$. Since the frequency of the wave is $\omega = \omega_0(\partial\Delta\phi/\partial t)$, the phase modulation $\Delta\phi(t)$ leads to a frequency modulation

$$\begin{aligned}\Delta\omega(t) &= -\partial(\Delta\phi)/\partial t \\ &= -\frac{2\pi\omega_0^2}{c^2k_0}\chi^{(3)}\frac{\partial|\mathscr{E}|^2}{\partial t}z.\end{aligned} \tag{14}$$

The spectrum of the self-phase-modulated field is, therefore, expected to be broadened. It can be calculated from the Fourier transformation

$$|E(\omega)|^2 = \left|\frac{1}{2\pi}\int_{-\infty}^{\infty}\mathscr{E}(t)e^{-i\omega_0 t}e^{i\omega t}\,dt\right|^2. \tag{15}$$

An example is shown in Figure 1.1. We assume here a 4.5-ps full width at half-maximum (FWHM) Gaussian laser pulse propagating in a nonlinear medium that yields an SPM output with a maximum phase modulation of

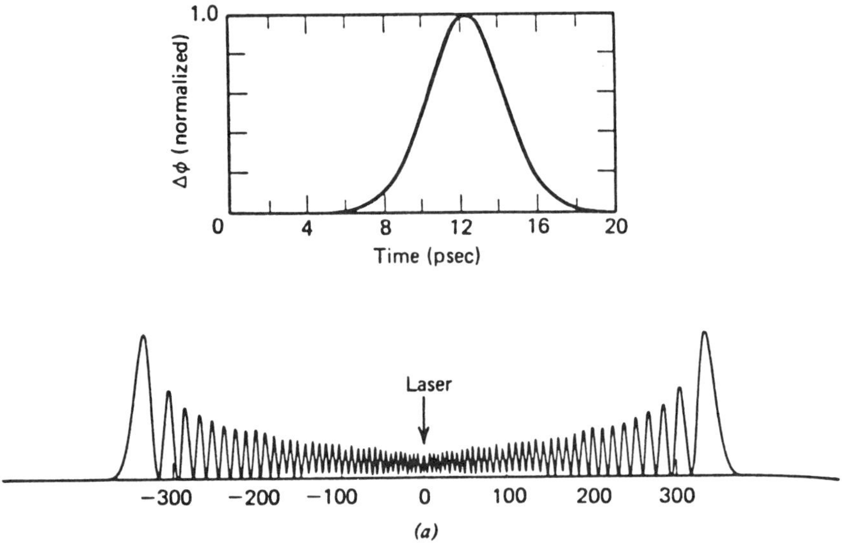

FIGURE 1.1. Theoretical power spectrum obtained by assuming an instantaneous response of Δn to the intensity variation $|E(t)|^2$, so that the phase modulation $\Delta\phi(t)$ is proportional to $|E(t)|^2$. (a) $\Delta\phi$ versus t and (b) power spectrum of the phase-modulated pulse.

$\Delta\phi_{\max} \simeq 72\pi$ rad. The spectrum of the output shows a broadening of several hundred cm^{-1} with a quasi-periodic oscillation. It is symmetric with respect to the incoming laser frequency because the SPM pulse is symmetric. The leading half of the $\Delta\phi$ pulse is responsible for the Stokes broadening and the lagging half for the anti-Stokes broadening. The structure of the spectrum can be understood roughly as follows. As shown in Figure 1.1, the $\Delta\phi$ curve following the laser pulse takes on a bell shape. For each point on such a curve, one can always find another point with the same slope, except, of course, the inflection points. Since $\partial\phi/\partial t = -\omega$, these two points describe radiated waves of the same frequency but different phases. These two waves will interfere with each other. They interfere constructively if the phase difference $\Delta\phi_{12}$ is an integer of 2π and destructively if $\Delta\phi_{12}$ is an odd integer of π. Such interference then gives rise to the peaks and valleys in the spectrum. The inflection points that have the largest slope on the curve naturally lead to the two outermost peaks with $|\omega_{\max}| \sim |\partial\phi/\partial t|_{\max}$. To find how many peaks we should expect in the spectrum, we need only to know $\phi_{\max}$, as the number of pairs of constructive and destructive interferences is simply $N \sim \phi_{\max}/2\pi$ on each side of the spectrum. The broadened spectrum has Stokes–anti-Stokes symmetry because $\Delta\phi(t)$ is directly proportional to $|E(t)|^2$ and is a symmetric pulse.

With the above qualitative picture in mind, we can now generalize our discussion of SPM somewhat. The response of the medium to the laser pulse

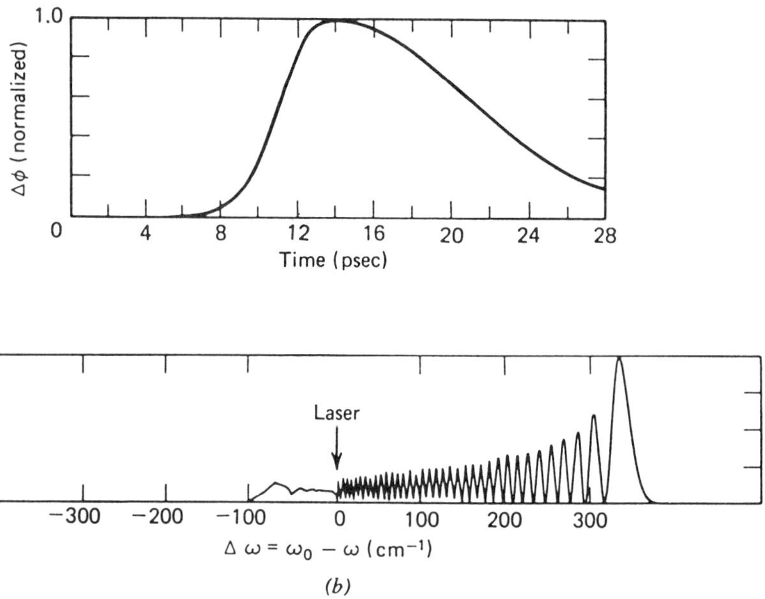

FIGURE 1.2. Theoretical power spectrum obtained by assuming a transient response of Δn to the intensity variation $|E(t)|^2$ so that $\Delta\phi(t)$ is no longer proportional to $|E(t)|^2$. (a) $\Delta\phi$ versus t and (b) power spectrum of the phase-modulated pulse.

is generally not instantaneous. One therefore expects

$$\Delta\phi(z,t) = (\omega/c)\int_0^z \Delta n(z,t)\,dz,$$

$$\Delta n(z,t) = \int_{-\infty}^{t} n_2(z, t-t')|E(z,t')|^2\,dt'. \tag{16}$$

Then, even if $|E(t)|^2$ is symmetric, $\Delta\phi(t)$ is asymmetric and is no longer proportional to $|E(t)|^2$. The consequence is a Stokes–anti-Stokes asymmetry. An example is given in Figure 1.2. Because of the finite response time of the medium, the leading part of the $\Delta\phi(t)$ curve always sees a larger portion of the intensity pulse $|E(t)|^2$, and therefore the Stokes side of the spectrum is always stronger. This Stokes–anti-Stokes asymmetry can be drastic if the response time becomes comparable to or smaller than the laser pulse width.

In the more rigorous theory, one should also expect a distortion of the pulse shape as the pulse propagates on in the nonlinear medium. Self-steepening of the pulse, for example, is possible and may also affect the spectral broadening (DeMartini et al., 1967; Gustafson et al., 1969; see Chapter 6). However, the above qualitative discussion still applies since the $\Delta\phi(t)$ curve should still take on an asymmetric bell shape in general.

The experimental situation is usually not as ideal as the simple theory describes. The laser beam has a finite cross section and will diffract. The transverse intensity variation also leads to a $\Delta n(\mathbf{r})$ that varies in the transverse directions. This causes self-focusing of the beam and complicates the simultaneously occurring SPM of the beam (Shen, 1975; Marburger, 1975). Moreover, stimulated light scattering could also occur simultaneously in the medium, in most cases initiated by self-focusing (Shen, 1975; Marburger, 1975). All these make the analysis of SPM extremely difficult.

One experimental case is, however, close to ideal, namely SPM of a laser pulse in an optical fiber. The transverse beam profile of a guided wave remains unchanged along the fiber. As long as the laser intensity is not too strong, self-focusing and stimulated scattering of light in the fiber can be neglected. For a sufficiently short pulse, the nonlinearity of the fiber is dominated by the electronic contribution and therefore has a nearly instantaneous response. Then if the pulse is not too short and the spectral broadening is not excessive, the slowly varying amplitude approximation is valid and $\partial^2 P^{(3)}/\partial t^2$ in the wave equation can be well approximated by $-\omega_0^2 P^{(3)}$. The only modification of the simple theory of SPM we have discussed is to take into account the fact that we now have a wave in a waveguide with a confined transverse dimension instead of an infinite plane wave in an open space. Thus the quantitative analysis can easily be worked out. Indeed, Stolen and Lin (1978) found excellent agreement between theory and experiment.

The above discussion of SPM in time can also be used to describe SPM in space. As we already mentioned, the transverse intensity variation of a laser beam can induce a spatial variation of Δn in the transverse directions. Let us consider here, for simplicity, a continuous-wave (cw) laser beam with a Gaussian transverse profile. The phase increment $\Delta\phi(r, z)$ varying with the transverse coordinate r is given by

$$\Delta\phi(r, z) = (\omega/c)\Delta n(r)z. \tag{17}$$

This leads to a distortion of the wave front. Since the beam energy should propagate along the ray path perpendicular to the wave front, this distortion of the wave front would cause the beam to self-focus. If the propagation length is sufficiently long, the beam will actually self-focus and drastically modify the beam cross section. However, if the length of the medium is much shorter than the self-focusing distance, then the self-focusing effect in the medium can be neglected and we are left with only the SPM effect on the beam. The results of Figure 1.1 can describe the spatial SPM equally well if we simply replace t by r and ω by $k_\perp$, where $k_\perp$ is the transverse component of the wave vector of the beam. We realize that $k_\perp$ defines the deflection angle θ of a beam by the relation $k_\perp = (\omega n_0/c)\sin\theta$. Therefore, the quasi-periodic spectrum in the $k_\perp$ space actually corresponds to a diffraction pattern with multiple bright and dark rings. This has indeed been observed experimentally (Durbin et al., 1981; Santamato and Shen, 1984). An example is shown in Figure 1.3. Self-

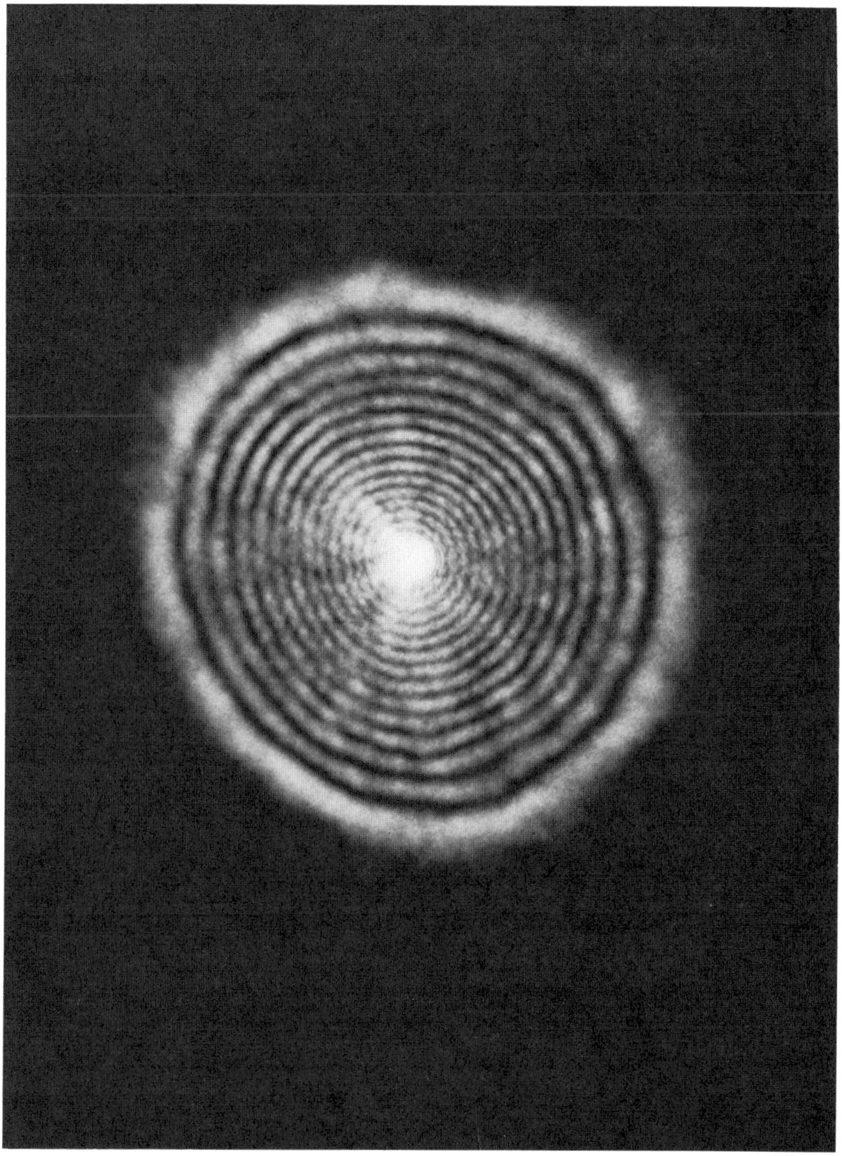

FIGURE 1.3. Diffraction ring pattern arising from spatial self-phase modulation of a CW Ar^+ laser beam passing through a 300-μm nematic liquid crystal film. (After Durbin et al., 1981; Santamato and Shen, 1984.)

focusing or diffraction in the medium can modify the spatial SPM through its modification of the beam profile. This is analogous to the self-steepening effect on the temporal SPM through its modification of the pulse shape.

We now return to the discussion of temporal SPM and spectral broadening. In the next section we consider the case where the incoming laser pulse is very short and spectral broadening is very extensive so that the approximations used in the simple theory of SPM need improvement.

4. More Rigorous Theory of Self-Phase Modulation and Spectral Superbroadening

Another experimental case of SPM that could avoid complications arising from self-focusing, stimulated scattering, or other nonlinear optical effects involves the propagation of an ultrashort laser pulse through a thin nonlinear medium. In this case, the medium is thin enough so that the self-focusing effect on SPM in the medium can be ignored. The pulse is short enough so that the transient stimulated light-scattering processes are effectively suppressed. Yet the pulse intensity can still be so high as to induce a very strong SPM, but not high enough to result in appreciable multiphoton absorption or optical breakdown. This is the case first studied by Fork et al. (1987). Using an 80-fs pulse at 627 nm focused to an intensity of 10^{13} to 10^{14} W/cm on a 500-μm film of ethylene glycol, they observed in the output a huge spectral broadening that appears as a white continuum. Unlike the spectral broadening discussed in the previous section, the present case shows a Stokes–anti-Stokes asymmetry that emphasizes the anti-Stokes side instead. Such a spectral superbroadening was observed earlier by Alfano and Shapiro (1970) in much longer media with picosecond pulses, but SPM in those cases was definitely affected by self-focusing. Obviously, the results of Fork et al. cannot be explained by the simple theory of SPM. We must resort to a more rigorous analysis.

We first notice that the self-steepening effect on the pulse is not included in the simple theory. This means that the approximations neglecting the $\partial^2\mathscr{E}/\partial t^2$ term and the terms involving the time derivatives of $|P^{(3)}|$ in the wave equation are not quite appropriate. They become worse for shorter pulses. In the more rigorous analysis of SPM, we must improve on these approximations. Let us now go back to Eq. (10). Without any approximation, we can transform it into an equation for the field amplitude (Yang and Shen, 1984):

$$\left(\frac{\partial}{\partial z}+\frac{n_0}{c}\frac{\partial}{\partial t}\right)\mathscr{E}+\frac{1}{i2k_0}\left(\frac{\partial^2}{\partial z^2}-\frac{n_0^2}{c^2}\frac{\partial^2}{\partial t^2}\right)\mathscr{E}$$
$$=-\frac{4\pi\omega_0^2}{i2k_0c^2}\chi^{(3)}|\mathscr{E}|^2\mathscr{E}+\frac{2\pi}{ik_0c^2}\left(-i2\omega_0\frac{\partial}{\partial t}+\frac{\partial^2}{\partial t^2}\right)\chi^{(3)}|\mathscr{E}|^2\mathscr{E}. \tag{18}$$

The last term on both sides of the equation has been neglected in the simple theory of SPM. By defining the differential operators $D_{\pm}\equiv(\partial/\partial z)\pm$

$(n_0/c)(\partial/\partial t)$, Eq. (18) can be written as

$$D_+\mathscr{E} + \frac{1}{i2k_0}D_-D_+\mathscr{E} = \frac{1}{i2k_0}\Pi,$$

$$\Pi = -\frac{4\pi\omega_0^2}{c^2}\left[1 + \frac{2i}{\omega_0}\frac{\partial}{\partial t} - \frac{1}{\omega_0^2}\frac{\partial^2}{\partial t^2}\right]\chi^{(3)}|\mathscr{E}|^2\mathscr{E}. \quad (19)$$

Since $D_- = -(2n_0/c)(\partial/\partial t) + D_+$, we have from Eq. (19)

$$\begin{aligned} D_+\mathscr{E} &= \frac{1}{i2k_0}\left[\left(-\frac{2n_0}{c}\frac{\partial}{\partial t} + D_+\right)D_+\mathscr{E} + \Pi\right] \\ &= \frac{1}{i2k_0}\sum_{m=0}^{\infty}\left[\left(-\frac{2n_0}{c}\frac{\partial}{\partial t} + D_+\right)\Big/ i2k_0\right]^m \Pi. \end{aligned} \quad (20)$$

It is then simply a question of how many terms in the power series expansion we need to include to better describe the SPM.

The zeroth-order approximation corresponds to neglecting all derivatives of $\chi^{(3)}|\mathscr{E}|^2\mathscr{E}$ and yields

$$D_+\mathscr{E} = -(2\pi\omega_0^2/ic^2k_0)\chi^{(3)}|\mathscr{E}|^2\mathscr{E}, \quad (21)$$

which is identical to Eq. (11) used as the basis for the simple theory of SPM. We recognize that under this lowest-order approximation,

$$\begin{aligned} D_+(\chi^{(3)}|\mathscr{E}|^2\mathscr{E}) &\propto (\chi^{(3)}|\mathscr{E}|^2)^2\mathscr{E} \\ &\ll \chi^{(3)}|\mathscr{E}|^2\mathscr{E} \quad \text{if } \chi^{(3)}|\mathscr{E}|^2 \ll 1. \end{aligned} \quad (22)$$

Therefore, we can use D_+ as an expansion parameter in the higher-order calculations. For the first-order approximation, we neglect terms involving $D_+^m(\chi^{(3)}|\mathscr{E}|^2\mathscr{E})$ with $m \geq 1$ in Eq. (20) and obtain

$$\begin{aligned} D_+\mathscr{E} &= \frac{1}{i2k_0}\sum_{m=0}^{\infty}\left[\left(\frac{1}{i\omega_0}\frac{\partial}{\partial t}\right)^m \Pi\right] \\ &= \left(1 + \frac{i}{\omega_0}\frac{\partial}{\partial t}\right)\left[-\frac{2\pi\omega_0^2}{ik_0c^2}\chi^{(3)}|\mathscr{E}|^2\mathscr{E}\right]. \end{aligned} \quad (23)$$

The calculation here has in a sense used $\chi^{(3)}|\mathscr{E}|^2$ as the expansion parameter. In the above first-order approximation, we have kept the $(\chi^{(3)}|\mathscr{E}|^2)^n\mathscr{E}$ terms with $n \leq 1$ including all their time derivatives. In ordinary cases, this is a very good approximation because usually $\chi^{(3)}|\mathscr{E}|^2 \ll 1$ and therefore the higher-order terms involving $(\chi^{(3)}|\mathscr{E}|^2)^n\mathscr{E}$ with $n \geq 2$ are not very significant. For example, in the ultrashort pulse case, we have $\chi^{(3)} \sim 10^{-14}$ esu (or $n_2 \sim 10^{-13}$ esu) for a condensed medium; even if the laser pulse intensity is $I \sim 10^{14}$ W/cm^2, we find $\chi^{(3)}|\mathscr{E}|^2 \sim 4 \times 10^{-3} \ll 1$. For larger $\chi^{(3)}|\mathscr{E}|^2$, one may need to include higher-order terms in the calculations. The next-order correction includes the $D_+(\chi^{(3)}|\mathscr{E}|^2\mathscr{E})$ term and all its time derivatives. They yield additional terms proportional to $(\chi^{(3)}|\mathscr{E}|^2)^2\mathscr{E}$ in the wave equation.

If $\chi^{(3)}|\mathscr{E}|^2 \gtrsim 1$, then the approach with series expansion will not be useful and we have to go back to the original nonlinear wave equation (19).

In the following discussion, we consider only cases with $\chi^{(3)}|\mathscr{E}|^2 \ll 1$. We are therefore interested in the solution of Eq. (23), which, with $n_2 = (2\pi\omega_0/k_0 c)\chi^{(3)}$, takes the form

$$\frac{\partial \mathscr{E}}{\partial z} + \frac{1}{c}\frac{\partial}{\partial t}(n_0 \mathscr{E} + n_2|\mathscr{E}|^2) = i\frac{n_2\omega_0}{c}|\mathscr{E}|^2\mathscr{E}. \tag{24}$$

For simplicity, we now neglect the dispersion of the response of the medium. This, as we mentioned earlier, is equivalent to assuming an instantaneous response. Insertion of $\mathscr{E} = |\mathscr{E}|\exp(i\phi)$ into Eq. (24) yields two separate equations for the amplitude and phase:

$$\left[\frac{\partial}{\partial z} + \frac{n_0}{c}\left(1 + \frac{3n_2}{n_0}|\mathscr{E}|^2\right)\frac{\partial}{\partial t}\right]|\mathscr{E}| = 0, \tag{25a}$$

$$\left[\frac{\partial}{\partial z} + \frac{n_0}{c}\left(1 + \frac{n_2}{n_0}|\mathscr{E}|^2\right)\frac{\partial}{\partial t}\right]\phi = \frac{n_2\omega_0}{c}|\mathscr{E}|^2. \tag{25b}$$

In comparison with Eq. (12) for the simple theory of SPM, the only difference is the addition of the $n_2|\mathscr{E}|^2$ ($= \Delta n$) terms on the left-hand sides of Eqs. (25). Its effect is obvious in causing a pulse shape deformation during the pulse propagation. With $\Delta n > 0$, we expect a pulse steepening in the lagging edge. This is because the peak of the pulse then propagates at a lower velocity than either the leading or the lagging part of the pulse (DeMartini et al., 1967).

Let us first neglect the self-steepening effect on the amplitude pulse. Clearly, self-steepening in the lagging part of the ϕ pulse should lead to a spectral broadening with Stokes–anti-Stokes asymmetry emphasizing the anti-Stokes side, because it is the lagging part of the phase modulation that gives rise to the broadening on the anti-Stokes side. To be more quantitative, we assume an input laser pulse with $|\mathscr{E}(0,t)|^2 = A^2/\cosh(t/\tau)$, whose shape remains unchanged in propagating through the medium so that $|\mathscr{E}(z,t)|^2 = A^2/\cosh[(t - n_0/c)/\tau]$. The solution of Eq. (25b) can then be found analytically as

$$\phi = \omega_0\tau\{x - \sinh^{-1}[\sinh x - (n_2/c\tau)A^2 z]\} \tag{26}$$

with $x = [t - (n_0/c)z]/\tau$. The corresponding frequency modulation is given by

$$\Delta\omega/\omega_0 = [1 + (Q^2 - 2Q\sinh x)/\cosh x]^{-1/2} - 1. \tag{27}$$

Here we have defined

$$Q = n_2 A^2 z/c\tau \tag{28}$$

as a characteristic parameter for spectral broadening. For $Q \ll 1$, we have

$$\Delta\omega/\omega_0 \simeq -Q\sinh x/\cosh^2 x, \tag{29}$$

which is identical to the result one would find from the simple theory of SPM. For $Q \gtrsim 1$, we expect spectral superbroadening with appreciable Stokes–

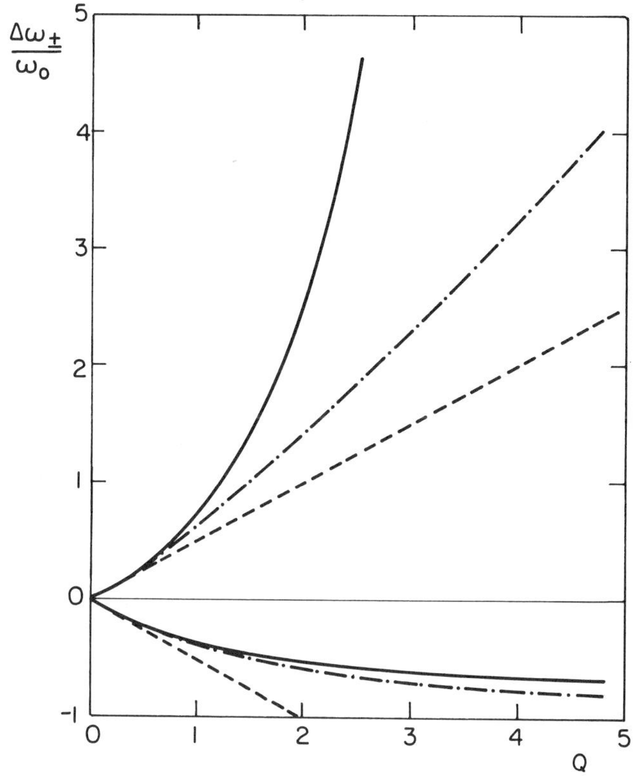

FIGURE 1.4. Maximum Stokes ($\Delta\omega_- < 0$) and anti-Stokes ($\Delta\omega_+ > 0$) shifts calculated with different models: simple theory of self-phase modulation (---); more rigorous theory without the self-steepening effect on the intensity pulse (-·-); more rigorous theory with the self-steepening effect (——). (After Yang and Shen, 1984.)

anti-Stokes asymmetry and a maximum anti-Stokes shift $\Delta\omega_+ \gtrsim \omega_0$. The maximum Stokes and anti-Stokes shifts, $\Delta\omega_-$ and $\Delta\omega_+$, respectively, can be directly obtained from Eq. (27):

$$\Delta\omega_\pm/\omega_0 = \tfrac{1}{2}[(Q^2 + 4)^{1/2} \pm |Q|] - 1. \tag{30}$$

This is plotted in Figure 1.4 in comparison with the result calculated from the simple theory of SPM. For $|Q| \ll 1$ we have $\Delta\omega_\pm \simeq \pm(1/2)\omega_0|Q|$, and for $|Q| \gg 1$ we have $\Delta\omega_+ \simeq \omega_0|Q|$ and $\Delta\omega_- \simeq \omega_0(1/|Q| - 1)$. It is seen that Stokes broadening is always limited by $|\Delta\omega_-| < \omega_0$, as it should be.

We now include the effect of self-steepening on the amplitude pulse. It can be shown that with $|\mathscr{E}(0,t)|^2 = A^2/\cosh(t/\tau)$, the solution of Eq. (25a) must satisfy the implicit algebraic equation

$$|\mathscr{E}(z,t)|^2 = A^2/\cosh[x - Q|\mathscr{E}(z,t)|^2/A^2]. \tag{31}$$

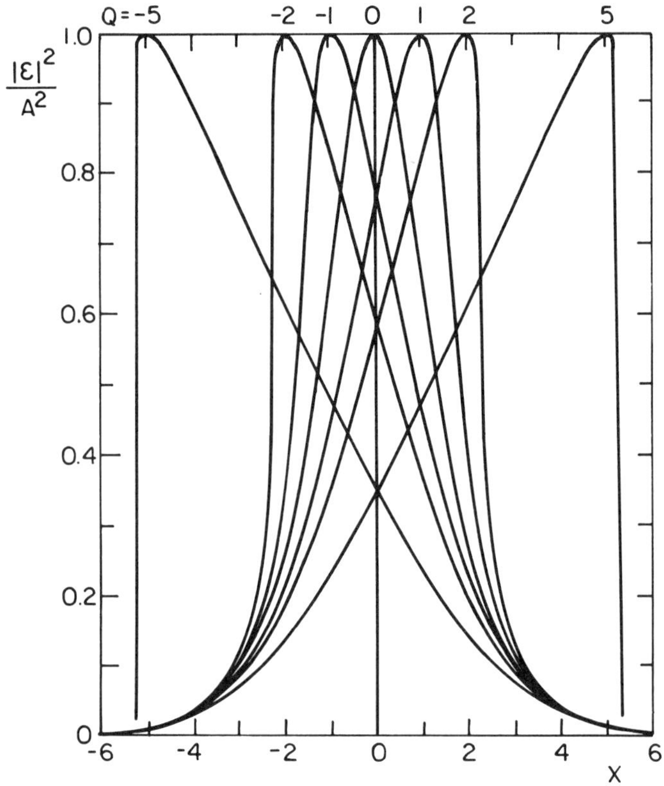

FIGURE 1.5. Self-steepening effect on the intensity pulse during its propagation in a nonlinear medium at various values of $Q = n_2 A^2 zc/\tau$, with $x = (t - n_0 z/c)/\tau$. (After Yang and Shen, 1984.)

A simple numerical calculation then allows us to find $|\mathscr{E}|^2$ as a function of x for a given Q. The results are shown in Figure 1.5. For $|Q| \gtrsim 1$, the self-steepening effect is apparent. Knowing $|\mathscr{E}(z,t)|^2$, we can again solve for $\phi(z,t)$ from Eq. (25b) and find $\Delta\omega(z,t)$ and $\Delta\omega_{\pm}$. This can be done numerically; the results are also presented in Figure 1.4. It is seen that for $Q > 1$, self-steepening of the pulse amplitude has increased the spectral broadening on the anti-Stokes side quite significantly. The additional spectral broadening comes in because the steepening of the amplitude pulse enhances the steepening of the ϕ pulse.

The spectral broadening actually results from frequency chirping since $\partial\phi(t)/\partial t = -\omega(t)$. This is shown in Figure 1.6 for the numerical example discussed above. As expected, the Stokes and anti-Stokes shifts appear, respectively, in the leading and lagging parts of the self-steepened pulse. The $\Delta\omega = 0$ point appears at larger x for larger values of Q because self-steepening shifts the peak of the pulse to larger x (see Figure 1.5).

We can compare the calculation with the experiment of Fork et al. In their

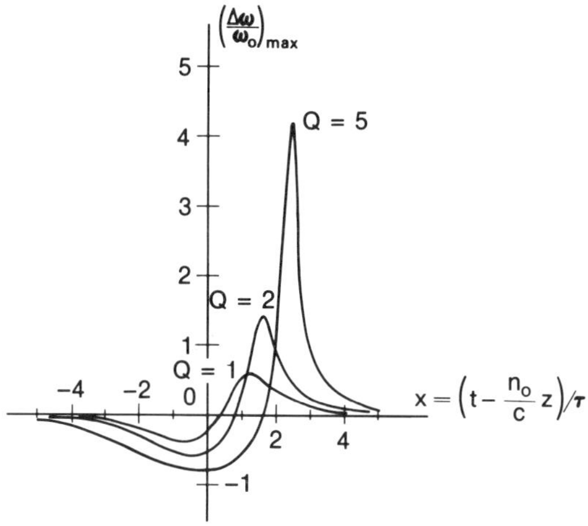

FIGURE 1.6. Frequency shift due to phase modulation as a function of $x = (t - n_0 z/c)/\tau$ at various values of $Q = n_2 A^2 zc/\tau$.

experiment, the relevant parameters are $n_2 \simeq 10^{-13}$ esu, $z = 0.05$ cm, $I \sim 10^{14}$ W/cm^2, and τ (pulse width) $\simeq 8 \times 10^{-14}$ s. The corresponding value of Q is about 2.3. They observed a Stokes broadening $\Delta\omega_-/\omega_0 \simeq -0.6$ and an anti-Stokes broadening $\Delta\omega_+/\omega_0 \simeq 2.3$. Our calculation gives $\Delta\omega_-/\omega_0 \simeq -0.54$ and $\Delta\omega_+/\omega_0 \simeq 3.5$. Considering the uncertainty in the experimental parameters, we can regard the agreement between theory and experiment as reasonable.

Manassah et al. (1985, 1986) used the method of multiple scales to solve Eq. (18) (neglecting dispersion). They also took $n_2|\mathscr{E}|^2/n_0$ as the expansion parameter in their series expansion and therefore necessarily obtained the same results as we discussed above.*

We have neglected in the above calculation the dispersion of the medium response. Normal dispersion may also reshape the pulse (Fisher and Bischel, 1975; Fisher et al., 1983), but in the present case the length of the medium is so short that this effect is not likely to be important. Anomalous dispersion with resonances in $\chi^{(3)}$ or n_2 could, however, give rise to resonant structure in the broadened spectrum. The calculation including the dispersion of $\chi^{(3)}$ is much more complicated and, in general, must resort to numerical solution (Fisher and Bischel, 1975; Fisher et al., 1983). In obtaining the time-dependent solution of the wave equation with the third-order nonlinearity, we have already taken all the four-wave mixing contributions into account. By adding a noise term with a blackbody spectrum in the nonlinear wave equation, the four-wave parametric generation process proposed by Penzkofer et al. (1973, 1975) could also be inclued in the calculation.

* A factor of 3 in front of n_2 is mistakenly left out in Eq. (3a) of Yang and Shen (1984).

5. Self-Focusing and Self-Phase Modulation

For pulsed laser beam propagation in a nonlinear medium, SPM in time and SPM in space necessarily appear together. SPM in time causes self-steepening of the pulse, which in turn enhances SPM in time. Similarly, SPM in the transverse beam profile causes self-focusing of the beam, which in turn enhances the transverse SPM. If the propagation distance in the medium is sufficiently long, these effects can build up to a catastrophic stage, namely self-steepening to a shock front and self-focusing to a spot limited in dimensions only by higher-order nonlinear processes and diffraction. SPM in time and SPM in space are then tightly coupled and strongly influenced by each other. In many experiments, the observed strong temporal SPM and extensive spectral broadening are actually initiated by self-focusing. In such cases, the input laser pulse is so weak that without self-focusing in the nonlinear medium, SPM would not be very significant. Self-focusing to a limiting diameter greatly enhances the beam intensity, and hence SPM can appear several orders of magnitude stronger. A quantitative description of such cases is unfortunately very difficult, mainly because the quantitative theory for self-focusing is not yet available. We must therefore restrict ourselves to a more qualitative discussion of the problem.

5.1 Self-Phase Modulation with Quasi- Steady-State Self-Focusing

In the early experiments on self-focusing of single-mode nanosecond laser pulses, it was found that the output of the self-focused light had a spectral broadening of several hundred cm^{-1} (Shen, 1975; Marburger, 1975). This was rather surprising because from the simple theory of SPM, picosecond pulses would be needed to create such a spectral broadening (Cheung et al., 1968; Gustafson et al., 1969; Shimizu, 1967). Later, the observation was explained by SPM of light trailing behind a moving focus (Shen and Loy, 1971; Wong and Shen, 1972). We briefly review the picture here and then use it to interpret the recently observed SPM and spectral superbroadening of ultrashort pulses in gases (Corkum et al., 1986; Glownia et al., 1986).

Figure 1.7 depicts the quasi-steady-state self-focusing of a laser pulse leading to a moving focus along the axis ($\hat{z}$) with a U-shaped trajectory described by the equation (Shen, 1975; Marburger, 1975)

$$z_f(t) = \frac{K}{\sqrt{P(t_R)} - \sqrt{P_0}}, \tag{32}$$

where P_0 is the critical power for self-focusing, $\sqrt{P(t_R)}$ is the laser power at the retarded time $t_R = t - z_f n_0/c$, and K is a constant that can be determined from experiment. This equation assumes instantaneous response of Δn to the applied field, which is a good approximation as long as the response time τ is

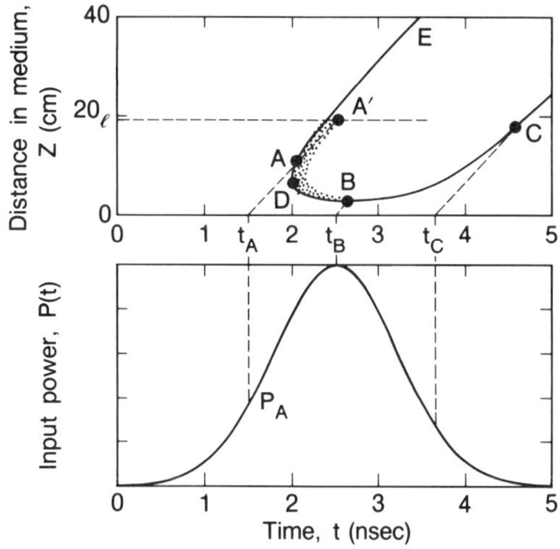

FIGURE 1.7. Self-focusing for an input laser pulse in (a) leading to the trajectory of a moving focus in the form of U curve in (b). The dashed lines in (b), with slope equal to the light velocity, depict how light propagates in the medium along the z axis at various times. The shaded region around the U curve has appreciably larger Δn. Light traversing the medium along the dashed lines through the shaded region should acquire a phase increment $\Delta\phi$ that varies with time.

much shorter than the laser pulse width. Here, we are interested only in the upper branch of the U curve, along which the focus has a forward velocity faster than light. Because of the high laser intensity in the focal spot, the locally induced Δn should be appreciable and should last for a duration not shorter than the relaxation time τ. Thus one can imagine that the moving focus creates in the medium a channel of Δn at least $\tau\, dz_f/dt$ long trailing after the focus. Consider now the defocused light from a local focal spot. Since it lags behind the moving focus (which travels faster than light), it experiences the Δn dielectric channel created by the focus over a certain distance and will diffract only weakly. In other words, the defocused light from the focus is partially trapped in the Δn channel. This partial trapping of light in turn helps to maintain the Δn channel and make it last longer. The emission from the focal spot at the end of the medium then takes the form of an asymmetric pulse (with a pulse width of the order of a few τ) with a longer trailing edge.

The above picture is also illustrated in Figure 1.7. We use the shaded area around the U curve to denote the region with appreciable Δn. The laser input at t_A focuses at A, but defocuses more gradually because of the existing Δn channel in front of it. The partially trapped light then propagates along the axis from A to the end of the medium at A', crossing the shaded region with appreciable Δn. It therefore acquires a significant phase increment $\Delta\phi$. From the figure, one may visualize that $\Delta\phi$ can be strongly phase modulated in time, varying from nearly zero to a maximum and back to zero in a few relaxation times. This could yield appreciable spectral broadening in the output of the self-focused light.

To be more quantitative, we realize that the light pulse emitted from a focus in the medium must be asymmetric and must have a pulse width of several τ. The shaded area in Figure 1.7 has a somewhat larger width since Δn is induced by the focused light. Knowing the trajectory of the moving focus, the beam intensity in the focal region, and how Δn responds to the intensity, we can calculate $\Delta\phi(t)$ and hence $\Delta\omega(t)$ and the broadened spectrum (Shen and Loy, 1971; Wong and Shen, 1972). As an example, consider the case of a 1.2-ns laser pulse propagating into a 22.5-cm CS_2 cell. The trajectory of the moving focus (focal diameter $\simeq 5\ \mu$m) is described by Eq. (32) with $K = 5.7\ (\text{kW})^{1/2}$-cm and $P_0 = 8$ kW. In this case, Δn is dominated by molecular reorientation; it obeys the dynamic equation

$$\left(\frac{\partial}{\partial t} + \frac{1}{\tau}\right)\Delta n = \frac{n_2}{\tau}|E(z,t)|^2. \tag{33}$$

For CS_2, $\tau = 2$ ps and $n_2 = 10^{-11}$ esu. The phase increment experienced by light waves traversing the cell along the axis is given by

$$\Delta\phi = \int_0^l (\omega/c)\Delta n[z, t' = t - (l - z)n_0/c]\,dz, \tag{34}$$

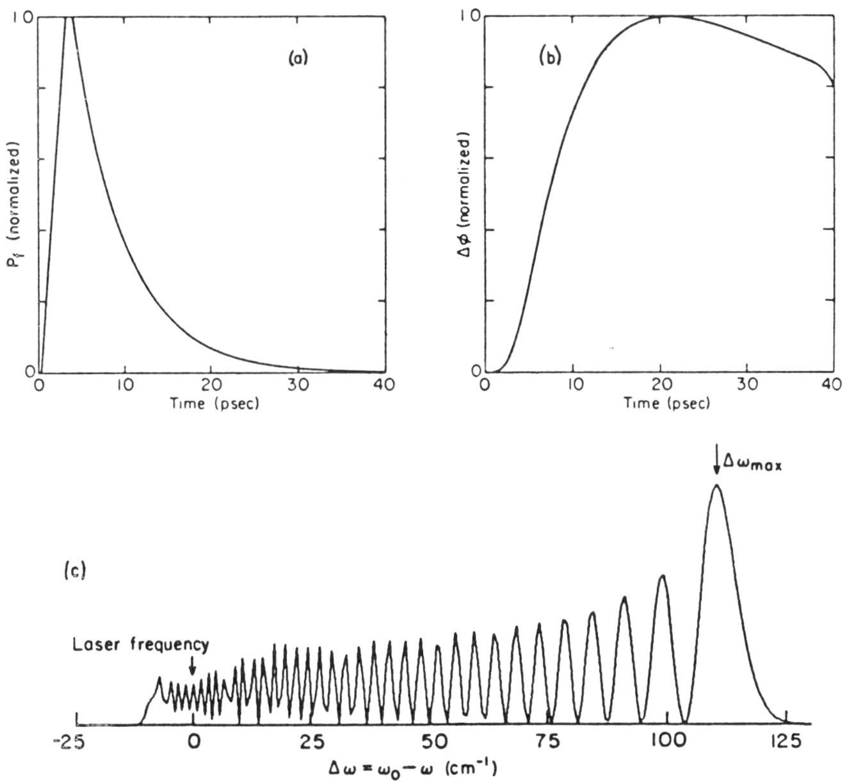

FIGURE 1.8. Theoretical power spectrum of a light pulse emitted from the focal region of a moving focus at the end of a CS_2 cell. (a) The intensity pulse; (b) $\Delta\phi$ versus t; (c) the power spectrum (see text for details). (After Shen and Loy, 1971; Wong and Shen, 1972.)

where l is the cell length. (For this illustrative example, we have neglected the diffraction effect on $\Delta\phi$.) We now simply assume that $|E(z,t)|^2$ in the focal region resulting from self-focusing has a pulse width of $\sim 3\tau$ and a pulse shape as shown in Figure 1.8a. Equations (33) and (34) then allow us to find $\Delta n(z,t)$ and $\Delta\phi(t)$. Knowing $\Delta\phi(t)$ and $E(l,t)$, we can then calculate the spectrum of the output from the focal spot at the end of the cell, as shown in Figure 1.8c. The experimentally observed spectrum has in fact the predicted spectral broadening (Shen and Loy, 1971; Wong and Shen, 1972), but it often has a strong central peak (Figure 1.9). This is presumably because in the calculation

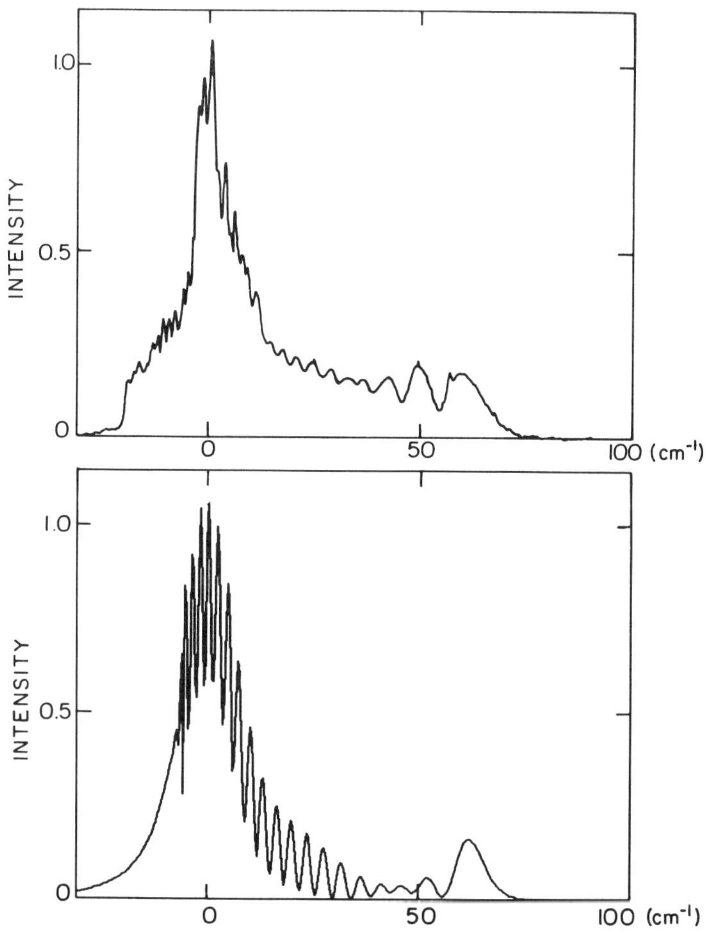

FIGURE 1.9. (a) Experimentally observed power spectrum of light emitted from the focal region at the end of a 10-cm CS_2 cell; the input pulse has a pulse width of 1.2 ns and a peak power of 27 kW. (b) Theoretical power spectrum using the moving focus model. (After Shen and Loy, 1971; Wong and Shen, 1972.)

we have neglected a significant portion of the beam that self-focuses from the periphery and experiences little phase modulation. For shorter input pulses or longer cells, self-focusing of the beam toward the end of the cell is more gradual; accordingly, the weakly phase-modulated part is less and the central peak in the spectrum is reduced. We also note that in Figures 1.8 and 1.9 the anti-Stokes broadening is much weaker. This is because the negatively phase-modulated part of the pulse has little intensity, as seen in Figure 1.8.

Using the picture sketched in Figure 1.7, we can actually predict the Stokes broadening with the correct order of magnitude by the following rough estimate. We approximate the upper branch of the U curve toward the end of the medium by a straight line with a slope equal to the end velocity of the moving focus. If Δn is the induced refractive index in the shaded area, then the phase modulation of the emitted light is given by

$$\Delta\phi(t) \simeq \left(\frac{\omega_0}{c}\right)\left(\frac{n_0}{c} - \frac{1}{v}\right)^{-1} \int_{t_0}^{t} \Delta n(l, t')\, dt', \tag{35}$$

where l is the length of the medium and t_0 is the time when $\Delta n(l, t)$ starts to become appreciable. The extent of Stokes broadening is readily obtained from

$$\Delta\omega_- = (\partial\Delta\phi/\partial t)_{\max} \simeq \omega_0 Q \tag{36}$$

with $Q = \Delta n_{\max} l_{\text{eff}}/cT$ where $T \sim 2\tau$ and $l_{\text{eff}} = T/[(n_0/c) - (1/v)]$. For the above example with CS_2, we have $\Delta n_{\max} \sim 10^{-3}$, $T \sim 4$ ps, $l_{\text{eff}} \sim 1$ cm, and $Q \sim 0.01$. The resultant Stokes broadening should be $\Delta\omega_- \sim 150\ \text{cm}^{-1}$. The experimentally observed broadening is about 120 cm^{-1}.

5.2 Spectral Superbroadening of Ultrashort Pulses in Gases

The above discussion can be used to explain qualitatively the recently observed spectral superbroadening of ultrashort pulses in gas media (Corkum et al., 1986; Glownia et al., 1986). In those experiments, picosecond or femtosecond laser pulses with energies of several hundred microjoules were weakly focused into a high-pressure gas cell. Spectral superbroadening with $\Delta\omega \sim 10^4\ \text{cm}^{-1}$ was observed. A few examples are shown in Figure 1.10. Self-focusing was apparently present in the experiment. We therefore use the above simple model for SPM with quasi-steady-state self-focusing to estimate the spectral broadening (Loy and Shen, 1973), assuming that Δn from the electronic contribution in the medium has a response time $\tau \sim 10$ fs. In this case, the position of the moving focus is given by (Shen, 1975; Marburger, 1975)

$$(z_f^{-1}(t) - f^{-1})^{-1} = \frac{K}{\sqrt{P(t_R)} - \sqrt{P_0}} \tag{37}$$

instead of Eq. (32), where f is the focal length of the external focusing lens. Let us consider, for example, external focusing of a 250-μJ, 100-fs pulse to a nominally 100-μm focal spot in a 3-atm, 100-cm Xe cell. Self-focusing yields

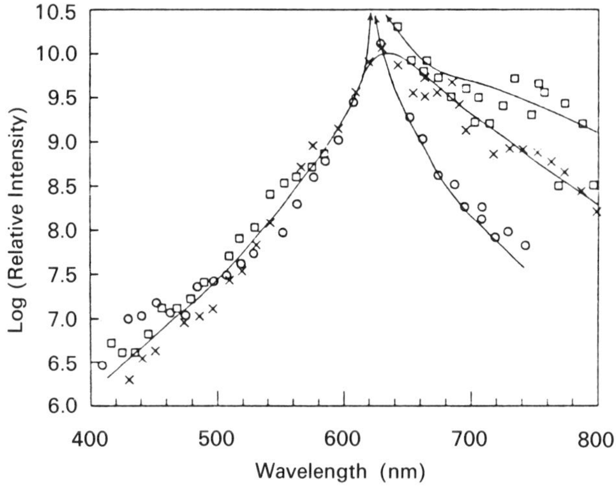

FIGURE 1.10. Continuum spectra of self-phase-modulated light from 70-fs pulses in 30-atm xenon (crosses), 2-ps pulses in 15-atm xenon (circles), and 2-ps pulses in 40-atm nitrogen (squares). The cell length is 90 cm. (After Corkum et al., 1986.)

a smaller focus, assumed to be 50 μm. We then use the values $n_2 \sim 10^{-16}$ esu, $I \sim 10^{14}$ W/cm^2, $\Delta n_{\max} \sim 4 \times 10^{-5}$, $l_{\text{eff}} \sim 10$ cm, and $T \sim 2\tau \sim 20$ fs; we find $Q \sim 1$ and hence $\Delta\omega_{-} \sim 10^4$ cm^{-1}. The above estimate is admittedly very crude because of uncertainties in the experimental parameters, but it does give a spectral superbroadening in order-of-magnitude agreement with the experiments.

Appreciable anti-Stokes broadening was also observed in the superbroadened spectrum of the SPM light from a gas medium. This seems to be characteristically different from what we have concluded from the discussion in the previous subsection. However, we realize that in the present case the moving focus terminates at $z = f$ instead of the end of the cell, and the total transmitted light is detected and spectrally analyzed. Thus the detected output pulse has essentially the same intensity envelope as the input pulse if we neglect the self-steepening effect, and the negatively phase-modulated part (the trailing edge) of the $\Delta\phi(t)$ pulse will overlap with the major part of the intensity pulse. Consequently, the spectral intensity of the anti-Stokes side should be nearly as strong as that of the Stokes side. The extent of the anti-Stokes broadening is expected to be somewhat less than that of the Stokes broadening because of the longer trailing edge of the $\Delta\phi$ pulse, unless the self-steepening effect becomes important.

Self-focusing in a gas medium should be more gradual than in a liquid cell. With weak external focusing, the focal dimensions resulting from combined external and self-focusing may not be very different from those resulting from external focusing alone. Thus, even with self-focusing, the SPM output from the gas medium may not have a much larger diffraction angle than the linearly transmitted output, as was observed in the experiments.

5.3 Self-Phase Modulation with Transient Self-Focusing

We have used the picture of a moving focus with a trailing dielectric channel to describe SPM initiated by quasi-steady-state self-focusing. For shorter input pulses, the velocity of the forward moving focal spot is closer to the light velocity, and consequently more light is expected to be trapped in the dielectric channel for a longer distance. In fact, when the pulse width is comparable to or shorter than the relaxation time τ, the entire self-focusing process becomes transient, and the input pulse will evolve into a dynamic trapping state (Loy and Shen, 1973).

The dynamic trapping model for transient self-focusing is an extension of the moving focus model for quasi-steady-state self-focusing. Consider the case where Δn is governed by Eq. (33) or, more explicitly,

$$\Delta n(z, \xi) = \frac{1}{\tau} \int_{-\infty}^{\xi} n_2 |E(z, \xi')|^2 \exp[-(\xi - \xi')/\tau] \, d\xi', \tag{38}$$

where $\xi = t - zn_0/c$. Because of this transient response of Δn, the later part of the pulse propagating in the medium may see a larger Δn than the earlier part. As a result, different parts of the pulse will propagate in the medium differently, as sketched in Figure 1.11 (Loy and Shen, 1973). The transient Δn makes the very leading edge of the pulse diffract, the middle part self-focus weakly, and the lagging part self-focus to a limiting diameter. The result is

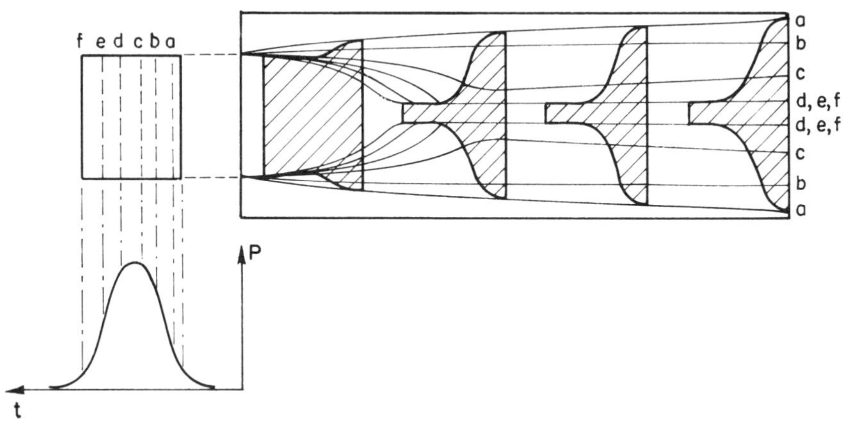

FIGURE 1.11. Sketch showing self-focusing of an ultrashort pulse in a medium with a transient response of Δn. Different parts (a, b, c, etc.) of the pulse focus and defocus along different ray paths. The pulse is first deformed into a horn shape and then propagates on without much further change. (After Loy and Shen, 1973.)

that in propagating through the medium, the pulse is first deformed into a horn-shaped pulse and then the horn-shaped pulse propagates on with only a slight change of the pulse shape due to diffraction of the front edge. In a long medium, the front-edge diffraction finally could blow up the deformed pulse. Note that this picture comes in because in transient self-focusing, both focusing and diffraction are much more gradual, leading to a long longitudinal focal dimension and hence the rather stable horn-shaped propagating pulse. Such a stable form of self-focused pulse propagation is known as dynamic trapping.

Since the various parts of the light pulse see different Δn's along their paths, the phase increments $\Delta\phi$ they acquire are also different. This means phase modulation and hence spectral broadening. As an approximation, we can assume that the overall phase modulation is dominated by that of a stable horn-shaped pulse propagating in the nonlinear medium over a finite length l_{eff}. For illustration, let us take an example in which the horn-shaped pulse can be described by

$$|E(r,\xi)|^2 = A_0^2 \exp[-\xi^2 t_p^2 - 2r^2/r_0^2(\xi)] \tag{39}$$

with

$$\begin{aligned} r_0(\xi)/a &= 1 && \text{if } \xi \le \xi_1 \\ &= (1-\Delta)e^{-(\xi-\xi_1)/\tau_1} + \Delta && \text{if } \xi_1 \le \xi \le \xi_2 \\ &= (1-\Delta)e^{-(\xi_2-\xi_1)/\tau_1} + \Delta && \text{if } \xi \ge \xi_2 \end{aligned}$$

where $\xi = t - zn_0/c$. We have picosecond pulse propagation in Kerr liquids in mind and therefore choose $t_p = 1.25\tau$, $\xi_1 = 2.5\tau$, $\xi_2 = 2\tau$, $\tau_1 = \tau$, $\Delta = 0.05$, and $\tau = 2$ ps (for CS_2). We also choose the pulse intensity as $A_0^2 = 80(n_2/n_0)k^2a^2$, where k is the wave vector and the effective pulse propagation distance $l_{\text{eff}} = 0.15ka^2$. From the three-dimensional wave equation, it can be shown that the phase modulation obeys the equation

$$\partial(\Delta\phi)/\partial z = k(\Delta n/n_0) - 2/kr_0^2(\xi). \tag{40}$$

The second term on the right of the equation is the diffractive contribution to $\Delta\phi$, which can be appreciable when r_0 is small. Knowing $|E(r,\xi)|^2$, we can find $\Delta n(z,\xi)$ from Eq. (38), and hence $\Delta\phi(z,\xi)$ from Eq. (40), and finally the broadened spectrum from $|E(0,\xi)|^2$ and $\Delta\phi(z,\xi)$ for $z = l_{\text{eff}}$, as shown in Figure 1.12.

The main qualitative result of the above calculation is that the spectrum has the quasi-periodic structure with nearly equal Stokes and anti-Stokes broadening, although the Stokes side is more intense. This agrees with the experimental observation (Cubbedu and Zagara, 1971; Cubbedu et al., 1971) and the more detailed numerical calculation of Shimizu and Courtens (1973). The reason is as follows. The Stokes–anti-Stokes symmetry results from a symmetric $\Delta\phi(t)$ pulse that overlaps well with the intensity pulse. The neck portion of the horn-shaped pulse with $\partial(\Delta\phi)/\partial t < 0$ contributes to the anti-Stokes broadening. As seen in Eq. (40), the time dependence of $\Delta\phi$ comes solely

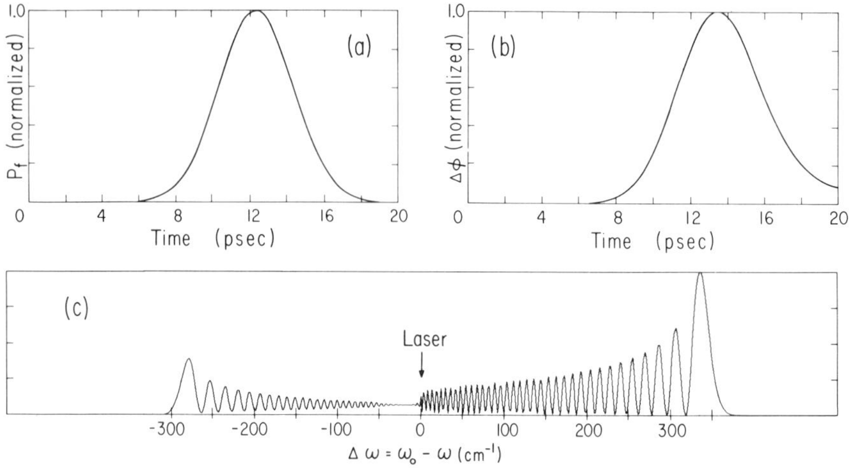

FIGURE 1.12. Theoretical power spectrum obtained by assuming a horn-shaped pulse propagating for a certain distance in a nonlinear medium without any change in its shape. (a) Normalized intensity output pulse; (b) $\Delta\phi$ versus t; (c) power spectrum of the output. (After Loy and Shen, 1973.)

from the time dependence in $(\Delta n/n_0 - 2/k^2 r_0^2)$. Without the diffraction term $(-1/k^2 r_0^2)$, the $\Delta\phi$ pulse would have a longer trailing edge because of the relaxation of Δn. With the diffraction term, the rapid reduction of r_0 toward the neck makes the $\Delta\phi$ pulse more symmetric. Thus it appears that the dynamic trapping model explains fairly well the qualitative features of SPM and spectral broadening initiated by transient self-focusing. The broadening is more extensive with more intense input pulses and longer propagation lengths. In a long medium, dynamic trapping may exist only over a limit section of the medium. In that case, the spectrally broadened light may seem to have originated from a source inside the medium. This has also been observed experimentally (Ho et al., 1987).

6. Conclusion

We have seen that the temporal SPM and the concomitant spectral broadening of light arise because an intense optical field can induce an appreciable refractive index change Δn in a medium. The theory of pure SPM is, in principle, quite straightforward. If the input pulse intensity is not very strong, the zeroth-order approximation taking into account only the direct contribution of Δn to the induced phase change $\Delta\phi$ should already give a fairly good description. The next-order approximation including the self-steepening effect on both $\Delta\phi$ and the amplitude pulse should satisfactorily cover the cases of strong SPM with spectral superbroadening.

Unfortunately, the temporal SPM is often complicated by the spatial SPM. The latter can lead to self-focusing, which dramatically alters the intensity distributions of the laser pulse in space and time and therefore drastically modifies the temporal SPM. In fact, in most practical cases, self-focusing occurs long before the temporal SPM becomes appreciable; it is actually self-focusing that increases the beam intensity in the medium and thus initiates a strong SPM in time. Only by using an optical waveguide or a very thin nonlinear medium can self-focusing be avoided. These are then the only experimental cases where a pure temporal SPM has been realized.

In the pure SPM case, the theoretical difficulty is in the description of Δn: it is not easy, in general, to predict quantitatively the nonlinear response of a medium from first principles; one must rely on experimental measurements. Quantitative measurements of $\Delta n(t)$ in the picosecond and femtosecond domains are still rare. In particular, measurements of Δn with femtosecond time resolution are still rather difficult. The various low-frequency resonances could make the time dependence of Δn very complex. Inclusion of the transient response of Δn (or the dispersion of Δn) in the theory complicates the calculation; one may have to resort to numerical solution of the problem. Experimentally, SPM of laser pulses in optical fibers has been well studied; SPM of ultrashort pulses in thin nonlinear media is, however, still not well documented. More careful quantitative measurements are needed in order to have a more detailed comparison with theory.

With self-focusing, the theory of SPM becomes extremely complex. The main difficulty lies in the fact that a quantitative theory capable of describing the details of self-focusing is not yet available. We must then rely on the more qualitative physical pictures for self-focusing. Therefore, the discussion of the subsequently induced SPM can be at most semiquantitative. Thus, we find it quite satisfying that the predicted spectral broadening from such theoretical discussions can give order-of-magnitude agreement with the experimental observations in a number of very different cases: nanosecond pulse propagation in liquids to pico- or femtosecond pulse propagation in gases.

Spectral superbroadening is often observed with the propagation of ultrashort pulses in condensed media and is apparently initiated by self-focusing. In most cases, the details of the self-focusing process have not been measured; in some cases, even the quantitative information on Δn is not available. The measurements on spectral broadening also tend to show strong fluctuations. All these made even an order-of-magnitude comparison between theory and experiment rather difficult.

A complete theory of SPM with self-focusing requires the solution of a time-dependent three-dimensional wave equation. With self-focusing modifying the laser pulse rapidly in both space and time, such a solution, even on the largest computer, is a formidable task. In our opinion, the best way to tackle the problem is to try to simplify the calculation by reasonable approximations derived from the physical picture that has already been established for self-focusing.

Acknowledgment. This work was supported by the Director, Office of Energy Research, Office of Basic Energy Sciences, Materials Sciences Division of the U.S. Department of Energy under Contract No. DE-AC03-76SF00098.

References

Alfano, R.R. and S.L. Shapiro (1970) Emission in the region 4000–7000 A via four-photon coupling in glass; Observation of self-phase modulation and small scale filaments in crystals and glasses; Direct distortion of electronic clouds of rare-gas atoms in intense electric fields. Phys. Rev. Lett. **24**, 584, 592, 1219.

Alfano, R.R. and S.L. Shapiro (1971) Picosecond spectroscopy using the inverse Raman effect. Chem. Phys. Lett. **8**, 631.

Alfano, R.R., Q. Li, T. Jimbo, J.T. Manassah, and P.P. Ho (1986) Induced spectral broadening of a weak picosecond pulse in glass produced by an intense picosecond pulse. Opt. Lett. **11**, 626.

Alfano, R.R., Q. Wang, T. Jimbo, and P.P. Ho (1987) Induced spectral broadening about a second harmonic generated by an intense primary ultrashort laser pulse in ZnSe. Phys. Rev. A **35**, 459.

Bloembergen N. and P. Lallemand (1966) Complex intensity-dependent index of refraction, frequency broadening of stimulated Raman lines, and stimulated Rayleigh scattering. Phys. Rev. Lett. **16**, 81.

Brewer, R.G. (1967) Frequency shifts in self-focusing light. Phys. Rev. Lett. **19**, 8.

Busch, G.E., R.P. Jones, and P.M. Rentzepis (1973) Picosecond spectroscopy using a picosecond continuum. Chem. Phys. Lett. **18**, 178.

Cheung, A.C., D.M. Rank, R.Y. Chiao, and C.H. Townes (1968) Phase modulation of Q-switched laser beams in small-scale filaments. Phys. Rev. Lett. **20**, 786.

Corkum, P.B., P.P. Ho, R.R. Alfano, and J.T. Manassah (1985) Generation of infrared supercontinuum covering 3–14 μm in dielectrics and semiconductors. Opt. Lett. **10**, 624.

Corkum, P.B., C. Rolland, and T. Rao (1986) Supercontinuum generation in gases. Phys. Rev. Lett. **57**, 2268.

Cubbedu, R. and F. Zagara (1971) Nonlinear refractive index of CS_2 in small-scale filaments. Opt. Commun. **3**, 310.

Cubbedu, R., R. Polloni, C.A. Sacchi, O. Svelto, and F. Zagara (1971) Study of small-scale filaments of light in CS_2 under picosecond excitation. Phys. Rev. Lett. **26**, 1009.

DeMartini, F., C.H. Townes, T.K. Gustafson, and P.L. Kelley (1967) Self-steepening of light pulses. Phys. Rev. **164**, 312.

Durbin, S.D., S.M. Arakelian, and Y.R. Shen (1981) Laser-induced diffraction rings from a nematic liquid crystal film. Opt. Lett. **6**, 411.

Fabellinski, I.L. (1967) *Molecular Scattering of Light*. Plenum, New York, Chapter VIII.

Fisher, R.A. and W. Bischel (1975) Numerical studies of the interplay between self-phase modulation and dispersion for intense plane-wave light pulses. J. App. Phys. **46**, 4921.

Fisher, R.A., B. Suydam, and D. Yevich (1983) Optical phase conjugation for time domain undoing of dispersive self-phase modulation effects. Opt. Lett. **8**, 611.

Fork, R.L., C.V. Shank, C. Hirliman, and R. Yen (1983) Femtosecond white-light continuum pulses. Opt. Lett. **8**, 1.

Fork, R.L., C.H. Brito Cruz, P.C. Becker, and C.V. Shank (1987). Compression of optical pulses to six femtoseconds by using cubic phase compensation. Opt. Lett. **12**, 483.

Glownia, J., G. Arjavalingam, P. Sorokin, and J. Rothenberg (1986) Amplification of 350-fs pulses in XeCl excimer gain modules. Opt. Lett. **11**, 79.

Gustafson, T.K., J.P. Taran, H.A. Haus, J.R. Lifsitz, and P.L. Kelley (1969) Self-phase modulation, self steepening, and spectral development of light in small-scale trapped filaments. Phys. Rev. **177**, 306.

Hellwarth, R.W. (1970) Theory of molecular light scattering spectra using the linear-dipole approximation. J. Chem. Phys. **52**, 2128.

Hellwarth, R.W. (1977) Third-order optical susceptibilities of liquids and solids. Prog. Quantum Electron. **5**, 1.

Ho, P.P., Q.X. Li, T. Jimbo, Y.L. Ku, and R.R. Alfano (1987) Supercontinuum pulse generation and propagation in a liquid CCl_4. Appl. Opt. **26**, 2700.

Ippen, E.P. and C.V. Shank (1975) Dynamic spectroscopy and subpicosecond pulse compression. Appl. Phys. Lett. **27**, 488.

Jones, W.J. and B.P. Stoicheff (1964) Induced absorption at optical frequencies. Phys. Rev. Lett. **13**, 657.

Lallemand, P. (1966) Temperature variation of the width of stimulated Raman lines in liquids. Appl. Phys. Lett. **8**, 276.

Loy, M.M.T. and Y.R. Shen (1973) Study of self-focusing and small-scale filaments of light in nonlinear media. IEEE J. Quantum Electron. **QE-9**, 409.

Manassah, J.T., R.R. Shapiro, and M. Mustafa (1985) Spectral distribution of an ultrashort supercontinuum laser source. Phys. Lett. **A107**, 305.

Manassah, J.T., M.A. Mustafa, R.R. Alfano, and P.P. Ho (1986) Spectral extent and pulse shape of the supercontinuum for ultrashort laser pulse. IEEE J. Quantum Electron. **QE-22**, 197.

Marburger, J.H. (1975) Self-focusing: theory. Prog. Quantum Electron. **4**, 35.

Nakatsuka, H. and D. Grischkowsky (1981) Recompression of optical pulses broadened by passage through optical fibers. Opt. Lett. **6**, 13.

Nakatsuka, H., D. Grischkowsky, and A.C. Balant (1981) Nonlinear picosecond pulse propagation through optical fibers with positive group velocity dispersion. Phys. Rev. Lett. **47**, 910.

Nikolaus, B. and D. Grischkowsky (1983a) 12 × pulse compression using optical fibers. Appl. Phys. Lett. **42**, 1.

Nikolaus, B. and D. Grischkowsky (1983b) 90-fs tunable optical pulses obtained by two-state pulse compression. Appl. Phys. Lett. **43**, 228.

Penzkofer, A. (1974) Parametrically generated spectra and optical breakdown in H_2O and NaCl. Opt. Commun. **11**, 265.

Penzkofer, A., A. Laubereau, and W. Kaiser (1973) Stimulated short-wave radiation due to single-frequency resonances of $\chi^{(3)}$. Phys. Rev. Lett. **31**, 863.

Penzkofer, A., A. Seilmeier, and W. Kaiser (1975) Parametric four-photon generation of picosecond light at high conversion efficiency. Opt. Commun. **14**, 363.

Santamato, E. and Y.R. Shen (1984) Field curvature effect on the diffraction ring pattern of a laser beam dressed by spatial self-phase modulation in a nematic film. Opt. Lett. **9**, 564.

Shen, Y.R. (1966) Electrostriction, optical Kerr effect, and self-focusing of laser beams. Phys. Lett., **20**, 378.
Shen Y.R. (1975) Self-focusing: experimental. Prog. Quantum Electron. **4**, 1.
Shen, Y.R. (1984) *The Principles of Nonlinear Optics.* Wiley, New York, Chapters 1 and 16.
Shen, Y.R. and M.M.T. Loy (1971) Theoretical investigation of small-scale filaments of light originating from moving focal spots. Phys. Rev. A **3**, 2099.
Shimizu, F. (1967) Frequency broadening in liquids by a short light pulse. Phys. Rev. Lett. **19**, 1097.
Shimizu, F. and E. Courtens (1973) Recent results on self-focusing and trapping. In *Fundamental and Applied Laser Physics*, M.S. Feld, A. Javan, and N.A. Kurnit, eds. Wiley, New York, p. 67.
Stoicheff, B.P. (1963) Characteristics of stimulated Raman radiation generated by coherent light. Phys. Lett. **7**, 186.
Stolen, R. and C. Lin (1978) Self-phase modulation in silica optical fibers. Phys. Rev. A **17**, 1448.
Topp, M.R. and P.M. Rentzepis (1971) Time-resolved absorption spectroscopy in the 10^{-12}-sec range. J. Appl. Phys. **42**, 3415.
Treacy, E.P. (1968) Compression of picosecond light pulses. Phys. Lett. **28A**, 34.
Treacy, E.P. (1969a) Measurements of picosecond pulse substructure using compression techniques. Appl. Phys. Lett. **14**, 112.
Treacy, E.P. (1969b) Optical pulse compression with diffraction gratings. IEEE J. Quantum Electron. **5**, 454.
Wong, G.K.L. and Y.R. Shen (1972) Study of spectral broadening in a filament of light. Appl. Phys. Lett. **21**, 163.
Yang, G. and Y.R. Shen (1984) Spectral broadening of ultrashort pulses in a nonlinear medium. Opt. Lett. **9**, 510.

2 Supercontinuum Generation in Condensed Matter

Q.Z. WANG, P.P. HO, and R.R. ALFANO

1. Introduction

Supercontinuum generation, the production of intense ultrafast broadband "white light" pulses, arises from the propagation of intense picosecond or shorter laser pulses through condensed or gaseous media. Various processes are responsible for continuum generation. These are called self-, induced-, and cross-phase modulations and four-photon parametric generation. Whenever an intense laser pulse propagates through a medium, it changes the refractive index, which in turn changes the phase, amplitude, and frequency of the incident laser pulse. A phase change can cause a frequency sweep within the pulse envelope. This process has been called *self-phase modulation* (*SPM*) (Alfano and Shapiro, 1970a). Nondegenerate *four-photon parametric generation* (*FPPG*) usually occurs simultaneously with the SPM process (Alfano and Shapiro, 1970a). Photons at the laser frequency parametrically generate photons to be emitted at Stokes and anti-Stokes frequencies in an angular pattern due to the required phase-matching condition. When a coherent vibrational mode is excited by a laser, stimulated Raman scattering (SRS) occurs. SRS is an important process that competes and couples with SPM. The interference between SRS and SPM causes a change in the emission spectrum resulting in *stimulated Raman scattering cross-phase modulation* (*SRS-XPM*) (Gersten et al., 1980). A process similar to SRS-XPM occurs when an intense laser pulse propagates through a medium possessing a large second-order χ^2 and third-order χ^3 susceptibility. Both second harmonic generation (SHG) and SPM occur and can be coupled together. The interference between SHG and SPM alters the emission spectrum and is called *second harmonic generation cross-phase modulation* (*SHG-XPM*) (Alfano et al., 1987). A process closely related to XPM, called *induced phase modulation* (*IPM*) (Alfano, 1986), occurs when a weak pulse at a different frequency propagates through a disrupted medium whose index of refraction is changed by an intense laser pulse. The phase of the weak optical field can be modulated by the time variation of the index of refraction originating from the primary intense pulse.

TABLE 2.1. Brief history of experimental continuum generation.

Investigator	Year	Material	Laser wavelength/ pulsewidth	Spectrum	Frequency (cm^{-1})	Process
Alfano, Shapiro	1968–1973	Liquids and solids	530 nm/ 8 ps or 1060 nm/ 8 ps	Visible and near IR	6,000	SPM
Stolen et al.	1974–1976	Fibers	530 nm/ns	Visible	500	SPM
Shank, Fork et al.	1983	Glycerol	620 nm/100 fs	UV, visible, near IR	10,000	SPM
Corkum, Ho, Alfano	1985	Semiconductors dielectrics	10 μm/6 ps	IR	1,000	SPM
Corkum, Sorokin	1986	Gases	600 nm/2 ps 300 nm/0.5 ps	Visible and UV	5,000	SPM
Alfano, Ho, Manassah, Jimbo	1986	Glass	1060 nm/ 530 nm/8 ps	Visible	1,000	IPM (XPM)
Alfano, Ho, Wang, Jimbo	1986	ZnSe	1060 nm/8 ps	Visible	1,000	SHG-XPM (ISB)
Alfano, Ho, Baldeck	1987	Fibers	530 nm/30 ps	Visible	1,000	SRS-XPM

The first study of the generation and mechanisms of the ultrafast supercontinuum dates back to the years 1968 to 1972, when Alfano and Shapiro first observed the "white" picosecond continuum in liquids and solids (Alfano and Shapiro, 1970a). Spectra extending over $\sim$6000 cm^{-1} in the visible and infrared wavelength region were observed. Over the years, improvements in the generation of ultrashort pulses from mode-locked lasers led to the production of wider supercontinua in the visible, ultraviolet, and infrared wavelength regions using various materials. Table 2.1 highlights the major accomplishments in this field over the past 20 years.

In this chapter we focus on the picosecond supercontinuum generation in liquids, solids, and crystals. Supercontinuum generation in gases, XPM, and IPM are discussed by Corkum and Rolland (Chapter 7), Glownia et al. (Chapter 8), Baldeck et al. (Chapter 4), Agrawal (Chapter 3), and Manassah (Chapter 5), respectively.

2. Simplified Model

Before we go further, let us first examine the nonlinear wave equation to describe the self-phase modulation mechanism. A thorough theoretical study of supercontinuum generation has been dealt with in Chapters 1, 3, and 5.

The optical electromagnetic field of a supercontinuum pulse satisfies Maxwell's equations:

$$\begin{aligned} \nabla \times \mathbf{E} &= -\frac{1}{c}\frac{\partial \mathbf{B}}{\partial t}, \\ \nabla \times \mathbf{H} &= \frac{1}{c}\frac{\partial \mathbf{D}}{\partial t} + \frac{4\pi}{c}\mathbf{J}, \\ \nabla \cdot \mathbf{D} &= 4\pi\rho, \\ \nabla \cdot \mathbf{B} &= 0. \end{aligned} \tag{1}$$

Equations (1) can be reduced to (see Appendix)

$$\frac{\partial A}{\partial z} + \frac{1}{v_g}\frac{\partial A}{\partial t} = i\frac{\omega_0 n_2}{2c}|A|^2 A, \tag{2}$$

where $A(z, t)$ is the complex envelope of the electric field and $v_g = 1\bigg/\left(\frac{\partial k}{\partial \omega}\right)_{\omega_0}$ is the group velocity. The total refractive index n is defined by $n^2 = n_0^2 + 2n_0 n_2 |A(t)|^2$, where n_2 is the key parameter called the nonlinear refractive index. This coefficient is responsible for a host of nonlinear effects: self- and cross-phase modulation, self-focusing, and the optical Kerr effect, to name the important effects. Equation (2) was derived using the following approximations: (1) linearly polarized electric field, (2) homogeneous radial fields, (3) slowly varying envelope, (4) isotropic and nonmagnetic medium, (5) negligible Raman effect, (6) frequency-independent nonlinear susceptibility $\chi^{(3)}$, and (7) neglect of group velocity dispersion, absorption, self-steepening, and self-frequency shift.

Denoting by a and α the amplitude and phase of the electric field envelope $A = ae^{i\alpha}$, Eq. (2) reduces to

$$\frac{\partial a}{\partial z} + \frac{1}{v_g}\frac{\partial a}{\partial t} = 0 \tag{3a}$$

and

$$\frac{\partial \alpha}{\partial z} + \frac{1}{v_g}\frac{\partial \alpha}{\partial t} = \frac{\omega_0 n_2}{2c}a^2. \tag{3b}$$

The analytical solutions for the amplitude and phase are

$$a(\tau) = a_0 F(\tau) \tag{4a}$$

and

$$\alpha(z, \tau) = \frac{\omega_0 n_2}{2c}\int_0^z a^2\, dz' = \frac{\omega_0 n_2}{2c}a_0^2 F^2(\tau) z, \tag{4b}$$

where a_0 is the amplitude, $F(\tau)$ the pulse envelope, and τ the local time $\tau = t - z/v_g$. For materials whose response time is slower than pure electronic but faster than molecular orientation (i.e., coupled electronic, molecular redistribution, libratory motion) the envelope is just the optical pulse shape. For a "pure" electronic response, the envelope should also include the optical cycles in the pulse shape.

The electric field envelope solution of Eq. (2) is given by

$$A(z,\tau) = a(\tau)\exp\left[i\frac{\omega_0 n_2}{2c}a_0^2 F^2(\tau)z\right]. \tag{5}$$

The main physics behind the supercontinuum generation by self-phase modulation is contained in Eq. (5) and is displayed in Figure 2.1. As shown in Figure 2.1a, the index change becomes time dependent and, therefore, the phase of a pulse propagating in a distorted medium becomes time dependent, resulting in self-phase modulation. The electric field frequency is continuously shifted (Figure 2.1c) in time. This process is most important in the generation of femtosecond pulses (see Chapter 10 by Johnson and Shank).

Since the pulse duration is much larger than the optical period $2\pi/\omega_0$ (slowly varying approximation), the electric field at each position τ within the pulse has a specific *local* and *instantaneous frequency* at given time that is

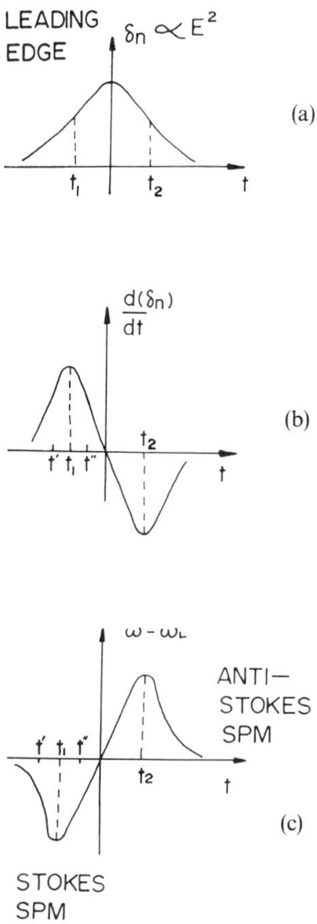

FIGURE 2.1. A simple mechanism for SPM for a non-linear index following the envelope of a symmetrical laser pulse: (a) time-dependent nonlinear index change; (b) time rate of change of index change; (c) time distribution of SPM-shifted frequencies $\omega(t) - \omega_0$.

given by

$$\omega(\tau) = \omega_0 + \delta\omega(\tau), \tag{6a}$$

where

$$\delta\omega(\tau) = -\frac{\partial\alpha}{\partial\tau} = -\frac{\omega_0}{2c} n_2 a_0^2 z \frac{\partial F^2(\tau)}{\partial\tau}. \tag{6b}$$

The $\delta\omega(\tau)$ is the frequency shift generated at a particular time location τ within the pulse shape. This frequency shift is proportional to the derivative of the pulse envelope, which corresponds to the generation of new frequencies resulting in wider spectra.

Pulses shorter than the excitation pulse can be produced at given frequencies. It was suggested by Y.R. Shen many years ago that Alfano and Shapiro in 1970 most likely produced femtosecond pulses via supercontinuum generation. Figure 2.1c shows the frequency distribution within the pulse shape. The leading edge, the pulse peak, and the trailing edge are red shifted, nonshifted, and blue shifted, respectively.

The spectrum of SPM pulses is obtained by taking the Fourier transform of the complex temporal envelope $A(z, \tau)$:

$$A(\Omega) = \frac{1}{2\pi} \int_{-\infty}^{\infty} A(z, \tau) \exp[i\Omega\tau]\, d\tau, \tag{7}$$

where $\Omega = \omega - \omega_0$. The intensity spectrum is given by

$$S(\Omega z) = \frac{c}{4\pi} |A(\Omega, z)|^2. \tag{8}$$

In practical cases, the phase of $A(z, \tau)$ is large compare with π, and the stationary phase method leads to

$$\Delta\omega(z)_{\max} = \frac{\omega_0}{2c} n_2 a_0^2 \left[\left.\frac{\partial F^2}{\partial\tau}\right|_{\tau_1} - \left.\frac{\partial F^2}{\partial\tau}\right|_{\tau_2} \right] z. \tag{9}$$

The intensity

$$\begin{aligned} S(\Omega, z) = \left(\frac{c}{4\pi}\right)\left(\frac{4\pi c}{\omega_0 n_2 z}\right) \Bigg\{ & F^2(\tau') \Big/ \left.\frac{\partial F^2}{\partial\tau^2}\right|_{\tau'} + F^2(\tau'') \Big/ \left.\frac{\partial^2 F^2}{\partial\tau^2}\right|_{\tau''} \\ & + 2 \frac{F(\tau')F(\tau'')}{[\partial^2 F^2/\partial\tau^2|_{\tau'} \partial^2 F^2/\partial\tau^2|_{\tau''}]^{1/2}} \\ & \times \cos\left[\Omega(\tau' - \tau'') + \frac{\omega_0}{2c} n_2 a_0^2 z (F^2(\tau') - F^2(\tau'')) \right] \Bigg\}, \end{aligned} \tag{10}$$

where $\Delta\omega_{\max}$ is the maximum frequency spread, τ_1 and τ_2 are the pulse envelope inflection points, and τ' and τ'' are the points of the pulse shape that have the same frequency.

An estimate of the modulation frequency $\delta\omega_M$ can be made by calculating the maximum number of interference minima and dividing this number into the maximum frequency broadening. A straightforward calculation leads to

$$\delta\omega_M \approx 2\pi \frac{\partial F^2}{\partial \tau}\bigg|_{\tau_1} \bigg/ F^2(\tau_1) \approx 2\pi \frac{\partial F^2}{\partial \tau}\bigg|_{\tau_2} \bigg/ F^2(\tau_2). \tag{11}$$

For a Gaussian laser pulse given by

$$F(\tau) = \exp[-\tau^2/2\tau_0^2], \tag{12}$$

the modulation frequency of the SPM spectrum is (Alfano, 1972)

$$\delta\omega_M = \frac{4\pi}{\tau_0} \quad \text{or} \quad \delta\bar{\nu}_M = \frac{2}{\tau_0 c}. \tag{13}$$

Using this relation, the average modulation period of 13 cm^{-1} corresponds to an initial pulse duration of 5 ps emitted from mode-locked Nd:glass laser. The maximum frequency extent in this case is (Alfano, 1972)

$$\Delta\omega_{\max} \sim \frac{\omega_0 n_2 a_0^2 z}{c\tau_0}. \tag{14}$$

The maximum frequency shift (Eq. (14)) indicates the following salient points:

- The frequency extent is inversely proportional to the pumping pulse duration. The shorter the incoming pulse, the greater the frequency extent. The first white light band supercontinuum pulses were generated using picosecond laser pulses (Alfano and Shapiro, 1970a, b).
- The spectral broadening is proportional to n_2. The supercontinuum generation can be enhanced by increasing the nonlinear refractive index. This is discussed in detail in Section 6.
- The spectral broadening is linearly proportional to amplitude a_0^2. Therefore, multiple-excitation laser beams of different wavelengths may be used to increase the supercontinuum generation. This leads to the basic principle behind IPM and XPM. These processes are described by Baldeck et al. (Chapter 4) and Manassah (Chapter 5).
- The spectral broadening is proportional to ω_0 and z.

The chirp—the temporal distribution of frequency in the pulse shape—is an important characteristic of SPM broadened pulse. In the linear chirp approximation, the chirp coefficient C is usually defined by the phase relation

$$\alpha = C\tau^2. \tag{15}$$

For a Gaussian electric field envelope and linear approximation, the envelope reduces to

$$F^2(\tau) = \exp[-\tau^2/\tau_0^2] \approx 1 - \tau^2/\tau_0^2. \tag{16}$$

The linear chirp coefficient derived from Eqs. (5) and (16) becomes

$$C = \left(\frac{\omega_0}{2c}\right)\left(\frac{n_2 a_0^2 z}{\tau_0^2}\right). \tag{17}$$

Typical calculated SPM spectra are displayed in Figure 2.2. The spectral

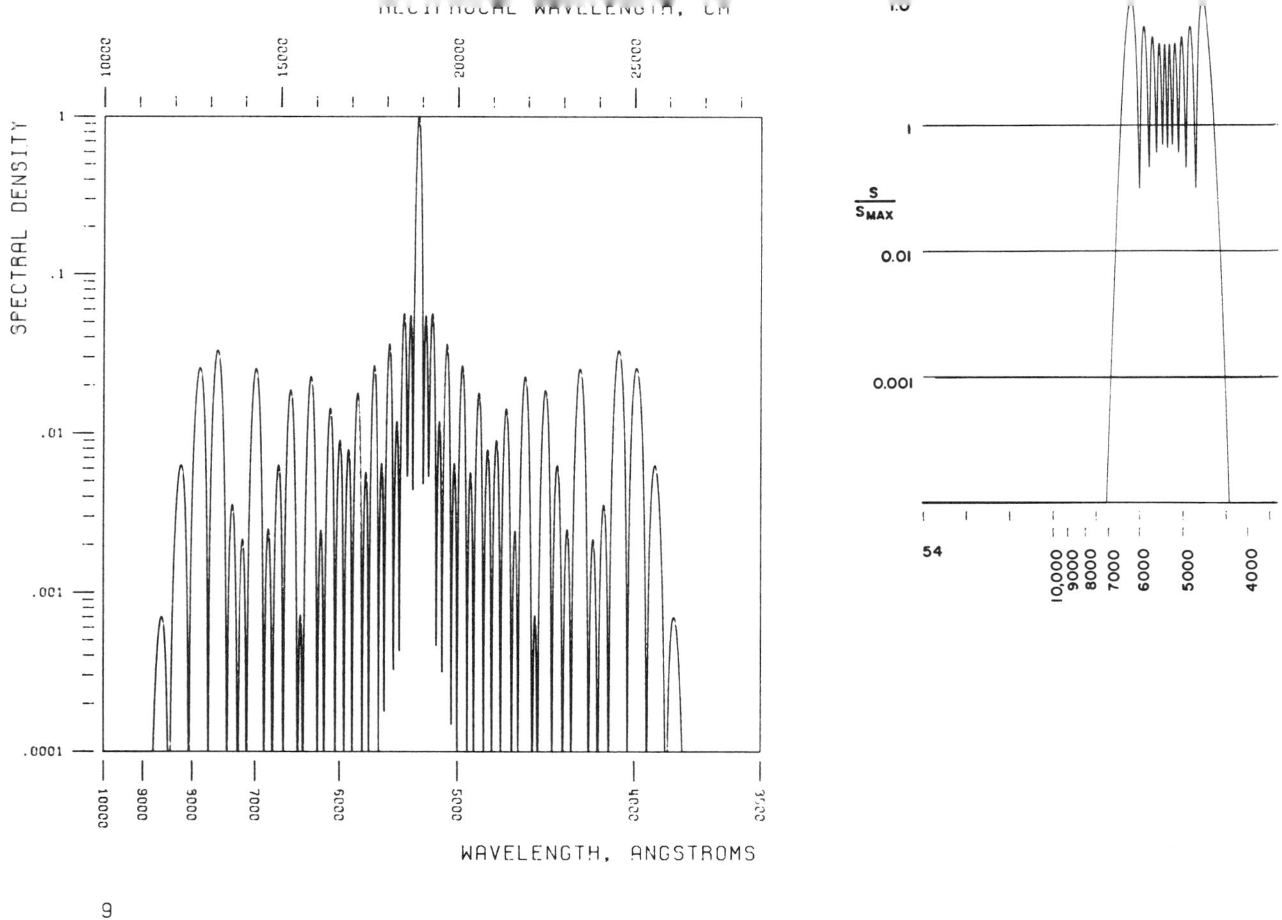

FIGURE 2.2. Calculated SPM spectrum: (a) for response time slower than pure electronic but faster than molecular orientation: $\beta = 30$ and $\tau = 0.1$ ps; (b) for pure electronic response: $\beta = 30$ and $\tau = 0.1$ ps. (From Alfano, 1972.)

densities of the SPM light are normalized and β is defined as $\beta = (n_2 a_0^2 \omega_0 z)/2c$, which measures the strength of the broadening process. Figure 2.2a shows the spectrum for a material response time slower than pure electronic but faster than molecular orientation for $\beta = 30$ and $\tau = 0.1$ ps. The extent of the spectrum is about 7000 cm^{-1}. Figure 2.2b shows the SPM spectrum for a quasi-pure electronic response for $\beta = 30$ and $\tau = 0.1$ ps. Typical SPM spectral characteristics are apparent in these spectra.

3. Experimental Arrangement for SPM Generation

To produce the supercontinuum, an ultrafast laser pulse is essential with a pulse duration in the picosecond and femtosecond time region. A mode-locked laser is used to generate picosecond and femtosecond light pulses. Table 2.2 lists the available mode-locked lasers that can produce picosecond and femtosecond laser pulses. Measurements performed in the 1970s used a mode-locked Nd: glass laser with output at 1.06 μm with power of $\sim 5 \times 10^9$ W and the second harmonic (SHG) at 530 nm with power of 2×10^8 W. Typically, one needs at least a few microjoulis of 100-fs pulse passing through a 1-mm sample to produce continuum.

A typical experimental setup for ultrafast supercontinuum generation is shown in Figure 2.3. Both spectral and spatial distributions are measured. The 8-ps SHG pulse of 5 mJ is reduced in size to a collimated 1.2-mm-diameter beam across the sample by an inverted telescope. For weaker excitation pulses, the beam is focused into the sample using a 10- to 25- cm focal lens. The typical sample length used is 10 to 15 cm for picosecond pulses and 0.1 to 1 cm for 100-f pulses. The intensity distribution of the light at the exit face of the sample was magnified 10 times and imaged on the slit of a spectrograph. The spectrum of each individual filament within the slit was displayed. Usually there were 5 to 20 filaments. A thin quartz-wedge beam splitter was used to photograph

TABLE 2.2. Available ultrafast mode-locked lasers.

Oscillator	Wavelength (nm)	Pulse duration
Ruby	694.3	30 ps
YAG	1064	30 ps
Silicate glass	1060	8 ps
Phosphate glass	1054	6 ps
Dye	Tunable (SYNC or flash lamp)	5–10 ps
Dye + CO_2 + semiconductor switches	9300	1–10 ps
Dye (CPM)	610–630	100 fs
Dye + pulse compression (SYNC)	Tunable	300 fs
Dye + CPM (prisms in cavity)	620	27–60 fs
Dye + SPM + pulse compression (prisms and grating pairs)	620	6–10 fs

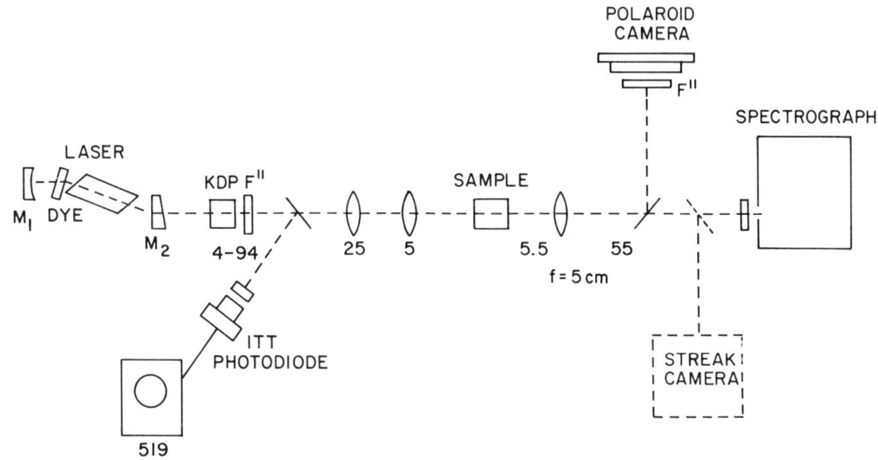

FIGURE 2.3. Experimental arrangement for generating and observing supercontinuum and self-focusing. (From Alfano, 1972.)

filament formation of the Stokes (anti-Stokes) side of the spectra; three type 3-68 and three type 3-67 (two type 5-60) Corning filters were used to prevent the 530-nm direct laser light from entering the spectrograph. To reduce nonfilament light, a wire 2 mm in diameter was sometimes placed at the focal point of the imaging lens. Previously, spectra were taken on Polaroid type 57 film. At present, video systems such as an Silicon-intensified target (SIT) camera together with a PC computer are commonly used to display the spectra. Today, to obtain temporal information about the supercontinuum, a streak camera is added to the experimental system.

4. Generation of Supercontinuum in Solids

In the following sections, we review the experimental measurements of supercontinuum generated in condensed matter. Topics discussed include supercontinuum generation in various kinds of solids and liquids, optical glass fibers, liquid argon, liquid and solid krypton, magnetic crystals, and dielectric crystals.

The mechanisms behind SPM are discussed in Chapter 1 by Shen and Yang. In general, various mechanisms are responsible for SPM in condensed matter and give rise to the coefficient of the intensity-dependent refractive index n_2. These are the orientational Kerr effect, electrostriction, molecular redistribution, librations, and electronic distribution. In suitably chosen media (central-symmetric molecules) these frequency-broadening mechanisms may be distinguished from the electronic mechanism through their different time responses

(Lallemand, 1966). The relation times for these mechanisms are given approximately by (Brewer and Lee, 1968)

$$\tau(\text{orientation}) = \frac{4\pi}{3}\eta a^3/kT > 10^{-12}\ \text{s}, \tag{18}$$

$$\tau(\text{molecular types}) = \frac{\langle x^2 \rangle}{D} = \frac{6\pi\eta a\langle x^2 \rangle}{kT} > 10^{-12}\ \text{s}, \tag{19}$$

$$\tau(\text{libration about field}) = \sqrt{\frac{2I}{\alpha E_0^2}} > 10^{-12}\ \text{s}, \tag{20}$$

$$\tau(\text{electronic}) = \frac{2\pi a_0 \hbar}{e^2} > 1.5 \times 10^{-16}\ \text{s}, \tag{21}$$

where η is the viscosity ($\eta = 0.4$ cp for liquids and $\eta = 10^6$ cp for glasses); a is the molecular radius; D is the diffusion coefficient ($\geq 10^{-5}$ cm/s for liquids) and x is the diffusion distance of the clustering, $\sim 10^{-8}$ cm; I is the moment of inertia, $I_{\text{argon}} = 9.3 \times 10^{-38}$ esu and $I_{\text{CCl}_4} = 1.75 \times 10^{-38}$ esu; α is the polarizability, $\alpha_{\text{argon}} = 1.6 \times 10^{-24}$ esu and $\alpha_{\text{CCl}_4} = 1.026 \times 10^{-24}$ esu; and E_0 is the amplitude of the electric field, taken as 10^5 esu, which is close to the atomic field. The response time for an electron distortion is about the period of a Bohr orbit, $\sim 1.5 \times 10^{-16}$ s. Thus, typical calculated relaxation time responses for diffusional motions are $> 10^{-12}$ s, while the electronic distortion response time is ~ 150 as.

With picosecond light pulses Brewer and Lee (1968) showed that the dominant mechanism for filament formation should be electronic in very viscous liquids. Molecular rocking has been suggested as the cause of the broadening and self-focusing in CS_2. The molecules are driven by the laser field to rock about the equilibrium position of a potential well that has been set up by the neighboring molecules. This mechanism is characteristized by a relaxation time:

$$\tau_1 = \frac{\eta}{G} = 2.3 \times 10^{-13}\ \text{s}, \tag{22}$$

where G is the shear modulus $\sim 1.5 \times 10^{10}$ dynes/cm and viscosity $\eta = 3.7 \times 10^{-3}$ p for CS_2.

In solids, mechanisms giving rise to the coefficient of the intensity-dependent refractive index n_2 for picosecond pulse excitation are either direct distortion of electronic clouds around nuclei or one of several coupled electronic mechanisms: librational distortion, where electronic structure is distorted as the molecule rocks; electron-lattice distortion, where the electron cloud distorts as the lattice vibrates; and molecular distortion, where electronic shells are altered as the nuclei redistribute spatially. The electrostriction mechanism is rejected because it exhibits a negligible effect for picosecond and femtosecond pulses.

Typical supercontinuum spectra generated in solids and liquids using 8-ps pulses at 530 nm are displayed in Figure 2.4. All continuum spectra are similar despite the different materials.

4.1 Supercontinuum in Glasses

Spectra from the glass samples show modulation (see Figure 2.4a). The spectral modulation ranged from as small as a few wave numbers to hundreds of wave numbers. The filament size was approximately 5 to 50 μm. Typically, 5 to 20 small-scale filaments were observed. Occasionally, some laser output pulses from the samples did not show modulation or had no regular modulation pattern. Typical Stokes sweeps from these filaments were 1100 cm^{-1} in extra-dense flint glass of length 7.55 cm and 4200 cm^{-1} in both borosilicate crown (BK-7) and light barium (LBC-1) glass of length 8.9 cm. Sweeps on the anti-Stokes side were typically 7400 cm^{-1} in BK and LBC glasses. The sweep is polarized in the direction of the incident laser polarization for unstrained glasses.

4.2 Supercontinuum in Quartz

SPM spectra from quartz using an 8-ps pulse at 530 nm are similar to the spectra from glasses displayed in Figure 2.4a. Typical Stokes sweeps from the filaments were 3900 cm^{-1} in a quartz crystal of length 4.5 cm, and the anti-Stokes sweeps were 5500 cm^{-1}.

4.3 Supercontinuum in NaCl

Sweeps of 3900 cm^{-1} in NaCl of length 4.7 cm to the red side of 530 nm were observed. Sweeps on the anti-Stokes side were about 7300 cm^{-1}. Some of the spectra show modulation with ranges from a few wave numbers to hundreds of wave numbers. Some laser shots showed no modulation or no regular modulation pattern. For unstrained NaCl, the supercontinuum light is polarized in the direction of the incident laser polarization.

4.4 Supercontinuum in Calcite

Sweeps of 4400 cm^{-1} and 6100 cm^{-1} to the Stokes and anti-Stokes sides of 530 nm were observed in a calcite crystal of length 4.5 cm (see Figure 2.4b). Some spectra showed modulation structure within the broadened spectra; some showed no modulation or no regular modulation pattern. The exit supercontinuum light has same polarization as the incident laser. The SRS threshold is lower for laser light traveling as an O-wave than an E-wave. SPM dominates the E-wave spectra.

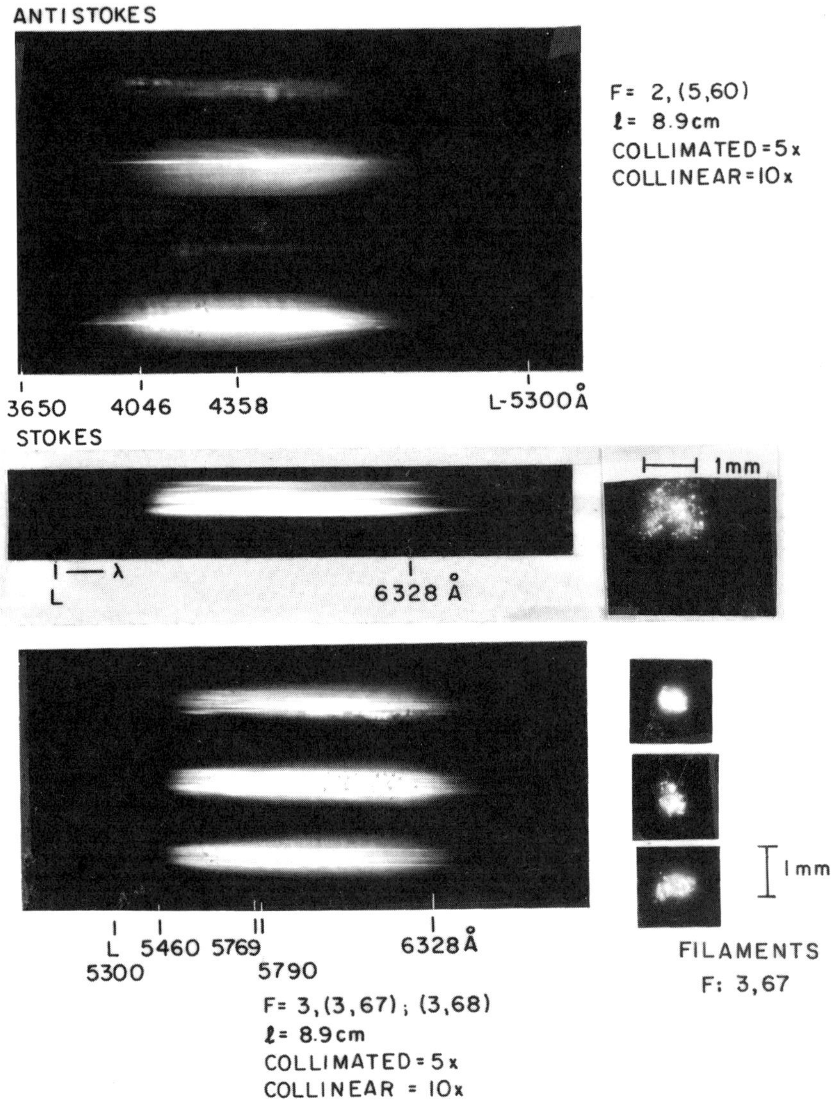

(a)

FIGURE 2.4. Supercontinuum spectra from various kind of solids and liquids. (a) Stokes and anti-Stokes SPM from BK-7 glass and filament formation for different laser shots. The filaments are viewed through Corning 3-67 filters. (b) Stokes and anti-Stokes SPM from calcite for different laser shots. The laser beam propagates as an O-wave through the sample. (c) Stokes and anti-Stokes SPM spectra from calcite for different laser shots. The laser beam propagates as an E-wave. (From Alfano, 1972.)

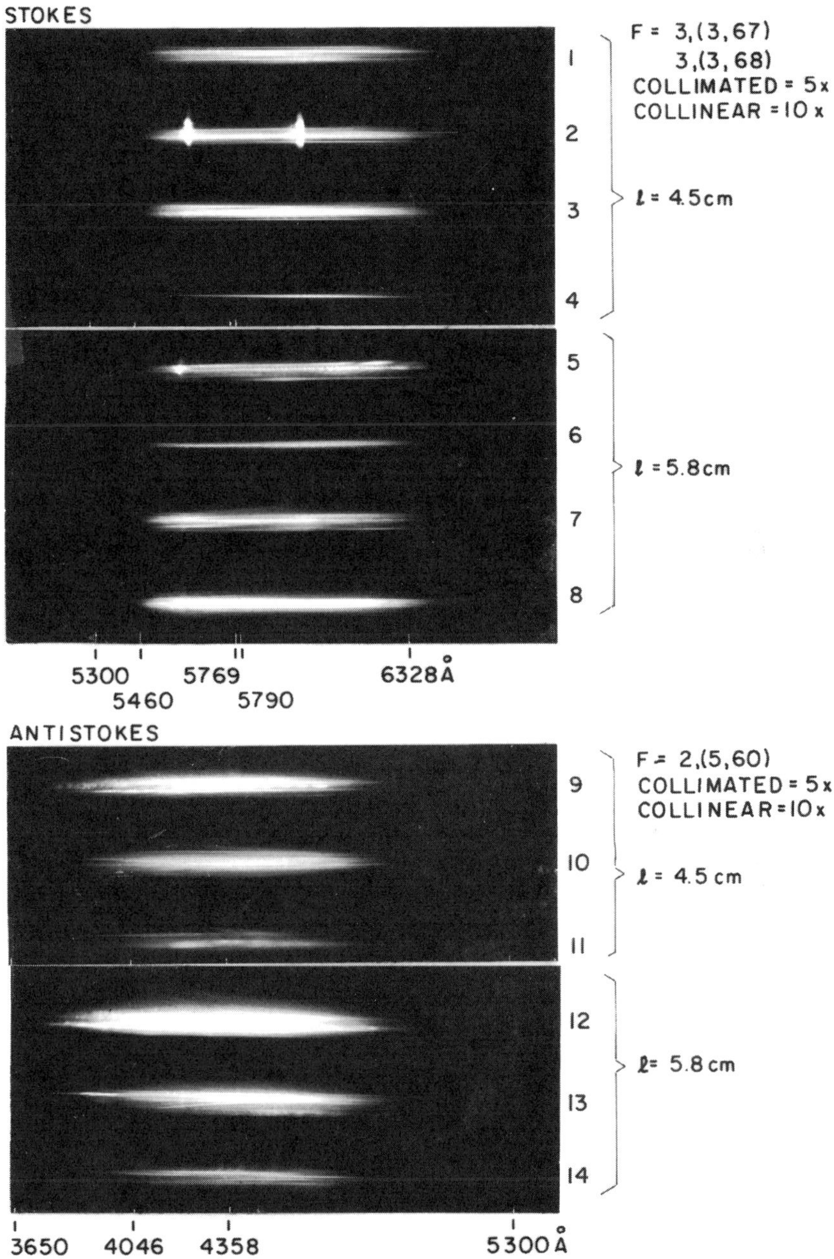

(b)

FIGURE 2.4 (*continued*)

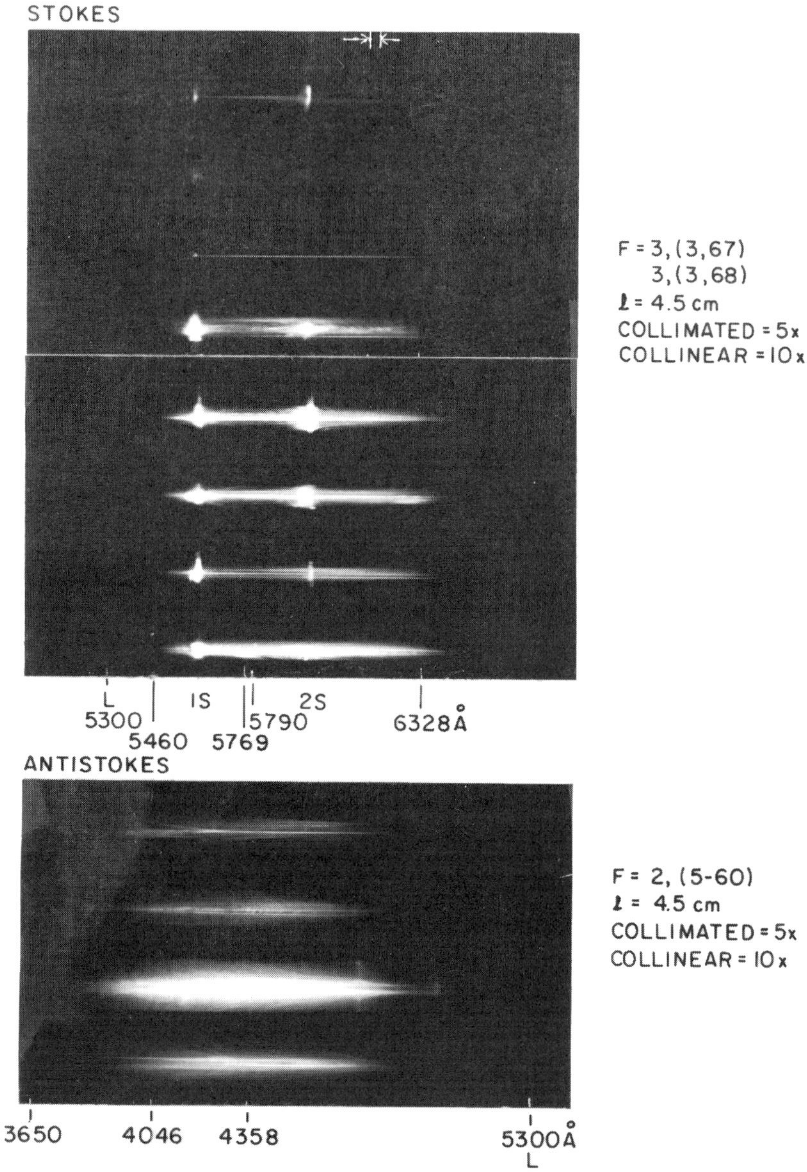

(c)

FIGURE 2.4 (*continued*)

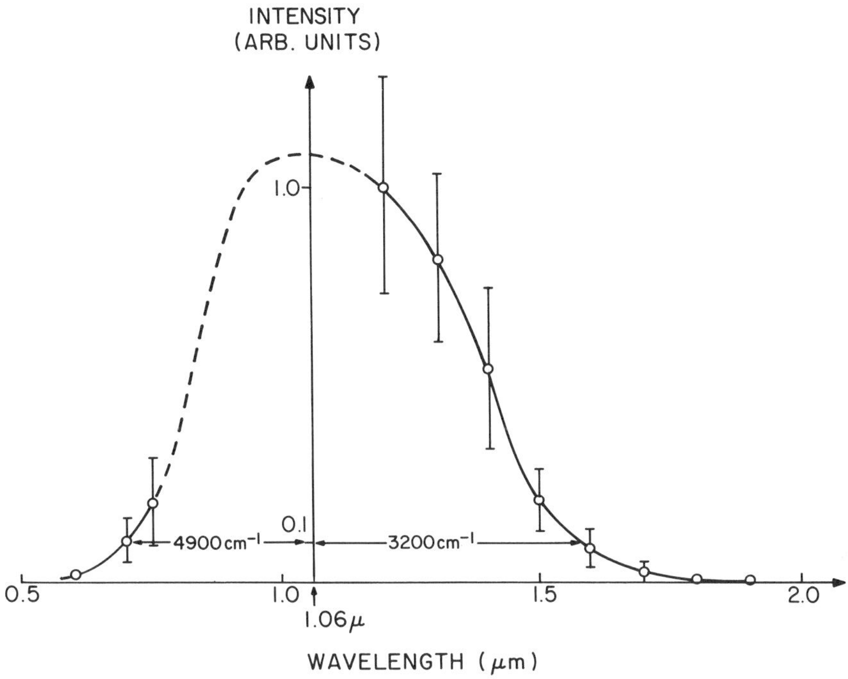

FIGURE 2.5. Relative emission intensity versus emission wavelength for KBr. Exciting wavelength = 1.06 μm. (From Yu et al., 1975.)

4.5 Supercontinuum in KBr

A high-power broadband coherent source in the near- and medium-infrared region can be realized by passing an intense 1.06-μm picosecond pulse through a KBr crystal. Figure 2.5 shows the spectra from a 10-cm-long KBr crystal with excitation of a 9-ps, 10^{11} W/cm^2 pulse at 1.06 μm. On the Stokes side the maximum intensity occurs at 1.2 μm. When the signal drops to 10^{-1} the span of the spectral broadening is $\Delta v_s = 3200$ cm^{-1} on the Stokes side and $\Delta v_a = 4900$ cm^{-1} on the anti-Stokes side. Beyond 1.6 μm the signal level falls off rapidly. At 1.8 μm the signal is 10^{-2} and at 2 μm no detectable signal can be observed (Yu et al., 1975).

4.6 Supercontinuum in Semiconductors

Infrared supercontinuum spanning the range 3 to 14 μm can be obtained when an intense picosecond pulse generated from a CO_2 laser is passed into GaAs, AgBr, ZnSe, and CdS crystals (Corkum et al., 1985).

The supercontinuum spectra measured from a 6-cm-long Cr-doped GaAs crystal and a 3.8-cm AgBr crystal for different laser pulse durations and

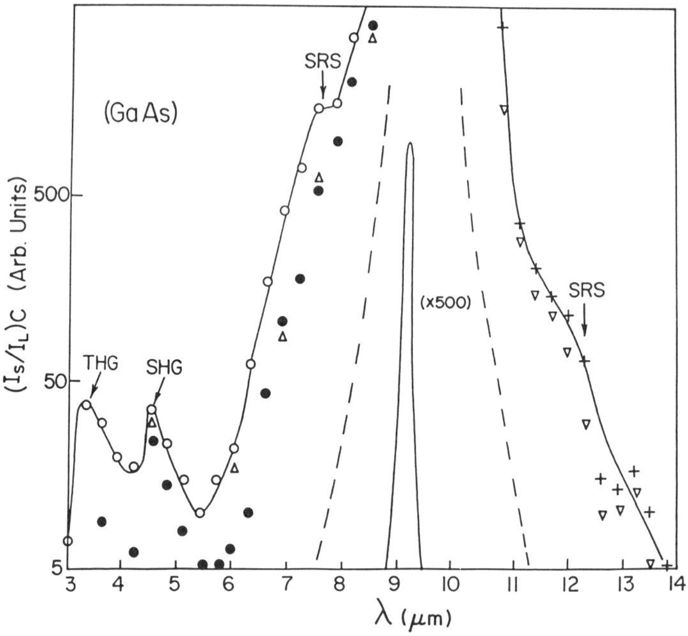

FIGURE 2.6. Supercontinuum spectra from a 6-cm-long Cr-doped GaAs crystal. (From Corkum et al., 1985.)

intensities are plotted in Figures 2.6 and 2.7, respectively. The signals were normalized for the input laser energy and the spectral sensitivity of filters, grating, and detectors. Each point represents the average of three shots. The salient feature of the curves displayed in Figures 2.6 and 2.7 is that the spectral broadening spans the wavelength region from 3 to 14 μm. The wave number spread on the anti-Stokes side is much greater than that on the Stokes side. From data displayed in Figure 2.6, the maximum anti-Stokes spectral broadening is $\Delta\omega_a = 793$ cm^{-1}. Including second and third harmonic generation (SHG and THG), it spans 2000 cm^{-1}. On the Stokes side, $\Delta\omega_s = 360$ cm^{-1}, yielding a value of $\delta\omega_a/\delta\omega_s \sim 2.2$. For AgBr, Figure 2.7 shows that $\Delta\omega_a = 743$ cm^{-1} and $\Delta\omega_s = 242$ cm^{-1}, yielding $\Delta\omega_a/\Delta\omega_s \sim 3$.

The spectral broadening mechanism for the supercontinuum can originate from several nonlinear optical processes. These include self-phase modulation, the four-wave parametric effect, higher-order harmonic generation, and stimulated Raman scattering. In Figure 2.6 the supercontinuum from the GaAs has two small peaks at 4.5 and 3.3 μm. These arise from the SHG and THG, respectively. Small plateaus are located at 7.5 and 12 μm. These arise from the first-order anti-Stokes and Stokes stimulated Raman scattering combined with SPM about these wavelengths. The SPM is attributed to an electronic mechanism.

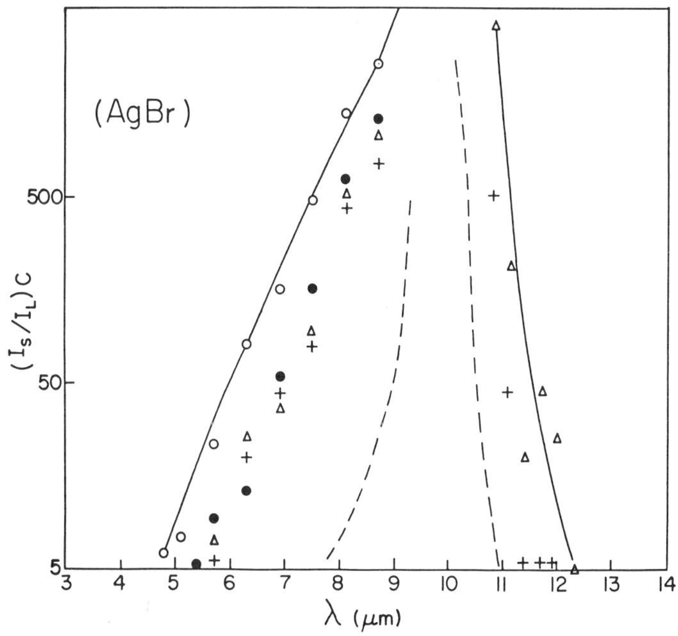

FIGURE 2.7. Supercontinuum spectra from a 3.8-cm-long AgBr crystal. (From Corkum et al., 1985.)

Summarizing the important experimental aspects of the spectra in condensed matter: The spectra are characterized by very large spectral widths and a nonperiodic or random substructure. Occasionally, a periodic structure interference minimum and maximum are observed. The modulation frequencies range from a few cm^{-1} to hundreds of cm^{-1}, and some modulation progressively increases away from the central frequency. The Stokes and anti-Stokes spectra are approximately equal in intensity and roughly uniform. The extents on the Stokes and anti-Stokes sides are not symmetric. The peak intensity at the central frequency is 10^2–10^3 the intensity of the SPM spectra at a given frequency.

5. Generation of Supercontinuum in Liquids

Nonlinear optical effects in solids are very effective; however, damage generated in solid media often limits their usefulness for ultrashort high-power effects. Various kinds of inorganic and organic liquids are useful media for generating picosecond or femtosecond supercontinuum light pulses since they are self-healing media. The supercontinuum spectra produced in liquids (Alfano, 1972) are similar to the spectra displayed in Figure 2.4 (Alfano, 1972). The following highlights the supercontinuum phenomena in the varous favorite liquid media of the authors. These liquids give the most intense and uniform supercontinuum spectral distributions.

5.1 Supercontinuum in H_2O and D_2O

The supercontinuum generated in H_2O and D_2O by the second harmonic of a mode-locked neodymium glass laser spanned several thousand wave numbers. The time duration was equal to or less than the picosecond pulse that generated it (Busch et al., 1973). The continuum extended to below 310 nm on the anti-Stokes side and to the near-IR region on the Stokes side. There were sharp absorptions at 450 nm in the H_2O continuum and at 470 nm in the D_2O continuum resulting from the inverse Raman effect (Alfano and Shapiro, 1970b; Jones and Stoicheff, 1964). Focusing a 12-mJ, 1060-nm single pulse 14 ps in duration into 25 cm of liquid D_2O resulted in a continuum that showed practically no structure, extending from 380 to at least 800 nm and highly directional and polarized (Sharma et al., 1976). Enhancing the supercontinuum intensity using water with ions is discussed in Section 10.

5.2 Supercontinuum in CCl_4

Another favorite liquid for producing a supercontinuum is CCl_4, in which the spectra produced are similar to the spectra displayed in Figure 2.4. A typical flat white supercontinuum extending from 430 nm through the visible and near infrared could be produced by focusing an 8-ps pulse at 1060 nm with about 15 mJ pulse energy into a cell containing CCl_4 (Magde and Windsor, 1974).

5.3 Supercontinuum in Phosphoric Acid

Orthophosphoric acid was found to be a useful medium for generating picosecond continuum light pulses ranging from the near UV to the near IR. By focusing a pulse train from a mode-locked ruby laser into a 10-cm-long cell containing phosphoric acid (60% by weight) solution in water by an 8-cm focal lens, a supercontinuum from near 450 nm to the near IR was obtained. The supercontinuum spectra contain structure arising from Raman lines (Kobayashi, 1979).

5.4 Supercontinuum in Polyphosphoric Acid

The supercontinuum from polyphosphoric acid was generated by focusing an optical pulse at 694.3 nm with 100 mJ pulse energy and a pulse width of 28 ps into a cell of any length from 2 to 20 cm containing polyphosphoric acid. It reaches 350 nm on the anti-Stokes side, being limited by the absorption of polyphosphoric acid, and 925 nm on the Stokes side, being cut off by limitations of IR film sensitivity (Nakashima and Mataga, 1975).

6. Supercontinuum Generated in Optical Fibers

The peak power and the interaction length can be controlled better in optical fibers than in bulk materials. Optical fibers are particularly interesting material for nonlinear optical experiments. In this section, we discuss supercontinuum generation in glass optical fibers. Details of the use of SPM for pulse compression are discussed in other chapters.

The generation of continua in glass optical fibers was performed by Stolen et al. in 1974. Continua covering $\sim 500\,\text{cm}^{-1}$ were obtained. Shank et al. (1982) compressed 90-fs optical pulses to 30-fs pulses using SPM in an optical fiber followed by a grating compressor. Using the SPM in an optical fiber with a combination of prisms and diffraction gratings, they were able to compress 30 fs to 6 fs (Fork et al., 1987; also see Chapter 10 by Johnson and Shank).

A typical sequence of spectral broadening versus input peak power using 500-fs pulses (Baldeck et al., 1987b) is shown in Figure 2.8. The spectra show

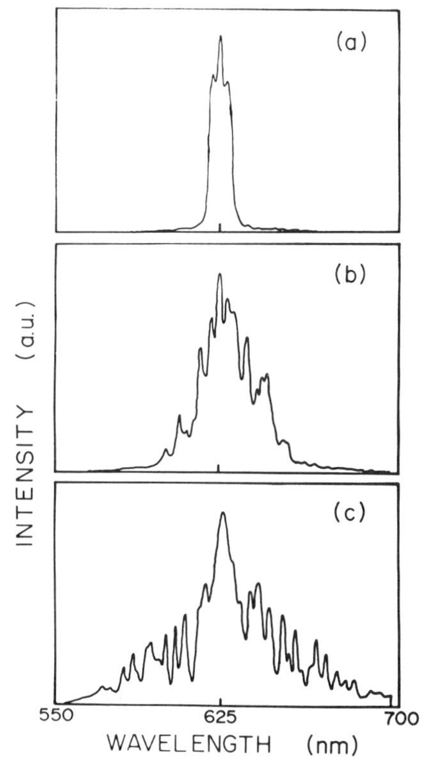

FIGURE 2.8. Sequence of spectral broadening versus increasing input energy in a single-mode optical fiber (length = 30 cm). The intensity of the 500-fs pulse was increased from (a) to (c). (From Baldeck et al., 1987b.)

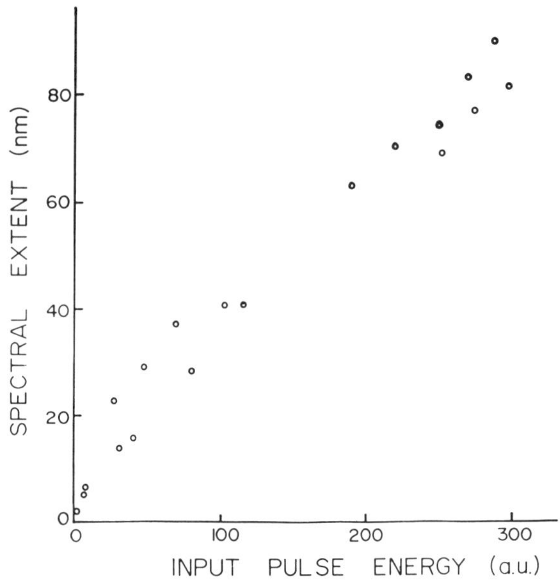

FIGURE 2.9. Supercontinuum spectra versus input pulse energy in a single-mode optical fiber (length = 30 cm) for a 500-fs pulse. (From Baldeck et al., 1987b.)

SPM characteristic of heavy modulation. The spectral extent is plotted against the energy in Figure 2.9 for 500-fs pulses (Baldeck et al., 1987b). The relative energy of each pulse was calculated by integrating its total broadened spectral distribution. The supercontinuum extent increased linearly with the pulse intensity. The fiber length dependence of the spectral broadening is plotted in Figure 2.10. The broadening was found to be independent of the length of the optical fiber for $l > 20$ cm. This is due to group velocity dispersion. The SPM spectral broadening occurs in the first few centimeters of the fiber for such short pulses (Baldeck et al., 1987b).

In multimode optical fibers, the mode dispersion is dominant and causes pulse distortion. Neglecting the detailed transverse distribution of each mode, the light field can be expressed by

$$E(t) = \sum_i a_i A_i(t) \exp[i\omega_0 t - ik_i z], \tag{23}$$

where ω_0 is the incident laser frequency; a_i, $A_i(t)$, and $k_i = n_i\omega_0/c$ are the effective amplitude, electric field envelope function at the local time $\tau = t - z/v_{gi}$, and wave number of mode i, respectively; and v_{gi} is the group velocity of mode i. The effective refractive index of mode i is denoted by n_i and

$$n_i = n_{0i} + n_{2i}|E(t)|^2, \tag{24}$$

where n_{0i} and n_{2i} are the linear refractive index and the nonlinear coefficient of the ith mode, respectively. The nonlinearities of different modes are as-

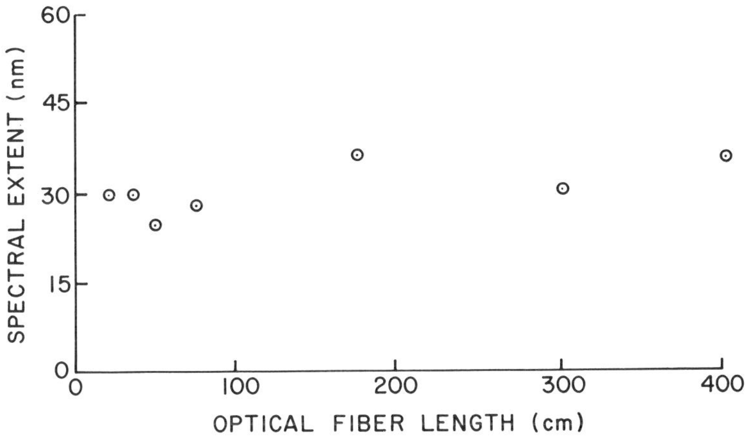

FIGURE 2.10. Supercontinuum spectra versus optical fiber length for a 500-fs pulse. (From Baldeck et al., 1987b.)

sumed to be the same, that is, $n_{2i} = n_2$. Substituting Eq. (21) into Eq. (20), we obtain

$$E(t) = \sum_i a_i A_i(t) \exp[i\omega_0 t - in_{0i}\omega_0 z/c - i\Delta\phi(t)], \tag{25}$$

where

$$\Delta\phi(t) = (n_2\omega_0 z/c)|E(t)|^2. \tag{26}$$

After inserting Eq. (23) into Eq. (26), the time-dependent phase factor $\Delta\phi(t)$ can be expanded in terms of $E_i(t)$:

$$\Delta\phi(t) = \sum_i \sum_j (n_2\omega_0 z/c) a_i a_j A_i(t) A_j(t) \exp[i(n_{0i} - n_{0j})\omega_0 z/c]. \tag{27}$$

In the picosecond time envelope, the terms of $i \neq j$ oscillate rapidly. Their contributions to the time-dependent phase factor are washed out. The approximate $\Delta\phi(t)$ has the form

$$\Delta\phi = \sum_i (n_2\omega_0 z/c) a_i^2 A_i^2(t). \tag{28}$$

The pulse shape changes due to the different group velocities of various modes. When most of the incident energy is coupled into the lower modes, the pulse will have a fast rising edge and a slow decay tail since the group velocity is faster for lower-order mode. This feature was observed using a streak camera. Therefore, the $\Delta\phi(t, z)$ of Eq. (28) will also have a fast rising edge and a slow decay tail. The time derivative of the phase $\Delta\phi(t, z)$ yields an asymmetric frequency broadening.

Figure 2.11 shows the spectral continuum generated from multimode glass optical fibers using 8-ps pulses at 530 nm. The spectral broadening is asymmetric about the incident laser frequency. It is shifted much more to the Stokes side than to the anti-Stokes side. The observed spectra did not show a

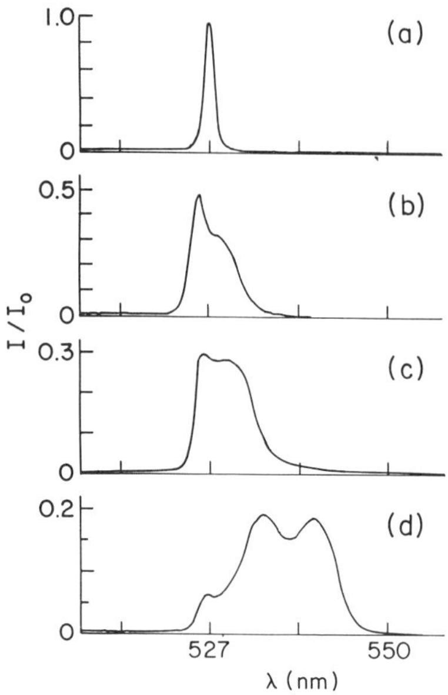

FIGURE 2.11. Output spectra for 8-ps laser pulses at 527 nm propagating through different lengths of multimode optical fibers: (a) no fiber; (b) 22 cm; (c) 42 cm; (d) 84 cm. (From Wang et al., 1988.)

modulation. This can be explained by the spectral resolution of the measurement system. The calculated modulation period is about 0.13 nm, which is much smaller than the resolution of the measurement system (about 1 nm) (Wang et al., 1988).

7. Supercontinuum Generation in Rare-Gas Liquids and Solids

Continuum generation is a general phenomenon that occurs in all states of matter. A system for testing the role of the electronic mechanism is rare-gas liquids and solids (Alfano and Shapiro, 1970a). Rare-gas liquids are composed of atoms possessing spherical symmetry. Thus, there are no orientational, librational, or electron-lattice contributions to the nonlinear refractive index n_2. However, interrupted rocking of argon can occur in which a distorted atom can rock about an equilibrium value before it collides with other atoms. Contributions to the nonlinear refractive index might be expected from electrostriction, molecular redistribution, interrupted rocking, and a distortion of the electron clouds:

$$n_2 = n_{2_{\text{ELECTRONIC}}} + n_{2_{\text{MR}}} + n_{2_{\text{LIBRATION}}} + n_{2_{\text{ELECTROSTR.}}}. \tag{29}$$

Electrostriction is ruled out because picosecond exciting pulses are too short. Molecular redistribution arises from fluctuations in the local positional arrangement of molecules and can contribute significantly to n_2. However, n_2 due to all mechanisms except electronic was estimated to be $\sim 2 \times 10^{-14}$ esu for liquid argon from depolarized inelastic-scattering data. Electronic distortion ($n_2 = 6 \times 10^{-14}$ esu) slightly dominates all nonlinear index contributions (Alfano and Shapiro, 1970a; Alfano, 1972). Furthermore, the depolarized inelastic light-scattering wing vanishes in solid xenon, implying that the molecular redistribution contribution to n_2 vanishes in rare-gas solids. Observations of self-focusing and SPM in rare-gas solids appear to provide a direct

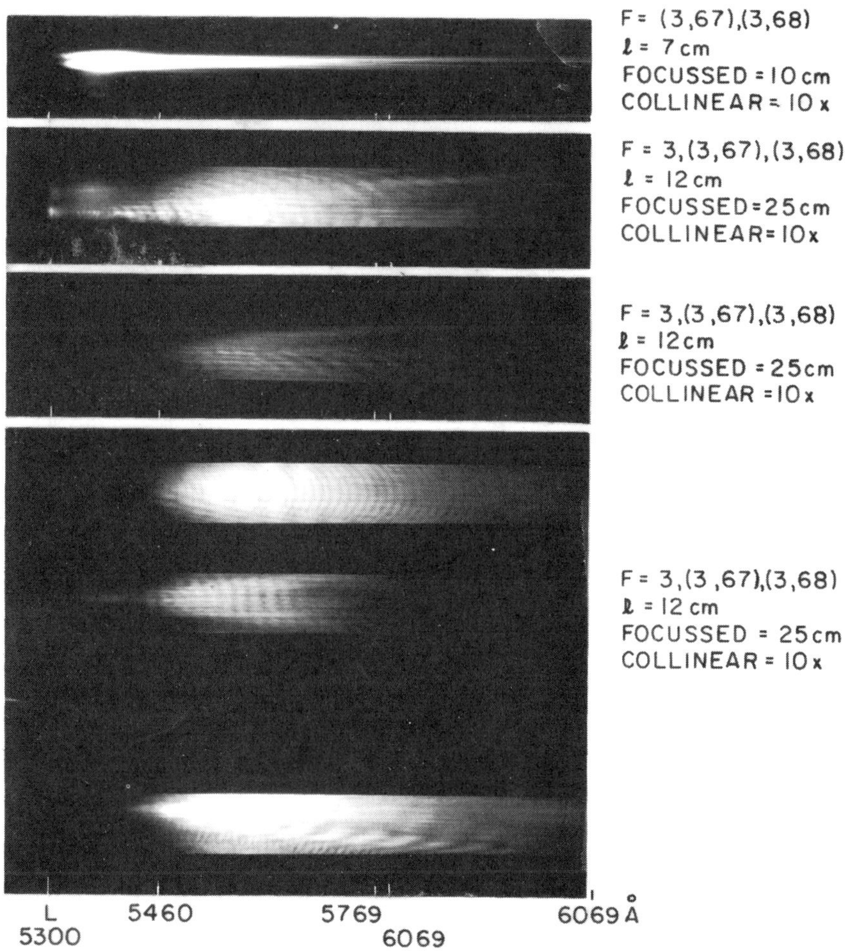

(a)

FIGURE 2.12. Supercontinuum spectra for picosecond laser pulses at 530 nm passing through rare-gas liquids and solids; (a) Stokes SPM from liquid argon for different laser shots; (b) anti-Stokes SPM for liquid argon for different laser shots; (c) Stokes SPM for liquid and solid krypton for different laser shots. (From Alfano, 1972.)

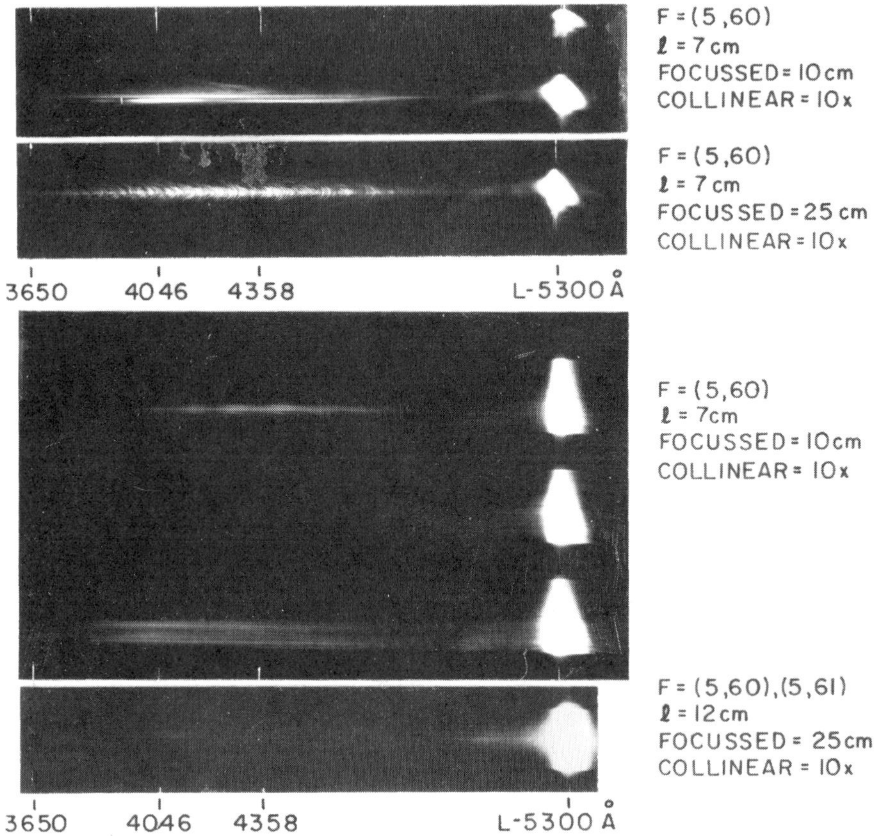

(b)

FIGURE 2.12 (*continued*)

proof that atomic electronic shells are distorted from their spherical symmetry under the action of the applied field. However, both pure electronic and molecular redistribution mechanisms contribute to n_2 in rare-gas liquids. The response time of the system for a combination of both of these mechanisms lies between 10^{-15} and 10^{-12} s. For femtosecond and subpicosecond pulses, the dominant mechanism for n_2 and SPM is electronic in origin.

The experimental setup used to generate and detect a supercontinuum in rare-gas liquids and solids is the same as that shown in Figure 2.3 with the exception that the samples are placed in an optical dewar.

Typical supercontinuum spectra from rare-gas liquids and solids are displayed in Figure 2.12. Sweeps of 1000 to 6000 cm^{-1} were observed to both the Stokes and anti-Stokes sides of 530 nm in liquid argon. Modulation ranges from a few cm^{-1} to hundreds of cm^{-1}. Similar spectral sweeps were observed in liquid and solid krypton.

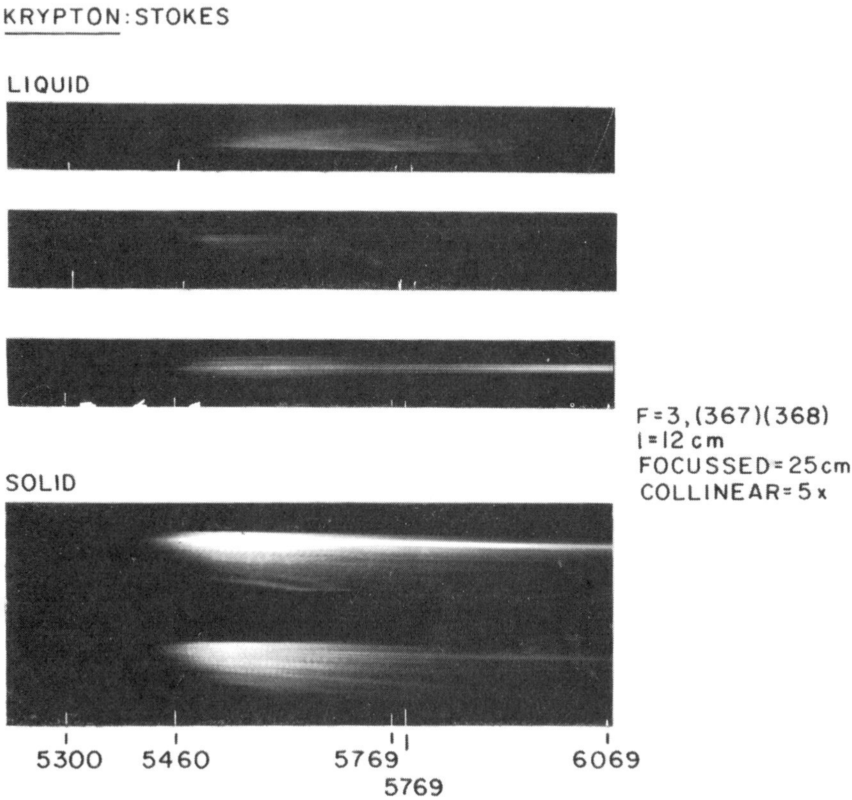

(c)

FIGURE 2.12 (*continued*)

A most important point is that the threshold for observing SPM in liquid krypton is 0.64 ± 0.12 that in liquid argon. The SPM threshold ratio of solid and liquid krypton is 0.86 ± 0.15. In liquid argon, SPM spectra appear at a threshold power of ~0.5 GW focused in a 12-cm sample. The swept light is also collimated, polarized, and modulated. These observations rule out dielectric breakdown.

The refractive index in rare-gas liquids is given by $n_\parallel = n_0 + n_2\langle E^2\rangle$, where $n_\parallel$ is the refractive index parallel to the field. $\langle E^2\rangle^{1/2}$ is the rms value of the electric field. The electronic nonlinear refractive index in rare gas liquids is given by

$$n_2 = [(n_0^2 + 2)^4/81n_0]\pi N\rho, \tag{30}$$

where n_0 is the linear refractive index, ρ is the second-order hyperpolarizability, and N is the number of atoms per unit volume. The term $n_2 = 0.6 \times 10^{-13}$ esu in liquid argon and $\simeq 1.36 \times 10^{-13}$ in liquid krypton. For liquid argon and liquid and solid krypton, the refractive indices are taken as 1.23, 1.30, and 1.35, respectively (McTague et al., 1969). Intense electric fields distort atoms and produce a birefringence. The anisotropy in refractive index between light traveling with the wave vector parallel and perpendicular to the applied electric field is given by (Alfano, 1972)

$$\delta n_{\parallel} - \delta n_{\perp} = \tfrac{1}{3} n_2 E_0^2, \tag{31}$$

where $\delta n_{\parallel}$ and $\delta n_{\perp}$ are the changes in refractive indices parallel and perpendicular to the field. The value of $n_2 E_0^2$ is $\sim 5 \times 10^{-5}$ V/m in liquid argon when $E_0 \sim 1.5 \times 10^7$ V/m ($\sim 4 \times 10^{11}$ W/cm^2). This change in index explains the self-focusing and SPM described above which was observed by Alfano and Shapiro in 1970. Similar SPM effects occur in organic and inorganic liquids, often accompanied by SRS and inverse Raman effects.

8. Supercontinuum Generation in Antiferromagnetic $KNiF_3$ Crystals

The influence of magnetic processes on nonlinear optical effects is an interesting topic. In this section, we discuss the supercontinuum generation associated with the onset of magnetic order in a $KNiF_3$ crystal (Alfano et al., 1976). Light at 530 nm is well suited for the excitation pulse because $KNiF_3$ exhibits a broad minimum in its absorption (Knox et al., 1963) between 480 and 610 nm.

Typical spectra from an unoriented 5-cm-long $KNiF_3$ single crystal are displayed in Figure 2.13 for 530-nm picosecond excitation (Alfano et al., 1976). The spectra are characterized by extensive spectral broadening ranging up to ~ 3000 cm^{-1} to either side of the laser frequency. The intensity, although not the spectral broadening, of the output exhibited the large temperature dependence illustrated in Figure 2.14. There is no sharp feature at 552 nm, the position expected for stimulated Raman scattering by the 746-cm^{-1} magnon pair excitation. Usually, the spectra were smooth; however, occasionally structure was observed. A periodic structure with a modulation frequency of tens to hundreds of wave numbers was evident. The frequency broadening light is also polarized in the same direction as the incident 530-nm pulse. This property is the same observed in glass, crystals, and liquids (see Sections 3–6). Self-focusing was also observed, usually in the form of 10 to 40 small self-focused spots 5 to 20 μm in diameter at the exit face of the crystal. Using a focused beam, optical damage could also be produced. It should be emphasized that spectral broadening was always observed even in the absence of self-focusing, damage, or periodic spectral intensity modulation.

Figure 2.15 shows the output intensity at 570 nm as a function of input intensity for two temperatures: above and below the Néel temperature. The output intensity is approximately exponential in the input intensity at both temperatures. However, the slope is more than a factor of two larger at 77 K than at 300 K. The rapid rise in conversion efficiency of four orders of magnitude within a small interval of input intensity is indicative of an amplifi-

FIGURE 2.13. Spectra for picosecond laser pulse at 530 nm passing through 5-cm-long $KNiF_3$. (From Alfano et al., 1976.)

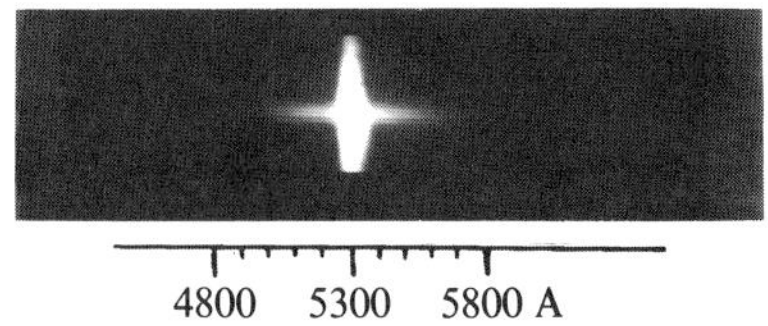

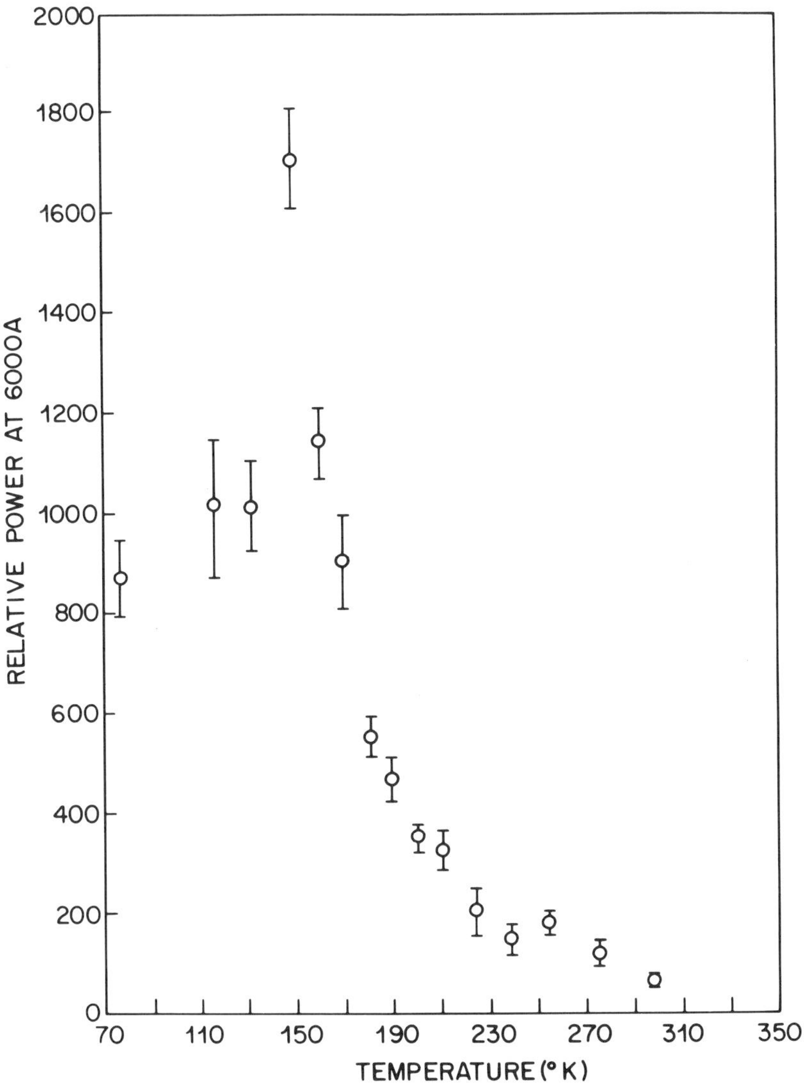

FIGURE 2.14. Intensity of the frequency-broadening emission from $KNiF_3$ as a function of temperature at fixed pump intensity at 552 nm. (From Alfano et al., 1976.)

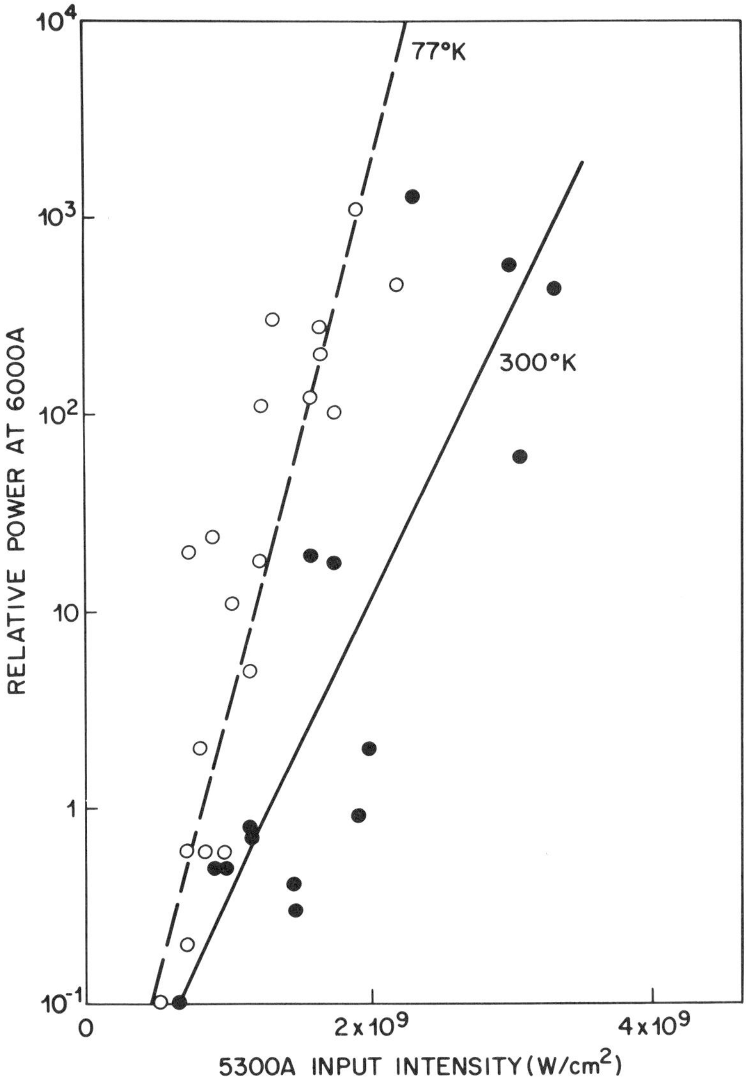

FIGURE 2.15. Intensity dependence of continuum spectra at 570 nm from $KNiF_3$ as a function of pumping laser intensity at fixed lattice temperature. (From Alfano et al., 1976.)

cation process with very large gain. Identical curves were obtained at 552 and 600 nm output wavelengths. The similarity in results for several output frequencies shows that simple stimulated magnon pair scattering is not the dominant process. If it were, one would expect the behavior at 552 nm to differ considerably from that at other wavelengths.

The most novel experimental results in $KNiF_3$ are the large ($\sim 20\times$) intensity increases below T_N. Spectra at 552, 570, and 600 nm behave identically—within experimental error—consistent with the observations in Figure 2.15. The temperature dependence of the relative **peak** intensity for the spontaneous magnon pair scattering in the $KNiF_3$ sample (using 514.5 nm laser light) was measured and is plotted in Figure 2.14. For $KNiF_3$ the magnon pair scattering accounts for the entire inelastic light scattering and therefore for the non-σ electronic contribution to χ^3 (Hellwarth et al., 1975). The temperature dependence is compelling evidence for the magnetic origin of the low-temperature-enhanced nonlinear optical spectral broadened intensity.

The observation can be semiquantitatively accounted for in terms of a temperature-dependent spin contribution to the overall nonlinear susceptibility $\chi^{(3)}_{ijkl}$ that governs four-photon parametric mixing as the primary process. In general $\chi^{(3)}$ may be written as a sum of electronic and Raman contributions (Levenson and Bloembergen, 1974). For $KNiF_3$ we may consider the latter to consist solely of the magnon pair Raman scattering contribution (Chinn et al., 1971; Fleury et al., 1975), which we can approximate as a Lorentzian:

$$\chi^{(3)}_{ijkl}(-\omega_3, \omega_1, \omega_1, -\omega_2) = \chi^{(3)}_E + K \frac{\alpha^m_{ij}\alpha^m_{kl} + \alpha^m_{il}\alpha^m_{jk}}{\omega_m - (\omega_1 - \omega_2) + i\Gamma_m}. \tag{32}$$

Here ω_m and Γ_m denote the temperature-dependent frequency and linewidth, respectively, of the magnon pair excitations, α^m_{ij} is the magnon pair polarizability, and $\chi^{(3)}_E$ is the usual nonresonant, temperature-independent "electronic" contribution from nonlinear distortion of the electronic orbits. The second term in Eq. (32) is called magnetic $\chi^{(3)}_M$. Since the **integrated intensity** of the spontaneous magnon pair Raman spectrum, which is $\sim|\alpha^m|^2$, has been measured and found to be essentially temperature independent (Chinn et al., 1971; Fleury et al., 1975), the only quantities in Eq. (32) that vary significantly with temperature are ω_m and Γ_m. The observed temperature independence of the extent of spectral broadening, $\delta\omega$, may be explained by noting that $\delta\omega \sim 2\Delta\omega n_2 k E_1^2 l$ due to self-phase modulation. Here $\Delta\omega$ is the spectral width of the input pulse, k is its propagation constant, E_1 is the field amplitude, and l is the path length. n_2 is the nonlinear refractive index, which contains a purely electronic contribution, σ, and a contribution proportional to the integrated Raman scattering cross section (Hellwarth et al., 1975). Since neither σ nor $|\alpha^m|^2$ is temperature dependent in $KNiF_3$, n_2 and therefore $\Delta\omega$ should not vary either, in agreement with observations.

The observed strong temperature dependence of the intensity of the frequency-broadened spectrum (see Figure 2.14) arises from the resonant term in Eq. (32) through the primary process $2\omega_1 \rightarrow \omega_2 + \omega_3$, which is strongest when $\omega_2 = \omega_1 + \omega_m$ and increases as Γ_m decreases (on cooling below the Néel temperature). That is, the resonant contribution to $\chi^{(3)}$ in Eq. (32) varies with temperature in the same way as the peak spontaneous magnon pair cross section: $\Gamma_m^{-1}(T)$. However, the individual contribution to $\chi^{(3)}$ cannot be directly inferred from the dependence of the broadened spectrum. This is because the latter receives significant contributions from secondary processes of the form $\omega_1 + \omega_2' \rightarrow \omega_3' + \omega_4$ etc., in which products of the primary process interact with the pump to smooth the spectral distribution and wash out the sharp features that the resonant spin nonlinearity produces in the primary process. The large values of pump intensity and source spectral width make possible strong amplification in spite of imprecise phase matching in the forward direction. Such behavior (washing out of stimulated Raman features by the spectral broadening process) has frequently been observed in both liquids and crystals. Thus a full quantitative description of the nonlinear optical processes in $KNiF_3$ is not yet possible.

9. Generation of Supercontinuum near Electronic Resonances in Crystals

Since the active medium of a laser possesses well-defined electronic energy levels, knowledge of SPM near electronic levels is of paramount importance. SPM near electronic levels of a PrF_3 crystal has been investigated experimentally and theoretically to gain additional information on the SPM process —in particular, on the role played by the electronic levels and on how the continuum spectrum evolves through and beyond the electronic absorption levels (Alfano et al., 1974).

Experimentally, the Stokes and anti-Stokes spectrum and filament formation from the PrF_3 crystal are investigated under intense picosecond pulse excitation at the wavelength of 530 nm. The *c* axis of the crystal is oriented along the optical axis. The intensity distribution at the exit face of the crystal is magnified by $10\times$ and imaged on the slit of a Jarrell-Ash $\frac{1}{2}$-m-grating spectrograph so that the spectrum of each filament can be displayed. The spectra are recorded on Polaroid type 57 film. No visible damage occurred in the PrF_3 crystal.

The PrF_3 crystal was chosen for the experiment because its electronic levels are suitably located on the Stokes and anti-Stokes sides of the 530-nm excitation wavelength. The absorption spectra of a $\frac{1}{2}$-mm-thick PrF_3 crystal and the energy level scheme of Pr^{3+} ions are shown in Figure 2.16. The fluorides of Pr have the structure of the naturally occurring mineral tysonite with D_{34}^4 symmetry.

Typical spectra of frequency broadening from PrF_3 about 530 nm are shown in Figure 2.17 for different laser shots. Because of the absorption associated with the electronic level, it is necessary to display the spectrum over different wavelength ranges at different intensity levels. In this manner,

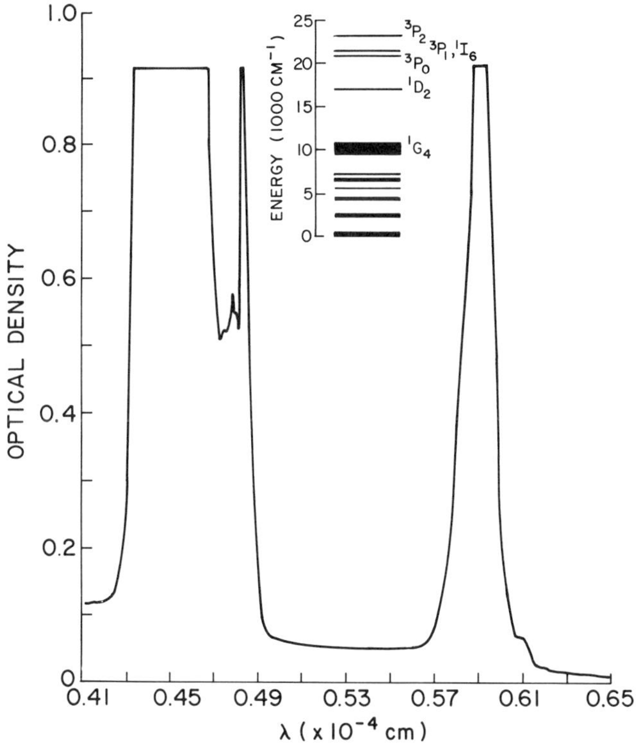

FIGURE 2.16. Absorption spectra of 0.5-mm-thick PrF_3 crystal; insert is the level scheme of Pr^{3+} ions. (From Alfano et al., 1974.)

the development of the SPM spectrum through the electronic absorption levels can be investigated. Using appropriate filters, different spectral ranges are studied and displayed in the following figures: in Figure 2.18a the Stokes side for frequency broadening $\bar{v}_B > 100\ \text{cm}^{-1}$ at an intensity level (I_{SPM}) of $\sim 10^{-2}$ of the laser intensity (I_L), in Figure 2.18b the Stokes side for $\bar{v}_B > 1500\ \text{cm}^{-1}$ at $I_{SPM} \sim 10^{-4} I_L$, in Figure 2.18c the anti-Stokes side for $\bar{v}_B > 100\ \text{cm}^{-1}$ at $I_{SPM} \sim 10^{-2} I_L$, and in Figure 2.18d the anti-Stokes side for $\bar{v}_B > 1500\ \text{cm}^{-1}$ at $I_{SPM} \sim 10^{-4} I_L$. Usually 50 to 100 small-scale filaments 5 to 50 μm in diameter are observed.

Several salient features are evident in the spectra displayed in Figures 2.17 and 2.18. In Figure 2.17 the Stokes and anti-Stokes spectra are approximately equal in intensity and frequency extent. The peak intensity at the central frequency is ~ 100 times the intensity of the SPM at a given frequency. The extent of the frequency broadening is $\sim 1500\ \text{cm}^{-1}$, ending approximately at the absorption lines. Occasionally a periodic structure of minima and maxima is observed that ranges from a few cm^{-1} to $100\ \text{cm}^{-1}$, and for some observations no modulation is observed. Occasionally an absorption band appears

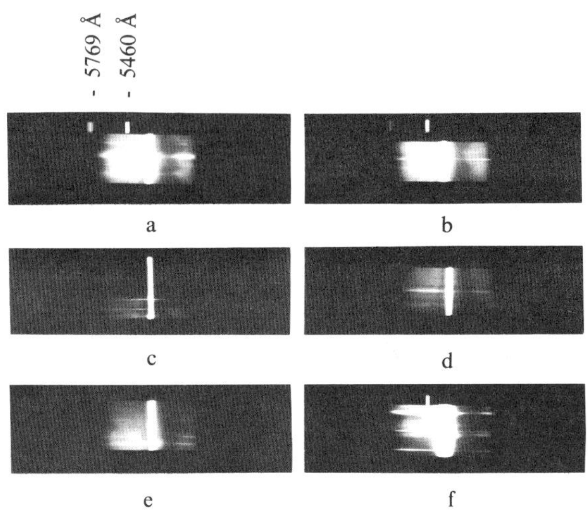

FIGURE 2.17. Spectra from PrF_3 excited by 4-ps laser pulses at 530 nm; neutral density (ND) filters: (a) ND = 1.5; (b) ND = 1.5; (c) ND = 2.0; (d) ND = 2.0; (e) ND = 1.7; (f) ND = 1.4. A wire is positioned after the collection lens at the focal length. (From Alfano et al., 1974.)

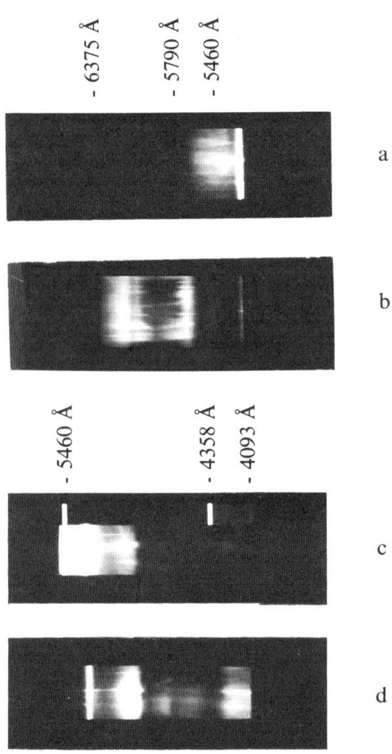

FIGURE 2.18. Spectra on the Stokes and anti-Stokes sides of the 530-nm excitation: (a) Stokes side, Corning 3-68 filter, wire inserted, ND = 2.0; (b) Stokes side, Corning 3-66 filter, wire inserted; (c) ant-Stokes side, wire inserted, ND = 1.0; (d) anti-Stokes side, Corning 5-61, wire inserted. (From Alfano et al., 1974.)

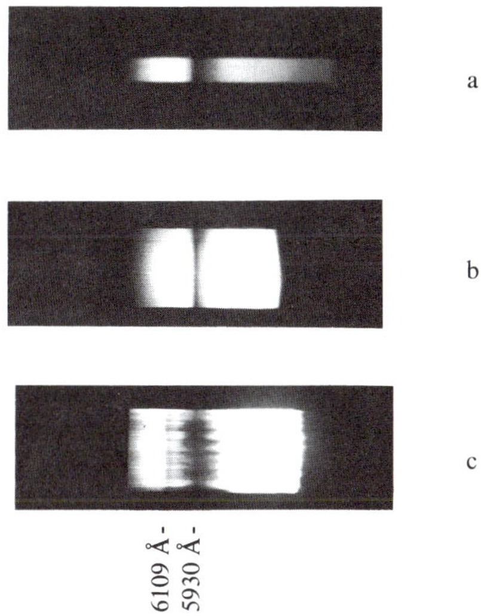

FIGURE 2.19. Comparison of the Stokes absorption spectra of PrF_3 photographed with different light sources: (a) light emitted from a tungsten lamp is passed through 0.5-cm-thick crystal; (b) SPM light emitted from BK-7 glass is passed through 0.5-mm-thick crystal; (c) SPM light is generated within the 5-cm PrF_3. (From Alfano et al., 1974.)

on the anti-Stokes side of the 530-nm line whose displacement is 430 cm^{-1}. In Figure 2.18 the main feature is the presence of a much weaker super-broadband continuum whose frequency extends through and past the well-defined absorption lines of the Pr^{3+} ion to a maximum frequency of >3000 cm^{-1} on the Stokes side (end of film sensitivity) and >6000 cm^{-1} on the anti-Stokes side. The intensity of the continuum at a given frequency outside absorption lines is $\sim 10^{-4}$ the laser intensity.

The observed absorption lines on the anti-Stokes side of 530 nm are located at 441.5, 465.3, and 484.5 nm and on the Stokes side at 593 and 610.9 nm. These lines correspond within ± 0.7 nm to the absorption lines measured with a Cary 14. The absorption lines measured from the Cary spectra are ~ 3 cm^{-1} at 611.2 nm, 62 cm^{-1} at 5938.8 nm, 46 cm^{-1} at 485.2 nm, and >100 cm^{-1} at 441.2 nm. Figure 2.19 compares the Stokes absorption spectra of a PrF_3 crystal photographed with a $\frac{1}{2}$-m Jarrell-Ash spectrograph with different broadband light sources. Figure 2.19a was obtained with light emitted from a tungsten lamp passing through a $\frac{1}{2}$-mm PrF_3 crystal, Figure 2.19b was obtained with the Stokes side of the broadband picosecond continuum generated in BK-7 glass passing through a $\frac{1}{2}$-mm PrF_3, and Figure 2.19c was obtained with the broadband light generated in a 5-cm PrF_3 crystal. Notice that the absorption line at 611.2 nm is very pronounced in the spectra obtained with the continuum generated in PrF_3, whereas with conventional absorption techniques it is barely visible. The anti-Stokes spectrum obtained with light emitted from a tungsten filament lamp passing through a $\frac{1}{2}$-mm PrF_3 crystal is shown in Figure 2.20a. This is compared with the spectrum obtained with broadband light generated in a 5-cm PrF_3 crystal shown in Figure 2.20b.

The angular variation of the anti-Stokes and Stokes spectral emission from

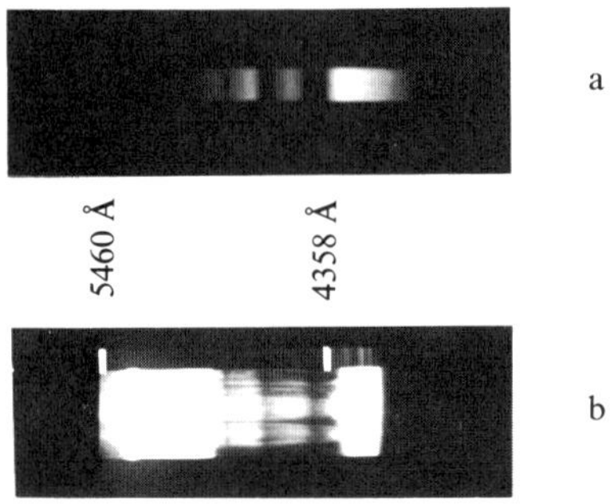

FIGURE 2.20. Comparison of the anti-Stokes absorption spectrum of PrF_3 photographed with (a) light emitted from a tungsten lamp passing through 0.5-mm-thick crystal and (b) SPM light generated within the 5-cm PrF_3. (From Alfano et al., 1974.)

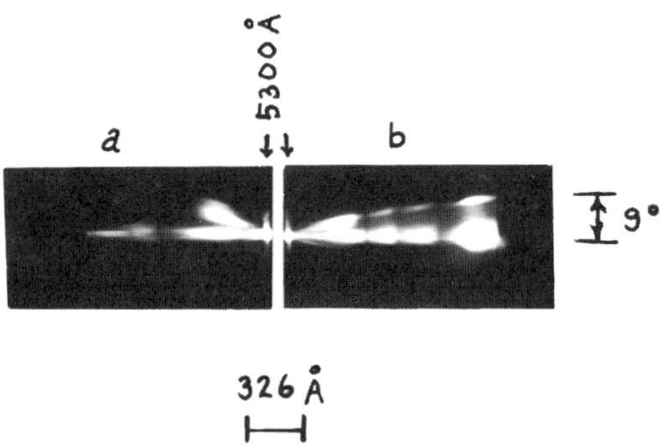

FIGURE 2.21. Angular variation of the (a) Stokes and (b) anti-Stokes spectral patterns emitted from PrF_3 crystal: (a) Corning 4(3-67) filters, ND = 1.0; (b) Corning 2(5-60) filters. (From Alfano et al., 1974.)

PrF_3 is displayed in Figure 2.21. The light emitted from the sample is focused on the slit of a $\frac{1}{2}$-m Jarrell-Ash spectrograph with a 5-cm focal length lens with the laser beam positioned near the bottom of the slit so that only the upper half of the angular spectrum curve is displayed. In this fashion, a larger angular variation of the spectrum is displayed. Emission angles $>9^0$ go off slit and are not displayed. This spectrum is similar to four-photon emission patterns observed from glass and liquids under picosecond excitation.

The experimental results show that a discontinuity in intensity occurs when the self-phase modulation frequency extends beyond the absorption line frequency. This is due to almost total suppression of the signal beyond the absorption resonance (Alfano et al., 1974). A similar argument and conclusion hold for the blue side of the laser line. The residual weak intensity that exists beyond the absorption line is not due to SPM. It can arise, however, from three-wave mixing. Since there was a continuum of frequencies created by SPM, it might be possible for three such frequencies, ω_1, ω_2, and ω_3, to mix to create a signal at frequency $\omega_1 + \omega_2 - \omega_3$ that lies beyond the absorption line. Since the frequencies are chosen from a continuum, it is also possible for phase matching to be achieved. For the spectrum in the domain between the laser frequency and the absorption line, the extent of self-broadening is proportional to the intensity. Since the energy in the pulse is proportional to the product of the frequency extent and the intensity spectrum, the intensity spectrum remains approximately constant. The observed absorption band in the continuum on the anti-Stokes side about 400 cm^{-1} away from the excitation frequency (see Figure 2.17) is probably due to the inverse Raman effect (Jones and Stoicheff, 1964). The observed absorption band is located in the vicinity of strong Raman bands: 401, 370, and 321 cm^{-1}.

A curious feature of the associated weak broadband spectrum is the existence of a pronounced absorption line at a position (611.2 nm) where the linear absorption would be expected to be rather weak. A possible explanation for this is as follows: Imagine tracing the spatial development of the phase modulation spectrum. At a short distance, where the bounds of the spectrum have not yet intersected a strong absorption line, the spectrum is reasonably flat. On intersecting the absorption line, the spectrum abruptly drops (Alfano et al., 1974). The mechanism of FFPG is presumably responsible for the appearance of the signal beyond the absorption line limit. This explanation is also supported by the appearance of the angular emission pattern (see Figure 2.21). As the spectrum continues to develop, one reaches a point where the limit of the regenerated spectrum crosses a weak absorption line. One can again expect a drastic drop in the spectrum at the position of this line. At still greater distances renewed four-photon parametric regeneration accounts for the feeble signal. A continuum is generated behind absorption bands due to contributions from SPM, three-wave mixing (TWM), and FFPG.

10. Enhancement of Supercontinuum in Water by Addition of Ions

The most common liquids used to generate a continuum for various applications are CCl_4, H_2O, and D_2O. In most applications of the ultrafast supercontinuum, it is necessary to increase the conversion efficiency of laser excitation energy to the supercontinuum. One method for accomplishing this is based on the induced- or cross-phase modulation. Another way is to increase n_2 in materials. In this section, chemical means are used to obtain a tenfold enhancement of the ultrafast supercontinuum in water by adding Zn^{2+} or K^+ ions (Jimbo et al., 1987) for 8-ps pulse generation.

The optical Kerr gate (OKG) (Ho and Alfano, 1979) was used to measure the nonlinear refractive index of the salt solutions. The primary and second harmonic light beams were separated by a dichroic mirror and then focused into a 1-cm-long sample cell filled with the same salt solutions that produced the ultrafast supercontinuum pulse enhancements. The size of the nonlinear index of refraction, n_2, was determined from the transmission of the probe beam through the OKG.

Three different two-component salt solutions of various concentrations were tested. The solutes were KCl, $ZnCl_2$, and K_2ZnCl_4. All measurements were performed at $20 \pm 1°C$. Typical spectra of ultrafast supercontinuum pulses exhibited both SPM and FPPG features. The collinear profile arising from SPM has nearly the same spatial distribution as the incident 8-ps, 530-nm laser pulse. The two wings correspond to FPPG pulse propagation. The angle arises from the phase-matching condition of the generated wavelength emitted at different angles from the incident laser beam direction. FPPG spectra sometimes appear as multiple cones and sometimes show modulated features. SPM spectra also show modulated patterns. These features can be explained by multiple filaments.

Typical ultrafast supercontinuum pulse spectra on the Stokes side for different aqueous solutions and neat water, measured with the optical multichannel analyzer, are shown in Figure 2.22. The salient features in Figure 2.22 are a wideband SPM spectrum together with the stimulated Raman scattering of the OH stretching vibration around 645 nm. The addition of salts causes the SRS signal to shift toward the longer-wavelength region and sometimes causes the SRS to be weak (Figure 2.22a). The SRS signal of pure water and dilute solution appears in the hydrogen-bonded OH stretching region (~ 3400 cm^{-1}). In a high-concentration solution, it appears in the non-hydrogen-bonded OH stretching region (~ 3600 cm^{-1}). The latter features of SRS were observed in an aqueous solution of $NaClO_4$ by Walrafen (1972).

To evaluate quantitatively the effect of cations on ultrafast supercontinuum generation, the ultrafast supercontinuum signal intensity for various samples at a fixed wavelength were measured and compared. Figure 2.23 shows the dependence of the supercontinuum (mainly from the SPM contribution) signal intensity on salt concentration for aqueous solutions of K_2ZnCl_4, $ZnCl_2$, and KCl at 570 nm (Figure 2.23a) and 500 nm (Figure 2.23b). The data were normalized with respect to the average ultrafast supercontinuum signal intensity obtained from neat water. These data indicated that the supercontinuum pulse intensity was highly dependent on salt concentration and that both the Stokes and the anti-Stokes sides of the supercontinuum signals from a saturated K_2ZnCl_4 solution were about 10 times larger than from neat water. The insets in Figure 2.23 are the same data plotted as a function of K^+ ion concentration for KCl and K_2ZnCl_4 aqueous solutions. Solutions of KCl and K_2ZnCl_4 generate almost the same amount of supercontinuum if the K^+ cation concentration is same, even though they contain different amounts of Cl^- anions. This indicates that the Cl^- anion has little effect on generation of the supercontinuum. The Zn^{2+} cations also enhanced the supercontinuum, though to a lesser extent than the K^+ cations.

The measurements of the optical Kerr effect and the ultrafast supercontinuum in salt-saturated aqueous solutions are summarized in Table 2.3. The measured n_2 (pure H_2O) is about 220 times smaller than n_2 (CS_2). The value $G_{SPM}(\lambda)$ represents the ratio of the SPM signal intensity from a particular salt solution to that from neat water at wavelength λ. G_{Kerr} is defined as the ratio of the transmitted intensity caused by a polarization change of the probe beam in a particular salt solution to that in neat water; G_{Kerr} is equal to $[n_2\ (\text{particular solution})/n_2\ (\text{water})]^2$. Table 2.3 shows that, at saturation,

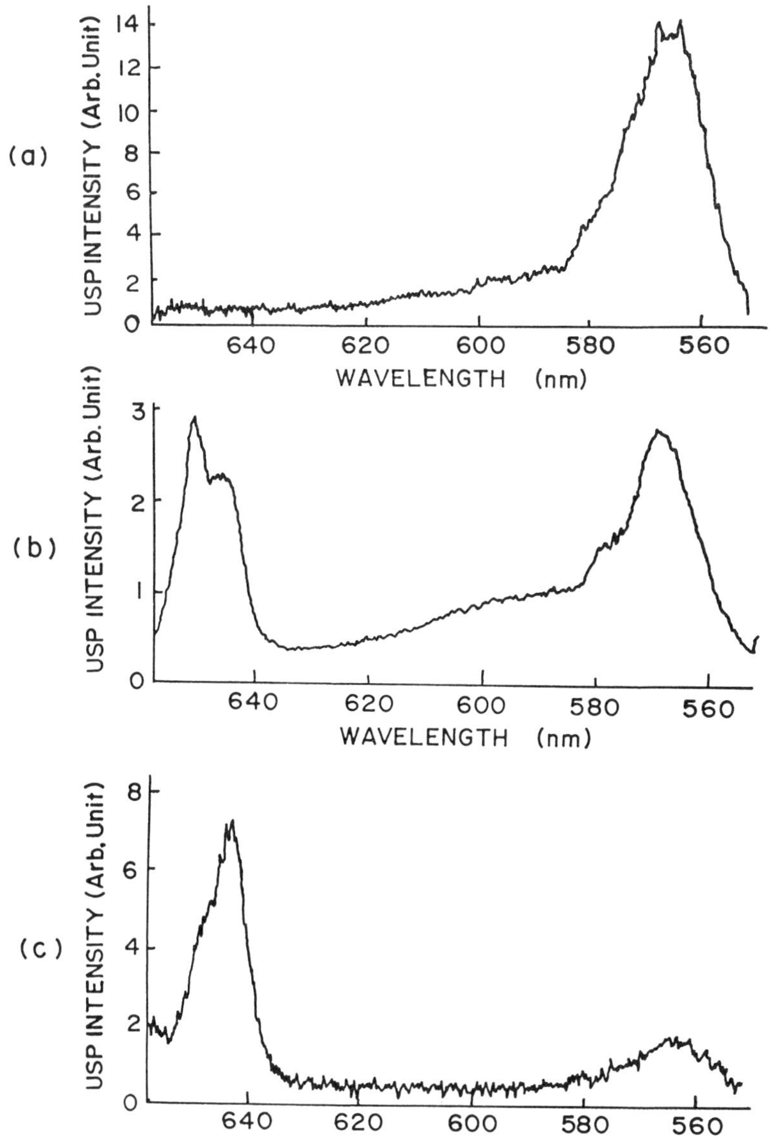

FIGURE 2.22. SPM spectrum of (a) saturated K_2ZnCl_4 solution, (b) 0.6-m K_2ZnCl_4, and (c) pure water. The SRS signal (645 nm) is stronger in pure water, and it disappears in high-concentration solution. (From Jimbo et al., 1987.)

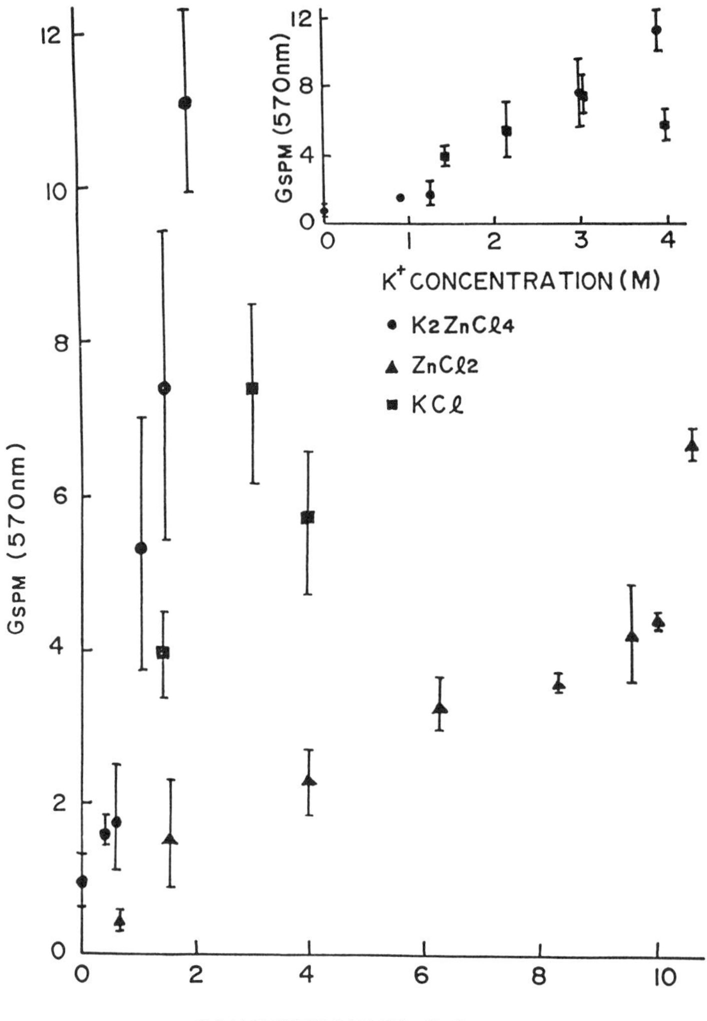

(a)

FIGURE 2.23. Salt concentration dependence of the SPM signal (a) on the Stokes side and (b) on the anti-Stokes side 20°C. Each data point is the average of about 10 laser shots. The inserts are the same data plotted as a function of K^+ ion concentration for KCl and K_2ZnCl_4 aqueous solutions. (From Jimbo et al., 1987.)

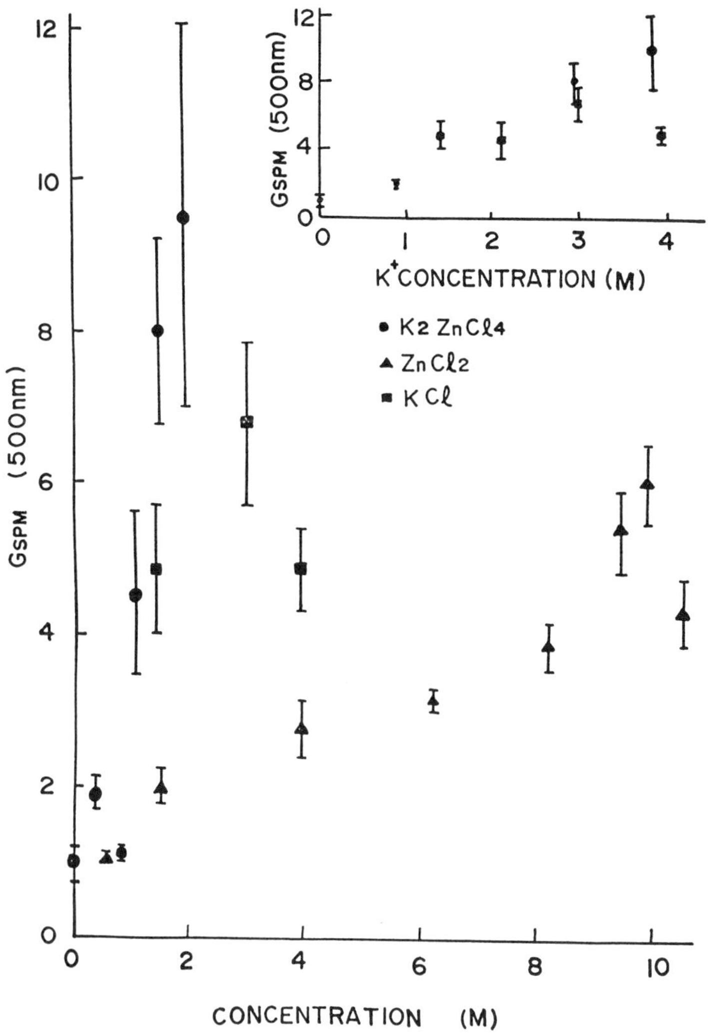

(b)

FIGURE 2.23 (*continued*)

TABLE 2.3. Enhancement of the supercontinuum and optical Kerr effects signals in saturated aqueous solutions at 20°C.[a]

Signal	K_2ZnCl_4 (1.9 M)	KCl (4.0 M)	$ZnCl_2$ (10.6 M)
$G_{SPM}(570)$	11 ± 1	5.6 ± 0.9	6.6 ± 0.4
$G_{SPM}(500)$	9.5 ± 2.5	4.9 ± 0.2	4.3 ± 0.5
G_{Kerr}	16 ± 1	6.1 ± 1.4	35 ± 9

[a] $G_{SPM}(\lambda) = [I_{SPM}(\lambda)/I_{laser}(530 \text{ nm})]_{solution}/[I_{SPM}(\lambda)/I_{laser}(530 \text{ nm})]_{water}$ and $G_{Kerr} = [I_{Kerr}(\text{solution})]/[I_{Kerr}(\text{water})]$.

K_2ZnCl_4 produced the greatest increase in the supercontinuum. Although $ZnCl_2$ generated the largest enhancement of the optical Kerr effect, it did not play an important role in the enhancement of the ultrafast supercontinuum (a possible reason for this is discussed below). The optical Kerr effect signal from saturated solutions of $ZnCl_2$ was about 2 to 3 times greater than that from saturated solutions of K_2ZnCl_4.

The enhancement of the optical nonlinearity of water by the addition of cations can be explained by the cations' disruption of the tetrahedral hydrogen-bonded water structures and their formation of hydrated units (Walrafen, 1972). Since the nonlinear index n_2 is proportional to the number density of molecules, hydration increases the number density of water molecules and thereby increases n_2. The ratio of the hydration numbers of Zn^{2+} and K^+ has been estimated from measurements of G_{Kerr} and compared with their values based on ionic mobility measurements. At the same concentration of KCl and $ZnCl_2$ acqueous solution, (G_{Kerr} generated by $ZnCl_2$ solution)/(G_{Kerr} generated by KCl solution) $= [N(Zn^{2+})/N(K^+)]^2 \sim 2.6$, where $N(Zn^{2+}) \sim 11.2 \pm 1.3$ and $N(K^+) \sim 7 \pm 1$ represent the hydration numbers for the Zn^{2+} and K^+ cations, respectively. The calculation of the hydration number of $N(Zn^{2+})/N(K^+) \sim 1.5$ is in good agreement with the Kerr nonlinearity measurements displayed in Table 2.3.

In addition, from our previous measurements and discussions of nonlinear processes in mixed binary liquids (Ho and Alfano, 1978), the total optical nonlinearity of a mixture modeled from a generalized Langevin equation was determined by the coupled interactions of solute-solute, solute-solvent, and solvent-solvent molecules. The high salt solution concentration may contribute additional optical nonlinearity to the water owing to the distortion from the salt ions and the salt-water molecular interactions.

The finding that Zn^{2+} cations increased G_{Kerr} more than G_{SPM} is consistent with the hydration picture. The transmitted signal of the OKG depends on Δn, while the ultrafast supercontinuum signal is determined by $\partial n/\partial t$. The ultrafast supercontinuum also depends on the response time of the hydrated units. Since the Zn^{2+} hydrated units are larger than those of K^+, the response time will be longer. These effects will be reduced for longer pulses. Two additional factors may contribute to part of the small discrepancy between G_{SPM} and G_{Kerr} for $ZnCl_2$. The first one is related to the mechanism of δn generation in which χ_{1111} is involved in the generation of SPM while the difference $\chi_{1111} - \chi_{1112}$ is responsible for the optical Kerr effect. The second is the possible dispersion of n_2 because of the difference in wavelength between the exciting beams of the ultrafast supercontinuum and the optical Kerr effects.

The optical Kerr effect is enhanced 35 times by using $ZnCl_2$ as a solute, and the ultrafast supercontinuum is enhanced about 10 times by using K_2ZnCl_4 as a solute. The enhancement of the optical nonlinearity has been attributed to an increase in the number density of water molecules owing to hydration and the coupled interactions of solute and solvent molecules. Addition of ions can be used to increase n_2 for SPM generation and gating.

11. Temporal Behavior of SPM

In addition to spectral features, the temporal properties of the supercontinuum light source are important for understanding the generation and compression processes. In this section, the local generation, propagation, and pulse duration reduction of SPM are discussed.

11.1 Temporal Distribution of SPM

In Section 2, using the stationary phase method, it was described theoretically that the Stokes and anti-Stokes frequencies should appear at well-defined locations in time within leading and trailing edges of the pump pulse profile (Alfano, 1972). Theoretical analyses by Stolen and Lin (1978) and Yang and Shen (1984) obtained similar conclusions.

Passing an 80-fs laser pulse through a 500-μm-thick ethylene glycol jet stream, the pulse duration of the spectrum in time was measured by the autocorrelation method (Fork et al., 1983). These results supported the SPM mechanism for supercontinuum generation. In the following, the measurements of the distribution of various wavelengths for the supercontinuum generated in CCl_4 by intense 8-ps laser pulses (Li et al., 1986) are presented. Reduction of the pulse duration using the SPM principle is discussed in Section 10.3.

The incident 530-nm laser pulse temporal profile is shown in Figure 2.24. The pulse shape can be fitted with a Gaussian distribution with duration $\tau(\text{FWHM}) = 8$ ps. The spectral and temporal distributions of the supercontinuum pulse were obtained by measuring the time difference using a streak camera. The measured results are shown as circles in Figure 2.25. Each data point corresponds to an average of about six laser shots. The observation is consistent with the SPM and group velocity dispersion. To determine the temporal distribution of the wavelengths generated within a supercontinuum, the group velocity dispersion effect (Topp and Orner, 1975) in CCl_4 was corrected. Results corrected for both the optical delay in the added filters and the group velocity are displayed as triangles in Figure 2.25. The salient feature

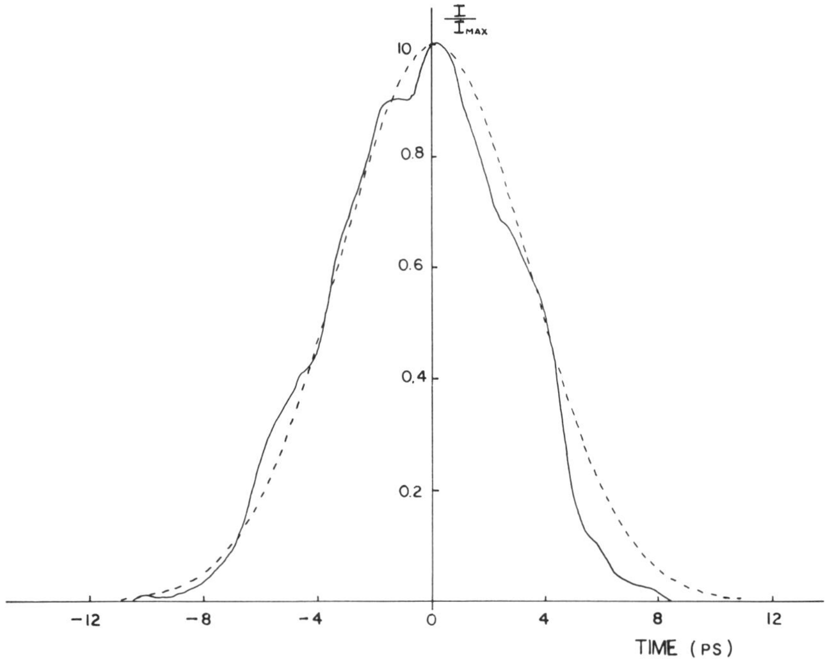

FIGURE 2.24. Temporal profile of a 530-nm incident laser pulse measured by a 2-ps-resolution streak camera. The dashed line is a theoretical fit to an 8-ps EWHM Gaussian pulse. (From Li et al., 1986.)

of Figure 2.25 indicates that the Stokes wavelengths of the continuum lead the anti-Stokes wavelengths.

Using the stationary phase SPM method [Eq. (6)], the generated instantaneous frequency ω of the supercontinuum can be expressed by

$$\omega(t) - \omega_L = -(\omega_L l/c)\,\partial(\Delta n)/\partial t. \tag{33}$$

where ω_L is the incident laser angular frequency, l is the length of the sample, and Δn is the induced nonlinear refractive index $n_2 E^2$. A theoretical calculated curve for the sweep is displayed in Figure 2.26 by choosing appropriate parameters to fit the experimental data of Figure 2.25. An excellent fit using a stationary phase model up to maximum sweep demonstrates that the generation mechanism of the temporal distribution of the supercontinuum arises from the SPM. During the SPM process, a wavelength occurs at a well-defined time within the pulse. The above analysis will be supported by the additional experimental evidence for SPM described in Section 10.3 (see Figure 2.29).

11.2 Local Generation and Propagation

The dominant mechanisms responsible for the generation of the ultrafast supercontinuum as mentioned in Sections 1 and 2 are SPM, FPPG, XPM,

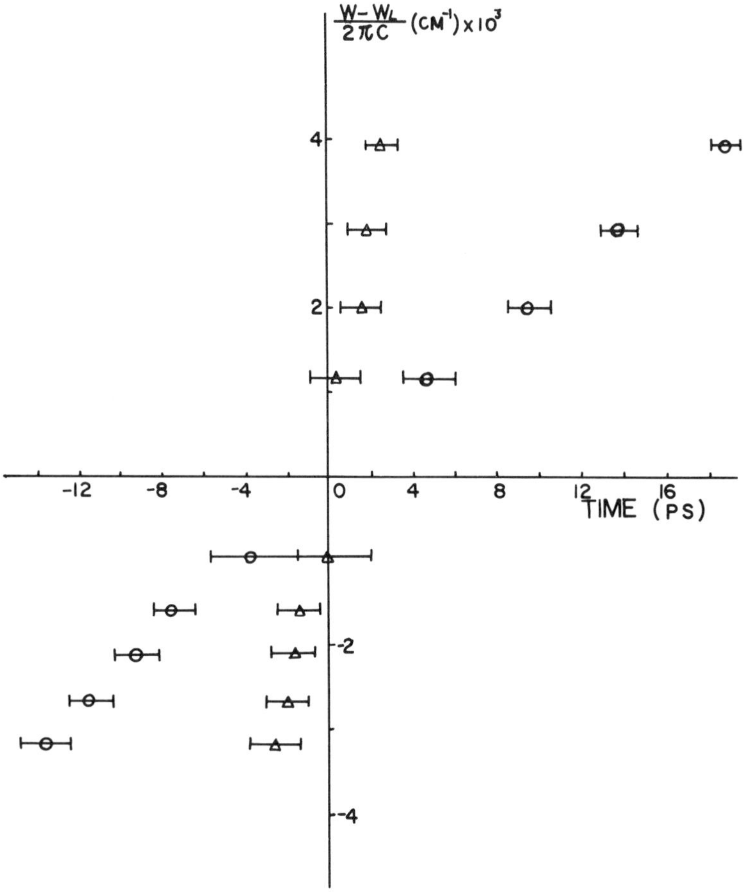

FIGURE 2.25. Measured supercontinuum temporal distribution at different wavelengths: (o) data points with correction of the optical path in filters; (Δ) data points with correction of both the optical path in filters and group velocity dispersion in liquid. (From Li et al., 1986.)

and SRS. In the SPM process, a newly generated wavelength could have bandwidth-limited duration at a well-defined time location (Alfano, 1972) in the pulse envelope. In the FPPG and SRS processes, the duration of the supercontinuum pulse could be shorter than the pump pulse duration due to the high gain about the peak of the pulse. In either case, the supercontinuum pulse will be shorter than the incident pulse at the local spatial point of generation. These pulses will be broadened in time due to the group velocity dispersion in condensed matter (Ho et al., 1987).

Typical data on the time delay of 10-nm-bandwith pulses centered at 530, 650, and 450 nm wavelengths of the supercontinuum generated from a 20-cm-long cell filled with CCl_4 are displayed in Figure 2.27. The peak locations of

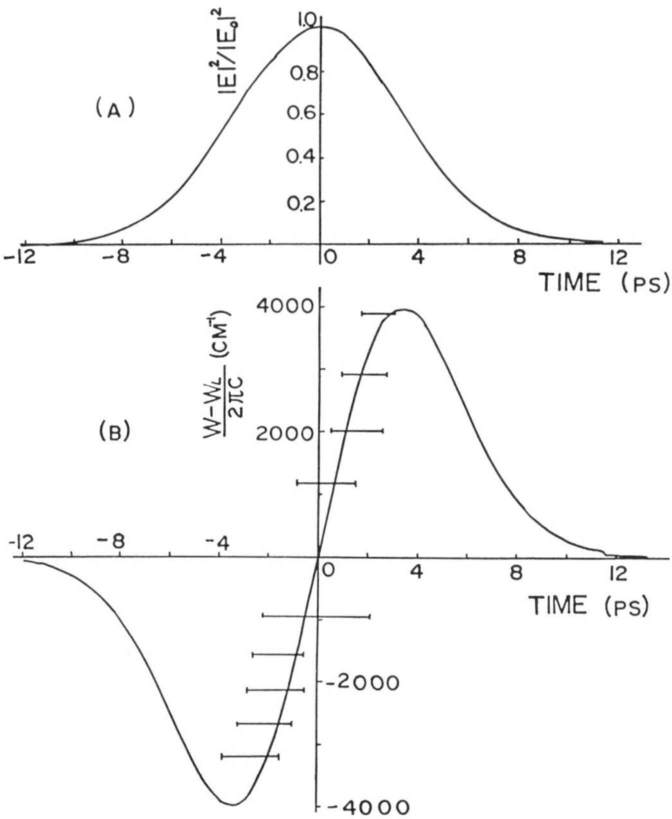

FIGURE 2.26. Comparison of the measured temporal distribution of supercontinuum with the SPM model. (From Li et al., 1986.)

530, 650, and 450 nm are -49, -63, and -30 ps, respectively. The salient features in Figure 2.27 (Ho et al., 1987) indicate that the duration of all 10-nm-band supercontinuum pulses is only 6 ps, which is shorter than the incident pulse of 8 ps, the Stokes side (650 nm) of the supercontinuum pulse travels ahead of the pumping 530 nm by 14 ps, and the anti-Stokes side (450 nm) of the supercontinuum pulse lags the 530 nm by 10 ps.

If the supercontinuum could be generated throughout the entire length of the sample, the Stokes side supercontinuum pulse generated by the 530-nm incident laser pulse at $z = 0$ cm of the sample would be ahead of the 530-nm incident pulse after propagating through the length of the sample. Over this path, 530 nm could continuously generate the supercontinuum pulse. Thus, the Stokes side supercontinuum generated at the end of the sample coincides in time with the 530-nm incident pulse. In this manner, a supercontinuum pulse centered at a particular Stokes frequency could have a pulse greater than the incident pulse extending in time from the energing of the 530-nm pulse to the position where the Stokes frequency was originally produced at $z \sim 0$ cm.

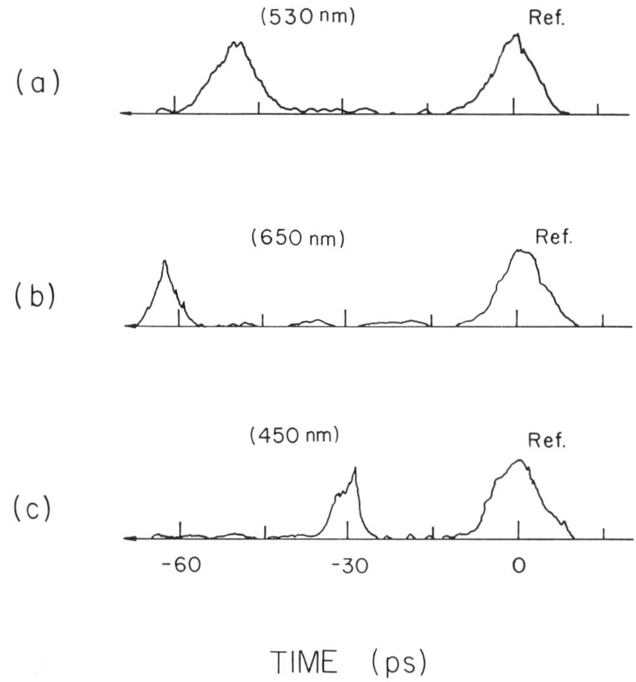

FIGURE 2.27. Temporal profiles and pulse locations of a selected 10-nm band of a supercontinuum pulse at different wavelengths propagated through a 20-cm-long CCl_4 cell: (a) $\lambda = 530$ nm; (b) $\lambda = 650$ nm; (c) $\lambda = 450$ nm. Filter effects were compensated. (From Ho et al., 1987.)

From a similar consideration, the anti-Stokes side supercontinuum pulse would also be broadened. However, no slow asymmetric tail for the Stokes pulse or rise for the anti-Stokes pulse is displayed in Figure 2.27. These observation suggest the local generation of supercontinuum pulses.

A model to describe the generation and propagation features of the supercontinuum pulse has been formulated based on local generation. The time delay of Stokes and anti-Stokes supercontinuum pulses relative to the 530-nm pump pulse is accounted for by the filaments formed ~5 cm from the sample cell entrance window. The 5-cm location is calculated from data in Figure 2.27 by using the equation

$$T_{530} - T_{\text{supercon.}} = \Delta x\left(\frac{1}{v_{530}} - \frac{1}{v_{\text{supercon.}}}\right), \tag{34}$$

where Δx is the total length of supercontinuum pulse travel in CCl_4 after the generation, T_{350} and $T_{\text{supercon.}}$ are the 530-nm and supercontinuum pulse peak time locations in Figure 2.27, and v_{530} and $v_{\text{supercon.}}$ are the group velocities of the 530-nm and supercontinuum pulses, respectively.

The duration of the supercontinuum pulse right at the generation location is either limited by the bandwidth of the measurement from the SPM process or shortened by the parametric generation process. In either cases, a 10-nm-bandwidth supercontinuum pulse will have a shorter duration than the incident pulse. After being generated, each of these 10-nm-bandwidth supercontinuum pulses will travel through the rest of the sample and will continuously generated by the incident 530 nm over a certain interaction length before these two pulses walk off. The interaction length can be calculated as (Alfano, 1972).

$$l = \tau \frac{v_{530} v_{\text{supercon.}}}{v_{530} - v_{\text{supercon.}}}, \tag{35}$$

where l is the interaction length over the pump and the supercontinuum pulses stay spatially coincident by less than the duration (FWHM) of the incident pump pulse, and τ is the duration of the supercontinuum pulse envelope. From Eq. (35), one can estimate the interaction length from the measured τ of the supercontinuum pulse. Using parameters $\tau = 6$ ps, $v_{530} = c/1.4868$, and $v_{\text{supercon.}} = c/1.4656$, the interaction length $l = 8.45$ cm is calculated. This length agrees well with the measured beam waist length of 8 cm for the pump pulse in CCl_4.

Since no long tails were observed from the supercontinuum pulses to the dispersion delay times of the Stokes and anti-Stokes supercontinuum pulses, the supercontinuum was not generated over the entire length of 20 cm but only over 1 to 9 cm. This length is equivalent to the beam waist length of the laser in CCl_4. The length of the local SPM generation over a distance of 8.45 cm yields a possible explanation for the 6-ps supercontinuum pulse duration. In addition, a pulse broadening of 0.3 ps calculated from the group velocity dispersion of a 10-nm band at 650-nm supercontinuum traveling over 20 cm of liquid CCl_4 is negligible in this case.

Therefore, the SPM pulses have shorter durations than the pump pulse and were generated over local spatial domains in the liquid cell.

11.3 SPM Pulse Duration Reduction

The principle behind the pulse narrowing based on the spectral temporal distribution of the SPM spectrally broadened in time within the pulse is described in Sections 2 and 10.1. At each time t within the pulse there is a frequency $\omega(t)$. When a pulse undergoes SPM, the changes in the optical carrier frequency within the temporal profile are greatest on the rising and falling edges, where the frequency is decreased and increased, respectively. Near the peak of the profile, and in the far leading and trailing wings, the carrier frequency structure is essentially unchanged. The maximum frequency shift is proportional to the intensity gradient on the sides of the pulse, and this determines the position of the outer lobes of the power spectrum. If these are then attenuated by a spectral window of suitably chosen width, the wings of the profile where the high- and low-frequency components are chiefly concentrated will be depressed, while the central peak will be largely unaffected.

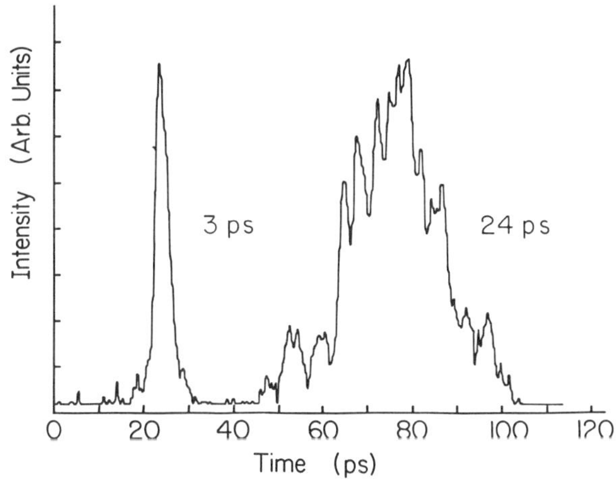

FIGURE 2.28. Streak camera temporal profile of the 25-ps, 530-nm incident laser pulse and 10-nm-bandwidth pulse at 580 nm. The 3-ps pulse was obtained by spectral filtering a SPM frequency continuum generated in D_2O. (From Dorsinville et al., 1987.)

The overall effect is to create a pulse that is significantly narrower in time than the original pulse duration. A file can be used to select a narrow portion of the pulse, giving rise to a narrower pulse in time.

A threefold shortening of 80-ps pulses to 30 ps from an Nd:YAG laser broadened from 0.3 to 4 Å after propagation through 125 m of optical fiber with a monochromator as a spectral window was demonstrated using this technique (Gomes et al., 1986). The measurements of pulses at different wavelengths of the frequency sweep of supercontinuum pulses generated by 8-ps laser pulses propagating in CCl_4 show that the continuum pulses have a shorter duration ($\sim$6 ps) than the pumping pulses (Li et al., 1986).

A major advance occurred when a 25-ps laser pulse was focused into a 5-cm-long cell filled with D_2O. A continuum was produced. Using 10-nm-bandwidth narrowband filters, tunable pulses of less than 3 ps in the spectral range from 480 to 590 nm (Figure 2.28) were produced (Dorsinville et al., 1987).

To identify the SPM generation mechanism, the temporal distribution of the continuum spectrum was determined by measuring the time delay between the continuum and a reference beam at different wavelengths using a streak camera. The results are displayed in Figure 2.29, which is similar to data displayed in Figure 2.25. The time delay was $\sim$22 ps for a 140-nm change in wavelength; as predicted by the SPM mechanism, the Stokes wavelength led the anti-Stokes wavelength (Alfano, 1972). The delay due to group velocity over a 5-cm D_2O cell for the 140-nm wavelength change is less than 3 ps. The remaining 18 ps is well accounted for by the SPM mechanism using a 25-ps (FWHM) pulse and the stationary phase method (Alfano, 1972). Furthermore, a 10-nm selected region in the temporal distribution curve corresponds to an

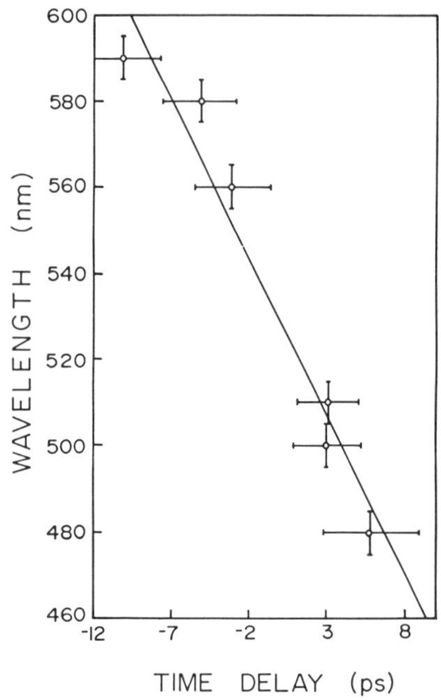

FIGURE 2.29. Continuum temporal distribution at different wavelengths. Horizontal error bars correspond to 10-nm bandwidths of the filter. (From Dorsinville et al., 1987.)

$\sim$2.6-ps width matching the measured pulse duration (Figure 2.28). This observation suggests that by using narrower bandwidth filters the pulse duration can be shortened to the uncertainty limit.

12. Higher-Order Effects on Self-Phase Modulation

A complete description of SPM-generated spectral broadening should take into account higher-order effects such as self-focusing, group velocity dispersion, self-steepening, and initial pulse chirping. Some of these effects are described by Suydam (Chapter 6), Shen (Chapter 1), and Agrawal (Chapter 3). These effects will influence the observed spectral profiles.

12.1 Self-Focusing

In the earliest experiments using picosecond pulses, the supercontinuum pulses were often generated in small-scale filaments resulting from the self-focusing of intense laser beams (Alfano, 1972). Self-focusing arises from the radial dependence of the nonlinear refractive index $n(r) = n_0 + n_2 E^2(r)$ (Shen, 1984; Auston, 1977). It has been observed in many liquids, bulk materials (Shen, 1984), and optical fibers (Baldeck et al., 1987a). It effects on the continuum pulse generation can be viewed as good and bad. On the one hand, it facilitates the spectral broadening by concentrating the laser beam energy. On the other hand, self-focusing is a random and unstable phenomenon that is not controllable. Femtosecond supercontinua are generated with thinner samples than picosecond supercontinua, so it can reduce but not totally eliminate self-focusing effects.

12.2 Dispersion

Group velocity dispersion (GVD) arises from the frequency dependence of the refractive index. These effects are described by Agrawal (Chapter 3). The first-order GVD term leads to a symmetric temporal broadening (Marcuse, 1980). A typical value for the broadening rate arising from $\partial^2 k/\partial\omega^2$ is 500 fs/m · nm (in silica at 532 nm). In the case of supercontinuum generation, spectral widths are generally large (several hundred nanometers), but interaction lengths are usually small (<1 cm). Therefore, the temporal broadening arising from GVD is often negligible for picosecond pulses but is important for femtosecond pulses. Limitations on the spectral extent of supercontinuum generation are also related to GVD. Although the spectral broadening should increase linearly with the medium length (i.e., $\Delta\omega(z)_{\max} = \omega_0 n_2 a^2 z/c\,\Delta\tau$), it quickly reaches a maximum as shown in Figure 2.10. This is because GVD, which is large for pulses having SPM-broadened spectra, reduces the pulse peak power a^2 and broadens the pulse duration $\Delta\tau$. As shown in Figure 2.25, the linear chirp parameter is decreased by the GVD chirp in the normal dispersion regime. This effect is used to linearize chirp in the pulse compression technique.

The second-order term $\partial^3 k/\partial\omega^3$ has been found to be responsible for asymmetric distortion of temporal shapes and modulation of pulse propagation in the lower region of the optical fiber (Agrawal and Potasek, 1986). Since the spectra of supercontinuum pulses are exceptionally broad, this term should also lead to asymmetric distortions of temporal and spectral shapes of supercontinuum pulses generated in thick samples. These effects have been observed.

In multimode optical fibers, the mode dispersion dominates and causes distortion of the temporal shapes. This in turn yields asymmetric spectral broadening (Wang et al., 1988).

12.3 Self-Steepening

Pulse shapes and spectra of intense supercontinuum pulses have been found to be asymmetric (De Martini et al., 1967). There are two potential sources of asymmetric broadening in supercontinuum generation. The first one is the second-order GVD term. The second one is self-steepening, which is intrinsic to the SPM process and occurs even in nondispersion media. Details of the effects of self-steepening can be found in Suydam (Chapter 6), Shen and Yang (Chapter 1), and Manassah (Chapter 5).

Because of the intensity and time dependence of the refractive index, $n = n_0 + n_2 E(t)^2$, the supercontinuum pulse peak sees a higher refractive index than its edges. Because $v = c/n$, the pulse peak travels slower than the leading and trailing edges. This results in a sharpened trailing edge. Self-steepening

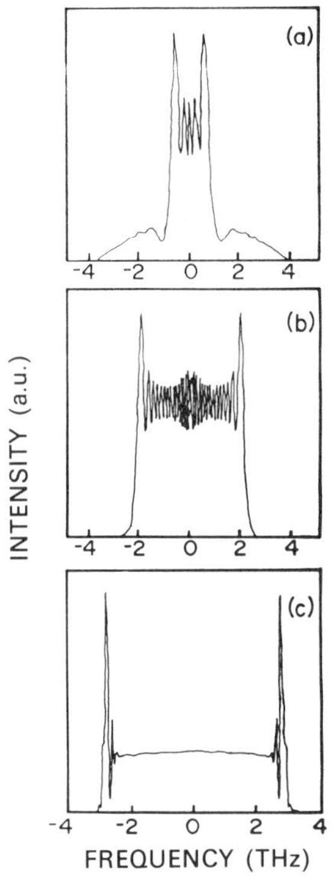

FIGURE 2.30. Influence of initial pulse chirping on SPM-broadened spectra in optical fibers. Peak power = 1000 W. (a) $C = 50$; (b) $C = 0$; (c) $C = -50$ [see Eq. (17)]. (From Baldeck et al., 1987b.)

occurs and more blue-shifted frequencies (sharp trailing edge) are generated than red frequencies. Several theoretical approaches have given approximate solutions for the electric field envelope distorted by self-steepening and asymmetric spectral extent. Actual self-steepening effects have not been observed in the time domain.

12.4 Initial Pulse Chirping

Most femtosecond and picosecond pulses are generated with initial chirps. Chirps arise mainly from GVD and SPM in the laser cavity. As shown in Figure 2.30, the spectral broadening is reduced for positive chirps and enhanced for negative chirps in the normal dispersion regime. The spectral distribution of SPM is also affected by the initial chirp.

13. Overview

Supercontinuum generation is the generation of bursts of "white" light, which can be obtained by passing intense picosecond or femtosecond pulses through various materials. Because of the nonlinear response of the medium, the pulse envelope yields a phase modulation that initiates the wide frequency broadening (up to 10,000 cm^{-1}). The phase modulation can be generated by the pulse itself, a copropagating pump pulse, or the copropagating stronger pulse. These different configurations are called self-phase modulation (SPM), induced-phase modulation (IPM), and cross-phase modulation (XPM), respectively. The SPM process for supercontinuum generation in various materials was reviewed in this chapter. The latter two processes are closely related to each other and are described by Baldeck et al. (Chapter 4), Agrawal (Chapter 3), and Manassah (Chapter 5).

Using an 8-ps laser pulse at 530 nm, typical Stokes sweeps were 4400 cm^{-1} in a calcite crystal of length 4 cm, 3900 cm^{-1} in a quartz crystal of length 4.5 cm, 1100 cm^{-1} in extra-dense flint glass of length 7.55 cm, 3900 cm^{-1} in NaCl of length 4.7 cm, and 4200 cm^{-1} in both BK-7 and LBC-1 glasses of length 8.9 cm. Sweeps on the anti-Stokes side were typically 6100 cm^{-1} in calcite, 5500 cm^{-1} in quartz, 7300 cm^{-1} in NaCl, and 7400 cm^{-1} in BK-7 and LBC-1 glasses. An infrared supercontinuum spanning the range from 3 to 14 μm can be obtained by passing an intense laser pulse generated from a CO_2 laser through GaAs, AgBr, ZnSe, and CdS crystals. Near- and medium-infrared spectral sweeps of 3200 cm^{-1} on the Stokes side and 4900 cm^{-1} on the anti-Stokes side can be realized by passing a strong 1.06-μm pulse through a KBr crystal of length 10 cm. Sweeps on the order of 1000 cm^{-1} are observed to both the red and blue sides of 530 nm in liquid argon. Similar spectral sweeps are observed in liquid and solid krypton arising from electronic mechanism for SPM. Using a picosecond laser train of wavelength 530 nm, the spectra were broadened up to 3000 cm^{-1} to either side of the laser frequency in a 5-cm-long magnetic $KNiF_3$ single crystal. Production of SPM near electronic levels of PrF_3 crystal and enhancement of supercontinuum in water by addition of Zn^{2+} and K^+ cations have been also discussed. The temporal properties of supercontinuum pulses have been described. Higher-order effects on SPM arising from dispersion, self-focusing, self-steepening, and initial pulse chirping were briefly described.

SPM will continue to be an important nonlinear process in science and technology and has been one of the most important ultrafast nonlinear optical processes for more than 20 years since the advent of ultrashort laser pulses!

References

Agrawal, G.P. and M.J. Potasek (1986) Nonlinear pulse distortion in single-mode optical fibers at the zero-dispersion wavelength. Phys. Rev. A **33**, 1765–1776.

Alfano, R.R. (1972) Interaction of picosecond laser pulses with matter. GTE Technical Report TR 72–330. Published as Ph.D. thesis at New York University, 1972.

Alfano, R.R. (1986) The ultrafast superdcontinuum laser source. *Proc. International Conference Laser* '85. STS Press, McLean, Virigina, pp. 110–122.

Alfano, R.R. and P.P. Ho (1988) Self-, cross-, and induced-phase modulations of ultrashort laser pulse propagation. IEEE J. Quantum Electron. **QE-24**, 351–363.

Alfano, R.R. and S.L. Shapiro (1970a) Emission in the region 4000–7000 A via four-photon coupling in glass. Phys. Rev. Lett. **24**, 584–587; Observation of self-phase modulation and small scale filaments in crystals and glasses. Phys. Rev. Lett. **24**, 592–594; Direct distortion of electronic clouds of rare-gas atoms in intense electric fields. Phys. Rev. Lett. **24**, 1219–1222.

Alfano, R.R. and S.L. Shapiro (1970b) Picosecond spectroscopy using the inverse Raman effect. Chem. Phys. Lett. **8**, 631–633.

Alfano, R.R., L. Hope, and S. Shapiro (1972) Electronic mechanism for production of self-phase modulation. Phys. Rev. A **6**, 433–438.

Alfano, R.R., J. Gersten, G. Zawadzkas, and N. Tzoar (1974) Self-phase-modulation near the electronic resonances of a crystal. Phys. Rev. A **10**, 698–708.

Alfano, R.R., P. Ho, P. Fleury, and H. Guggeneheim (1976) Nonlinear optical effects in antiferromagnetic $KNiF_3$. Opt. Commun. **19**, 261–264.

Alfano, R.R., Q. Li, T. Jimbo, J. Manassah, and P. Ho (1986) Induced spectral broadening of a weak picosecond pulse in glass produced by an intense ps pulse. Opt. Lett. **11**, 626–628.

Alfano, R.R., Q.Z. Wang, T. Jimbo, and P.P. Ho (1987) Induced spectral broadening about a second harmonic generated by an intense primary ultrashort laser pulse in ZnSe crystals. Phys. Rev. A **35**, 459–462.

Alfano, R.R., Q.Z. Wang, D. Ji, and P.P. Ho (1989) Harmonic cross-phase-modulation in ZnSe. App. Phys. Lett. **54**, 111–113.

Anderson, D. and M. Lisak (1983) Nonlinear asymmetric self-phase modulation and self-steepening of pulses in long optical waveguides. Phys. Rev. A **27**, 1393–1398.

Auston, D.H. (1977) In *Ultrafast Light Pulses*, S.L. Shapiro, ed., Springer,Verlag, New York.

Baldeck, P.L., F. Raccah, and R.R. Alfano (1987a) Observation of self-focusing in optical fibers with picosecond pulses. Opt. Lett. **12**, 588–589.

Baldeck, P.L., P.P. Ho, and R.R. Alfano (1987b) Effects of self-, induced-, and cross-phase modulations on the generation of ps and fs white light supercontinuum. Rev. Phys. Appl. **22**, 1877–1894.

Bloembergen, N. and P. Lallemand (1966) Complex intensity dependent index of refraction frequency broadening of stimulated Raman lines and stimulated Rayleigh scattering. Phys. Rev. Lett. **16**, 81–84.

Bourkoff, E., W. Zhao, and R.I. Joseph (1987) Evolution of femtosecond pulses in single-mode fibers having higher-order nonlinearity and dispersion. Opt. Lett. **12**, 272–274.

Brewer, R.G. (1967) Frequency shifts in self-focused light. Phys. Rev. Lett. **19**, 8–10.

Brewer, R.G. and C.H. Lee (1968) Self-trapping with picosecond light pulses. Phys. Rev. Lett. **21**, 267–270.

Busch, G.E., R.P. Jones, and P.M. Rentzepis (1973) Picosecond spectroscopy using a picosecond continuum. Chem. Phys. Lett. **18**, 178–185.

Chinn, S.R., H. Zeiger, and J. O'Connor (1971) Two-magnon Raman scattering and exchange interactions in antiferromagnetic $KNiF_3$ and K_2NiF_4 and ferrimagnetic $RbNiF_3$. Phys. Rev. **B3**, 1709–1735.

Corkum, P., P. Ho, R. Alfano, and J. Manassah (1985) Generation of infrared supercontinuum covering 3–14 μm in dielectrics and semiconductors. Opt. Lett. **10**, 624–626.

Corkum, P.B., C. Rolland, and T. Rao (1986) Supercontinuum generation in gases. Phys. Rev. Lett. **57**, 2268–2271.

Cornelius, P. and L. Harris (1981) Role of self-phase modulation in stimulated Raman scattering from more than one mode. Opt. Lett. **6**, 129–131.

DeMartini, F., C.H. Townes, T.K. Gustafson, and P.L. Kelly (1967) Self-steepening of light pulses. Phys. Rev. **164**, 312–322.

Dorsinville, R., P. Delfyett, and R.R. Alfano (1987) Generation of 3 ps pulses by spectral selection of the supercontinuum generated by a 30 ps second harmonic Nd:YAG laser pulse in a liquid. Appl. Opt. **27**, 16–18.

Fisher, R.A. and W. Bischel (1975) Numerical studies of the interplay wave laser pulse. J. Appl. Phys. **46**, 4921–4934.

Fisher, R.A., B. Suydam, and D. Yevich (1983) Optical phase conjugation for time domain undoing of dispersion self-phase modulation effects. Opt. Lett. **8**, 611–613.

Fleury, P.A., W. Hayes, and H.J. Guggenheim (1975) Magnetic scattering of light in $K(NiMg)F_3$. J. Phys. C: Solid State **8**, 2183–2189.

Fork, R.L., C.V. Shank, C. Hirliman, R. Yen, and J. Tomlinson (1983) Femtosecond white-light continuum pulse. Opt. Lett. **8**, 1–3.

Fork, R.L., C.H. Brito Cruz, P.C. Becker, and C.V. Shank (1987) Compression of optical pulses to six femtoseconds by using cubic phase compensation. Opt. Lett. **12**, 483–485.

Gersten, J., R. Alfano, and M. Belic (1980) Combined stimulated Raman scattering in fibers. Phys. Rev. A **21**, 1222–1224.

Girodmaine, J.A. (1962) Mixing of light beams in crystals. Phys. Rev. Lett. **8**, 19–20.

Glownia, J., G. Arjavalingam, P. Sorokin, and J. Rothenberg (1986) Amplification of 350-fs pulses in XeCl excimer gain modules. Opt. Lett. **11**, 79–81.

Goldberg, L. (1982) Broadband CARS probe using the picosecond continua. In *Ultrafast Phenomena III*. Springer-Verlag, New York, pp. 94–97.

Gomes, A.S.L., A.S. Gouveia-Neto, J.R. Taylor, H. Avramopoulos, and G.H.C. New (1986) Optical pulse narrowing by the spectral windowing of self-phase modulated picosecond pulses. Opt. Commun. **59**, 399.

Gustafson, T.K., I.P. Taran, H.A. Haus, J.R. Lifisitz, and P.L. Kelly (1969) Self-modulation, self-steepening, and spectral development of light in small-scale trapped filaments. Phys. Rev **177**, 306–313.

Gustafson, T.K., J. Taran, P. Kelley, and R. Chiao (1970) Self-modulation of picosecond pulse in electro-optical crystals. Opt. Commun. **2**, 17–21.

Hellwarth, R.W. (1970) Theory of molecular light scattering spectra using linear-dipole approximation. J. Chem. Phys. **52**, 2128–2138.

Hellwarth, R.W., J. Cherlow, and T.T. Yang (1975) Origin and frequency dependence of nonlinear optical susceptibilities of glasses. Phys. Rev. B **11**, 964–967.

Heritage, J., A. Weiner, and P. Thurston (1985) Picasecond pulse shaping by spectral phase and amplitude manipulation. Opt. Lett. **10**, 609–611.

Ho, P.P. and R.R. Alfano (1978) Coupled molecular reorientational relaxation kinetics in mixed binary liquids directly measured by picosecond laser techniques. J. Chem. Phys. **68**, 4551–4563.

Ho, P.P. and R.R. Alfano (1979) Optical Kerr effect in liquids. Phys. Rev. A **20**, 2170–4564.

Ho, P.P., Q.X. Li, T. Jimbo, Y.L. Ku, and R.R. Alfano (1987) Supercontinuum pulse generation and propagation in a liqud carbon tetrachloride. Appl. Opt. **26**, 2700–2702.

Ishida, Y., K. Naganuma, T. Yagima, and C. Lin (1984) Ultrafast self-phase modulation in a colliding pulse mode-locking ring dye laser. In *Ultrafast Phenomena IV*. Springer-Verlag, New York, pp. 69–71.

Jimbo, T., V.L. Caplan, Q.X. Li, Q.Z. Wang, P.P. Ho, and R.R. Alfano (1987) Enhancement of ultrafast supercontinuum generation in water by addition of Zn^{2+} and K^+ cations. Opt. Lett. **12**, 477–479.

Johnson, A., R. Stolen, and W. Simpson (1986) The observation of chirped stimulated Raman scattering light in fibers. In *Ultrafast Phenomena* V.G.R. Fleming and A.E. Siegman ed. Springer-Verlag, New York, pp. 160–163.

Jones, W.J. and B.P. Stoicheff (1964) Inverse Raman spectra: induced absorption at optical frequencies. Phys. Rev. Lett. **13**, 657–659.

Knox, K., R.G. Shulman, and S. Sugano (1963) Covalency effects in $KNiF_3$. II. Optical studies. Phys. Rev. **130**, 512–516.

Knox, W., R. Fork, M. Dower, R. Stolen, and C. Shank (1985) Optical pulse compression to 8-fs at 5-kHz repetition rate. Appl. Phys. Lett. **46**, 1120–1121.

Kobayashi, T. (1979) Broadband picosecond light generation in phosphoric acid by a mode-locked laser. Opt. Commun. **28**, 147–149.

Lallemand, P. (1966) Temperature variation of the width of stimulated Raman lines in liquids. Appl. Phys. Lett. **8**, 276–277.

Levenson, M.D. and N. Bloembergen (1974) Dispersion of the nonlinear optical susceptibility tensor in centrosymmetric media. Phys. Rev. B **10**, 4447–4463.

L, Q.X., T. Jimbo, P.P. Ho, and R.R. Alfano (1986) Temporal distribution of picosecond super-continuum generated in a liquid measured by a streak camera. Appl. Opt. **25**, 1869–1871.

Lozobkin, V., A. Malytin, and A. Prohorov (1970) Phase self-modulation of Nd:glass radiation with mode-locking. JETP Lett. **12**, 150–152.

Magde, D. and M.W. Windsor (1974) Picosecond flash photolysis and spectroscopy: 3,3′-diethyloxadicarbocyanine iodide (DODCI). Chem. Phys. Lett. **27**, 31–36.

Manassah, J.T., P.P. Ho, A. Katz, and R.R. Alfano (1984) Ultrafast supercontinuum laser source. Photonics Spectra **18** November, 53–59.

Manassah, J.T., R.R., Alfano and M. Mustafa (1985a) Spectral distribution of an ultrashort supercontinuum laser source. Phys. Lett. A **107**, 305–309.

Manassah, J.T., M. Mustafa, R. Alfano, and P. Ho (1985b) Induced supercontinuum and steepening of an ultrafast laser pulse. Phys. Lett. **113A**, 242–247.

Manassah, J.T., M. Mustafa, R.R. Alfano, and P.P. Ho (1986) Spectral extent and pulse shape of the supercontinuum for ultrashort laser pulse. IEEE J. Quantum Electron. **QE-22**, 197–204.

Marcuse, D. (1980) Pulse distortion in single-mode optical fibers. Appl. Opt. **19**, 1653–1660.

Masuhara, H., H. Miyasaka, A. Karen, N. Mataga, and Y. Tsuchiya (1983) Temporal characteristics of picosecond continuum as revealed by two-dimensional analysis of streak images. Opt. Commun. **4**, 426.

McTague, J., P. Fleury, and D. DuPre (1969) Intermolecular light scattering in liquids. Phys. Rev. **188**, 303–308.

Nakashima, N. and N. Mataga (1975) Picosecond flash photolysis and transient spectral measurements over the entire visible, near ultraviolet and near infrared regions. Chem. Phys. Lett. **35**, 487–492.

Patel, C.K.N. and E.D. Shaw (1971) Tunable stimulated Raman scattering from mobile carriers in semiconductors. Phys. Rev. B **3**, 1279–1295.

Penzokfer, A., A. Laubereau, and W. Kasier (1973) Stimulated short-wave radiation due to single frequency resonances of $\chi^{(3)}$. Phys. Rev. Lett. **31**, 863–866.

Potasek, M.J., G.P. Agrawal, and S.C. Pinault (1986) Analytical and numerical study of pulse broadening in nonlinear dispersive optical fibers. J. Opt. Soc. Am. B **3**, 205–211.

Shank, C. (1983) Measurement of ultrafast phenomena in the femtosecond domain. Science **219**, 1027.

Shank, C.V., R.L. Fork, R. Yen, and R.H. Stolen (1982) Compression of femtosecond optical pulses. Appl. Phys. Lett. **40**, 761–763.

Sharma, D.K., R.W. Yid, D.F. Williams, S.E. Sugamori, and L.L.T. Bradley (1976) Generation of an intense picosecond continuum in D_2O by a single picosecond 1.06 μ pulse. Chem. Phys. Lett. **41**, 460–465.

Shen, Y.R. (1966) Electrostriction, optical Kerr effect and self-focusing of laser beams. Phys. Lett. **20**, 378.

Shen, Y.R. (1984) *The Principles of Nonlinear Optics*. Wiley, New York.

Shimizu, F. (1967) Frequency broadening in liquids by a short light pulse. Phys. Rev. Lett. **19**, 1097–1100.

Stolen, R.H. and A.M. Johnson (1986) The effect of pulse walkoff on stimulated Raman scattering in optical fibers. IEEE J. Quantum Electron. **QE-22**, 2154–2160.

Stolen, R.H. and C. Lin (1978) Self-phase modulation in silica optical fibers. Phys. Rev. A **17**, 1448–1453.

Topp, M.R. and G.C. Orner (1975) Group velocity dispersion effects in picosecond spectroscopy. Opt. Commun. **13**, 276.

Tzoar, N. and M. Jain (1981) Self-phase modulation in long-geometry waveguide. Phys. Rev. A **23**, 1266–1270.

Walrafen, G.E. (1972) Stimulated Raman scattering and the mixture model of water structure. Adv. Mol. Relaxation Processes **3**, 43–49.

Wang, Q.Z., D. Ji, Lina Yang, P.P. Ho, and R.R. Alfano (1989) Self-phase modulation in multimode optical fibers with modest high power. Produced by moderately high power picosecond pulses. *Opt. Lett.*, in press.

Yablonovitch, E. and N. Bloembergen (1972) Avalanche ionization of the limiting diameter of filaments induced by light pulses in transparent media. Phys. Rev. Lett. **29**, 907–910.

Yang, G. and Y.R. Shen (1984) Spectral broadening of ultrashort pulse in a nonlinear medium. Opt. Lett. **9**, 510–512.

Yu, W., R. Alfano, C.L. Sam, and R.J. Seymour (1975) Spectral broadening of picosecond 1.06 μm pulse in KBr. Opt. Commun. **14**, 344.

Appendix: Nonlinear Wave Equation with Group Velocity Dispersion

We start with Maxwell equations for the electric and magnetic fields **E** and **H** in Gaussian units

$$\begin{aligned} \nabla \times \mathbf{E} &= -\frac{1}{c}\frac{\partial \mathbf{B}}{\partial t}, \\ \nabla \times \mathbf{H} &= \frac{1}{c}\frac{\partial \mathbf{D}}{\partial t} + \frac{4\pi}{c}\mathbf{J}, \\ \nabla \cdot \mathbf{D} &= 4\pi\rho, \\ \nabla \cdot \mathbf{B} &= 0. \end{aligned} \tag{A.1}$$

The helping equations are $\mathbf{D} = \varepsilon\mathbf{E}$ and $\mathbf{B} = \mu\mathbf{H}$, and $\mathbf{J}$ and ρ are the current and charge densities, respectively. For nonmagnetic material, $\mathbf{B} \approx \mathbf{H}$. The refractive index of an isotropic material possessing nonlinearity can be written as

$$n(\omega) = [\varepsilon(\omega)]^{1/2} = n_0(\omega) + n_2|\mathbf{E}|^2, \tag{A.2}$$

where $n_0(\omega)$ is the linear refractive index and n_2 the nonlinear refractive index. In the absence of sources, from Maxwell equations one can readily obtain the wave equation

$$\nabla^2\mathbf{E}(\mathbf{r},t) - \frac{1}{c^2}\frac{\partial^2}{\partial t^2}\mathbf{D}_L(\mathbf{r},t) = \frac{2n_0 n_2}{c^2}\frac{\partial^2}{\partial t^2}(|\mathbf{E}|^2\mathbf{E}(\mathbf{r},t)), \tag{A.3}$$

where $\mathbf{D}_L(\mathbf{r},t)$ is the linear electric displacement vector. In obtaining the equation, we have used $\nabla \times (\nabla \times \mathbf{E}) = \nabla(\nabla \cdot \mathbf{E}) - \nabla^2\mathbf{E} \approx -\nabla^2\mathbf{E}$ and neglected the $(n_2)^2$ term.

The electric field can be written as

$$\mathbf{E}(\mathbf{r},t) = \Phi(x,y)\mathbf{E}(z,t), \tag{A.4}$$

where $\Phi(x,y)$ is the transverse distribition function. Substitute Eq. (A.4) into the wave equation and averaging over transverse coordinates, we have

$$\frac{\partial^2}{\partial z^2}\mathbf{E}(z,t) - \frac{1}{c}\frac{\partial^2}{\partial t^2}\mathbf{D}_L(z,t) = \frac{2n_0\bar{n}_2}{c^2}\frac{\partial^2}{\partial t^2}|\mathbf{E}(z,t)|^2\mathbf{E}(z,t). \tag{A.5}$$

We have neglected the $\partial^2/\partial x^2$ and $\partial^2/\partial y^2$ terms. The effective nonlinear refractive index $\bar{n}_2$ is

$$\bar{n}_2 = \frac{\int n_2\Phi^2(x,y)\,dx\,dy}{\int \Phi^2(x,y)\,dx\,dy} \approx \frac{1}{2}n_2. \tag{A.6}$$

Using a plane wave approximation with $(k_0 z - \omega_0 t)$ representation, a linearly polarized electric field propagating along z direction can be written as

$$\mathbf{E}(z,t) = \hat{\mathbf{e}}A(z,t)\exp[i(k_0 z - \omega_0 t)], \tag{A.7}$$

where $\hat{\mathbf{e}}$ is the unit vector of polarization of electric field, ω_0 the carrier frequency, k_0 the carrier wave number, and $A(z,t)$ the pulse envelope function. The form of $\mathbf{D}_L(z,t)$ becomes

$$\mathbf{D}_L(z,t) = \int_{-\infty}^{+\infty} n_0^2(\omega)\tilde{E}(z,\omega)\exp(-i\omega t)\,d\omega, \tag{A.8}$$

where

$$\tilde{E}(z,\omega) = \frac{1}{2\pi}\int_{-\infty}^{+\infty} E(z,t)\exp(i\omega t)\,dt. \tag{A.9}$$

If the $(\omega_0 t - k_0 z)$ representation is used, one obtains sign changes in the final reduced wave equation.

Using the foregoing equations, we can write the linear polarization term on the left-hand side of the one-dimensional wave equation as

$$
\begin{aligned}
-\frac{1}{c^2}\frac{\partial^2}{\partial t^2}\mathbf{D}_L(z,t) \\
&= -\frac{1}{c^2}\int_{-\infty}^{+\infty} n_0^2(\omega)\tilde{E}(z,\omega)\exp(-i\omega t)d\omega \\
&= \frac{1}{2\pi}\int_{-\infty}^{\infty}\int_{-\infty}^{\infty} k^2(\omega)A(z,t')\exp[i\omega(t'-t)]\exp[i(k_0 z - \omega_0 t')]d\omega\, dt'.
\end{aligned} \tag{A.10}
$$

The derivation of the wave equation then proceeds by expanding $k^2(\omega)$ about the carrier frequency ω_0 in the form:

$$
k^2(\omega) \approx k_0^2 + 2k_0 k_0^{(1)}(\omega - \omega_0) + k_0 k_p^{(2)}(\omega - \omega_0)^2 + \cdots, \tag{A.11}
$$

where $k_0 = k(\omega_0)$ is the propagation constant, $k_0^{(1)} = \left.\frac{\partial k}{\partial \omega}\right|_{\omega=\omega_0}$ is the inverse of group velocity, and $k_0^{(2)} = \left.\frac{\partial^2 k}{\partial \omega^2}\right|_{\omega=\omega_0}$ is the inverse of group velocity of dispersion. It is then possible to evaluate the integral of Eq. (A.10) by using the convenient delta function identities

$$
\frac{1}{2\pi}\int_{-\infty}^{\infty} \exp[i(\omega - \omega_0)(t'-t)]d\omega = \delta(t'-t) \tag{A.12}
$$

as well as

$$
\frac{1}{2\pi}\int_{-\infty}^{\infty} (\omega - \omega_0)\exp[i(\omega - \omega_0)(t'-t)]d\omega = i\delta^{(1)}(t'-t), \tag{A.13}
$$

and

$$
\frac{1}{2\pi}\int_{-\infty}^{\infty} (\omega - \omega_0)^2\exp[i(\omega - \omega_0)(t'-t)]d\omega = -\delta^{(2)}(t'-t). \tag{A.14}
$$

In these relations, $\delta^{(n)}(t)$ is an nth-order derivative of the Dirac delta function, with the property that

$$
\int_{-\infty}^{\infty} \delta^{(n)}(t - t_0)f(t)\,dt = \left.\frac{d^n f(t)}{\partial t^n}\right|_{t=t_0} \tag{A.15}
$$

when applied to a function $f(t)$. Substitute Eq. (A.11) into Eq. (A.10) and use Eq. (A.12) to (A.15), the second term on the left-hand side of Eq. (1.5) becomes

$$
\begin{aligned}
&-\frac{1}{c^2}\frac{\partial^2}{\partial t^2}\mathbf{D}_L(z,t)\\
&=\frac{1}{2\pi}\int_{-\infty}^{\infty}\int_{-\infty}^{\infty}[k_0^2+2k_0k_0^{(1)}(\omega-\omega_0)+k_0k_0^{(2)}(\omega-\omega_0)^2]A(z,t')\\
&\quad\times\exp[i\omega(t'-t)]\exp[i(k_0z-\omega_0t)]\,d\omega\,dt'\\
&=\int_{-\infty}^{\infty}[k_0^2\delta(t'-t)+i2k_0k_0^{(1)}\delta^{(1)}(t'-t)-k_0k_0^{(2)}\delta^{(2)}(t'-t)]A(z,t')\\
&\quad\times\exp[i(k_0z-\omega_0t)]\,dt'\\
&=\left[k_0^2A+i2k_0k_0^{(2)}\frac{\partial A}{\partial t}-k_0k_0^{(2)}\frac{\partial^2 A}{\partial t^2}\right]\exp[i(k_0z-\omega_0t)]. \qquad \text{(A.16)}
\end{aligned}
$$

Neglecting the second derivative of $A(z,t)$ with respect to z and $\frac{\partial^2}{\partial t^2}|A(z,t)|^2\cdot A(z,t)$, the first term on the left-hand side and the term on the right-hand side of Eq. (1.5) are simply

$$\frac{\partial^2 E(z,t)}{\partial z^2}\approx\left[-k_0^2A(z,t)+i2k_0\frac{\partial A(z,t)}{\partial z}\right]\exp[i(k_0z-\omega_0t), \qquad \text{(A.17)}$$

and

$$\frac{2n_0\bar{n}_2}{c^2}\frac{\partial^2}{\partial t^2}|E|^2E\approx-\frac{n_0n_2\omega_0^2}{c^2}|A|^2A\exp[i(k_0z-\omega_0t)], \qquad \text{(A.18)}$$

respectively.

Inserting Eqs. (A.16) to (A.18) into Eq. (1.5), the wave equation for electric field reduces to the wave equation for the pulse envelope

$$i\left(\frac{\partial A}{\partial z}+\frac{1}{v_g}\frac{\partial A}{\partial t}\right)-\frac{1}{2}k_0^{(2)}\frac{\partial^2 A}{\partial t^2}+\frac{\omega_0}{2c}n_2|A|^2A=0, \qquad \text{(A.19)}$$

where $v_g\equiv 1/k_0^{(1)}$ is the group velocity. In Eq. (A.19), the first two terms describe the envelope propagation at the group velocity v_g; the third term determines the temporal pulse broadening due to group velocity dispersion; the fourth characterizes the second order of the nonlinear polarization, which is responsible for the self-phase modulation effect and spectral broadening. Neglecting the group velocity dispersion term in Eq. (A.19), we obtain

$$\frac{\partial A}{\partial z}+\frac{1}{v_g}\frac{\partial A}{\partial t}=i\frac{\omega_0}{2c}n_2|A|^2A. \qquad \text{(A.20)}$$

This is Eq. (2.2).

3 Ultrashort Pulse Propagation in Nonlinear Dispersive Fibers

GOVIND P. AGRAWAL

1. Introduction

The use of silica fibers for transmission of optical pulses has become widespread, as is evident from the recent advances in optical fiber communications (Basch, 1986; Miller and Kaminow, 1988). For pulses not too short (pulse width >1 ns) and not too intense (peak power <10 mW), the fiber plays a passive role (except for energy loss) and acts as a transporter of optical pulses from one place to another without significantly affecting their shape or spectrum. However, as pulses become shorter and more intense, two physical mechanisms, chromatic dispersion and index nonlinearity, both intrinsic to the silica material, start to affect the pulse shape and spectrum during propagation.

The fiber has found many novel and interesting applications in this active role. It has been used for pulse compression (Shank et al., 1982; Nikolaus and Grischkowsky, 1983), and pulses with durations as short as 6 fs have been produced (Fork et al., 1987). Such short pulses provide the capability for ultrafast continuum spectroscopy with a time resolution approximating that allowed by the transform-limited bandwidth (Shank et al., 1986). In the anomalous dispersion regime the fiber supports optical solitons resulting from a balance between the dispersive and nonlinear effects (Hasegawa and Tappert, 1973; Mollenauer et al., 1980). The solitons may be useful for high-speed optical communications over long distances (Hasegawa, 1983; Mollenauer et al., 1986). The soliton formation capacity of optical fibers has also been exploited to develop the soliton laser (Mollenauer and Stolen, 1984). An important application of optical fibers is in the field of ultrafast supercontinuum generation (Alfano, 1985; Shank et al., 1986). Several nonlinear effects such as self-phase modulation (SPM), cross-phase modulation (XPM), and stimulated Raman scattering (SRS) can lead to an extensive spectral broadening of the incident pulse, resulting in an almost "white" spectrum (Alfano and Shapiro, 1970). Although SPM in optical fibers was studied nearly a decade ago (Stolen and Lin, 1978), it is only recently that the effect of XPM on propagation of ultrashort pulses in optical fibers has attracted attention

(Alfano et al., 1987; Islam et al., 1987; Agrawal, 1987; Schadt and Jaskorzynska, 1987; Baldeck et al., 1987).

This chapter reviews how the nonlinear and dispersive effects in optical fibers influence the propagation characteristics of ultrashort pulses with widths in the picosecond range. In Section 2 we outline the derivation of the basic propagation equation satisfied by the amplitude of the pulse envelope. In the presence of SRS it becomes necessary to consider the coupled amplitude equations satisfied by the pump and Raman pulses. These equations include the effect of group velocity mismatch, group velocity dispersion, SPM, XPM, SRS gain, and pump depletion. Section 3 considers the propagation of a single pulse below the SRS threshold and identifies various propagation regimes. Sections 4 and 5 then consider the cases of normal dispersion and anomalous dispersion regimes separately. The effect of SRS on pulse propagation is studied in Section 6, where the results of numerical simulations are presented and compared with the experiments. Particular attention is paid to the XPM effects in Section 7, where the case of two incident pulses is investigated. Finally, Section 8 gives a brief summary of the main conclusions of the chapter.

2. Propagation Equation

As is the case for all electromagnetic phenomena, pulse propagation in optical fibers is governed by Maxwell's equations. For a nonmagnetic medium, this amounts to solving the wave equation

$$\nabla \times \nabla \times \mathbf{E} = -\mu_0 \frac{\partial^2 \mathbf{D}}{\partial t^2}, \tag{1}$$

where $\mathbf{E}$ is the electric field, μ_0 is the vacuum permeability, and $\mathbf{D}$ is the electric flux density. The constitutive relation between $\mathbf{D}$ and $\mathbf{E}$ takes a simple form in the frequency domain and can be written as

$$\tilde{\mathbf{D}}(\mathbf{r}, \omega) = \varepsilon_0 \tilde{n}^2 \tilde{\mathbf{E}}(\mathbf{r}, \omega), \tag{2}$$

where ε_0 is the vacuum permittivity, $\tilde{n}$ is the refractive index, and $\tilde{\mathbf{E}}$ and $\tilde{\mathbf{D}}$ are the Fourier transforms of $\mathbf{E}$ and $\mathbf{D}$ respectively, i.e.,

$$\tilde{\mathbf{E}}(\mathbf{r}, \omega) = \frac{1}{2\pi} \int_{-\infty}^{\infty} \mathbf{E}(\mathbf{r}, t) \exp(i\omega t)\, dt \tag{3}$$

with a similar relation for $\tilde{\mathbf{D}}(\mathbf{r}, \omega)$. Using Eqs. (1)–(3), we obtain

$$\nabla^2 \tilde{\mathbf{E}} + (\tilde{n}\omega/c)^2 \tilde{\mathbf{E}} = 0, \tag{4}$$

where we have assumed that $\nabla \cdot \mathbf{E} \cong 0$. A suitable form for the refractive index in optical fibers is

$$\tilde{n} = n(\mathbf{r}, \omega) + i\alpha c/2\omega + n_2 |\mathbf{E}|^2, \tag{5}$$

where the linear index $n(\mathbf{r}, \omega)$ is inhomogeneous (in the transverse dimensions

x and y) to account for dielectric waveguiding and is frequency dependent to account for chromatic dispersion in silica fibers. The fiber loss is taken into account by the absorption coefficient α. Finally, the last term in Eq. (5) accounts for the fiber nonlinearity. The nonlinear coefficient $n_2 = 3.2 \times 10^{-16}$ cm^2/W for silica fibers and is nearly frequency independent in the frequency range of interest. It is implicitly assumed that the nonlinear response is instantaneous, an assumption that becomes questionable for ultrashort optical pulses with widths <0.1 ps.

Since the refractive index does not vary appreciably along the fiber length (z direction), Eq. (4) can be solved by assuming that

$$\tilde{\mathbf{E}}(x, y, z, \omega) = \hat{e}\tilde{a}(z, \omega)\psi(x, y) = \hat{e}E_0\psi(x, y)\exp(i\tilde{\beta}z), \tag{6}$$

where $\hat{e}$ is the polarization unit vector, $\tilde{\beta}$ is the wave number, and $\psi(x, y)$ is obtained by solving

$$\frac{\partial^2\psi}{\partial x^2} + \frac{\partial^2\psi}{\partial y^2} + \left(\frac{\tilde{n}^2\omega^2}{c^2} - \tilde{\beta}^2\right)\psi = 0. \tag{7}$$

Since the last two terms in Eq. (5) are much smaller than the linear index $n(x, y, \omega)$, they can be treated by first-order perturbation theory. The complex wave number $\tilde{\beta}$ is thus well approximated by

$$\tilde{\beta} \cong \beta(\omega) + i\alpha/2 + \Delta\beta_{\text{NL}}, \tag{8}$$

where

$$\Delta\beta_{\text{NL}} = \frac{n_2\omega}{c}\iint_\sigma |\psi(x, y)|^2|\mathbf{E}|^2\,dx\,dy \tag{9}$$

and $\beta(\omega)$ is obtained by solving the eigenvalue equation

$$\frac{\partial^2\psi}{\partial x^2} + \frac{\partial^2\psi}{\partial y^2} + \left(n(x, y, \omega)\frac{\omega^2}{c^2} - \beta^2(\omega)\right)\psi = 0 \tag{10}$$

for a given index distribution of the fiber. For a single-mode fiber, $\psi(x, y)$ is the field distribution of the fundamental fiber mode and has been taken to satisfy the normalization condition

$$\iint_\sigma |\psi(x, y)|^2\,dx\,dy = 1, \tag{11}$$

where the integration is carried over the fiber cross section σ.

The dispersion relation (8) can be used to obtain the envelope propagation equation by noting from Eq. (6) that

$$\frac{\partial\tilde{a}}{\partial z} = i[\beta(\omega) + i\alpha/2 + \Delta\beta_{\text{NL}}]\tilde{a}. \tag{12}$$

Since the pulse spectrum is centered at the carrier frequency ω_0, it is useful to expand $\beta(\omega)$ in a Taylor series

$$\beta(\omega) = \beta_0 + (\omega - \omega_0)\beta_1 + \tfrac{1}{2}(\omega - \omega_0)^2\beta_2 + \tfrac{1}{6}(\omega - \omega_0)^3\beta_3 + \cdots, \quad (13)$$

where $\beta_n = d^n\beta/d\omega^n$ is evaluated at ω_0. The cubic and higher-order terms are generally neglected by assuming that the spectral width $\Delta\omega \ll \omega_0$. If $\beta_2 \simeq 0$ for some specific values of ω_0 or if the pulses are ultrashort (<0.1 ps), it may be necessary to include the cubic term. It is useful to define the slowly varying pulse amplitude $A(z, t)$ by using

$$a(z, t) = A(z, t)\exp[i(\beta_0 z - \omega_0 t)], \quad (14)$$

where $a(z, t)$ is the inverse Fourier transform of $\tilde{a}(z, \omega)$. During the Fourier transform operation $\omega - \omega_0$ is replaced by the differential operator $i(\partial/\partial t)$. The use of Eqs. (12)–(14) then leads to the following propagation equation:

$$\frac{\partial A}{\partial z} + \beta_1 \frac{\partial A}{\partial t} + \frac{\alpha}{2}A + \frac{i}{2}\beta_2 \frac{\partial^2 A}{\partial t^2} = i\gamma|A|^2 A, \quad (15)$$

where the nonlinearity coefficient

$$\gamma = n_2\omega_0/cA_{\text{eff}}. \quad (16)$$

The parameter A_{eff} is known as the effective fiber core area and from Eq. (9) is given by (noting that $E = A\psi$)

$$A_{\text{eff}}^{-1} = \iint_\sigma |\psi(x, y)|^4\,dx\,dy. \quad (17)$$

Equation (15) describes the propagation of an optical pulse in single-mode fibers. The pulse envelope moves at the group velocity $v_g = 1/\beta_1$. The group velocity dispersion (GVD) is included through the parameter β_2 while the parameter γ takes into account the fiber nonlinearity responsible for SPM. Both material dispersion and waveguide dispersion contribute to β_2. However, the contribution of material dispersion generally dominates. In that case β_2 can be approximated by

$$\beta_2 \simeq (\lambda/2\pi c^2)D(\lambda), \quad (18)$$

where $\lambda = 2\pi c/\omega_0$ is the carrier wavelength and $D(\lambda) = \lambda^2(d^2n/d\lambda^2)$. For silica fibers $D(\lambda) = 0.066$ at $\lambda = 0.53$ μm, and therefore $\beta_2 \simeq 0.06$ ps^2/m. The nonlinearity parameter γ depends on the fiber parameters through A_{eff}. Its estimated values are in the range 0.01 to 0.02 W^{-1} m^{-1}. The GVD parameter β_2 becomes negative for $\lambda > \lambda_{ZD}$, where λ_{ZD} is sometimes referred to as the zero-dispersion wavelength. In the anomalous dispersion regime ($\beta_2 < 0$), the fiber can support optical solitons. The wavelength $\lambda_{ZD} \cong 1.3$ μm but can be changed in the range 1.2 to 1.6 μm by modifying the fiber design that changes the waveguide dispersion contribution to β_2.

Depending on the experimental conditions, Eq. (15) may need modification. For example, this equation does not include the effect of SRS that becomes important at high power levels. SRS transfers the pulse energy to the Stokes pulse (shifted downward in frequency by about 13 THz) as the pump pulse propagates inside the fiber. Equation (15) can be generalized to include SRS if we include the Raman gain and consider a coupled set of equations describing the interaction between the pump and Stokes pulses with amplitudes A_1 and A_2, respectively;

$$\frac{\partial A_1}{\partial z} + \frac{1}{v_{g1}}\frac{\partial A_1}{\partial t} + \frac{1}{2}(\alpha_1 + g_1|A_2|^2)A_1 + \frac{i}{2}\beta_{21}\frac{\partial^2 A_1}{\partial t^2} - i\gamma_1(|A_1|^2 + 2|A_2|^2)A_1 = 0, \tag{19}$$

$$\frac{\partial A_2}{\partial z} + \frac{1}{v_{g2}}\frac{\partial A_2}{\partial t} + \frac{1}{2}(\alpha_2 - g_2|A_1|^2)A_2 + \frac{i}{2}\beta_{22}\frac{\partial^2 A_2}{\partial t^2} - i\gamma_2(|A_2|^2 + 2|A_1|^2)A_2 = 0. \tag{20}$$

Here $v_{gj} = \beta_j^{-1}$ is the group velocity, g_j is the Raman gain coefficient, β_{2j} is the GVD coefficient, and γ_j is the nonlinear coefficient ($j = 1$ and 2 for pump and Stokes pulses, respectively). It is important to include both SPM and XPM in the nonlinear term (Gersten et al., 1980; Schadt and Jaskorzynska, 1987; Agrawal, 1987).

Equation (15) should also be modified for ultrashort optical pulses with widths $\lesssim 0.1$ ps. The spectral width $\Delta\omega$ of such pulses becomes comparable to the carrier frequency ω_0, and several approximations made in the derivation of Eq. (15) become questionable. Furthermore, the pulse spectrum is wide enough that the low-frequency components get amplified from the Raman gain at the expense of the high-frequency components. As a result, the spectrum shifts toward shorter frequencies as the pulse propagates inside the fiber, a phenomenon referred to as self-frequency shift (Mitschke and Mollenauer, 1986; Gordon, 1986). One can no longer consider the pump and Stokes pulses separately. Rather than using Eqs. (19) and (20) to describe SRS and related nonlinear effects, a generalization of Eq. (15) is needed. Kodama and Hasegawa (1987) have carried out such a generalization. The resulting propagation equation is

$$\frac{\partial A}{\partial z} + \beta_1\frac{\partial A}{\partial t} + \frac{\alpha}{2}A = -\frac{i}{2}\beta_2\frac{\partial^2 A}{\partial t^2} + i\gamma|A|^2 A + \frac{1}{6}\beta_3\frac{\partial^3 A}{\partial t^3} - a_1\frac{\partial}{\partial t}(|A|^2 A) - a_2 A\frac{\partial |A|^2}{\partial t} \tag{21}$$

This equation has three additional terms compared with Eq. (15). The term proportional to β_3 results from including the cubic term in the expansion (13). The term proportional to a_1 is responsible for self-steepening; a_1 is generally approximated by $a_1 \simeq 2\gamma/\omega_0$. The term proportional to a_2 includes the effect of SRS among other things. In particular, the imaginary part of a_2 is responsible for the self-frequency shift. The physical origin of this effect is related to the retarded nonlinear response of the fiber. In this chapter we do not consider these effects and only refer to the recent work (Zysset et al., 1987; Bourkoff et al., 1987; Golovchenko et al., 1987; Vysloukh and Matveeva, 1987; Serkin, 1987).

3. Propagation Regimes

We first consider the case of pulse propagation below the SRS threshold such that Eq. (15) is applicable. It is useful to define the normalized variables

$$\tau = (t - \beta_1 z)/T_0, \qquad A = \sqrt{P_0}\exp(-\alpha z/2)U, \tag{22}$$

where T_0 and P_0 are related to the width and the peak power of the incident pulse. Equation (15) then takes the form

$$i\frac{\partial U}{\partial z} = \pm\frac{1}{2L_D}\frac{\partial^2 U}{\partial \tau^2} - \frac{1}{L_{NL}}e^{-\alpha z}|U|^2 U, \tag{23}$$

where

$$L_D = T_0^2/|\beta_2| \quad \text{and} \quad L_{NL} = (\gamma P_0)^{-1}. \tag{24}$$

The choice of the sign in the dispersion term depends on the sign of β_2; a plus sign is chosen for normal GVD ($\beta_2 > 0$) and a minus sign is chosen for anomalous GVD ($\beta_2 < 0$). The dispersion length L_D and the nonlinear length L_{NL} provide the length scales over which dispersive and nonlinear effects become important. Depending on their relative magnitudes, pulse propagation in optical fibers can be classified in three different categories.

3.1 Dispersion-Dominant Regime

If the fiber length L is such that $L \ll L_{NL}$ while $L \sim L_D$, the nonlinear term in Eq. (23) plays a relatively minor role and can be neglected. The resulting linear equation is readily solved to yield

$$U(z,\tau) = \int_{-\infty}^{\infty} \tilde{U}(0,\omega)\exp[-i\omega\tau \pm i\omega^2(z/2L_D)]\,d\omega, \tag{25}$$

where $\tilde{U}(0,\omega)$ is the Fourier transform of the incident pulse amplitude $U(0,\tau)$. Equation (25) has been widely used to study pulse broadening in the dispersion-dominant regime. In particular, a simple analytic expression for the pulse width can be obtained for Gaussian pulses (Marcuse, 1981). For an unchirped Gaussian pulse, $U(0,\tau) = \exp(-\tau^2/2)$, and the pulse broadens by a factor of $(1 + z^2/L_D^2)^{1/2}$ as it propagates along the fiber. The dispersion length L_D corresponds to the fiber length over which the pulse width increases by about 40% (in the absence of nonlinear effects). The dispersion-dominant regime or Eq. (25) is applicable when the fiber and pulse parameters are such that

$$\frac{L_D}{L_{NL}} = \frac{\gamma P_0 T_0^2}{|\beta_2|} \ll 1 \tag{26}$$

As a rough estimate, P_0 should be less than 10 mW for 10-ps pulses.

3.2 Nonlinearity-Dominant Regime

If the fiber length L is such that $L \ll L_D$ while $L \gtrsim L_{NL}$, the dispersion term in Eq. (23) plays a relatively minor role and can be neglected as long as the pulse shape is relatively smooth so that $|\partial^2 U/\partial\tau^2|$ is not too large. The resulting equation can be solved analytically with the result

$$U(z,\tau) = U(0,\tau)\exp[i\phi_{\text{SPM}}(z,\tau)], \tag{27}$$

where

$$\phi_{\text{SPM}}(z,\tau) = \left(\frac{1 - e^{-\alpha z}}{\alpha L_{NL}}\right)|U(0,\tau)|^2. \tag{28}$$

Equation (27) shows that the nonlinearity gives rise to SPM while it does not affect the pulse shape. The effect of SPM is to broaden the pulse spectrum with considerable internal structure. Figure 3.1 shows the calculated SPM spectra for a Gaussian pulse obtained by taking the Fourier transform of Eq. (27) with $\alpha = 0$ and $z = 40L_{NL}$. Quite generally, the number of internal peaks increases linearly with the fiber length (for $\alpha = 0$), and the dominant peaks occur near the spectral boundaries. The spectral range is determined by $\Delta\omega_{\max} = |\partial\phi_{\text{SPM}}/\partial t|_{\max}$, which is the maximum chirp induced by the nonlinearity. The SPM-induced spectral features have been observed in optical fibers and were used to estimate n_2 (Stolen and Lin, 1978).

3.3 Dispersive Nonlinear Regime

When the fiber length L is longer than or comparable to both L_D and L_{NL}, dispersion and nonlinearity act together as the pulse propagates along the fiber. Their mutual interaction plays an important role and has been used to generate solitons and to compress optical pulses. To understand this interaction theoretically, Eq. (23) is solved numerically and has been found to be extremely helpful in interpreting the experimental results (Mollenauer et al., 1980; Grischowsky and Balant, 1982; Tomlinson et al., 1984). It is useful to normalize the fiber length by introducing $\xi = z/L_D$ and to write Eq. (23) in the following normalized form:

$$i\frac{\partial U}{\partial \xi} = \pm\frac{1}{2}\frac{\partial^2 U}{\partial \tau^2} - N^2 e^{-\alpha z}|U|^2 U, \tag{29}$$

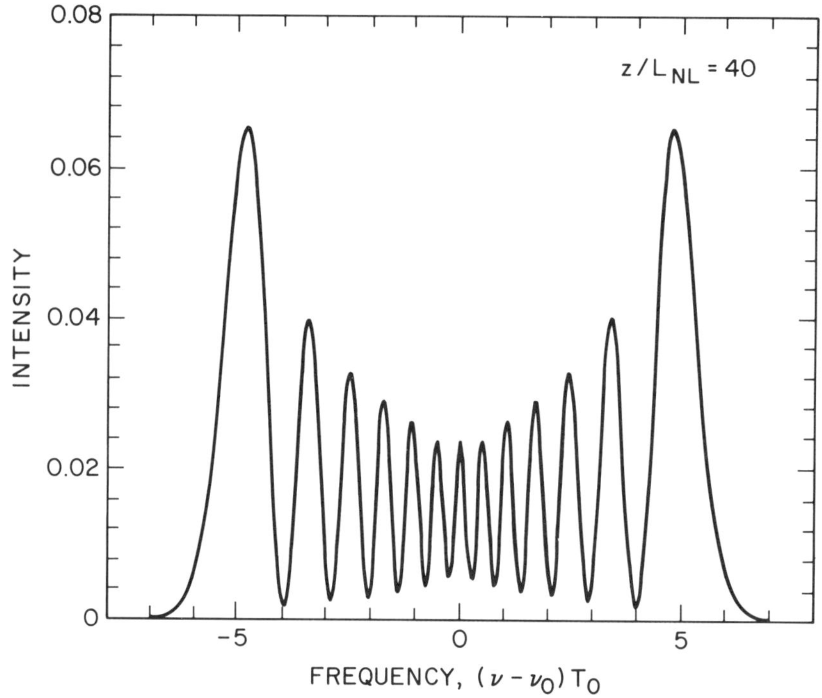

FIGURE 3.1. Calculated SPM spectrum of a Gaussian pulse (in the absence of dispersion) at a distance $z = 40L_{NL}$, where L_{NL} is the nonlinear length defined by Eq. (24).

where

$$N^2 = \frac{L_D}{L_{NL}} = \frac{\gamma P_0 T_0^2}{|\beta_2|}. \tag{30}$$

In the lossless case ($\alpha = 0$), Eq. (29) is often referred to as the nonlinear Schrödinger equation (NSE). In the anomalous dispersion regime (corresponding to the choice of minus sign), the NSE is known to have exact solutions, known as solitons (Zakharov and Shabat, 1972; Hasegawa and Tappert, 1973). For an initial pulse shape $U(0, \tau) = \text{sech}(\tau)$ and integer values of the parameter N, the solitons follow a periodic evolution pattern with the period $\xi_0 = \pi/2$. In the unnormalized units, the soliton period z_0 becomes

$$z_0 = \frac{\pi}{2} L_D = \frac{\pi T_0^2}{2|\beta_2|}. \tag{31}$$

The fundamental soliton corresponds to $N = 1$ and propagates without change in its shape. From Eq. (30), the peak power necessary to excite the fundamental soliton is $P_1 = |\beta_2|/\gamma T_0^2$. For a hyperbolic secant pulse, the pulse width T_p (FWHM) is related to T_0 by $T_p \simeq 1.76T_0$. This relation should be used for comparison with experiments. As a rough estimate, for 1.55-μm solitons in silica fibers, $z_0 \sim 25$ m and $P_1 \sim 1$ W when $T_p = 1$ ps.

In the next two sections we consider the pulse propagation characteristics in the normal and anomalous GVD regimes based on the numerical solution of Eq. (29). A fast Fourier transform (FFT)-based beam propagation method (Fleck et al., 1976; Lax et al., 1981; Agrawal and Potasek, 1986) was used for the numerical solution. Although a finite-difference scheme can be used, the FFT method is considerably faster for a given accuracy (typically by more than a factor of 10). In the FFT method, Eq. (29) is formally written as

$$\frac{\partial U}{\partial \xi} = (D + Q)U, \tag{32}$$

where

$$D = \pm \frac{i}{2} \frac{\partial^2}{\partial \tau^2} \quad \text{and} \quad Q = iN^2 e^{-\alpha z} |U|^2. \tag{33}$$

The field is propagated inside the fiber by a small distance δ using the following prescription:

$$U(\xi + \delta, \tau) = \left[\exp\left(\frac{\delta D}{2}\right)\exp\left(\int_{\xi}^{\xi+\delta} Q(\xi')\,d\xi'\right)\exp\left(\frac{\delta D}{2}\right)\right]U(\xi, \tau). \tag{34}$$

The numerical procedure consists of propagating the field by a distance $\delta/2$ with dispersion only, multiplying the result by a nonlinear term that represents the effect of nonlinearity over the whole step length δ, and then propagating the field for the remaining distance $\delta/2$ with dispersion only. In effect, the nonlinearity is assumed to be lumped at the midplane of each segment. This technique is a generalization of the split-step method (Hasegawa and Tappert, 1973; Fisher and Bischel, 1975).

The propagation in a linear dispersive medium indicated by the exponential operator $\exp(\delta D/2)$ in Eq. (34) can be accomplished using the Fourier transform method. The use of the FFT algorithm makes this step relatively fast. The integral in Eq. (34) is well approximated using the trapezoidal rule:

$$\int_{\xi}^{\xi+\delta} Q(\xi')\,d\xi' \simeq \tfrac{1}{2}\delta[Q(\xi) + Q(\xi + \delta)]. \tag{35}$$

However, $Q(\xi + \delta)$ cannot be evaluated since $U(\xi + \delta, \tau)$ is not known while evaluating Eq. (35) at the midsegment located at $\xi + \delta/2$. One can approximate $Q(\xi + \delta)$ by $Q(\xi)$ if $|U|^2$ has not significantly changed in going from ξ to $\xi + \delta$. This approximation can, however, restrict the step size δ to relatively small values. Another approach is to follow an iterative procedure in which Eq. (35) is evaluated several times with increasingly accurate values of $Q(\xi + \delta)$. In practice, two iterations are often sufficient to obtain the desired accuracy.

4. Normal Dispersion

In the normal dispersion regime, the GVD parameter $\beta_2 > 0$. For silica fibers, this regime corresponds to $\lambda < 1.3$ μm and has been extensively studied because of the availability of solid-state and dye lasers in the visible region. At $\lambda = 0.53$ μm, $\beta_2 \simeq 0.06$ ps^{2}m and $\gamma \simeq 0.01$ W^{-1} m^{-1} for typical fiber parameters. From Eq. (24), $L_D \sim 100$ m for $T_0 \simeq 3$–5 ps and $L_{NL} \sim 1$ m for $P_0 \sim 100$ W. Thus the parameter $N \sim 10$ under typical experimental conditions, although $N \sim 100$ is feasible for wider and more intense pulses. Figure 3.2 shows the evolution of pulse shape along the fiber for $N = 30$ after assuming $\alpha = 0$ and a Gaussian pulse with $U(0, \tau) = \exp(-\tau^2/2)$. As the pulse propagates, it broadens and develops a nearly rectangular profile with sharp leading and trailing edges. The combination of rapidly varying intensity and SPM in these steep-slope regions broadens the pulse spectrum. Because the new frequency components are mainly generated near the edges, the pulse develops a nearly linear frequency chirp across its entire width (Grischkowsky and Balant, 1982; Tomlinson et al., 1984). This linear chirp helps to compress the pulse by passing it subsequently through a dispersive delay line (often a grating pair).

An interesting feature of Figure 3.2 is the presence of rapid oscillations in the wings of the pulse. Tomlinson et al. (1985) have interpreted these oscillations in terms of optical wave breaking resulting from a mixing of the SPM-induced frequency-shifted components with the unshifted light in the wings. This phenomenon can also be understood as a four-wave mixing process and should be manifest through the presence of sidelobes in the pulse spectrum. Figure 3.3 shows the pulse shape and spectrum at $\xi = 0.08$. As seen there, oscillations near the pulse edges are accompanied by the sidelobes in the pulse spectrum. The central structure in the spectrum is due to SPM (see Figure 3.1).

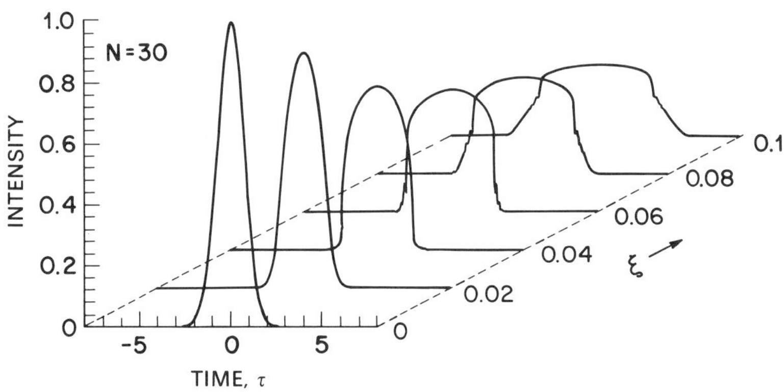

FIGURE 3.2. Evolution of the pulse shape of an unchirped Gaussian pulse along the fiber length ($\xi = z/L_D$) for $N = 30$ in the normal dispersion regime. The peak power is related to N through Eq. (30).

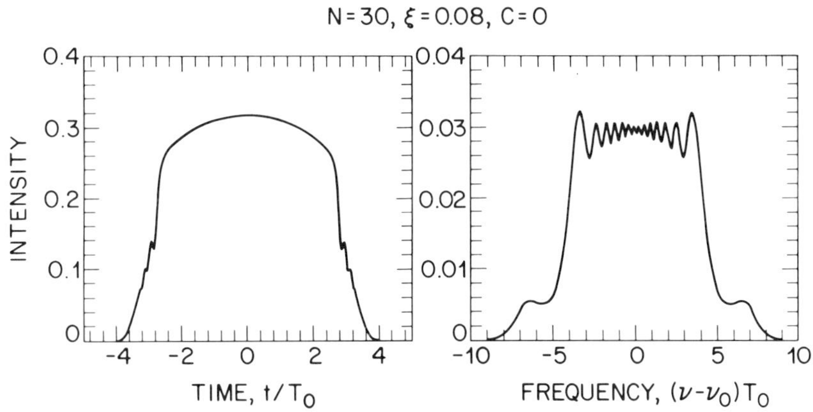

FIGURE 3.3. Pulse shape and spectrum of an unchirped Gaussian pulse at $\xi = 0.08$ for $N = 30$. Sidelobes in the spectrum and oscillations near the pulse edges are due to optical wave breaking.

The results shown in Figures 3.2 and 3.3 are for an initially unchirped pulse. The pulses emitted by practical laser sources are often chirped and may follow quite a different evolution pattern than that shown in Figure 3.2 (Lassen et al., 1985). To include the effect of initial chirp, the incident amplitude is taken to be

$$U(0, \tau) = \exp[-\tfrac{1}{2}(1 + iC)\tau^2], \tag{36}$$

where C is a measure of chirp. C can be estimated by noting that the spectral width of the chirped pulse is larger by a factor of $(1 + C^2)^{1/2}$ than the transform-limited value for $C = 0$. Figure 3.4 shows the pulse shape and spectrum under conditions identical to those of Figure 3.3 except for the chirp parameter that has a value of $C = 20$. A comparison of Figures 3.3 and 3.4 shows how much an initial chirp can modify the propagation behavior. For example, the shape for the initially chirped pulse is nearly triangular rather than rectangular. The pulse evolution is also sensitive to the fiber loss (Lassen et al., 1985). Thus, for an actual comparison between theory and experiment it is necessary to include both the chirp and the loss in numerical simulations.

5. Anomalous Dispersion

In the anomalous dispersion regime, the GVD parameter $\beta_2 < 0$. For silica fibers, this regime corresponds to $\lambda > 1.3\ \mu$m and has been extensively studied, using color-center lasers, for example (Mollenauer et al., 1980). At $\lambda = 1.55\ \mu$m, $\beta_2 \simeq -0.02$ ps^2/m and $\gamma \simeq 0.01$ W^{-1} m^{-1} for typical fiber parameters. For $T_0 \simeq 1$ ps, the dispersion length $L_D \simeq 50$ m; the peak power corresponding

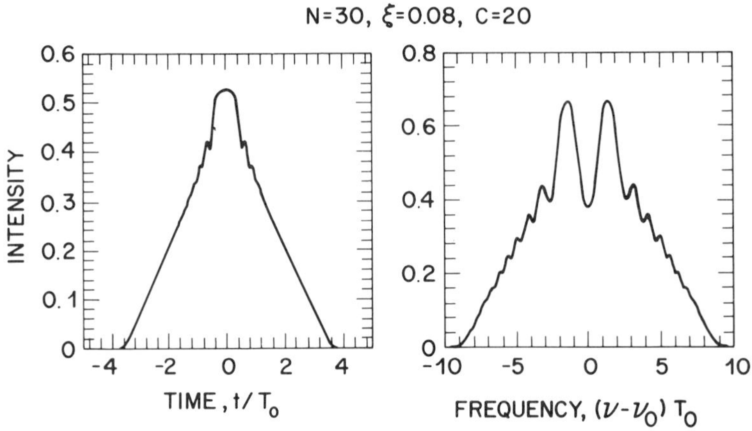

FIGURE 3.4. Pulse shape and spectrum for parameters identical to those of Figure 3.3 except for the chirp parameter, which has a value $C = 20$ instead of 0.

to the fundamental soliton ($N = 1$) can be estimated from Eq. (30) and is $P_0 \sim 1$ W. In the absence of fiber loss, the fundamental soliton can propagate undistorted for arbitrarily long distances. This is the reason behind the current interest in soliton-based communication systems (Hasegawa and Tappert, 1973). In practice, however, the fiber loss leads to a gradual broadening of the soliton. Periodic compensation of the loss by Raman gain has been proposed to restore the original soliton (Mollenauer et al., 1986).

The higher-order solitons ($N > 1$) follow a periodic evolution pattern along the fiber with a period $z_0 = (\pi/2)L_D$. Figure 3.5 shows, as an example, the evolution pattern of an $N = 3$ soliton over one period obtained by solving Eq. (29) with $N = 3$, $\alpha = 0$, and $U(0, \tau) = \text{sech}(\tau)$. The pulse initially narrows, develops a two-peak structure, and then reverses its propagation behavior beyond $z/z_0 = 0.5$ such that the original pulse is restored at $z = z_0$. Both pulse narrowing and pulse restoration at the soliton period have been observed experimentally (Stolen et al., 1983). In particular, initial narrowing of the higher-order soliton has been used to compress the optical pulses by suitably selecting the peak power and the fiber length. A combination of a normal dispersion fiber with a grating pair followed by an anomalous dispersion fiber has been used to compress optical pulses by more than three orders of magnitude (Tai and Tomita, 1986) and to generate pulses as short as 33 fs (Gouveia-Neto et al., 1987).

The soliton formation capacity of optical fibers has led to the development of the soliton laser (Mollenauer and Stolen, 1984). A piece of single-mode fiber inside the cavity is used to shape the pulses. The pulse width can be controlled by adjusting the fiber length. Pulses as short as 50 fs have been generated directly from a soliton laser and compressed to about 19 fs using the

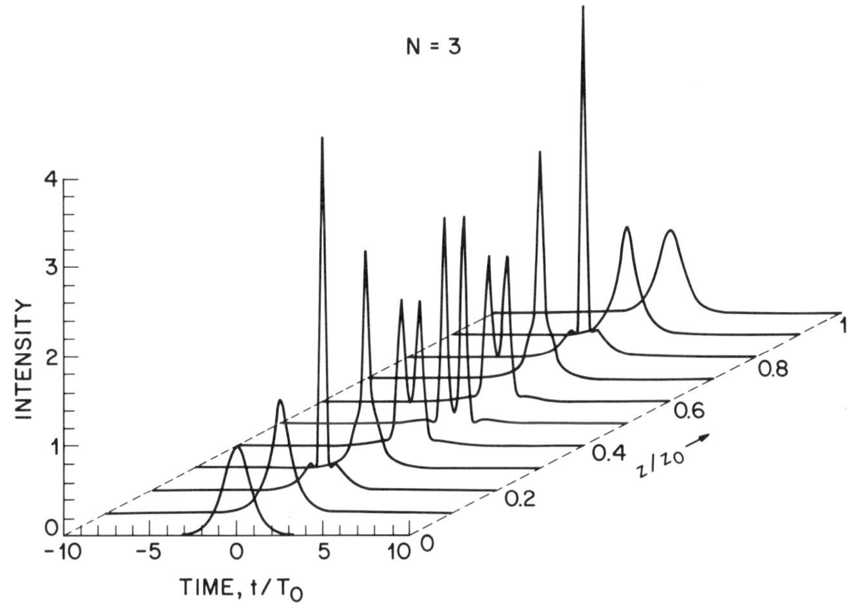

FIGURE 3.5. Evolution of $N = 3$ soliton over one period. Note the initial pulse narrowing in the anomalous dispersion regime.

high-order soliton effect in an external fiber (Mitschke and Mollenauer, 1987).

Equation (29) has proved to be very useful in understanding the propagation behavior in the anomalous dispersion regime of optical fibers. However, as discussed in Section 2, for ultrashort pulses ($T_0 \lesssim 0.1$ ps) Eq. (29) is no longer adequate, and Eq. (21) should be used in its place. The necessity of including the higher-order nonlinear terms in Eq. (21) was established in a recent experiment where a new effect known as the self-frequency shift was observed (Mitschke and Mollenauer, 1986). In this phenomenon the center frequency of the soliton shifts toward red as the pulse propagates inside the fiber. Physically, the low-frequency components are amplified from the Raman gain at the expense of the high-frequency components. Equation (21) includes the Raman gain through the last term and can be used to study such self-frequency shifts (Gordon, 1986; Kodama and Hasegawa, 1987).

The effect of higher-order dispersion on the soliton behavior [governed by the term proportional to β_3 in Eq. (21)] has also attracted attention. In general, such effects become significant only when the pulse wavelength nearly coincides with the zero-dispersion wavelength of the fiber and can lead to considerable pulse distortion (Agrawal and Potasek, 1986). It has been shown that the fiber can support a soliton even at the zero-dispersion wavelength (Wai et al., 1987) by shifting the center wavelength toward the anomalous dispersion regime.

6. Stimulated Raman Scattering

In preceding sections the peak power of the incident pulse has been assumed to be considerably below the SRS threshold (Smith, 1972). When this condition is not satisfied, the Raman gain provided by the fiber leads to a rapid buildup of the Stokes pulse. Figure 3.6 shows the Raman gain spectrum for silica fibers (Stolen, 1980). The large spectral width results from the amorphous nature of fused silica. Since the largest gain occurs at a frequency shifted downward by about 13 THz from the pump frequency, the Stokes pulse appears red-shifted by that amount. If the incident pulse is so short ($\lesssim 0.1$ ps) that its spectral width exceeds the Raman shift, the distinction between the pump and Stokes pulses is not sharp since their spectra overlap. In that case pulse evolution is governed by Eq. (21). However, for pulse widths $\gtrsim 1$ ps, the pump and Stokes pulses can be treated separately, and their propagation in optical fibers is described by the coupled set of Eqs. (19) and (20). In this section we consider numerical solution of these equations and compare the results with the experimental work on SRS. The numerical method is a generalization of that used to solve the NSE, described in Section 3.3.

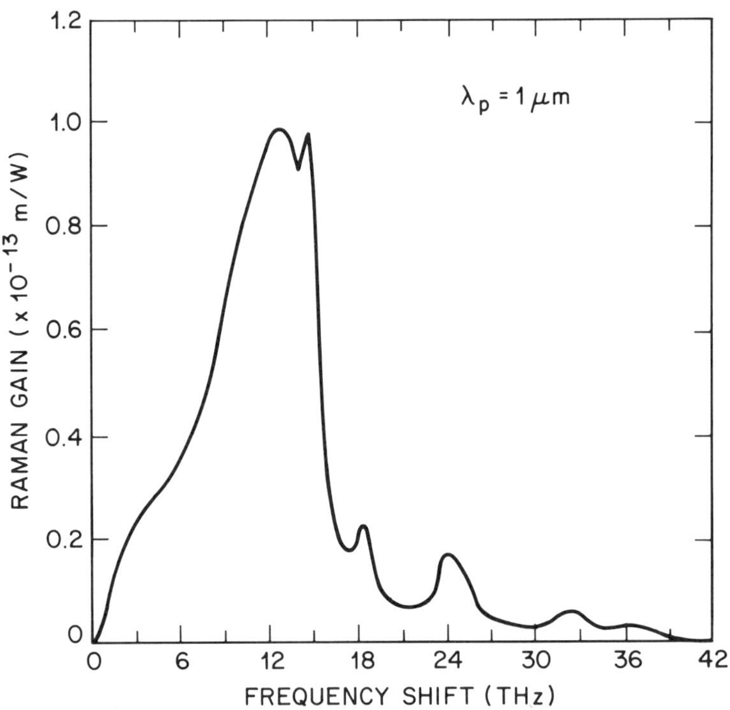

FIGURE 3.6. Raman gain spectrum for silica fibers (after Stolen, 1980). The peak gain is about 1×10^{-13} m/W for a 1-μm pump and scales as λ_p^{-1} with change in the wavelength.

For the purpose of computations, it is convenient to define the reduced time

$$T = t - z/v_{g1} \tag{37}$$

and write Eqs. (19) and (20) in the form

$$\frac{\partial A_p}{\partial z} + \tfrac{1}{2}(\alpha_p + g_p|A_s|^2)A_p + \frac{i}{2}\beta_p \frac{\partial^2 A_p}{\partial T^2} - i\gamma_p(|A_p|^2 + 2|A_s|^2)A_s = 0, \tag{38}$$

$$\frac{\partial A_s}{\partial z} + d\frac{\partial A_s}{\partial T} + \tfrac{1}{2}(\alpha_s - g_s|A_p|^2)A_s + \frac{i}{2}\beta_s \frac{\partial^2 A_s}{\partial T^2} - i\gamma_s(|A_s|^2 + 2|A_p|^2)A_p = 0, \tag{39}$$

where we have used the subscripts p and s to identify the pump and Stokes variables. The parameter

$$d = \frac{v_{gp} - v_{gs}}{v_{gp}v_{gs}} \tag{40}$$

accounts for the group velocity mismatch between the pump and Raman pulses. Because of the Stokes shift of about 13 THz, the parameters β_j, γ_j, and g_j are slightly different for $j = p$ and s. In particular, form Eqs. (16) and (18)

$$\beta_s = (\lambda_s/\lambda_p)\beta_p, \qquad \gamma_s = (\lambda_p/\lambda_s)\gamma_p, \tag{41}$$

where λ_p and λ_s are the center wavelengths for the pump and Raman pulses, respectively. Similarly, the gains are related by

$$g_s = (\lambda_p/\lambda_s)g_p. \tag{42}$$

For definiteness, the numerical results for SRS are presented here for $\lambda_p \simeq 0.53\ \mu$m; qualitatively similar behavior is expected to occur for other wavelengths. Using typical values for the fiber parameters, we estimate that $\alpha_p = 0.0012\ \text{m}^{-1}$, $\beta_p = 0.06\ \text{ps}^2\ \text{m}^{-1}$, $\gamma_p = 0.015\ \text{W}^{-1}\ \text{m}^{-1}$, and $g_p = 0.015\ \text{W}^{-1}\ \text{m}^{-1}$. Assuming a Gaussian unchirped pump pulse, the incident field $A_p(0, \tau)$ is taken to be

$$A_p(0, \tau) = \sqrt{P_0}\exp(-T^2/2T_0^2). \tag{43}$$

We choose $T_0 = 10$ ps, which corresponds to a full width at half-maximum (FWHM) of about 16 ps. The peak pump power is taken to be $P_0 = 500$ W. Numerical simulations require the use of a seed Raman pulse. It was taken to be a Gaussian pulse with a width identical to that of the pump pulse but with a power of 10^{-6} of the pump pulse. The two pulses overlap initially but the Raman pulse travels faster than the pump pulse inside the fiber because of the group velocity mismatch. The walk-off effect is included through the parameter d in Eq. (39); its value is estimated to be about 5 ps/m. Because of the use of a specific seed pulse, the results presented here are applicable more directly to the case of Raman amplification rather than noise-induced SRS. Most of the qualitative features are, however, expected to apply to SRS as well.

Figure 3.7 shows the evolution of the pump and Raman pulses as they copropagate inside a 5-m-long fiber. As the Raman pulse moves ahead of the

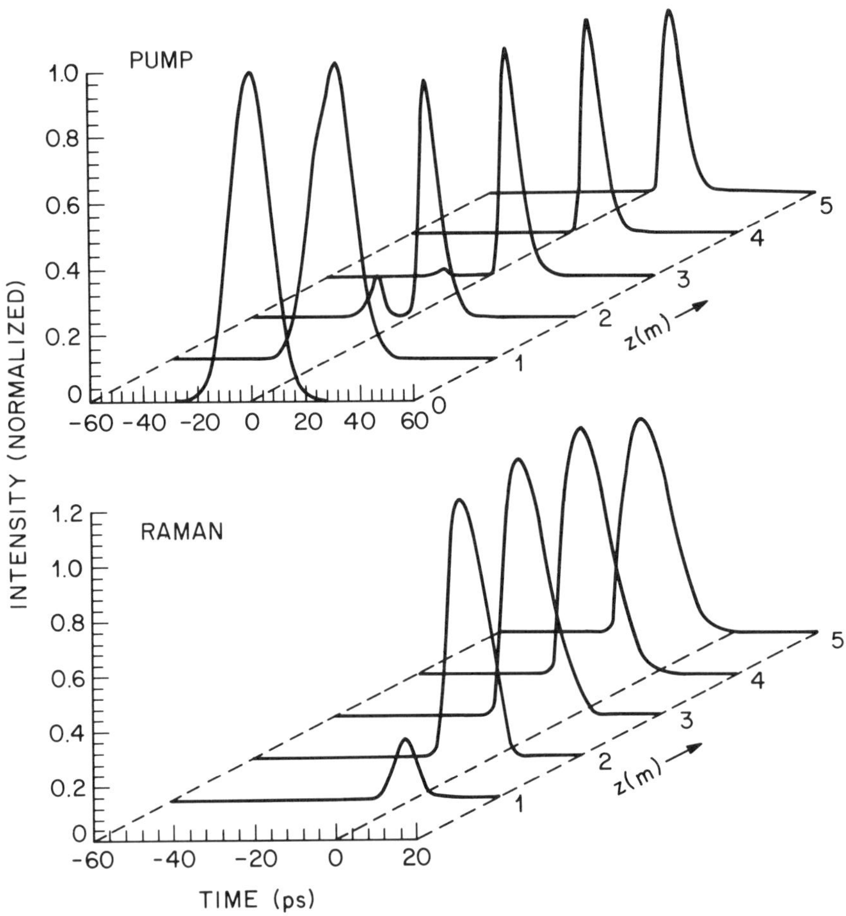

FIGURE 3.7. Evolution of pump pulse and Raman pulse along the fiber length. Pump pulse parameters are $\lambda_p = 0.53\ \mu\text{m}$, $T_0 = 10$ ps, and $P_0 = 500$ W. Pulse walk-off is included by taking $d = -5$ ps/m.

pump pulse, the energy for the generation of the Raman pulse comes from the leading edge of the pump pulse. This is seen clearly in Figure 3.8, where the pulse shapes are shown overlapped at $z = 2$ m and 5 m. At $z = 2$ m, the creation of the Raman pulse has led to a two-peak structure for the pump pulse as a result of pump depletion. The peak near the leading edge disappears with further propagation as the Raman pulse walks through it. At $z = 5$ m, the two pulses are completely separated. Because of the walk-off effect, both pulses are asymmetric and narrower than the incident pulse. In particular, the pump pulse consists of the trailing portion of the incident pulse that remained undepleted because of the walk-off of the Raman pulse.

The spectral evolution of the two pulses along the fiber length is shown in Figure 3.9. Both pulses exhibit asymmetric spectral broadening with con-

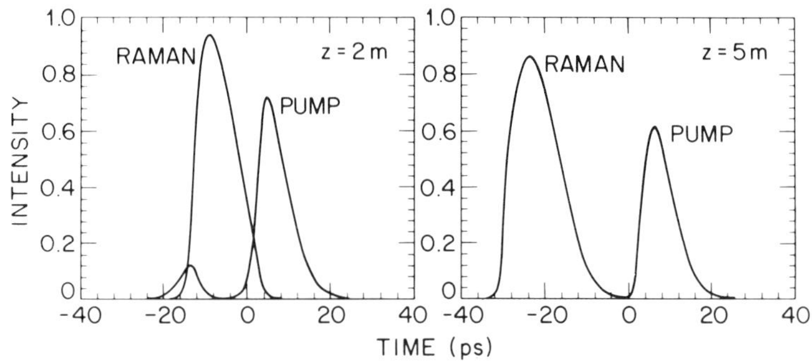

FIGURE 3.8. Comparison of pump and Raman pulse shapes at $z = 2$ m and $z = 5$ m with parameters of Figure 3.7. At $z = 2$ m, the pump pulse develops a two-peak structure because of pump depletion occurring at the location of the Raman pulse. Two pulses are completely separated after propagating a distance of 5 m.

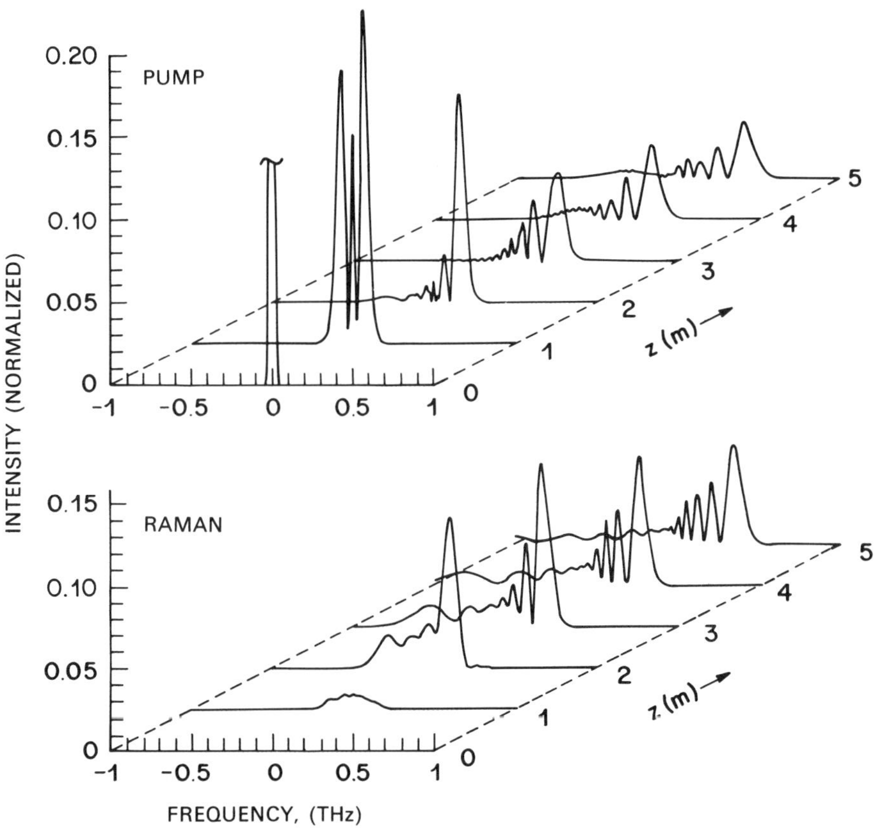

FIGURE 3.9. Evolution of pump and Raman pulse spectra along the fiber length. The parameter values are identical to those of Figure 3.7.

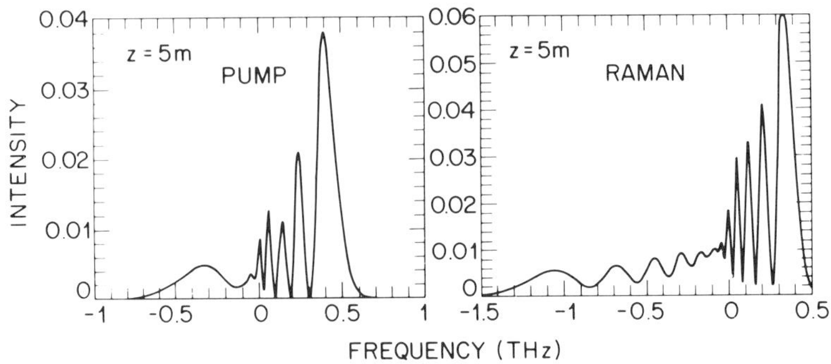

FIGURE 3.10. Comparison of pump and Raman pulse spectra at $z = 5$ m. Note the spectral broadening of the Raman pulse by more than a factor of 2 compared with that of the pump pulse.

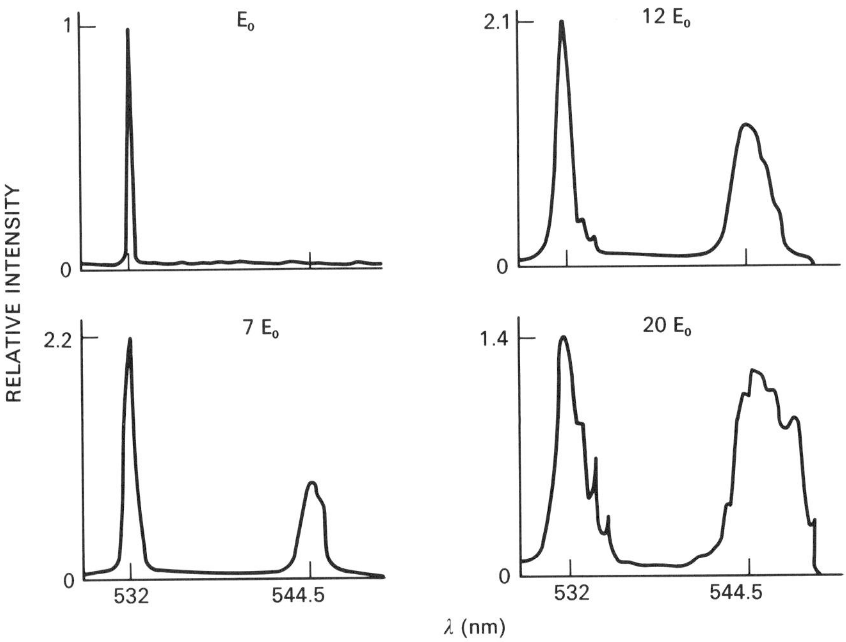

FIGURE 3.11. Experimental spectra of pump and Raman pulses after a 25-ps pump pulse at 532 nm has propagated 10 m inside a fiber. Four spectra correspond to different pump pulse energies normalized to E_0, an energy below the Raman threshold. (After Alfano et al., 1987.)

siderable internal structure. In general, the spectral width of the Raman pulse exceeds that of the pump pulse by a factor of 2 or more. These features are more clearly seen in Figure 3.10, where the spectra at $z = 5$ m are shown on an expanded scale. Note that the spectral range of the Raman pulse is about 2.5 times that of the pump pulse. This feature was predicted by Gersten et al. (1980) and has been verified by Alfano et al. (1987). Figure 3.11 shows their experimentally observed spectra at the output end of a 10-m-long fiber for a 25-ps pump pulse with increasing pump pulse energies. The Raman line is broader by about a factor of 2.8. The spectral asymmetry of the pump line is also evident. Similar qualitative features have been observed at other pump wavelengths. Figure 3.12 shows, as an example, the observed pump spectra at various peak power levels when 140-ps-wide pump pulses at 1.06 μm are propagated along a 150-m-long fiber (Kean et al., 1987). For peak powers below 100 W, the observed spectra are due to SPM (see Figure 3.1). The Raman threshold is reached near 100 W. The spectrum at 148 W shows the asymmetric spectral broadening resulting from the Raman interaction and should be compared with that shown in Figure 3.10. The physical origin of

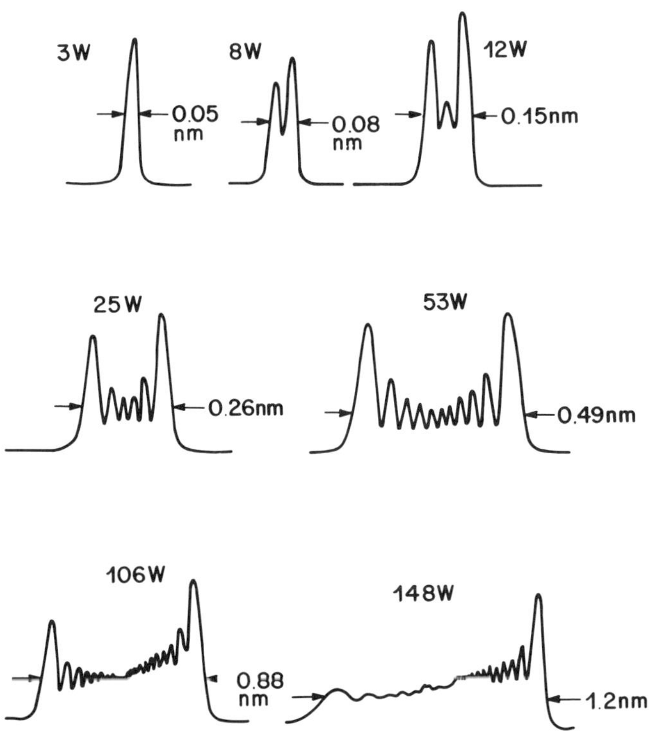

FIGURE 3.12. Experimental spectra of the pump pulse at different peak powers after a 140-ps pulse at 1.06 μm has propagated inside a 150-m-long fiber. The Raman threshold is reached at about 100 W. (After Kean et al., 1987.)

spectral broadening is related to XPM, while the physical origin of spectral asymmetry is related to the pulse walk-off. These features are discussed further in the following section.

7. Cross-Phase Modulation

Although the major qualitative features seen in the pulse spectra of Figures 3.9 and 3.10 are due to XPM, it is difficult to isolate the XPM-induced features. For this reason, we consider in this section the interaction of two incident pulses copropagating along the fiber without SRS; that is, $g_p = 0$ and $g_s = 0$ in Eqs. (38) and (39). The other parameters are identical to those used in Section 6 except for the incident peak powers, which we take to be $P_1 = P_2 =$ 200 W. For these power levels both pulses are below the Raman threshold for $L = 5$ m; that is, $g_s PL < 16$ (Smith, 1972; Stolen, 1980). Both incident pulses are assumed to be Gaussian with $T_0 = 10$ ps; they overlap completely at the input end of the fiber. The pulse walk-off effects are included by taking $d = -2$ ps/m, a value appropriate when the center wavelengths of two pulses are spaced by about 5 nm. Similar to the SRS case shown in Figures 3.7 and 3.9, the evolution of the pulse shapes and spectra along a 5-m-long fiber was studied. It was found that the shapes of both pulses remain largely unaffected during XPM interaction except for the walk-off effect; that is, the peaks of the Gaussian pulses drifted apart at a rate of about 2 ps/m, as one would have expected.

The spectral evolution is, however, governed by the combined effect of SPM and XPM. Figure 3.13 shows the pulse spectra at $z = 5$ m for the two pulses. Similar to the case of SPM (see Figure 3.1), the spectra exhibit internal

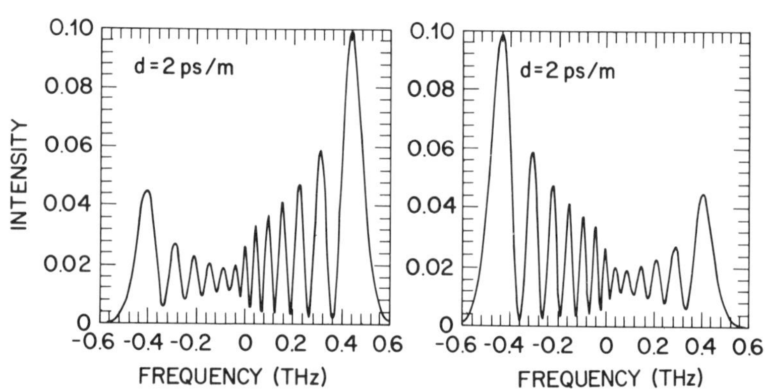

FIGURE 3.13. Calculated spectra of two pulses at $z = 5$ m. Both pulses are identical (except for their center wavelengths) and launched simultaneously inside the fiber with parameters $T_1 = T_2 = 10$ ps and $P_1 = P_2 = 200$ W. Group velocity mismatch is included by taking $d = -2$ ps/m.

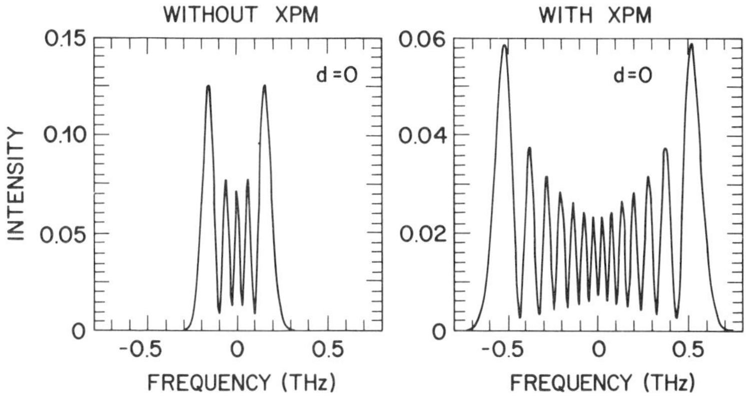

FIGURE 3.14. Same as in Figure 3.13 except that the group velocity mismatch is ignored by setting $d = 0$. Since the spectra are identical, only the spectrum of one pulse is shown. The right and left plots show the spectrum without and with XPM, respectively. Note the XPM-induced broadening by a factor of about 3.

structure with multiple peaks. The noticeable new features are (1) that the spectrum of each pulse is asymmetric and (2) that the spectrum of one pulse is the mirror image of the other (shifted by the initial wavelength separation of the two pulses). In general, blue-shifted components are more intense for the slower pulse while the opposite occurs for the fast-moving pulse.

In order to isolate the features related to XPM and group velocity msmatch, Figure 3.14 shows the spectrum for the case $d = 0$. Also included for comparison is the spectrum expected in the absence of XPM. Since the spectra of both pulses are found to be identical for $d = 0$, the spectra of only one pulse are shown in Figure 3.15. A comparison of the two spectra in Figure 3.14 shows that in the absence of group velocity mismatch ($d = 0$), the effect of XPM is to broaden the spectrum by nearly a factor of 3 over that expected by SPM alone. By comparing Figures 3.13 and 3.14 we conclude that it is the pulse walk-off that is responsible for the spectral asymmetry seen in Figure 3.13. It should, however, be stressed that XPM is also necessary for spectral asymmetry since it is XPM that provides the physical mechanism for two pulses to interact with each other.

A simple analytic model for the XPM interaction between two copropagating pulses can be developed to understand the origin of spectral features seen in Figures 3.13 and 3.14. The model makes the following three assumptions to simplify the coupled-amplitude Eqs. (19) and (20). (1) The effects of GVD are negligible so that $\beta_{21} = \beta_{22} = 0$. This approximation is justified if the fiber length is much shorter than the dispersion length L_D associated with each pulse [see Eq. (23)]. (2) The peak powers of the incident pulses are below the Raman threshold so that $g_1 = g_2 = 0$. (3) The fiber loss is negligible so that $\alpha_1 = \alpha_2 = 0$. The last assumption is not necessary but is made to simplify the treatment. The inclusion of fiber loss does not affect the conclusions drawn here.

It can readily be verified by using Eqs. (19) and (20) with these simplifications that the pulse shapes remain unchanged during propagation. The effect of SPM and XPM is to generate a nonlinear phase shift accumulated over the fiber length L (Islam et al., 1987). The phase shift for pulse 1 is given by

$$\Phi_1(T) = \gamma_1 \int_0^L [|A_1(0, T)|^2 + 2|A_2(0, T - zd)|^2]\, dz, \tag{44}$$

where T is the reduced time given by Eq. (37) and d is the group velocity mismatch defined in a way similar to Eq. (40). If the two pulses are Gaussian in shape and are launched simultaneously with complete overlap, the incident fields are given by

$$A_m(0, T) = \sqrt{P_m} \exp(-T^2/2T_m^2), \tag{45}$$

where $m = 1$ or 2, P_m is the peak power, and T_m is the $1/e$ half-width. We assume equal pulse widths, $T_1 = T_2 = T_0$, but allow for unequal peak powers. By using Eq. (45) in Eq. (44) and performing the integration, we obtain

$$\Phi_1(\tau) = \gamma_1 L \left\{ P_1 \exp(-\tau^2) + P_2 \frac{\sqrt{\pi}}{\delta} [\mathrm{erf}(\tau) - \mathrm{erf}(\tau - \delta)] \right\}, \tag{46}$$

where erf stands for the error function, $\tau = T/T_0$ is the normalized time coordinate, and

$$\delta = Ld/T_0 \tag{47}$$

is the normalized walk-off parameter.

The nonlinearity-induced chirp is obtained by using $\Delta\nu_1 = -(1/2\pi)\,\partial\Phi_1/\partial T$. By differentiating Eq. (46), it is given by

$$\Delta\nu_1(\tau) = \frac{\gamma_1 L}{\pi T_0} e^{-\tau^2} \left\{ P_1 \tau + \frac{P_2}{\delta} [\exp(2\tau\delta - \delta^2) - 1] \right\}. \tag{48}$$

To the lowest order in δ, the chirp is given by the simple relation

$$\Delta\nu_1(\tau) \simeq \frac{\gamma_1 L}{\pi T_0} e^{-\tau^2} [P_1 \tau + P_2(2\tau - \delta)]. \tag{49}$$

The chirp $\Delta\nu_2$ can be obtained by interchanging the subscripts 1 and 2 and replacing δ by $-\delta$ in Eq. (49) and is given by

$$\Delta\nu_2(\tau) = \frac{\gamma_2 L}{\pi T_0} e^{-\tau^2} [P_2 \tau + P_1(2\tau + \delta)]. \tag{50}$$

Equations (49) and (50) can be used to understand all the spectral features seen in Figures 3.13 and 3.14. Consider first the case $\delta = 0$ (no walk-off), for which the numerical pulse spectra are shown in Figure 3.14. Since $P_1 = P_2$ was assumed in numerical simulations, Eq. (49) indicates that the effect of

XPM is to broaden the pulse spectrum by a factor of 3 over that expected from the SPM alone. This is in agreement with the results of Figure 3.14. The fact that the broadening factor is slightly smaller than 3 in Figure 3.14 is related to the GVD, which reduces the peak powers slightly by broadening the pulse during its propagation inside the fiber.

In the presence of group velocity mismatch ($\delta \neq 0$), Eq. (49) shows that the chirp can be larger or smaller depending on the relative signs of τ and δ. For the results shown in Figure 3.13, $\delta < 0$. As a consequence, the chirp is larger near the trailing edge ($\tau > 0$) and smaller near the leading edge ($\tau < 0$) for pulse 1 while the opposite occurs for pulse 2. Since blue components are generated near the trailing edge, they are expected to be more intense for pulse 1. For the same reason, the spectral spread should be larger toward the blue side than the red side for pulse 1. This feature is clearly seen in Figure 3.13. Using a similar argument, Eq. (50) predicts that the spectrum of pulse 2 will have a larger spread on the red side, and the red components will be more intense. For the specific case of $\gamma_i \simeq \gamma_2$ and $P_1 = P_2$, the two spectra should be mirror images of each other since $\Delta\nu_2(\tau) = -\Delta\nu_1(-\tau)$. This feature would, of course, be absent for unequal peak powers.

Equations (49) and (50) can be used to predict the relative spectral widths of two spectra for the case of $P_1 \neq P_2$. Since the spectral range is relatively unaffected by pulse walk-off (compare Figures 3.13 and 3.14), we set $\delta = 0$. The ratio of the spectral widths is then given by

$$\frac{\Delta\nu_2}{\Delta\nu_1} = \frac{\lambda_1}{\lambda_2}\left(\frac{P_2 + 2P_1}{P_1 + 2P_2}\right), \tag{51}$$

where we used Eq. (16) to show the dependence on wavelengths. For $\lambda_1 \simeq \lambda_2$, the width ratio can vary from 0.5 to 2 depending on the power ratio P_2/P_1. In particular, when a weak probe pulse copropagates with an intense pulse ($P_2 \ll P_1$), its spectrum is broadened by a factor of more than 2 over that of the pump pulse because of XPM (Gersten et al., 1980; Alfano et al., 1987). Note, however, that the width ratio can be quite different if λ_1 and λ_2 differ significantly. For example, the spectral broadening would be by a factor of 4 if the probe pulse is at the second harmonic of the pump pulse.

8. Conclusion

This chapter has reviewed the propagation characteristics of ultrashort optical pulses in single-mode fibers influenced by the dispersive and nonlinear effects. When the pulse peak power is below the Raman threshold, the propagation behavior is modeled well by the nonlinear Schrödinger equation (29). New qualitative features arise depending on whether the propagation occurs in the normal or the anomalous dispersion regime. In the latter case, the fiber supports optical solitons and has found applications in the design of the soliton laser. When the pulse peak power exceeds the Raman threshold, the evolution of the pump and Raman pulses is governed by the coupled set of Eqs. (38) and (39). In this case it is important to take into account both the group velocity mismatch and XPM. In particular, XPM is responsible for new spectral features such as an additional broadening of the Raman pulse spectrum by a factor of 2 or more over that of the pump pulse. Of course, SRS is not a prerequisite for the observation of these novel qualitative features; similar propagation characteristics can be observed by launching simultaneously two optical pulses inside the fiber. Clearly, optical fibers are extremely versatile for studying nonlinear phenomena and should find new applications, including one in the design of ultrafast supercontinuum laser sources.

References

Agrawal, G.P. (1987) Modulation instability induced by cross-phase modulation. Phys. Rev. Lett. **59**, 880–883.

Agrawal, G.P. and M.J. Potasek (1986) Nonlinear pulse distortion in single-mode optical fibers at the zero-dispersion wavelength. Phys. Rev. A **33**, 1765–1776.

Alfano, R.R. (1985) Ultrafast supercontinuum laser source. *Proc. Lasers '85*, pp. 110–122.

Alfano, R.R., P.L. Baldeck, F. Raccah, and P.P. Ho (1987) Cross-phase modulation measured in optical fibers. Appl. Opt. **26**, 3491–3492.

Alfano R.R. and S.L. Shapiro (1970) Observation of self-phase modulation in crystals and glasses. Phys. Rev. Lett. **24**, 592–594.

Baldeck, P.L., P.P. Ho, and R.R. Alfano (1987) Effects of self, induced and cross phase modulations on the generation of picosecond and femtosecond white light supercontinua. Rev. Phys. Appl. **22**, 1677–1694.

Basch, E.E., ed. (1986) *Optical-Fiber Transmission.* Sams, Indianapolis, Indiana.

Bourkoff, E., W. Zhao, R.L. Joseph, and D.N. Christodoulides (1987) Evolution of femtosecond pulses in single-mode fibers having higher-order nonlinearity and dispersion. Opt. Lett. **12**, 272–274.

Fisher, R.A. and W.K. Bischel (1975) Numerical studies of the interplay between self-phase modulation and dispersion for intense plane-wave laser pulses. J. Appl. Phys. **46**, 4921–4934.

Fleck, J.A., J.R. Morris, and M.D. Feit (1976) Time-dependent propagation of high-energy laser beams through the atmosphere. Appl. Phys. **10**, 129–160.

Fork, R.L., C.H. Brito Cruz, P.C. Becker, and C.V. Shank (1987) Compression of optical pulses to six femtoseconds by using cubic phase compensation. Opt. Lett. **12**, 483–485.

Gersten, J.I., R.R. Alfano, and M. Belic (1980) Combined stimulated Raman scattering and continuum self-phase modulations. Phys. Rev. A **21**, 1222–1224.

Golovchenko, E.A., E.M. Dianov, A.N. Pilipetskii, A.M. Prokhorov, and V.N. Serkin (1987) Self-effect and maximum contraction of optical femtosecond wave packets in a nonlinear dispersive medium. JETP Lett. **45**, 91–95.

Gordon, J.P. (1986) Theory of the soliton self-frequency shift. Opt. Lett. **11**, 662–664.

Gouveia-Neto, A.S., A.S.L. Gomes, and J.R. Taylor (1987) Generation of 33-fsec pulses at 1.32 μm through a high-order soliton effect in a single-mode optical fiber. Opt. Lett. **12**, 395–397.

Grischkowsky, D. and A.C. Balant (1982) Optical pulse compression based on enhanced frequency chirping. Appl. Phys. Lett. **41**, 1–3.

Hasegawa, A. (1983) Amplification and reshaping of optical solitons in glass fiber—IV. Opt. Lett. **8**, 650–652.

Hasegawa, A. and F. Tappert (1973) Transmission of stationary nonlinear optical pulses in dispersive dielectric fibers. I. Anomalous dispersion. Appl. Phys. Lett. **23**, 142–144.

Islam, M.N., L.F. Mollenauer, R.H. Stolen, J.R. Simpson, and H.T. Shang (1987) Cross-phase modulation in optical fibers. Opt. Lett. **12**, 625–627.

Kean, P.N., K. Smith, and W. Sibbett (1987) Spectral and temporal investigation of self-phase modulation and stimulated Raman scattering in a single-mode optical fiber. IEE Proc. **134**, 163–170.

Kodama, Y. and A. Hasegawa (1987). Nonlinear pulse propagation in a monomode dielectric guide. IEEE J. Quantum Electron. **QE-23**, 510–524.

Lassen, H.E., F. Mengel, B. Tromborg, N.C. Albertsen, and P.L. Christiansen (1985). Evolution of chirped pulses in nonlinear single-mode fibers. Opt. Lett. **10**, 34–36.

Lax, M., J.H. Batteh, and G.P. Agrawal (1981) Chenneling of intense electromagnetic beams. J. Appl. Phys. **52**, 109–125.

Marcuse, D. (1981) Pulse distortion in single-mode fibers. Appl. Opt. **19**, 1653–1660.

Miller, S.E. and I.P. Kaminow, eds. (1988) *Optical Fiber Telecommunications II*. Academic Press, Boston, Massachusetts.

Mitschke, F.M. and L.F. Mollenauer (1986). Discovery of the soliton self-frequency shift. Opt. Lett. **11**, 659–661.

Mitschke, F.M. and L.F. Mollenauer (1987). Ultrashort pulses from the soliton laser. Opt. Lett. **12**, 407–409.

Mollenauer, L.F. and R.H. Stolen (1984) The soliton laser. Opt. Lett. **9**, 13–15.

Mollenauer, L.F., R.H. Stolen, and J.P. Gordon (1980) Experimental observation of picosecond pulse narrowing and solitons in optical fibers. Phys. Rev. Lett. **45**, 1095–1097.

Mollenauer, L.F., J.P. Gordon, and M.N. Islam (1986) Soliton propagation in long fibers with periodically compensated loss. IEEE J. Quantum Electron. **QE-22**, 157–173.

Nikolaus, B. and D. Grischkowsky (1983) 12 × pulse compression using optical fibers. Appl. Phys. Lett. **42**, 1–2.

Schadt, D. and B. Jaskorzynska (1987) Frequency chirp and stimulated Raman scattering influenced by pulse walk-off in optical fibers. J. Opt. Soc. Am. **4**, 856–862.

Serkin, V.N. (1987) Colored envelope solitons in optical fibers. Sov. Tech. Phys. Lett. **13**, 320–321.

Shank, C.V., R.L. Fork, R. Yen, R.H. Stolen, and W.J. Tomlinson (1982) Compression of femtosecond optical pulses. Appl. Phys. Lett. **40**, 761–763.

Shank, C.V., R.L. Fork, C.H. Brito Cruz, and W. Knox (1986). In *Ultrafast Phenomena V*, G.R. Fleming and A.E. Siegman, eds. Springer-Verlag, Heidelberg.

Smith, R.G. (1972) Optical power handling capacity of low loss optical fibers as determined by stimulated Raman and Brillouin scattering. Appl. Opt. **11**, 2489–2494.

Stolen, R.H. (1980) Nonlinearity in fiber transmission. Proc. IEEE **68**, 1232–1236.

Stolen, R.H. and C. Lin (1978) Self-phase modulation in silica optical fibers. Phys. Rev. A **17**, 1448–1453.

Stolen, R.H., L.F. Mollenauer, and W.J. Tomlinson (1983) Observation of pulse restoration at the soliton period in optical fibers. Opt. Lett. **8**, 186–188.

Tai, K. and A. Tomita (1986) 1100 × optical fiber pulse compression using grating pair and soliton effect at 1.319 μm. Appl. Phys. Lett. **48**, 1033–1035.

Tomlinson, W.J., R.H. Stolen, and C.V. Shank (1984) Compression of optical pulses chirped by self-phase modulation in fibers. J. Opt. Soc. Am. B **1**, 139–149.

Tomlinson, W.J., R.H. Stolen, and A.M. Johnson (1985) Optical wave breaking in nonlinear optical fibers. Opt. Lett. **10**, 457–459.

Vysloukh, V.A. and T.A. Matveeva (1987) Influence of inertia of nonlinear response on compression of femtosecond pulses. Sov. J. Quantum Electron. **17**, 498–500.

Wai, P.K., C.R. Menyuk, H.H. Chen, and Y.C. Lee (1987) Soliton at the zero-group-dispersion wavelength of a single-mode fiber. Opt. Lett. **12**, 628–630.

Zakharov, V.E. and A.B. Shabat (1972) Exact theory of two-dimensional self-focusing and one-dimensional self-modulation of waves in nonlinear media. Sov. Phys. JETP **34**, 62–69.

Zysset, B., P. Beaud, and W. Hodel (1987) Generation of optical solitons in the wavelength region 1.37–1.49 μm. Appl. Phys. Lett. **50**, 1027–1029.

4
Cross-Phase Modulation: A New Technique for Controlling the Spectral, Temporal, and Spatial Properties of Ultrashort Pulses

P.L. BALDECK, P.P. HO, and R.R. ALFANO

1. Introduction

Self-phase modulation (SPM) is the principal mechanism responsible for the generation of picosecond and femtosecond white-light supercontinua. When an intense ultrashort pulse propagates through a medium, it distorts the atomic configuration of the material, which changes the refractive index. The pulse phase is time modulated, which causes the generation of new frequencies. This phase modulation originates from the pulse itself (*self*-phase modulation). It can also be generated by a copropagating pulse (*cross*-phase modulation).

Several schemes of nonlinear interaction between optical pulses can lead to cross-phase modulation (XPM). For example, XPM is intrinsic to the generation processes of stimulated Raman scattering (SRS) pulses, second harmonic generation (SHG) pulses, and stimulated four-photon mixing (SFPM) pulses. More important, the XPM generated by pump pulses can be used to control, with femtosecond time response, the spectral, temporal, and spatial properties of ultrashort probe pulses.

Early studies on XPM characterized induced polarization effects (optical Kerr effect) and induced phase changes, but did not investigate spectral, temporal and spatial effects on the properties of ultrashort pulses. In 1980, Gersten, Alfano, and Belic predicted that Raman spectra of ultrashort pulses would be broadened by XPM (Gersten et al., 1980). The first experimental observation of XPM spectral effects dates to early 1986, when it was reported that intense picosecond pulses could be used to enhance the spectral broadening of weaker pulses copropagating in bulk glasses (Alfano et al., 1986). Since then, several groups have been studying XPM effects generated by ultrashort pump pulses on copropagating Raman pulses (Schadt et al., 1986; Schadt and Jaskorzynska, 1987a; Islam et al. 1987a; Alfano et al., 1987b; Baldeck et al., 1987b–d; Manassah, 1987a, b; Hook et al., 1988), second harmonic pulses (Alfano et al., 1987a; Manassah, 1987c; Manassah and Cockings, 1987; Ho et al., 1988), stimulated four-photon mixing pulses (Baldeck and Alfano, 1987), and probe pulses (Manassah et al., 1985; Agrawal et al., 1989a; Baldeck et al., 1988a, c). Recently, it has been shown that XPM leads to the generation of modulation instability (Agrawal, 1987; Agrawal et al., 1989b; Schadt and Jaskorzynska, 1987b; Baldeck et al., 1988b, 1988d; Gouveia-Neto et al., 1988a, b), solitary waves (Islam et al., 1987b; Trillo et al., 1988), and pulse compression (Jaskorzynska and Schadt, 1988; Manassah, 1988; Agrawal et al. 1988). Finally, XPM effects on ultrashort pulses have been proposed to tune the frequency of probe pulses (Baldeck et al., 1988a), to eliminate the soliton self-frequency shift effect (Schadt and Jaskorzynska, 1988), and to control the spatial distribution of light in large core optical fibers (Baldeck et al., 1987a).

This chapter reviews some of the key theoretical and experimental works that have predicted and described spectral, temporal, and spatial effects attributed to XPM. In Section 2, the basis of the XPM theory is outlined. The nonlinear polarizations, XPM phases, and spectral distributions of copropagating pulses are computed. The effects of pulse walk-off, input time delay, and group velocity dispersion broadening are particularly discussed. (Additional work on XPM and on SPM theories can be found in Manassah (Chapter 5) and Agrawal (Chapter 3).) Experimental evidence for spectral broadening enhancement, induced-frequency shift, and XPM-induced optical amplification is presented in Section 3. Sections 4, 5, and 6 consider the effects of XPM on Raman pulses, second harmonic pulses, and stimulated four-photon mixing pulses, respectively. Section 7 shows how induced focusing can be initiated by XPM in optical fibers. Section 8 presents measurements of modulation instability induced by cross-phase modulation in the normal dispersion region of optical fibers. Section 9 describes XPM-based devices that could be developed for the optical processing of ultrashort pulses with terahertz repetition rates. Finally, Section 10 summarizes the chapter and highlights future trends.

2. Cross-Phase Modulation Theory

2.1 Coupled Nonlinear Equations of Copropagating Pulses

The methods of multiple scales and slowly varying amplitude (SVA) are the two independent approximations used to derive the coupled nonlinear equations of copropagating pulses. The multiple scale method, which has been used for the first theoretical study on induced-phase modulation, is described in Manassah (Chapter 5). The following derivation is based on the SVA approximation.

The optical electromagnetic field of two copropagating pulses must ultimately satisfy Maxwell's vector equation:

$$\nabla \times \nabla \times \mathbf{E} = -\mu_0 \frac{\partial \mathbf{D}}{\partial t} \tag{1a}$$

and

$$\mathbf{D} = \varepsilon \mathbf{E} + \mathbf{P}^{NL}, \tag{1b}$$

where ε is the medium permitivity at low intensity and $\mathbf{P}^{NL}$ is the nonlinear polarization vector.

Assuming a pulse duration much longer than the response time of the medium, an isotropic medium, the same linear polarization for the copropagating fields, and no frequency dependence for the nonlinear susceptibility $\chi^{(3)}$, the nonlinear polarization reduces to

$$P^{NL}(r, z, t) = \chi^{(3)} E^3(r, z, t), \tag{2}$$

where the transverse component of the total electric field can be approximated by

$$E(r, z, t) = \tfrac{1}{2}\{A_1(r, z, t)e^{i(\omega_1 t - \beta_1 z)} + A_2(r, z, t)e^{i(\omega_2 t - \beta_2 z)} + \text{c.c.}\}. \tag{3}$$

A_1 and A_2 refer to the envelopes of copropagating pulses of carrier frequencies ω_1 and ω_2, and β_1 and β_2 are the corresponding propagation constants, respectively.

Substituting Eq. (3) into Eq. (2) and keeping only the terms synchronized with ω_1 and ω_2, one obtains

$$P^{NL}(r, z, t) = P_1^{NL}(r, z, t) + P_2^{NL}(r, z, t), \tag{4a}$$

$$P_1^{NL}(r, z, t) = \tfrac{3}{8}\chi^{(3)}(|A_1|^2 + 2|A_2|^2)A_1 e^{i(\omega_1 t - \beta_1 z)}, \tag{4b}$$

$$P_2^{NL}(r, z, t) = \tfrac{3}{8}\chi^{(3)}(|A_2|^2 + 2|A_1|^2)A_2 e^{i(\omega_2 t - \beta_2 z)}, \tag{4c}$$

where P_1^{NL} and P_2^{NL} are the nonlinear polarizations at frequencies ω_1 and ω_2, respectively. The second terms in the right sides of Eqs. (4b) and (4c) are cross-phase modulations terms. Note the factor of 2.

Combining Eqs. (1)–(4) and using the slowly varying envelope approximation (at the first order for the nonlinearity), one obtains the coupled nonlinear wave equations:

$$\frac{\partial A_1}{\partial z} + \frac{1}{v_{g1}}\frac{\partial A_1}{\partial t} + \frac{i}{2}\beta_1^{(2)}\frac{\partial^2 A_1}{\partial t^2} = i\frac{\omega_1}{c}n_2[|A_1|^2 + 2|A_2|^2]A_1, \tag{5a}$$

$$\frac{\partial A_2}{\partial z} + \frac{1}{v_{g2}}\frac{\partial A_2}{\partial t} + \frac{i}{2}\beta_2^{(2)}\frac{\partial^2 A_2}{\partial t^2} = i\frac{\omega_2}{c}n_2[|A_2|^2 + 2|A_1|^2]A_2, \tag{5b}$$

where v_{gi} is the group velocity for the wave i, $\beta_i^{(2)}$ is the group velocity dispersion for the wave i, and $n_2 = 3\chi^{(3)}/8n$ is the nonlinear refractive index.

In the most general case, numerical methods are used to solve Eqs. (5). However, they have analytical solutions when the group velocity dispersion temporal broadening can be neglected.

Denoting the amplitude and phase of the pulse envelope by a and α, that is,

$$A_1(\tau, z) = a_1(\tau, z)e^{i\alpha_1(\tau, z)} \quad \text{and} \quad A_2(\tau, z) = a_2(\tau, z)e^{i\alpha_2(\tau, z)}, \tag{6}$$

and assuming $\beta_1^{(2)} \approx \beta_2^{(2)} \approx 0$, Eqs. (5a) and (5b) reduce to

$$\frac{\partial a_1}{\partial z} = 0, \tag{7a}$$

$$\frac{\partial \alpha_1}{\partial z} = i\frac{\omega_1}{c} n_2 [a_1^2 + 2a_2^2], \tag{7b}$$

$$\frac{\partial a_2}{\partial z} + \left(\frac{1}{v_{g2}} - \frac{1}{v_{g1}}\right)\frac{\partial a_2}{\partial \tau} = 0, \tag{7c}$$

$$\frac{\partial \alpha_2}{\partial z} = i\frac{\omega_2}{c} n_2 [a_2^2 + 2a_1^2], \tag{7d}$$

where $\tau = (t - z/v_{g1})/T_0$ and T_0 is the $1/e$ pulse duration.

In addition, Gaussian pulses are chosen at $z = 0$:

$$A_1(\tau, z = 0) = \sqrt{\frac{P_1}{A_{\text{eff}}}}\, e^{-\tau^2/2}, \tag{8a}$$

$$A_2(\tau, z = 0) = \sqrt{\frac{P_2}{A_{\text{eff}}}}\, e^{-(\tau-\tau_d)^2/2}, \tag{8b}$$

where P is the pulse peak power, A_{eff} is the effective cross-sectional area, and $\tau_d = T_d/T_0$ is the normalized time delay between pulses at $z = 0$. With the initial conditions defined by Eqs. (8), Eqs. (5) have analytical solutions when temporal broadenings are neglected:

$$A_1(\tau, z) = \sqrt{\frac{P_1}{A_{\text{eff}}}}\, e^{-\tau^2/2} e^{i\alpha_1(t,z)}, \tag{9a}$$

$$A_2(\tau, z) = \sqrt{\frac{P_2}{A_{\text{eff}}}}\, e^{-(\tau-\tau_d-z/L_w)^2/2} e^{i\alpha_2(\tau,z)}, \tag{9b}$$

$$\alpha_1(\tau, z) = \frac{\omega_1}{c} n_2 \frac{P_1}{A_{\text{eff}}} z e^{-\tau^2} + \sqrt{\pi}\frac{\omega_1}{c} n_2 \frac{P_2}{A_{\text{eff}}} L_w \left[\operatorname{erf}(\tau - \tau_d) - \operatorname{erf}\left(\tau - \tau_d - \frac{z}{L_w}\right)\right], \tag{9c}$$

$$\alpha_2(\tau, z) = \frac{\omega_2}{c} n_2 \frac{P_2}{A_{\text{eff}}} z e^{-(\tau-\tau_d-z/L_w)^2} + \sqrt{\pi}\frac{\omega_2}{c} n_2 \frac{P_1}{A_{\text{eff}}} L_w \left[\operatorname{erf}(\tau) - \operatorname{erf}\left(\tau - \frac{z}{L_w}\right)\right], \tag{9d}$$

where $L_w = T_0/(1/v_{g1} - 1/v_{g2})$ is defined as the walk-off length.

Equations (9c) and (9d) show that the phases $\alpha_i(\tau, z)$ of copropagating pulses that overlap in a nonlinear Kerr medium are modified by a cross-phase modulation via the peak power $P_{j \neq i}$. In the case of ultrashort pulses this cross-phase modulation gives rise to the generation of new frequencies, as does self-phase modulation.

The instantaneous XPM-induced frequency chirps are obtained by differentiating Eqs. (9c) and (9d) according to the instantaneous frequency formula $\Delta\omega = -\partial\alpha/\partial\tau$. These are

$$\Delta\omega_1(\tau, z) = 2\frac{\omega_1}{c} n_2 \frac{P_1}{A_{\text{eff}}} \frac{z}{T_0} \tau e^{-\tau^2} 2\frac{\omega_1}{c} n_2 \frac{P_2}{A_{\text{eff}}} \frac{L_w}{T_0} [e^{-(\tau-\tau_d-2/L_w)^2} - e^{-(\tau-\tau_d-z/L_w)2}], \tag{10a}$$

$$\Delta\omega_2(\tau, z) = 2\frac{\omega_2}{c} n_2 \frac{P_2}{A_{\text{eff}}} \frac{z}{T_0} (\tau - \tau_d - 2/L_w) e^{-(\tau-\tau_d-z/L_w)^2}$$

$$+ 2\frac{\omega_2}{c} n_2 \frac{P_1}{A_{\text{eff}}} \frac{L_w}{T_0} [e^{-\tau^2} - e^{-(\tau-2/L_w)^2}], \tag{10b}$$

where $\Delta\omega_1 = \omega - \omega_1$ and $\Delta\omega_2 = \omega - \omega_2$. The first and second terms on the right sides of Eqs. (10a) and (10b) are contributions arising from SPM and XPM, respectively. It is interesting to notice in Eq. (10) than the maximum frequency chirp arising from XPM is inversely proportional to the group velocity mismatch $L_w/T_0 = 1/(1/v_{g1} - 1/v_{g2})$ rather than the pump pulse time duration or distance traveled z as for ZPM. Therefore, the time duration of pump pulses does not have to be as short as the time duration of probe pulses for XPM applications.

More generally, spectral profiles affected by XPM can be studied by computing the Fourier transform:

$$S(\omega - \omega_0, z) = \frac{1}{2\pi} \int_{-\infty}^{+\infty} a(\tau, z) e^{i\alpha(\tau, z)} e^{i(\omega-\omega_0)\tau}\, d\tau \tag{11}$$

where $|S(\omega - \omega_0, z)|^2$ represents the spectral intensity distribution of the pulse. Equation (10) is readily evaluated numerically using fast Fourier transform algorithms.

Analytical results of Eqs. (9) take in account XPM, SPM, and group velocity mismatch. These results are used in the Section 2.2 to isolate the specific spectral features arising from the nonlinear interaction of copropagating pulses. Higher-order effects due to group velocity dispersion broadening are discussed in Section 2.3.

2.2 Spectral Broadening Enhancement

The spectral evolution of ultrashort pulses interacting in a nonlinear Kerr medium is affected by the combined effects of XPM, SPM, and pulse walk-off.

For a negligible group velocity mismatch, XPM causes the pulse spectrum to broaden more than expected from SPM alone. The pulse phase of Eqs. (9c) and (9d) reduces to

$$\alpha_i(\tau, z) = \frac{\omega_i}{c} n_2 \frac{(P_i + 2P_j)}{A_{\text{eff}}} z e^{-\tau^2}. \tag{12}$$

The maximum spectral broadening of Gaussian pulses, computed using Eqs.

(12), is given by

$$\Delta\omega_i(z) \approx \frac{\omega_i}{c} n_2 \frac{(P_i + 2P_j)}{A_{\text{eff}}} \frac{z}{T_0}. \tag{13}$$

Thus, the spectral broadening enhancement arising from XPM is given by

$$\frac{\Delta\omega_{i\text{SPM+XPM}}}{\Delta\omega_{i\text{SPM}}} = 1 + \frac{2P_j}{P_i}. \tag{14}$$

Therefore, XPM can be used to control the spectral broadening of probe pulses using strong command pulses. This spectral control is important, for it is based on the electronic response of the interacting medium. It could be turned on and off in a few femtoseconds, which could lead to applications such as the pulse compression of weak probe pulses, frequency-based optical computation schemes, and the frequency multiplexing of ultrashort optical pulses with terahertz repetition rates.

The effect of pulse walk-off on XPM-induced spectral broadening can be neglected when wavelengths of pulses are in the low dispersion region of the nonlinear material, the wavelength difference or/and the sample length are small, and the time duration of pulses is not too short. For other physical situations, the group velocity mismatch and initial time delay between pulses affect strongly the spectral shape of interacting pulses (Islam et al., 1987a; Manassah 1987a; Agrawal et al., 1988, 1989a; Baldeck et al., 1988a).

Figure 4.1 shows how the spectrum of a weak probe pulse can be affected by the XPM generated by a strong copropagating pulse. The wavelength of the pump pulse was chosen where the pump pulse travels faster than the probe pulse. Initial time delays between pulses at the entrance of the nonlinear medium were selected to display the most characteristic interaction schemes. Figures 4.1a and 4.1b are displayed for reference. They show the probe pulse spectrum without XPM interaction (Figure 4.1a) and after the XPM interaction but for negligible group velocity mismatch (Figure 4.1b). Figure 4.1c is for the case of no initial time delay and total walk-off. The probe spectrum is shifted and broadened by XPM. The anti-Stokes shift is characteristic of the probe and pump pulse walk-off. The probe pulse is blue shifted because it is modulated only by the back of the faster pump pulse. When the time delay is chosen such that the pump pulse enters the nonlinear medium after the probe and has just time to catch up with the probe pulse, one obtains a broadening similar to that in Figure 4.1c but with a reverse Stokes shift (Figure 4.1d). The XPM broadening becomes symmetrical when the input time delay allows the pump pulse not only to catch up with but also to pass partially through the probe pulse (Figure 4.1e). However, if the interaction length is long enough to allow the pump pulse to completely overcome the probe pulse, there is no XPM-induced broadening (Figure 4.1f).

The diversity of spectral features displayed in Figure 4.1 can easily be understood by computing the phase and frequency chirp given by Eqs. (9) and (10) (Figure 4.2). For reference, Figure 4.2a shows the locations of the pump pulse (solid line) and the probe pulse (dotted line) at the output of the nonlinear

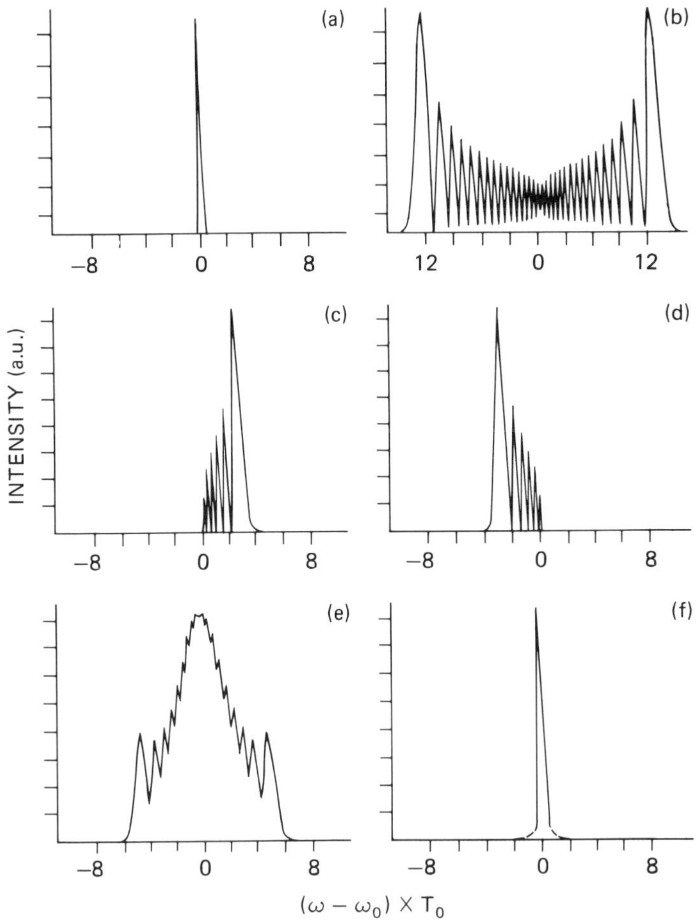

FIGURE 4.1. Influence of cross-phase modulation, walk-off, and input time delay on the spectrum of a probe pulse from Eqs. (9) and (11) with $P_1 \ll P_2$. $\phi = 2(\omega_1/c)n_2 P_2 L_w$, $\delta = z/L_w$, and τ_d are the XPM, walk-off, and input time delay parameters, respectively. (a) Reference spectrum with no XPM; i.e. $\phi = 0$. (b) XPM in the absence of walk-off; i.e., $\phi = 50$ and $\delta = 0$. (c) XPM, total walk-off, and no initial time delay; i.e., $\phi = 50$, $\delta = -5$, and $\tau_d = 0$. (d) XPM and initial time delay to compensate the walk-off; i.e., $\phi = 50$, $\delta = -5$, and $\tau_d = 5$. (e) XPM and symmetrical *partial* walk-off; i.e., $\phi = 50$, $\delta = -3$, and $\tau_d = 1.5$. (f) XPM and symmetrical *total* walk-off; i.e., $\phi = 50$, $\delta = -5$, and $\tau_d = 2.5$.

sample (case of no initial delay and total walk-off). In this case the XPM phase, which is integrated over the fiber length, has the characteristic shape of an error function whose maximum corresponds to neither the probe pulse maximum nor the pump pulse maximum (Figure 4.2b). The probe pulse (dotted line in Figure 4.2c) sees only the blue part of the frequency chirp (solid line in Figure 4.2c) generated by the pump pulse. As a result, the probe spectrum is

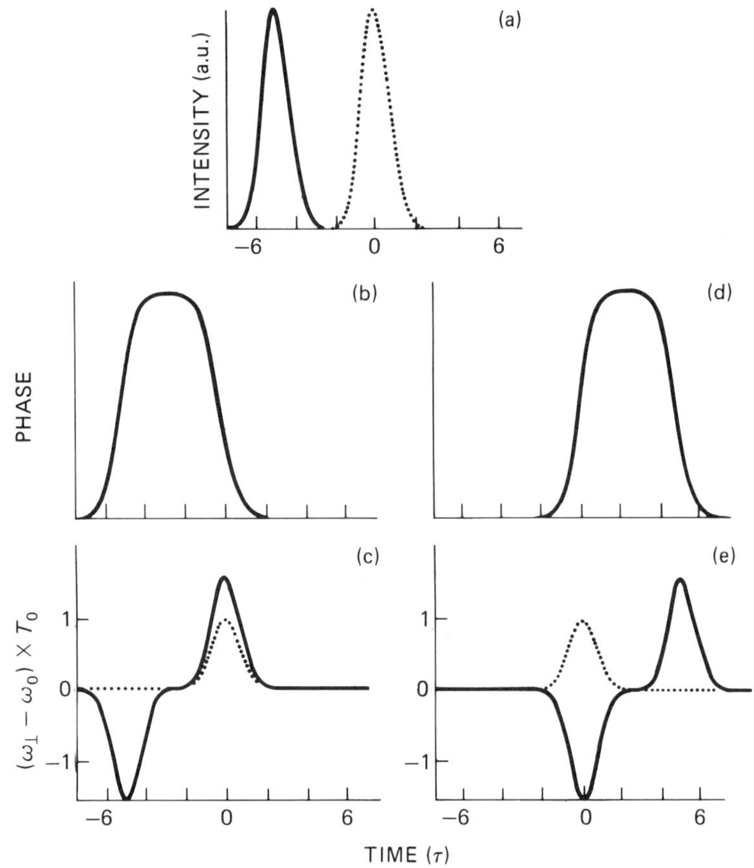

(a)

FIGURE 4.2. Influence of cross-phase modulation, walk-off, and input time delay on the phase and frequency chirp of a probe pulse. (a) Locations of pump (solid line) and probe (dotted line) at the output of the nonlinear medium for total walk-off and no initial time-delay; i.e., $\delta = -5$ and $\tau_d = 0$. (b) XPM phase with a total walk-off and no initial time delay; i.e., $\phi = 50$, $\delta = -5$, and $\tau_d = 0$. (c) XPM-induced chirp (solid line) with total walk-off and no initial time delay. (Dotted line) Probe pulse intensity. (d) XPM phase with an initial time delay to compensate the walk-off; i.e., $\phi = 50$, $\delta = -5$, and $\tau_d = 5$. (e) XPM-induced chirp (solid line) with an initial time delay to compensate the walk-off. (Dotted line) Probe pulse intensity. (f) XPM phase and symmetrical *partial* walk-off; i.e., $\phi = 50$, $\delta = -3$, and $\tau_d = 1.5$. (g) XPM-induced chirp (solid line) and symmetrical *partial* walk-off. (Dotted line) Probe pulse intensity. (h) XPM phase and symmetrical *total* walk-off; i.e., $\phi = 50$, $\delta = -5$, and $\tau_d = 2.5$. (i) XPM-induced chirp (solid line) and symmetrical *total* walk-off. (Dotted line) Probe pulse intensity.

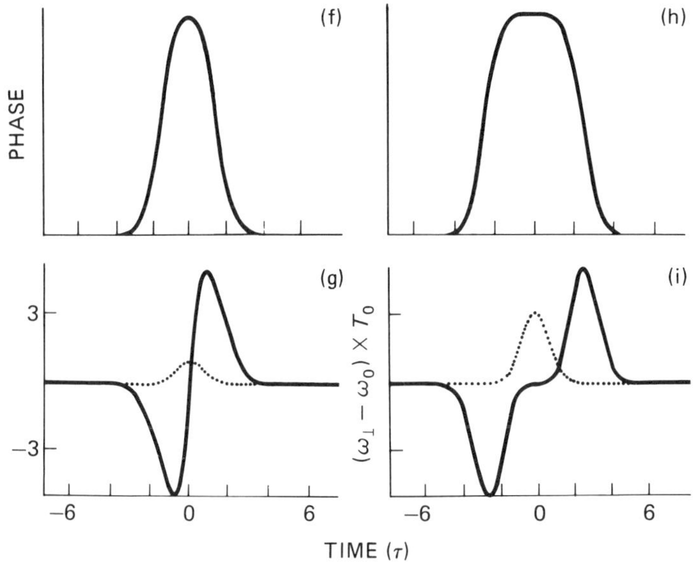

(b)

FIGURE 4.2 (*continued*)

simultaneously broadened and shifted toward the highest frequencies (Figure 4.1c). One should notice that opposite to the SPM frequency chirp, the XPM chirp in Figure 4.2c is not monotonic. The pulse leading edge and trailing edge have a positive chirp and negative chirp, respectively. As a result, dispersive effects (GVD, grating pair, ...) are different for the pulse front and the pulse back. In the regime of normal dispersion ($\beta^{(2)} > 0$), the pulse front would be broadened by GVD while the pulse back would be sharpened. Figures 4.2d and 4.2e show XPM-induced phase and frequency chirp for the mirror image case of Figures 4.2b and 4.2c. The probe spectrum is now shifted toward the smallest frequencies. Its leading edge has a negative frequency chirp, while the trailing edge has a positive one. A positive GVD would compress the pulse front and broaden the pulse back. The case of a partial symmetrical walk-off is displayed in Figures 4.2f and 4.2g. In first approximation, the time dependence of the XPM phase associated with the probe pulse energy is parabolic (Figure 4.2f), and the frequency chirp is quasi-linear (4.2g). This is the prime quality needed for the compression of a weak pulse by following the XPM interaction by a grating pair compressor (Manassah, 1988). Figures 4.2h and 4.2i show why there is almost no spectral broadening enhancement when the pump pulse passes completely through the probe pulse (Figure 4.1f): the part of XPM associated with the probe pulse energy is constant (Figure 4.2h). The probe pulse is phase modulated, but *the phase shift is time independent*. Therefore, there is neither frequency chirp (Figure 4.2i) nor spectral broadening enhancement by XPM.

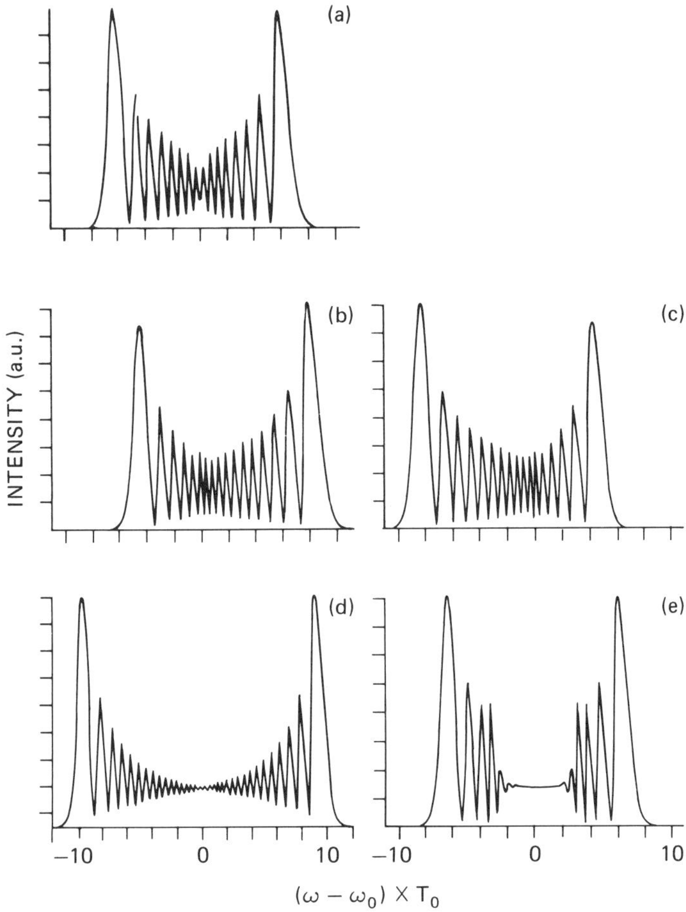

FIGURE 4.3. Influence of self-phase modulation, cross-phase modulation, walk-off, and input time delay on the spectrum of a probe pulse from Eqs. (9) and (11) with $P_1 = P_2$. The parameter values in Figure 4.1 are used.

The combined effects of XPM and walk-off on the spectra of weak probe pulses (negligible SPM) have been shown in Figure 4.1 and 4.2. When the group velocity mismatch is large, the spectral broadening is not significant and the above spectral features reduce to a tunable induced-frequency shift of the probe pulse frequency (see Section 3.2). When strong probe pulses are used, the SPM contribution has to be included in the analysis. Figure 4.3 show how the results of Figure 4.1 are modified when the probe power is the same as the pump power, that is, the SPM has to be taken in account. Figure 4.3a shows the spectral broadening arising from the SPM alone. Combined effects of SPM and XPM are displayed in Figures 4.3b to 4.3e with the same initial delays as in Figure 4.1. The SPM contribution to the spectral broadening is larger than the XPM contribution because the XPM interaction length is limited by the walk-off between pump and probe pulses.

The XPM spectral features described in this section have been obtained using first-order approximation of the nonlinear polarization, propagation constant, and nonlinearity in the nonlinear wave equation (Eq. 1). Moreover, plane wave solutions and peak powers below the stimulated Raman scattering threshold have been assumed. For practical purposes it is often necessary to include the effects of (1) first- and second-order group velocity dispersion broadening, $\beta^{(2)}$ and $\beta^{(3)}$, (2) induced- and self-steepening, (3) four-wave mixing occurring when pump and probe pulses are coupled through $\chi^{(3)}$, (4) stimulated Raman scattering generation, (5) the finite time response of the nonlinearity, and (6) the spatial distribution of interacting fields (i.e., induced- and self-focusing, diffraction, Gaussian profile of beams, ...). In Section 2.3 the combined effect of XPM and group velocity dispersion broadening $\beta^{(2)}$ is shown to lead to new kinds of optical wave breaking and pulse compression. Some other effects that lead to additional spectral, temporal, and spatial features of XPM are discussed by Agrawal (Chapter 3) and Manassah (Chapter 5).

2.3 Optical Wave Breaking and Pulse Compression due to Cross-Phase Modulation in Optical Fibers

When an ultrashort light pulse propagates through an optical fiber, its shape and spectrum change considerably as a result of the combined effect of group velocity dispersion $\beta^{(2)}$ and self-phase modulation. In the normal dispersion regime of the fiber ($\lambda \leq 1.3\ \mu$m), the pulse can develop rapid oscillations in the wings together with spectral sidelobes as a result of a phenomenon known as optical wave breaking (Tomlinson et al., 1985). In this section it is shown that a similar phenomenon can lead to rapid oscillations near one edge of a weak pulse that copropagates with a strong pulse (Agrawal et al., 1988).

To isolate the effects of XPM from those of SPM, a pump-probe configuration is chosen ($P_2 \ll P_1$) so that pulse 1 plays the role of the pump pulse and propagates without being affected by the copropagating probe pulse. The probe pulse, however, interacts with the pump pulse through XPM. To study how XPM affects the probe evolution along the fiber, Eqs. (5a) and (5b) have been solved numerically using a generalization of the beam propagation or the split-step method (Agrawal and Potasek, 1986). The numerical results depend strongly on the relative magnitudes of the length scales L_d and L_w, where $L_d = T_0^2/|\beta_2|$ is the dispersion length and $L_w = v_{g1}v_{g2}T_0/|v_{g1} - v_{g2}|$ is the walk-off length. If $L_w \ll L_d$, the pulses walk off from each other before GVD has an opportunity to influence the pulse evolution. However, if L_w and L_d become comparable, XPM and GVD can act together and modify the pulse shape and spectra with new features.

To show these features as simply as possible, a specific case is considered in which $L_w/L_d = 0.1$ and $\lambda_1/\lambda_2 = 1.2$. Both pulses are assumed to propagate in the normal GVD regime with $\beta_1 = \beta_2 > 0$. It is assumed that the pump pulse

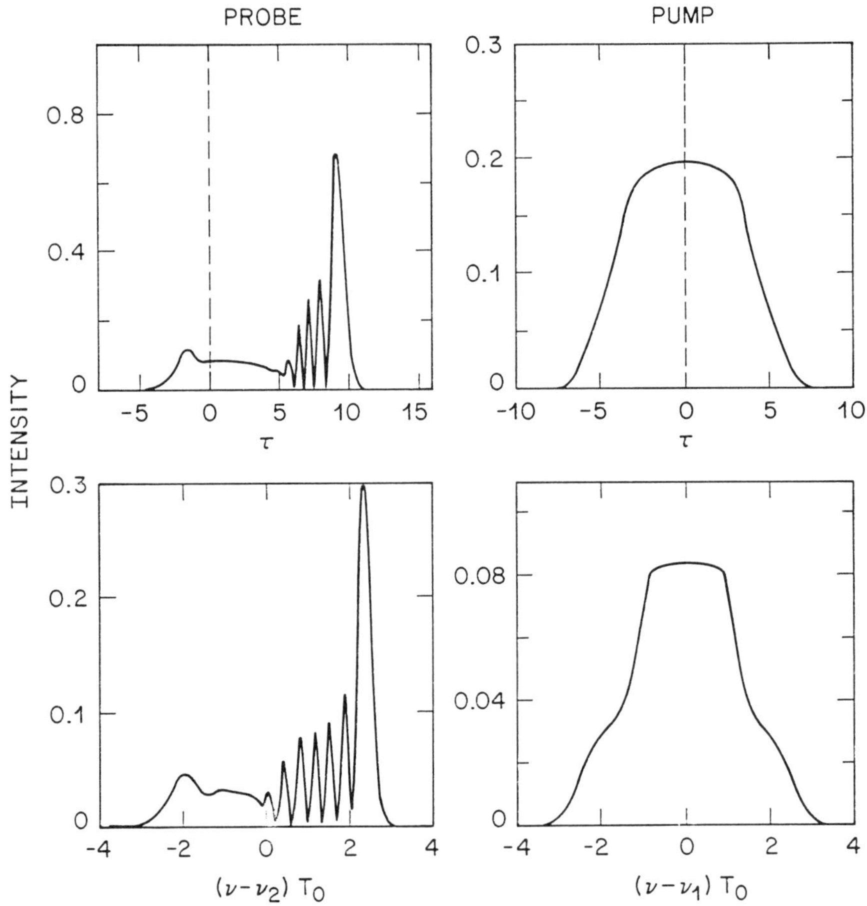

FIGURE 4.4. Shape and spectrum of probe pulse (left) and pump pulse (right) at $z/L_d =$ 0.4 when the two pulses copropagate in the normal dispersion regime of a single-mode fiber. The parameters are $N = 10$, $L_w/L_d = 0.1$, $\lambda_1/\lambda_2 = 1.2$, and $\tau_d = 0$. Oscillations near the trailing edge (positive time) of the probe pulse are due to XPM-induced optical wave breaking. (From Agrawal et al., 1988.)

goes faster than the probe pulse ($v_{g1} > v_{g2}$). At the fiber input both pulses are taken to be a Gaussian of the same width with an initial delay τ_d between them. First, the case $\tau_d = 0$ is considered, so the two pulses overlap completely at $z = 0$. Figure 4.4 shows the shapes and spectra of the pump and probe pulses at $z/L_d = 0.4$ obtained by solving Eqs. (5a) and (5b) numerically with $N = (\gamma_1 P_1 L_d)^{0.5} = 10$. For comparison, Figure 4.5 shows the probe and pump spectra under identical conditions but without GVD effects ($\beta_1 = \beta_2 = 0$). The pulse shapes are not shown since they remain unchanged when the GVD effects are excluded.

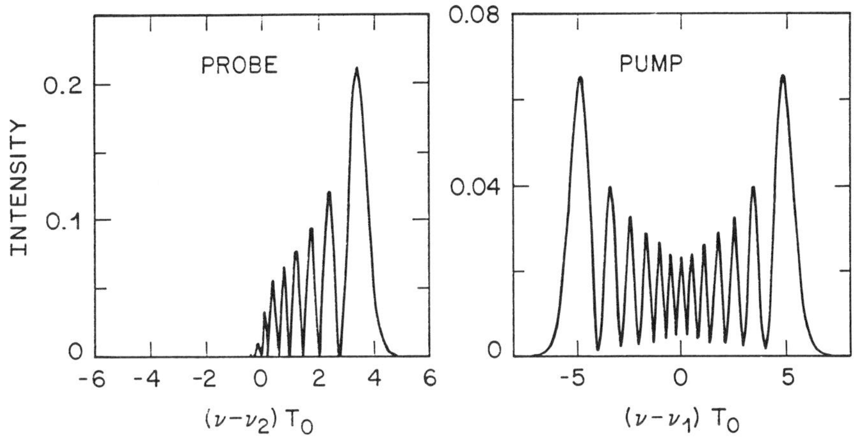

FIGURE 4.5. Spectra of probe and pump pulses under conditions identical to those of Figure 4.4 but without the GVD effects ($\beta_1 = \beta_2 = 0$). Pulse shapes are not shown as they remain unchanged. (From Agrawal et al., 1988.)

From a comparison of Figures 4.4 and 4.5 it is evident that GVD can substantially affect the evolution of features expected from SPM or XPM alone. Consider first the pump pulse for which XPM effects are absent. The expected from dispersive SPM for $N = 10$. With further propagation, the pump pulse eventually develops rapid oscillations in the wings as a result of conventional SPM-induced optical wave breaking. Consider now the probe pulse for which SPM effects are absent and probe pulse evolution is governed by dispersive XPM. In absence of GVD, the pulse shape would be a narrow Gaussian centered at $\tau = 4$ (the relative delay at the fiber output because of group velocity mismatch). The GVD effects not only broaden the pulse considerably but also induce rapid oscillations near the trailing edge of the probe pulse. These oscillations are due to XPM-induced optical wave breaking.

To understand the origin of XPM-induced optical wave breaking, it is useful to consider the frequency chirp imposed on the probe pulse by the copropagating pulse. As there is total walk-off and no initial delay, maximum chirp occurs at the center of the probe pulse. Since the chirp is positive, blue-shifted components are generated by XPM near the pulse center. As a result of the normal GVD, the peak of the probe pulse moves slower than its tails. Since the peak lags behind as the probe pulse propagates, it interferes with the trailing edge. Oscillations seen near the trailing edge of the probe pulse in Figure 4.4 result from such an interference. Since the basic mechanism is analogous to the optical wave-breaking phenomenon occurring in the case of dispersive XPM, we call it XPM-induced optical wave breaking.

In spite of the identical nature of the underlying physical mechanism, optical wave breaking exhibits different qualitative features in the XPM case compared with the SPM case. The most striking difference is that the pulse shape is asymmetric with only one edge developing oscillations. For the case shown in Figure 4.4 oscillations occur near the trailing edge. If the probe and pump wavelengths were reversed so that the pump pulse moved slower than the probe pulse, oscillations would occur near the leading edge since the pump pulse would interact mainly with that edge. In fact, in that case the shape and the spectrum of the probe pulse are just the mirror images of those shown in Figures 4.4 and 4.5.

The effect of initial delay between probe and pump pulses is now investigated. The effect of initial delay on XPM-induced spectral broadening has been discussed in the dispersionless limit ($\beta_1 = \beta_2 = 0$) in Section 2.2. For example, if the pump pulse is delayed by the right amount so that it catches

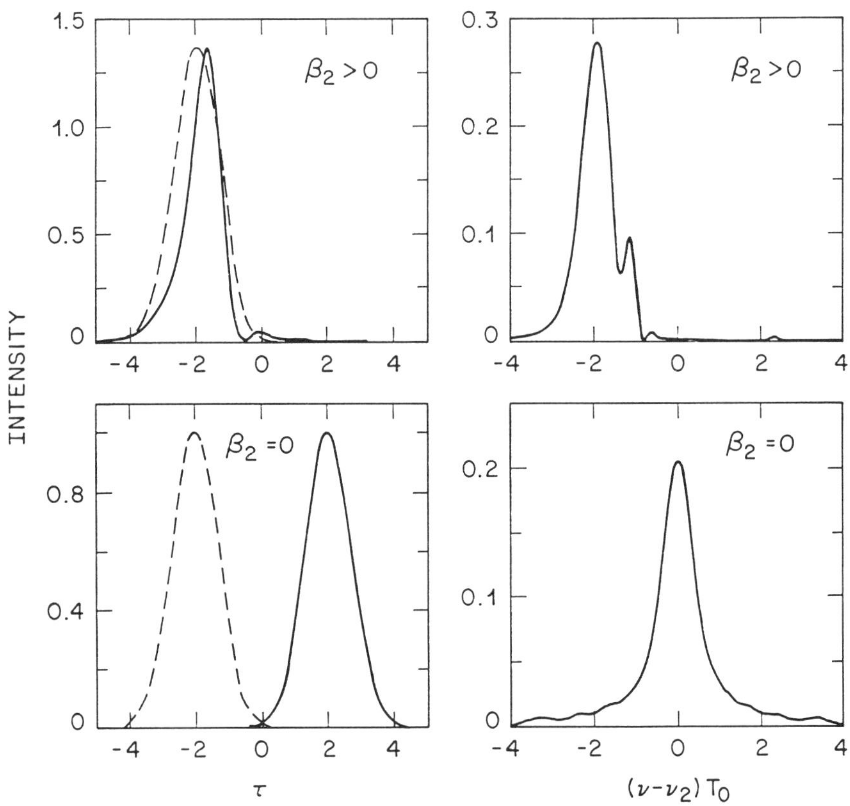

FIGURE 4.6. Probe shape and spectrum with (top) and without (bottom) the GVD effects under conditions identical to those of Figure 4.5 except that $\tau_d = -2$. Note the important effect on pulse evolution of the initial time delay between the pump and probe pulses. (From Agrawal et al., 1988.)

up with the probe pulse at the fiber output, the probe spectrum is just the mirror image of that shown in Figure 4.4, exhibiting a red shift rather than a blue shift. Futhermore, if τ_d is adjusted such that the pump pulse catches up with the probe pulse halfway through the fiber, the probe spectrum is symmetrically broadened since the pump walks through the probe in a symmetric manner. Our numerical results show that the inclusion of GVD completely alters this behavior. Figure 4.6 shows the probe shape and spectrum under conditions identical to those of Figure 4.4 except that the probe pulse is advanced ($\tau_d = -2$) such that the pump pulse would catch it halfway through the fiber in the absence of GVD effects. The lower row shows the expected behavior in the dispersionless limit, showing the symmetrical spectral broadening in this case of symmetrical walk-off. A direct comparison reveals how much the presence of GVD can affect the SPM effects on the pulse evolution. In particular, both the pulse shape and spectra are asymmetric. More interestingly, the probe pulse is compressed, in sharp contrast to the case of Figure 4.4, where GVD led to a huge broadening. This can be understood qualitatively from Eq. (10). For the case shown in Figure 4.6, the XPM-induced chirp is negative and nearly linear across the trailing part of the probe pulse. Because of this chirp, the traveling part is compressed as the probe pulse propagates inside the fiber.

Experimental observation of XPM-induced optical wave breaking would require the use of femtosecond pulses. This can be seen by noting that for picosecond pulses with $T_0 = 5$–10 ps, typically $L_d \approx 1$ km while $L_w \approx 1$ m even if the pump-probe wavelengths differ by as little as 10 nm. By contrast, if $T_0 = 100$ fs, both L_d and L_w become comparable (≈ 10 cm), and the temporal changes in the probe shape discussed here can occur in a fiber less than a meter long. Pulses much shorter than 100 fs should also not be used since higher-order nonlinear effects such as self-steepening and a delayed nonlinear response then become increasingly important. Although these effects are not expected to eliminate the phenomenon of XPM-induced optical wave breaking, they may interfere with the interpretation of experimental data.

3. Pump-Probe Cross-Phase Modulation Experiments

Cross-phase modulation is intrinsic to numerous schemes of ultrashort pulse interaction. The first observation of spectral effects arising from XPM was reported using a pump-probe scheme (Alfano et al., 1986). The phase modulation generated by the infrared pulse at the probe wavelength was referred to as an induced-phase modulation (PM). More recently, the induced-frequency shift and spectral broadening enhancement of picosecond probe pulses have been observed using optical fibers as nonlinear media (Baldeck et al., 1988a; Islam et al., 1987a, b). Pump-probe experiments on XPM are of prime importance for they could lead to applications for pulse compression, optical communication, and optical computation purposes. Results of the pump-probe experiments on XPM are discussed in this section.

3.1 Spectral Broadening Enhancement by Cross-Phase Modulation in BK-7 Glass

The possibility of enhancing the spectral broadening of a probe pulse using a copropagating pump pulse was first observed experimentally in early 1986 (Alfano et al., 1986). The spectral broadening of a weak 80-μJ picosecond 530-nm laser in BK-7 glass was enhanced over the entire spectral band by the presence of an intense millijoule picosecond 1060-nm laser pulse. The spectral distributions of the self-phase modulation and the cross-phase modulation signals were found to be similar. The dominant enhancement mechanism for the induced supercontinuum was determined to be a cross-phase modulation process, not stimulated four-photon scattering.

The experimental setup is shown in Figure 4.7. A single 8-ps laser pulse at 1060 nm generated from a mode-locked glass laser system was used as the pump beam. Its second harmonic was used as the probe beam. These pulses at the primary 1060-nm and the second harmonic 530-nm wavelengths were weakly focused into a 9-cm-long BK-7 glass. A weak supercontinuum signal was observed when both 530- and 1060-nm laser pulses were sent through the sample at the same time. This signal could arise from the IPM process and/or stimulated four-photon parametric generation (FPPG).

In this induced supercontinuum experiment, the 530-nm laser pulse intensity was kept nearly constant with a pulse energy of about 80 μJ. The primary 1060-nm laser pulse energy was a controlled variable changing from 0 to 2 mJ. Filters were used to adjust the 1060-nm pump-laser pump intensity. The output beam was separated into three paths for diagnosis.

The output beam along path 1 was imaged onto the slit of a 0.5-m Jarrel-Ash spectrograph to separate the contributions from the possible different mechanisms for the supercontinuum by analyzing the spatial distribution of the spectrum from phase modulation and stimulated four-photon scattering processes. In this spectrograph measurement, films were used to measure the spatial distribution of the supercontinuum spectrum and a photomultiplier tube was used to obtain quantitative readings. To distinguish different contributions from either phase modulation or stimulated four-photon scattering, geometric blocks were arranged in the path for the selection of a particular process. An aperture of 6 mm diameter was placed in front of the entrance slit of the spectrograph to measure the signal contributed phase modulation, while an aluminum plate of 7 mm width was placed in front of the spectrograph entrance slit to measure the $\lambda = 570$ nm contribution.

The beam along path 2 was directed into a spectrometer with an optical multichannel analyzer to measure the supercontinuum spectral intensity distribution. The spectrum was digitized, displayed, and stored in 500 channels

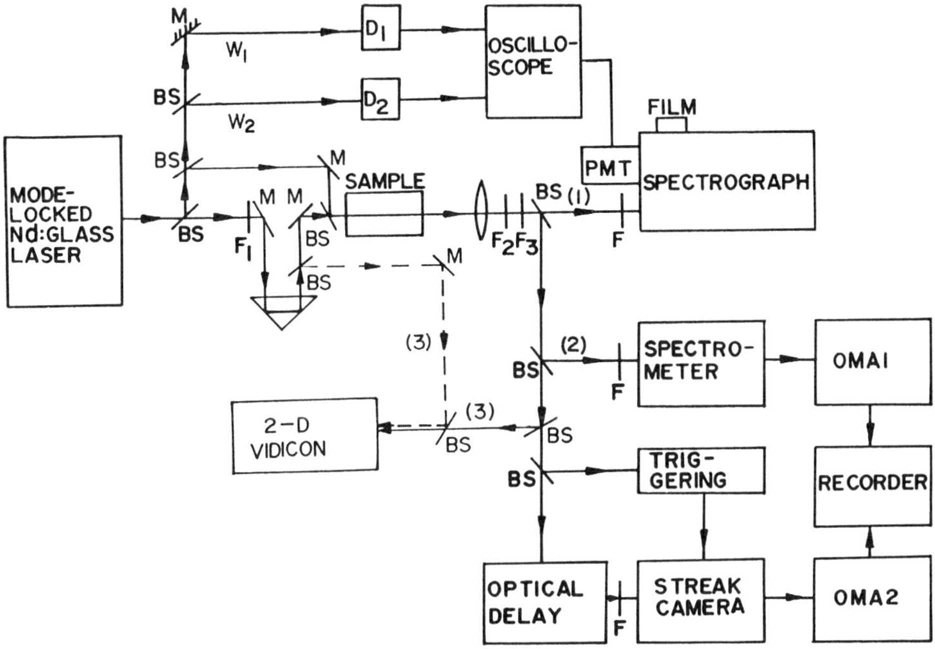

FIGURE 4.7. Schematic diagram of the experimental arrangment for measuring the spectral broadening enhancement of probe pulses by induced-phase modulation. F_1: Hoya HA30 (0.03%), R72 (82%), Corning 1-75 (1%), 1-59 (15%), 0-51 (69%), 3-75 (80%). The numbers in parentheses correspond to the transmittivity at 1054 nm. All these color filters have about 82% transmittivity at 527 nm. F_2: 1-75 + 3-67 for Stokes side measurements; F_2: 1-75 + 2 (5-57) for anti-Stokes side measurements; F_3: neutral density filters; F: ND3 + 1-75; D1, D2: detectors; M: dielectric-coated mirror; BS: beam splitter. (From Alfano et al., 1986.)

as a function of wavelength. The beam along path 3 was delayed and directed into a Hamamatsu Model C1587 streak camera to measure the temporal distribution of the laser pulse and induced supercontinuum. The duration of the induced supercontinuum with a selected 10-nm bandwidth was measured to be about the same as the incident laser pulse duration.

Experimental results for the spectral distribution of induced supercontinuum and supercontinuum are displayed in Figure 4.8. More than 20 laser shots for each data point in each instance have been normalized and smoothed. The average gain of the induced supercontinuum in a BK-7 glass from 410- to 660-nm wavelength was about 11 times that of the supercontinuum. In this instance, both the 530- and 1060-nm laser pulse energies were maintained nearly constant: 80 μJ for 530 nm and 2 mJ for 1060 nm. In this experiment, the 530-nm laser pulse generated a weak supercontinuum and the intense 1060-nm laser pulse served as a catalyst to enhance the supercontinuum in

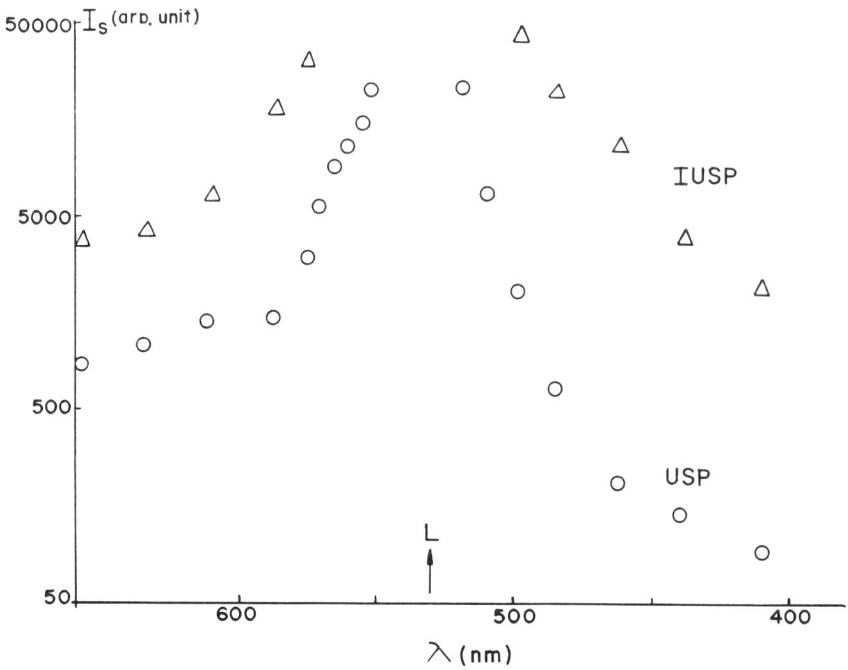

FIGURE 4.8. Intensities of the induced ultrafast supercontinuum pulse (IUSP) and the ultrafast supercontinuum pulse (USP). Each data point was an average of about 20 laser shots and was corrected for the detector, filter, and spectrometer spectral sensitivity. (Δ) IUSP (F_1: 3-75); (○) USP from 527 nm (F_1: HA30). USP from 1054 nm, which is not shown here was ≈1% of the IUSP signal. The measured 527-nm probe pulse was about 5 × 10 counts on this arbitrary unit scale. The error bar of each data point is about ±20%. (From Alfano et al., 1986.)

the 530-nm pulse. The supercontinuum generated by the 1060-nm pulse alone in this spectral region was less than 1% of the total induced supercontinuum. The spectral shapes of the induced supercontinuum pulse and the supercontinuum pulse in Figure 4.8 are similar. Use of several liquid samples such as water, nitrobenzene, CS_2, and CCl_4 has also been attempted to obtain the induced supercontinuum. There was no significant (twofold) enhancement from all other samples that we tested.

A plot of the intensity dependence of the induced supercontinuum is displayed in Figure 4.9 as a function of the 1060-nm pump pulse energy. The wavelengths plotted in Figure 4.9 were $\lambda = 570$ nm for the Stokes side and $\lambda = 498$ nm for the anti-Stokes side. The 530-nm pulse energy was set at 80 ± 15 μJ. The induced supercontinuum increased linearly as the added 1060-nm laser pulse energy was increased from 0 to 200 μJ. When the 1060-nm pump pulse was over 1 mJ, the supercontinuum enhancement reached a plateau and saturated at a gain factor of about 11 times over the supercon-

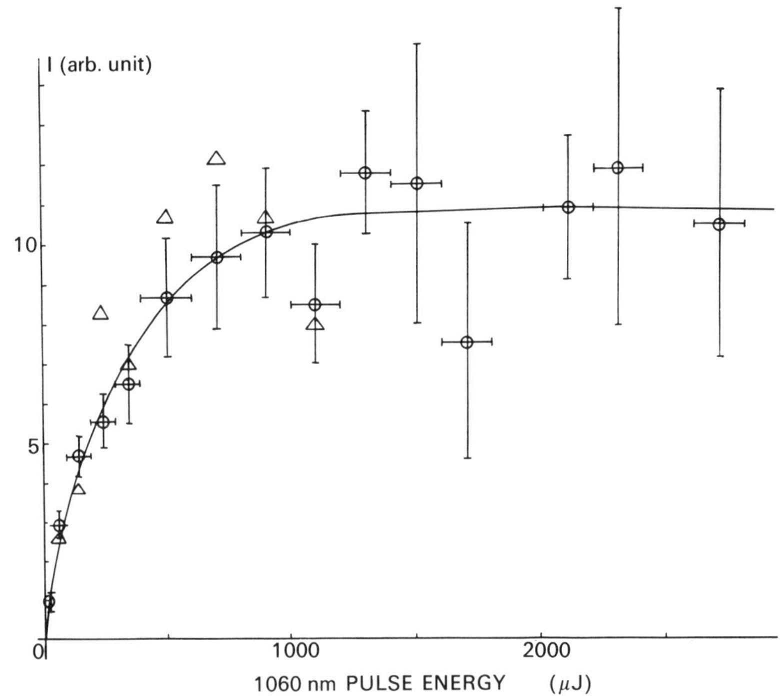

FIGURE 4.9. Dependence of the IUSP signal on the intensity of the 1.06-μm pump pulse. (○) Stokes side at $\lambda = 570$ nm; (Δ) anti-Stokes side at $\lambda = 498$ nm. The error bars of the anti-Stokes side were similar to those of the Stokes side. The solid line is a guide for the eye. The vertical axis is the normalized $I_{\text{IUSP}}/I_{527\text{ nm}}$. (From Alfano et al., 1986.)

tinuum intensity generated by only the 530-nm pulse. This gain saturation may be due to the trailing edge of the pulse shape function being maximally distorted when the primary pulse intensity reaches a certain critical value. This implies a saturation of the PM spectral distribution intensity when the pumped primary pulse energy is above 1 mJ, as shown in Figure 4.9.

Since the supercontinuum generation can be due to the phase modulation and/or the stimulated four-photon scattering processes, it is important to distinguish between these two different contributions to the induced supercontinuum signal. Spatial filtering of the signal was used to separate the two main contributions. The induced supercontinuum spectrum shows a spatial spectral distribution similar to that of the conventional supercontinuum. The collinear profile that is due to the phase modulation has nearly the same spatial distribution as the incident laser pulse. Two emission wings at noncollinear angles correspond to the stimulated four-photon scattering continuum arising from the phase-matching condition of the generated wavelengths emitted at different angles from the incident laser beam direction.

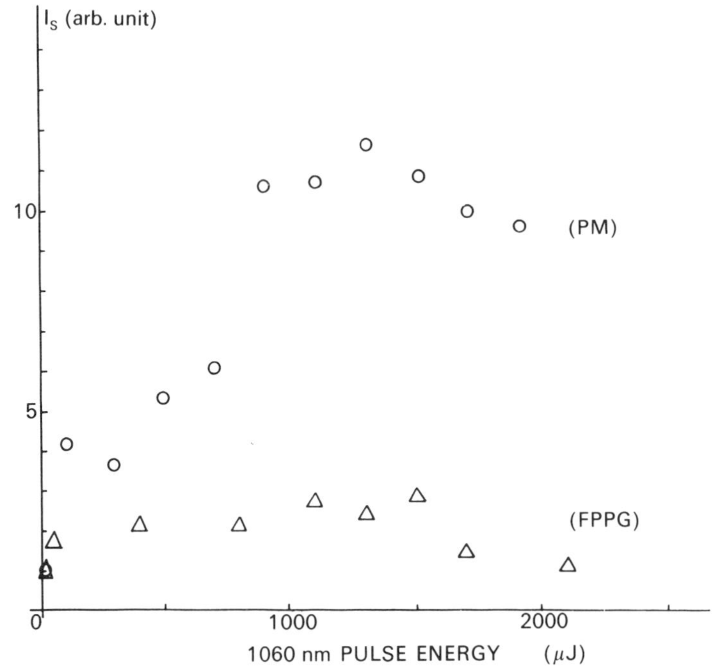

FIGURE 4.10. Dependence of I_s(PM) and I_s(FPPG) at $\lambda = 570$ nm or the intensity of the 1054-nm pump laser pulse. (○) PM; (△) FPPG. The measured signal has been normalized with the incident 527-nm pulse energy. The error bar of each data point is about ±20% of the average value. (From Alfano et al., 1986.)

Using a photomultiplier system and spatial filtering, quantitative measurements of the induced supercontinuum contributions from the collinear PM and the noncollinear stimulated four-photon scattering parts were obtained (Figure 4.10). These signals, measured at $\lambda = 570$ nm from the collinear PM and the noncollinear parts of the induced supercontinuum, are plotted as a function of the pump pulse energy. There was little gain from the contribution of the stimulated four-photon scattering process over the entire pulse-energy-dependent measurement as shown in Figure 4.10. The main enhancement of the induced supercontinuum generation is consequently attributed to the PM mechanism, which corresponds to the collinear geometry. Another possible mechanism for the observed induced supercontinuum could be associated with the enhanced self-focusing of the second harmonic pulse induced by the primary pulse. There was no significant difference in the spatial intensity distribution of the 530-nm probe beam with and without the added intense 1060-nm pulse.

In this experiment the spectral broadening of 530-nm pulses was enhanced by nonlinear interaction with copropagating strong infrared pulses in a BK-7 glass sample. The spectral change has been found to arise from a phase modulation process rather than a stimulated four-photon mixing process. It is in good agreement with predictions of the induced-phase modulation theory. This experiment showed the first clear evidence of a cross-phase modulation spectral effect.

3.2 Induced-Frequency Shift of Copropagating Pulses

Optical fibers are convenient for the study of nonlinear optical processes. The optical energy is concentrated into small cross sections (typically 10^{-7} cm^2) for long interaction lengths. Thus, large nonlinear effects are possible with moderate peak powers (10–10^4 W). Optical fibers appear to be an ideal medium in which to investigate XPM effects. The first pump probe experiment using picosecond pulses propagating in optical fibers demonstrated the importance of the pulse walk-off in XPM spectral effects (Baldeck et al., 1988a). It was shown that ultrashort pulses that overlap in a nonlinear and highly dispersive medium undergo a substantial shift of their carrier frequencies. This new coherent effect, which was referred to as an induced-frequency shift, resulted from the combined effect of cross-phase modulation and pulse walk-off. In the experiment, the induced-frequency shift was observed by using strong infrared pulses that shifted the frequency of weak picosecond green pulses copropagating in a 1-m-long single-mode optical fiber. Tunable red and blue shifts were obtained at the fiber output by changing the time delay between infrared and green pulses at the fiber input.

A schematic of the experimental setup is shown in Figure 4.11. A mode-locked Nd:YAG laser with a second harmonic crystal was used to produce

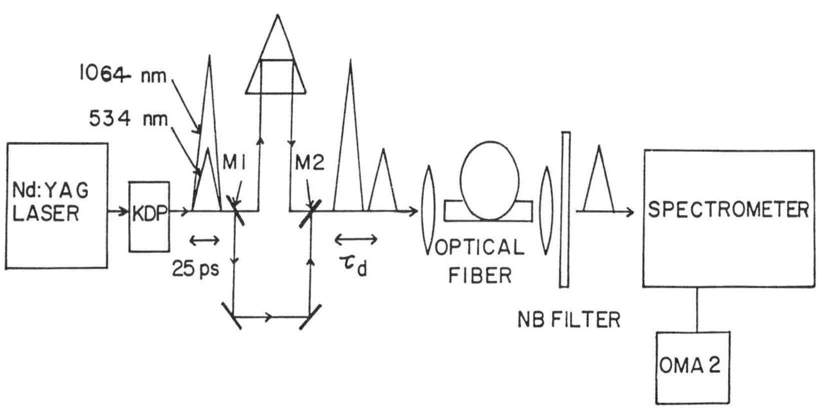

FIGURE 4.11. Experimental setup used to measure the induced-frequency shift of 532-nm pulses as a function of the time delay between pump and probe pulses at the optical fiber input. Mirrors M_1 and M_2 are wavelength selective; i.e., they reflect 532-nm pulses and transmit 1064-nm pulses. (From Baldeck et al., 1988a.)

33-ps infrared pulses and 25-ps green pulses. These pulses were separated using a Mach-Zehnder interferometer delay scheme with wavelength-selective mirrors. The infrared and green pulses propagated in different interferometer arms. The optical path of each pulse was controlled using variable optical delays. The energy of infrared pulses was adjusted with neutral density filters in the range 1 to 100 nJ while the energy of green pulses was set to about 1 nJ. The nonlinear dispersive medium was a 1-m-long single-mode optical fiber (Corguide of Corning Glass). This length was chosen to allow for total walk-off without losing control of the pulse delay at the fiber output. The group velocity mismatch between 532 and 1064-nm pulses was calculated to be about 76 ps/m in fused silica. The spectrum of green pulses was measured using a grating spectrometer (1 meter, 1200 lines/mm) and an optical multichannel analyzer (OMA2).

The spectra of green pulses propagating with and without infrared pulses are plotted in Figure 4.12. The dashed spectrum corresponds to the case of green pulses propagating alone. The blue-shifted and red-shifted spectra are those of green pulses copropagating with infrared pulses after the input delays were set at 0 and 80 ps, respectively. The main effect of the nonlinear interaction was to shift the carrier frequency of green pulses. The induced-wavelength shift versus the input delay between infrared and green pulses is plotted in

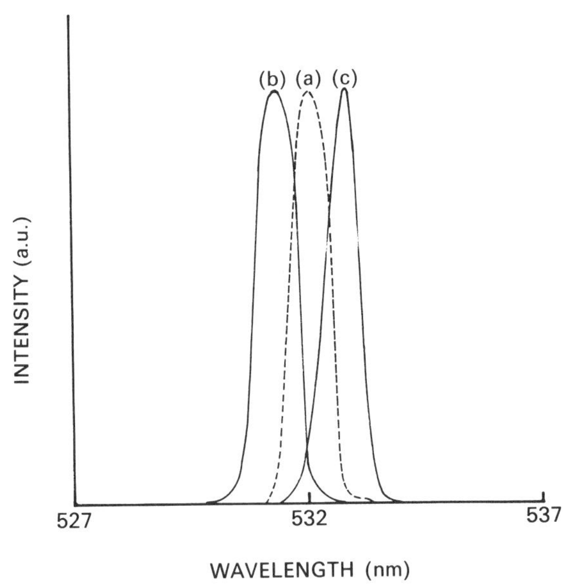

FIGURE 4.12. Cross-phase modulation effects on spectra of green 532-nm pulses. (a) Reference spectrum (no copropagating infrared pulse). (b) Infrared and green pulses overlapped at the fiber input. (c) Infrared pulse delayed by 80 ps at the fiber input. (From Baldeck et al., 1988a.)

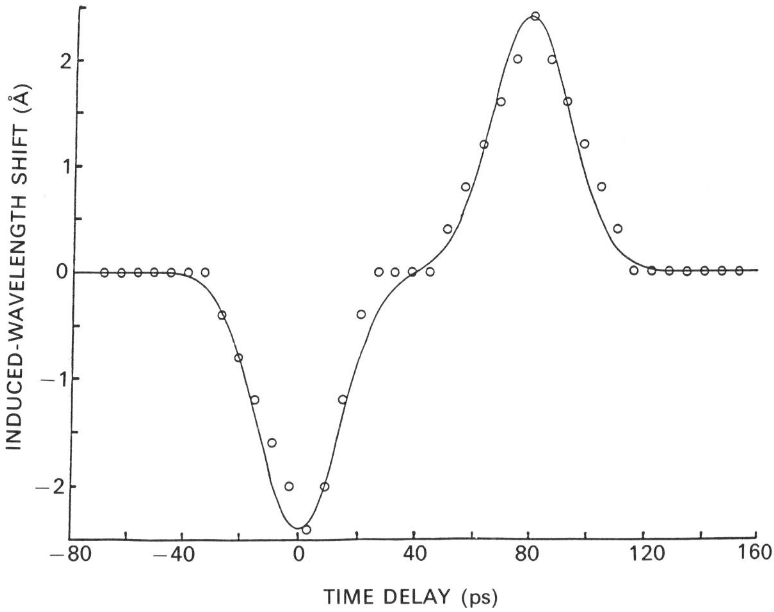

FIGURE 4.13. Induced wavelength shift of green 532-nm pulses as a function of the input time delay between 532-nm pulses and infrared 1064-nm pulses at the input of a 1-m-long optical fiber. (o) Experimental points. The solid line is the theoretical prediction from Eq. (3.3). (From Baldeck et al., 1988a.)

Figure 4.13. The maximum induced-wavelength shift increased linearly with the infrared pulse peak power (Figure 4.14). Hence, the carrier wavelength of green pulses could be tuned up to 4 Å toward both the red and blue sides by varying the time delay between infrared and green pulses at the fiber input. The solid curves in Figure 4.13 and 4.14 are from theory.

When weak probe pulses are used the SPM contribution can be neglected in Eqs. (9) and (10). Thus, nonlinear phase shifts and frequency chirps are given by

$$\alpha_1(\tau, z) \approx \sqrt{\pi}\frac{\omega_1}{c} n_2 \frac{P_2}{A_{\text{eff}}} L_w \left[\text{erf}(\tau - \tau_d) - \text{erf}\left(\tau - \tau_d + \frac{z}{L_w}\right)\right], \quad (15)$$

$$\delta\omega_1(\tau, z) \approx -2\frac{\omega_1}{c} n_2 \frac{P_2}{A_{\text{eff}}} \frac{L_w}{T_0} [e^{-(\tau-\tau_d)^2} - e^{-(\tau-\tau_d-z/L_w)^2}]. \quad (16)$$

When the pulses coincide at the fiber entrance ($t_d = 0$) the point of maximum phase is generated ahead of the green pulse peak because of the group velocity mismatch (Eq. 15). The green pulse sees only the trailing part of the XPM profile because it travels slower than the pump pulse. This leads to a blue induced-frequency shift (Eq. 16). Similarly, when the initial delay is set

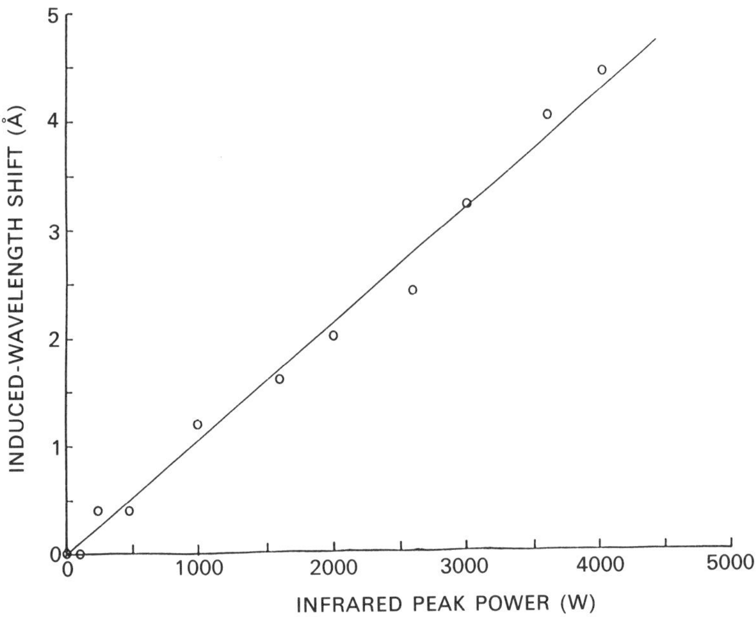

FIGURE 4.14. Maximum induced wavelength shift of 532-nm pulses versus the peak power of infrared pump pulses. (o) Experimental points. The solid line is the theoretical prediction from Eq. (3.4). (From Baldeck et al. 1988a.)

at 80 ps, the infrared pulse has just sufficient time to catch up with the green pulse. The green pulse sees only the leading part of the XPM phase shift, which gives rise to a red induced-frequency shift. When the initial delay is about 40 ps, the infrared pulse has time to pass entirely through the green pulse. The pulse envelope sees a constant dephasing and there is no shift of the green spectrum (Figure 4.13).

Equations (15) and (16) can be used to fit our experimental data shown in Figures 4.13 and 4.14. Assuming that the central part of the pump pulses provides the dominant contribution to XPM, we set $t = 0$ in Eq. (16) and obtain

$$\delta\omega_1(\tau, z) = \frac{\omega_1}{c} n_2 \frac{P_2}{A_{\text{eff}}} \frac{L_w}{T_0} [e^{-(\tau-\tau_d)^2} - e^{-(\tau-\tau_d+z/L_w)^2}]. \tag{17}$$

The maximum induced-frequency shift occurs at $t_d = d = z/L_w$ and is given by

$$|\Delta\omega_{\max}| = \frac{\omega_1}{c} n_2 \frac{P_2}{A_{\text{eff}}} \frac{L_w}{T_0}. \tag{18}$$

Equations (17) and (18) are plotted in Figures 4.13 and 4.14, respectively. There is very good agreement between this simple analytical model and experimental data. It should be noted that only a simple parameter (i.e., the infrared peak power at the maximum induced-frequency shift) has been adjusted to fit the data. Experimental parameters were $\lambda = 532$ nm, $T_0 = 19.8$ ps (33 ps FWHM), $L_w = 26$ cm, and $\delta = 4$.

We have shown experimentally and theoretically that ultrashort optical pulses that overlap in a nonlinear and highly dispersive medium can undergo a substantial shift of their carrier frequency. This induced-frequency shift has been demonstrated using strong infrared pulses to shift the frequency of copropagating green pulses. The results are well explained by an analytical model that includes the effect of cross-phase modulation and pulse walk-off. This experiment led to a conclusive observation of XPM spectral effects.

3.3 XPM-Induced Spectral Broadening and Optical Amplification in Optical Fibers

This section presents additional features that can arise from the XPM interaction between a pump pulse at 630 nm and a probe pulse at 532 nm. With this choice of wavelengths, the group velocity dispersion between the pump pulse and the probe pulse is reduced and the XPM interaction enhanced. The spectral width and the energy of the probe pulse were found to increase in the presence of the copropagating pump pulse (Baldeck et al., 1988c).

A schematic of the experimental setup is shown in Figure 4.15. A mode-locked Nd:YAG laser with a second harmonic crystal was used to produce pulses of 25-ps duration at 532 nm. Pump pulses were obtained through stimulated Raman scattering by focusing 90% of the 532-nm pulse energy into a 1-cm cell filled with ethanol and using a narrowband filter centered at 630 nm. Resulting pump pulses at 630 nm were recombined with probe pulses and coupled into a 3-m-long single-mode optical fiber. Spectra of probe pulses

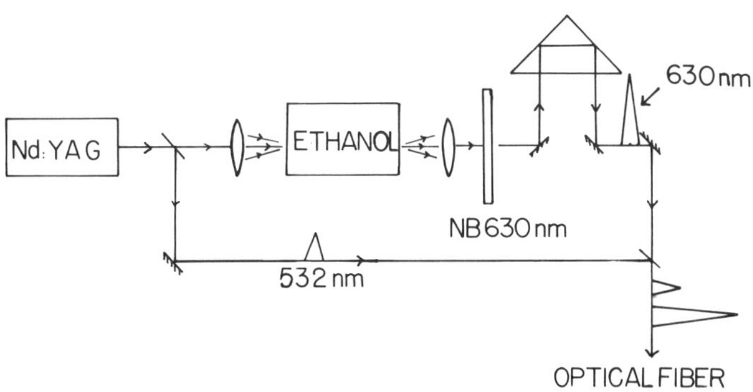

FIGURE 4.15. Experimental setup for generating copropagating picosecond pulses at 630 and 532 nm. (From Baldeck et al., 1988c.)

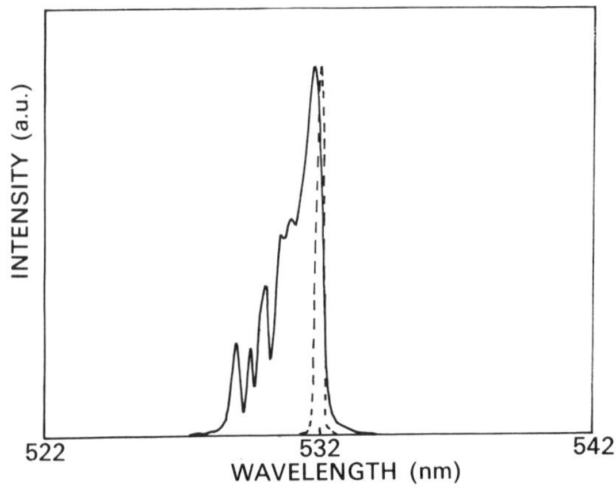

FIGURE 4.16. Cross-phase modulation effects on the spectrum of a probe picosecond pulse. (Dashed line) Reference spectrum without XPM. (Solid line) With XPM and no time delay between pump and probe pulses at the optical fiber input. (From Baldeck et al., 1988c.)

were recorded for increasing pump intensities and varying input time delays between pump and probe pulses.

With negative delays (late pump at the optical fiber input), the spectrum of the probe pulse was red shifted as in the 1064 nm/532 nm experiment (Figure 4.12). A new XPM effect was obtained when both pulses entered the fiber simultaneously. The spectrum of the probe pulse not only shifted toward blue frequencies as expected but also broadened (Figure 4.16). An spectral broadening as wide as 10 nm could be induced, which was, surprisingly, at least one order of magnitude larger than predicted by the XPM theory. As shown in Figure 4.16, the probe spectrum extended toward the blue-shifted frequencies with periodic resonant lines. These lines could be related to modulation instability sidelobes that have been predicted theoretically to occur with cross-phase modulation (see Section 8).

The optical amplification of the probe pulse is another new and unexpected feature arising from the XPM interaction. Pump power-dependent gain factors of 3 or 7 were measured using probe pulses at 532 nm and pump pulses at 630 or 1064 nm, respectively. Figure 4.17 shows the dependence of the XPM-induced gain for the probe pulse at 532 nm with the input time delay between the probe pulse and the pump pulse at 630 nm. The shape of the gain curve corresponds to the overlap function of pump and probe pulses. Figure 4.18 shows the dependence of the gain factor on the intensity of pump pulses at 1064 nm. This curve is typical of a parametric amplification with pump depletion. The physical origin of this XPM-induced gain is still under investigation. It could originate from an XPM-phase-matched four-wave mixing process.

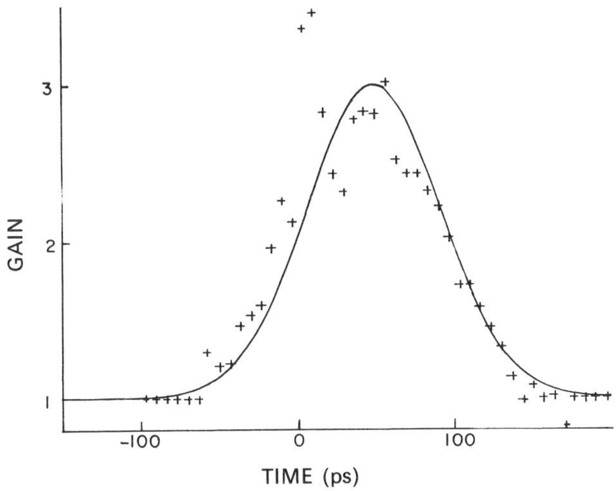

FIGURE 4.17. XPM-induced optical gain $I_{532}(\mathrm{out})/I_{532}(\mathrm{in})$ versus input time delay between pump pulses at 630 nm and probe pulses at 532 nm. (Crosses) Experimental data; (solid line) fit obtained by taking the convolution of pump and probe pulses. (From Baldeck and Alfano, 1988c.)

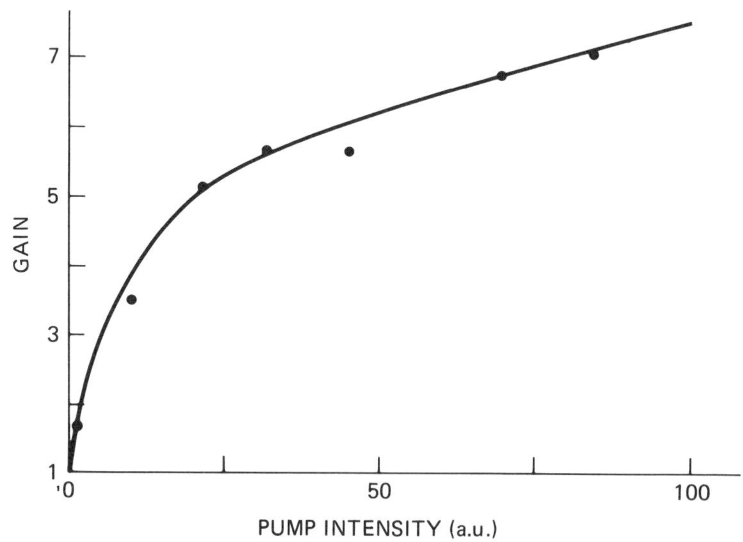

FIGURE 4.18. XPM-induced optical gain $I_{532}(\mathrm{out})/I_{532}(\mathrm{in})$ versus intensity of pump pulses at 1064 nm. (From Baldeck et al., 1987c-d.)

The spectral distribution of probe pulses can be significantly affected by the XPM generated by a copropagating pulse. In real time, the probe pulse frequency can be tuned, its spectrum broadened, and its energy increased. XPM appears as a new technique for controlling the spectral properties and regenerating ultrashort optical pulses with terahertz repetition rates.

4. Cross-Phase Modulation with Stimulated Raman Scattering

When long samples are studied optically, stimulated Raman scattering (SRS) contributes to the formation of ultrafast supercontinua. In 1980, Gersten, Alfano, and Belic predicted that ultrashort pulses should generate broad Raman lines due to the coupling among laser photons and vibrational phonons (Gersten et al., 1980). This phenomenon was called cross-pulse modulation (XPM). It characterized the phase modulation of the Raman pulse by the intense pump laser pulse. Cornelius and Harris (1981) stressed the role of SPM in SRS from more than one mode. Recently, a great deal of attention has been focused on the combined effects of SRS, SPM, and group velocity dispersion for the purposes of pulse compression and soliton generation (Dianov et al., 1984; Lu Hian-Hua et al., 1985; Stolen and Johnson, 1986; French et al., 1986; Nakashima et al., 1987; Johnson et al., 1986; Gomes et al., 1988; Weiner et al., 1986–1988, to name a few). Schadt et al. numerically simulated the coupled wave equations describing the changes of pump and Stokes envelopes (Schadt et al., 1986) and the effect of XPM on pump and Stokes spectra (Schadt and Jaskorzynska, 1987a) in nonlinear and dispersive optical fibers. Manassah (1987a, b) obtained analytical solutions for the phase and shape of a weak Raman pulse amplified during the pump and Raman pulse walk-off. The spectral effects of XPM on picosecond Raman pulses propagating in optical fibers were measured and characterized (Islam et al., 1987a, b; Alfano et al., 1987b; Baldeck et al., 1987b, d). In this section we review (1) Schadt and Jaskorzynska theoretical analysis of stimulated Raman scattering in optical fibers and (2) measurements of XPM and SPM effects on stimulated Raman scattering.

4.1 Theory of XPM with SRS

The following theoretical study of stimulated Raman scattering generation of picosecond pulses in optical fibers is from excerpts from Schadt et al. (1986) and Schadt and Jaskorzynska (1987a).

In the presence of copropagating Raman and pump pulses the nonlinear polarization can be approximated in the same way as in Section 2.1 by

$$P^{NL}(r,z,t) = \chi^{(3)}E^3(r,z,t), \tag{19}$$

where the total electric field $E^3(r,z,t)$ is given by

$$E(r, z, t) = \tfrac{1}{2}\{A_p(r, z, t)e^{i(\omega_{pt} - \beta_{Pz})} + A_s(r, z, t)e^{i(\omega_{st} - \beta_{Sz})} + \text{c.c.}\}. \tag{20}$$

In this case $A_1 = A_p$ and $A_2 = A_s$.

The subscripts P and S refer to the pump and Stokes Raman pulses, respectively. The anti-Stokes Raman is neglected. Substituting Eq. (20) into Eq. (19) and keeping only terms synchronized with either pump or Stokes carrier frequency, the nonlinear polarization becomes

$$P_P^{NL}(z, t) = \tfrac{3}{8}\{i2\chi_R^{(3)}|A_S|^2 + \chi_{PM}^{(3)}[|A_P|^2 + 2|A_S|^2]\}A_P e^{i(\omega_{Pt} - \beta_{Pz})} + \text{c.c.}, \tag{21a}$$

$$P_S^{NL}(z, t) = \tfrac{3}{8}\{-i2\chi_R^{(3)}|A_P|^2 + \chi_{PM}^{(3)}[|A_S|^2 + 2|A_P|^2]\}A_S e^{i(\omega_{St} - \beta_{Sz})} + \text{c.c.}, \tag{21b}$$

where $\chi^{(3)} = \chi_{PM}^{(3)} + i\chi_R^{(3)}$, $\chi_R^{(3)}$ gives rise to the Raman gain (or depletion) of the probe (or pump), and $\chi_{PM}^{(3)}$ leads to self- and cross-phase modulations. Note the factor 2 associated with XPM.

As in the pump-probe case, the phase shift contribution of the nonlinear polarization at the pump (or Raman) frequency depends not only on the pump (or Raman) peak power but also on the Raman (or pump) peak power. This gives rise to cross-phase modulation during the Raman scattering process.

Using the expressions for P_P^{NL} and P_S^{NL} in the nonlinear wave equation, leads to the coupled nonlinear dispersive equations for Raman and pump pulses:

$$\frac{\partial A_P}{\partial Z} + \frac{z_K}{z_W}\frac{\partial A_P}{\partial T} + \frac{i}{2}\frac{z_K}{z_D}\frac{\partial^2 A_P}{\partial T^2} = -\frac{1}{2}\frac{\Omega_P}{\Omega_S}\frac{z_K}{z_A}|A_S|^2 A_P + \frac{i}{2}[|A_P|^2 + 2|A_S|^2]A_P - \frac{z_K}{z_L}A_P, \tag{22a}$$

$$\frac{\partial A_S}{\partial Z} + \frac{i}{2}\frac{k_R''}{k_P''}\frac{z_K}{z_D}\frac{\partial^2 A_S}{\partial T^2} = \frac{1}{2}\frac{z_K}{z_A}|A_P|^2 A_S + \frac{i}{2}\frac{\Omega_S}{\Omega_P}[|A_S|^2 + 2|A_P|^2]A_S - \frac{\Gamma_S}{\Gamma_P}\frac{z_K}{z_L}A_S, \tag{22b}$$

where $A_1 = a_1/|a_{0P}|$ are the complex amplitudes a_1 normalized with respect to the initial peak amplitude $|a_{0P}|$ of the pump pulse. The index $1 = P$ refers to the pump, whereas $1 = S$ refers to the Stokes pulse. $Z = z/z_K$ and $T = (t - z/v_s)/\tau_0$ are the normalized propagation distance and the retarded time normalized with respect to the duration of the initial pump pulse. $\Omega = \omega/(1/\tau_0)$ is a normalized frequency. Moreover, the following quantities were introduced:

$z_K = 1/\gamma_P|a_{0P}|^2 = 1/(|a_{0P}|^2 n_2\omega_P/c)$is the Kerr distance, with the PM coefficient γ_P, the Kerr coefficient n_2, and ω_P as the carrier frequency of the pump pulse; c is the velocity of light.

$z_W = \tau_0/(v_p^{-1} - v_s^{-1})$ is the walk-off distance; v_p and v_s are the group velocities at the pump and Stokes frequencies, respectively.

$z_D = \tau_0^2/k_P''$ is the dispersion length; $k_P'' = \partial^2 k_P/\partial\omega^2$, where k_P is the propagation constant of the pump.

$z_D = 1/\alpha_s|a_{0P}|^2 = 1/\gamma|a_{0P}|^2$ is the amplification length, with g the Raman gain coefficient.

$z_L = 1/\Gamma_P$ is the pump loss distance, where Γ_P is the attenuation coefficient at the pump frequency.

The derivation of Eqs (22) assumes that a quasi-steady-state approximation holds. Thus, it restricts the model to pulses much longer than the vibrational dephasing time (~ 100 fs) of fused silica. The Raman gain or loss is assumed to be constant over the spectral regions occupied by the Stokes and pump pulses, respectively. Furthermore, the quasi-monochromatic approximation is used, which is justified as long as the spectral widths of the pulses are much smaller than their carrier frequencies. As a consequence of these simplifications, the considered spectral broadening of the pulses is a result only of phase modulations and pulse reshaping. The direct transfer of the chirp from the pump to the Stokes pulse by SRS is not described by the model. The frequency dependence of the linear refractive index is included to a second-order term, so both the walk-off arising from a group velocity mismatch between the pump and Stokes pulses and the temporal broadening of the pulses are considered.

Using Eqs. (22a) and (22b), Schadt and Jaskorzynska numerically simulated the generation of picosecond Raman pulses in optical fibers. They particularly investigated the influence of walk-off on the symmetry properties of pulse spectra and temporal shapes and the contributions from SPM and XPM to the chirp of the pulses.

4.1.1 Influence of Walk-Off on the Symmetry Properties of the Pulse Spectra

Results obtained in absence of walk-off are shown in Figure 4.19 (Schadt and Jaskorzynska, 1987a). The pump spectrum, broadened and modulated by SPM, is slightly depleted at its center due the energy transfer toward the Raman pulse (Figure 4.19a). The Raman spectrum is almost as wide as the pump spectrum, but without modulations (Figure 4.19b). The spectral broadening of the Raman spectrum arises mainly from XPM. The modulationless feature appears because the Raman pulse, being much shorter than the pump pulse, picks up only the linear part of the XPM-induced chirp. Such a linearly chirped Raman pulse could be efficiently compressed using a grating-pair pulse compressor.

The influence of the walk-off on the Raman process is displayed in Figure 4.20. The pronounced asymmetry of the spectra in Figure 4.20a and 4.20b is connected with the presence of the pulse walk-off in two different ways. When the Stokes pulse has grown strong enough to deplete the pump pulse visibly, it has also moved toward the leading edge of the pump (it is referred only to regions of normal dispersion). The leading edge has in the meantime been downshifted in frequency as a result of SPM. Consequently, the pump pulse loses energy from the lower-frequency side. On the other hand, the asymmetric depletion of the pump gives rise to the asymmetric depletion buildup of the frequency shift itself, as can be seen from Figures 4.20c and 4.20d. Theoretical

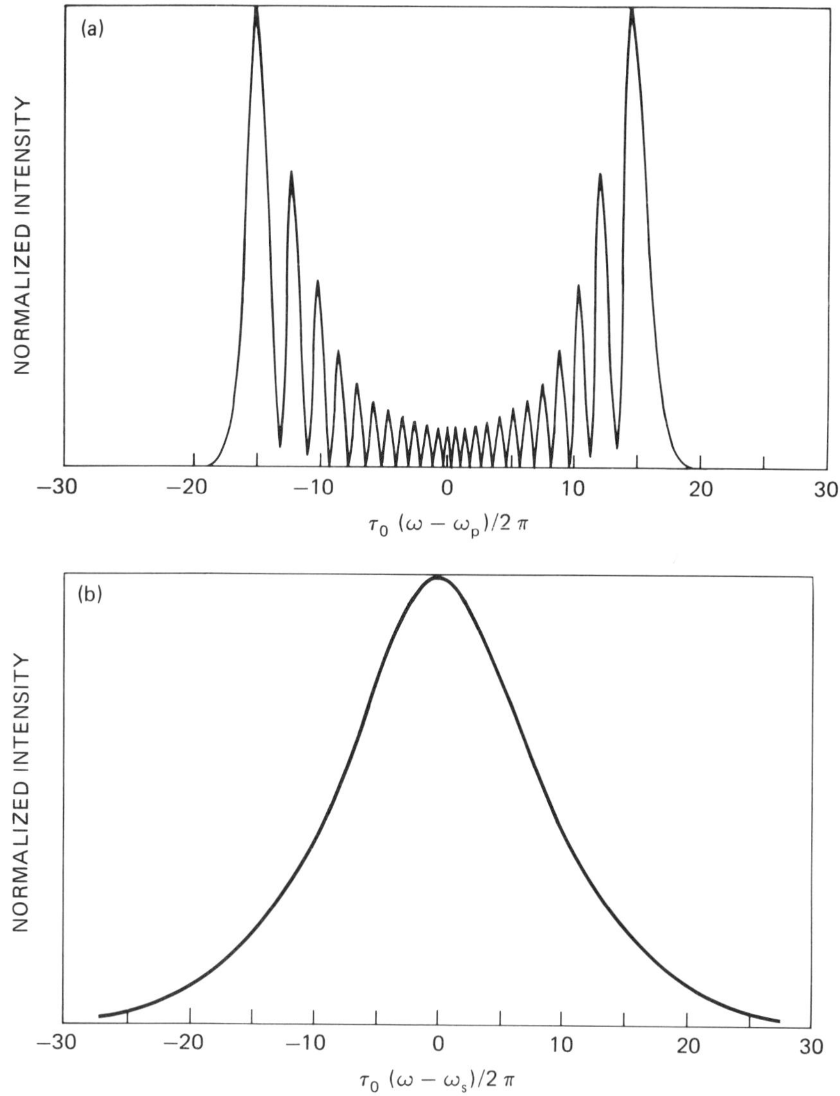

FIGURE 4.19. Spectra of pump and Stokes Raman pulses in the absence of walk-off. (a) Spectrum of the pump pulse. (b) Spectral broadening of the Stokes pulse because of phase modulations. (From Schadt and Jaskorzynska, 1987a.)

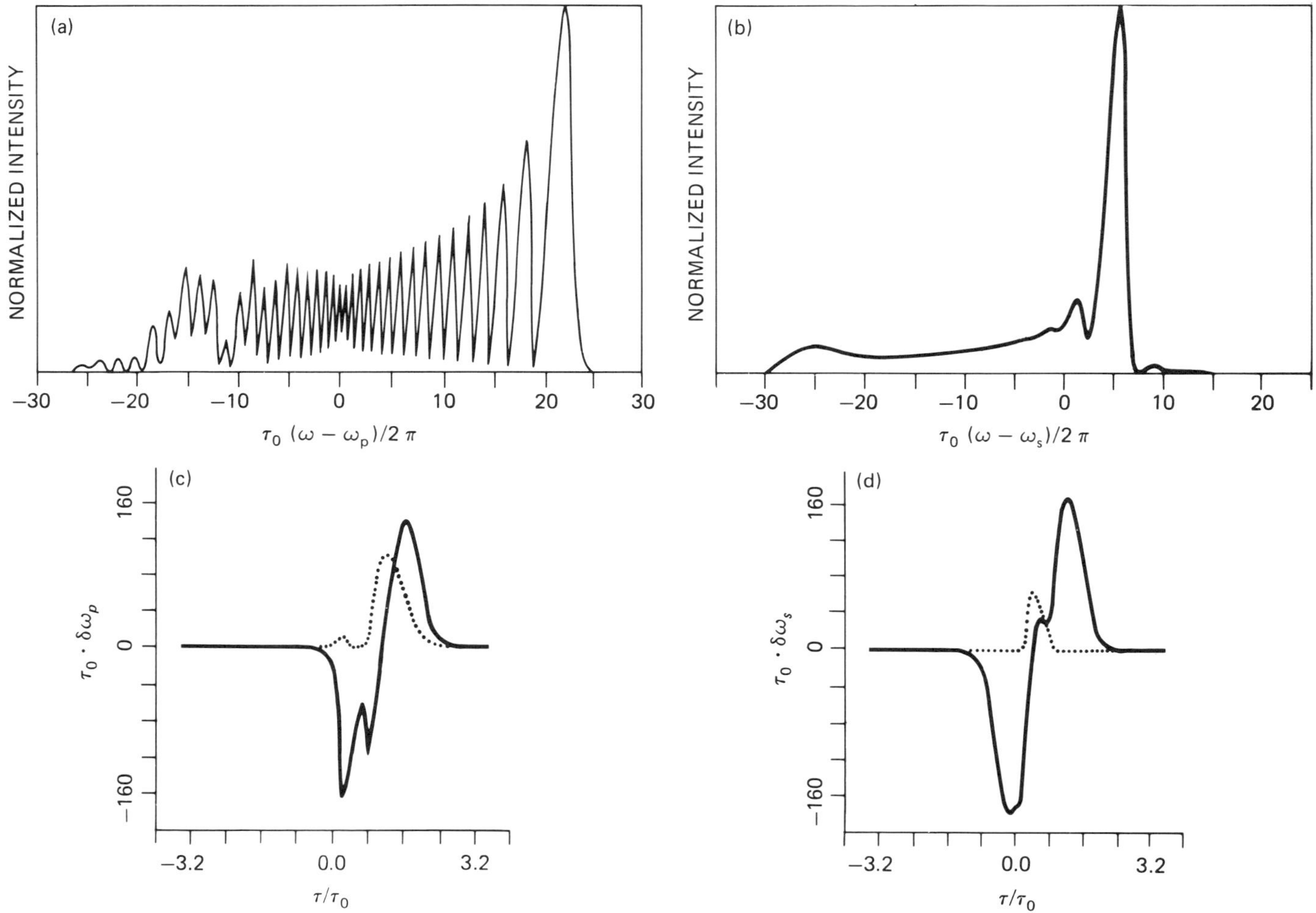

FIGURE 4.20. Influence of walk-off on spectra and chirps of pump and Stokes Raman pulses. (a) Spectrum of the pump pulse. (b) Spectral broadening of the Stokes pulse because of phase modulations. (c) Chirp (solid line) and shape (dashed line) of the pump pulse. (d) Chirp (solid

spectra in Figure 4.20 agree very well with measured spectra (Gomes et al., 1986; Weiner et al., 1986; Zysset and Weber, 1986).

4.1.2 Contributions from Self-Phase Modulation and Cross-Phase Modulation to the Chirp of Pulses

The chirps of Raman and pump pulses originate from SPM and XPM. The contributions from SPM and XPM are independent as long as the effect of second-order dispersion is negligible. In Figure 4.21 are plotted the contributions to the pump and Stokes chirps coming from either SPM only (Figure

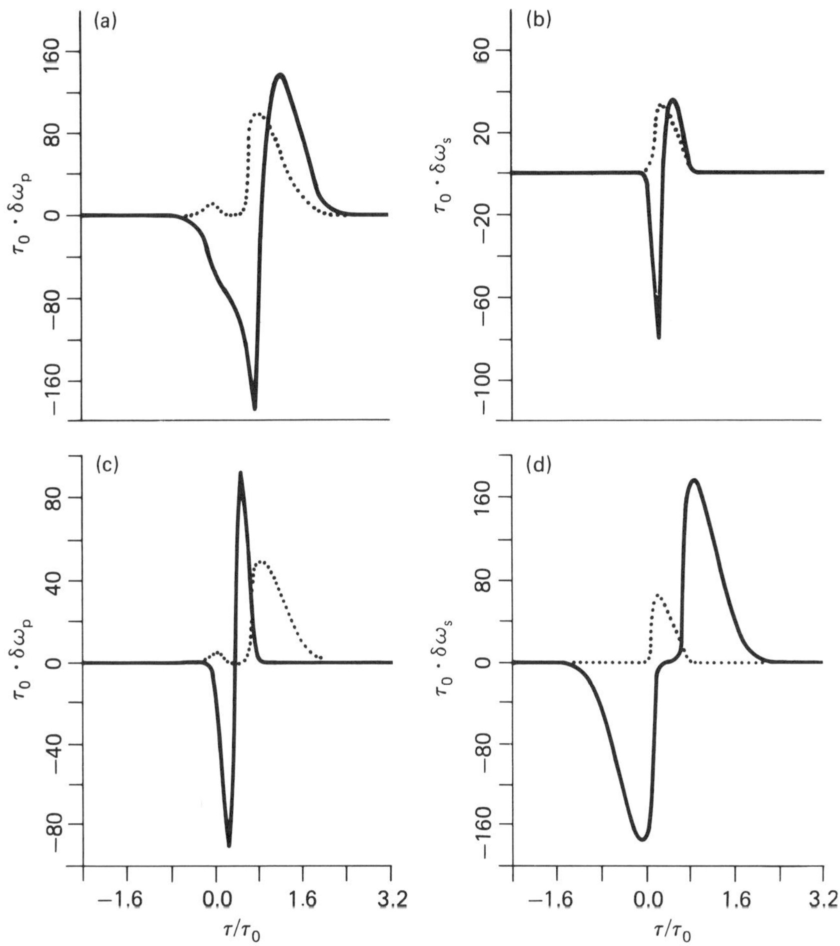

Figure 4.21. Chirp components that are due to SPM and XPM for the case of walk-off. (a) Pump chirp due to SPM only. (b) Stokes chirp due to SPM only. (c) Pump chirp due to XPM only. (d) Stokes chirp due to XPM only. (From Schadt and Jaskorzynska, 1987a.)

4.21a, 4.21b) or XPM only (Figures 4.21c, 4.21d). The shapes of SPM contributions shown in Figures 4.21a and 4.21b apparently reflect the history of their buildup according to the changes of pulse shapes during the propagation. Their strong asymmetry is a result of an asymmetric development of the pulse shapes that is due to walk-off. The XPM affecting the pump pulse in the initial stage of the Raman process plays a lesser role as the pump depletion becomes larger. This constituent of the chirp, associated with the Stokes pulse is built up just in the region where most of the pump energy is scattered to the Stokes frequency if the Raman process goes fast enough. However, if for a fixed walk-off SRS is slow, as in the case illustrated by Figure 4.21c, the leading part of the pump pulse will remain affected by the XPM.

The most characteristic feature of the XPM-induced part of the Stokes chirp, shown in Figure 4.21d, is a plateau on the central part of the Stokes pulse. In the case of the lower input power (Figure 4.21d) this plateau can be attributed mainly to the effect of walk-off. Since pump depletion becomes considerable only close to the end of the propagation distance, it has little influence on the buildup of the chirp. For higher input powers the range over which the chirp vanishes is wider. Consequently, after the walk-off distance the effect of XPM on the Stokes chirp is negligible for a severely depleted pump, whereas in the case of insignificant pump depletion the leading part of the Stokes pulse will remain influenced by XPM.

Schadt et al. have developed a numerical model to describe combined effects of SRS, SPM, XPM, and walk-off in single-mode optical fibers. They explained the influence of the above effects on pump and Stokes spectra and chirps. They separately studied the contributions of SPM and XPM to the chirps and found that both walk-off and pump depletion tend to cancel the effect of XPM on the chirp in the interesting pulse regions. However, for more conclusive results an investigation of the direct transfer of the pump chirp and consideration of the finite width of the Raman gain curve are needed.

4.2 Experiments

In the late 1970s and early 1980s, numerous experimental studies investigated the possibility of using SRS to generate and amplify Raman pulses in optical fibers (Stolen, 1979). However, most of these studies involved "long" nanosecond pulses and/or neglected to evaluate SPM and XPM contributions to the pump and Raman spectral broadenings. It was not until 1987, after the success of the first spectral broadening enhancement experiment (Alfano et al., 1986), that measurements of XPM effects on Raman pulses were reported (Islam et al., 1987a; Alfano et al., 1987b). In this section, research work at AT&T Bell Laboratories and at the City College of New York is reported.

4.2.1 XPM Measurements with the Fiber Raman Amplification Soliton Laser

Islam et al. showed the effects of pulse walk-off on XPM experimentally in the Fiber Raman Amplification Soliton Laser (FRASL) (Islam et al., 1986).

They proved that XPM prevents a fiber Raman laser from producing pedestal-free, transform limited pulses except under restrictive conditions (Islam et al., 1987b). The following simple picture of walk-off effects and experimental evidence is excerpted from reference (Islam et al., 1987a).

The spectral features and broadening resulting from XPM depend on the walk-off between the pump and signal pulses. These spectral features can be confusing and complicated, but Islam et al. show that they can be understood both qualitatively and quantitively and quantitatively by concentrating on the phase change as a function of walk-off. XPM is most pronounced when the pump and signal are of comparable pulse widths and when they track each other. The phase change $\Delta\phi$ induced on the signal is proportional to the pump intensity, and the signal spectrum (Figure 4.22a) looks like that obtained from self-phase-modulation (SPM).

The opposite extreme occurs when the phase shift is uniform over the width of the signal pulse. This may happen in the absence of pump depletion or spreading if the pump walks completely through the signal, or if the signal is much narrower than the pump and precisely tracks the pump. XPM is canceled in this limit, and the original spectral width of the signal (much narrower than any shown in Figure 4.22) results.

A third simple limit exists when the pump and signal coincide at first, but then the pump walks off. This is most characteristic of stimulated amplification processes (i.e., starting from noise), and may occur also in synchronously-pumped systems such as the FRASL. The net phase change turns out to be proportional to the integral of the initial pump pulse, and, as Figure 4.22b shows, the signal spectrum is asymmetric and has "wiggles." Figure 4.22c treats the intermediate case where the pump starts at the trailing edge of the signal, and in the fiber walks through to the leading edge. A symmetric spectrum results if the walk-off is symmetric.

A FRASL consists of a optical fiber ring cavity that is synchronously pumped by picosecond pulses and designed to lase at the stimulated Raman scattering Stokes wavelength (Figure 4.23). To obtain the generation of soliton Raman pulses the pump wavelength is chosen in the positive group velocity dispersion region of the optical fiber, whereas the Raman wavelength is in the negative group velocity dispersion region. Inserting a narrowband tunable etalon in the resonant ring, Islam et al. turned their laser in a pump-probe configuration in which they could control the seed feedback into the fiber and observe the spectral broadening in a single pass. The effect of walk-off on XPM could be studied by changing the fiber length in the cavity. Output Raman signals were passed through a bandpass filter to eliminate the pump and then sent to a scanning Fabry-Perot and an autocorrelator.

When a 50-m fiber is used in the FRASL ($l < l_w$), the signal remains with the pump throughout the fiber. With no etalon in the cavity, the signal spectrum is wider than the 300-cm^{-1} free spectral range of the Fabry-Perot. Even with the narrow-passband etalon introduced into the cavity, the spectral width remains greater than 300 cm^{-1} (Figure 4.24a). Therefore, more or less independent of the seed, the pump in a single pass severely broad-

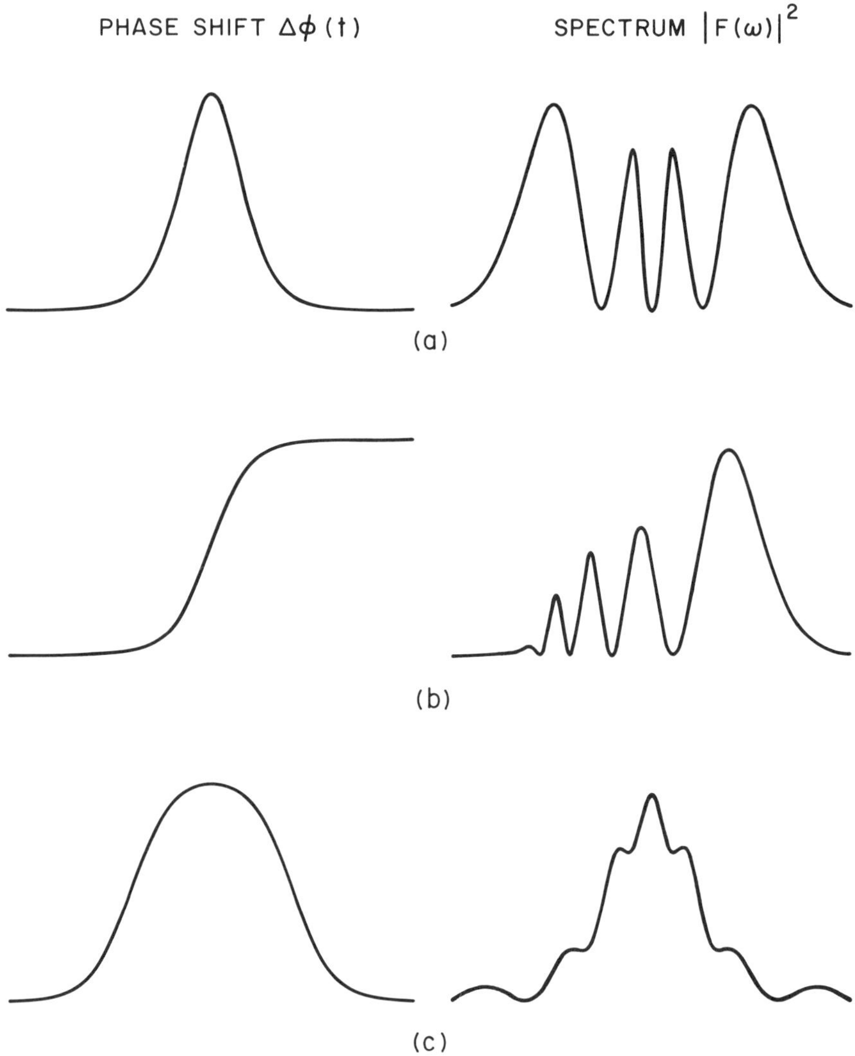

FIGURE 4.22. Phase shifts and spectra corresponding to various degrees of walk-off between pump and signal pulses. (a) Perfect tracking case ($t_0 = \beta = 0$, $2A^2l = 3.5\pi$, $\alpha = 1$); (b) pump and signal coincide initially, and then pump walks off ($t_0 = 0$, $\beta l = 4$, $2A^2/\beta = 3.5\pi$, $\alpha = 1$); and (c) pump walks from trailing edge of signal to the leading edge ($t_0 = -2, \beta l = 4, 2A^2/\beta = 3.5\pi, \alpha = 1$) (from Islam et al., 1987a).

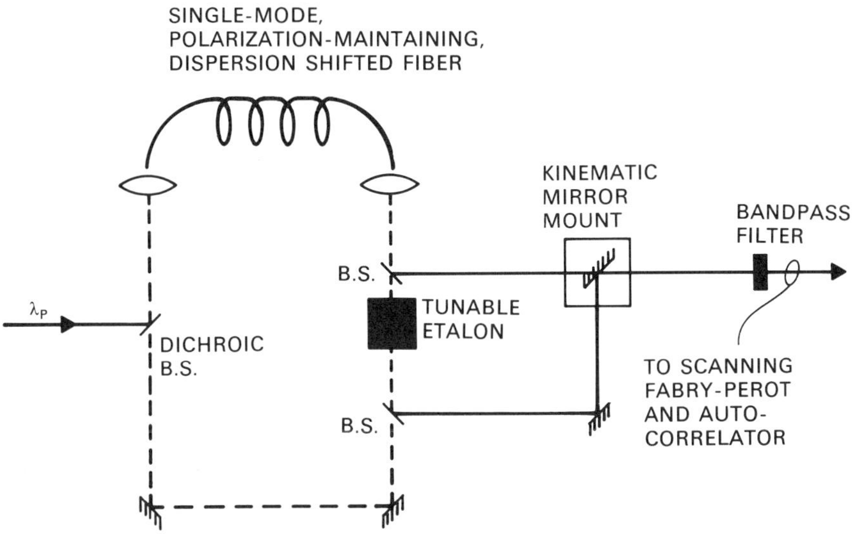

FIGURE 4.23. Modified fiber Raman amplification soliton laser (FRASL). B.S., beam splitter. (From Islam et al., 1987a.)

ens the signal spectrum. As expected from theory, the Raman spectrum is featureless.

If the fiber length is increased to 100 m ($l \approx l_w$), there is partial walk-off between the pump and signal and XPM again dominates the spectral features. Without an etalon in the FEASL cavity, the emerging spectrum is wide and has wiggles (Figure 4.24b). By time dispersion tuning the FRASL, thus varying the amount of walk-off, the details of the spectrum can be changed as shown in Figure 4.24c. Even after the etalon is inserted and the cavity length appropriately adjusted, the spectrum remained qualitatively the same (Figure 4.24d).

When there is complete walk-off between pump and signal ($l = 400$ m $\gg l_w$), without an etalon the spectrum is symmetric and secant-hyperboliclike, althoug still broad (Figure 4.24e). The effects of XPM are reduced considerably, but they are not canceleled completely because the walk-off is asymmetrized by pump depletion. As Figure 4.24f shows, the addition of the etalon narrows the spectrum (the narrow peak mimics the seed spectrum). However, XPM still produces a broad spectral feature (at the base of the peak), which is comparable in width to the spectrum without the filter (Figure 4.24e). In autocorrelation, it was found that the low-level wider feature corresponded to a $\tau \approx 250$ fs peak, while the narrow spectral peak results in a broader $\tau \approx 2.5$ ps pulse.

With these experimental results, Islam et. al have conclusively assessed the effects of walk-off on Raman XPM. It should be noted that, despite the long nonlinear interaction lengths, spectral broadenings were small and the SPM

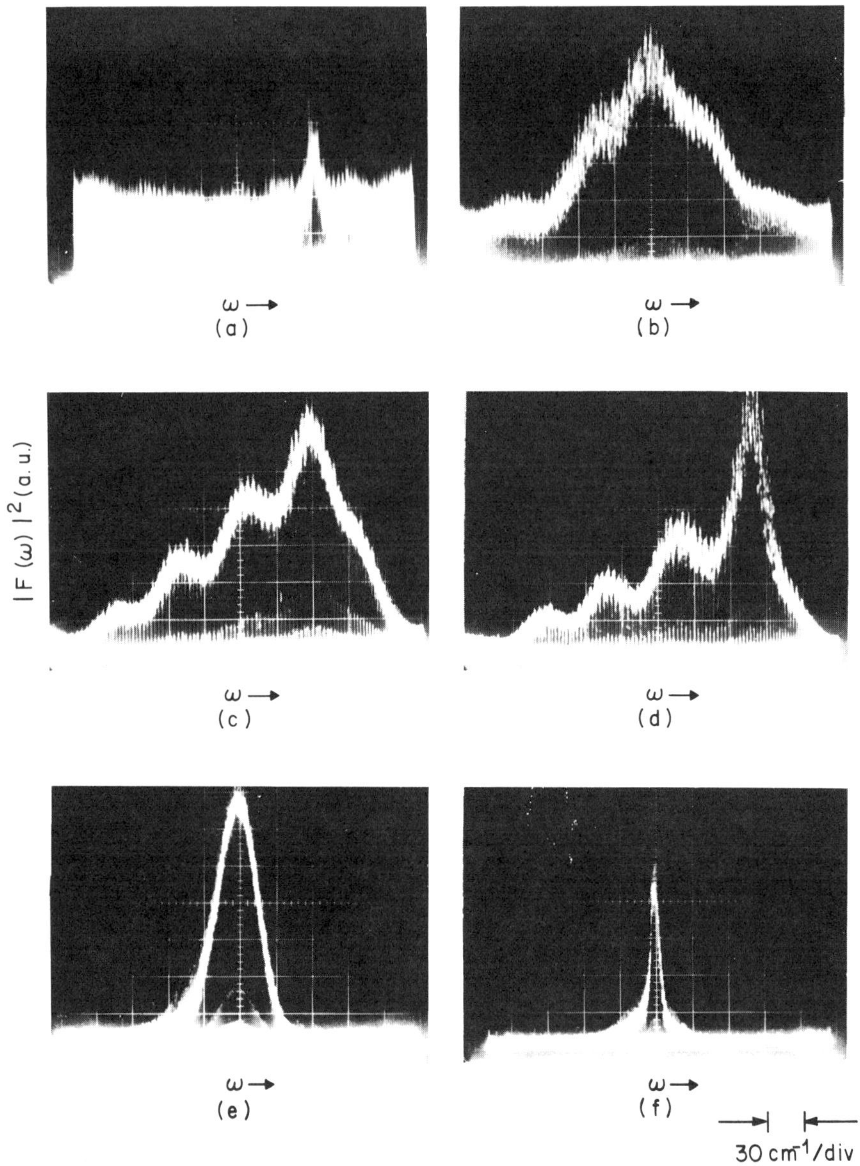

FIGURE 4.24. Experimental spectra for various fiber lengths (l) with and without the tunable etalon in the FRASL cavity. (a) $l = 50$ m with etalon in cavity. (b) $l = 100$ m, no etalon. (c) $l = 100$ m, no etalon, but different FRASL cavity length than in (b). (d) Same as (c), except with etalon inserted. (e) $l = 400$ m, no etalon. (f) Same as (e), except with etalon inserted. Here, except for the wings, the spectrum is nearly that of the etalon. The vertical scales are in arbitrary units, and the signal strength increases for increasing fiber lengths. (From Islam et al., 1987a.)

generated by the Raman pulse itself was negligible. Furthermore, measured spectral features were characteristic of XPM for the Raman amplification scheme, as expected for the injection of Raman seed pulses in the optical fiber loop.

4.2.2 Generation of Picosecond Raman Pulses in Optical Fibers

Stimulated Raman scattering of ultrashort pulses in optical fibers attracts a great deal of interest because of its potential applications for tunable fiber lasers and all-optical amplifiers. XPM effects on weak Raman pulses propagating in long low-dispersive optical fibers were characterized in the preceding section. Temporal and spectral modifications of pump and Raman pulses are more complex to analyze when Raman pulses are generated in short lengths (i.e., high Raman threshold) of very dispersive optical fibers. In addtion to XPM and walk-off, one has to take into account pump depletion, SPM of the Raman pulse, Raman-induced XPM of the pump pulse, group velocity dispersion broadening, higher-order SRS, and XPM-induced modulation instability. This section presents measurements of the generation of Raman picosecond pulses from the noise using short lengths of a single-mode optical fiber (Alfano et al., 1987b; Baldeck et al., 1987b–d).

A mode-locked Nd:YAG laser was used to generate 25-ps time duration pulses at $\lambda = 532$ nm with a repetition rate of 10 Hz. The optical fiber was custom-made by Corning Glass. It has a 3-μm core diameter, a 0.24% refractive index difference, and a single-mode cutoff at $\lambda = 462$ nm. Spectra of output pulses were measured using a grating spectrometer (1 m, 600 lines/mm) and recorded with an optical multichannel analyzer OMA2. Temporal profiles of pump and Raman pulses were measured using a 2-ps resolution Hamamatsu streak camera.

Spectra of pump and Raman pulses, which were measured for increasing pump energy at the output of short fiber lengths, are plotted in Figure 4.25. The dashed line in Figure 4.25a is the reference laser spectrum at low intensity. Figures 4.25a (solid line) and 4.25b show spectra measured at the Raman threshold at the output of 1- and 6-m-long optical fibers, respectively. The Raman line appears at $\lambda = 544.5$ nm (about 440 cm^{-1}). The laser line is broadened by SPM and shows XPM-induced sidebands, which are discussed in Section 8. For moderate pump intensities above the stimulated Raman scattering threshold, spectra of Raman pulses are broad, modulated, and symmetrical in both cases (Figures 4.25c and d). For these pump intensities, the pulse walk-off (6 m corresponds to two walk-off lengths) does not lead to asymmetric spectral broadening. For higher pump intensities, Raman spectra become much wider (Figures 4.25e and f). In addition, spectra of Raman pulses generated in the long fiber are highly asymmetric (Figure 4.25f). The intensity-dependent features observed in Figure 4.24 are characteristic of spectral broadenings arising from nonlinear phase modulations such as SPM and XPM as predicted by the theory (Section 4.1). At the lowest intensities XPM dominates, while at the highest intensities the SPM generated by the Raman

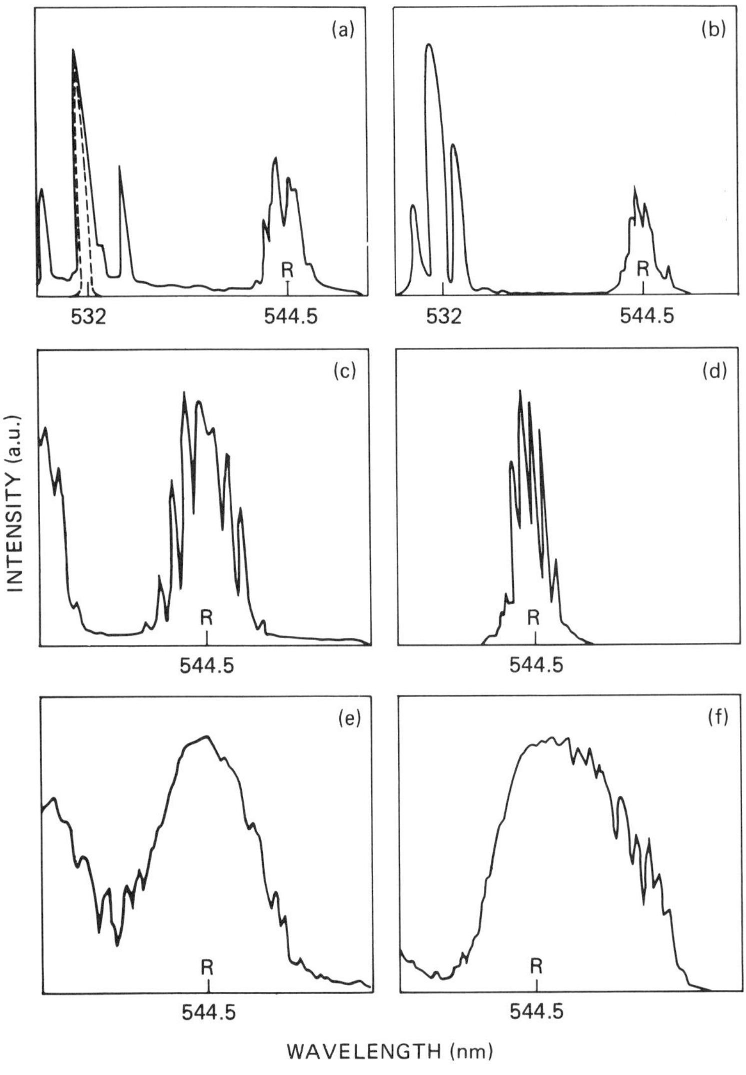

FIGURE 4.25. Spectra of picosecond Raman pulses generated in short lengths of a single-mode optical fiber. The laser and Raman lines are at 532 and 544.5 nm, respectively. Results in the left column and right column were obtained with 1- and 6-m-long single-mode optical fibers, respectively. (a and b) Dashed line: referenced of laser spectrum at low intensity; solid line: pump and Raman lines near the stimulated Raman scattering threshold. Frequency sidebands about the laser line are XPM-induced modulation instability sidebands (see Section 8). (c and d) Raman spectra for moderate pump peak powers above threshold. (e and f) Raman spectra for higher pump peak powers. (From Baldeck et al., 1987c-d.)

pulse itself is the most important. However, it should be noted that the widths of Raman spectra shown in Figure 4.25 are one order of magnitude larger than expected from the theory. Modulation instability induced by pump pulses could explain such a discrepancy between measurements and theory (Section 8).

Temporal measurements of the generation process were performed to test whether the spectral asymmetry originated from the pump depletion reshaping as in the case of longer pulses (Schadt et al., 1986). Pump and Raman profiles were measured at the output of a 17-m-long fiber (Figure 4.26). The dotted line is for a pump intensity at the SRS threshold and the solid line for a higher pump intensity. The leading edge of the pump pulse is partially "eaten" but is not completely emptied because of the quick walk-off between pump and Raman pulses. Thus, the leading edge of the pump pulse does not become very sharp, and the contribution of pump depletion effects to the spectral asymmetry of pump and Raman pulses does not seem to be significant.

Figure 4.26 shows a typical sequence of temporal profiles measured for input pump intensities strong enough to generate higher-order stimulated Raman scattering lines. The temporal peaks are the maxima of high-order SRS scatterings that satisfy the group velocity dispersion delay of 6 ps/m for each frequency shift of 440 cm^{-1}. These measurements show that (1) the Raman process clamps the peak power of pulses propagating into an optical fiber to a maximum value and (2) high-order stimulated Raman scatterings occur in cascade during the laser pulse propagation.

4.2.3 Generation of Femtosecond Raman Pulses in Ethanol

Nonlinear phenomena such as supercontinuum generation and stimulated Raman scattering were first produced in unstable self-focusing filaments generated by intense ultrashort pulses in many liquids and solids. Optical fibers are convenient media for studying such nonlinear phenomena without the catastrophic features of collapsing beams. However, optical fibers are not suitable for certain applications such as high-power experiments, the generation of large Raman shifts (>1000 cm^{-1}), and Raman pulses having high peak powers (>1 MW). In this section, spectral measurements of SRS generation in ethanol are presented. Spectral shapes are shown to result from the combined effects of XPM, SPM, and walk-off.

Spectral measurements of SRS in ethanol have been performed using the output from a CPM ring dye amplifier system (Baldeck et al., 1987b). Pulses of 500 fs duration at 625 nm were amplified to an energy of about 1 mJ at a repetition rate of 20 Hz. Pulses were weakly focused into a 20-cm-long cell filled with ethanol. Output pulses were imaged on the slit of a $\frac{1}{2}$-m Jarrell-Ash spectrometer and spectra were recorded using an optical multichannel analyzer OMA2.

Ethanol has a Raman line shifted by 2928 cm^{-1}. Figure 4.27 shows how the Stokes spectrum of the Raman line changes as a function of the pump intensity. Results are comparable to those obtained using optical fibers. At low

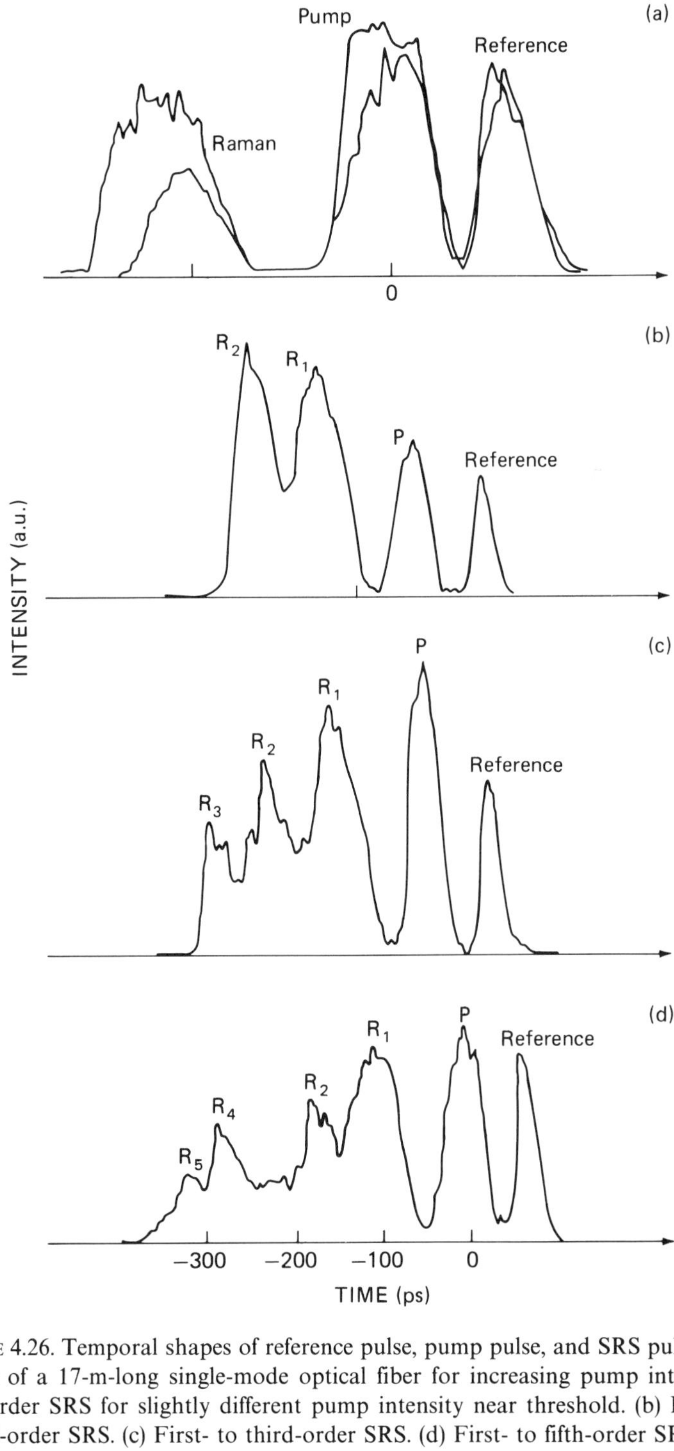

FIGURE 4.26. Temporal shapes of reference pulse, pump pulse, and SRS pulses at the output of a 17-m-long single-mode optical fiber for increasing pump intensity. (a) First-order SRS for slightly different pump intensity near threshold. (b) First- and second-order SRS. (c) First- to third-order SRS. (d) First- to fifth-order SRS. (From Baldeck et al., 1987d.)

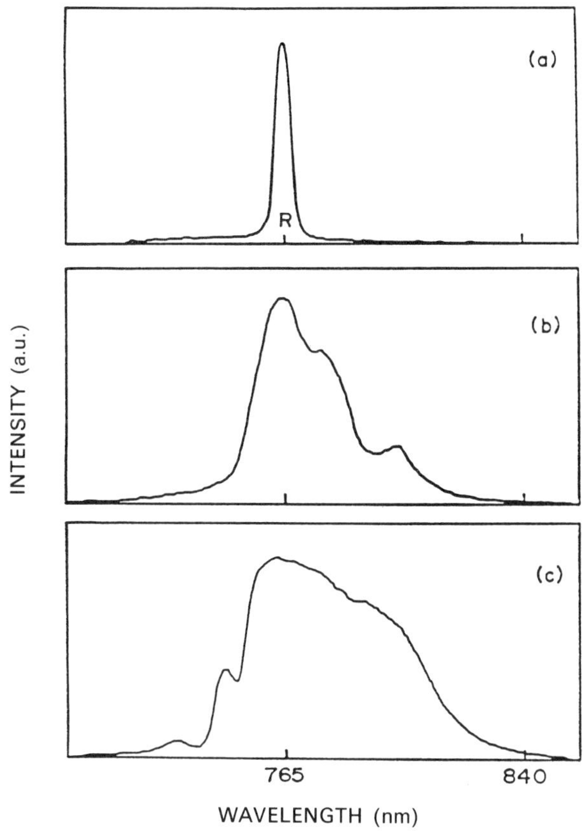

FIGURE 4.27. Effects of cross- and self-phase modulations on the Stokes-shifted Raman line generated by 500-fs pulses in ethanol. (a to c) Increasing laser intensity. (From Baldeck et al., 1987b.)

intensity the Stokes spectrum is narrow and symmetrical (Figure 4.27). As the pump intensity increases the Raman spectrum broadens asymmetrically with a long tail pointing toward the longer wavelengths. Spectra of the anti-Stokes Raman line were also measured (Baldeck et al., 1987b). They were as wide as Stokes spectra but with tails pointing toward the shortest wavelengths, as predicted by the sign of the walk-off parameter.

5. Harmonic Cross-Phase Modulation Generation in ZnSe

Like stimulated Raman scattering, the second harmonic generation (SHG) process involves the copropagation of a weak generated-from-the-noise pulse with an intense pump pulse. The SHG of ultrashort pulses occurs simul-

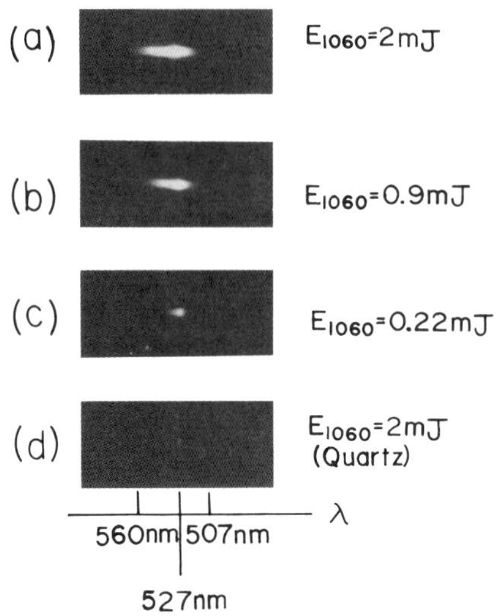

FIGURE 4.28. Induced-spectral-broadening spectra in ZnSe crystal excited by an intense 1060-nm laser pump. In (d) the ZnSe crystal was replaced by a 3.7-cm-long quartz crystal. (From Alfano et al., 1987a.)

taneously with cross-phase modulation, which affects both the temporal and spectral properties of second harmonic pulses. In this section, measurements of XPM on the second harmonic generated by an intense primary picosecond pulse in ZnSe crystals are reported (Alfano et al., 1987a; Ho et al., 1988).

The laser system consisted of a mode-locked Nd:galss laser with single-pulse selector and amplifier. The output laser pulse had about 2 mJ energy and 8 ps duration at a wavelength of 1054 nm. The 1054-nm laser pulse was weakly focused into the sample. The spot size at the sample was about 1.5 mm in diameter. The second harmonic produced in this sample was about 10 nJ. The incident laser energy was controlled using neutral density filters. The output light was sent through a $\frac{1}{2}$-m Jarrell-Ash spectrometer to measure the spectral distribution of the signal light. The 1054-nm incident laser light was filtered out before detection. A 2-ps time resolution Hamamatsu streak camera system was used to measure the temporal characteristics of the signal pulse. Polycrystalline ZnSe samples 2, 5, 10, 22, and 50 mm thick were purchased from Janos, Inc. and a single crystal of ZnSe 16 mm thick was grown at Philips.

Typical spectra of non-phase-matched SHG pulses generated in a ZnSe crystal by 1054-nm laser pulses of various pulse energies are displayed in Figure 4.28. The spectrum from a quartz sample is included in Figure 4.28d for reference. The salient features of the ZnSe spectra indicate that the extent of the spectral broadening about the second harmonic line at 527 nm depends on the intensity of the 1054-nm laser pulse. When the incident laser pulse energy was 2 mJ, there was significant spectral broadening of about 1100 cm^{-1} on the Stokes side and 770 cm^{-1} on the anti-Stokes side (Figure 4.29). There was no significant difference in the spectral broadening distribution measured in the single and polycrystalline materials. The spectral width of the SHG

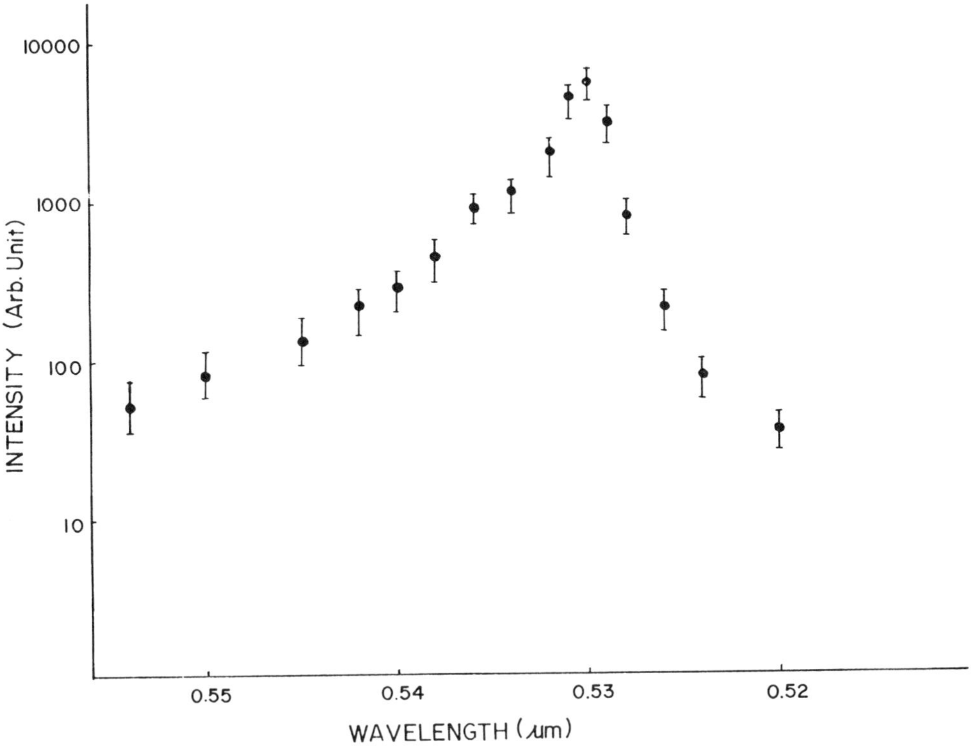

FIGURE 4.29. Spectral measurement of the induced spectrally broadened pulse about $\lambda = 527$ nm by sending a 1054-nm pulse through 22-mm ZnSe. (From Alfano et al., 1988.)

signal is plotted for the Stokes and anti-Stokes sides as a function of the incident pulse energy in Figure 4.30. The salient feature of Figure 4.30 is that the Stokes side of the spectrum is broader than the anti-Stokes side. When the incident pulse energy was less than 1 mJ, the spectral broadening was found to be monotonically increasing on the pulse energy of 1054 nm. The spectral broadening generated by sending an intense 80-μJ, 527-nm, 8-ps laser pulse alone through these ZnSe crystals was also measured for comparison with the ± 1000 cm^{-1} induced spectral broadening. The observed spectral broadening was only 200 cm^{-1} when the energy of the 527-nm pulse was over 0.2 mJ. This measurement suggests that the self-phase modulation process from the 10-nJ SHG pulse in ZnSe is too insignificant to explain the observed 1000 cm^{-1}. Most likely, the broad spectral width of the SHG signal arises from the XPM generated by the pump during the generation process.

The temporal profile and propagation time of the intense 1054-nm pump pulse and the second harmonic pulse propagating through a 22-mm ZnSe polycrystalline sample is shown in Figure 4.31. A pulse delay of ~ 189 ps at

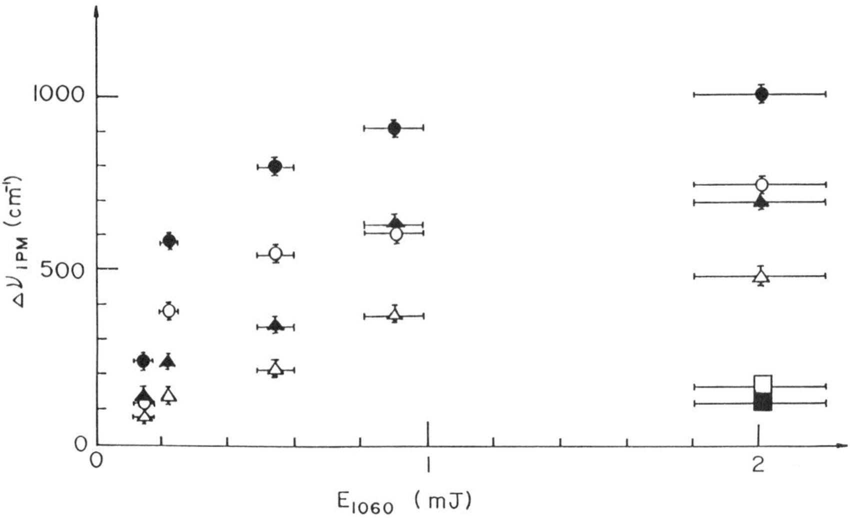

FIGURE 4.30. Intensity dependence of induced spectral width about 530 nm in ZnSe pumped by a 1060-nm laser pulse. The horizontal axis is the incident laser pulse energy. (○) 2.2-cm-long polycrystalline ZnSe anti-Stokes broadening. (●) 2.2-cm-long polycrystalline ZnSe Stokes broadening. (▲) 1.6-cm-long single-crystal ZnSe anti-Stokes broadening. (△) 1.6-cm-long single-crystal ZnSe Stokes broadening. (□) 3.7-cm-long quartz crystal anti-Stokes broadening. (■) 3.7-cm-long quartz crystal Stokes broadening. The measured Δn is defined as the frequency spread from 527 nm to the farthest detectable wavelengths measured either photographically or by an optical multichannel analyzer. (From Alfano and Ho, 1988.)

1054 nm was observed (Figure 4.31a) when an intense 1054-nm pulse passed through the crystal. The second harmonic signal, which spread from 500 to 570 nm, indicated a sharp spike at 189 ps and a long plateau from 189 to 249 ps (Figure 4.31b). Using 10-nm bandwidth narrowband filters, pulses of selected wavelengths from the second harmonic signal were also measured. For example, time delays corresponding to the propagation of two pulses with wavelengths centered at 530 and 550 nm are displayed in Figures 4.31c and d, respectively. All traces from Figure 4.31 indicated that the induced spectrally broadened pulses have one major component emitted at nearly the same time as the incident pulse (Figure 4.31a). The selected wavelength shifted 10 nm from the second harmonic wavelength has shown a dominant pulse distribution generated at the end of the crystal. Furthermore, when a weak 3-nJ, 527-nm calibration pulse propagated alone through the 22-mm ZnSe, a propagation time of about 249 ps was observed, as expected from the group velocity.

The difference in the propagation times of a weak 527-nm calibration pulse and a 1054-nm pump pulse through a ZnSe crystal can be predicted perfectly

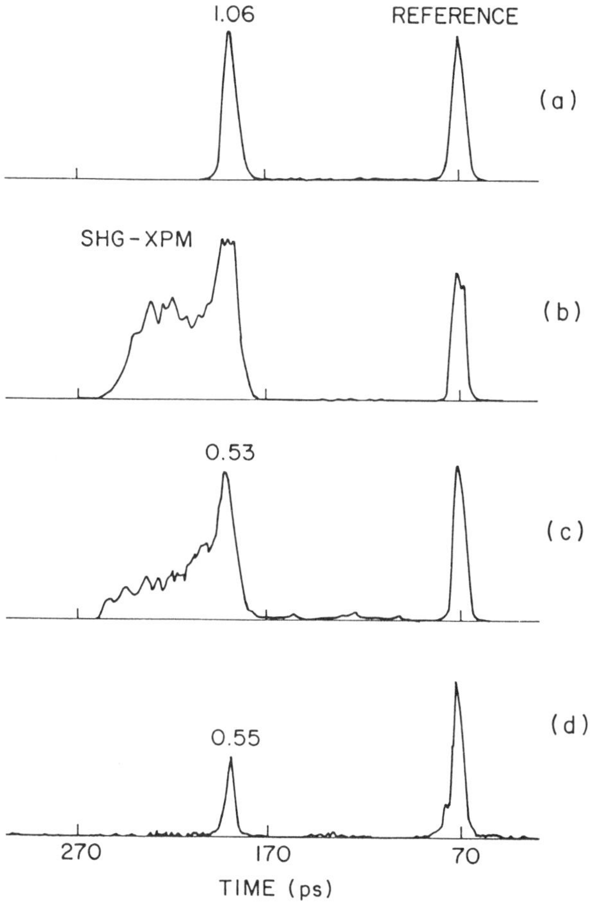

FIGURE 4.31. Temporal profile and propagation delay time of (a) incident 1054 nm, (b) SHG-XPM signal of all visible spectra, and (c) selected 530 nm from SHG-XPM of a 22-nm-long ZnSe crystal measured by a 2-ps resolution streak camera system. (d) same as (c) for a signal selected at 550 nm. The reference time corresponds to a laser pulse traveling through air without the crystal. The right side of the time scale is the leading time. The vertical scale is an arbitrary intensity scale. (From Alfano and Ho, 1988.)

by the difference in group velocities. The measured group refractive indices of ZnSe can be fitted to $n_{g,1054} = 3.39$ and $n_{g,1054} = 2.57$, respectively. These values are in agreement with the calculated values.

The sharp spike and plateau of the second harmonic pulse can be explained using the XPM model of second harmonic generation (Ho et al., 1980). Because of lack of phase matching, i.e., destructive interferences, the energy of the second harmonic pulse cannot build up along the crystal length. As a result, most of the second harmonic power is generated at the exit face of the crystal, which explains the observed spike. However, since very intense pump pulses are involved there is a partial phase matching due to the cross-phase modulation and two photon absorption effects at the second harmonic wavelength. Some second harmonic energy can build up between the entrance and exit faces of the sample, which explains the plateau feature.

6. Cross-Phase Modulation and Stimulated Four-Photon Mixing in Optical Fibers

Stimulated four-photon mixing (SFPM) is an ideal process for designing parametric optical amplifiers and frequency converters. SFPM is produced when two high-intensity pump photons are coupled by the third-order susceptibility $\chi^{(3)}$ to generate a Stokes photon and an anti-Stokes photon. The frequency shifts of the SFPM waves are determined by the phase-matching conditions, which depend on the optical geometry. SFPM was produced in glass by Alfano and Shapiro (1970) using picosecond pulses. Later, SFPM was successfully demonstrated by a number of investigators in few mode, birefringent, and single-mode optical fibers (Stolen, 1975; Stolen et al., 1981; Washio et al., 1980). Most of the earlier experiments using optical fibers were performed with nanosecond pulses. Lin and Bosch (1981) obtained large-frequency shifts; however, the spectral dependence on the input intensity was not investigated. In the following, measurements of the intensity dependence of SFPM spectra generated by 25-ps pulses in an optical fiber are reported (Baldeck and Alfano, 1987). For such short pulses, spectra are influenced by the combined effects of SPM and XPM. The broadening of SFPM lines and the formation of frequency continua are investigated.

The experimental method is as follows. A Quantel frequency-doubled mode-locked Nd:YAG laser produced 25-ps pulses. An X20 microscope lens was used to couple the laser beam into the optical fiber. The spectra of the output pulses were measured using a 1-m, 1200 lines/mm grating spectrometer. Spectra were recorded on photographic film and with an optical multichannel analyzer OMA2. Average powers coupled in the fiber were measured with a power meter at the optical fiber output. The 15-m-long optical fiber had a core diameter of 8 mm and a normalized frequency $V = 4.44$ at 532 nm. At this wavelength, the four first LP modes (LP_{01}, LP_{11}, LP_{21}, and LP_{02}) were allowed to propagate.

Typical intensity-dependent spectra are displayed in Figures 4.32, 4.33, and 4.34. At low intensity, $I < 10^8$ W/cm^2, the output spectrum contains only the pump wavelength $\lambda = 532$ nm (Figure 4.32a). At approximately 5×10^8 W/cm^2 three sets of symmetrical SFPM lines (at $\Omega = 50$, 160, and 210 cm^{-1}) and the first SRS Stokes line (at 440 cm^{-1}) appear (Figures 4.32b and c). As the intensity increases the SFPM and SRS lines broaden, and a Stokes frequency continuum is generated (Figures 4.32d and e). Above an intensity threshold of 20×10^8 W/cm^2, new sets of SFPM lines appear on the Stokes

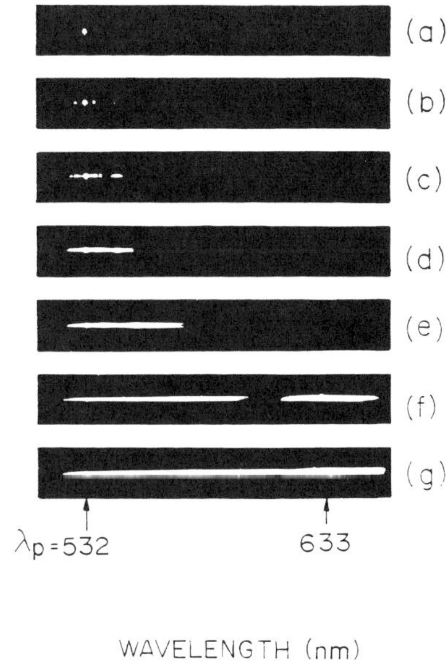

FIGURE 4.32. Evolution of a stimulated four-photon spectrum with increasing pulse intensity. (a) $I < 10^8$ W/cm^2; (b and c) $I = 5 \times 10^8$ W/cm^2; (d) $I = 10 \times 10^8$ W/cm^2; (e) $I = 15 \times 10^8$ W/cm^2; (f) $I = 30 \times 10^8$ W/cm^2; (g) $I = 35 \times 10^8$ W/cm^2. (From Baldeck and Alfano, 1987.)

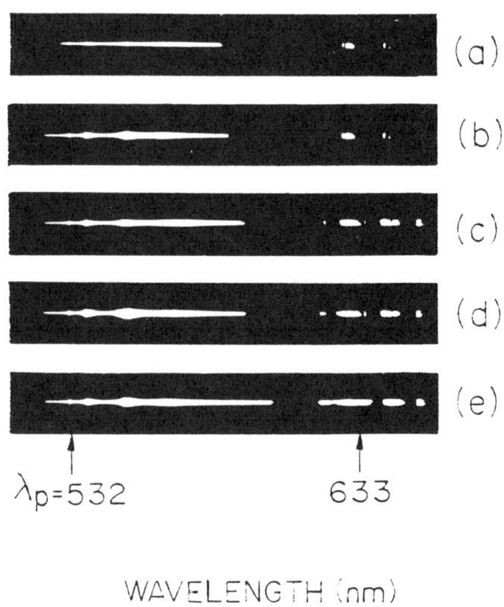

FIGURE 4.33. (a to e) Sequence of the large-shift SFPM line broadening. The pulse peak intensity increases from $I = 20 \times 10^8$ W/cm^2 in (a) to $I = 30 \times 10^8$ W/cm^2 in (e) in steps of 2.5×10^8 W/cm^2. (From Baldeck and Alfano, 1987.)

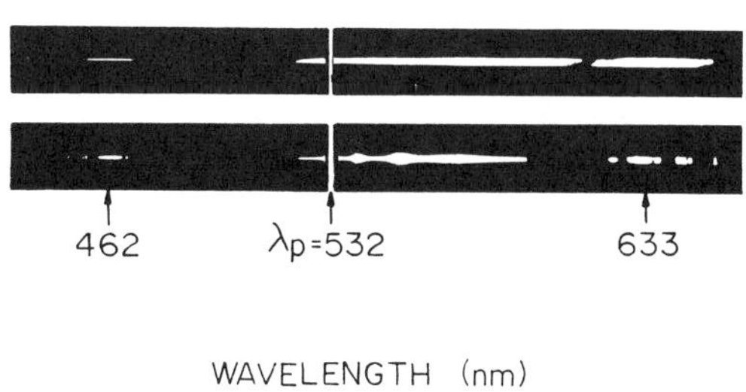

FIGURE 4.34. Examples of large-shift Stokes lines with their corresponding anti-Stokes lines. Photographs of the Stokes and anti-Stokes regions were spliced together. (From Baldeck and Alfano, 1987.)

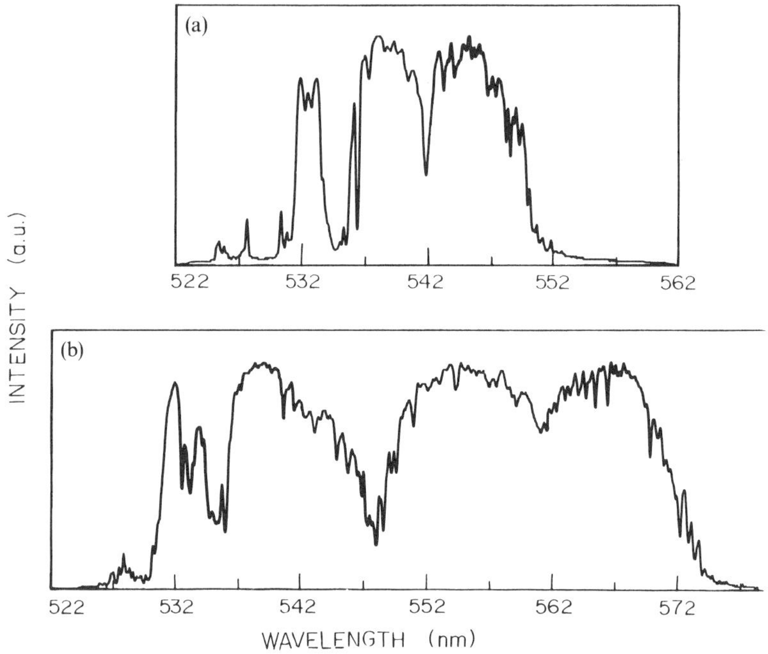

FIGURE 4.35. Supercontinuum generation. (a) The pump, SFPM, and first SRS Stokes lines are broadened at $I = 10 \times 10^8$ W/cm^2. (b) The broadened second and third SRS Stokes lines appear and extend the spectrum toward the Stokes wavelengths at $I = 15 \times 10^8$ W/cm^2. (From Baldeck and Alfano, 1987.)

and anti-Stokes sides with frequency ranging from 2700 to 3865 cm^{-1}. Finally, the large shifts merge (Figure 4.32f) and contribute to the formation of a 4000 cm^{-1} frequency continuum (Figure 4.32g). Figure 4.33 shows how the large Stokes shift SFPM lines are generated and broaden when the pump intensity increases from 20×10^8 to 30×10^8 W/cm^2. Figure 4.34 gives two examples of complete spectra including the large-shift anti-Stokes and Stokes lines. The measured SFPM shifts correspond well with the phase-matching condition of SFPM in optical fibers.

Figure 4.35 shows the development of a Stokes continuum from the combined effects of SFPM, SRS, SPM, and XPM. As the pump intensity is increased, the pump, SFPM, and first SRS lines broaden and merge (Figure 4.35a). For stronger pump intensities, the continuum is duplicated by stimulated Raman scattering, and the continuum expands toward the lowest optical frequencies (Figure 4.35b). As shown, the maximum intensities of new frequencies are self-limited.

The broadening of the SFPM and SRS lines arises from self- and cross-phase modulation effects. It is established that spectral broadenings generated by SPM are inversely proportional to the pulse duration and linearly propor-

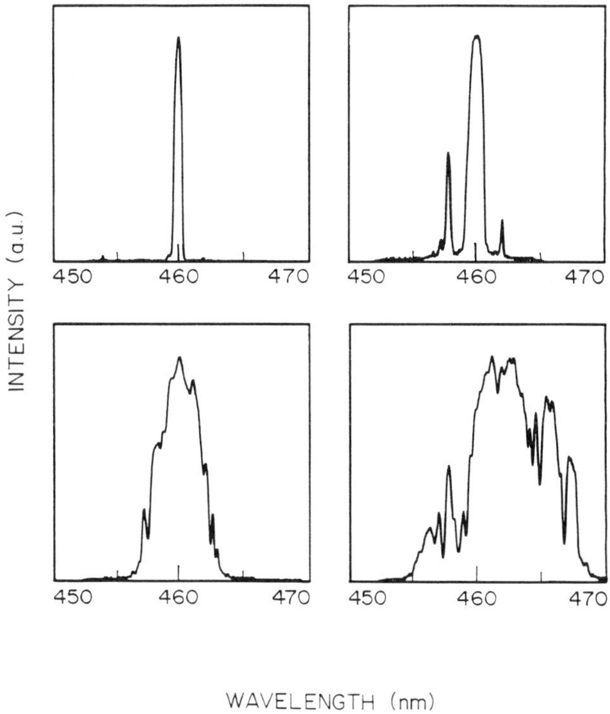

FIGURE 4.36. (a to d) Spectral broadening of the anti-Stokes SFPM line generated at 460 nm. The pulse peak intensity increases from $I = 20 \times 10^8$ W/cm^2 in (a) to $I = 30 \times 10^8$ W/cm^2 in (e) in steps of 2.5×10^8 W/cm^2. (From Baldeck and Alfano, 1987.)

tional to the pump intensity. In this experiment, SPM effects are important because of the pump pulse shortness (25 ps) and intensity (10^9 W/cm^2). Furthermore, the modulation that is seen in the continuum spectrum fits well with the spectrum modulation predicted by phase modulation theories.

Figure 4.36 shows the spectral broadening of the anti-Stokes SFPM line of $\lambda = 460$ nm ($\Omega = 2990$ cm^{-1}). This line is a large-shift SFPM anti-Stokes line generated simultaneously with the $\lambda = 633$ nm SFPM Stokes line by the laser pump of $\lambda = 532$ nm (see Figure 4.34). the corresponding frequency shift and mode distribution are $W = 2990$ cm^{-1} and LP_{01} (pump)–LP_{11} (Stokes and anti-Stokes), respectively. From Figures 4.36a to d, the peak intensity of the $\lambda = 460$ nm line increases from approximately 20×10^8 to 30×10^8 W/cm^2 in steps of 2.5×10^8 W/cm^2. In Figure 4.36a, the spectrum contains only the 460-nm SFPM line generated by the laser pump ($\lambda = 532$ nm). In Figure 4.36b, the line begins to broaden and two symmetrical lines appear with a frequency shift of 100 cm^{-1}. This set of lines could be a new set of small-shift SFPM lines generated by the 460-nm SFPM line acting as a new pump wavelength.

Figures 4.36c and d show significant broadening, by a combined action of SFPM, SPM, and XPM, of the 460 nm into a frequency continuum. Similar effects were observed on the Stokes side as displayed in Figure 4.33.

The intensity effects on SFPM spectra generated by 25-ps pulses propagating in optical fibers have been investigated experimentally. In contrast to SFPM lines generated by nanosecond pulses, spectra were broadened by self-phase modulation and cross-phase modulation. Intensity-saturated wide frequency continua covering the whole visible spectrum were generated for increasing intensities. Applications are for the design of wideband amplifiers, the generation of "white" picosecond pulses, and the generation by pulse compression of femtosecond pulses at new wavelengths.

7. Induced Focusing by Cross-Phase Modulation in Optical Fibers

Cross-phase modulation originates from the nonlinear refractive index $\Delta n(r,t) = 2n_2 E_P^2(r,t)$ generated by the pump pulse at the wavelength of the probe pulse. Consequently, XPM has not only temporal and spectral effects but also spatial effects. Induced focusing is a spatial effect of XPM on the probe beam diameter. Induced focusing is the focusing of a probe beam because of the radial change of the refractive index induced by a copropagating pump beam. Induced focusing is similar to the self-focusing (Kelley, 1965) of intense lasers beams that has been observed in many liquids and solids. Overviews and references on self-focusing in condensed media are given by Auston (1977) and Shen (1984).

In 1987, Baldeck, Raccah, and Alfano reported on experimental evidence for focusing of picosecond pulses propagating in an optical fiber (Baldeck et al., 1987a). Focusing occurred at Raman frequencies for which the spatial effect of the nonlinear refractive index was enhanced by cross-phase modulation. Results of this experiment on induced focusing by cross-phase modulation in optical fibers are summarized in this section.

The experimental setup is shown in Figure 4.37. A Quantel frequency-doubled mode-locked Nd:YAG laser produced 25-ps pulses at 532 nm. The laser beam was coupled into the optical fiber with a 10 × microscope lens. A stable modal distribution was obtained with a Newport FM-1 mode scrambler. Images of the intensity distribution at the output face were magnified by 350 × and recorded on photographic film. Narrowband (NB) filters were used to select frequencies of the output pulses. The optical fiber was a commercial multimode step-index fiber (Newport F-MLD). Its core diameter was 100 μm, its numerical aperture 0.3, and its length 7.5 m.

Several magnified images of the intensity distributions that were observed

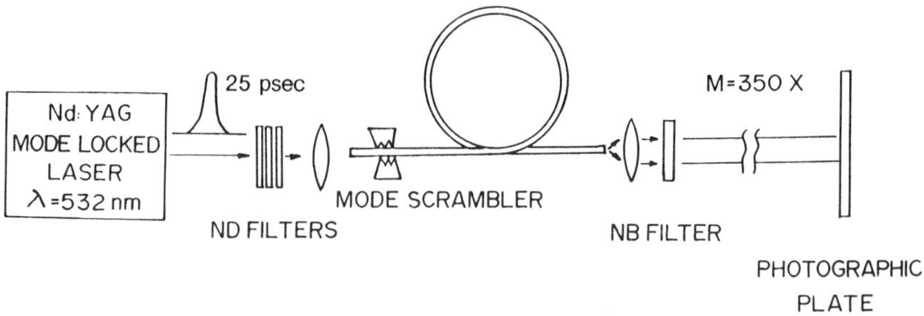

FIGURE 4.37. Experimental setup for the observation of Raman focusing in a large-core optical fiber. (From Baldeck et al., 1987.)

at the output face of the fiber for different input pulse energies are shown in Figure 4.38. The intensity distribution obtained for low pulse energies ($E < 1$ nJ) is shown in Figure 4.38a. It consists of a disk profile with a speckle pattern. The intensity distribution of the disk covers the entire fiber core area. The disk diameter, measured by comparison with images of calibrated slits, is 100 μm, which corresponds to the core diameter. The characteristics of this fiber allow for the excitation of about 200,000 modes. The mode scrambler distributed the input energy to most of the different modes. The speckle pattern is due to the interference of these modes on the output face. Figure 4.37b shows the intensity distribution in the core for intense pulses ($E > 10$ nJ). At the center of the 100-mm-diameter disk image, there is an intense smaller (11-mm) ring of a Stokes-shifted frequency continuum of light. About 50% of the input energy propagated in this small-ring pattern. The corresponding intensities and nonlinear refractive indices are in the ranges of gigawatts per square centimeter and 10^{-6}, respectively. For such intensities, there is a combined effect of stimulated Raman scattering, self-phase modulation, and cross-phase modulation that generates the observed frequency continuum. In Figure 4.37c, an NB filter selected the output light pattern at 550 nm. This clearly shows the ring distribution of the Stokes-shifted wavelengths. Such a ring distribution was observed for a continuum of Stokes-shifted wavelengths up to 620 nm for the highest input energy before damage.

The small-ring intensity profile is a signature of induced focusing at the Raman wavelengths. First, the small ring is speckleless, which is characteristic of single-mode propagation. This single-mode propagation means that the guiding properties of the fiber are dramatically changed by the incoming pulses. Second, SRS, SPM, and XPM occur only in the ring structure, i.e., where the maximum input energy has been concentrated. Our experimental results may be explained by an induced-gradient-index model for induced focusing. For high input energies, the Gaussian beam induces a radial change of the refractive index in the optical fiber core. The step-index fiber becomes

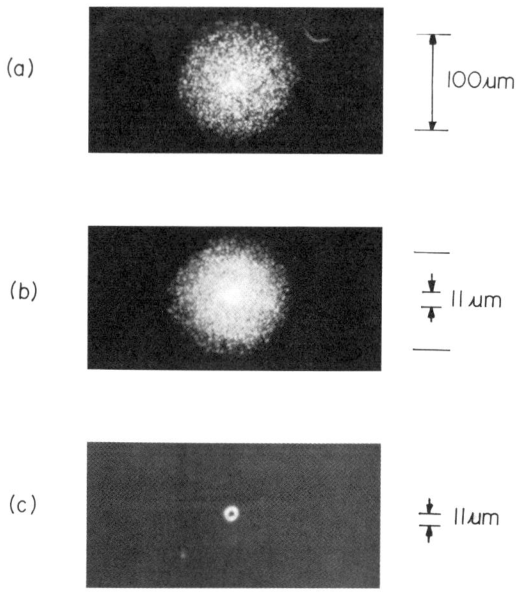

FIGURE 4.38. Images of the intensity distributions at the optical fiber output: (a) input pulses of low energies ($E < 1$ nJ); (b) input pulses of high energies ($E > 10$ nJ); (c) same as (b) with an additional narrowband filter centered at $\lambda = 550$ nm. ($M = 350\chi$). (From Baldeck et al., 1987a.)

a gradient-index fiber, which modifies its light-guiding properties. There is further enhancement of the nonlinear refractive index at Raman frequencies because of XPM. Thus, Stokes-shifted light propagates in a well-marked induced-gradient-index fiber. The ray propagation characteristics of a gradient-index fiber are shown schematically in Figure 4.39 (Keiser, 1983). The cross-sectional view of a skew-ray trajectory in a graded-index fiber is shown. For a given mode u, there are two values for the radii, r_1 and r_2, between which the mode is guided. The path followed by the corresponding ray lies completely within the boundaries of two coaxial cylindrical surfaces that form a well-defined ring. These surfaces are known as the caustic surfaces. They have inner and outer radii r_1 and r_2, respectively. Hence, Figure 4.39 shows that skew rays propagate in a ring structure comparable to the one shown in Figure 4.38c. This seems to support the induced-gradient-index model for induced focusing in optical fibers.

Induced focusing of Raman picosecond pulses has been observed in optical fibers. Experimental results may be explained by an induced-gradient-index model of induced focusing. An immediate application of this observation

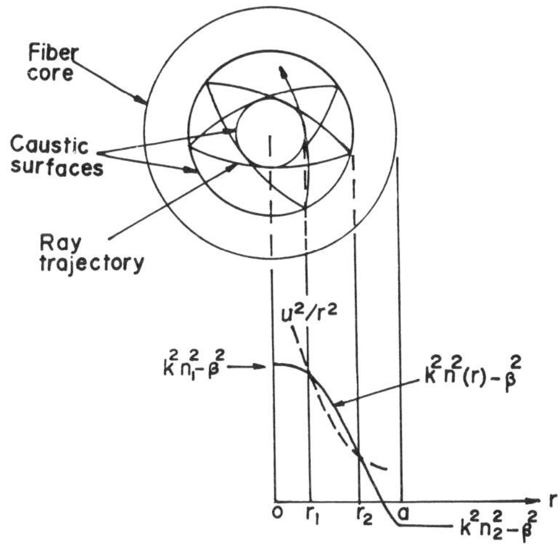

FIGURE 4.39. Cross-sectional projection of a skew ray in a gradient-index fiber and the graphical representation of its mode solution from the WBK method. The field is oscillatory between the turning points r_1 and r_2 and is evanescent outside this region.

could be the single-mode propagation of high-bit-rate optical signals in large-core optical fibers.

8. Modulation Instability Induced by Cross-Phase Modulation in Optical Fibers

Modulation instability refers to the sudden breakup in time of waves propagating in nonlinear dispersive media. It is a common nonlinear phenomenon studied in several branches of physics (an overview on modulation instability can be found in Hasegawa, 1975). Modulation instabilities occur when the steady state becomes unstable as a result of an interplay between the dispersive and nonlinear effects. Tai, Hasegawa, and Tomita have observed the modulation instability in the anomalous dispersion regime of silica fibers, i.e., for wavelengths greater than 1.3 μm (Tai et al., 1986). Most recently, Agrawal (1987) has suggested that a new kind of modulation instability can occur even in the normal dispersion regime when two copropagating fields interact with each other through the nonlinearity-induced cross-phase modulation. This section summarizes the first observation by Baldeck, Alfano, and Agrawal of such a modulation instability initiated by cross-phase modulation in the normal dispersion regime of silica optical fibers (Baldeck et al. 1988b, 1989d).

Optical pulses at 532 nm were generated by either a mode-locked Nd:YAG laser or a Q-switched Nd:YAG laser with widths of 25 ps or 10 ns, respectively. In both cases the repetition rate of pulses was 10 Hz. Pulses were coupled into a single-mode optical fiber using a microscope lens with a magnification of 40. The peak power of pulses into the fiber could be adjusted in the range 1 to 10^4 W by changing the coupling conditions and by using neutral density

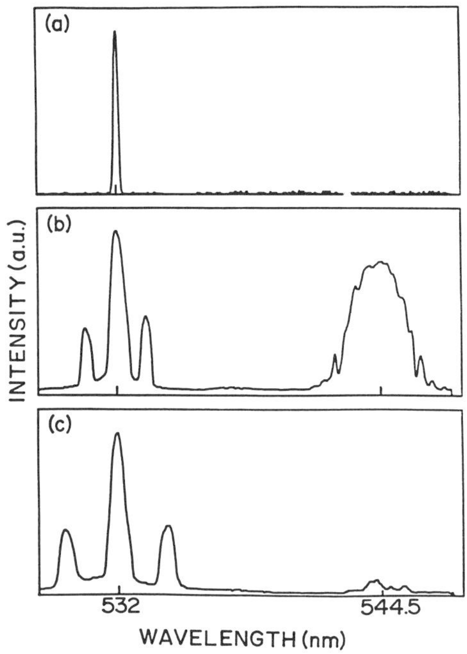

FIGURE 4.40. Characteristic frequency sidebands of modulation instability resulting from cross-phase modulation induced by the simultaneously generated Raman pulses in lengths L of a single-mode optical fiber. The laser line is at $\lambda = 532$ nm and the Raman line at $\lambda = 544.5$ nm. The time duration of input pulses is 25 ps. (a) Reference spectrum at low intensity; (b) Spectrum at about the modulation instability threshold and $L = 3$ m; (c) same as (b) for $L = 0.8$ m. (From Baldeck et al., 1988d-1989.)

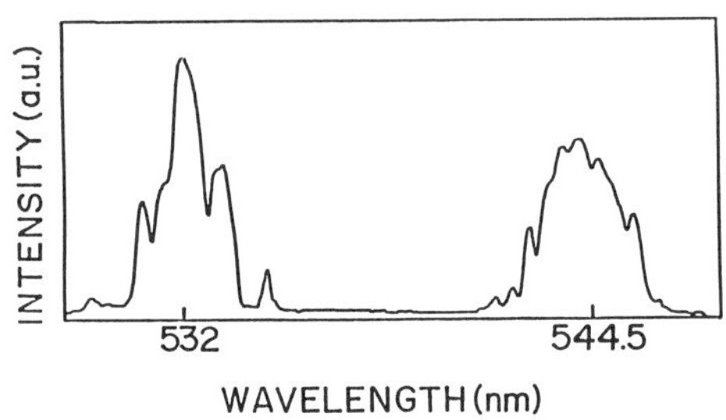

FIGURE 4.41. Secondary sidebands observed for pulse energies well above the modulation instability threshold. (From Baldeck et al., 1988d-1989.)

filters. The optical fiber was custom-made by Corning Glass. It has a 3-μm core diameter, a 0.24% refractive index difference, and a single-mode cutoff at $\lambda = 462$ nm. Spectra of output pulses were measured using a grating spectrometer (1 m, 600 lines/mm) and recorded with an optical multichannel analyzer OMA2.

Figures 4.40 and 4.41 show spectra of intense 25-ps pulses recorded for different peak powers and fiber lengths. Figure 4.40a is the reference spectrum of low-intensity pulses. Figures 4.40b and c show spectra measured at about the modulation instability threshold for fiber lengths of 3 and 0.8 m, respectively. They show modulation instability sidebands on both sides of the laser wavelength at 532 nm and the first-order stimulated Raman scattering line at 544.5 nm. Notice that the frequency shift of sidebands is larger for the shorter fiber. Secondary sidebands were also observed for pulse energy well above the modulation instability threshold and longer optical fibers as shown in Figure 4.41.

Similar to spectra in the experiment of Tai et al., spectra shown in Figures 4.40 and 4.41 are undoubtedly signatures of modulation instability. A major salient difference in the spectra in Figures 4.40 and 4.41 is that they show modulation instability about 532 nm, a wavelength in the *normal dispersion regime* of the fiber. According to the theory, modulation instability at this wavelength is possible only if there is a cross-phase modulation interaction (Agrawal, 1987). As shown in Figure 4.40, modulation instability sidebands were observed only in the presence of stimulated Raman scattering light. It has recently been demonstrated that cross-phase modulation is intrinsic to the stimulated Raman scattering process (see Section 4). Therefore, sideband features observed in Figures 4.40 and 4.41 are conclusively a result of the cross-phase modulation induced by the simultaneously generated Raman pulses. To rule out the possibility of a multimode or single-mode stimulated four-photon mixing process as the origin of the sidebands, Baldeck et al. note that the fiber is truly single-mode (cutoff wavelength at 462 nm) and that the sideband separation changes with the fiber length.

The strengthen the conclusion that the sidebands are due to modulation instability induced by cross-phase modulation, Baldeck et al. measured and compared with theory the dependence of sideband shifts on the fiber lengths. For this measurement, they used 10-ns pulses from the Q-switched Nd:YAG laser to ensure quasi-CW operation. The spectra were similar to those obtained with 25-ps pulses (Figure 4.40). As shown in Figure 4.42, the sidelobe separation, defined as the half-distance between sideband maxima, varied from 1.5 to 8.5 nm for fiber lengths ranging from 4 to 0.1 m, respectively. The energy of input pulses was set at approximately the modulation instability threshold for each fiber length. The solid line in Figure 4.42 corresponds to

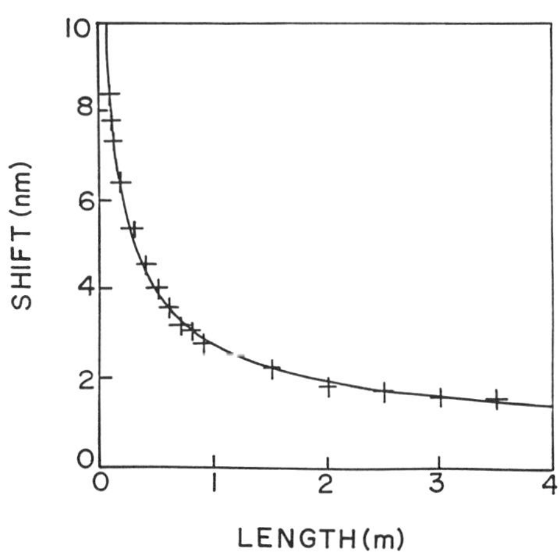

FIGURE 4.42. Sideband shifts versus fiber length near the modulation instability threshold. The time duration of input pulses is 10 ns. Crosses are experimental points. The solid line is the theoretical fit from Eq. (25). (From Baldeck et al. 1988d-1989.)

the theoretical fit. As discussed in Agrawal (1987), the maximum gain of modulation instability sidebands is given by $g_{\max} = k''\Omega_m^2$, where $\Omega_m = 2\pi f_m$ is the sideband shift. Thus, the power of a sideband for an optical fiber length L is given by

$$P(\Omega_m, L) = P_{\text{noise}} \exp(k''\Omega_m^2 L), \tag{23}$$

where P_{noise} is the initial spontaneous noise and $k'' = \partial(v_g)^{-1}/\partial\omega$ is the group velocity dispersion at the laser frequency.

For such amplified spontaneous emission, it is common to define a threshold gain g_{th} by

$$P_{\text{th}}(L) = P_{\text{noise}} \exp(g_{\text{th}}), \tag{24}$$

where P_{th} is the sideband power near threshold such that each sideband contains about 10% of the input energy. A typical value for g_{th} is 16 (Tai et al., 1986).

From Eqs. (23) and (24) the dependence of the sideband shift on the fiber length near threshold is given by

$$\Omega_m = (g_{\text{th}}/k''L)^{1/2}. \tag{25}$$

At $\lambda = 532$ nm, the group velocity dispersion is $k'' \approx 0.06\ \text{ps}^2/\text{m}$. The theoretical fit shown in Figure 4.42 (solid line) is obtained using this value and $g_{\text{th}} = 18.1$ in Eq. (25). The good agreement between the experimental data and the theory of modulation instability supports the belief of Baldeck et al. that they have observed cross-phase modulation-induced modulation instability, as predicted in Agrawal (1987).

Tai et al. have shown that modulation instability leads to the breakup of long quasi-CW pulses in trains of picosecond subpulses. The data in Figure 4.42 show that the maximum sideband shift is $\Delta\lambda_{\max} \approx 8.5$ nm or 8.5 THz, which corresponds to the generation of femtosecond subpulses within the envelope of the 10-ns input pulses with a repetition time of 120 fs. Even though autocorrelation measurements were not possible because of the low repetition rate (10 Hz) needed to generate pulses with kilowatt peak powers, Baldeck et al. believe they have generated for the first time modulation instability subpulses shorter than 100 fs.

Baldeck et al. (1988b) observed modulation instability in the normal dispersion regime of optical fibers. Modulation instability sidebands appear about the pump frequency as a result of cross-phase modulation induced by the simultaneously generated Raman pulses. Sideband frequency shifts were measured for many fiber lengths and found to be in good agreement with theory. In this experiment, cross-phase modulation originated from an optical wave generated inside the nonlinear medium, but similar results are expected when both waves are incident externally. Modulation instability induced by cross-phase modulation represents a new kind of modulation instability that not only occurs in normally dispersive materials but also, most important, has the potential to be controlled in real time by switching on or off the copropagating pulse responsible for the cross-phase modulation. Using optical fibers, such modulation instabilities could lead to the design of a novel source of femtosecond pulses at visible wavelengths.

9. Applications of Cross-Phase Modulation for Ultrashort Pulse Technology

Over the last 20 years, picosecond and femtosecond laser sources have been developed. Researchers are now investigating new applications of the unique properties of these ultrashort pulses. The main efforts are toward the design of communication networks and optical computers with data streams in, eventually, the tens of terahertz. For these high repetition rates, electronic componants are too slow and all-optical schemes are needed. The discovery of cross-phase modulation effects on ultrashort pulses appears to be a major breakthrough toward the real-time all-optical coding/decoding of such short pulses. As examples, this section describes the original schemes for a frequency shifter, a pulse compression switch, and a spatial light deflector. These all-optical devices are based on spectral, temporal, and spatial effects of cross-phase modulation on ultrashort pulses.

The first XPM-based technique to control ultrashort pulses was developed in the early 1970s. It is the well-known optical Kerr gate, which is shown in Figure 4.43. A probe pulse can be transmitted through a pair of cross-polarizers only when a pump pulse induces the (cross-) phase (modulation) needed for the change of polarization of the probe pulse. The principle of the optical Kerr gate was demonstrated using nonlinear liquids (Shimizu and Stoicheff, 1969; Duguay and Hansen, 1969) and optical fibers (Stolen and Ashkin, 1972; Dziedzic et al., 1981; Ayral et al., 1984). In optical fibers, induced-phase effects can be generated with milliwatt peak powers because of their long interaction lengths and small cross sections (White et al., 1988). XPM effects in optical fibers have been shown to alter the transmission of

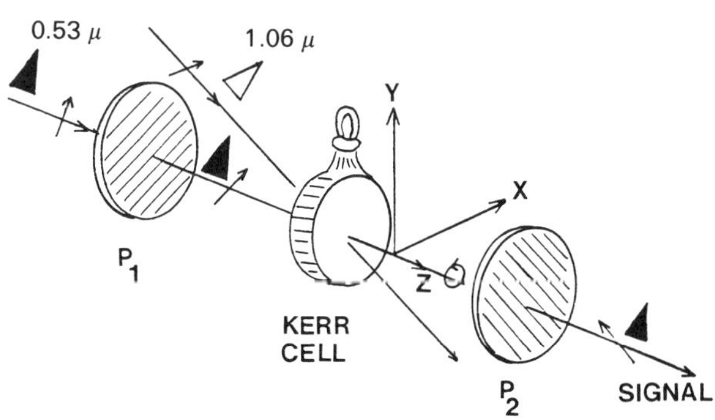

FIGURE 4.43. Schematic diagram of an optical Kerr gate.

frequency multiplexed signals (Chraplyvy et al., 1984) and also to allow quantum nondemolition measurements (Levenson et al., 1986; Imoto et al., 1987). In addition, phase effects arising from XPM have been used to make all-fiber logic gates (Kitayama et al., 1985a), ultrafast optical multi/demultiplexers (Morioka et al., 1987), and nonlinear interferometers (Monerie and Durteste, 1987).

The novelty of our most recent work was to show that XPM leads not only to phase effects but also to spectral, temporal, and spatial effects on ultrashort pulses. New schemes for XPM-based optical signal processors are proposed in Figure 4.44. The design of an ultrafast frequency shifter is shown in Figure 4.44a. It is based on spectral changes that occur when pulses copropagate in a nonlinear dispersive medium. In the absence of a pump pulse, the weak signal pulse passes undistorted through the nonlinear medium. When the signal pulse copropagates in the nonlinear medium with a pump pulse, its carrier wavelength can be changed by an amount $\Delta\lambda$ that is linearly proportional to the peak power of the pump pulse (see Section 3.2). Thus, in Figure 4.44 the signal pulses S1 and S2 have their frequencies shifted by $\Delta\lambda_1$ and $\Delta\lambda_2$ by the pump pulses P1 and P2, while S3 is not affected by the stream of pump pulses.

The design of a pulse-compression switch is proposed in Figure 4.44b. It is a modified version of the usual optical fiber/grating-pair pulse compression scheme (see Chapter 9 by Dorsinville et al. and Chapter 10 by Johnson and Shank). First, the probe pulse is spectrally broadened by a copropagating pump pulse in the nonlinear medium (case of negligible group velocity mismatch; see Sections 2.2 and 3.1). Then, or simultaneously, it is compressed in time by a dispersive element. Thus, in the presence of the pump pulse, the signal pulse is compressed ("on" state), while in its absence, the signal pulse is widely broadened ("off" state) by the device.

An example of an all-optical spatial light deflector based on spatial effects of XPM is shown in Figure 4.44c. In this scheme, the pump pulse profile leads to an induced focusing of the signal pulse through the induced nonlinear refractive index (Section 7). The key point in Figure 4.43c is that half of the pump pulse profile is cut by a mask, which leads to an asymmetric induced-focusing effect and a spatial deflection of the signal pulse. This effect is very similar to the self-deflection of asymmetric optical beams (Swartlander and Kaplan, 1988). In the proposed device, pump pulses originate from either path P1 or path P2, which have, respectively, their left side or right side blocked. Thus, if a signal pulse copropagates with a pump pulse from P1 or P2, it is deflected on, respectively, the right or left side of the nondeflected signal pulse.

The prime property of future XPM-based optical devices will be their switching speed. They will be controlled by ultrashort pulses that will turn on or off the induced nonlinearity responsible for XPM effects. With short pulses, the nonlinearity originates from the fast electronic response of the interacting material. As an example, the time response of electronic nonlinearity in optical fibers is about 2 to 4 fs (Grudinin et al., 1987). With such a response time, one

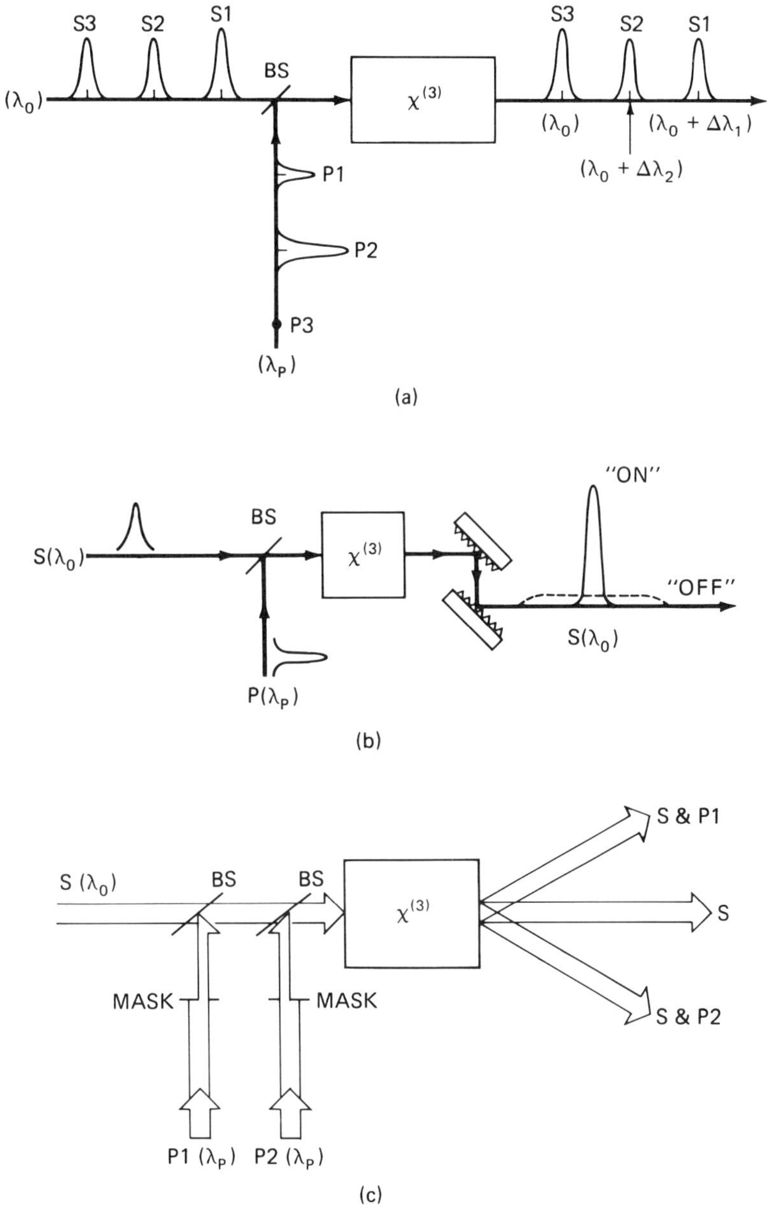

FIGURE 4.44. Schematic diagrams of ultrafast optical processors based on cross-phase modulation effects. (a) Ultrafast frequency shifter; (b) all-optical pulse compression switch; (c) all-optical spatial light deflector.

can envision the optical processing of femtosecond pulses with repetition rates up to 100 THz.

10. Conclusion

This chapter reviewed cross-phase modulation effects on ultrashort optical pulses. It presented XPM measurements that were obtained during the years 1986 to 1988. XPM is a newly identified physical phenomenon with important potential applications based on the picosecond and femtosecond pulse technology. XPM is similar to SPM but corresponds to the phase modulation caused by the nonlinear refractive index *induced by a copropagating pulse*. As for SPM, the time and space dependences of XPM lead to spectral, temporal, and spatial changes of ultrashort pulses.

Experimental investigations of cross-phase modulation effects began in 1986, when the spectral broadening enhancement of a probe pulse was reported for the first time. Subsequently, spectra of Raman, second harmonic, and stimulated four-photon mixing picosecond pulses were found to broaden with increasing pump intensities. Moreover, it was demonstrated that the spectral shape of Raman pulses was affected by the pulse walk-off, that the frequency of copropagating pulses could be tuned by changing the input time delay between probe and pump pulses, and that modulation instability could be obtained in the normal dispersion regime of optical fibers. All these results are well understood in terms of the XPM theory. Furthermore, induced focusing of Raman pulses, which was recently observed in optical fibers, was explained as a spatial effect of XPM.

The research trends are now toward more quantitative comparisons between measurement and theory and the development of XPM-based applications. Future experiments should clarify the relative contributions of SPM, XPM, and modulation instability to the spectral broadening of Raman, second harmonic, and stimulated four-photon mixing pulses. As, XPM appears to be a new tool for controlling (with the fast femtosecond time response of electronic nonlinearities) the spectral, temporal, and spatial properties of ultrashort pulses. Applications could include the frequency tuning in real time of picosecond pulses, the compression of weak pulses, the generation of femtosecond pulse trains from CW beams by XPM-induced modulation instability, and the spatial scanning of ultrashort pulses. The unique controllability of XPM should open up a broad range of new applications for the supercontinuum laser source.

Experiments on induced- and cross-phase modulations have been performed by the authors in close collaboration with T. Jimbo, Z. Li, Q.Z. Wang, D. Ji, and F. Raccah. Theoretical studies were undertaken in collaboration with J. Gersten and Jamal Manassah of the City College of New York and, most recently, with Govind P. Agrawal of AT&T Bell Laboratories.

We gratefully acknowledge partial support from Hamamatsu Photonics K.K.

11. Addendum

This chapter was written during the spring of 1988. Since then many more of new theoretical and experimental results on XPM effects have been or are being published by various research groups. The reference list in the introduction section of this chapter has been updated. The interested readers should refer themselves to original reports in the most recent issues of optics and applied physics publications.

References

Agrawal, G.P. (1987) Modulation instability induced by cross-phase modulation. Phys. Rev. Lett. **59**, 880–883.

Agrawal, G.P. and M.J. Potasek (1986) Nonlinear pulse distortion in single-mode optical fibers at the zero-dispersion wavelength. Phys. Rev. **3**, 1765–1776.

Agrawal, G.P., P.L. Baldeck, and R.R. Alfano (1988) Optical wave breaking and pulse compression due to cross-phase modulation in optical fibers. Conference abstract #MW3, in *Digest of the 1988 OSA annual meeting.* Optical Society of America, Washington, D.C. Opt. Lett. **14**, 137–139 (1989).

Agrawal, G.P., P.L. Baldeck, and R.R. Alfano (1989a) Temporal and spectral effects of cross-phase modulation on copropagating ultrashort pulses in optical fibers. Submitted for publication in Phys. Rev. A.

Agrawal, G.P., P.L. Baldeck, and R.R. Alfano (1989b) Modulation instability induced by cross-phase modulation in optical fibers. Phys. Rev. A (April 1989).

Alfano, R.R. and P.P. Ho (1988) Self-, cross-, and induced-phase modulations of ultrashort laser pulse propagation. IEEE J. Quantum Electron. **24**, 351–364.

Alfano, R.R., and S.L. Shapiro (1970) Emission in the region 4000–7000 Å via four-photon coupling in glass. Phys. Rev. Lett. **24**, 584–587. Observation of self-phase modulation and small scale filaments in crystals and glasses. Phys. Rev. Lett. **24**, 592–594.

Alfano, R.R., Q. Li, T. Jimbo, J.T. Manassah, and P.P. Ho (1986) Induced spectral broadening of a weak picosecond pulse in glass produced by an intense ps pulse. Opt. Lett. **11**, 626–628.

Alfano, R.R., Q.Z. Wang, T. Jimbo, and P.P. Ho (1987a) Induced spectral broadening about a second harmonic generated by an intense primary ultrafast laser pulse in ZnSe crystals. Phys. Rev. **A35**, 459–462.

Alfano, R.R., P.L. Baldeck, F. Raccah, and P.P. Ho (1987b) Cross-phase modulation measured in optical fibers. Appl. Opt. **26**, 3491–3492.

Alfano, R.R., P.L. Baldeck, and P.P. Ho (1988) Cross-phase modulation and induced-focusing of optical nonlinearities in optical fibers and bulk materials. Conference abstract #ThA3, In *Digest of the OSA topical meeting on nonlinear optical properties of materials.* Optical Society of America, Washington, D.C.

Auston, D.H. (1977) In *Ultrafast Light Pulses* S.L. Shapiro, ed. Springer-Verlag, Berlin, 1977.

Ayral, J.L., J.P. Pochelle, J. Raffy, and M. Papuchon (1984) Optical Kerr coefficient measurement at 1.15 μm in single-mode optical fibers. Opt. Commun. **49**, 405–408.

Baldeck, P.L. and R.R. Alfano (1987) Intensity effects on the stimulated four-photon spectra generated by picosecond pulses in optical fibers. Conference abstract #FQ7,

March meeting of the American Physical Society, New York, New York, 1987; J. Lightwave Technol. L.T-5, 1712–1715.

Baldeck, P.L., F. Raccah, and Alfano R.R. (1987a) Observation of self-focusing in optical fibers with picosecond pulses. Opt. Lett. **12**, 588–589.

Baldeck, P.L., P.P. Ho, and R.R. Alfano (1987b) Effects of self, induced-, and cross-phase modulations on the generation of picosecond and femtosecond white light supercontinua. Rev. Phys. Appl. **22**, 1677–1694.

Baldeck, P.L., P.P. Ho, and R.R. Alfano (1987c) Experimental evidences for cross-phase modulation, induced-phase modulation and self-focusing on picosecond pulses in optical fibers. Conference abstract #TuV4, in *Digest of the 1987 OSA annual meeting*. Optical Society of America, Washington, D.C.

Baldeck, P.L., F. Raccah, R. Garuthara, and R.R. Alfano (1987d) Spectral and temporal investigation of cross-phase modulation effects on picosecond pulses in singlemode optical fibers. Proceeding paper #TuC4, International Laser Science conference ILS-III, Atlantic City, New Jersey, 1987.

Baldeck, P.L., R.R. Alfano, and G.P. Agrawal (1988a) Induced-frequency shift of copropagating pulses. Appl. Phys. Lett. **52**, 1939–1941.

Baldeck, P.L., R.R. Alfano, and G.P. Agrawal (1988b) Observation of modulation instability in the normal dispersion regime of optical fibers. Conference abstract #MBB7, in *Digest of the 1988 OSA annual meeting*. Optical Society of America, Washington, D.C.

Baldeck P.L., R.R. Alfano, and G.P. Agrawal (1988c) Induced-frequency shift, induced spectral broadening and optical amplification of picosecond pulses in a single-mode optical fiber. Proceeding paper #624, Electrochemical Society symposium on non-linear optics and ultrafast phenomena, Chicago, Illinois, 1988.

Baldeck, P.L., R.R. Alfano, and G.P. Agrawal (1988d) Generation of sub-100-fsec pulses at 532 nm from modulation instability induced by cross-phase modulation in single-mode optical fibers. Proceeding paper #PD2, in *Ultrafast Phenomena 6*. Springer-Verlag, Berlin.

Baldeck, P.L., R.R. Alfano (1989) Cross-phase modulation: a new technique for controlling the spectral, temporal and spatial properties of ultrashort pulses. SPIE Proceedings of the 1989 Optical Science Engineering conference, Paris, France.

Chraplyvy, A.R. and J. Stone (1984) Measurement of cross-phase modulation in coherent wavelength-division multiplexing using injection lasers. Electron. Lett. **20**, 996–997.

Chraplyvy, A.R., D. Marcuse and P.S. Henry (1984) Carrier-induced phase noise in angle-modulated optical-fiber systems. J. Lightwave Technol. **LT-2**, 6–10.

Cornelius, P. and L. Harris (1981) Role of self-phase modulation in stimulated Raman scattering from more than one mode. Opt. Lett. **6**, 129–131.

Dianov, E.M., A.Y. Karasik, P.V. Mamyshev, G.I. Onishchukov, A.M. Prokhorov, M.F. Stel'Makh, and A.A. Formichev (1984) Picosecond structure of the pump pulse in stimulated Raman scattering in optical fibers. Opt. Quantum Electron. **17**, 187.

Duguay, M.A. and J.W. Hansen (1969) An ultrafast light gate. Appl. Phys. Lett. **15**, 192–194.

Dziedzic, J.M., R.H. Stolen, and A. Ashkin (1981) Optical Kerr effect in long fibers. Appl. Opt. **20**, 1403–1406.

French, P.M.W., A.S.L. Gomes, A.S. Gouveia-Neto, and J.R. Taylor (1986) Picosecond stimulated Raman generation, pump pulse fragmentation, and fragment compression in single-mode optical fibers. IEEE J. Quantum Electron. **QE-22**, 2230.

Gersten J., R.R. Alfano, and M. Belic (1980) Combined stimulated Raman scattering and continuum self-phase modulation. Phys. Rev. **A # 21**, 1222–1224.

Gomes, A.S.L., W. Sibbet, and J.R. Taylor (1986) Spectral and temporal study of picosecond-pulse propagation in a single-mode optical fibers. Appl. Phys. **B # 39**, 44–46.

Gomes, A.S.L., V.L. da Silva, and J.R. Taylor (1988) Direct measurement of nonlinear frequency chirp of Raman radiation in single-mode optical fibers using a spectral window method. J. Opt. Soc. Am. **B # 5**, 373–380.

Gouveia-Neto, A.S., M.E. Faldon, A.S.B. Sombra, P.G.J. Wigley, and J.R. Taylor (1988a) Subpicosecond-pulse generation through cross-phase modulation-induced modulation instability in optical fibers. Opt. Lett. **12**, 901–906.

Gouveia-Neto, A.S., M.E. Faldon, and J.R. Taylor (1988b) Raman amplification of modulation instability and solitary-wave formation. Opt. Lett. **12**, 1029–1031.

Grudinin, A.B., E.M. Dianov, D.V. Korobkin, A.M. Prokhorov, V.N. Serkinand, and D.V. Khaidarov (1987) Decay of femtosecond pulses in single-mode optical fibers. Pis'ma Zh. Eksp. Teor. Fiz. **46**, 175–177. [Sov. Phys. JETP Lett. **46**, 221, 225.]

Hasegawa, A. (1975). *Plasma Instabilities and Nonlinear Effects.* Springer-Verlag, Heidelberg.

Ho, P.P., Q.Z. Wang, D. Ji, and R.R. Alfano (1988) Propagation of harmonic cross-phase-modulation pulses in ZnSe. Appl. Phys. Lett. 111–113.

Hook, A.D. Anderson, and M. Lisak (1988) Soliton-like pulses in stimulated Raman scattering. Opt. Lett. **12**, 114–116.

Imoto, N., S. Watkins, and Y. Sasaki (1987) A nonlinear optical-fiber interferometer for nondemolition measurement of photon number. Optics Commun. **61**, 159–163.

Islam, M.N., L.F. Mollenauer, R.H. Stolen (1986) Fiber Raman amplification soliton laser, in *Ultrafast Phenomena 5*. Springer-Verlag, Berlin.

Islam, M.N., L.F. Mollenauer, R.H. Stolen, J.R. Simson, and H.T. Shang (1987a) Cross-phase modulation in optical fibers. Opt. Lett. **12**, 625–627.

Islam, M.N., L.F. Mollenauer, R.H. Stolen, J.R. Simson, and H.T. Shang (1987b) Amplifier/compressor fiber Raman lasers. Opt. Lett. **12**, 814–816.

Jaskorzynska, B. and D. Schadt (1988) All-fiber distributed compression of weak pulses in the regime of negative group-velocity dispersion. IEEE J. Quantum Electron. **QE-24**, 2117–2120.

Johnson, A.M., R.H. Stolen, and W.M. Simpson (1986) The observation of chirped stimulated Raman scattered light in fibers. In *Ultrafast Phenomena 5*. Springer-Verlag, Berlin.

Keiser, G. (1983) In *Optical Fiber Communications.* McGraw-Hill, New York.

Kelley, P.L. (1965) Self-focusing of optical beams. Phys. Rev. Lett. **15**, 1085.

Kimura, Y., K.I. Kitayama, N. Shibata, and S. Seikai (1986) All-fibre-optic logic "AND" gate. Electron. Lett. **22**, 277–278.

Kitayama, K.I., Y. Kimura, and S. Seikai (1985a) Fiber-optic logic gate. Appl. Phys. Lett. **46**, 317–319.

Kitayama, K.I., Y. Kimura, K. Okamoto, and S. Seikai (1985) Optical sampling using an all-fiber optical Kerr shutter. Appl. Phys. Lett. **46**, 623–625.

Levenson, M.D., R.M. Shelby, M. Reid, and D.F. Walls (1986) Quantum nondemolition detection of optical quadrature amplitudes. Phys. Rev. Lett. **57**, 2473–2476.

Lin, C. and M.A. Bosh (1981) Large Stokes-shift stimulated four-photon mixing in optical fibers. Appl. Phys. Lett. **38**, 479–481.

Lu, Hian-Hua, Yu-Lin Li, and Jia-Lin Jiang (1985) On combined self-phase modulation and stimulated Raman scattering in fibers. Opt. Quantum Electron. **17**, 187.

Manassah, J.T. (1987a) Induced phase modulation of the Raman pulse in optical fibers. Appl. Opt. **26**, 3747–3749.

Manassah, J.T. (1987b) Time-domain characteristics of a Raman pulse in the presence of a pump. Appl. Opt. **26**, 3750–3751.

Manassah, J.T. (1987c) Amplitude and phase of a pulsed second-harmonic signal. J. Opt. Soc. Am. **B # 4**, 1235–1240.

Manassah, J.T. (1988) Pulse compression of an induced-phase modulated weak signal. Opt. Lett. **13**, 752–755.

Manassah, J.T. and O.R. Cockings (1987) Induced phase modulation of a generated second-harmonic signal. Opt. Lett. **12**, 1005–1007.

Manassah, J.T., M. Mustafa, R.R. Alfano, and P.P. Ho (1985) Induced supercontinuum and steepening of an ultrafast laser pulse. Phys. Lett. **113A**, 242–247.

Monerie, M. and Y. Durteste (1987) Direct interferometric measurement of nonlinear refractive index of optical fibers by cross-phase modulation. Electron. Lett. **23**, 961–962.

Morioka, T., M. Saruwatari, and A. Takada (1987) Ultrafast optical multi/demultiplexer utilising optical Kerr effect in polarisation-maintaining single-mode optical fibers. Electron. Lett. **23**, 453–454.

Nakashima, T., M. Nakazawa, K. Nishi, and H. Kubuta (1987) Effect of stimulated Raman scattering on pulse-compression characteristics. Opt. Lett. **12**, 404–406.

Schadt, D., B. Jaskorzynska, and U. Osterberg (1986) Numerical study on combined stimulated Raman scattering and self-phase modulation in optical fibers influenced by walk-off between pump and Stokes pulses. J. Opt. Soc. Am. **B # 3**, 1257–1260.

Schadt, D. and B. Jaskorzynska (1987a) Frequency chirp and spectra due to self-phase modulation and stimulated Raman scattering influenced by walk-off in optical fibers. J. Opt. Soc. Am. **B # 4**, 856–862.

Schadt, D. and B. Jaskorzynska (1987b) Generation of short pulses from CW light by influence of cross-phase modulation in optical fibres. Electron. Lett. **23**, 1091–1092.

Schadt, D. and B. Jaskorzynska (1988) Suppression of the Raman self-frequency shift by cross-phase modulation. J. Opt. Soc. Am. **B # 5**, 2374–2378.

Shen, Y.R. (1984) In *The Principles of Nonlinear Optics*. Wiley, New York.

Shimizu, F. and B.P. Stoicheff (1969) Study of the duration and birefringence of self-trapped filaments in CS_2. IEEE J. Quantum Electron. **QE-5**, 544.

Stolen, R.H. (1975) Phase-matched stimulated four-photon mixing. IEEE J. Quantum Electron. **QE-11**, 213–215.

Stolen, R.H. (1979) In *Nonlinear properties of Optical fibers*, S.E. Miller and A.G. Chynoweth, eds. Academic Press, New York, Chapter 5.

Stolen, R.H. and A. Ashkin (1972) Optical Kerr effect in glass waveguide. Appl. Phys. Lett. **22**, 294–296.

Stolen, R.H., M.A. Bosh, and C. Lin (1981) Phase matching in birefringent fibers. Opt. Lett. **6**, 213–215.

Stolen, R.H. and A.M. Johnson (1986) The effect of pulse walk-off on stimulated Raman scattering in optical fibers. IEEE J. Quantum Electron. **QE-22**, 2230.

Swartzlander, G.A., Jr., and A.E. Kaplan (1988) Self-deflection of laser beams in a thin nonlinear film. J. Opt. Soc. Am. **B5**, 765–768.

Tai, K., A. Hasegawa, and A. Tomita (1986) Observation of modulation instability in optical fibers. Phys. Rev. Lett. **56**, 135–138.

Tomlinson, W.J., R.H. Stolen, and A.M. Johnson (1985) Optical wave breaking of pulses in nonlinear optical fibers. Opt. Lett. **10**, 457–459.

Trillo, S., S. Wabnitz, E.M. Wright, and G.I. Stegeman (1988) Optical solitary waves induced by cross-phase modulation. Opt. Lett. **13**, 871–873.

Washio, K., K. Inoue, and T. Tanigawa (1980) Efficient generation near-IR stimulated light scattering in optical fibers pumped in low-dispersion region at 1.3 mm. Electron. Lett. **16**, 331–333.

Weiner, A.M., J.P. Heritage, and R.H. Stolen (1986) Effect of stimulated Raman scattering and pulse walk-off on self-phase modulation in optical fibers. In *Digest of the Conference on Lasers and Electro-Optics.* Optical Society of America, Washington, D.C., p. 246.

Weiner, A.M., J.P. Heritage, and R.H. Stolen (1988) Self-phase modulation and optical pulse compression influenced by stimulated Raman scattering in fibers. J. Opt. Soc. Am. **B5**, 364–372.

White, I.H., R.V. Penty, and R.E. Epworth (1988) Demonstration of the optical Kerr effect in an all-fibre Mach-Zehnder interferometer at laser diode powers. Electron. Lett. **24**, 172–173.

Zysset B. and H.P. Weber (1986) Temporal and spectral investigation of Nd:YAG pulse compression in optical fibers and its application to pulse compression. In *Digest of the Conference on Lasers and Electro-Optics.* Optical Society of America, Washington, D.C., p. 182.

5
Simple Models of Self-Phase and Induced-Phase Modulation

JAMAL T. MANASSAH

Introduction

Supercontinuum (Alfano and Shapiro, 1970) generation is the production of nearly continuous spectra by propagating intense picosecond and subpicosecond laser pulses through nonlinear media. Induced supercontinuum (Manassah et al., 1985; Alfano et al., 1986) is the superbroadening of the spectrum of a weak pulse due to the presence of a strong pulse propagating simultaneously with it in a nonlinear medium. These observable physical effects form the motivation for the study of self-phase and induced-phase modulation. This chapter examines, for some idealized simple models of the nonlinear material and incoming pulse, the amplitude, phase, geometric shape, and spectral distribution of an outgoing pulse on exiting from the nonlinear material for cases of both absent and present pump. The chapter is limited in scope and extent; it is confined to some analytical and semianalytical cases developed by the author and co-workers. Effects related to group velocity dispersion (GVD) are not generally included among our models.

To define the nomenclature for the different models, we next write the general form of Maxwell's equation in a nonlinear medium. The propagation of a monochromatic electromagnetic wave in a linear medium is described by the equation

$$\left(\nabla_T^2 + \frac{\partial^2}{\partial z^2}\right)\tilde{E}(\mathbf{r}, \omega) + k^2(\omega)\tilde{E}(\mathbf{r}, \omega) = 0.$$

Denoting, $k(\omega_0) \equiv k_0$, the above Maxwell's equation can be described for any ω near ω_0 to second order in the difference $(\omega - \omega_0)$ by

$$\left(\nabla_T^2\tilde{E} + \frac{\partial^2 E}{\partial z^2}\right) + [k_0^2 + 2k_0k_0'(\omega - \omega_0) + (k_0'^2 + k_0k_0'')(\omega - \omega_b)^2]\tilde{E} = 0,$$

where

$$k_0' = \left.\frac{dk}{d\omega}\right|_{\omega=\omega_0} \quad \text{and} \quad k'' = \left.\frac{d^2k}{d\omega^2}\right|_{\omega=\omega_0},$$

where k' is the group velocity and k'' is the group velocity dispersion. Consequently, for a pulse with center frequency ω_0, the electric field then obeys the equation (Tzoar and Jain, 1979)

$$\nabla_T^2 E(\mathbf{r}, t) + \frac{\partial^2 E(\mathbf{r}, t)}{\partial z^2} + \exp(-i\omega_0 t)\left[k_0^2 + 2k_0 k_0' i\left(\frac{\partial}{\partial t}\right) - (k_0'^2 + k_0 k_0'')\frac{\partial^2}{\partial_t^2}\right]$$
$$\times \exp(i\omega_0 t)E(\mathbf{r}, t) = 0,$$

where ω in Fourier space has been replaced by $[i(\partial/\partial t)]$ in the time domain. In normalized coordinates defined by

$$E = E_0\Phi, \qquad T = t/\tau, \qquad \mathbf{R}' = \mathbf{r}/c\tau,$$
$$W_0 = \omega_0\tau, \qquad K_0 = (c\tau)k_0,$$

where E_0 is the maximum pulse amplitude and τ is its width, Maxwell's equation for the function Φ can be written as

$$\frac{E_0}{c^2\tau^2}\left\{\nabla_T'^2\Phi + \frac{\partial^2}{\partial Z'^2}\Phi + \exp(-iW_0 T)\left[K_0^2 + 2K_0 K_0'\left(i\frac{\partial}{\partial T}\right)\right.\right.$$
$$\left.\left. - (K_0'^2 + K_0 K_0'')\frac{\partial^2}{\partial T^2}\right]\Phi\exp(iW_0 T)\right\} = 0,$$

where

$$K_0' = \left.\frac{dK}{dW}\right|_{W_0} = ck_0' \quad \text{and} \quad K_0'' = \left.\frac{d^2K}{dW^2}\right|_{W_0} = \frac{c}{\tau}k_0''.$$

The source term for Maxwell's equation in a nonlinear medium is given by (Bloembergen, 1965)

$$\mu_0\frac{\partial^2}{\partial t}P_{NL}$$

where P_{NL} is the nonlinear polarization ($\chi^{(2)}$ medium means that P_{NL} is proportional to E^2; $\chi^{(3)}$ medium means that P_{NL} is proportional to $|E|^2E$, etc.). In all phenomenological treatment (i.e., Sections 1 through 4), we assume the instantaneous form for the polarization, which implies physically that the pulse duration is much longer than the material relaxation time.

In the following, plane wave refers to the case where the ∇_T^2 term of Maxwell's equation is neglected, finite beam size refers to the case where ∇_T^2 effects are incorporated in the analysis, dispersionless refers to the case where the k_0'' term in Maxwell's equation is neglected, steepened pulse refers to the case where the time derivative of the envelope of the nonlinear polarization is kept (see also Chapter 6, by Suydam), and low-intensity pulse refers to the case where this steepening is negligible.

1. Low-Intensity Pulse in $\chi^{(3)}$ Dispersionless Medium

In this section we review the conventional self-phase modulation theory (Alfano and Shapiro, 1970) and generalize its results to the case of finite beam size. We show that the effects of diffraction can combine with those of self-phase modulation and self-focusing to exhibit many new features both in the shapes of the pulse and phase and in the spectral distribution.

1.1 Conventional Self-Phase Modulation Theory

This topic is covered in greater detail by Shen and Yang (Chapter 1 and Wang et al., Chapter 2), but for completeness we will quickly review it. The simplified nonlinear wave equation for the electric field envelope is given by

$$\frac{\partial A}{\partial z} + \frac{1}{v_g}\frac{\partial A}{\partial t} = i\frac{\omega}{2c} n_2 |A|^2 A, \tag{1}$$

where v_g is the group velocity and n_2 is the Kerr index of refraction. In this equation we neglected group velocity dispersion, self-steepening (De Martini et al., 1967), and absorption and we assumed that $\chi^{(3)}$ is instantaneous (i.e., we neglect all effects associated with nonzero relaxation time). Denoting the amplitude and phase of the envelope A by a and α, respectively, their differential equations are given by

$$\frac{\partial a}{\partial z} + \frac{1}{v_g}\frac{\partial a}{\partial t} = 0, \tag{2}$$

$$\frac{\partial \alpha}{\partial z} + \frac{1}{v_g}\frac{\partial \alpha}{\partial t} = \frac{\omega}{2c} n_2 a^2, \tag{3}$$

and the solutions are

$$a(\bar{u}) = a_0 F(\bar{u}), \tag{4}$$

$$\alpha(z, \bar{u}) = \frac{\omega_0}{2c} n_2 a_0^2 F^2(\bar{u}) z, \tag{5}$$

where $\bar{u}$ is the pulse comoving coordinate defined by

$$\bar{u} = t - z/v_g, \tag{6}$$

a_0 is the amplitude at $z = 0$, and F is the pulse shape form function. The above solution assumes that $\alpha(z = 0) = 0$; otherwise this initial phase should be added to the solution of Eq. (5). It should be noted that in this approximation the amplitude of the pulse is unmodified and only the phase of the pulse is affected, hence the name self-phase modulation. At this point, we refer the reader to other articles (Alfano, 1972 and Chapter 2) for the analysis of the experimental consequences of the above solutions.

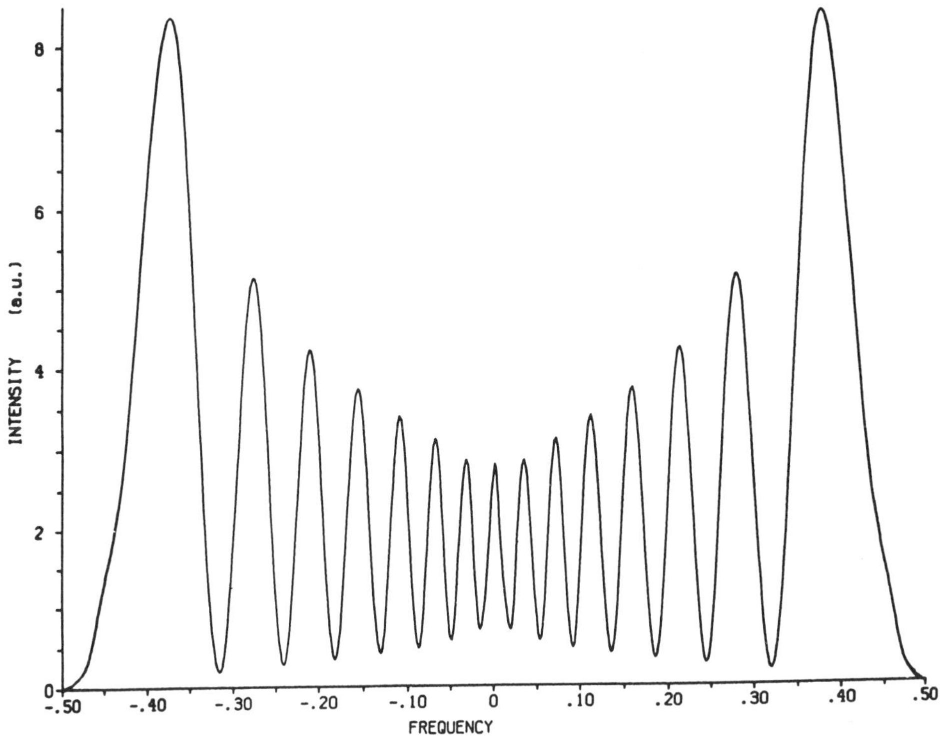

FIGURE 5.1. A typical spectral distribution predicted by the conventional self-phase modulation theory.

The spectral extent of a Gaussian pulse associated with the above supercontinuum is

$$\Delta\omega \approx \frac{\omega_0}{c}\frac{n_2 a_0^2 z}{\tau}, \tag{7}$$

where τ is the incoming pulse width. The spectral distribution is symmetric (see fig. 5.1) and its modulation period is

$$\delta\omega \approx 4\pi/\tau. \tag{8}$$

The linear chirp coefficient for the self-phase modulated pulse in the above approximation is given for a Gaussian pulse by

$$|m| = \frac{\omega_0}{2c}\frac{n_2 a_0^2 z}{\tau^2}, \tag{9}$$

where the coefficient m is related to the curvature of the phase at its extremum, that is,

$$\alpha|_{\text{extremum}} = m\bar{u}^2. \tag{10}$$

1.2 Grating Transform and Pulse Compression

The frequency sweep in the self-phase modulated pulse is used as the first step in the fiber-grating compressor (Tomlinson et al., 1984; Grischkowsky and Balant, 1986). In this section, we find the characteristics of the grating that optimizes the compressor for a self-phase modulated signal in the same approximations as in Section 1.1. Other details are given in Chapter 10 by Johnson and Shank. To find the grating transform of an incoming pulse we use the techniques of Fourier optics (Goodman, 1968). Essentially, in the Fourier domain, the outgoing signal is equal to the incoming signal multiplied by the optical element transfer function (Gaskill, 1978). The transfer function of a grating is described by (Treacy, 1969; Martinez et al., 1984)

$$H(\omega) = \exp[i\Phi(\omega)], \tag{11}$$

where

$$\Phi(\omega) = c(\omega - \omega_g)^2, \tag{12}$$

that is, the transfer function is a pure phase. Using the properties of the Fourier transform, the outgoing pulse can be written as the convolution of the incoming pulse with the time representation of the grating transfer function. Thus, the outgoing pulse electric field can be written as a function of the incoming pulse amplitude and phase as

$$\begin{aligned} E_{\text{out}}(\bar{U}) = \exp[iK\bar{U}]\frac{(1+i)}{(8\pi g)^{1/2}} \int_{-\infty}^{\infty} d\bar{U}'\, a(\bar{U} - \bar{U}')\exp[i\alpha(\bar{U} - \bar{U}')] \\ \times \exp\left[-\frac{i\bar{U}'^2}{4g}\right] d\bar{U}', \end{aligned} \tag{13}$$

where K is the incoming pulse center frequency normalized to the pulse duration (i.e., $K = \omega\tau$) and $g = c/\tau^2$.

For an incoming Gaussian pulse that is weakly self-phase modulated, the amplitude and the phase are given by (4) and (5) but the integration in Eq. (13) with these expressions cannot be performed except numerically. However, an approximation to this integral can be obtained by approximating the phase to a Gaussian as given by Eq. (10). Under these conditions the outgoing pulse can be approximated by

$$\begin{aligned} E_{\text{out}}(\bar{U}) = \exp[iK\bar{U}]\exp\left[-\frac{i\bar{U}^2}{4g}\right]\frac{(1+i)}{(8g)^{1/2}} \\ \times \left\{\frac{E_0}{[1/4 + (m - 1/4g)^2]^{1/4}} \exp\left[-\frac{\bar{U}^2}{2(4g)^2}\frac{1}{1/4 + (m - 1/4g)^2}\right]\right\}. \end{aligned} \tag{14}$$

The optimum compression condition then can be directly deduced to give

$$g = m/(1 + 4p^2). \tag{15}$$

The amplitude scaling factor and the compression ratio under this optimization condition are respectively related to the parameter given by

$$S = (1 + 4m^2)^{1/4}. \tag{16}$$

The outgoing pulse (amplitude)^2 is then written as

$$|E_{\text{out}}|^2 = (SE_0)^2 \exp[-S^4 \bar{U}^2]. \tag{17}$$

The numerically derived $|E_{\text{out}}|^2$ actually deviates on the wings from Eq. (18) results, because of the deviation of the phase from the approximate parabolic shape assumed in the derivation of (17). Experimentally, these deviations were compensated for by the addition of prisms (Fork et al., 1987) that correct for cubic terms in the exact phase expression.

1.3 Self-Focusing and Self-Phase Modulation in a Parabolic Graded Index Medium

In this section we examine the combined effects of Kerr nonlinearity, diffraction, and graded index waveguiding on the spatial and spectral profiles of an intense pulse propagating in a parabolic graded index material with the same axis of symmetry as the pulse (Manassah et al., 1988a). The beam transverse geometric shape, radius of curvature, phase, and spectrum are computed as functions of the material parameters and pulse peak powers. Approximate analytical results are derived for the beam waist radius of curvature and phase. This calculation is motivated by two considerations:

1. Graded-index optical fibers (Thomas et al., 1982) are closely approximated by such materials.
2. The results for homogeneous bulk material in the presence of heating due to absorption of the laser beam can be closely approximated by this model for thermal positive lensing materials.

The approximations that we make in the following calculation are that

1. the graded index profile is approximated by

$$n = n_0(1 - Ar^2/2n_0);$$

2. the GRIN material boundary with other materials is at a distance larger than the beam radius (in the language of optical fibers this translates into neglecting the core-cladding boundary conditions) and the beam radius a is much smaller than the core radius r_c;
3. effects of group velocity dispersion are neglected;
4. the self-steepening (De Martini et al., 1967) of the amplitude is neglected (i.e., the time derivative of the nonlinear polarization is neglected);
5. the quadratic index of refraction n_2 is not modified by the radial variation in the ordinary index of refraction;

6. one component of the electric field is kept (i.e., we are neglecting the vector nature of the electic field);
7. the graded index of refraction is of the order required to guide the beam within the material transverse dimension; and
8. the spatial changes in the physical quantities over the light wavelength are unimportant. Under these assumptions, ε, the envelope of the electric field, obeys the equation

$$\nabla_T^2 \varepsilon - 2ik\varepsilon' - kk_2 r^2 \varepsilon + \frac{n_2 k^2}{n_0} |\varepsilon|^2 \varepsilon = 0, \tag{18}$$

where ∇_T^2 is the transverse component of the Laplacian, $\varepsilon' = \partial\varepsilon/\partial z$, $k_2 = kA = 2kn_0 \Delta/r_c^2$, and Δ is the relative index difference between the core center and the cladding.

If the initial condition for the incoming pulse is given by

$$\varepsilon(r, 0, \bar{u}) = \varepsilon_0 \exp\left(-\frac{r^2}{a^2}\right) \exp\left(-\frac{\bar{u}^2}{2\tau^2}\right), \tag{19}$$

that is, the initial pulse is assumed to be Gaussian both in the transverse plane and in the comoving coordinate system, $\bar{u} = (z/v_g - t)$, where v_g is the pulse group velocity, a the initial beam radius, τ the pulse duration, and ε_0 the magnitude of the pulse amplitude.

In the following, the product $\varepsilon_0 \exp(-\bar{u}^2/2\tau^2)$ is denoted by $\tilde{\varepsilon}_0$. Then an approximate solution to Eq. (18) with the boundary condition given by Eq. (19), correct to order r^2/a^2, can be obtained through the trial solution (Marburger, 1975)

$$\varepsilon(r, z, \bar{u}) = \frac{\tilde{\varepsilon}_0}{\bar{\omega}(z, \bar{u})} \exp\left[-\frac{r^2}{a^2 \bar{\omega}^2(z, \bar{u})} - \frac{ik}{2} \rho(z, \bar{u}) r^2 + ik\alpha(z, \bar{u})\right], \tag{20}$$

where the different functions can be interpreted as follows: $\bar{\omega}$ is the normalized beam radius, ρ is the inverse of the beam radius of curvature, and $k\alpha$ is the longitudinal phase on the material axis. This approximation of self-similarity of the beam is well justified for powers smaller than the critical power (i.e., the power at which self-focusing becomes possible), in particular for instances where $\bar{\omega}_{\max}$ and $\bar{\omega}_{\min}$ are not too far apart.

The equations satisfied by the subsidiary functions $\bar{\omega}$, ρ, and α are given by

$$\rho = \frac{1}{\bar{\omega}} \frac{\partial \bar{\omega}}{\partial z}, \tag{21}$$

$$\frac{\partial \alpha}{\partial z} = \frac{a^2}{2\bar{\omega}^2} [B - C], \tag{22}$$

$$\frac{\partial^2 \bar{\omega}}{\partial z^2} + A\bar{\omega} + [2C - B] \frac{1}{\bar{\omega}^3} = 0, \tag{23}$$

where

$$A = \frac{k_2}{k} \equiv \frac{1}{L_w^2}, \tag{24}$$

$$B = \frac{4}{a^4 k^2} \equiv \frac{1}{L_d^2}, \tag{25}$$

$$C = \frac{n_2 \tilde{\varepsilon}_0^2}{n_0 a^2}, \tag{26}$$

and L_w and L_d are the characteristic lengths for waveguiding and diffraction, respectively.

The solutions to Eqs (21), (22), and (23) satisfying the boundary conditions $\bar{\omega} = 1$, $\rho = 0$, and $\alpha = 0$ at $z = 0$ are

1. the normalized beam radius:

$$\bar{\omega} = [\beta \cos(\gamma z) + \delta]^{1/2}; \tag{27}$$

2. the inverse of the radius of curvature:

$$\rho = -\frac{1}{2} \frac{\beta\gamma \sin(\gamma z)}{[\beta \cos(\gamma z) + \delta]}; \tag{28}$$

3. the longitudinal phase:

$$k\alpha = \frac{ka^2(B - C)}{2[B - 2C]^{1/2}} \operatorname{arctg}\left[\left(\frac{B - 2C}{A}\right)^{1/2} \operatorname{tg}(A^{1/2} z)\right] \tag{29}$$

for $B - 2C \geq 0$ and

$$k\alpha = \frac{ka^2}{4} \frac{B - C}{[2C - B]^{1/2}} \ln \left\{ \frac{1 + \left(\frac{2C - B}{A}\right)^{1/2} \operatorname{tg}(A^{1/2} z)}{1 - \left(\frac{2C - B}{A}\right)^{1/2} \operatorname{tg}(A^{1/2} z)} \right\} \tag{30}$$

for $B - 2C \leq 0$, where

$$\gamma = 2A^{1/2}, \tag{31}$$

$$\beta = \frac{2C - B + A}{2A}, \tag{32}$$

$$\delta = \frac{A - 2C + B}{2A}. \tag{33}$$

For negative n_2 (i.e., defocusing medium), $\beta < \delta$ is satisfied for all values of ε and Eq. (29) is the solution for the longitudinal phase everywhere. For positive n_2 (i.e., focusing medium), $\beta = \delta$ for the critical field ε_c, which is given by (Kelley, 1965)

$$\varepsilon_c^2 = 2n_0/a^2 k^2 n_2. \tag{34}$$

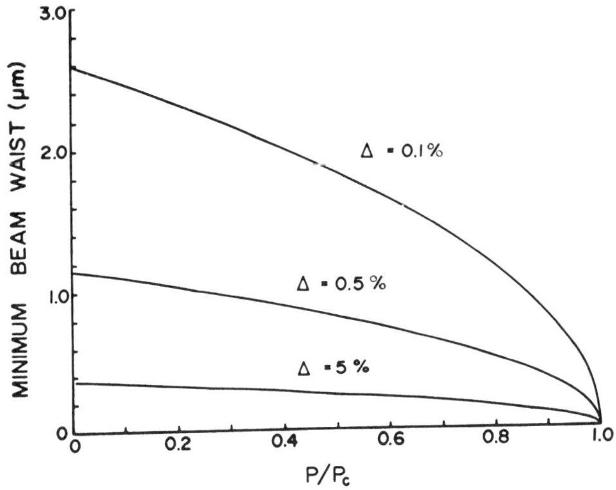

FIGURE 5.2. Minimum beam waist as a function of the normalized peak power P/P_c. Δ is the relative refraction index difference between the GRIN core center and the cladding. In silica fibers $P_c \approx 10^6$ W.

It should be noted that, in the low intensity limit, Eqs. (27), (28), and (29) lead to the usual results for Gaussian beams propagating in lenslike media (Kogelnik, 1965; Kogelnik and Li, 1966).

Equation (27) shows that the normalized beam radius $\bar{\omega}$ varies periodically along the optical axis length. The period of variation depends only on the waveguiding characteristic length L_w. The magnitude of $\bar{\omega}_{\text{min}}$, the minimum normalized beam radius, depends on all three characteristic lengths associated with A, B, and C. In Figure 5.2, the minimum beam radius is plotted as a function of the pulse normalized peak power P/P_c for different values of the graded index parameter Δ, where P_c is the critical power for self-focusing. The minimum beam radius decreases for increasing peak powers and collapses at $P = P_c$.

The inverse of the radius of curvature ρ is plotted in Figure 5.3 as a function of the normalized length $2z/\pi L_w$. For increasing peak powers, the curvature is clearly enhanced periodically by self-focusing at the beam waist (i.e., minimum beam diameter) locations.

The total phase of the electric field can be computed using Eqs. (20), (28), and (29). This phase, and consequently the spectral broadening arising from self-phase modulation, is radially dependent. It is worth noting that the time-dependent part of the phase ϕ', denoted ϕ'_t, which is equal to total phase $-\omega_0 t$, reduces to that of the conventional SPM theory in the case of a homogeneous medium ($k_2 = 0$) and for an incoming plane wave ($a \Rightarrow \infty$). Furthermore, for weak waveguiding ($k_2 \Rightarrow 0$), but finite initial beam diameter, the time-dependent phase ϕ'_t reduces to

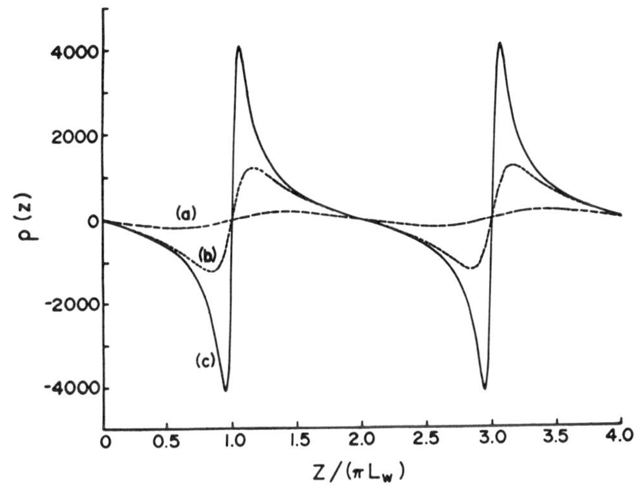

FIGURE 5.3. Inverse of radius of curvature as a function of normalized length $2z/\pi L_w$ for a graded-index fiber. ($a = 1.5$ μm, $\Delta = 0.48\%$, and $L_w = 15.3$ μm). (a) $P/P_c = 0.1$; (b) $P/P_c = 0.9$; (c) $P/P_c = 0.99$.

$$\lim_{k_2 \Rightarrow 0} \phi_t' = \phi_{\mathrm{SPM}}\left(1 - \frac{2r^2}{a^2}\right). \tag{35}$$

Thus, in this limit the spectral extent as a function of the radius varies as $(1 - 2r^2/a^2)$. This approximation is valid for $r/a \ll 1$ and reflects the $|\varepsilon|^2$ change.

In Figure 5.4, the longitudinal phase contribution $k\alpha(z, \bar{u} = 0)$, denoted the α-phase, is plotted as a function of the normalized material length for different power levels. As shown, the α-phase mostly increases by steps at z locations corresponding to the periodic positions of the minimum beam waist. As a result, the total amount of the longitudinal phase yielded by the pulse is often much larger than the usual SPM, and it depends strongly on the waveguiding, diffraction, and nonlinear parameters. It is worth noting that for $z \ll L_w$ the regularized α-phase, defined as the value of the α-phase at a certain power minus its value for zero intensity, has the same sign as ϕ_{SPM}; however, this sign changes for $z > L_w$. Physically, this result leads to the reverse of the red leading the blue in the supercontinuum and may have important consequences for soliton propagation in graded-index fibers and for pulse compression. Finally, one notices that the α-phase approaches a ladder function for values of ε that equalize B and $2C$ (i.e., $P \rightarrow P_c$).

The temporal distribution of the longitudinal phase, the α-phase, can be studied using Eq. (29). If we define the parameter p by

$$C = \frac{p}{2} B \exp\left(\frac{-\bar{u}^2}{\tau^2}\right) \tag{36}$$

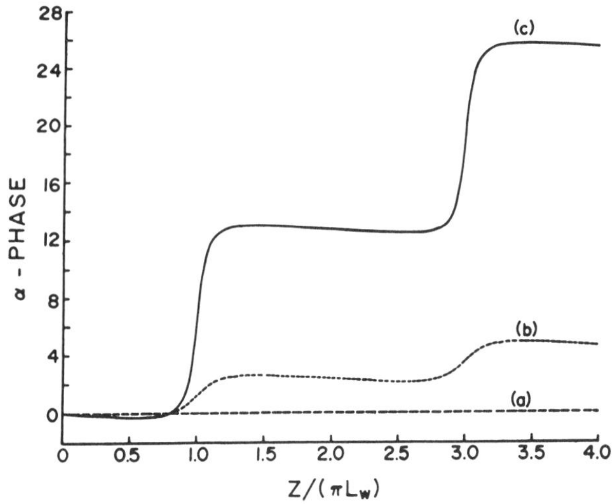

FIGURE 5.4. Longitudinal α-phase as a function of the material normalized length $2z/\pi L_w$ for a graded-index fiber. (a) $P/P_c = 0.1$; (b) $P/P_c = 0.9$; (c) $P/P_c = 0.99$.

where $p = 1$ corresponds to $P = P_c$, then

$$\frac{(\alpha\text{-phase})_{\text{reg.}}}{\phi_{\text{SPM}}(\bar{u}=0)} \approx \left(\frac{A}{B}\right)^{1/2} \frac{2}{p} \frac{[(1-p\exp(-\bar{u}^2/\tau^2))^{1/2} + (p/2)\exp(-\bar{u}^2/\tau^2) - 1]}{(1-p\exp(-\bar{u}^2/\tau^2))^{1/2}}. \tag{37}$$

It is worth noting that for small p, Eq. (37) can be approximated by

$$\lim_{p \ll 1} \frac{(\alpha\text{-phase})_{\text{reg.}}}{\phi_{\text{SPM}}(\bar{u}=0)} \approx -\frac{p}{4}\left(\frac{A}{B}\right)^{1/2} \exp\left(-\frac{2\bar{u}^2}{\tau^2}\right). \tag{38}$$

In Figure 5.5 the α-phase, normalized to its maximum value, is plotted as a function of time for different power levels. For small peak powers the width of the phase envelope is $\sqrt{2}$ smaller than the conventional SPM phase, as shown by Eq. (38). As the pulse peak power increases and tends to the critical power, the phase width significantly decreases, and the value of the phase at its maximum increases dramatically.

The spectral distribution of a pulse propagating in this medium can be obtained by taking the Fourier transform of the electric field and then taking its magnitude squared. It should be noted that all time dependence as given by Eq. (20) should be incorporated in the numerical calculation. Special care should be exercised at the beam waist, where $\bar{\omega}$ as a function of time features the same narrowing characteristics observed above for the longitudinal phase. At locations other than the waists, most of the pulse energy is not modulated. For small peak powers ($P \ll P_c$) the SPM spectra generated in parabolic GRIN material are similar to the conventional SPM broadened spectra, as observed in Figure 5.6a. On the other hand, for peak powers near the critical

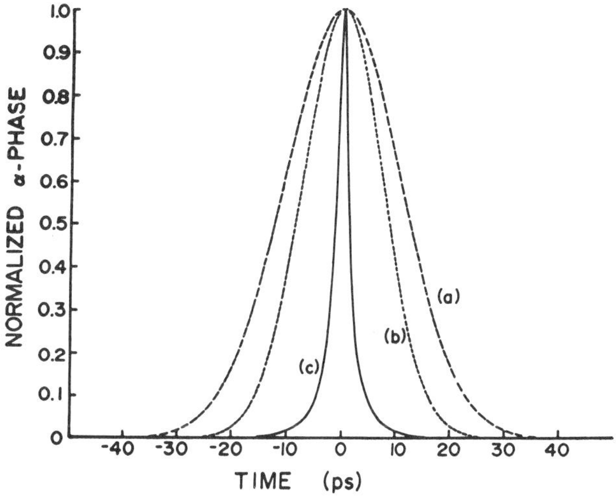

FIGURE 5.5. The α-phase normalized to its maximum value plotted versus time for a graded-index fiber. (a) Conventional SPM phase for $P/P_c = 0.1$; (b) $P/P_c = 0.1$; (c) $P/P_c = 0.992$. Maximum values of the phase are, respectively, 0.92, 0.55, and 1880 radians. ($a = 25\ \mu$m, $\Delta = 0.48\%_0$, $z = 10$ cm, and $\tau = 15$ ps.)

power for self-focusing ($P \approx P_c$) new SPM features appear; see Figure 5.6b. There is an intense peak at the laser wavelength over a much weaker background of white light. This effect is due to the narrowing of the α-phase and the limited modulation over the pulse duration.

In conclusion, the combined effects of self-focusing, diffraction, and waveguiding lead to novel features in both the time and frequency domains of the electric field of a pulse. Novel features in the SPM spectra, distinct from conventional SPM theory, appear.

1.4 Self-Focusing, Self-Phase Modulation, and Diffraction in Bulk Homogeneous Material

In this section we specialize the results of the previous section to $k_2 \Rightarrow 0$ (i.e., homogeneous bulk material) (Manassah et al., 1988b). We explore the simultaneous effects of self-focusing, self-phase modulation, and diffraction on the propagation of an ultrafast pulse in a homogeneous Kerr medium. The competing effects of self-focusing and diffraction are shown to modify the shape and magnitude of the pulse amplitude and phase. These modifications are shown to affect the spectral distribution of the supercontinuum compared to that predicted by conventional self-phase modulation theory.

In the case of a homogeneous medium ($A \Rightarrow 0$), the results of Section 1.3 can be written as

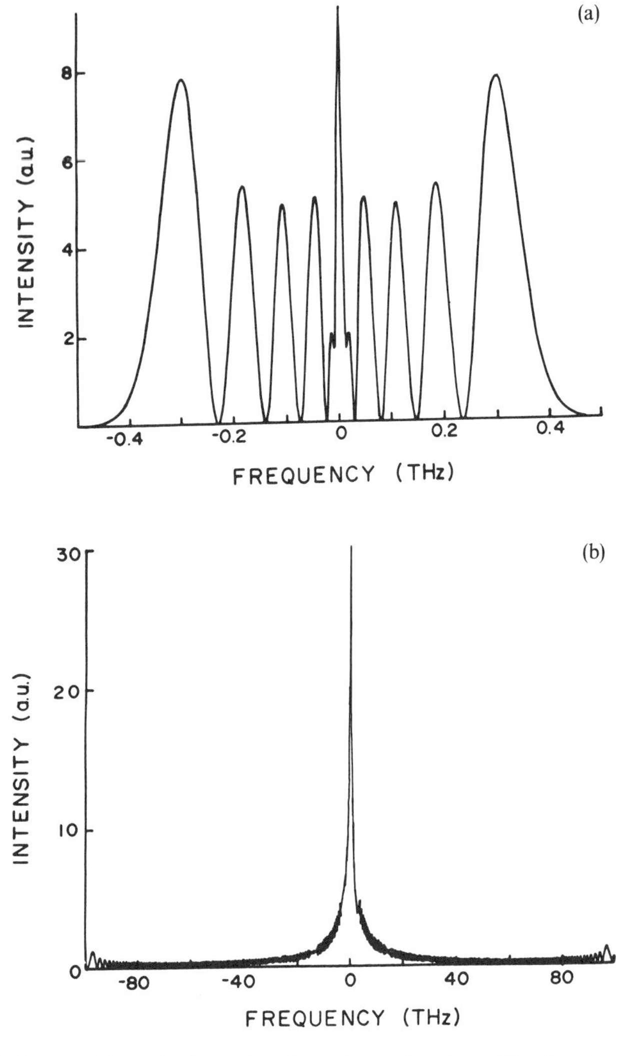

FIGURE 5.6. Spectral broadening of a Gaussian pulse ($\tau = 15$ ps) outgoing from a graded-index fiber ($a = 25\ \mu$m, $\Delta = 0.48\%$). (a) $P/P_c = 0.1$ and $z = 0.5$ m; (b) $P/P_c = 0.997$ and $z = 0.1$ m.

$$\bar{\omega} = [1 + y^2(1 - p')]^{1/2}, \tag{39}$$

$$L_d\rho = \frac{y(1 - p')}{[1 + y^2(1 - p')]}, \tag{40}$$

$$k\alpha = \frac{[1 - \frac{1}{2}p']}{[1 - p']^{1/2}} \operatorname{arctg}[(1 - p')^{1/2} y] \tag{41}$$

for $p' < 1$, and

$$k\alpha = \frac{[1 - \frac{1}{2}p']}{[p' - 1]^{1/2}} \ln \left\{ \frac{1 + y(p' - 1)^{1/2}}{1 - y(p' - 1)^{1/2}} \right\} \tag{42}$$

for $p' > 1$, where

$$p' = p \exp(-\bar{u}^2/\tau^2), \tag{43}$$

$p = \varepsilon_0^2/\varepsilon_c^2$, ε_c is the critical field for self-focusing, and y is the material length in units of the Rayleigh diffraction length L_d, that is,

$$y = z/L_d. \tag{44}$$

In the above notation, the normalized self-focusing distance (Kelley, 1965) is

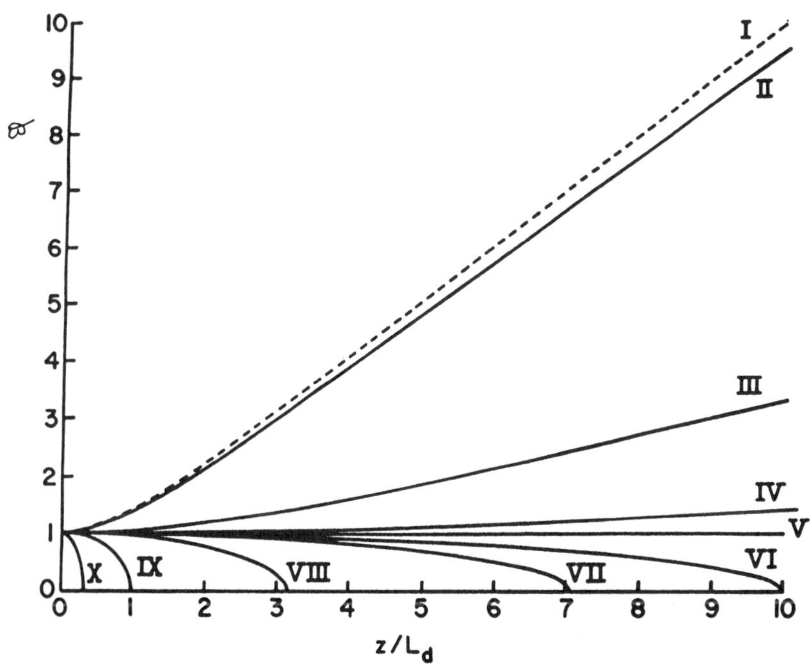

FIGURE 5.7. Normalized beam radius for a pulse propagating in a homogeneous medium at $u = 0$ as a function of the normalized length (z/L_d). (I) $p \ll 1$; (II) $p = 0.1$; (III) $p = 0.9$; (IV) $p = 0.99$; (V) $p = 1$; (VI) $p = 1.01$; (VII) $p = 1.02$; (VIII) $p = 1.1$; (IX) $p = 2$; (X) $p = 10$.

$$y_{\text{foc}} = 1/(p - 1)^{1/2}. \tag{45}$$

In the following, the analysis is limited to $y < y_{\text{foc}}$, where the above solutions are valid.

It should be noted that the solutions of $\bar{\omega}$, ρ, and α in the linear regime ($\varepsilon_0 \Rightarrow 0$) reduce to the standard formulas of a Gaussian pulse propagating in a homogeneous medium (Tien et al., 1965), specifically:

$$\bar{\omega} = (1 + y^2)^{1/2}, \tag{46}$$

$$\rho L_d = y/(1 + y^2), \tag{47}$$

$$k\alpha = \text{arctg}(y) \equiv \eta(z). \tag{48}$$

In Figure 5.7 the normalized beam radius is plotted for $\bar{u} = 0$ as a function of the normalized length (z/L_d) and for different electric field intensities [$p = \varepsilon_0^2/\varepsilon_c^2$]. For weak fields ($p < 1$) diffraction is the dominant effect, while for the most intense fields ($p > 1$) the nonlinearity is dominant, there is self-focusing, and $\bar{\omega}$ approaches zero as the length approaches the self-focusing distance.

In Figure 5.8 the normalized inverse radius of curvature ($L_d/R = \rho L_d$) is plotted for $\bar{u} = 0$ as a function of the normalized length. For $p < 1$, this

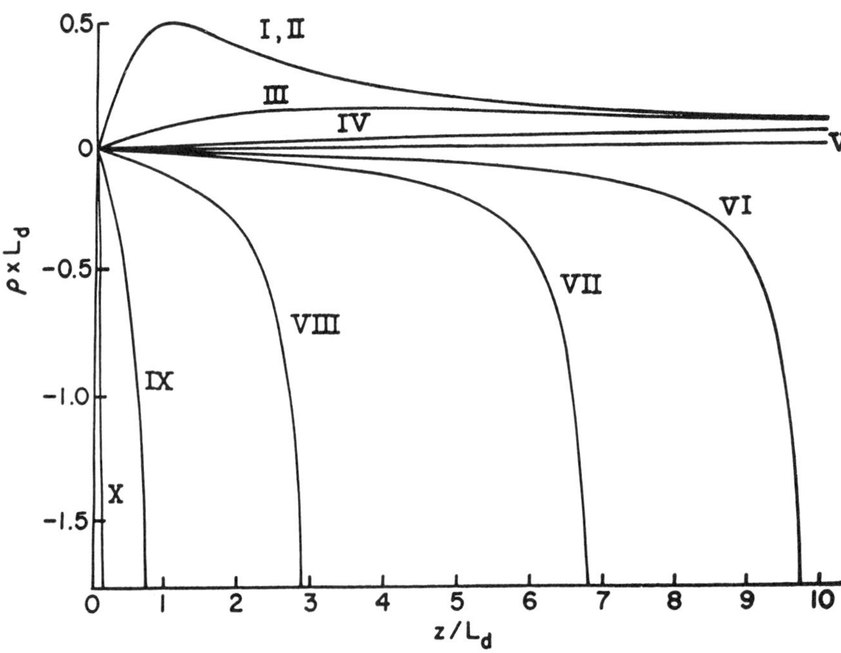

FIGURE 5.8. Normalized inverse radius of curvature (ρL_d) for a pulse propagating in a homogeneous medium at $u = 0$ as a function of the normalized length (z/L_d). (I) $p \ll 1$; (II) $p = 0.1$; (III) $p = 0.9$; (IV) $p = 0.99$; (V) $p = 1$; (VI) $p = 1.01$; (VII) $p = 1.02$; (VIII) $p = 1.1$; (IX) $p = 2$; (X) $p = 10$.

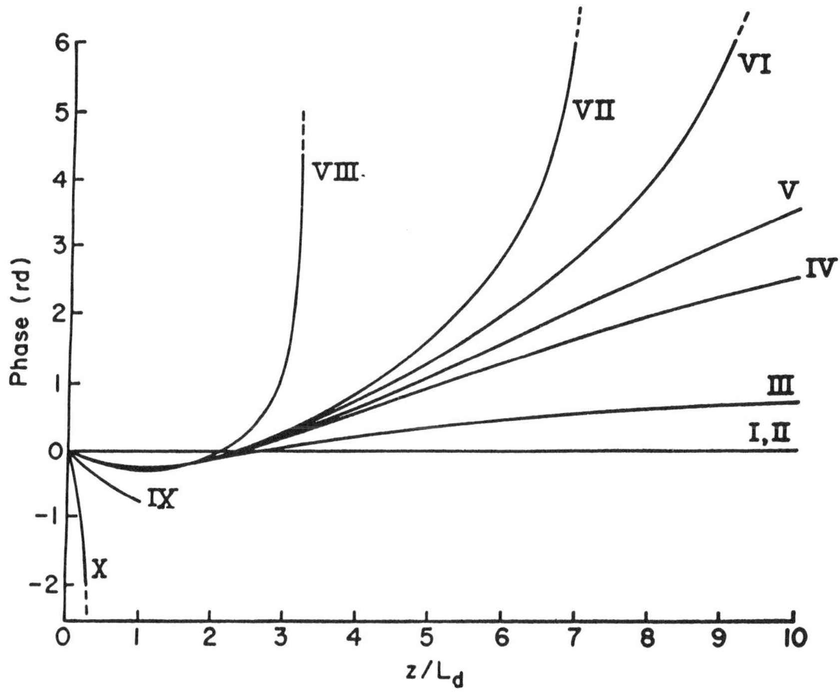

FIGURE 5.9. Regularized longitudinal phase for a pulse propagating in a homogeneous medium at $u = 0$ as a function of the normalized length (z/L_d). (I) $p \ll 1$; (II) $p = 0.1$; (III) $p = 0.9$; (IV) $p = 0.99$; (V) $p = 1$; (VI) $p = 1.01$; (VII) $p = 1.02$; (VIII) $p = 1.1$; (IX) $p = 2$; (X) $p = 10$.

quantity is positive, has a maximum for $y = 1/(1 - p)^{1/2}$, and goes to zero for very large distances; that is, diffraction is the dominant effect. For $p > 1$, this quantity is negative, monotonically decreasing, and approaches $-\infty$ at the self-focusing distance.

In Figure 5.9 the regularized longitudinal phase (i.e., its value for a specific p minus its value for $p = 0$) is plotted for $\bar{u} = 0$ as a function of the normalized length. For $0 < p < 2$ the longitudinal phase changes sign in the total length interval; when positive it is the reverse sign of that predicted by the conventional self-phase modulation theory. For $1 < p < 2$ this phase has a positive asymptote for the sample length equal to the self-focusing distance. For $p > 2$ this phase is everywhere negative, admits the conventional self-phase modulation curve as a tangent at the origin, but decreases much faster than the SPM result as the length increases until it reaches a negative asymptote at the self-focusing distance.

In Figure 5.10a the electric field intensity $(\varepsilon_0^2/\bar{\omega}^2)$ is plotted for $p \approx 1$ as a function of the normalized time $(\bar{u}/\tau)$ for different normalized lengths of the sample; as can be observed, the pulse compresses with increasing length.

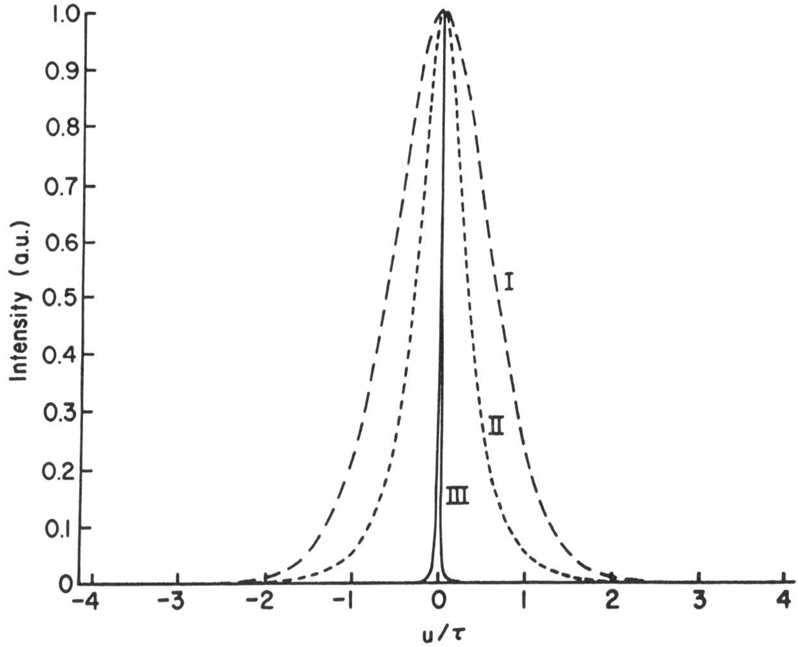

(a)

FIGURE 5.10. (a) Normalized intensity of the pulse as a function of time for $p = 0.9999$. Homogeneous material. (I) $z/L_d = 1$; (II) $z/L_d = 3$; (III) $z/L_d = 100$. (b) Regularized longitudinal phase as a function of time for $p = 0.9999$. Homogeneous material. (I) $z/L_d = 1$; (II) $z/L_d = 3$; (III) $z/L_d = 100$. [The values of the phase peaks are, respectively, 0.285, 0.25, and 46.87. The corresponding phases for SPM theory are, respectively, -0.5, -1.5, and -50.] (c) Spectral distribution intensity as a function of the normalized frequency difference. $p = 0.9999$ and $z/L_d = 100$.

Physically it should be noted that for very large distances the dispersion effects will impose a limit on the value of this compression. In Figure 5.10b the regularized longitudinal phase is plotted for $p < 1$ as function of the normalized time for different normalized lengths of the sample. As can be observed, for small length the phase has the same sign as in conventional SPM theory, for intermediate length the phase changes sign in its width interval, and for large length the phase is always positive, leading to the reverse of the red leading the blue in the supercontinuum. In Figure 5.10c the spectral distribution for $p \cong 1$ and large z is plotted; it should be noted that a central peak appears in the spectrum.

In Figure 5.11 the electric field intensity, phase, and spectral distribution are plotted as a function of $\bar{u}$ for length close to the self-focusing distance. Both the field intensity and the phase narrow, albeit at different rates. The magnitudes of the maxima of both intensity and phase increase dramatically

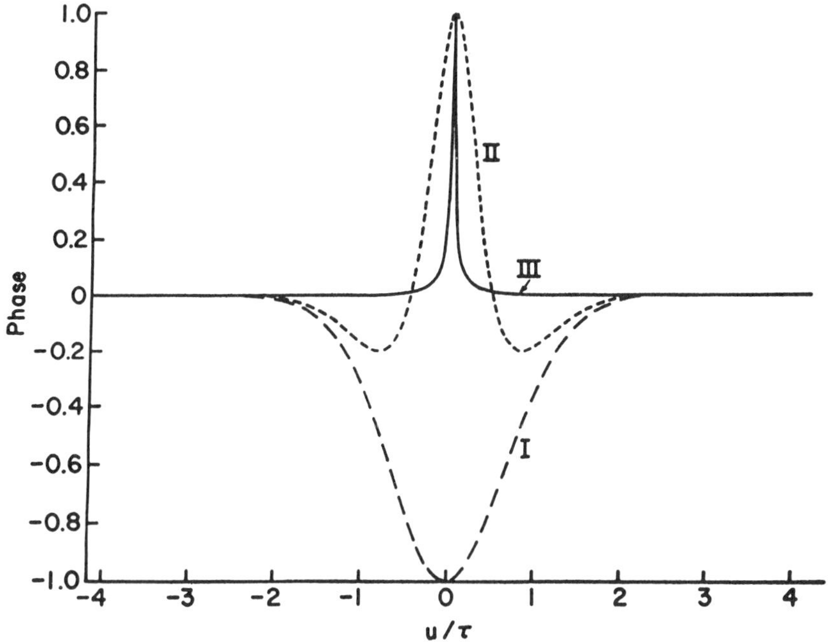

(b)

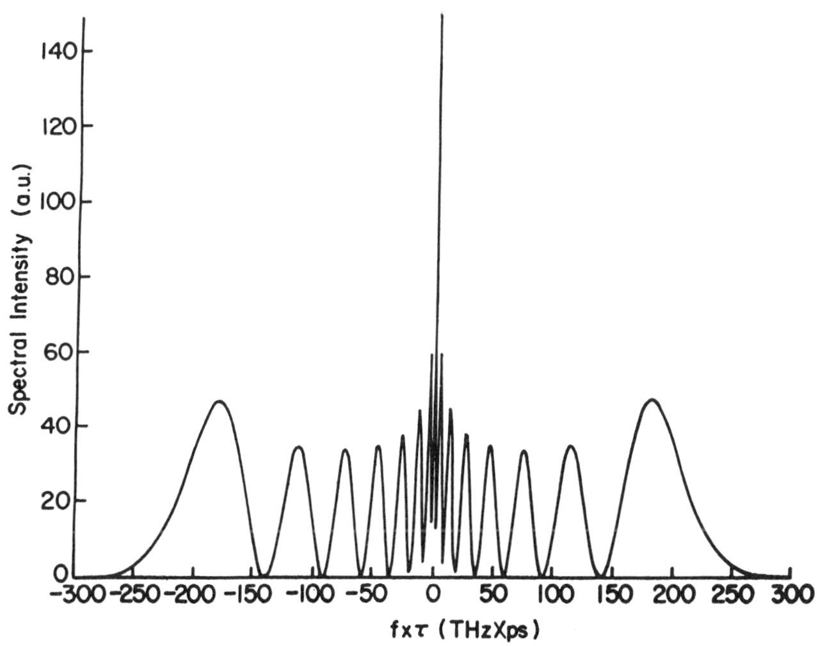

(c)

FIGURE 5.10 (*continued*)

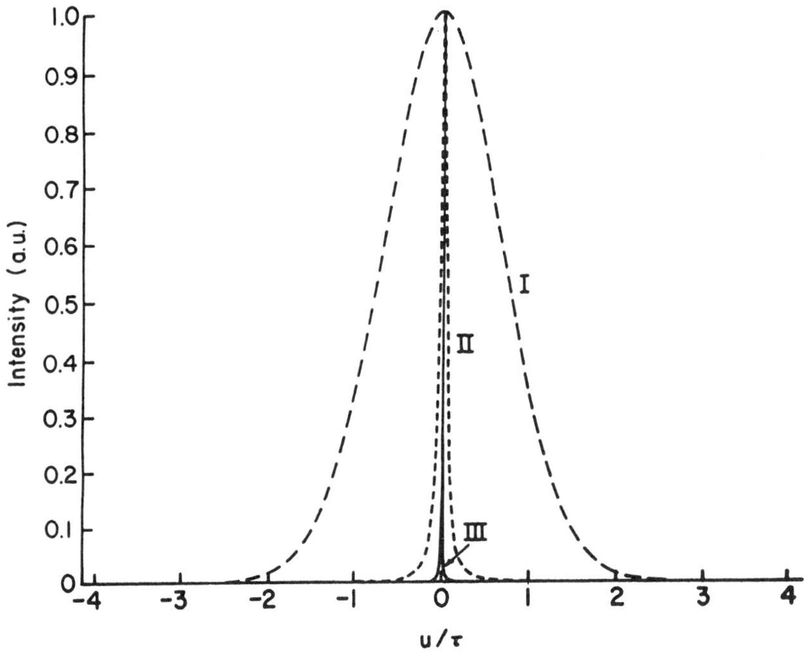

(a)

FIGURE 5.11. (a) Normalized intensity ($\varepsilon^2/\bar{\omega}^2$) of the pulse as a function of time. $p = 10$. (I) $z/L_d = 0$; (II) $z/L_d = 0.333$; (III) $z/L_d = 0.3333$. [The focusing distance for this case corresponds to 1/3.] (b) Regularized longitudinal phase as a function of time. $p = 10$. (I) $z/L_d = 0$; (II) $z/L_d = 0.333$; (III) $z/L_d = 0.3333$. [The values of the phase peaks are, respectively, 0.51, -5.4, and -8.4. The corresponding phases for SPM theory are, respectively, -0.5, -1.67, and -1.67.] (c) Spectral distribution intensity as a function of the normalized frequency difference. $p = 10$. (I) $z/L_d = 0.333$; (II) $z/L_d = 0.3333$. [$z/L_d = 1/3$ is the self-focusing length for $p = 10$.]

from those of conventional SPM theory because of the strong self-focusing effect. If the electric field can still be supported by the medium (i.e., no breakdown occurs), Figure 5.11c shows the dramatic change in the spectral extent over an infinitesimal change in the sample length.

1.5 Thermal Focusing Effects on the Supercontinuum

It has long been recognized (Gordon et al., 1965) that the propagation of a laser beam in a material produces local heating in its vicinity and that the temperature gradient in the material induces a transverse gradient of the refractive index, which leads to a lensing effect. In Section 1.3 we showed that the phase shape and spectral distribution of a pulse propagating in a parabolic graded-index material differ significantly from the conventional self-phase

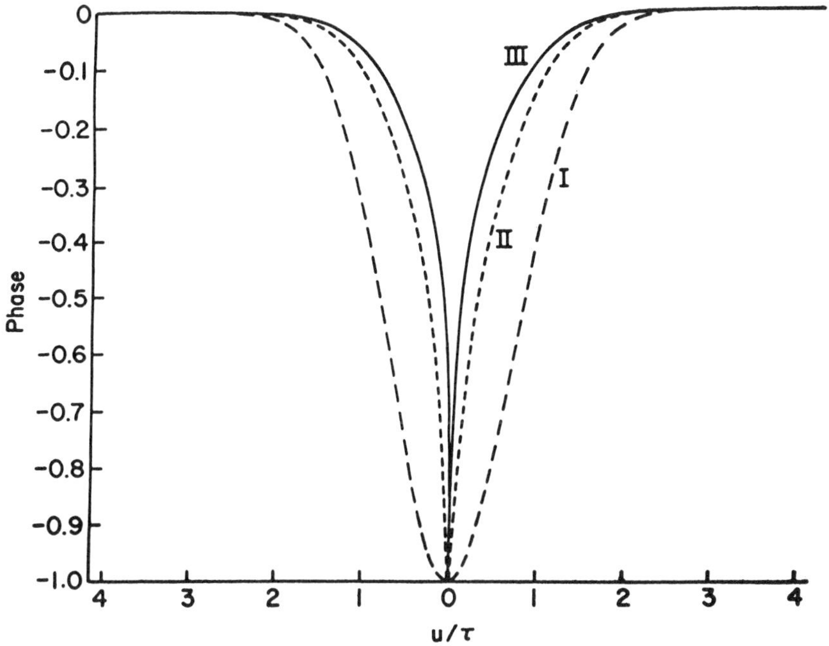

(b)

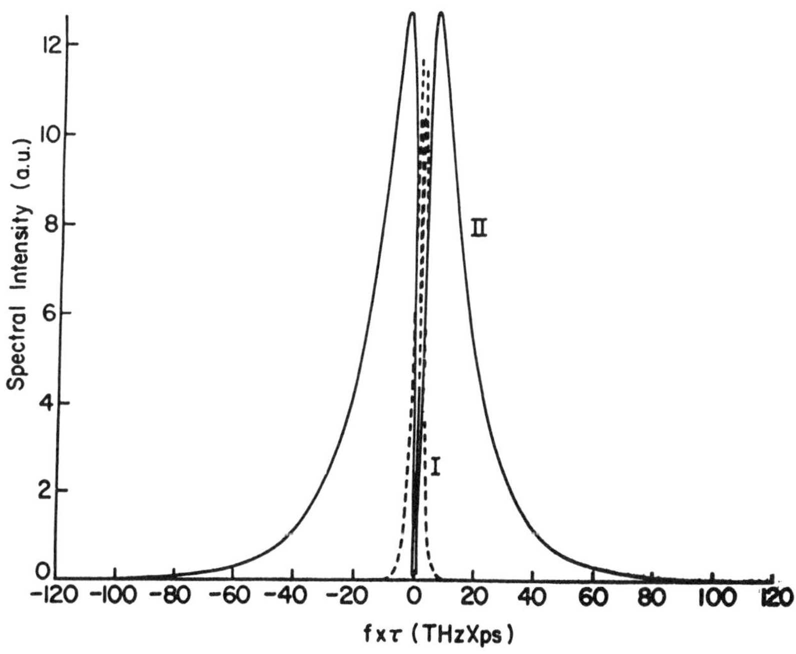

(c)

FIGURE 5.11 (*continued*)

modulation results of Section 1.1. The temperature gradient can, in the case of supercontinuum generation with high repetition rate, thus produce significant variations in the spectral extent and shape. Properly controlled (e.g., by using a CW heating beam), it is also possible to use this effect in optical fibers, lead glass, and certain semiconductors where $dn/dT > 0$ (Dabby and Whinnery, 1968) to control the sign of the pulse phase so as to reverse the red leading the blue in the supercontinuum and possibly to compress pulses within $\chi^{(3)}$ materials without the need for external gratings or prisms (Manassah et al., 1988c). In this section we review the effects of heating on the index of refraction profile and show how a homogeneous bulk material transforms in the vicinity of the heating beam into a graded-index material.

The following derivation is due to Gordon et al. (1965). If the field of the heating laser beam is given by

$$E(r) = E_0 \exp(-r^2/r_H^2), \tag{49}$$

the heat generated per unit length is proportional to the square of this field. The Green's function for the heat diffusion equation is given by

$$G(r, r', t) = \frac{1}{4\pi\kappa t} \exp\left[-\frac{(r^2 + r'^2)}{4Dt}\right] I_0\left[\frac{rr'}{2Dt}\right], \tag{50}$$

where

$$D = \kappa/\rho C_p, \tag{51}$$

κ is the thermal conductivity (cal/cm · s · K), ρ is the density (g/cm^3), and C_p is the specific heat (cal/g · K). The temperature distribution for the distributed source is

$$\Delta T(r, t) = \int_0^\infty \int_0^t 2\pi r' Q(r') G(r, r', t')\, dr'\, dt', \tag{52}$$

where $Q(r)$ is the heat generated per unit length. The solution for ΔT is given as function of the exponential integrals:

$$\Delta T(r, t) = \frac{A r_H^2}{8\kappa}\left\{Ei\left(\frac{-2r^2}{r_H^2}\right) - Ei\left(\frac{-2r^2}{8Dt + r_H^2}\right)\right\}, \tag{53}$$

where A can be related to the heating beam total power and the material rate of dissipation through

$$A r_H^2 = 0.48 bP/\pi, \tag{54}$$

where b is the fractional dissipation per centimeter.

For a temperature rise ΔT, the corresponding index of refraction is

$$n(r, t) = n_0 + \left(\frac{dn}{dT}\right)\Delta T(r, t) \tag{55}$$

where dn/dT measures the change in the material index of refraction for a change in its temperature. For small r, combining Eqs. (53) and (55), due to the temperature gradient, the index of refraction in the material is given by

$$n = n_0\left[1 + \frac{\delta'}{r_H^2} r^2\right], \tag{56}$$

where

$$\delta' = -\frac{0.12Pb}{n_0 \kappa \pi}\left(\frac{dn}{dT}\right)\frac{8Dt}{r_H^2 + 8Dt}; \tag{57}$$

that is, the problem reduces to a case similar to that treated in Section 1.3. The conclusions are as follows for dn/dT positive:

1. The normalized beam diameter is a periodic function in z. The wavelength of this periodicity is proportional to $|\delta'|^{-1/2}$. In Figure 5.12, $\bar{\omega}$ is plotted for selected values of $|\delta'|$.
2. The longitudinal phase, for $u = 0$, as function of z mostly increases by steps at z locations corresponding to the periodic positions of the minimum beam

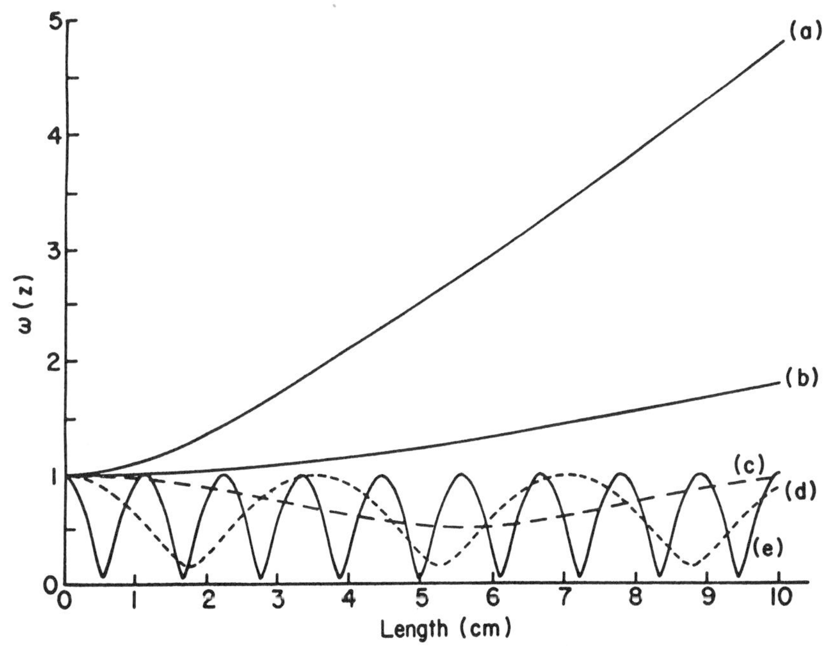

FIGURE 5.12. Normalized probe beam diameter as a function of the material thickness in the presence of thermal focusing. (a) $p = 0$, $|\delta'| = 0$; (b) $p = 0.9$, $|\delta'| = 0$; (c) $p = 0.9$, $|\delta'| = 2.5 \times 10^{-3}$; (d) $p = 0.9$, $|\delta'| = 7.8 \times 10^{-3}$; (e) $p = 0.9$, $|\delta'| = 2.5 \times 10^{-2}$.

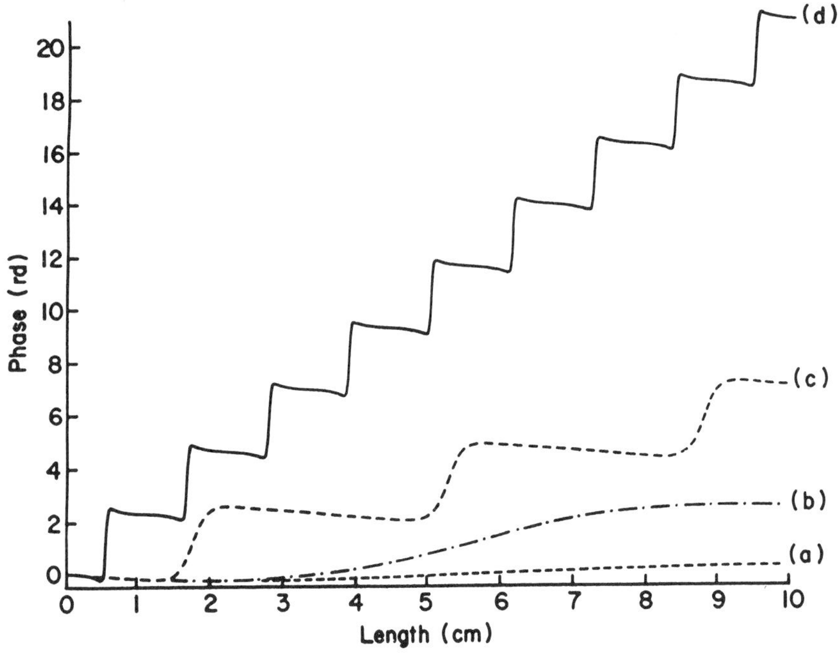

FIGURE 5.13. Longitudinal probe phase at $\bar{u} = 0$ as a function of the material thickness in the presence of thermal focusing. $p = 0.9$. (a) $|\delta'| = 0$; (b) $|\delta'|$ 2.5×10^{-3}; (c) $|\delta'| = 7.8 \times 10^{-3}$; (d) $|\delta'| = 2.5 \times 10^{-2}$.

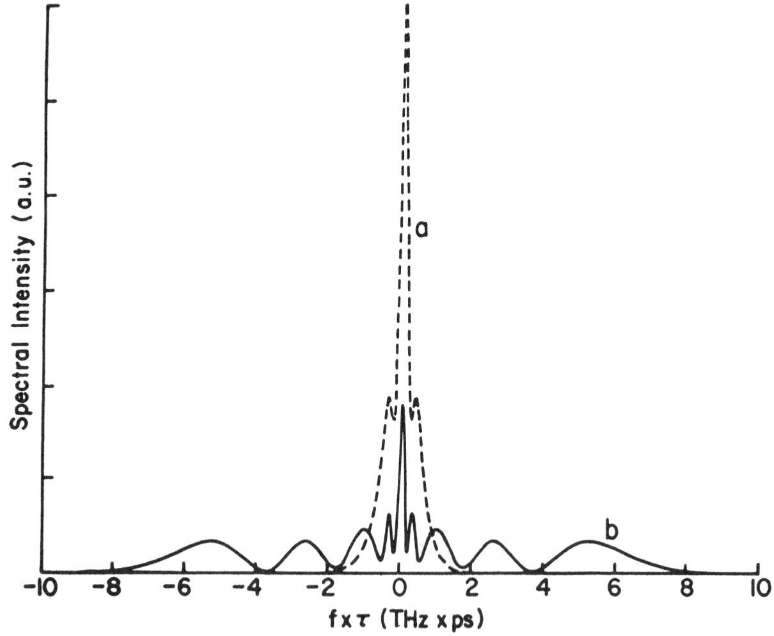

FIGURE 5.14. Probe spectral distribution in the presence of thermal focusing for $r = 0$, $z = 10$ cm, and $p = 0.9$. (a) $|\delta'| = 2.5 \times 10^{-3}$; (b) $|\delta'| = 2.5 \times 10^{-2}$.

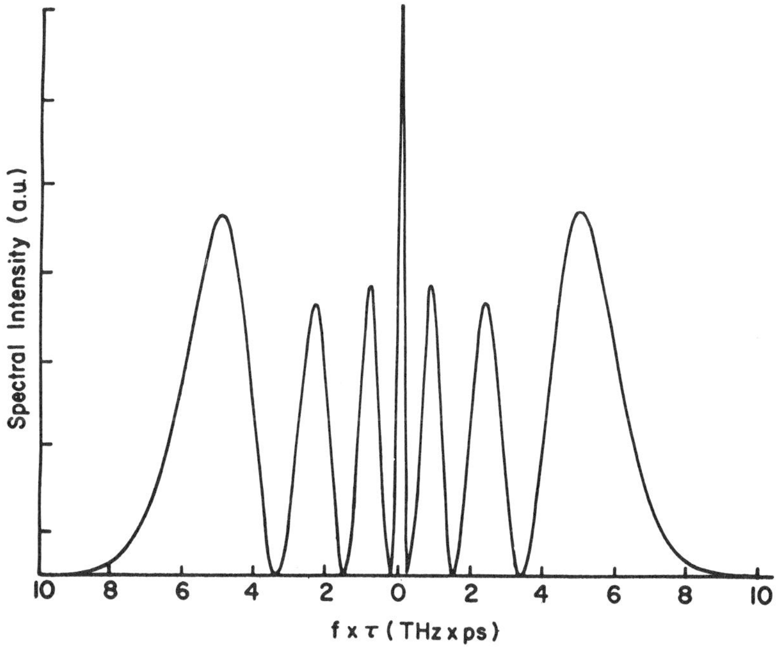

FIGURE 5.15. Probe spectral distribution in the presence of thermal focusing for $r = 0$, $z = 9.44$ cm, $p = 0.9$, and $|\delta'| = 2.5 \times 10^{-2}$.

diameter. As a result, this phase is not linear in z like the conventional self-phase modulation theory phase. Furthermore, its magnitude can be much larger. In Figure 5.13 this phase is plotted for different values of the δ' parameter.

3. The time-dependent portion of the longitudinal phase may have the reverse sign to that of conventional SPM. Physically, this result leads to the reverse of the red leading the blue in the supercontinuum. Furthermore, as the pulse peak magnitude increases and tends to the critical field of the nonlinear medium, the phase which decreases significantly. Therefore, for $\bar{\omega}$ close to its maximum, only a short period over the pulse duration is modulated; this portion increases as $\bar{\omega}$ approaches its minimum.

In Figure 5.14 the spectral intensity is plotted for the same length z but for different δ' parameters. For larger δ' the spectral extent is larger.

In Figure 5.15 the spectral intensity is plotted for the δ' of the curve in Figure 5.14b, but for a slightly different sample length. As can be observed, the spectral extent does not change significantly with a small variation in z; however, the spectral shape can be altered dramatically. Actually, the two lengths chosen correspond to successive maximum and minimum positions of the normalized beam diameter.

2. Pulse Steepening in a $\chi^{(3)}$ Dispersionless Medium. Plane-Wave Approximation

2.1 Nonlinear Wave Equation

Maxwell's equation, in a $\chi^{(3)}$ nonlinear medium where the dispersion of the linear index of refraction and its imaginary part are neglected, is given by

$$\nabla^2 \mathbf{E} - \frac{n^2}{c^2}\frac{\partial^2 \mathbf{E}}{\partial t^2} = \frac{2nn_2}{c^2}\frac{\partial^2}{\partial t^2}\langle \mathbf{E}\cdot\mathbf{E}\rangle \mathbf{E}, \tag{58}$$

where n is the linear index of refraction and n_2 is the nonlinear index of refraction. In the instance that the transverse variation of $\mathbf{E}$ is neglected (i.e., diffraction is neglected), one component of $\mathbf{E}$ is present and $\langle \mathbf{E}\cdot\mathbf{E}\rangle = |E|^2/2$, the wave equation reduces to

$$\frac{\partial^2 E}{\partial z^2} - \frac{1}{v_g^2}\frac{\partial^2 E}{\partial t^2} = \frac{nn_2}{c^2}\frac{\partial^2}{\partial t^2}|E|^2 E, \tag{59}$$

where v_g is the group velocity in the medium, assumed constant over the light bandwidth. To reduce the differential equation to a dimensionless form, we introduce the new dimensionless variables Φ, T, and Z defined as

$$T = t/\tau, \qquad Z = z/v_g\tau, \qquad E = E_0\Phi, \tag{60}$$

where τ is the pulse width and E_0 is the maximum amplitude of the electric field at the entrance plane of the $\chi^{(3)}$ medium. Introducing the nonlinear coupling constant ε, defined as

$$\varepsilon \equiv n_2|E_0|^2/n, \tag{61}$$

the wave equation in dimensionless form reduces to

$$\left(\frac{\partial^2}{\partial Z^2} - \frac{\partial^2}{\partial T^2}\right)\Phi = \varepsilon\frac{\partial^2}{\partial T^2}|\Phi|^2\Phi. \tag{62}$$

Typical values for the above parameters are

$$\tau \sim 10^{-14} \text{ to } 10^{-12} \text{ s},$$
$$n_2 \sim 10^{-22} \text{ to } 10^{-18} \text{ MKS},$$
$$v_g \sim 2\times 10^8 \text{ m/s}.$$

2.2 Method of Multiple Scales

The functional dependence of Φ on Z, T, and ε in the solution of (62) is not disjoint (Nayfeh, 1981). To first order in ε, Φ depends on the combinations εT and εZ as well as on the individual T, Z, and ε. Carrying the perturbation to higher orders, Φ additionally depends on $\varepsilon^2 T$, $\varepsilon^2 Z$, $\varepsilon^3 T$, $\varepsilon^3 Z$, ... Hence it is convenient to write Φ as

$$\Phi(Z, T; \varepsilon) = \Phi(Z_0, T_0, Z_1, T_1, Z_2, T_2, \ldots; \varepsilon), \tag{63}$$

where the new scaled variables Z_1, T_1, Z_2, T_2, etc. are defined as

$$T_0 = T, \qquad T_1 = \varepsilon T, \ldots, T_n = \varepsilon^n T, \tag{64}$$

$$Z_0 = Z, \qquad Z_1 = \varepsilon Z, \ldots, Z_n = \varepsilon^n Z. \tag{65}$$

The T_n's and Z_n's represent different time and distance scales. Φ is then determined as a function of the old and new variables. Next, we seek a uniform expansion solution to Φ in the form

$$\begin{aligned}\Phi = {} & \Phi_0(T_0, Z_0, T_1, Z_1, T_2, Z_2, \ldots) \\ & + \varepsilon\Phi_1(T_0, Z_0, T_1, Z_1, T_2, Z_2, \ldots) \\ & + \varepsilon^2\Phi_2(T_0, Z_0, T_1, Z_1, T_2, Z_2, \ldots) + \cdots.\end{aligned} \tag{66}$$

To express the derivatives in (62) as functions of the new variables, we use the chain rules for derivatives; then

$$\begin{aligned}\frac{\partial}{\partial T} &= \frac{\partial}{\partial T_0} + \varepsilon\frac{\partial}{\partial T_1} + \varepsilon^2\frac{\partial}{\partial T_2} + \cdots, \\ \frac{\partial}{\partial Z} &= \frac{\partial}{\partial Z_0} + \varepsilon\frac{\partial}{\partial Z_1} + \varepsilon^2\frac{\partial}{\partial Z_2} + \cdots, \\ \frac{\partial^2}{\partial T^2} &= \frac{\partial^2}{\partial T_0^2} + 2\varepsilon\frac{\partial^2}{\partial T_0 \partial T_1} + \varepsilon^2\left(2\frac{\partial^2}{\partial T_0 \partial T_2} + \frac{\partial^2}{\partial T_1^2}\right) + \cdots, \\ \frac{\partial^2}{\partial Z^2} &= \frac{\partial^2}{\partial Z_0^2} + 2\varepsilon\frac{\partial^2}{\partial Z_0 \partial Z_1} + \varepsilon^2\left(2\frac{\partial^2}{\partial Z_0 \partial Z_2} + \frac{\partial^2}{\partial Z_1^2}\right) + \cdots.\end{aligned}$$

Using expressions (66) and (67) in the partial differential equation (62) and equating the respective coefficients of ε^n, one obtains, respectively, for the terms multiplying ε^0, ε, and ε^2 the following equations:

$$\left(\frac{\partial^2}{\partial Z_0^2} - \frac{\partial^2}{\partial T_0^2}\right)\Phi_0 = 0, \tag{68}$$

$$\left(\frac{\partial^2}{\partial Z_0^2} - \frac{\partial^2}{\partial T_0^2}\right)\Phi_1 + 2\left(\frac{\partial^2}{\partial Z_1 \partial Z_0} - \frac{\partial^2}{\partial T_1 \partial T_0}\right)\Phi_0 = \frac{\partial^2}{\partial T_0^2}|\Phi_0|^2\Phi_0, \tag{69}$$

and

$$\begin{aligned}&\left(\frac{\partial^2}{\partial Z_0^2} - \frac{\partial^2}{\partial T_0^2}\right)\Phi_2 + 2\left(\frac{\partial^2}{\partial Z_1 \partial Z_0} - \frac{\partial^2}{\partial T_1 \partial T_0}\right)\Phi_1 \\ &\quad + \left(\frac{\partial^2}{\partial Z_1^2} - \frac{\partial^2}{\partial T_1^2} + 2\frac{\partial^2}{\partial Z_0 \partial Z_2} - 2\frac{\partial^2}{\partial T_0 \partial T_2}\right)\Phi_0 \\ &= 2\frac{\partial^2}{\partial T_0 \partial T_1}|\Phi_0|^2\Phi_0 + 2\frac{\partial^2}{\partial T_0^2}|\Phi_0|^2\Phi_1 + \frac{\partial^2}{\partial T_0^2}\Phi_0^2\Phi_1^*.\end{aligned} \tag{70}$$

Since the incoming pulse is described most easily by a comoving coordinate system (i.e., a new coordinate system that is moving with the pulse), we will define the U_n and V_n families of new coordinates as

$$
\begin{aligned}
U_0 &= Z_0 - T_0, & V_0 &= Z_0,\\
U_1 &= Z_1 - T_1, & V_1 &= Z_1,\\
U_2 &= Z_2 - T_2, & V_2 &= Z_2
\end{aligned}
\tag{71}
$$

The partial derivatives can be described in the new coordinates as

$$
\begin{aligned}
\frac{\partial}{\partial Z_n} &= \frac{\partial}{\partial U_n} + \frac{\partial}{\partial V_n},\\
\frac{\partial}{\partial T_n} &= -\frac{\partial}{\partial U_n},\\
\frac{\partial^2}{\partial Z_n^2} &= \frac{\partial^2}{\partial U_n^2} + \frac{\partial^2}{\partial V_n^2} + 2\frac{\partial^2}{\partial U_n \partial V_n},\\
\frac{\partial^2}{\partial T_n^2} &= \frac{\partial^2}{\partial U_n^2},\\
\frac{\partial^2}{\partial Z_n \partial Z_m} &= \frac{\partial^2}{\partial U_n \partial U_m} + \frac{\partial^2}{\partial V_n \partial V_m} + \frac{\partial^2}{\partial U_n \partial V_m} + \frac{\partial^2}{\partial U_m \partial V_n}.
\end{aligned}
\tag{72}
$$

The partial differential equations (68), (69), and (70) are then given by

$$
\left[\frac{\partial^2}{\partial V_0^2} + 2\frac{\partial}{\partial U_0}\frac{\partial}{\partial V_0}\right]\Phi_0 = 0,
\tag{73}
$$

$$
\left[\frac{\partial^2}{\partial V_0^2} + 2\frac{\partial}{\partial U_0}\frac{\partial}{\partial V_0}\right]\Phi_1 + 2\left[\frac{\partial}{\partial U_1}\frac{\partial}{\partial V_0} + \frac{\partial}{\partial U_0}\frac{\partial}{\partial V_1} + \frac{\partial}{\partial V_1}\frac{\partial}{\partial V_0}\right]\Phi_0 = \frac{\partial^2}{\partial U_0^2}|\Phi_0|^2\Phi_0,
\tag{74}
$$

and

$$
\begin{aligned}
&\left[\frac{\partial^2}{\partial V_0^2} + 2\frac{\partial}{\partial U_0}\frac{\partial}{\partial V_0}\right]\Phi_2 + 2\left[\frac{\partial}{\partial U_1}\frac{\partial}{\partial V_0} + \frac{\partial}{\partial U_0}\frac{\partial}{\partial V_1} + \frac{\partial}{\partial V_1}\frac{\partial}{\partial V_0}\right]\Phi_1\\
&\quad + \left[\frac{\partial^2}{\partial V_1^2} + 2\frac{\partial}{\partial U_1}\cdot\frac{\partial}{\partial V_1} + 2\frac{\partial}{\partial U_0}\cdot\frac{\partial}{\partial V_2} + 2\frac{\partial}{\partial U_2}\cdot\frac{\partial}{\partial V_0} + 2\frac{\partial}{\partial V_2}\cdot\frac{\partial}{\partial V_0}\right]\Phi_0\\
&= 2\frac{\partial}{\partial U_0}\cdot\frac{\partial}{\partial U_1}|\Phi_0|^2\Phi_0 + 2\frac{\partial^2}{\partial U_0^2}|\Phi_0|^2\Phi_1 + \frac{\partial^2}{\partial U_0^2}\Phi_0^2\Phi_1^*
\end{aligned}
\tag{75}
$$

An ansatz that solves the above partial differential equations correct to order ε in the (66) expansion is

$$
\begin{aligned}
\Phi_0 &= A(U_1, V_1, U_2, V_2)e^{iKU_0},\\
\Phi_1 &= 0,\\
\Phi_2 &= C(U_1, V_1, U_2, V_2,)e^{iKU_0},
\end{aligned}
\tag{76}
$$

where $K = \omega\tau = W$ and ω is the pulse center frequency. It is worth noting at this point that the solutions thus obtained are valid for $\varepsilon V \sim O(1)$ while those deduced from ordinary perturbation theory would have been valid only for $V \sim O(1)$. Furthermore, note that the specific form for C cannot be obtained from the above equations. With the above ansatz, the above system of partial differential equations reduces to (Manassah et al., 1986)

$$2iK\frac{\partial A}{\partial V_1} = -K^2|A|^2A, \tag{77}$$

$$\left[\frac{\partial^2}{\partial V_1^2} + 2\frac{\partial}{\partial U_1}\cdot\frac{\partial}{\partial V_1} + 2iK\frac{\partial}{\partial V_2}\right]A = 2iK\frac{\partial}{\partial U_1}|A|^2A. \tag{78}$$

2.3 Quasi-Linear Partial Differential Equations

Starting with Eqs. (77) and (78) we obtain the system of quasi-linear partial differential equations for the amplitude and phase of the pulse. Denote A by

$$A = ae^{i\alpha}. \tag{79}$$

Then the system of partial differential equations is given by

$$\frac{\partial a}{\partial V_1} = 0, \tag{80}$$

$$\frac{\partial \alpha}{\partial V_1} = \frac{K}{2}a^2, \tag{81}$$

$$3a^2\frac{\partial a}{\partial U_1} = 2\frac{\partial a}{\partial V_2}, \tag{82}$$

$$\frac{K}{4}a^4 + 2\frac{\partial \alpha}{\partial V_2} = a^2\frac{\partial \alpha}{\partial U_1}. \tag{83}$$

In variables U and V, the above equations lead to

$$\frac{\partial a}{\partial V} - \frac{3}{2}\varepsilon a^2\frac{\partial a}{\partial U} = 0, \tag{84}$$

$$\frac{\partial \alpha}{\partial V} - \frac{\varepsilon}{2}a^2\frac{\partial \alpha}{\partial U} = \frac{\varepsilon K}{2}a^2 - \frac{\varepsilon^2 K}{8}a^4. \tag{85}$$

In the same notation, the quasi-linear partial differential equations in other work are:

1. Conventional theory (Alfano and Shapiro, 1970):

$$\frac{\partial a}{\partial V} = 0, \tag{86}$$

$$\frac{\partial \alpha}{\partial V} = \frac{\varepsilon K}{2}a^2. \tag{87}$$

TABLE 5.1. Range of experimental parameters in supercontinuum generation experiments.[a]

	$n_2 = 10^{-22}$ (MKS) $= 0.9 \times 10^{-13}$ (esu)			$n_2 = 10^{-20}$ (MKS) $= 0.9 \times 10^{-11}$ (esu)		
P_c, critical power for self-focusing	2×10^6 W			2×10^4 W		
P_{in}, pulse input power, W	10^7	10^9	10^{11}	10^7	10^9	10^{11}
E_0, maximum amplitude of the incoming pulse, V/m	9×10^7	9×10^8	9×10^9	9×10^7	9×10^8	9×10^9
S_f, distance to focusing point; m	0.4	0.04	0.004	0.04	0.004	0.0004
$\varepsilon = n_2\|E_0\|^2/n$ ($n = 1.5$)	$\frac{2}{3} \times 10^{-6}$	$\frac{2}{3} \times 10^{-4}$	$\frac{2}{3} \times 10^{-2}$	$\frac{2}{3} \times 10^{-4}$	$\frac{2}{3} \times 10^{-2}$	$\frac{2}{3}$
$\varepsilon V = n_2\|E_0\|^2 z/c\tau$ ($z = 1\ \mu$m of sample thickness)	$\frac{1}{3} \times 10^{-7}$	$\frac{1}{3} \times 10^{-5}$	$\frac{1}{3} \times 10^{-3}$	$\frac{1}{3} \times 10^{-5}$	$\frac{1}{3} \times 10^{-3}$	$\frac{1}{3} \times 10^{-1}$

[a] Source wavelength = 1 μm, pulse width = 10^{-13} s, and beam diameter at input plane = 1 mm).

2. Slowly varying approximation (Anderson and Lisak, 1983):

$$\frac{\partial a}{\partial V} - 3\varepsilon a^2 \frac{\partial a}{\partial U} = 0, \tag{88}$$

$$\frac{\partial \alpha}{\partial V} - \varepsilon a^2 \frac{\partial \alpha}{\partial U} = \frac{\varepsilon K}{2} a^2. \tag{89}$$

3. Yang and Shen approximation (Yang and Shen, 1984)

$$\frac{\partial a}{\partial V} - \frac{\varepsilon}{2} a^2 \frac{\partial a}{\partial U} = 0, \tag{90}$$

$$\frac{\partial \alpha}{\partial V} - \frac{\varepsilon a^2}{2} \frac{\partial \alpha}{\partial U} = \frac{\varepsilon K}{2} a^2. \tag{91}$$

Next we will find the solution for a and α for a pulse whose initial shape is a sech pulse. In Manassah et al. (1986) the Gaussian pulse results are also given. In Table 5.1 some physical quantities in the present notation are given.

2.4 Pulse Amplitude

The solution of Eq. (84) with the boundary condition

$$a(U, 0) = \operatorname{sech}(U) \tag{92}$$

can be obtained by the usual technique of first-order partial differential equations; it is given by (Anderson and Lisak, 1983)

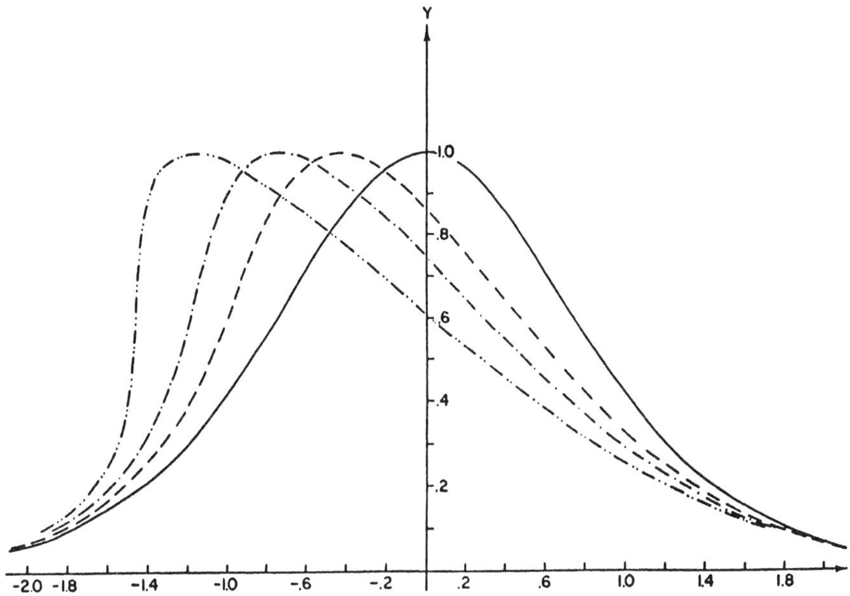

FIGURE 5.16. Steepened pulse amplitude as a function of U. —— $\varepsilon V = 0.0$; - - - $\varepsilon V = 0.3$; –·– $\varepsilon V = 0.5$; –··– $\varepsilon V = 0.8$.

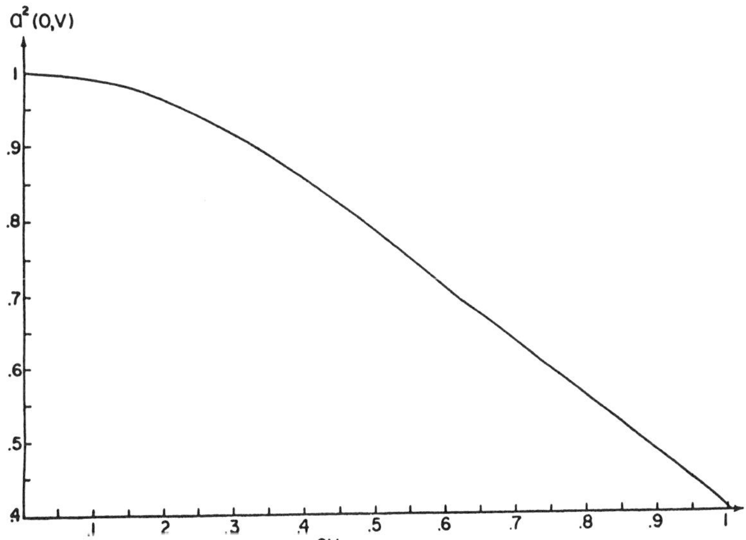

FIGURE 5.17. Magnitude of the steepened (amplitude)^2 at $U = 0$ as a function of εV.

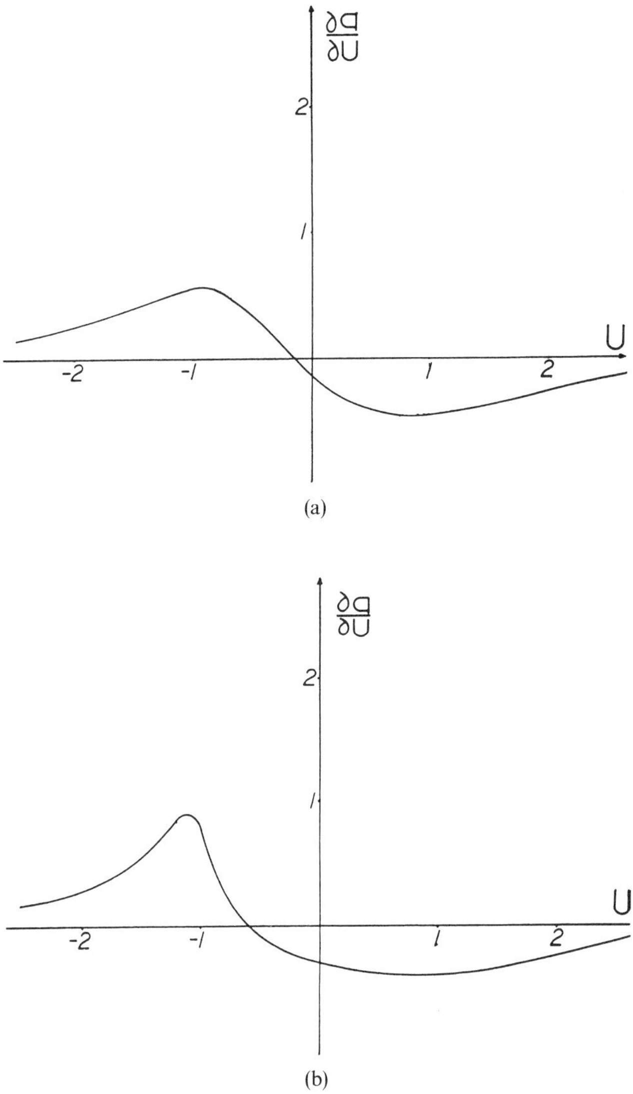

FIGURE 5.18. Slope of the steepened pulse as a function of U. (a) $\varepsilon V = 0.1$; (b) $\varepsilon V = 0.4$; (c) $\varepsilon V = 0.8$.

$$a^2 = \operatorname{sech}^2[U + \tfrac{3}{2}\varepsilon V a^2]. \tag{93}$$

The amplitude is maximum for

$$U_m = -\tfrac{3}{2}\varepsilon V. \tag{94}$$

The (amplitude)^2 of the pulse at $U = 0$ is given by the solution of

$$a_0^2 = \operatorname{sech}^2[\tfrac{3}{2}\varepsilon V a_0^2] \tag{95}$$

or

$$\varepsilon V = \frac{2}{3a_0^2} \ln\left[\frac{1}{a_0} + \left(\frac{1}{a_0^2} - 1\right)^{1/2}\right], \tag{96}$$

where $a_0 = a(0, V)$.

The partial derivative of the amplitude with respect to U is given by

$$\frac{\partial a}{\partial U} = \frac{-a(1 - a^2)^{1/2}}{3\varepsilon V a^2(1 - a^2)^{1/2} \pm 1}, \tag{97}$$

where $(-)$ refers to $U < -\frac{3}{2}\varepsilon V$ and $(+)$ otherwise.

The present model produces an amplitude that is asymmetric; this distortion is called self-steepening. A shock in the amplitude can develop for $\varepsilon V > \sqrt{\frac{3}{2}}$.

In Figure 5.16 the amplitude a is plotted as a function of U for different values of εV, in Figure 5.17 a_0 is plotted as a function of εV, and in Figure 5.18 $\partial a_0/\partial U$ is plotted as a function of U.

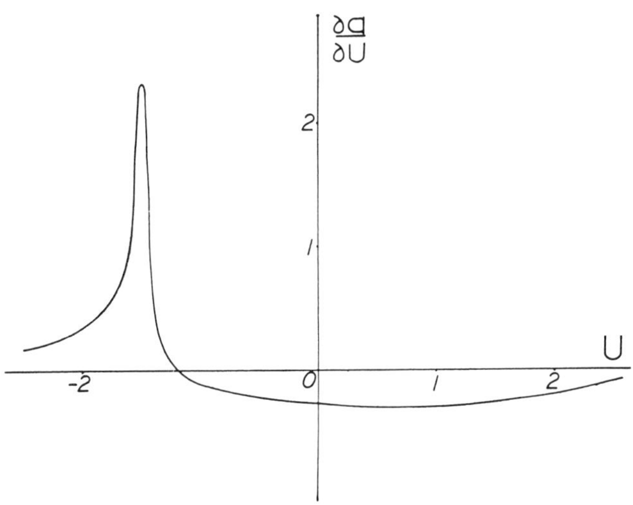

(c)

FIGURE 5.18 (*continued*)

2.5 Pulse Phase

The general solution of Eq. (85) is given by (Manassah et al., 1986)

$$\alpha = -KU - \frac{K\varepsilon}{2}\int_0^U a^2(p,V)\,dp - \frac{3K}{8}\varepsilon^2\int_0^V a^4(0,q)\,dq + F(U,V), \tag{98}$$

where the function $F(U, V)$ is a solution of the equation

$$\frac{\partial F}{\partial V} - \frac{\varepsilon}{2}a^2\frac{\partial F}{\partial U} = 0. \tag{99}$$

The boundary condition imposed by the physical condition of no initial chirp is that at the input plane (i.e., $V = 0$), α should be zero for all values of U. This implies that, for an incoming sech pulse, we have

$$F(U,0) = KU + \frac{K\varepsilon}{2}\tanh U. \tag{100}$$

The general solution of Eq. (99) is given by

$$F(f) = F\left(\frac{\varepsilon}{2}\int_0^V a^3(U,q)\,dq + \int_0^U a(p,0)\,dp\right). \tag{101}$$

The special form of α satisfying the boundary condition (100) is given by

$$\begin{aligned}\bar{\alpha} \equiv \frac{\alpha}{K} = -U - \frac{\varepsilon}{2}\int_0^U a^2(p,V)\,dp - \frac{3\varepsilon^2}{8}\int_0^V a^4(0,q)\,dq \\ + \tanh^{-1}\sin f(U,V) + \frac{\varepsilon}{2}\sin f(U,V),\end{aligned} \tag{102}$$

where

$$f(U,V) = \frac{\varepsilon}{2}\int_0^V a^3(U,q)\,dq + \sin^{-1}\tanh U. \tag{103}$$

The maximum of α, denoted by α^M, and its position, denoted by U_α, are respectively given by (Manassah and Mustafa, 1988a)

$$\alpha^M \approx K\varepsilon V/2 \tag{104}$$

and

$$U_\alpha \approx -\varepsilon V. \tag{105}$$

We note that the positions of the maxima of a and α are shifted from each other. Furthermore, this shift is linear in εV. Also note that while the value of the maximum of the amplitude is equal to 1 for all εV, α^M, the value of the phase maximum, increases linearly with εV. In Figure 5.19, $\alpha(U, V)$ is plotted for different values of the parameter εV. As can be observed, for small values of εV, α is symmetric and is centered on the $U = 0$ axis. However, as εV increases the α curve becomes skewed and its maximum is shifted to the left

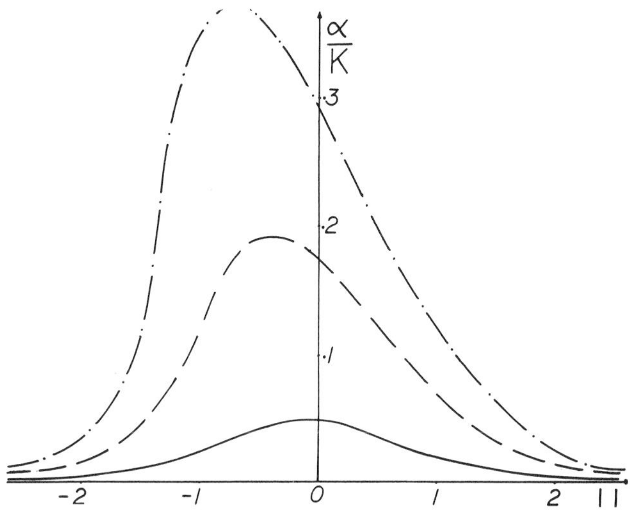

FIGURE 5.19. Phase of the steepened pulse as function of εV. (a) —— $\varepsilon V = 0.1$ (b) – – $\varepsilon V = 0.4$; (c) –·– $\varepsilon V = 0.8$.

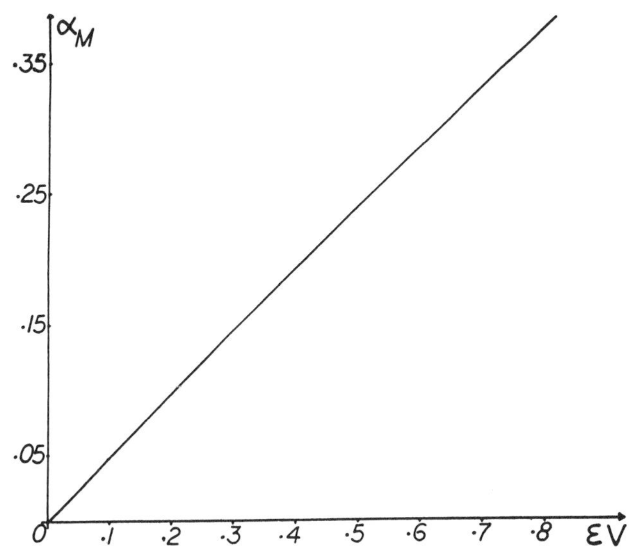

FIGURE 5.20. Computed magnitude of the maximum of the steepened pulse phase as function of εV.

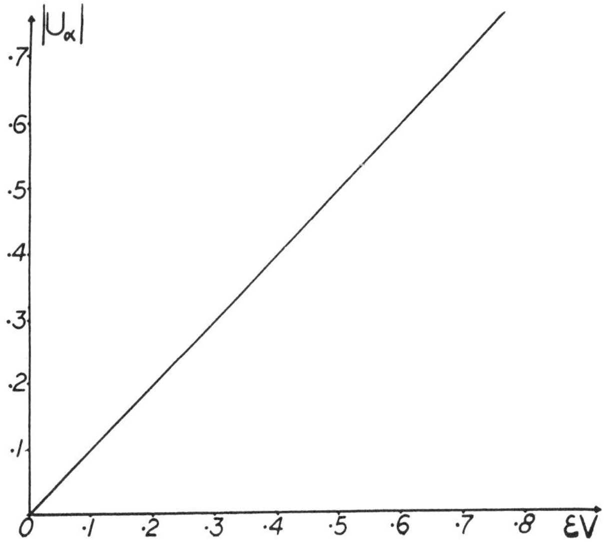

FIGURE 5.21. Computed position of the maximum of the steepened pulse phase as a function of εV.

of the $U = 0$ axis. In Figures 5.20 and 5.21 the computed values of α^M and U_α are plotted as functions of εV. As can be seen, the computed curves are well approximated by Eqs. (104) and (105).

The expression for the derivative of the phase with respect to U, also called the instantaneous frequency sweep, is given by

$$\frac{1}{K}\frac{\partial\alpha}{\partial U} = \bar{\alpha}' = -1 + a(U,V)[\cos f(U,V)]^{-1} + \frac{\varepsilon}{2}\{a(U,V)\cos[f(U,V)] - a^2(U,V)\}. \tag{106}$$

The second partial derivative of the phase is given by

$$\frac{1}{K}\frac{\partial^2\alpha}{\partial U^2} = \bar{\alpha}'' = \left[\frac{\partial a(U,V)}{\partial U} + a^2(U,V)\tan f(U,V)\right][\cos f(U,V)]^{-1} + \frac{\varepsilon}{2}\left[\frac{\partial a(U,V)}{\partial U}\cos f(U,V) - a^2(U,V)\sin f(U,V) - 2a(U,V)\frac{\partial a(U,V)}{\partial U}\right]. \tag{107}$$

In Figures 5.22 and 5.23 the phase partial derivatives are plotted for different values of the parameter εV. It should be observed that $\bar{\alpha}'$ and $\bar{\alpha}''$ for a self-phase modulated pulse are qualitatively different from those of a chirped Gaussian

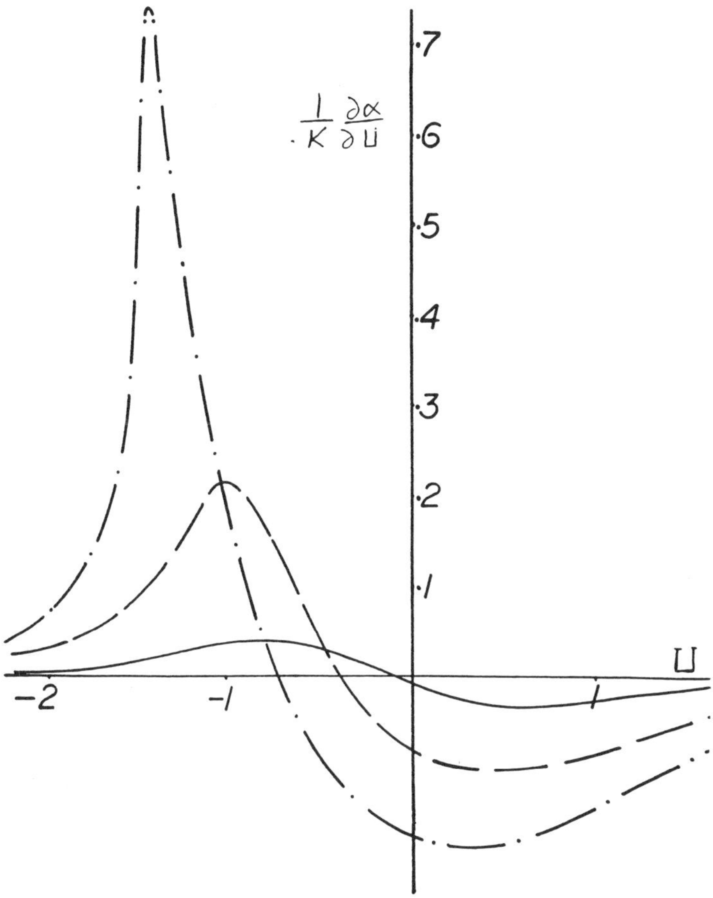

FIGURE 5.22. The U partial derivative of the steepened pulse phase as a function of εV. (a) —— $\varepsilon V = 0.1$; (b) -- $\varepsilon V = 0.4$; (c) -·- $\varepsilon V = 0.8$.

pulse, where α' is linear in U and α'' is a constant. Consequently, whereas in the time domain for small εV the phase can be approximated by a chirped Gaussian, in the frequency domain, where the values of $\bar{\alpha}'$ and $\bar{\alpha}''$ are critical, such an approximation is not valid. We will return to this point when we discuss the spectral distribution and the filter transform of this pulse. The critical results of this section are the asymmetries of the phase and its derivatives. In the following section, we discuss a direct method for measuring the asymmetry in the phase.

2.6 Direct Time Measurement of the Phase

The phase of the pulse can be directly measured using the Rothenberg-Grischkowsky interferometric technique (Rothenberg and Grischkowsky,

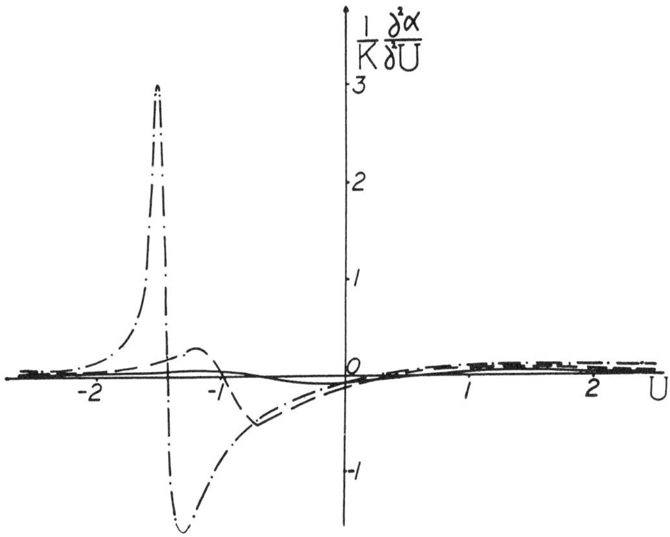

FIGURE 5.23. The U second partial derivative of the steepened pulse phase as a function of U. (a) —— $\varepsilon V = 0.1$; (b) -- $\varepsilon V = 0.4$; (c) -·- $\varepsilon V = 0.8$.

1987). Essentially, the measurement technique consists of adding two pulses, a reference pulse of known width and zero chirp and a signal pulse described by the above amplitude and phase, and then subtracting the sum of the intensities of the reference pulse and the signal pulse. The resultant is given by

$$I(U) = 2R(U)a(U)\cos[\alpha(U)]. \tag{108}$$

If the reference pulse $R(U)$ is obtained as a portion of the incoming pulse (i.e., sech U), then $I(U)$ is given by

$$I(U) = 2\,\mathrm{sech}\,U\,a(U)\cos[\alpha(U)]. \tag{109}$$

In Figure 5.24, $I(U)$ is plotted using the above theory and is compared to the results of conventional self-phase modulation theory. As can be observed (Manassah and Mustafa, 1988a), the asymmetry in the amplitude and phase are translated into an asymmetry in $I(U)$.

2.7 Interference Pattern of the Supercontinuum

In this section we compute the interference pattern generated by a self-phase-modulated pulse (Manassah and Mustafa, 1988b). We prove that the presence of the amplitude-phase time shift generates fringe position shifts. We also compute the Fourier transform of the interferometric intensity distribution and prove that the range of this transform is directly related to the range of $\overline{\alpha}'$.

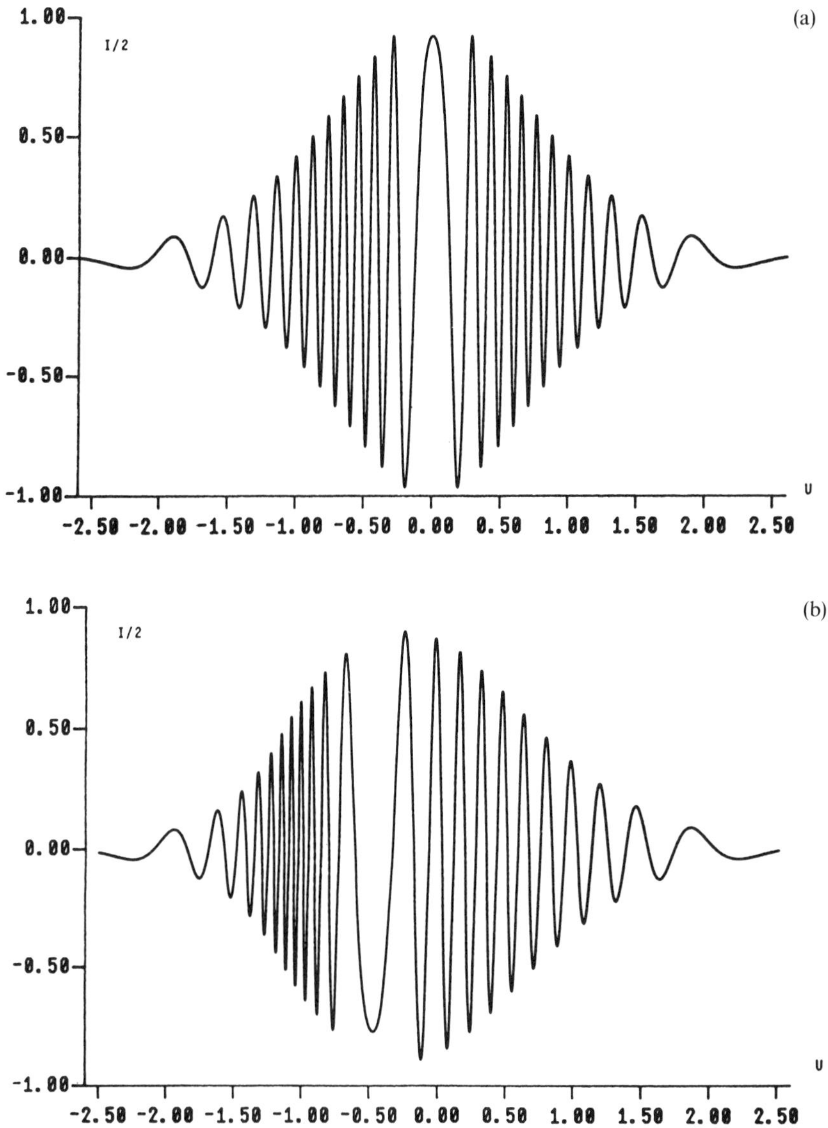

FIGURE 5.24. Difference signal between the output from a Mach-Zehnder interferometer and the sum of the input pulse to the interferometer (i.e., the SPM pulse) and the reference pulse (i.e., sech pulse). $K = 300$, $\varepsilon V = 0.5$. (a) Conventional SPM theory; (b) self-steepened theory.

For all interferometric problems (Born and Wolf, 1975), the method of analysis of nonmonochromatic light is first to find the intensity distribution for a specific frequency as a function of the path time delays of the specific physical setup and then to sum incoherently over the intensities of all frequency components. Therefore, if a light that is incident on a Young/Michelson interferometric system has an input spectral distribution given by $I_{\text{in}}(\omega)$, the output intensity I_{out} from this system is then given by (Manassah, 1987a)

$$I_{\text{out}}(\Delta) = \int I_{\text{in}}(\omega) H(\omega, \Delta)\, d\omega, \tag{110}$$

where Δ is the time delay associated with the two paths and $H(\omega, \Delta)$ is the response of the system to the incoming field of unit amplitude and frequency ω. Specifically, the monochromatic response function for the Young configuration is given by

$$H(\omega, \Delta) = 2\{1 + \cos(\omega\Delta)\}. \tag{111}$$

The relative output intensity for any Δ, normalized to the intensity for $\Delta = 0$, is then given by

$$\frac{I(\Delta)}{I(0)} = \frac{1}{2} + \frac{1}{2} \frac{\displaystyle\int_{-\infty}^{-\infty} dt\, a(t)a(t-\Delta)\cos\{\omega\Delta + \alpha(t) - \alpha(t-\Delta)\}}{\displaystyle\int_{-\infty}^{\infty} a^2(t)\, dt}. \tag{112}$$

For CW radiation, the minima of the interferometric intensity are located at

$$\Delta(n) = \frac{(2n+1)\pi}{\omega}.$$

In normalized time T, and using normalized frequency K, the normalized time delay Δ/τ is parametrized as y/K. In these units, the CW minima correspond to

$$y(n) = (2n+1)\pi \tag{113}$$

and the expression for the relative intensity of the interference pattern reduces to

$$I_R(y) = \frac{I(y)}{I(0)} = \frac{1}{2} + \frac{1}{2} \frac{\displaystyle\int_{-\infty}^{\infty} dT\, a(T)a(T-y/K)\cos\{y + \alpha(T) - \alpha(T-y/K)\}}{\displaystyle\int_{-\infty}^{\infty} dT\, a^2(T)}. \tag{114}$$

In Figure 5.25 we plot $I_R(y)$ for fixed K (i.e., fixed pulse width) but variable εV (i.e., changing pulse energy). As can be observed, both the values and positions of the extrema change with εV. Furthermore, for large n, the ratio of the magnitude of a maximum intensity to that of its neighboring minimum

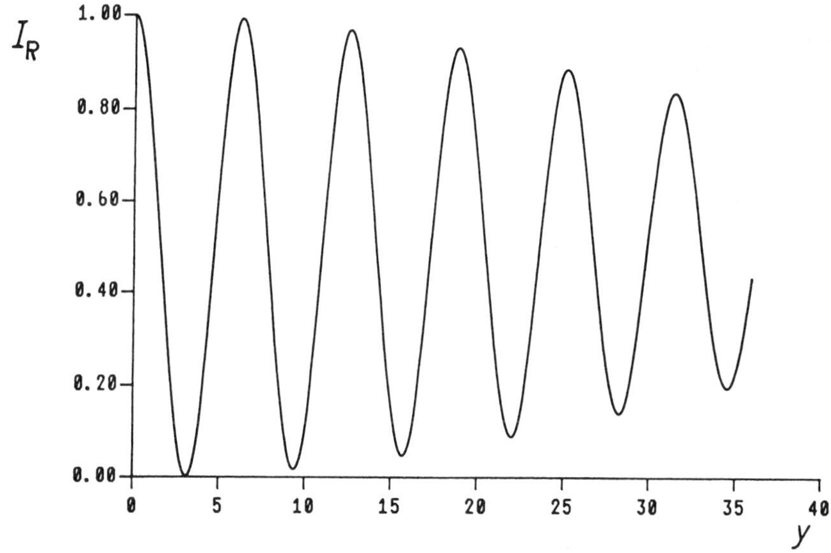

(a)

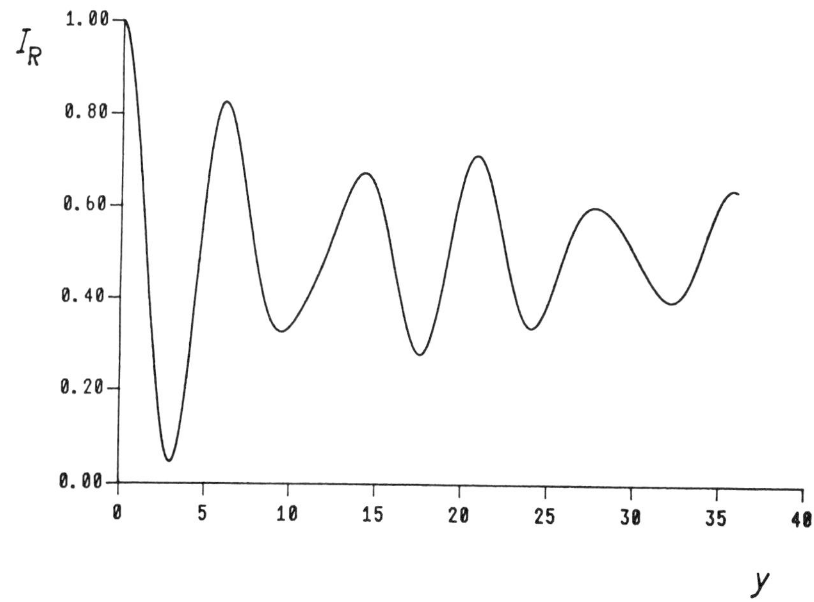

(b)

FIGURE 5.25. Young/Michelson normalized interferometric intensity distribution $I_R(y)$, for a steepened self-phase-modulated input, as a function of y $(= \omega_0 \Delta)$. $K = 50$. (a) $\varepsilon V = 0.1$; (b) $\varepsilon V = 0.5$; (c) $\varepsilon V = 0.8$.

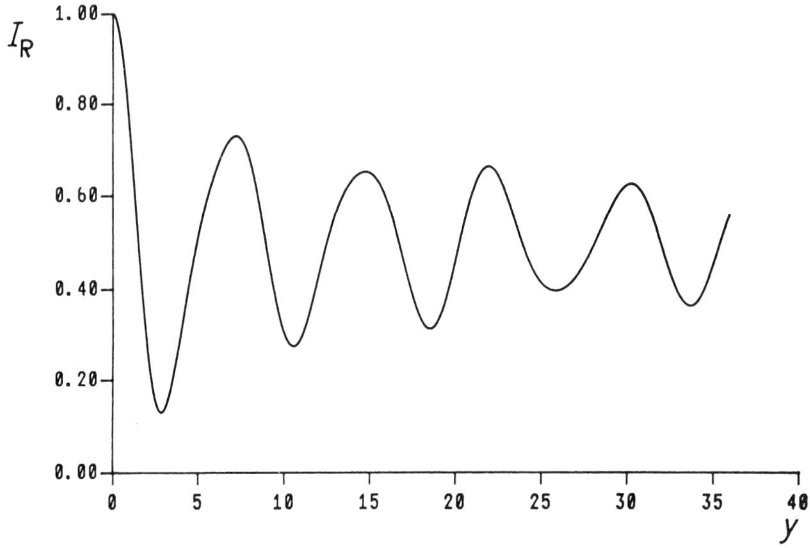

(c)

FIGURE 5.25 (*continued*)

intensity decreases with increasing εV. This ratio goes asymptotically, for large n, to 1.

In Figure 5.26, we plot the shifts in the position of the third minimum (i.e., the minimum that corresponds to $y(2) = 5\pi$ for CW radiation) as a function of K, but fixed εV. As can be observed, this shift depends only weakly on K. The approximate scaling—i.e., the strong dependence of the shift in fringe positions on the ratio εV, essentially the pulse intensity over the pulse duration, and not independently on the pulse width—will be discussed next. The phase of an SPM signal was shown in Section 2.5 to depend only on the parameter K through a multiplicative factor, $\alpha = K\bar{\alpha}$, where $\bar{\alpha}$ depends only on the parameter εV. Consequently, if for $y/K < 1$ we approximate Eq. (114) by the leading term of its Taylor series, we obtain

$$I_R^A(y) \approx \frac{1}{2} + \frac{1}{2} \frac{\displaystyle\int_{-\infty}^{\infty} dU\, a^2(U) \cos\{y(1 + \partial\bar{\alpha}/\partial U)\}}{\displaystyle\int_{-\infty}^{\infty} dU\, a^2(U)}. \qquad (115)$$

This approximate expression for I_R does not have an explicit dependence on K. We notice from Figure 5.26 that this asymptotic value for I_R is already within 2% of its exact value for $K \sim 90$ (i.e., a 30-fs pulse for $\lambda \sim 0.6\ \mu$m). Furthermore, by examining Eq. (114) it also becomes clear why the εV-dependent amplitude-phase time shift is responsible for the interferometric

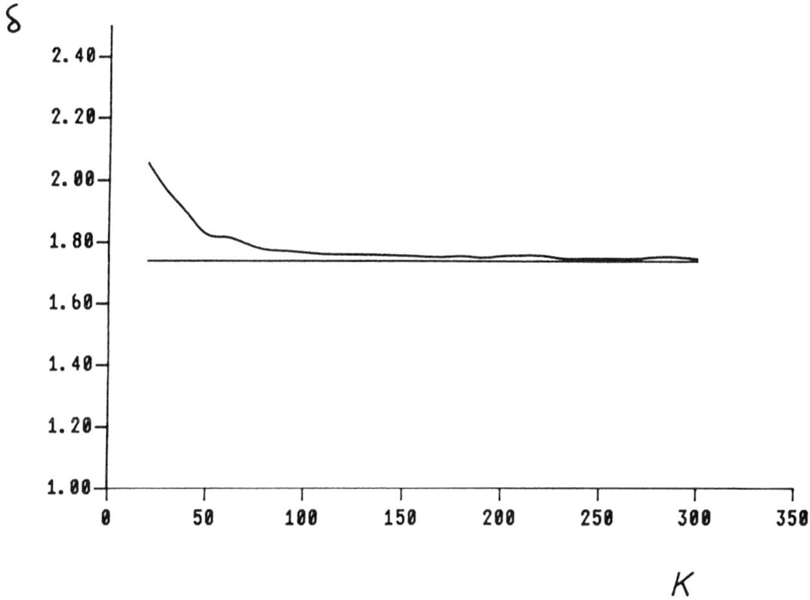

FIGURE 5.26. Shifts in the position of the Young interferometric third-order minimum, for a steepened self-phase-modulated input, $\varepsilon V = 0.5$, as a function of K $(= \omega_0 \tau)$.

TABLE 5.2. Shifts in the positions of the asymptotic interferometric intensity minima for various orders and εV values.

n / εV	0	1	2	3	4
0.1	−0.006	−0.016	−0.021	−0.029	−0.034
0.2	−0.018	−0.051	−0.073	−0.045	−0.343
0.3	−0.040	−0.101	−0.010	1.14	2.29
0.4	−0.072	−0.127	1.84	2.02	2.20
0.5	−0.113	.115	1.74	2.20	3.63
0.6	−0.161	1.29	1.59	3.15	4.59
0.7	−0.213	1.19	2.62	3.84	4.87
0.8	−0.254	1.04	2.51	4.01	5.67

minima shifts; in effect $\partial\bar{\alpha}/\partial U \neq 0$ for $U = U_a$ and the argument of the cosine function at that point is εV dependent.

In table 5.2 the shifts in the positions of the I_R^A minima are tabulated as functions of the order of the minima and of the parameter εV. A shift is defined as the difference between the actual minimum of $I_R^A(y)$ and the corresponding CW minimum as defined in Eq. (113). As can be observed, each εV has a distinct signature for the shifts.

A nonspectroscopic method for deducing εV is through an analysis of the Fourier transform of I_R^A. The expression for the Fourier transform can be directly deduced from Eq. (116) and the integral representation of the Dirac delta function, specifically:

$$F(x) \equiv \int_{-\infty}^{\infty} e^{ixy}\left[I_R^A(y) - \frac{1}{2}\right]dy$$
$$= \frac{\pi}{2N}\left\{\sum_i \left|\frac{a^2(U)}{\partial^2\bar{\alpha}/\partial U^2}\right|_{U=U_i} + \sum_j \left|\frac{a^2(U)}{\partial^2\bar{\alpha}/\partial U^2}\right|_{U=U_j}\right\}, \tag{116}$$

where U_i and U_j are respectively the solutions of

$$x + 1 + \partial\bar{\alpha}/\partial U = 0, \tag{117}$$

$$x - 1 - \partial\bar{\alpha}/\partial U = 0, \tag{118}$$

and N is the a^2 integral normalization factor. $F(x)$ is an even function of x. For $x > 0$, only Eq. (117) admits a solution since $|\partial\bar{\alpha}/\partial U| < 1$. The function $F(x)$ is not identically zero in the interval $[a, b]$, where

$$a = \min(\partial\bar{\alpha}/\partial U) + 1, \tag{119}$$

$$b = \max(\partial\bar{\alpha}/\partial U) + 1 \tag{120}$$

(i.e., a determination of the range of $F(x)$ specifies the value of the parameter εV). In Figure 5.27 the maximum and minimum of $(\partial\bar{\alpha}/\partial U)$ are plotted as functions of the parameter εV. It should also be observed here that, as we will see in Section 2.8, the spectral distribution extents on the Stokes and anti-Stokes sides are determined by the extrema of $(\partial\bar{\alpha}/\partial U)$. In Figure 5.28, the function $F(x)$ is plotted for selected values of εV.

2.8 Spectral Distribution

The spectral distribution for a signal is proportional to the magnitude squared of the time Fourier transform of the electric field, which is given by

$$\tilde{E}(K') = \int_{-\infty}^{\infty} \exp(-iK'U)E(U)\,dU. \tag{121}$$

In Figure 5.29, the spectral intensity (Manassah and Mustafa, 1988a) of the supercontinuum is computed from the specific values of the amplitude a and phase α, earlier computed. However, to gain an intuitive feeling for this spectrum structure, we observe that the stationary phase method can be applied to the integral of Eq. (121). Specifically, this equation can be approximated by

$$\tilde{E}(K) \approx \left(\frac{2\pi}{K}\right)^{1/2}\left\{\frac{a(U_1)\exp[iK(\delta + \bar{\alpha}(U_1)) + i\pi/4]}{[\alpha''(U_1)]^{1/2}} + \frac{a(U_2)\exp[iK(\delta + \bar{\alpha}(U_2)) - i\pi/4]}{|\bar{\alpha}''(U_2)|^{1/2}}\right\}, \tag{122}$$

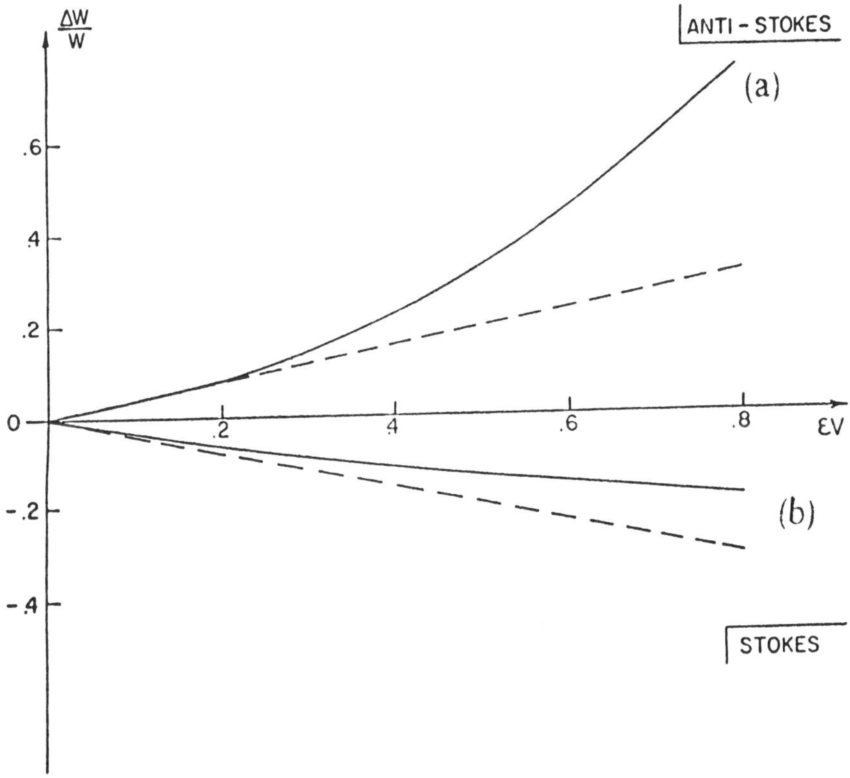

FIGURE 5.27. Values of the extrema of the U partial derivative of the steepened pulse phase as a function of εV. (a) Maxima; (b) minima. (Dashed line) Conventional SPM; (full line) steepened pulse.

where $K - K' = \delta K$ and U_1 and U_2 are the roots of the equation

$$\delta + \bar{\alpha}'(U) = 0, \tag{123}$$

(i.e., the stationary points), and U_1 and U_2 are chosen such that $\alpha''(U_1) > 0$ and $\alpha''(U_2) < 0$. Two stationary points exist for any $(-\delta)$ in the interval spanning the domain of α'. We observe the following features for the spectral distribution of the self-phase modulated supercontinuum:

1. The spectral extents are given by

$$K' - K|_{\text{anti-Stokes}} \approx \text{Max}(\partial\alpha/\partial U), \tag{124}$$

$$K' - K|_{\text{Stokes}} \approx \text{Min}(\partial\alpha/\partial U). \tag{125}$$

Since $\text{Max}(\partial\alpha/\partial U) > |\text{min}(\partial\alpha/\partial U)|$ (see Figure 5.27), the spectral extent on the anti-Stokes side ($\delta < 0$) is appreciably larger than the corresponding quantity on the Stokes side, a result clearly exhibited in Figure 5.29.

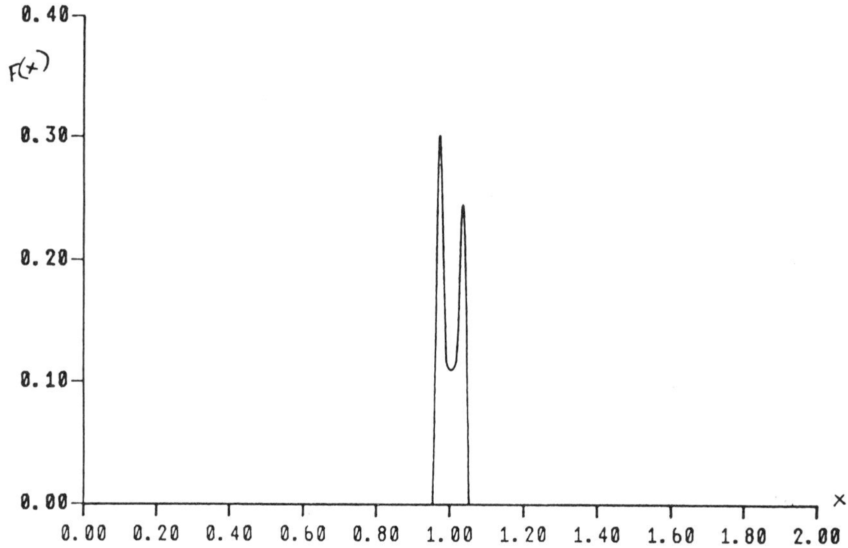

(a)

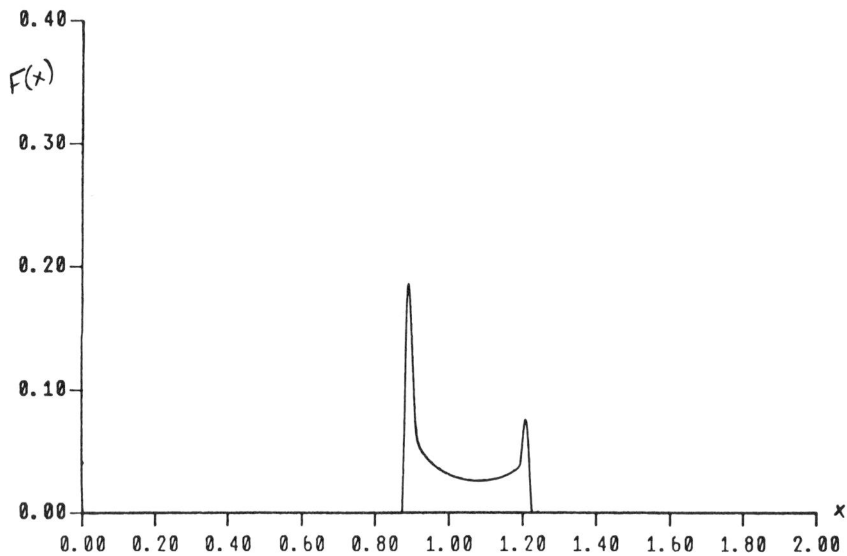

(b)

FIGURE 5.28. Fourier transform of the Young interferometer intensity distribution for a steepened SPM pulse as a function of x. (a) $\varepsilon V = 0.1$; (b) $\varepsilon V = 0.4$; (c) $\varepsilon V = 0.8$.

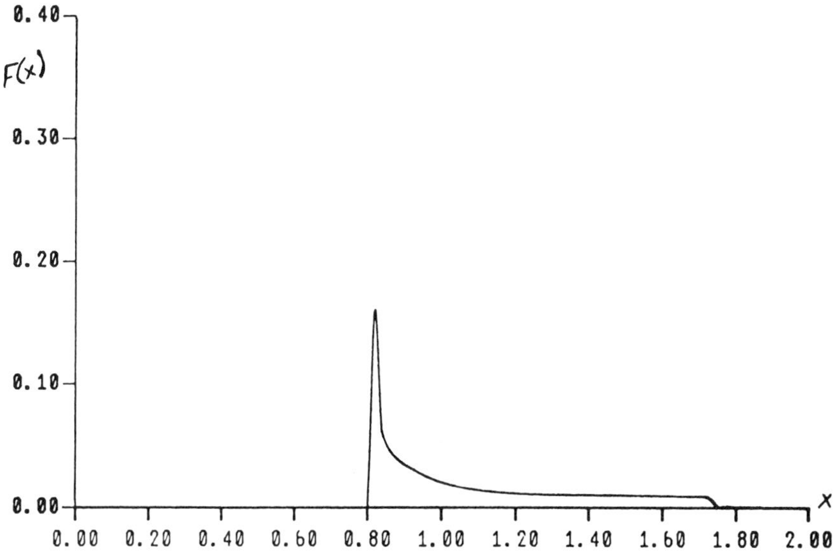

(c)

FIGURE 5.28 (*continued*)

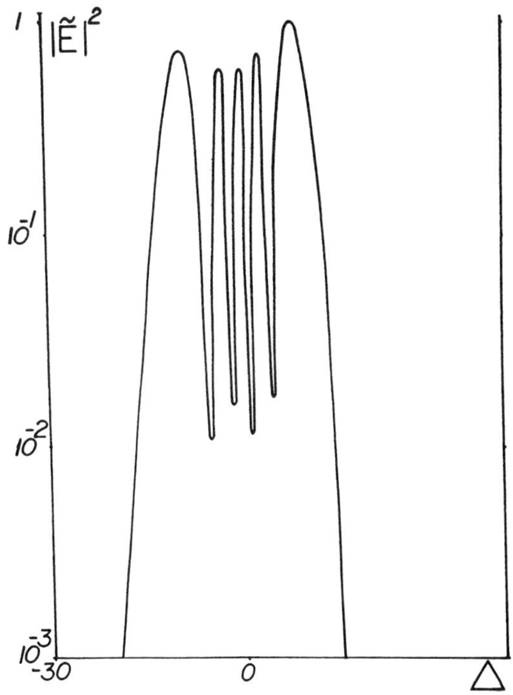

(a)

FIGURE 5.29. Normalized computed spectral distribution of the self-phase-modulated steepened pulse as a function of the frequency difference multiplied by the pulse duration. The pulse duration $\sim 10^{-13}$ s ($K = 300$). Left is anti-Stokes side. (a) $\varepsilon V = 0.1$; (b) $\varepsilon V = 0.4$; (c) $\varepsilon V = 0.8$.

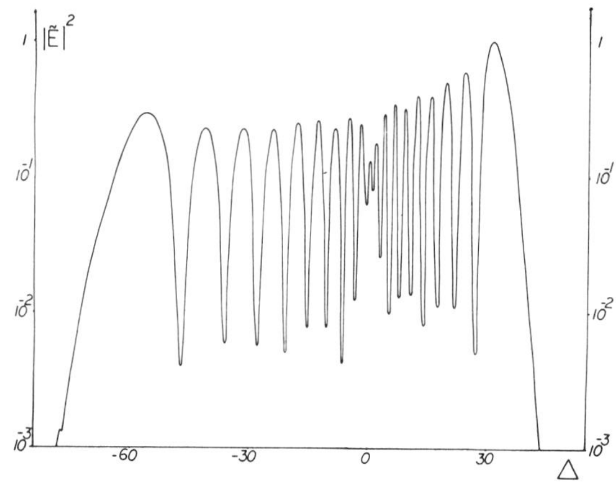

(b)

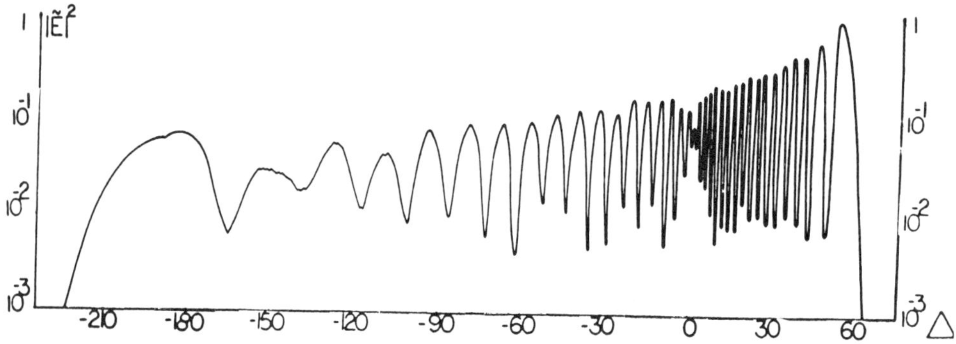

(c)

FIGURE 5.29 (*continued*)

2. The existence of two stationary points for all δ in the spectral distribution domain implies the existence of interference in the spectral intensity. M, the number of oscillations in the spectrum, is given by

$$M \approx \frac{\alpha^M}{\pi} \approx \frac{K\varepsilon V}{2\pi}, \tag{126}$$

where Eq. (104) has been used to derive the last relation. It is worth noting that M increases with both K (i.e., τ) and εV. The modulation frequency of these oscillations is approximately given by the frequency extent divided by the number of oscillations. Combining Eqs. (124), (125), and (126), we deduce that this quantity depends only on εV and is independent of K.

3. The envelope of the maxima of the spectral distribution is approximately given by

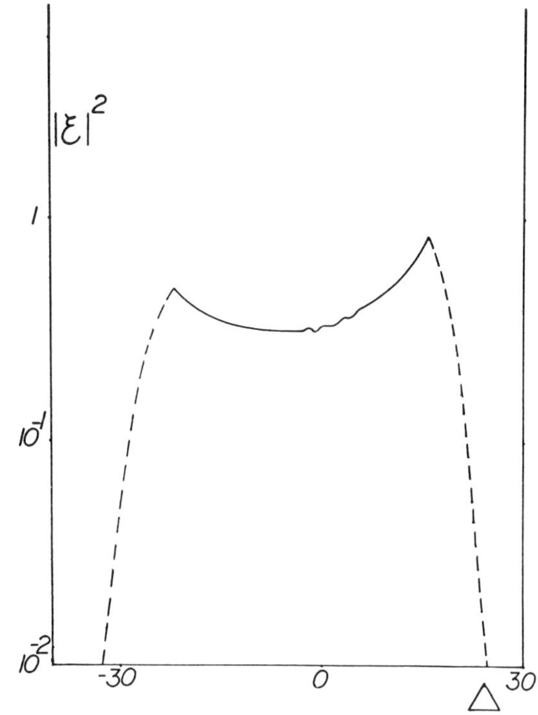

(a)

FIGURE 5.30. Approximate envelopes of the spectra of Figure 5.29 obtained by the method of the stationary phase approximation. (a) $\varepsilon V = 0.1$; (b) $\varepsilon V = 0.4$; (c) $\varepsilon V = 0.8$.

$$|\varepsilon|^2 \approx \frac{2\pi}{K}\left[\frac{a(U_1)}{(\alpha''(U_1))^{1/2}} + \frac{a(U_2)}{|\alpha''(U_2)|^{1/2}}\right]^2. \tag{127}$$

In Figure 5.30 this approximate envelope is plotted for selected εV. As can be observed, it mimics very well the exact envelope that can be deduced from Figure 5.29. The cutoff points for the envelope are, for $|\delta|$, equal to $(\max \bar{\alpha}')$ and $(\min \bar{\alpha}')$. This envelope maximum is in the shallow region of the α' curve. As can be observed from Figure 5.22, this is close to the region of $(\min \alpha')$. Physically, this translates into the spectrum having a sharp band edge close to its Stokes maximum extent.

2.9 The SPM-Spectral Maximum Shift

In Figure 5.31 we plot the position of the spectral distribution maximum as function of εV. We will refer to this displacement of the spectral maximum as the SPM-spectral maximum shift (Manassah and Mustafa, 1988a). For all practical purposes, the magnitude of this shift follows closely the Stokes

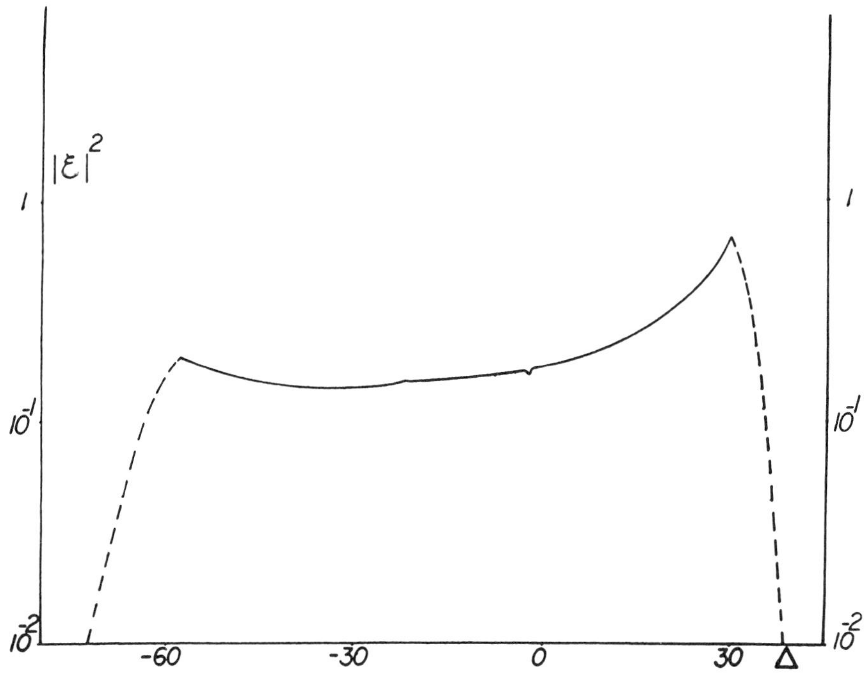

(b)

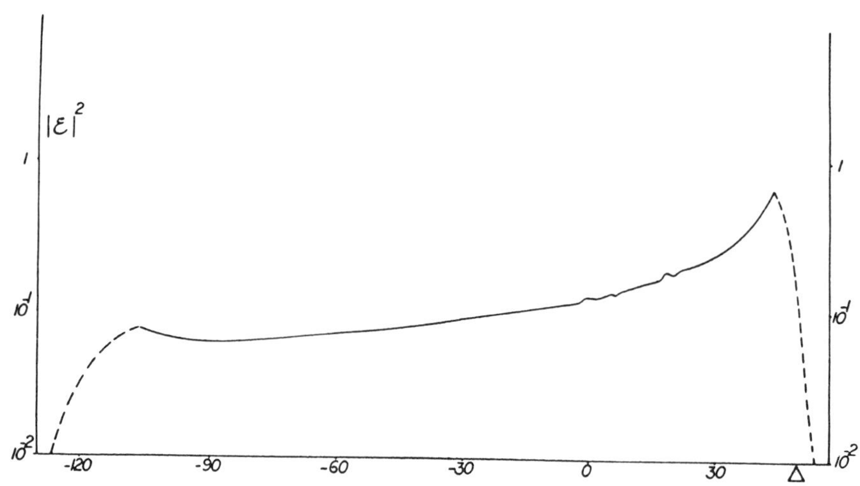

(c)

FIGURE 5.30 (*continued*)

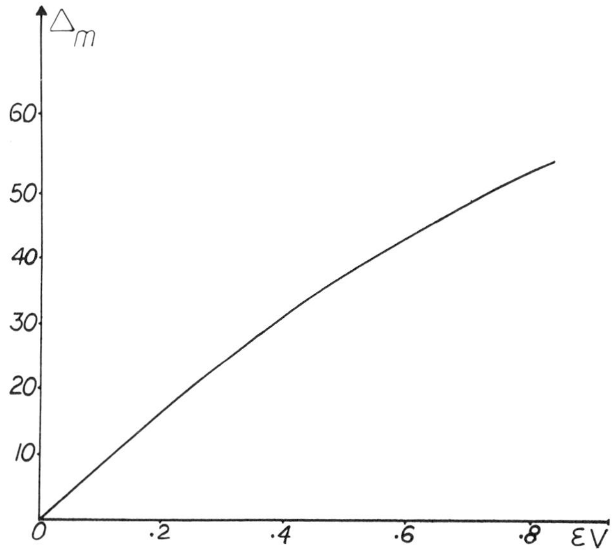

FIGURE 5.31. The SPM-spectral maximum shift, in the presence of pulse steepening, as a function of εV. ($K = 300$.)

frequency extent to within half a modulation cycle. As will be seen in Section 4, this shift may also be observed in induced phase modulation (IPM) due to the similarity relations in the spectral distributions of SPM and IPM proved to exist under certain conditions (Manassah et al., 1985). This similarity-induced shift should be distinguished from the induced frequency shift (Manassah, 1987b) that will be shown to exist because of the group velocity dispersion between pump and probe. Also, the SPM-spectral maximum shift whose origin is the self-steepening of the pulse shape should be distinguished from the self-frequency shift (Gordon, 1986) whose origin is the nonzero relaxation time of the medium.

Physically, the SPM-spectral maximum shift can be understood simply by noting that the $\chi^{(3)}$ nonlinearity does not change the total number of photons in the pulse. Thus, since the spectral anti-Stokes extent is larger than the Stokes extent, in order to conserve energy it is necessary that the peak of the spectral intensity be shifted to the Stokes side. Mathematically the SPM-spectral maximum shift can be approximated as a function of εV by

$$\Delta_m/K \approx \varepsilon V/4. \tag{128}$$

2.10 Filter Transform

In this section we compute the effects of amplitude filters on ultrafast self-phase-modulated pulses (Manassah, 1988c). We examine the dependence of the outgoing signal shape (i.e., width, amplitude, structure, and maxima positions) on the self-phase modulation parameters and filter characteristics. In particular, we show that amplitude filters can be used, under certain conditions, to compress pulses. The physical setup we are examining consists of an SPM pulse passing through an amplitude filter.

The amplitude filter transfer function is assumed to be Gaussian and is given by

$$H(K') = \exp\{-[(K' - K_f)/2\Delta_f\tau]^2\}, \tag{129}$$

where $K = \omega\tau$ and Δ_f and ω_f are, respectively, the filter spectral half-width and center frequency. The outgoing field from the filter is related to the respective ingoing field by

$$\widetilde{E_{\text{out}}}(K') = H(K')\widetilde{E_{\text{in}}}(K'). \tag{130}$$

Expressed as a function of the amplitude and phase of the SPM ingoing field, $E_{\text{out}}(U)$ is given by (Manassah, 1986a)

$$E^{\text{out}}(U) = \frac{E_0(\Delta_f\tau)}{\sqrt{\pi}}\exp(iK_f U)\int_{-\infty}^{\infty} dU'\, a(U')\exp[i\alpha(U')]$$
$$\times \exp[i(K - K_f)U']\exp[-(U' - U)^2(\Delta_f\tau)^2]. \tag{131}$$

In the case of very large $(\Delta_f\tau)$, specifically $(\Delta_f\tau)^2 \gg K(\varepsilon V)$, and using the following representation of the Dirac delta function:

$$\delta(U - U') = \lim_{\gamma\Rightarrow 0}\frac{1}{\gamma\sqrt{\pi}}\exp\left\{-\left[\frac{(U' - U)^2}{\gamma^2}\right]\right\}, \tag{132}$$

$E^{\text{out}} \Rightarrow E^{\text{in}}$. Physically, if the filter is transparent to all frequencies, there is no modification to the pulse.

On the other hand, if $1 \ll (\Delta_f\tau)^2 \ll K\varepsilon V$, we can evaluate Eq. (131) by the stationary phase method; specifically, we can approximate E^{out} by

$$E^{\text{out}}(U) \approx E_0(\Delta_f\tau)\left(\frac{2}{K}\right)^{1/2}\left\{\frac{a(U')\exp[iK(\delta + \bar{\alpha}(U')) + i\pi/4]}{[\alpha''(U')]^{1/2}}\right\}$$
$$\times \exp[-(U' - U)^2(\Delta_f\tau)^2] + \exp[-(U'' - U)^2(\Delta_f\tau)^2]$$
$$\times \left\{\frac{a(U'')\exp[iK(\delta + \bar{\alpha}(U'')) - i\pi/4]}{|\alpha''(U'')|^{1/2}}\right\}, \tag{133}$$

where U' and U'' are the roots of

$$\delta_f + \bar{\alpha}'(U) = 0 \tag{134}$$

(i.e., the stationary points of the integral),

$$K - K_f \equiv \delta_f K, \tag{135}$$

and U' and U'' are chosen such that $\alpha''(U') > 0$ and $\alpha''(U'') < 0$. Two stationary points exist for any $(-\delta_f)$ in the interval spanning the values of α'. Examining Eq. (133), we notice that E^{out} describes two pulses centered respectively at U' and U'' and of width $(\Delta_f \tau)$. Because $(\Delta_f \tau) \ll K\varepsilon V$, the interference term in $|E^{\text{out}}|^2$ is negligible. In the case that $(-\delta_f)$ is close to the maximum or minimum of $\bar{\alpha}'$, the two stationary points are narrowly separated and the two pulses merge into one. This limit is referred to as the one-daughter-pulse regime. It should be noted that the above results differ qualitatively from those associated with chirped Gaussian (CG) pulses. To understand these differences, we note that although the phase function α for SPM and CG pulses coincides in the region around the maximum of α, the respective derivative functions α' and α'' are quiet distinct. For CG pulses, α' is linear in U and its magnitude is symmetric with respect to an axis of symmetry and α'' is constant. For SPM pulses, as we have seen, α' is asymmetric and is bounded; it has two extremes. Consequently, whereas approximating α for an SPM pulse by the CG pulse may be acceptable in the time domain in certain specific instances, it is nearly never so in the frequency domain. The transformation by a filter is essentially a frequency, domain calculation and therefore we should expect major qualitative differences in features between the SPM and CG pulses going through the amplitude filters, namely:

1. For CG pulses Eq. (135) has a single solution, whereas for SPM pulses the same equation admits two, one, or zero solutions depending on the value of δ_f. Physically, this means that while the CG pulse produces only a single daughter pulse on passing through the filter, for an SPM pulse input, two pulses are the normal output from the filter. See Figure 5.32.
2. The outputs from two filters with center frequencies equidistant from the pulse center frequency have similar shapes for a CG input pulse; however, they do no for an SPM input pulse. This asymmetry is manifested through the amplitude, shape, and width of the outgoing pulses. Compare, for example, Figure 5.32 (iv) and (x).
3. In the CG case, the time of arrival of the pulse peak for different filter center frequencies is a linear function of the detuning with the pulse carrier frequency. For SPM pulses, this curve is given by the α' graph previously given (see Figure 5.22). In Figure 5.32, numerically evaluated values of $|E^{\text{out}}|^2$ for different values of the filter center frequency are shown. An efficient compression scheme will concentrate on values of K_f where only one narrow daughter pulse is generated (i.e., near the frequency corresponding to the spectral maximum extent). The computed values for the position of the pulse maximum and width agree to better than 10% with the approximate values of the stationary phase approximation.

For estimation purposes, in the intermediate region (i.e., $(\Delta_f \tau)^2 \sim K(\varepsilon V)$) the stationary phase approximation should be replaced by the method of steepest descent, which is applicable in case the argument of the exponent in the integrand is complex and the exponent is multiplied by a large number, in this instance K.

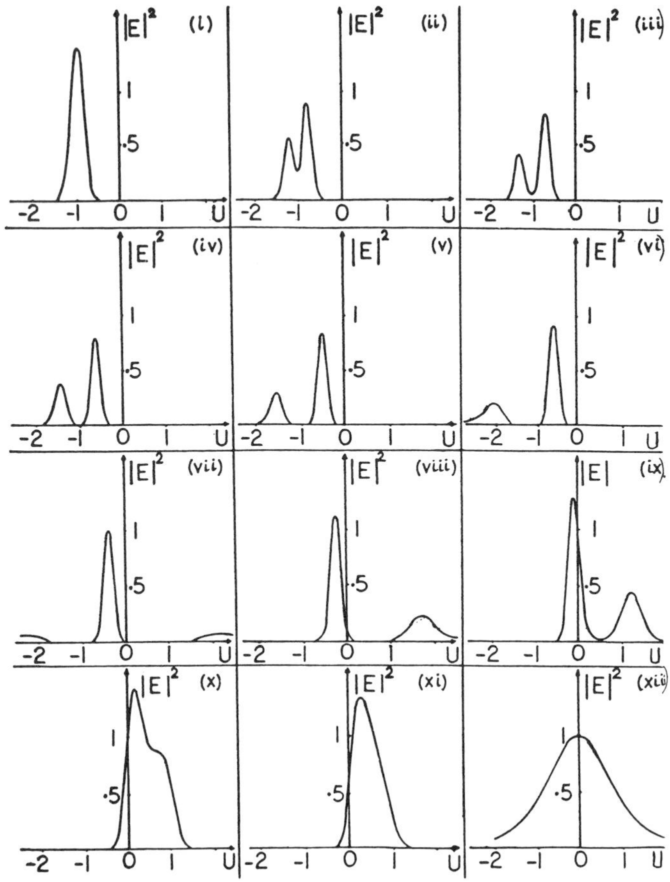

FIGURE 5.32. Field intensity outgoing from an amplitude filter, where the input is a steepened SPM signal ($K = 300$, $\varepsilon V = 0.4$, $\Delta_f \tau = 5$). All the intensities are normalized to the maximum value of the pulse resulting from a filter with the same center frequency as the incoming pulse. (i) $K - K_f = -60$; (ii) $K - K_f = -50$; (iii) $K - K_f = -40$; (iv) $K - K_f = -30$; (v) $K - K_f = -20$; (vi) $K - K_f = -10$; (vii) $K - K_f = 0$; (viii) $K - K_f = 10$; (ix) $K - K_f = 20$; (x) $K - K_f = 30$; (xi) $K - K_f = 34$; (xii) normalized sech^2 pulse.

In Figure 5.33 the computed magnitude of $|E^{(\text{out})}|^2$ and the compression ratio are plotted as functions of the parameter $(\Delta_f \tau)$ in the instance of optimal compression (i.e., one daughter pulse only). As can be observed, as the filter is broadened (more light passes through), the magnitude of $|E^{(\text{out})}|^2$ is increased. The compression ratio (the pulse width of $|E^{(\text{in})}|^2$ divided by the pulse width of $|E^{(\text{out})}|^2$) reaches a maximum as a function of $(\Delta_f \tau)$ and then decreases. As $(\Delta_f \tau)$ grows very large ($(\Delta_f \tau)^2 \gg K(\varepsilon V)$) the compression ratio approaches 1. An approximate value of the compression ratio can be

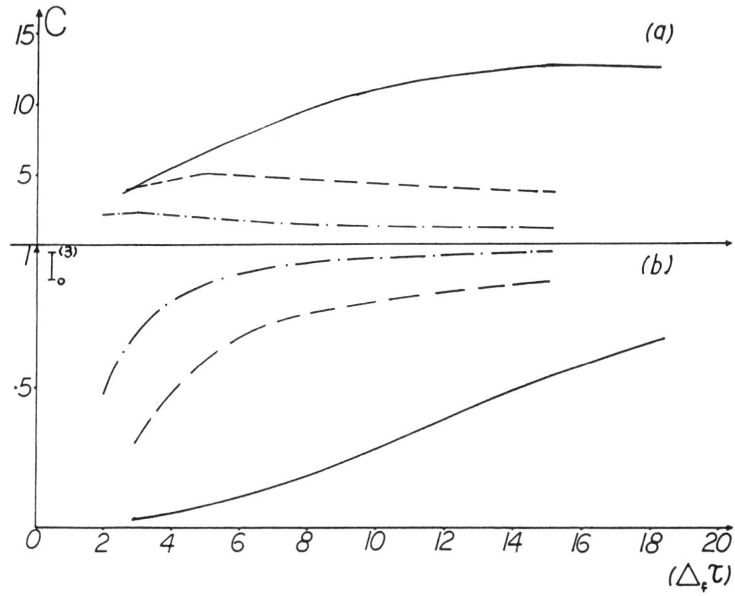

FIGURE 5.33. (a) Compression ratio and (b) intensity magnitude of the steepened SPM pulses outgoing from a filter as functions of $(\Delta_f\tau)$. (i) -- $K = 300$, $\varepsilon V = 0.4$, $K - K_f = -60$; (ii) -·-·- $K = 30$, $\varepsilon V = 0.4$, $K - K_f = -6$; (iii) —— $K = 300$, $\varepsilon V = 0.8$, $K - K_f = -210$.

estimated by the method of steepest descent, namely:

$$C^2 = (\Delta_f\tau)^2 - \frac{(\Delta_f\tau)^4[(\Delta_f\tau)^2 + 1]}{[(\Delta_f\tau)^2 + 1]^2 + g^2}, \tag{136}$$

where $g \approx K\alpha''(U_s)$, and U_s is the average value of the two collapsing stationary points. Comparison of this estimate for C with its computed values gives an error margin of less than 25% over the whole range of $(\Delta_f\tau)$.

Finally, we study the effect of laser source fluctuations on this compression scheme. In Figure 5.34 we show the effect of changing εV by $\pm 25\%$ which corresponds to $\pm 25\%$ in the ingoing laser intensity. While keeping $(\Delta_f\tau)$ constant, the effect of increasing εV is to create two daughter pulses. This can be understood by noting that the magnitude of the maximum of α' increases with εV; therefore an increase of εV means the resurgence of two well-separated stationary points. On the other hand, a decrease in εV drastically reduces the magnitude of $E^{(\text{out})}$ but with only slight variations in the shape. Therefore, in designing a setup for pulse compression through the above scheme, the nonlinear parameters should be selected to overcompensate the positive fluctuation.

It should be emphasized that the above compression scheme produces ultrashort pulses but with center frequency different from that of the incoming

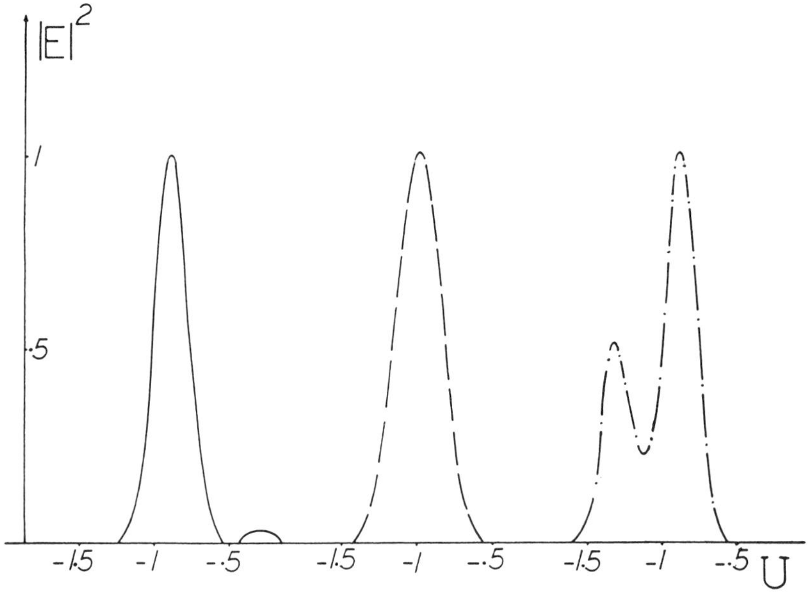

FIGURE 5.34. Effects of fluctuation in εV (i.e., laser intensity) on the shape of the filter output. ($K = 300$, $K - K_f = -60$, $\Delta_f \tau = 5$.) (i) —— $\varepsilon V = 0.3$; (ii) --- $\varepsilon V = 0.4$; (iii) -·-·- $\varepsilon V = 0.5$. The peak intensities are, respectively 8.27×10^{-3}, 6.03×10^{-1}, and 2.54×10^{-1}.

laser signal. This scheme permits, inter alia, the generation of femtosecond pulses in new regions (such as the ultraviolet) of the spectral domain using a laser source, with center frequency in the visible.

3. $\chi^{(5)}$ Dispersionless Medium

So far, our calculations assumed a $\chi^{(3)}$ nonlinear medium. Higher-order leading linearity is possible; for example, in cases where the nonlinearity is due to two-quantum photogeneration of nonequilibrium carriers in certain semiconductors the medium will be a $\chi^{(5)}$ medium. In this section we treat such a case. As we will observe, all calculations made for the $\chi^{(3)}$ medium case can be repeated for the $\chi^{(5)}$ medium case. We restrict ourselves here to some illustrative examples.

3.1 Self-Focusing in $\chi^{(5)}$ Material

In this section we examine the combined effects of the $\chi^{(5)}$ nonlinearity and diffraction on the spatial properties of a Gaussian pulse propagating in a $\chi^{(5)}$ medium. We make the same assumptions as in Sections 1.3 and 1.4.

The envelope of the electric field obeys the equation

$$\nabla_T^2 \varepsilon - 2ik\varepsilon' + \frac{n_4 k^2}{n_0} |\varepsilon|^4 \varepsilon = 0, \tag{137}$$

where ∇_T^2 is the transverse component of the Laplacian $\varepsilon' = \partial\varepsilon/\partial z$, n_4 is the quartic nonlinear index of refraction, and n_0 is the linear index of refraction of the medium.

We consider the initial condition

$$\varepsilon(r, 0) = \varepsilon_0 \exp(-r^2/a^2); \tag{138}$$

that is, we are considering a CW Gaussian beam (the generalization to a Gaussian pulse can be performed using the same steps as in Sections 1.3 and 1.4). An approximate solution to Eq. (137) with the boundary condition (138), correct to order r^2/a^2, can be obtained through the trial solution

$$\varepsilon(r, z) = \frac{\varepsilon_0}{\varpi(z)} \exp\left[-\frac{r^2}{a^2 \varpi^2(z)} - \frac{ik}{2} \rho(z) r^2 + ik\alpha(z) \right], \tag{139}$$

where the different functions have the same meaning as in Sections 1.3 and 1.4. The equations for ϖ, ρ, and α are given by

$$\rho = \varpi'/\varpi, \tag{140}$$

$$\alpha' = \frac{2}{k^2 a^2 \varpi^2} - \frac{n_4}{n_0} k^2 \frac{\varepsilon_0^4}{\varpi^4}, \tag{141}$$

$$\varpi'' - \frac{4}{a^4 k^2} \frac{1}{\varpi^3} + \frac{n_4}{n_0 a^2} \frac{4\varepsilon_0^4}{\varpi^5} = 0. \tag{142}$$

For the initial conditions $\varpi(z = 0) = 1$, and $\rho(z = 0) = 0$, the function ϖ obeys the equation

$$\pm \int_1^{\varpi} \frac{\varpi^2 \, d\varpi}{[(1 - \varpi^2)(\varpi^2 + F/2(D + \frac{1}{2}F))]^{1/2} [D + \frac{1}{2}F]^{1/2}} = z, \tag{143}$$

where

$$D = \frac{-4}{a^4 k^2} \quad \text{and} \quad F = \frac{4n_4}{n_0 a^2} \varepsilon_0^4. \tag{144}$$

The critical field is given by

$$\varepsilon_c = \left(\frac{2n_0}{a^2 k^2 n_4} \right)^{1/4}. \tag{145}$$

For $\varepsilon > \varepsilon_c$, the relation between ϖ and z is given by

$$y \equiv \frac{z}{L_d} = \frac{\sqrt{2}}{(s - 1)} \left\{ \frac{s}{2(s - \frac{1}{2})^{1/2}} \tilde{F}\left(\arccos \varpi, \left(\frac{s - 1}{2s - 1} \right)^{1/2} \right) - (s - \tfrac{1}{2})^{1/2} \tilde{E}\left(\arccos \varpi, \left(\frac{s - 1}{2s - 1} \right)^{1/2} \right) \right\}, \tag{146}$$

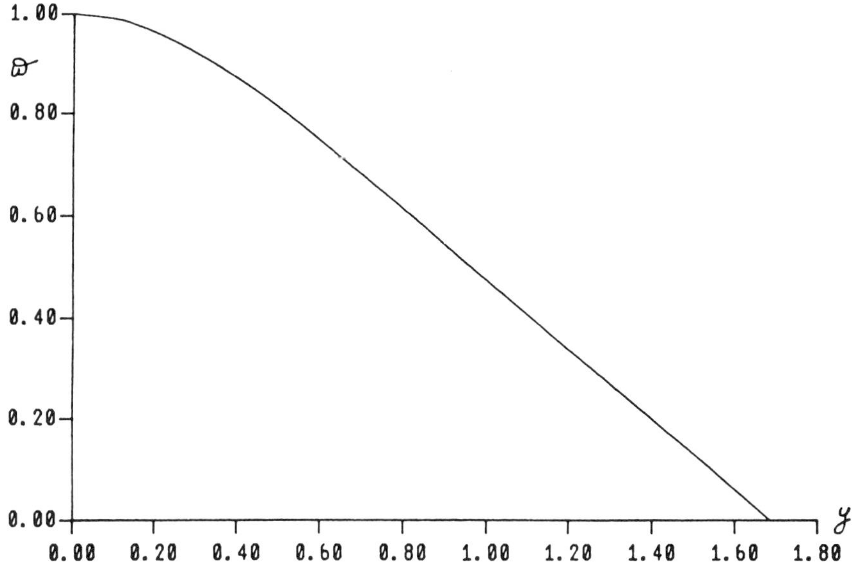

(a)

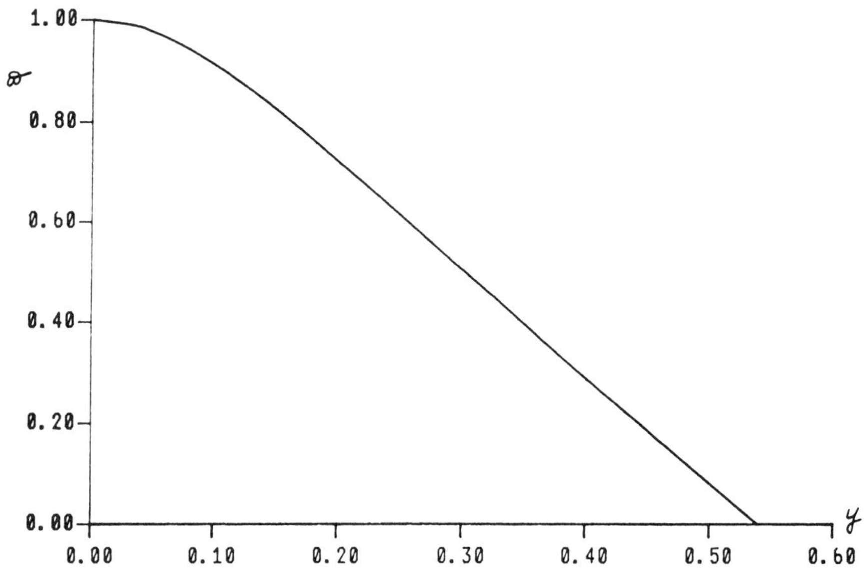

(b)

FIGURE 5.35. Normalized beam diameter of a pulse propagating in a dispersionless $\chi^{(5)}$ material, and with intensity larger than the critical intensity, plotted as a function of the distance expressed in units of the Rayleigh length. (a) $s = (1.1)^4$; (b) $s = 10$; (c) $s = (2)^4$.

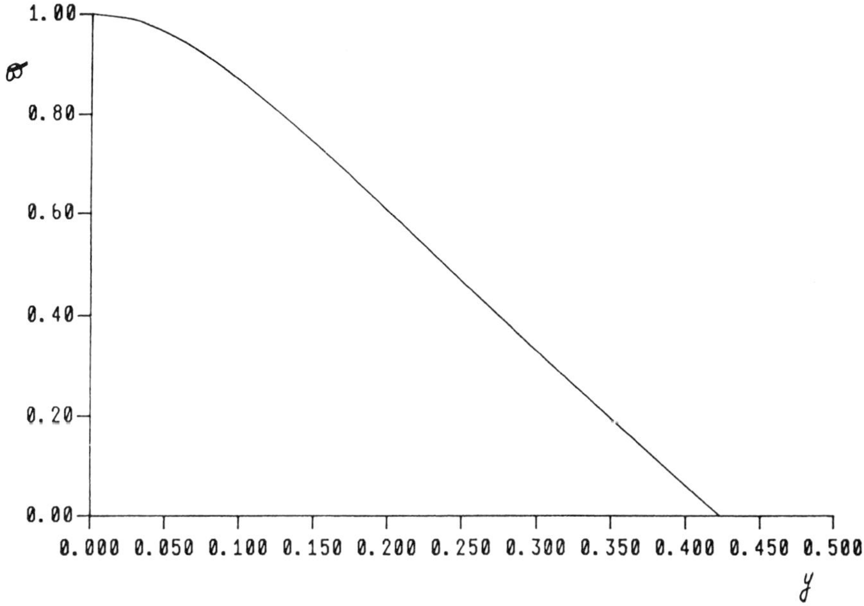

(c)

FIGURE 5.35 (*continued*)

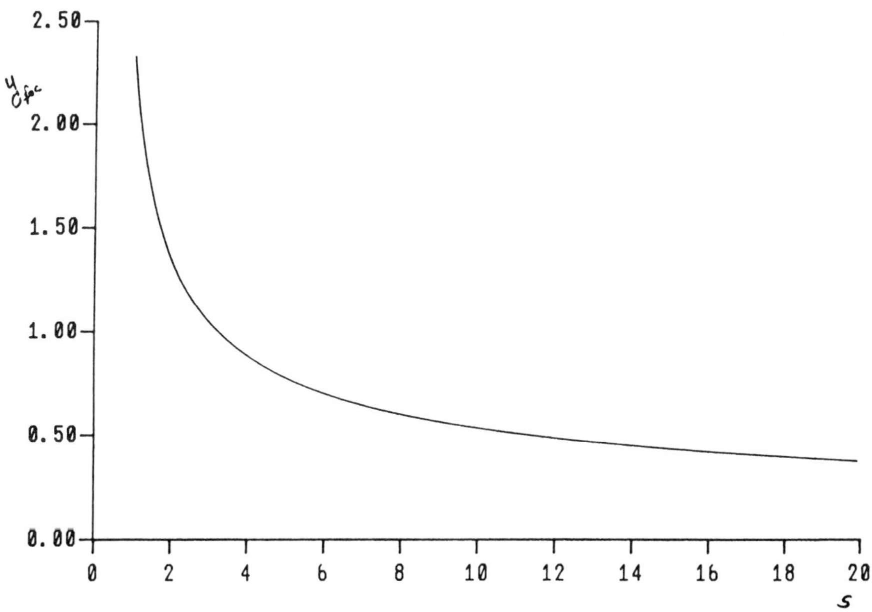

FIGURE 5.36. Normalized focusing distance in $\chi^{(5)}$ material as a function of the pulse (intensity)^2 measured in units of the (critical intensity)^2.

where L_d is the Rayleigh diffraction length,

$$s = \varepsilon_0^4/\varepsilon_c^4, \tag{147}$$

and $\tilde{F}$ and $\tilde{E}$ are the elliptic integrals of the first and second kinds.

In Figure 5.35 the function $\bar{\omega}$ as a function of the normalized length y is plotted for different values of s. In Figure 5.36 the normalized focusing distance as a function of s is plotted. As can be observed, the focusing distance decreases as $s^{-1/2}$ for large s, that is, as the normalized $(\text{energy})^{-1}$, whereas for a $\chi^{(3)}$ medium it decreases as $(\text{energy})^{-1/2}$.

3.2 Self-Steepened Pulse in $\chi^{(5)}$ Dispersionless Medium. Plane Wave Approximation

In this section we investigate the amplitude, phase, and spectral distribution of a high-intensity pulse propagation in a $\chi^{(5)}$ dispersionless medium (Manassah and Mustafa, 1988d). We consider the case where diffraction can be neglected. Our treatment follows exactly the same steps as in Section 2.1; consequently, here we give only the results.

The Maxwell wave equation is given by

$$\nabla^2 \mathbf{E} - \frac{n_0^2}{c^2}\frac{\partial^2}{\partial t^2}\mathbf{E} = \frac{4n_0 n_4}{c^2}\frac{\partial^2}{\partial t^2}\langle \mathbf{E}\cdot\mathbf{E}\rangle^2 \mathbf{E}, \tag{148}$$

where n is the linear index of refraction and n_4 is the quartic nonlinear index of refraction. This wave equation can be simplified under the assumption that one component of $\mathbf{E}$ is present, the transverse variation of $\mathbf{E}$ is neglected (i.e., diffraction effects are neglected), and $\langle \mathbf{E}\cdot\mathbf{E}\rangle = |E|^2/2$. The dimensionless variables Φ, Z, and T are defined in the same manner as in Section 2, and the dimensionless nonlinear coupling constant ε' is defined as

$$\varepsilon' = n_4|E_0|^4/n_0 \tag{149}$$

and the nonlinear wave equation form reduces to

$$\left(\frac{\partial^2}{\partial Z^2} - \frac{\partial^2}{\partial T^2}\right)\Phi = \varepsilon'\frac{\partial^2}{\partial T^2}|\Phi|^4\Phi. \tag{150}$$

Using the variables U, V, and K of Section 2 and the method of multiple scales, the quasi-linear partial differential equations for the amplitude a and the phase α are given by

$$\frac{\partial a}{\partial V} - \frac{5}{2}\varepsilon' a^4 \frac{\partial a}{\partial U} = 0, \tag{151}$$

$$\frac{\partial \alpha}{\partial V} - \frac{\varepsilon'}{2}a^4\frac{\partial \alpha}{\partial U} = \frac{K\varepsilon'}{2}a^4 - \frac{K\varepsilon'^2}{8}a^8. \tag{152}$$

For the initial boundary conditions

$$a(U,0) = \operatorname{sech}(U), \tag{153}$$

$$\alpha(U,0) = 0 \tag{154}$$

the solution for the amplitude is given by

$$a^2 = \operatorname{sech}^2(U + \tfrac{5}{2}\varepsilon' V a^4), \tag{155}$$

where $\varepsilon' V = n_4 |E_0|^4 z/c\tau$. The slope of the amplitude is given by

$$\frac{\partial a}{\partial U} = \frac{-a(1-a^2)^{1/2}}{10\varepsilon' V a^4 (1-a^2)^{1/2} \pm 1}, \tag{156}$$

where $(-, +)$ corresponds respectively to U (smaller, larger) than $-\frac{5}{2}\varepsilon' V$. Solution (155) derived for the amplitude under the condition of no dispersion and no absorption is valid for all values of $V < V_{\text{crit}}$, where V_{crit} is the critical value of V at which the optical amplitude shock develops (i.e., $\partial a/\partial U \Rightarrow \infty$). This solution is smoothed at the shock discontinuity by higher-order derivatives in Maxwell's equation. The value of V_{crit} is given by

$$\varepsilon' V_{\text{crit}} = 5\sqrt{5}/32 \approx 0.349. \tag{157}$$

In Figures 5.37 and 5.38 the amplitude and its first derivative are plotted for selected $\varepsilon' V$. The self-steepening effect is as expected, more pronounced as $\varepsilon' V$ increases, and so is the asymmetry. The position of the amplitude maximum,

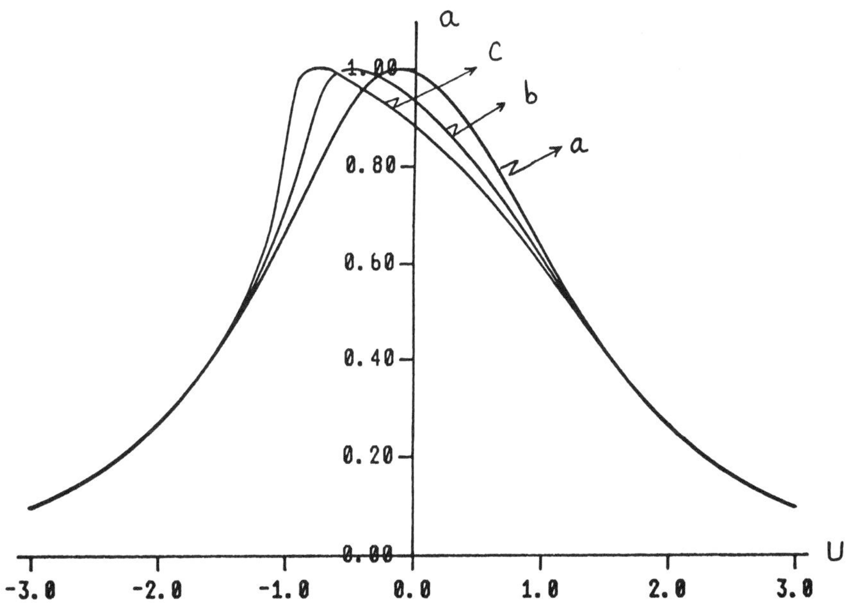

FIGURE 5.37. Steepened amplitude of a pulse propagating in dispersionless $\chi^{(5)}$ material as a function of U. (a) $\varepsilon' V = 0.05$; (b) $\varepsilon' V = 0.2$; (c) $\varepsilon' V = 0.34$.

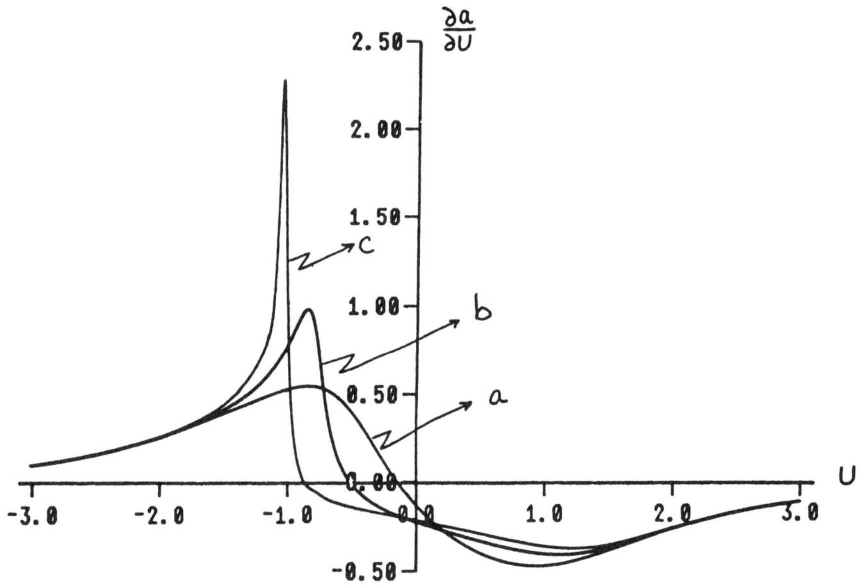

FIGURE 5.38. The U partial derivative of the steepened amplitude of a pulse propagating in dispersionless $\chi^{(5)}$ material as a function of U. (a) $\varepsilon' V = 0.05$; (b) $\varepsilon' V = 0.2$; (c) $\varepsilon' V = 0.34$.

denoted by U_a, is at

$$U_a = -\tfrac{5}{2}\varepsilon' V. \tag{158}$$

The solution of the phase equation (152) obeying (158) the boundary condition of Eq. (154) is given by

$$\bar{\alpha} \equiv \frac{\alpha}{K} = -U - \frac{\varepsilon'}{2}\int_0^U a^4(p, V)\,dp - \frac{5\varepsilon'^2}{24}\int_0^V a^8(0, q)\,dq$$

$$+ \tanh^{-1}[\sin g(U, V)] + \frac{\varepsilon'}{6}\sin[g(U, V)] - \frac{\varepsilon'}{18}\sin^3[g(U, V)] \tag{159}$$

where

$$g(U, V) = \frac{\varepsilon'}{2}\int_0^V a^5(U, q)\,dq + \sin^{-1}(\tanh U). \tag{160}$$

The first partial derivative of α with respect to U is given by

$$\frac{\partial\bar{\alpha}}{\partial U} = -1 - \frac{\varepsilon'}{6}a^4 + \frac{a}{\cos g} + \frac{\varepsilon'}{6}a(\cos g) - \frac{\varepsilon'}{6}a(\sin^2 g)(\cos g). \tag{161}$$

This quantity represents physically the normalized frequency sweep.

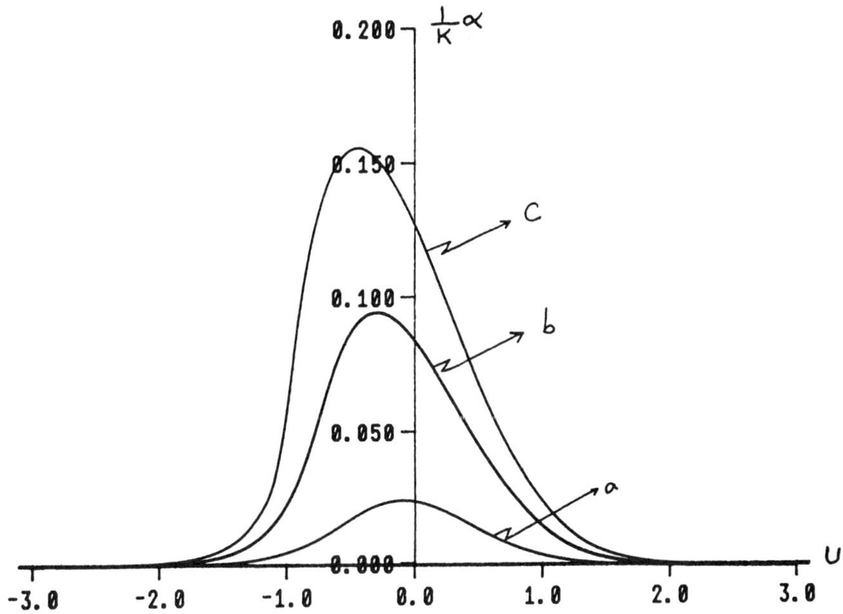

FIGURE 5.39. Normalized phase of a steepened pulse propagating in dispersionless $\chi^{(5)}$ material as a function of U. (a) $\varepsilon' V = 0.05$; (b) $\varepsilon' V = 0.2$; (c) $\varepsilon' V = 0.34$.

The second partial derivative of α with respect to U is given by

$$\frac{\partial \bar{\alpha}}{\partial U} = \frac{\partial a}{\partial U}\left[-\frac{2\varepsilon'}{3}a^3 + \frac{1}{\cos g} + \frac{\varepsilon}{6}\cos g - \frac{\varepsilon'}{6}\sin^2 g \cos g\right] + a^2 \frac{\sin g}{\cos^2 g}$$
$$- \frac{\varepsilon}{6}a^2 \sin g + \frac{\varepsilon}{6}a^2 \sin^3 g - \frac{\varepsilon}{3}a^2 \sin g \cos^2 g. \tag{162}$$

In Figures 5.39, 5.40, and 5.41, respectively, are plotted the phase and its derivatives $\partial\bar{\alpha}/\partial U$ and $\partial^2\bar{\alpha}/\partial U^2$. We note the following:

1. The phase is asymmetric with respect to the U axis. This is a result of the amplitude self-steepening.
2. The position and the value of the phase maximum, denoted respectively by U_α and α_M, are given by

$$U_\alpha \approx -\tfrac{3}{2}\varepsilon' V, \tag{163}$$

$$\alpha_M \approx \frac{K}{2}\varepsilon' V. \tag{164}$$

3. The maxima of the amplitude and the phase, as can be deduced by comparing Eqs. (158) and (163), are shifted with respect to each other. This

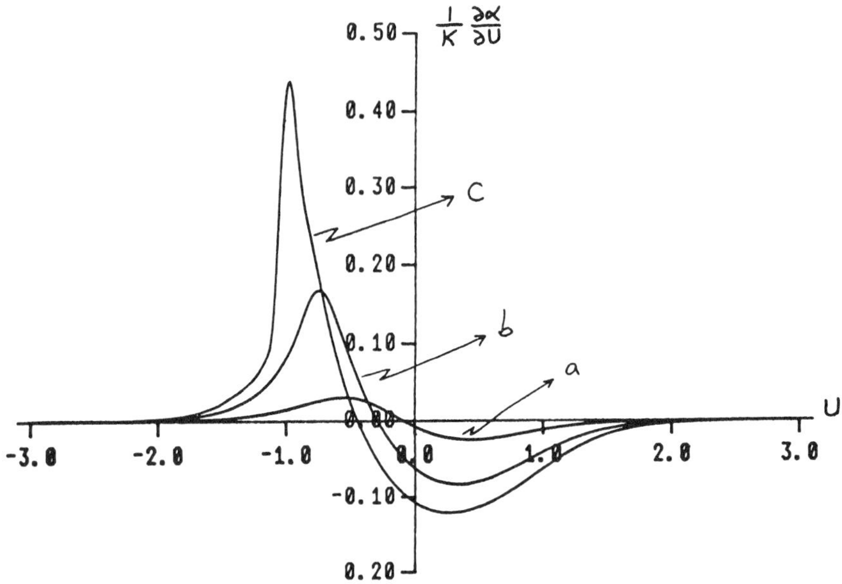

FIGURE 5.40. Normalized frequency sweep (slope of the phase) of a steepened pulse propagating in $\chi^{(5)}$ dispersionless medium as a function of U. (a) $\varepsilon' V = 0.05$; (b) $\varepsilon' V = 0.2$; (c) $\varepsilon' V = 0.34$.

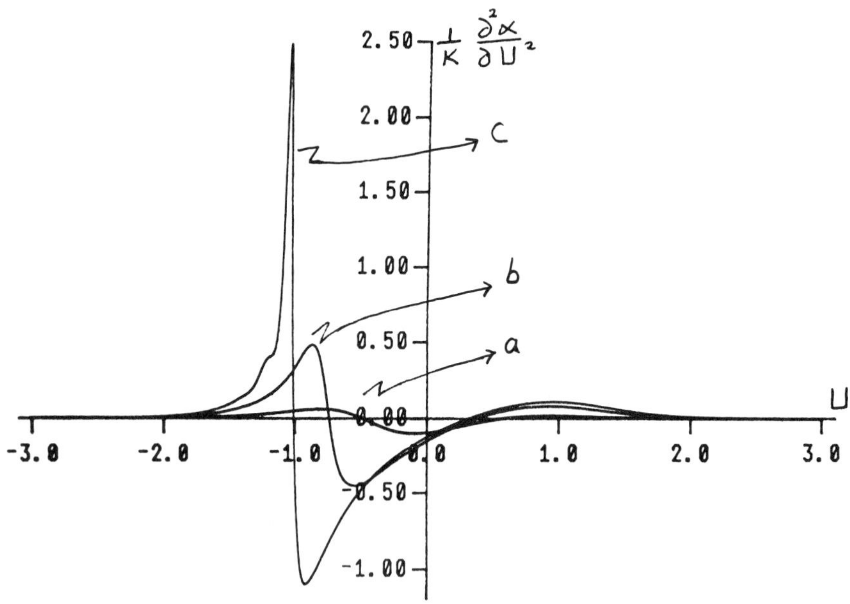

FIGURE 5.41. Normalized second partial derivative of the phase of a steepened pulse propagating in $\chi^{(5)}$ dispersionless medium as a function of U. (a) $\varepsilon' V = 0.05$; (b) $\varepsilon' V = 0.2$; (c) $\varepsilon' V = 0.34$.

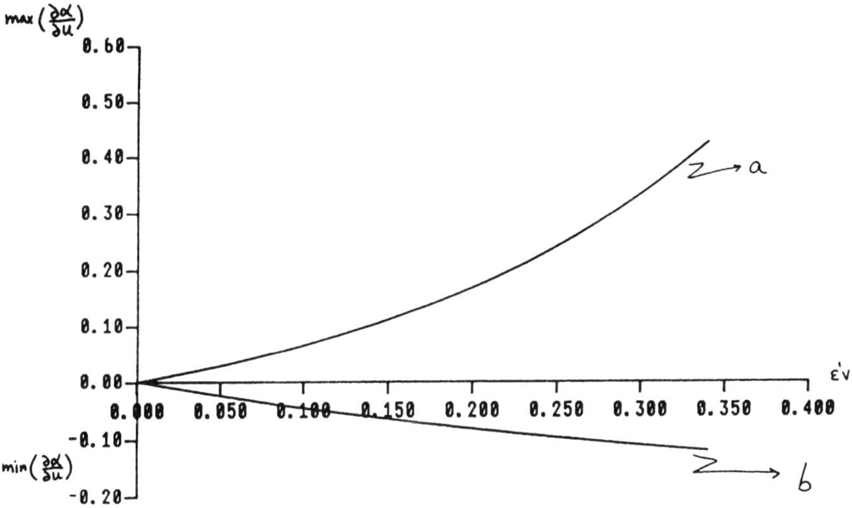

FIGURE 5.42. Extrema of the frequency sweep of a steepened pulse propagating in $\chi^{(5)}$ dispersionless medium as a function of $\varepsilon' V$. (a) Maxima; (b) minima.

results in a shift of the minima of the interference pattern as shown in Section 2.8.

4. The derivative of the phase is asymmetric with respect to the U axis. This results in an asymmetry of the spectral distribution between the Stokes and anti-Stokes portions of the spectrum.
5. The absolute value of the maximum of $\partial\alpha/\partial U$ is always bigger than that corresponding to its minimum value; therefore the anti-Stokes extent is larger than the Stokes extent of the spectrum. This inequality also leads to having for $x > 0$ the left peak of the Fourier transform of the interference pattern closer (in the notation of Section 2.8) to $x = 1$ than the right peak.

In Figure 5.42 we plot the values of $\max(\partial\alpha/\partial U)$ and $\min(\partial\alpha/\partial U)$ as functions of $\varepsilon' V$. The asymmetry in their values for the same $(\text{intensity})^2$ is clear. In Figure 5.43 we plot the function $F(x)$ for this case; the asymmetry in the values of the extrema is again exhibited. Notice that a peak appears at $x = 1$. This peak is associated with the spectral peak at the center frequency, as will be seen later.

The spectral distributions of the resulting supercontinuum obtained by taking the amplitude squared of the Fourier transform of the electric field are shown in Figure 5.44. We note that:

1. The detailed features of the calculated spectral extents conform very well with the estimated values deduced from the values of $\max(\partial\alpha/\partial U)$ and $\min(\partial\alpha/\partial U)$.

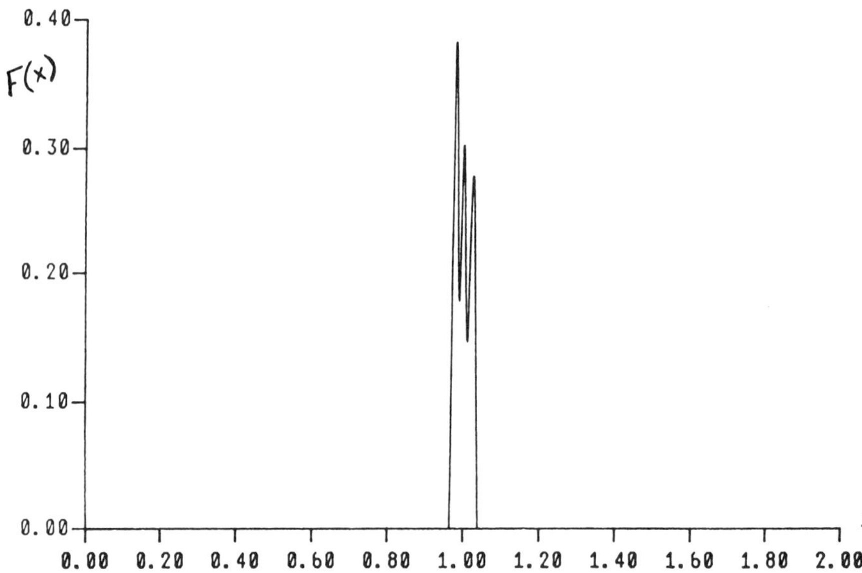

(a)

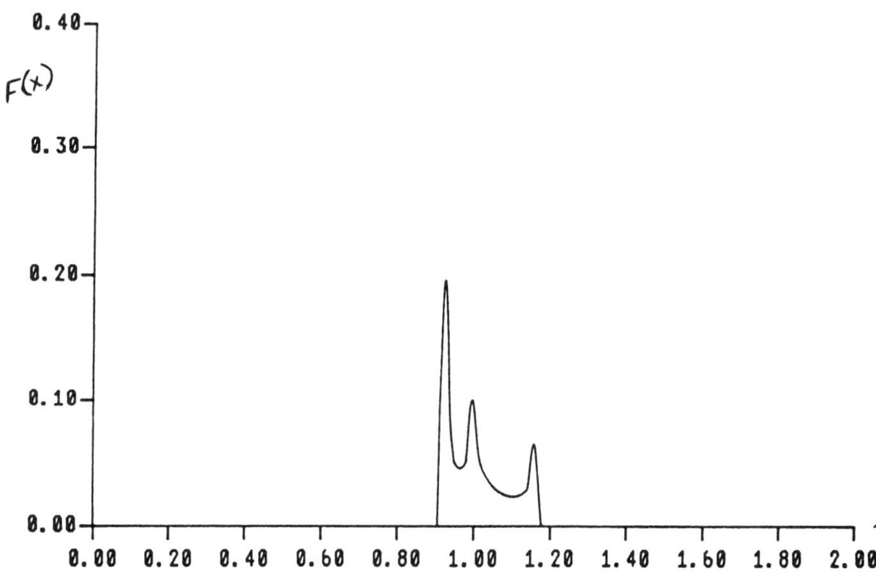

(b)

FIGURE 5.43. Envelope of the Fourier transform of the intensity distribution of a Young interferometer having as input a steepened pulse outgoing from a dispersionless $\chi^{(5)}$ material. (a) $\varepsilon' V = 0.05$; (b) $\varepsilon' V = 0.2$; (c) $\varepsilon' V = 0.34$.

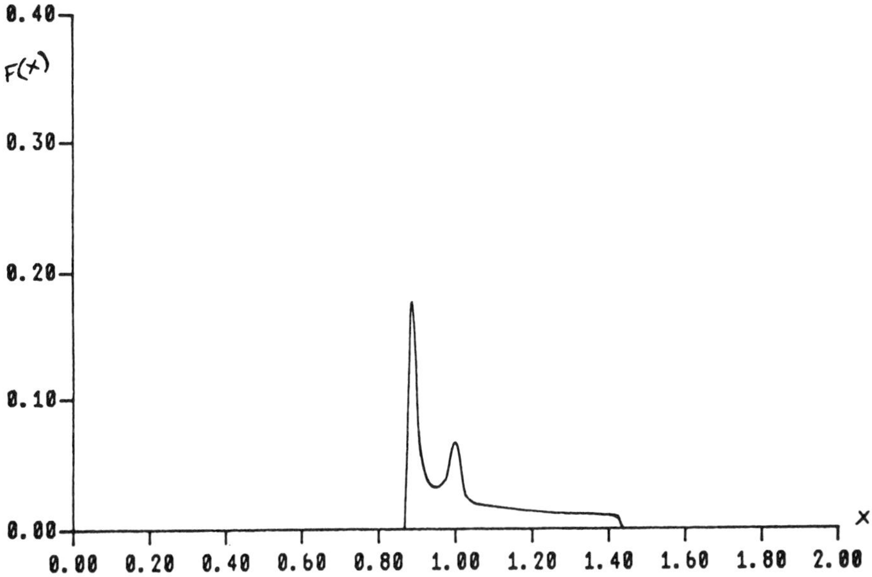

(c)

FIGURE 5.43 (*continued*)

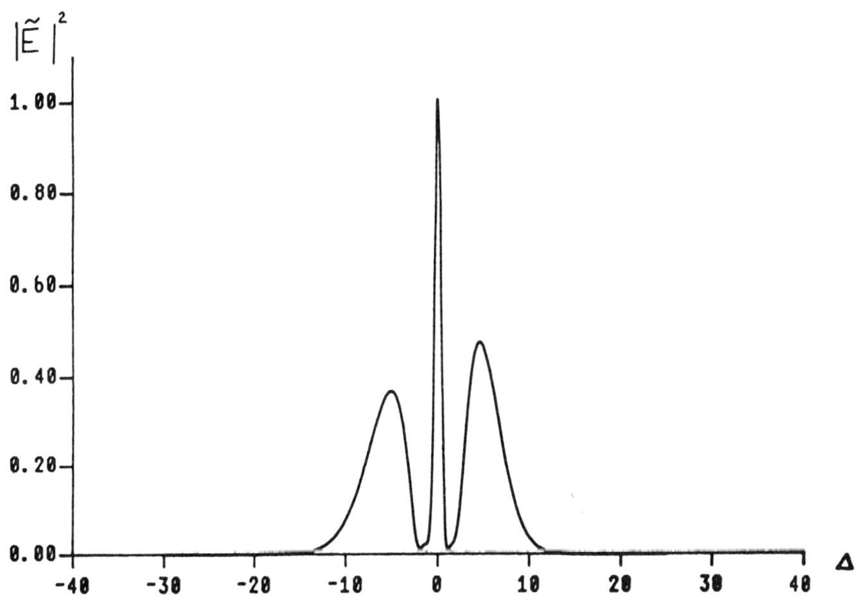

(a)

FIGURE 5.44. Spectral distribution of a steepened pulse outgoing from a dispersionless $\chi^{(5)}$ material. The zero of Δ is the center frequency of the original pulse. (a) $\varepsilon' V = 0.05$; (b) $\varepsilon' V = 0.2$; (c) $\varepsilon' V = 0.34$.

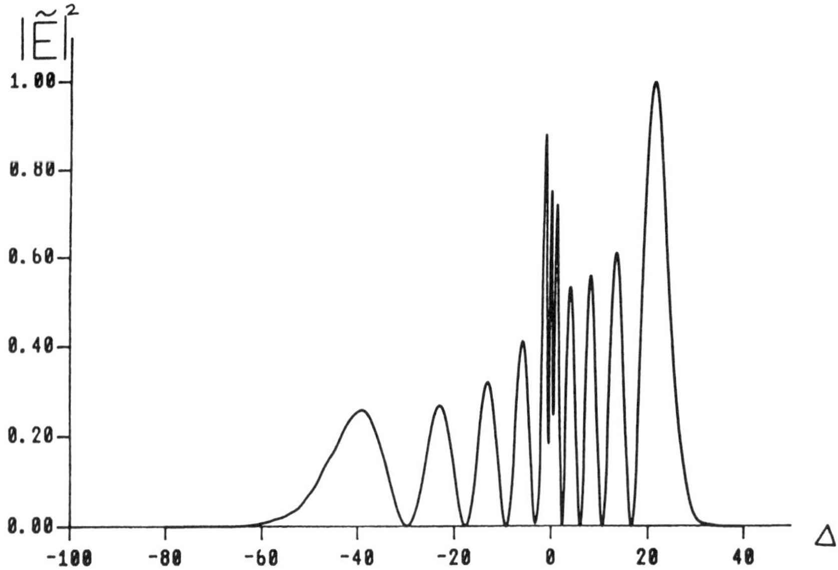

(b)

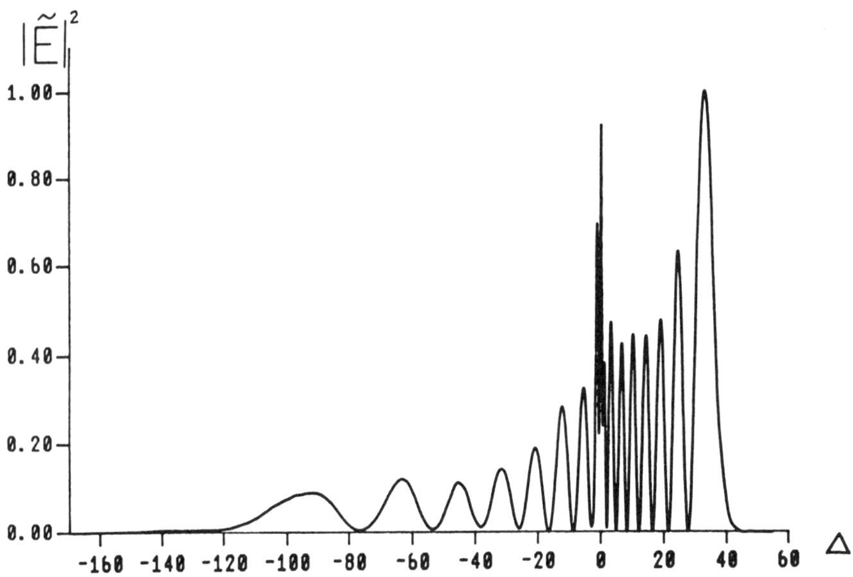

(c)

FIGURE 5.44 (*continued*)

2. The existence of a spectral peak on the edge of the Stokes portion of the spectrum is a result of the shallowness of the $\partial\alpha/\partial U$ curve near its minimum value.
3. The number of oscillations in the spectral distribution is estimated by the method of the stationary phase approximation to be given by

$$\alpha_M/\pi \approx K\varepsilon' V/2\pi \tag{165}$$

and is the close agreement with the numerically computed values.
4. The spectral shape has a peak at the center frequency, a new feature not present in the $\chi^{(3)}$ medium supercontinuum.

$\chi^{(5)}$ in the presence of self-steepening and material relaxation is treated in the appendix at the end of the chapter.

4. Induced Nonlinear Effects

In this section we study the geometric, time, and frequency domain effects on a pulse (probe) propagating in a $\chi^{(3)}$ medium due to the presence of a strong pump.

4.1 Induced-Phase Modulation

When a weak probe pulse is sent together with a pump, the phase of the probe pulse at different frequencies can be modulated by the time variation of the nonlinear index of refraction originating from the primary intense pulse (Manassah, 1987b). This process is defined as induced-phase modulation (IPM). If we denote by A and B the envelopes of the pump and probe, respectively, their differential equations are given for the lowest order of the quasi-linearized model (i.e., the pump shape distortion due to group velocity dispersion is neglected, higher derivatives beyond the first derivative of the index of refraction are neglected, the self-steepening effect is neglected, and the transverse variation of the envelope is neglected) by

$$\frac{\partial A}{\partial z} + \frac{1}{v_a}\frac{\partial A}{\partial t} = i\gamma_a[|A|^2 + 2|B|^2]A, \tag{166}$$

$$\frac{\partial B}{\partial z} + \frac{1}{v_b}\frac{\partial B}{\partial t} = i\gamma_b[|B|^2 + 2|A|^2]B, \tag{167}$$

where $\gamma_{a,b} = n_2 k_{a,b}/2n_0$, n_0 is the linear index of refraction, n_2 is the quadratic (Kerr) index of refraction, γ_a is the conventional SPM coefficient (see Section 1.1), and v_a and v_b are respectively the group velocities of the pump and the probe in the medium. We assume that the probe and pump durations are given by τ_b and τ_a and we introduce the normalized probe comoving coordinate $\bar{U}$ defined by

$$\bar{U} = \frac{1}{\tau_a}\left(t - \frac{z}{v_b}\right) \tag{168}$$

where the normalization is with respect to the pump duration. Denoting by F and G the pulse shape functions for the pump and probe, the initial conditions for the amplitude and phase of each pulse are given by

$$a(0,t) = a_0 F[(t - t_0)/\tau_a], \tag{169}$$

$$\alpha(0,t) = 0, \tag{170}$$

$$b(0,t) = b_0 G(t/\tau_b), \tag{171}$$

$$\beta(0,t) = 0, \tag{172}$$

where t_0 is the initial displacement of maxima between the pump and probe. The solutions of Eqs. (166) and (167), in the case that $a \gg b$, obeying the boundary conditions (169) through (172) are given by

$$a(z,t) = a_0 F\left[\left(t - t_0 - \frac{z}{v_a}\right)\Big/\tau_a\right], \tag{173}$$

$$\alpha(z,t) = \gamma_a a_0^2 F^2\left[\left(t - t_0 - \frac{z}{v_a}\right)\Big/\tau_a\right] z, \tag{174}$$

$$b(z,t) = b_0 G\left[\left(t - \frac{z}{v_b}\right)\Big/\tau_b\right], \tag{175}$$

$$\beta(z,t) = 2\gamma_b a_0^2 \psi(z,t), \tag{176}$$

where

$$\psi(z,t) = \int_0^z F^2\left\{\left[t - t_0 - \frac{z}{v_b} - z'\left(\frac{1}{v_a} - \frac{1}{v_b}\right)\right]\Big/\tau_a\right\} dz'. \tag{177}$$

If the pump form function is hyperbolic secant, then

$$\psi = \frac{1}{\eta}\left[\tanh\left(\frac{t - t_0}{\tau_a} - \frac{z}{\tau_a v_b}\right) - \tanh\left(\frac{t - t_0}{\tau_a} - \frac{z}{\tau_a v_b} - \eta z\right)\right], \tag{178}$$

where

$$\eta = \left(\frac{1}{v_a} - \frac{1}{v_b}\right)\Big/\tau_a. \tag{179}$$

η is the inverse walk-off distance between the pump and probe over the duration of the pump pulse. In the $\bar{U}$ notation, the functions a, α, b, and β reduce to

$$a = a_0 \operatorname{sech}[\bar{U} - \eta z - T_0], \tag{180}$$

$$\alpha = (\gamma_a a_0^2 z)\operatorname{sech}^2[\bar{U} - \eta z - T_0], \tag{181}$$

$$b = b_0 G(\sigma \bar{U}), \tag{182}$$

$$\beta = (2\gamma_b a_0^2 z)\psi = \frac{(2\gamma_b a_0^2 z)}{\eta z}[\tanh(\bar{U} - T_0) - \tanh(\bar{U} - T_0 - \eta z)], \tag{183}$$

where $T_0 = t_0/\tau_a$ and $\tau_a \equiv \sigma\tau_b$. It is worth noting that in this model the pump

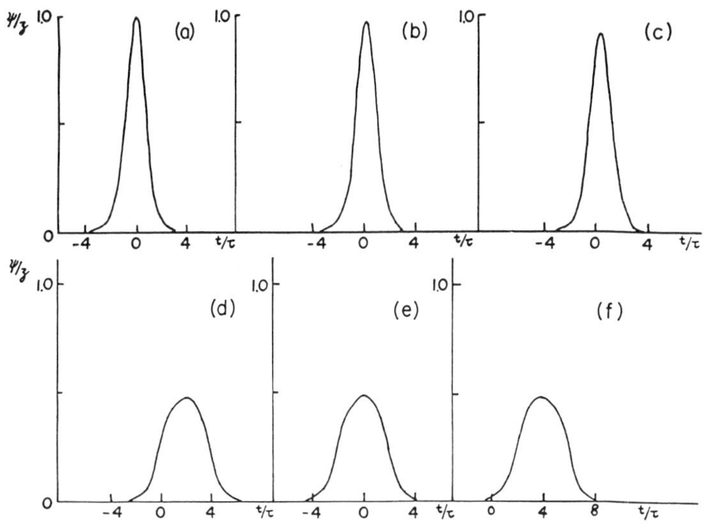

FIGURE 5.45. Induced phase function ψ. (a) $\eta = 0$, $T_0 = 0$; (b) $\eta z = 1/2$, $T_0 = 0$; (c) $\eta z = 1$, $T_0 = 0$; (d) $\eta z = 4$, $T_0 = 0$; (e) $\eta z = 4$, $T_0 = -2$; (f) $\eta z = 4$, $T_0 = 2$.

only modulates the probe; that is, there is no energy transfer between the two pulses. In Figure 5.45 the phase function ψ is plotted for different values of the parameters.

In the frequency domain, the electric field for the probe signal is proportional to

$$\tilde{E}_B(\Delta) = \frac{1}{2\pi}\int_{-\infty}^{\infty} b(\bar{U})\exp[i\beta(\bar{U}) + i\,\Delta\bar{U}]\,d\bar{U} \tag{184}$$

where $\Delta = (\omega' - \omega)\tau_a$ and ω is the probe center frequency. The spectral intensity is proportional to $|\tilde{E}_B(\Delta)|^2$. In Figure 5.46 the spectral intensity of the probe is plotted for different values of the parameters ηZ and T_0 in the case that $\sigma = 1$.

The spectral extent of the probe due to IPM can be obtained by the stationary phase method; in essence:

$$\bar{\Delta} \approx \left(\frac{\partial\beta}{\partial\bar{U}}\right)_{\max} - \left(\frac{\partial\beta}{\partial\bar{U}}\right)_{\min}, \tag{185}$$

where

$$\bar{\Delta} = \Delta_1 - \Delta_2 \tag{186}$$

and the values of Δ_1 and Δ_2 are given by

$$\Delta_{1,2} = \frac{2\gamma_b a_0}{\eta}[\operatorname{sech}^2(\bar{U}_{1,2} - \eta l) - \operatorname{sech}^2(U_{1,2})], \tag{187}$$

where l is the length of the material and $\bar{U}_{1,2}$ are the roots of the transcendental

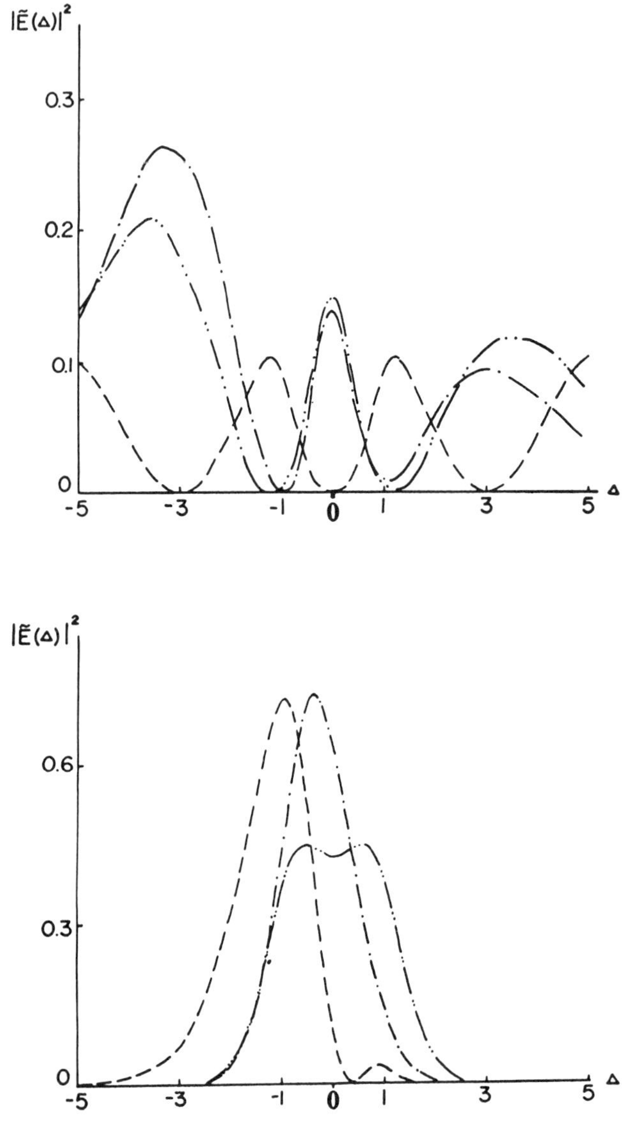

FIGURE 5.46. Spectral distribution of the probe. The probe and pump have the same initial width. *Top figure*: (a) -- $\eta z = 0$, $T_0 = 0$; (b) -···- $\eta z = 1/2$, $T_0 = 0$; (c) -·- $\eta z = 1$, $T_0 = 0$; *Bottom figure*: (a) -- $\eta z = 4$, $T_0 = 0$; (b) -···- $\eta z = 4$, $T_0 = -2$; (c) -·- $\eta z = 4$, $T_0 = 2$. $[a_0^2 \gamma z = 5\pi/2.]$

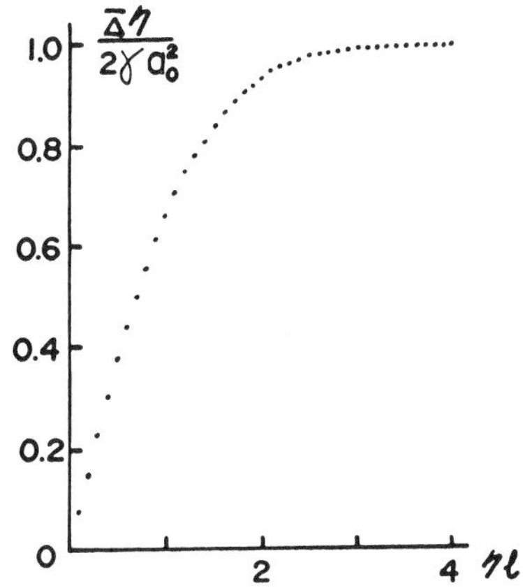

FIGURE 5.47. Induced frequency extent as a function of fiber length.

equation

$$\frac{\sinh(\bar{U} - \eta l)}{\cosh^3(\bar{U} - \eta l)} = \frac{\sinh(\bar{U})}{\cosh^3(\bar{U})}. \tag{188}$$

In Figure 5.47 the spectral extents are plotted as functions of ηl. We note that the spectral extent of the probe-induced broadening increases linearly with the length of the sample for small walk-off; however, it saturates for values of $\eta l > 2$. Physically, twice the walk-off distance measures the distance over which the pump and probe have some overlap; consequently it acts as a maximum effective length for material as far as induced-phase modulation is concerned.

In conclusion, the following features are observed in the induced super-continuum spectrum, in the presence of walk-off between pump and probe:

1. There is asymmetry in the spectral distribution between the Stokes and anti-Stokes sides.
2. The spectral intensity maximum is shifted to the Stokes side of the probe center frequency for $\eta > 0$ and to the anti-Stokes side for $\eta < 0$. This effect is called induced frequency shift.
3. The spectral distributions for equal magnitude but opposite sign η are symmetric with regard to the Δ axis.
4. The spectral distribution extent saturates for $l \gg \eta^{-1}$.

4.2 Raman Amplification and Induced-Phase Modulation

In optical fiber transmission one encounters the situation where Raman amplification and induced-phase modulation are simultaneously present (Manassah and Cockings, 1987a). In this section we study this combined effect to the lowest order approximation. The modifications to be made to Eqs. (166) and (167) are that an amplification term should be added to the probe amplitude equation and the γ's should be multiplied by 1/2 to take into account the value of the overlap integrals of the transverse distribution of the pump and signal over the fiber cross section. The differential equation for the probe amplitude is given by

$$\frac{\partial b}{\partial z} + \frac{1}{v_b}\frac{\partial b}{\partial t} = \frac{\delta}{2} a^2 b, \tag{189}$$

where δ is the Raman loss coefficient. The expressions for a, α, and β are the same as those given by Eqs. (180), (181), and (183) with $\gamma \Rightarrow \gamma/2$ and Eq. (182) is modified to read

$$b = b_0 G(\sigma \bar{U}) \exp\left[\frac{\delta a_0^2}{2}\psi\right]. \tag{190}$$

Here we treat the case $\sigma = 1$. The time of arrival of the peak of the pump at length l of the fiber is denoted by $\bar{U}_1$ and is given by

$$\bar{U}_1 = T_0 + \eta l. \tag{191}$$

With no pump present, the time of arrival of the probe peak is denoted by U_2' and is given by

$$\bar{U}_2' = 0. \tag{192}$$

In the presence of the pump, the coordinate of the probe peak, denoted by $\bar{U}_2$, is obtained by finding the maximum of b, which reduces to solving the transcendental equation

$$\tanh \bar{U} = \frac{\delta a_0^2}{2\eta}[\operatorname{sech}^2(\bar{U} - T_0) - \operatorname{sech}^2(\bar{U} - T_0 - \eta l)]. \tag{193}$$

We plot the values of $\bar{U}_1 - \bar{U}_2$ and of $b(\bar{U}_2)$ in the three following experimentally implementable cases:

1. ηl fixed, $T_0 = 0$, and a_0^2 (intensity of the pump) is changed—see Figure 5.48a.
2. a_0^2 fixed, $T_0 = 0$, and l is changed—see Figure 5.48b.
3. ηl fixed, pump intensity fixed, and T_0 is varied. In Figure 5.49 the pulse shape and position and in Figure 5.50 the maximum magnitude of the probe are plotted for this case.

The outstanding features observed are

1. a pull between the pump and probe pulses, that is, $|\bar{U}_1 - \bar{U}_2| < |\bar{U}_1 - \bar{U}_2'|$;
2. a time ordering with the pump pulse out last at the exit, if the initial time delay between the two pulses is smaller than t_{correl}, where $t_0 < t_{\text{correl}}$, and t_{correl} is the region of strong distortion of the signal;
3. a long time correlation (many pulse widths) in the pump-probe system;
4. an increase in the intensity of the pump increasing the pull;
5. the probe pulse perhaps splitting for $t_0 > t_{\text{correl}}$ and producing a daughter pulse, that is, a secondary pulse of weak amplitude;

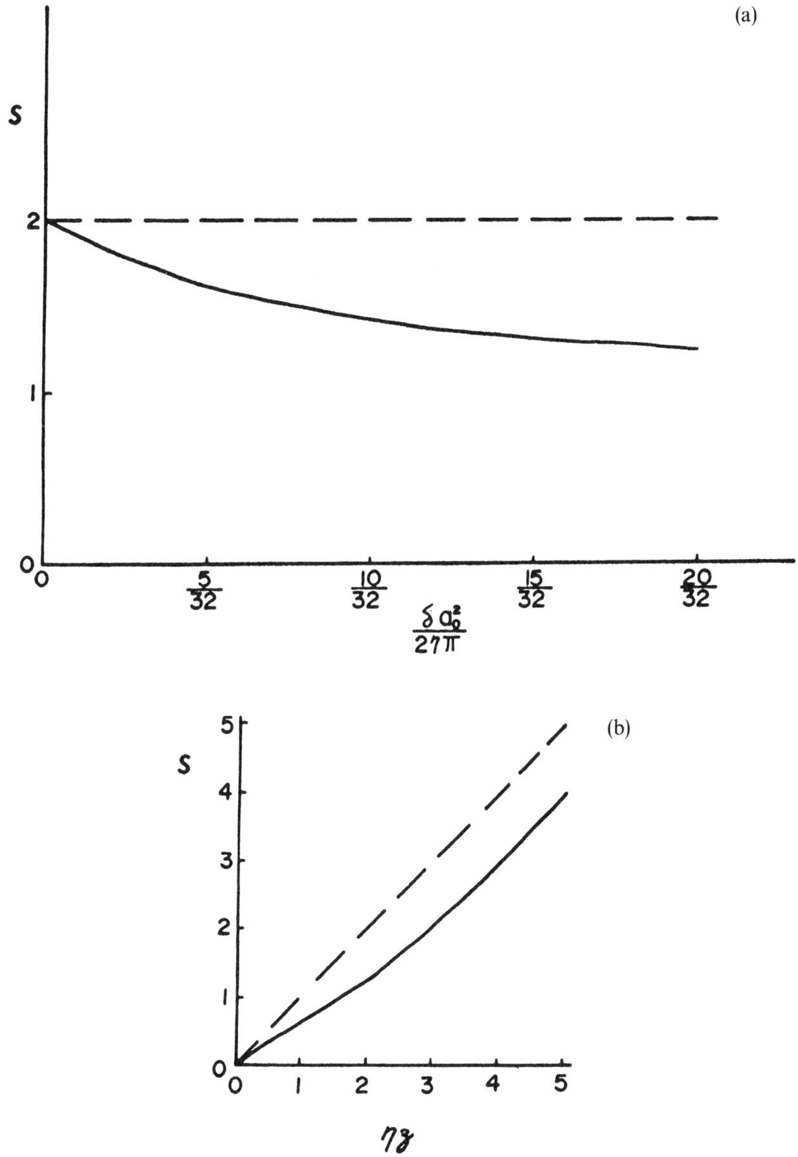

FIGURE 5.48. Time delay between pump and Raman pulses, at exit $S = \bar{U}_1 - \bar{U}_2$. (Solid line) Pump and signal system; (dashed line) noninteractive signal and pump. (a) $\eta z = 2$, $T_0 = 0$; (b) $\delta a_0/2\eta = (5\pi)/8$, $T_0 = 0$.

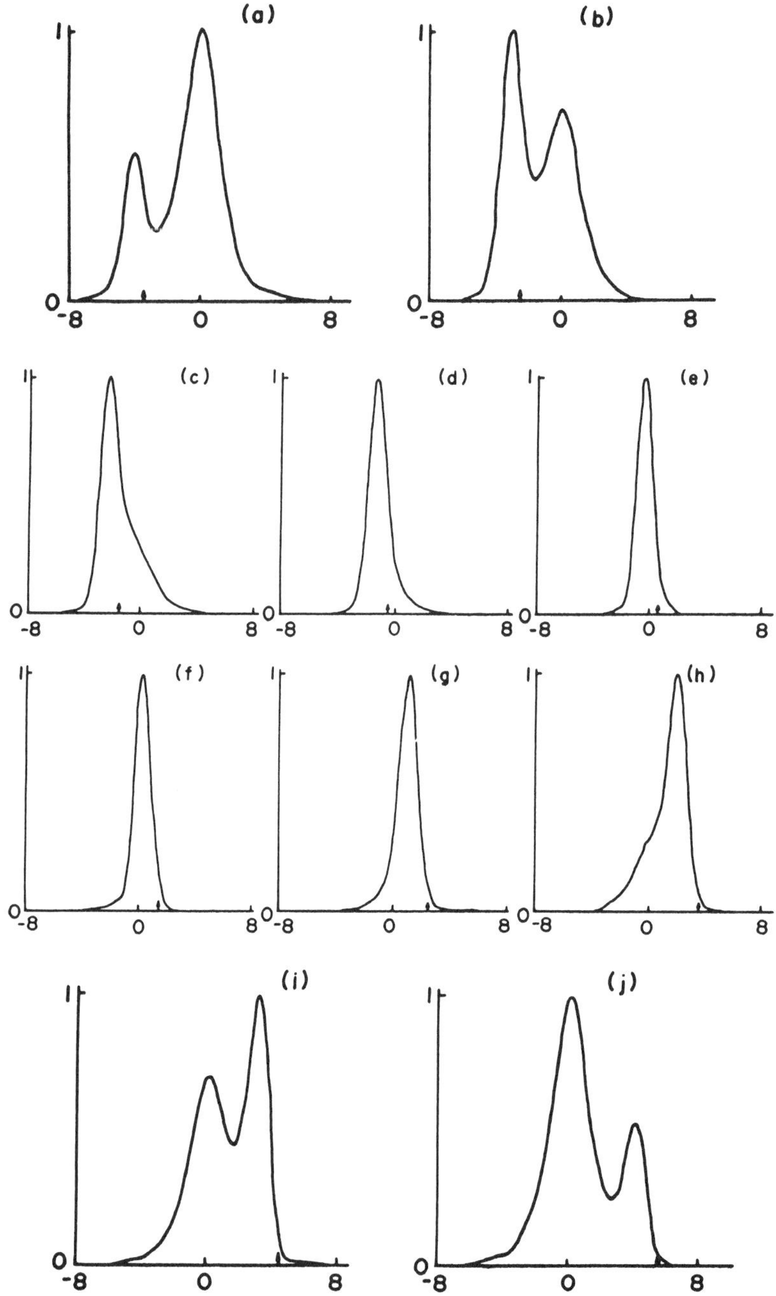

FIGURE 5.49. Normalized outgoing Raman probe amplitude (b/b_{max}) for different initial delay times between probe and pump. The arrow indicates the position of the outgoing pump maximum. In the absence of a pump, the probe maximum would have been in all these graphs at $T = 0$. Parameters for all cases are $\eta z = 2$, $\delta a_0^2/2\eta = 5\pi/8$. (a) $T_0 = -5.5$; (b) $T_0 = -4.5$; (c) $T_0 = -3.5$; (d) $T_0 = -2.5$; (e) $T_0 = -1.5$; (f) $T_0 = -0.5$; (g) $T_0 = 0.5$; (h) $T_0 = 1.5$; (i) $T_0 = 2.5$; (j) $T_0 = 3.5$.

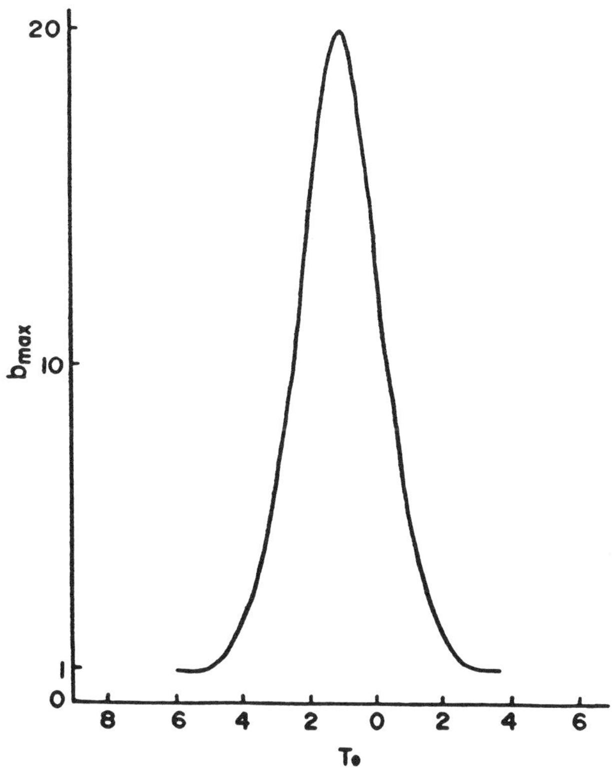

FIGURE 5.50. Maximum amplitude of the Raman signal as a function of initial delay time T_0. $\eta z = 2$; $\delta a_0^2/2\eta = 5\pi/8$.

6. the amplification curve being a narrow function of the initial time lag between pump and probe;
7. the amplification saturating as the length of the fiber exceeds a few walk-off distances.

4.3 Induced Pulse Compression

In Section 1.2 we discussed the fiber-grating compressor. In this section we study the possibility of using induced-phase modulation as the source of the chirp (Manassah, 1988a, 1988b). Walk-off effects between pump and probe, discussed in Section 4.1, are incorporated in this analysis. For computational simplicity we assume that the probe and pump pulse shapes are given, respectively, by a Gaussian and a hyperbolic secant.

In Figure 5.51A, the derivative of the probe normalized phase with respect to $\bar{U}$ is given for different values of ηz. The values of the extrema of this function determine the induced spectral extent of the probe. As can be observed, for a

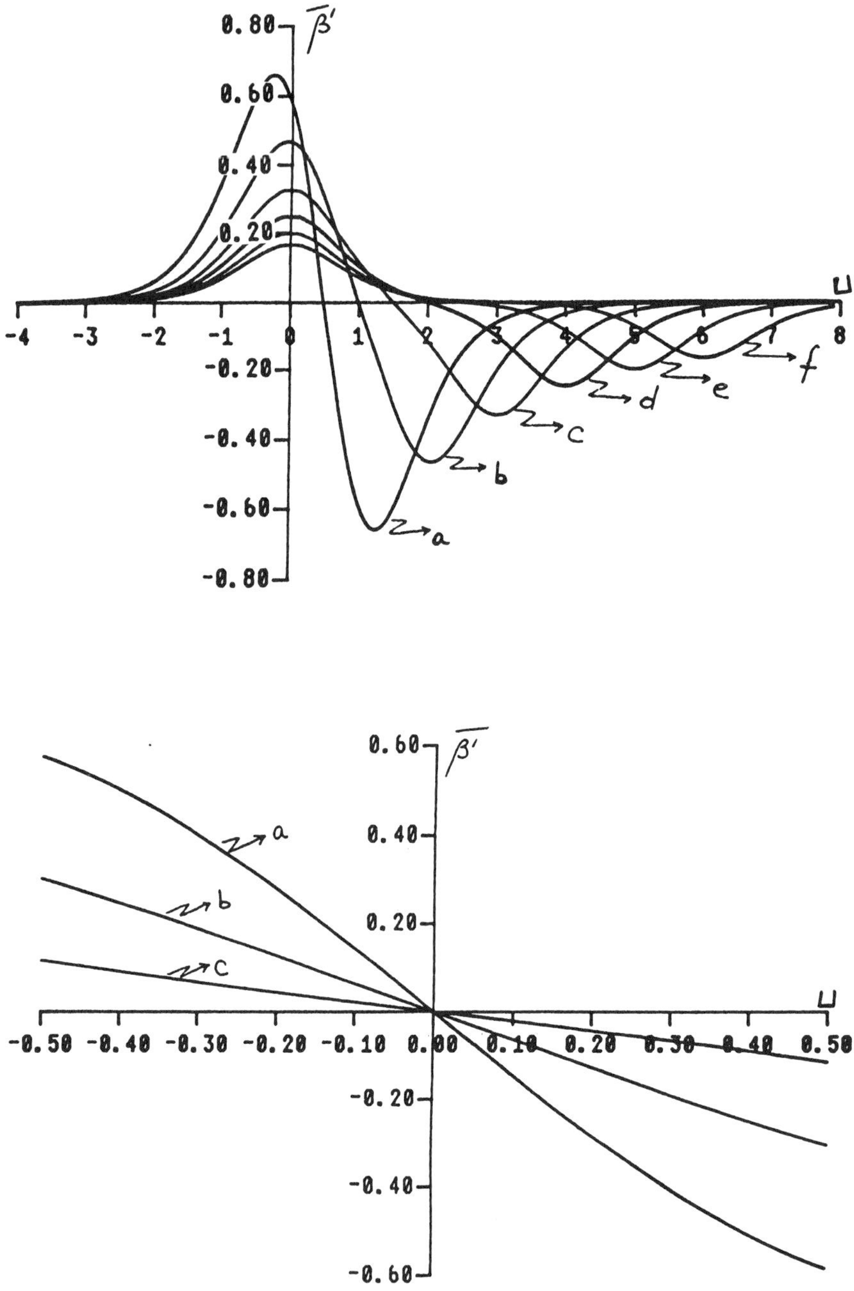

FIGURE 5.51. *Top*: Induced frequency sweep (i.e., derivative of the induced phase) for different ηz and $T_0 = 0$. (a) $\eta z = 1$; (b) $\eta z = 2$; (c) $\eta z = 3$; (e) $\eta z = 5$; (f) $\eta z = 6$. *Bottom*: Induced frequency sweep near $\bar{U} = 0$. (a) $\eta z = 1$; (b) $\eta z = 2$; (c) $\eta z = 3$. Best fits for the slopes are, respectively, -1.44, -0.64, and -0.22.

specific material length z, the induced spectral extent decreases with increasing η. Physically, an increase in the value of η corresponds to a shorter spatial overlap between probe and pulse and consequently to a smaller induced-phase modulation. It should also be noted that for $1 < \eta z < 3$ and for $T_0 = -\eta z/2$, ψ' is linear in $\bar{U}$ in the region around $\bar{U} = 0$ and is symmetric with respect to this point. In Figure 5.51B ψ' is plotted for such cases. Consequently, if the probe pulse duration is much shorter than the pump pulse duration (i.e., the probe amplitude range is entirely within the region of the linear chirp), it is reasonable to expect the induced-phase modulated probe pulse, under these conditions, to be efficiently compressed through the grating.

We assume that $T_0 = -\eta z/2$ and $\sigma \gg 1$ and we denote the induced phase by p, which is given by

$$|p| = (2\gamma_b a_0^2 z)\rho(\eta z), \tag{194}$$

where $\rho(\eta z)$ is a function of ηz equal to $(0.72, 0.32, 0.11)$, respectively, for $\eta z = (1, 2, 3)$.

If the incoming probe pulse intensity is given by

$$I_{\text{in}}^{(b)}(\bar{U}) = b_0^2 \exp(-2\bar{U}^2), \tag{195}$$

then the outgoing probe pulse from a pump-fiber-grating compressor, under the optimal phase condition, is given by

$$I_{\text{out}}^{(b)}(\bar{U}) = (\zeta b_0)^2 \exp(-2\lambda^2 \bar{U}^2), \tag{196}$$

where

$$\zeta^2 = (\sigma^4 + p^2)^{1/2}/\sigma^2, \tag{197}$$

$$\lambda^2 = (\sigma^4 + p^2)/\sigma^2, \tag{198}$$

and the optimal phase-matching condition is given by

$$g = p/4(\sigma^4 + p^2), \tag{199}$$

where the grating phase function is given by

$$\Phi_g = g(K' - K)^2 \tag{200}$$

where $K = \omega\tau_a$ is the probe central frequency normalized to the pump width.

In Figure 5.52 the numerically computed outgoing pulse intensity is plotted and is compared with the incoming pulse. The compression ratio agrees to better than 1% with the result of Eq. (198).

In Figure 5.53 the amplitude scale ζ and the relative width λ of the probe are plotted as function of p. It should be noted that in the choice of p, values for the SPM coefficient should be smaller than those at which pump distortion sets in (i.e., in the notation of Section 2, $\varepsilon V \approx 0.2$). As can be observed, the parameter λ is larger than σ; that is, the probe pulse is compressed even for large values of σ. Experimentally, for a 10-ps pump pulse, $\eta z = 3$; for $\gamma_a a_0^2 z = 1500$ (i.e., $\varepsilon V \approx 0.1$), $2\gamma_a = \gamma_b$, the value of $|p|$ is approximately 650.

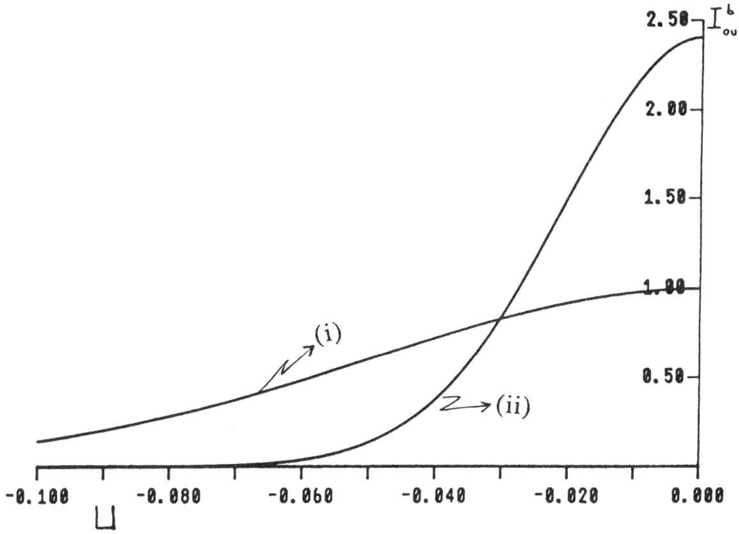

FIGURE 5.52. Induced-phase-modulated probe pulse at the input and output of the fiber-grating compressor as a function of $\bar{U}$. $\sigma = 10$; $\gamma_b a_0^2 z = 1000$; $\eta z = 3$. (i) Input; (ii) output.

Another method for pulse compression is through chirping-filtering, as discussed in Section 2.10. The disadvantage of this technique over the above one is the energy loss suffered by the probe in this process. The expression for the outgoing probe pulse, following its induced-phase modulation and filtering, is given by

$$E_{\text{out}}^{(b)}(\bar{U}) = \frac{[\Delta_f \tau_a]}{\sqrt{\pi}} \int_{-\infty}^{\infty} d\bar{U}' \exp[-\sigma^2 \bar{U}'^2]$$
$$\times \exp\left\{i(2\gamma_b a_0^2 z)\frac{1}{\eta z}\left(\tanh\left(\bar{U}' + \frac{\eta z}{2}\right) - \tanh\left(\bar{U}' - \frac{\eta z}{2}\right)\right)\right\}$$
$$\times \exp[i(K_b - K_f)\bar{U}'] \exp[-(\bar{U}' - \bar{U})^2(\Delta_f \tau)^2], \tag{201}$$

where we assumed that $T_0 = -\eta z/2$, Δ_f is the Gaussian filter width, $K_b = \omega_b \tau_a$, $K_f = \omega_f \tau_a$, and ω_f is the filter center frequency. The filter transform function is given by Eq. (129), and Eq. (201) can be analytically evaluated in the approximation of a linear chirp. The intensity of the outgoing pulse is then given by

$$I_{\text{out}}^{(b)}(\bar{U}) = (\xi b_0)^2 \exp[-2\chi^2 \bar{U}^2], \tag{202}$$

where

$$\xi^2 = \frac{\delta^2}{[(\sigma^2 + \delta^2)^2 + p^2]^{1/2}}, \tag{203}$$

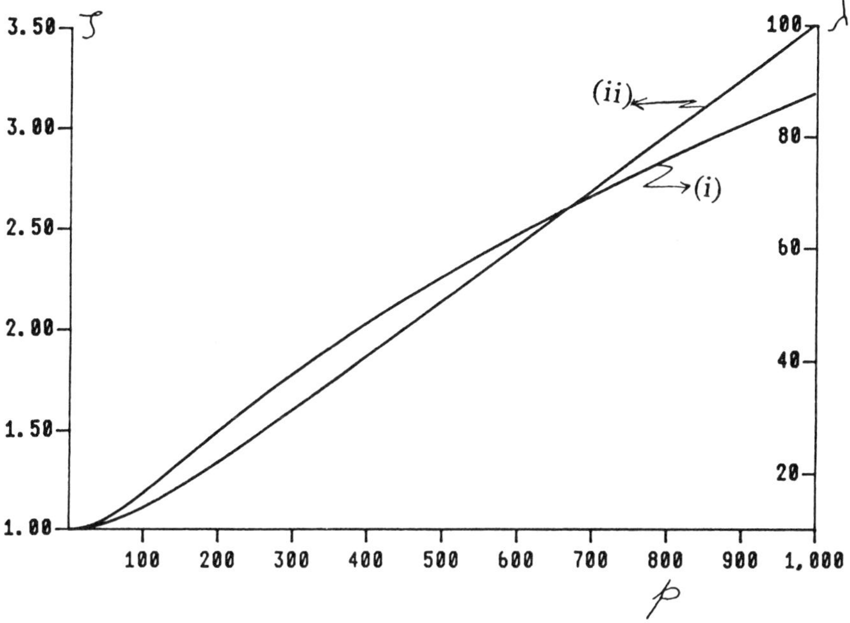

FIGURE 5.53. Fiber-grating outgoing induced-phase-modulated pulse's amplitude scale factor ζ and the pulse inverse width λ as functions of the induced-phase modulation parameter p. $\sigma = 10$. (i) Amplitude scale factor; (ii) inverse width.

$$\chi^2 = \delta^2 - \frac{(\sigma^2 + \delta^2)\delta^4}{(\sigma^2 + \delta^2)^2 + p^2}, \tag{204}$$

and

$$\delta = (\Delta_f \tau_a). \tag{205}$$

In Figure 5.54 the parameters ξ and χ are plotted as functions of δ for a value of the SPM coefficient just smaller than that at which pulse distortion sets in. As can be observed, the parameter χ can be larger than σ; that is, the probe pulse is compressed even for relatively large value of σ.

In conclusion, through combination with either a grating or amplitude filter, induced-phase modulation can lead to pulse compression.

4.4 Induced Focusing

In this section we compute the induced focusing effects (Bladeck et al., 1987; Manassah, 1988c) generated by a strong pump on the propagation of a probe Gaussian pulse in a $\chi^{(3)}$ medium. The differential equations for the probe pulse diameter, radius of curvature, and phase are expressed as functions of the pump pulse characteristics. We prove that waveguiding of the probe by the pump is possible.

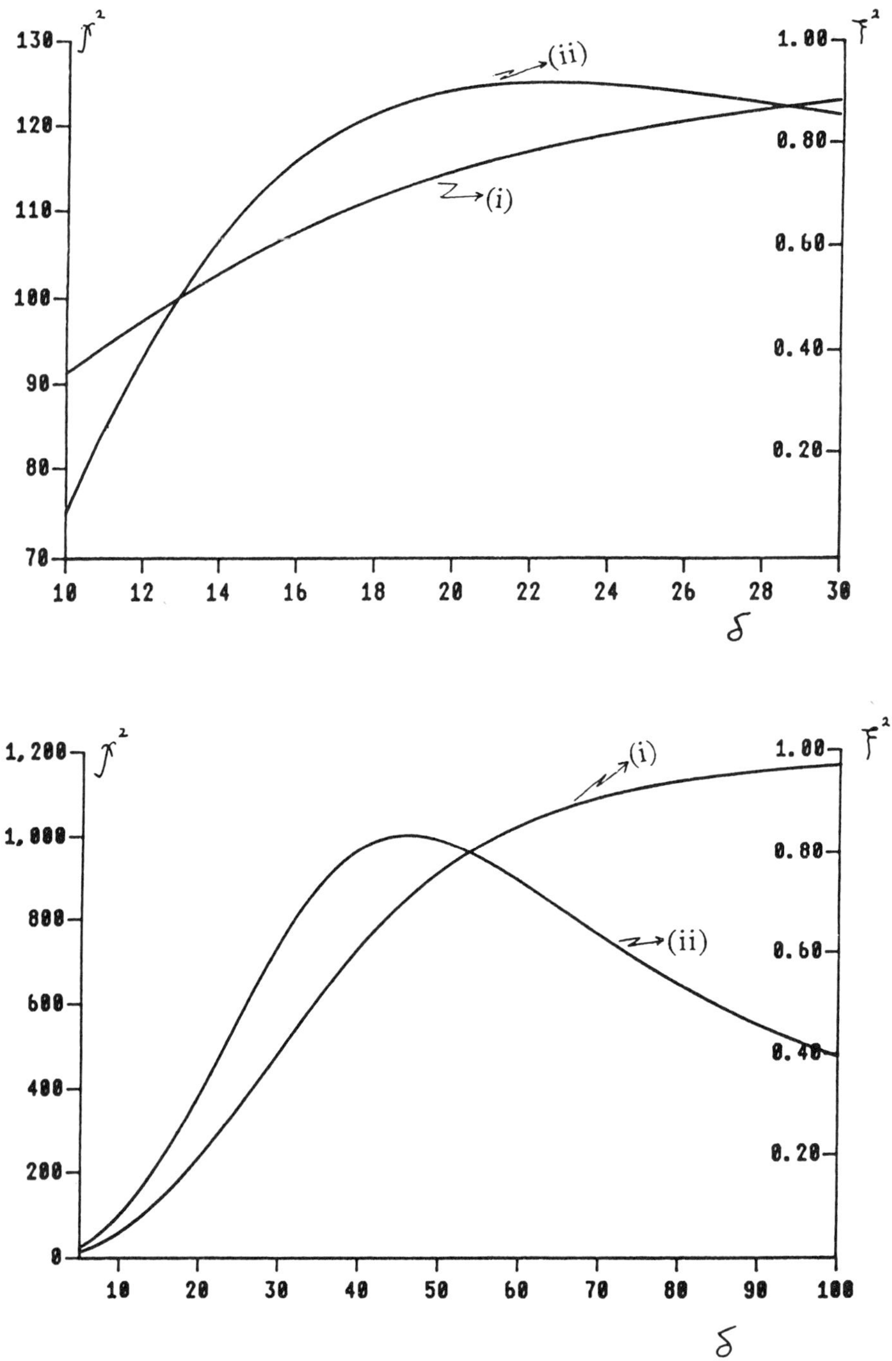

FIGURE 5.54. Fiber-filter outgoing induced-phase-modulated pulse's intensity amplitude scale factor and the pulse inverse width squared as functions of the normalized filter width ($\Delta_f \tau_a$). (i) Intensity scale factor; (ii) width squared (*top*) $\sigma = 10$, $p = 200$; (*bottom*) $\sigma = 10$, $p = 200$.

In the approximation where effects of group velocity dispersion and self-steepening are neglected and the general conditions for the slowly varying approximation are valid, the differential equations for the envelopes of the pump and probe pulses are given by

$$\nabla_T^2 A_1 - 2ik_1 A_1' + \frac{n_2}{n_0} k_1^2 [|A_1|^2 A_1 + 2|A_2|^2 A_1] = 0, \tag{206}$$

$$\nabla_T^2 A_2 - 2ik_2 A_2' + \frac{n_2}{n_0} k_2^2 [|A_2|^2 A_2 + 2|A_1|^2 A_2] = 0, \tag{207}$$

where A_1 and A_2 are respectively the pump and probe envelopes, $A' = \partial A/\partial z$, and ∇_T^2 is the transverse component of the Laplacian.

If the initial conditions for the envelopes are written as

$$A_{1,2}(r, 0, \bar{u}) = A_{1,2}^0 \exp\left(\frac{-r^2}{a_{1,2}^2}\right) \exp\left(\frac{-\bar{u}^2}{2\tau_{1,2}^2}\right), \tag{208}$$

that is, the initial pulses are assumed to be Gaussian both in the transverse plane and in the comoving coordinate system, $\bar{u} = (z/v_g - t)$ where v_g is the group velocity, a is the initial beam radius, τ is the pulse duration, A^0 is the magnitude of the pulse amplitude, and the subscripts (1, 2) refer respectively to the pump and probe.

In the following, the product $A^0 \exp(-\bar{u}^2/2\tau^2)$ is denoted by $\tilde{A}^0$. Furthermore, we will solve Eqs. (206) and (207) in the limit $A_1^0 \gg A_2^0$, and $\tau_1 \gg \tau_2$. For practical purposes, we are assuming that the pump pulse duration is long enough that any walk-off between pulse and pump because of a difference in group velocity can be neglected. Approximate solutions to Eqs. (206) and (207), correct to order r^2/a^2, can be obtained through the trial solutions:

$$A_{1,2}(r, z, \bar{u}) = \frac{\tilde{A}_{1,2}^0}{\bar{\omega}_{1,2}(z, \bar{u})} \exp\left[-\frac{r^2}{a_{1,2}^2 \bar{\omega}_{1,2}^2(z, \bar{u})} - \frac{ik_{1,2}}{2} \rho_{1,2}(z, \bar{u}) r^2 + ik_{1,2} \alpha_{1,2}(z, \bar{u})\right] \tag{209}$$

where the different functions have the same meaning as in Section 1.3; specifically, $\bar{\omega}$ is the normalized beam radius, ρ is the inverse of its radius of curvature, and $k\alpha$ is the longitudinal phase. The equations satisfied by these subsidiary functions are

$$\rho_{1,2} = \frac{1}{\bar{\omega}_{1,2}} \frac{\partial \bar{\omega}_{1,2}}{\partial z} \tag{210}$$

$$\frac{\partial \alpha_1}{\partial z} = \frac{a_1^2}{2\bar{\omega}_1^2} \left\{ \frac{4}{a_1^4 k_1^2} - \frac{n_2}{n} \frac{(\tilde{A}_1^0)^2}{a_1^2} \right\}, \tag{211}$$

$$\frac{\partial \alpha_2}{\partial z} = \frac{a_2^2}{2} \left\{ \frac{4}{a_2^4 k_2^2 \bar{\omega}_2^2} - \frac{2n_2}{n a_2^2} \frac{(\tilde{A}_1^0)^2}{\bar{\omega}_1^2} \right\}, \tag{212}$$

$$\frac{\partial^2 \bar{\omega}_1}{\partial z^2} = \left\{\frac{4}{a_1^4 k_1^2 \bar{\omega}_1^3} - \frac{2n_2}{na_1^2}\frac{(\tilde{A}_1^0)^2}{\bar{\omega}_1^3}\right\}, \tag{213}$$

$$\frac{\partial^2 \bar{\omega}_2}{\partial z^2} = \left\{\frac{4}{a_2^4 k_2^2 \bar{\omega}_2^3} - \frac{4n_2}{na_1^2}\frac{(\tilde{A}_1^0)^2 \bar{\omega}_2}{\bar{\omega}_1^4}\right\}. \tag{214}$$

Physically, the above factors are easily identifiable: $4/k^2a^4$ is the square of the inverse Rayleigh diffraction length (denoted L), and the Kerr phase for the pump pulse in the conventional SPM theory (see Section 1.1) is given by

$$\phi_{\text{SPM}} = -\frac{k_1 n_2 |\tilde{A}_1^0|^2 z}{2n} \tag{215}$$

where $n = n_0$.

The solutions for the pump subsidiary functions are the same as those derived in Section 1.4, specifically:

$$\bar{\omega}_1 = [1 + y_1^2(1 - p')]^{1/2}, \tag{216}$$

$$L_1 \rho_1 = \frac{y_1[1 - p']}{[1 + y_1^2(1 - p')]}, \tag{217}$$

$$k_1 \alpha_1 = \frac{[1 - \frac{1}{2}p']}{[(1 - p')^{1/2}]} \operatorname{arctg}[(1 - p')^{1/2} y_1], \tag{218}$$

where

$$p' = p \exp[-\bar{u}^2/\tau_1^2], \tag{219}$$

$$p = (A_1^0)^2/(A_1^c)^2, \tag{220}$$

$$y_1 = z/L_1, \tag{221}$$

$$(A_1^c)^2 = na_1^2/2n_2L_1^2. \tag{222}$$

$(A_1^c)^2$ is the critical field for self-focusing (i.e., the field for which the self-focusing distance is at $z = \infty$). In this notation, as noted in Section 1.4, the normalized self-focusing distance is $(y_1)_{\text{foc}} = 1/(p - 1)^{1/2}$ for $p > 1$. Here we will solve only for the probe subsidiary functions $\bar{\omega}_2$, ρ_2, and α_2 under the condition that $p \approx 1$ and that for the duration of the probe pulse we have $p \approx p'$. In this instance, $\bar{\omega}_1 = 1$ and the solutions of the probe subsidiary functions are given by

$$\bar{\omega}_2 = [\bar{\xi} \cos(\bar{\chi} z) + \bar{\eta}]^{1/2}, \tag{223}$$

$$\rho_2 = -\frac{1}{2}\frac{\bar{\xi}\bar{\chi} \sin(\bar{\chi} z)}{[\bar{\xi} \cos(\bar{\chi} z) + \bar{\eta}]}, \tag{224}$$

$$\alpha_2 = \left\{-\frac{a_1^2}{2L_1^2} z + \frac{a_2^2}{L_2^2}\frac{1}{\bar{\chi}(\bar{\eta} - \bar{\xi})^{1/2}} \operatorname{arctg}\left[(\bar{\eta} - \bar{\xi})^{1/2} \tan\frac{\bar{\chi} z}{2}\right]\right\}, \tag{225}$$

where

$$\bar{\xi} = \tfrac{1}{4}[2 - L_1^2/L_2^2], \tag{226}$$

$$\bar{\eta} = \tfrac{1}{4}[2 + L_1^2/L_2^2], \tag{227}$$

$$\bar{\chi} = 2\sqrt{2}/L_1, \tag{228}$$

and the range of $\bar{\omega}_2^2$ is between $L_1^2/2L_2^2$ and 1. In the same notation, the expressions for $\bar{\omega}_2$, ρ_2, and α_2 in the absence of a pump (i.e., $A_1^0 = 0$) are given by

$$\bar{\omega}_2 = (1 + y_2)^{1/2}, \tag{229}$$

$$\rho_2 L_2 = y_2[(1 + y_2^2)]^{-1}, \tag{230}$$

$$k_2\alpha_2 = \text{arctg}(y_2), \tag{231}$$

where $y_2 = z/L_2$.

In the figures, we consider the easily achievable experimental setup of a probe with a central frequency at twice that of the pump (i.e., $\lambda_2 = \lambda_1/2$) and for $a_1 = a_2$. In this case $L_2 = 2L_1$. In Figure 5.55, $\bar{\omega}_2$ is plotted in the presence and absence of the pump field; in Figure 5.56, ρ_2 is plotted; and in Figure 5.57, the longitudinal phase α_2 is plotted. The curves in the presence of the pump are reminiscent of those corresponding to the propagation of a Gaussian pulse

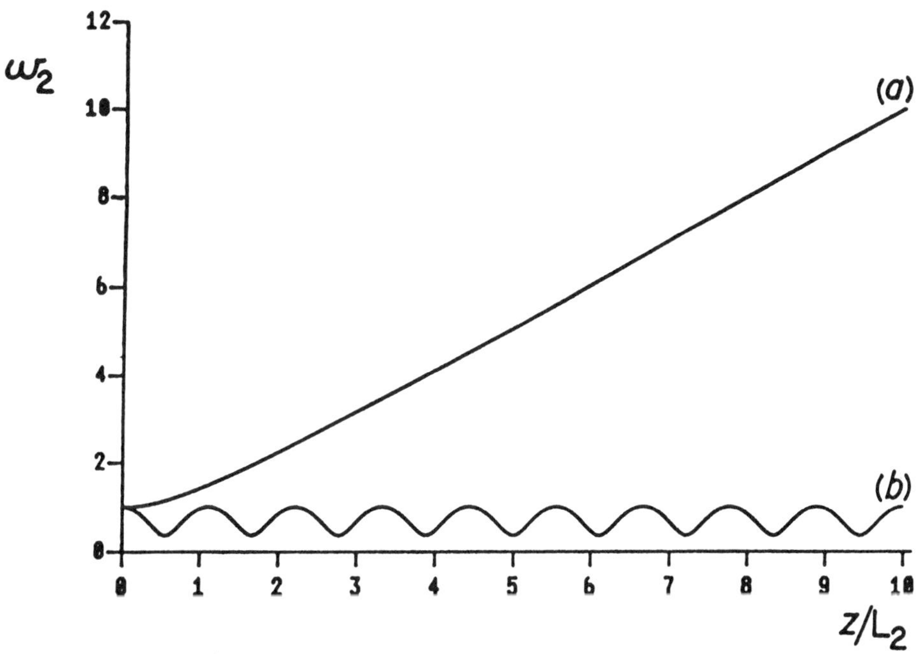

FIGURE 5.55. Normalized radius of an IPM probe as a function of distance. The pump intensity when present is at the self-focusing critical value. (a) No pump; (b) pump present.

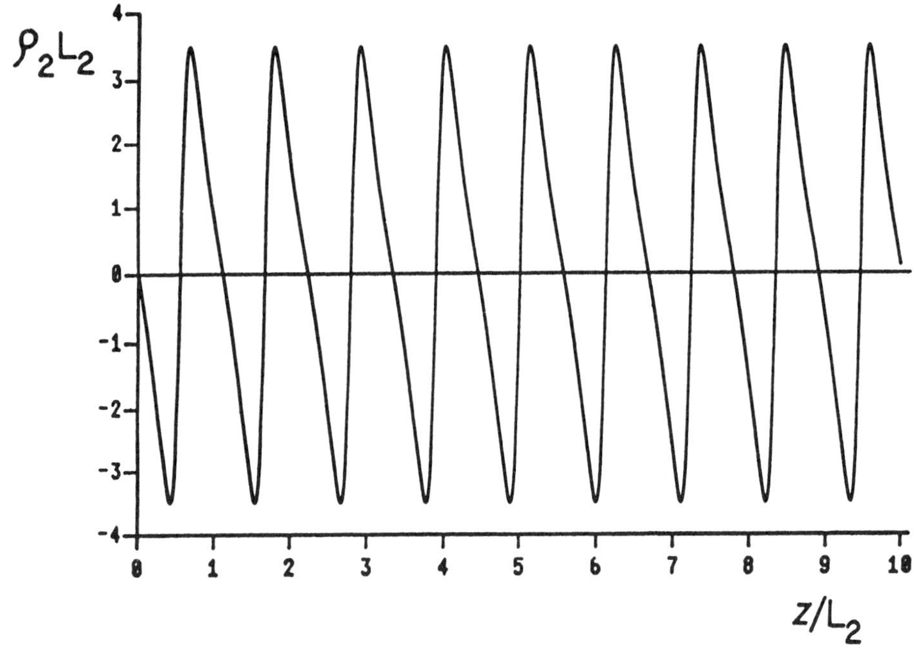

FIGURE 5.56. Inverse radius of curvature of an IPM probe as a function of distance. The pump intensity is at the self-focusing critical value.

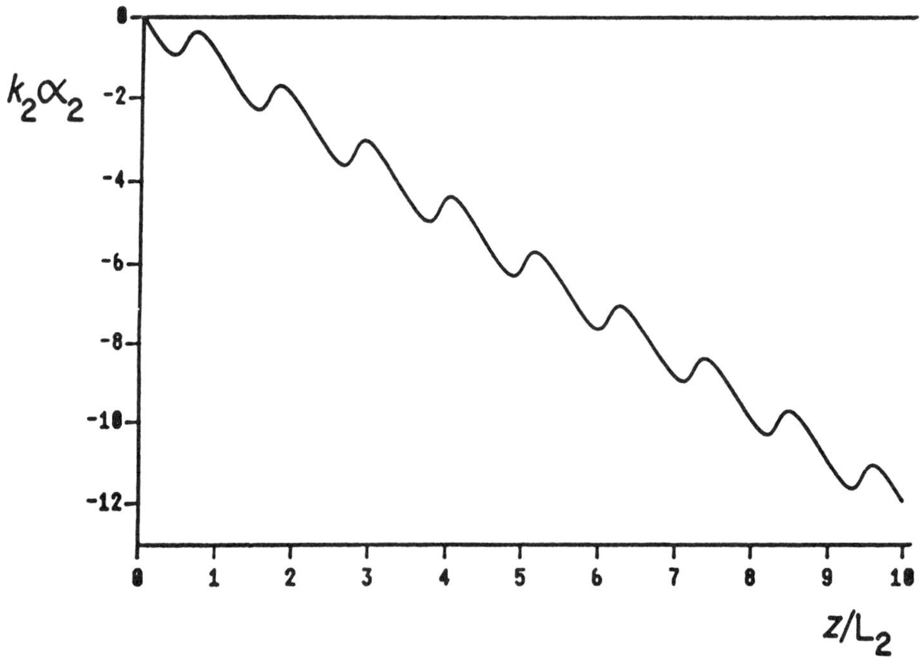

FIGURE 5.57. Longitudinal phase of an IPM probe as a function of distance. The pump intensity is at the self-focusing critical value.

in a quadratic graded-index material. This should come as no surprise, since Eq. (214) is equivalent to that case.

In conclusion, in this section we computed the geometric effects of induced-phase modulation of a pump on the propagation of a weak probe in a nonlinear $\chi^{(3)}$ material. We showed that, physically, the induced index of refraction can lead to a waveguiding of the probe. Furthermore the computed cumulative longitudinal phase [i.e., Eq. (225)] may lead to some interesting verification of the model through interference experiments.

4.5 Induced Self-Steepening

In this section we consider induced-phase modulation when the pump intensity produces self-steepening (Manassah et al., 1985). This case can be solved by the method of multiple scales used in Section 2. Using the notation of Section 2, for the case of the probe at the second harmonic frequency of the pump and assuming that the material has no $\chi^{(2)}$ (i.e., there is no conversion of two primary photons into a second harmonic photon) and that all conditions of Section 2.1 are satisfied, the total Φ_0 can be written as

$$\Phi_0 = Ae^{iKU} + \delta Be^{i2KU}. \tag{232}$$

this equation should be compared with Eq. (76). A corresponds to the pump and B to the probe and δ is the initial scaling factor between the probe and pump amplitudes. For the case of $\delta \ll 1$, that is, a weak probe, the equations that will correspond to Eqs. (84) and (85), for the nonlinear polarization $\langle|\Phi|^2\rangle\Phi$, are

$$\frac{\partial a}{\partial V} - \frac{3}{2}\varepsilon a^2 \frac{\partial a}{\partial U} = 0, \tag{233}$$

$$\frac{\partial b}{\partial V} - \frac{\varepsilon}{2} a^2 \frac{\partial b}{\partial U} - \varepsilon ab \frac{\partial a}{\partial U} = 0, \tag{234}$$

$$\frac{\partial \alpha}{\partial V} - \frac{\varepsilon}{2} a^2 \frac{\partial \alpha}{\partial U} = \frac{K\varepsilon}{2} a^2 - \frac{K\varepsilon^2}{8} a^4, \tag{235}$$

$$\frac{\partial \beta}{\partial V} - \frac{\varepsilon}{2} a^2 \frac{\partial \beta}{\partial U} = K\varepsilon a^2 - \frac{K\varepsilon^2}{4} a^4. \tag{236}$$

This system of equations indicates that once the pump pulse shape function is determined, b, α, and β can be deduced as functions of it. For the initial conditions, that the pump and probe have hyperbolic secant shape with the same width, the solutions of a and α are given by Eqs. (93) and (98), respectively, and those of b and β by

$$\beta = 2\alpha, \tag{237}$$

$$b = a. \tag{238}$$

Essentially the pump self-steepening is directly mirrored in the probe. The key results of the above analysis are as follows:

1. The ratio of the magnitudes of a and b is constant throughout the medium; that is, the ratio of the energy densities in the pump and the probe is preserved and thus the pump only modulates the probe.
2. For the case where the ratios of the pump and probe center frequencies are the same as the ratio of the phases (α, β), the supercontinuum and the induced supercontinuum spectral distributions are geometrically similar. This could provide direct experimental verification of the model.

The solutions of Eqs. (234) and (236), for pump and probe having the same initial shape and width, are given by Eqs. (237) and (238). The general solution for b is given by

$$b = aL, \tag{239}$$

where the function L can be written as $L(l)$, the function l satisfies the equation

$$\frac{\partial l}{\partial V} - \frac{\varepsilon}{2} a^2 \frac{\partial l}{\partial U} = 0, \tag{240}$$

its solution is given by

$$l = \frac{\varepsilon}{2} \int_0^V a^3(U, q)\, dq + \int_0^U a(p, 0)\, dp, \tag{241}$$

and the particular solution of L is then obtained by making it satisfy the initial condition. The β solution is given by Eq. (237) as long as $\beta(U, 0) = 0$.

The treatment in this section has so far considered a nonlinear index of refraction that has been averaged over time (the light cycle). If, on the other hand, we consider the time-dependent nonlinear index of refraction (i.e., the polarization is proportional to $|\Phi|^2 \Phi$), the equations of motion for the envelopes A and B are given by

$$\frac{\partial a}{\partial V} - \tfrac{1}{2}\varepsilon\left[(3a^2 + 2\delta^2 b^2)\frac{\partial a}{\partial U} + 4\delta^2 ab \frac{\partial b}{\partial U}\right] = 0, \tag{242}$$

$$\frac{\partial \alpha}{\partial V} - \tfrac{1}{2}\varepsilon(a^2 + 2\delta^2 b^2)\frac{\partial \alpha}{\partial U} = \tfrac{1}{2}K\varepsilon(a^2 + 2\delta^2 b^2) - \tfrac{1}{8}\varepsilon^2 K(a^2 + 2\delta^2 b^2)^2, \tag{243}$$

$$\frac{\partial b}{\partial V} - \tfrac{1}{2}\varepsilon\left[(2\delta^2 b^2 + 2a^2)\frac{\partial b}{\partial U} + 4ab\frac{\partial a}{\partial U}\right] = 0, \tag{244}$$

$$\frac{\partial \beta}{\partial V} - \tfrac{1}{2}\varepsilon(\delta^2 b^2 + 2a^2)\frac{\partial \beta}{\partial U} = K\varepsilon(2a^2 + \delta^2 b^2) - \tfrac{1}{4}\varepsilon^2 K(\delta^2 b^2 + 2a^2)^2. \tag{245}$$

In the limit considered earlier, $\delta^2 \ll 1$, the above equations reduce to

$$\frac{\partial a}{\partial V} - \tfrac{3}{2}\varepsilon a^2 \frac{\partial a}{\partial U} = 0, \tag{246}$$

$$\frac{\partial \alpha}{\partial V} - \tfrac{1}{2}\varepsilon a^2 \frac{\partial \alpha}{\partial U} = \tfrac{1}{2}K\varepsilon a^2 - \tfrac{1}{8}K\varepsilon^2 a^4, \tag{247}$$

$$\frac{\partial b}{\partial V} - \varepsilon a^2 \frac{\partial b}{\partial U} - 2\varepsilon ab \frac{\partial a}{\partial U} = 0, \tag{248}$$

$$\frac{\partial \beta}{\partial V} - \varepsilon a^2 \frac{\partial \beta}{\partial U} = 2K\varepsilon a^2 - \varepsilon^2 K a^4. \tag{249}$$

In this limit, the equations for a and α reduce to those of the SPM theory of Section 2 and the solutions for b and β are, respectively, given by

$$b(U,V) = L_1(U,V)a^4(U,V), \tag{250}$$

$$\begin{aligned}\beta(U,V) = {} & -2KU + 4\varepsilon K \int_0^U a^2(p,V)\,dp \\ & + 3K\varepsilon^2 \int_0^V a^4(0,q)\,dq + KL_2(U,V)\end{aligned} \tag{251}$$

where L_1 and L_2 satisfy the partial differential equation

$$\frac{\partial L_i}{\partial V} - \varepsilon a^2 \frac{\partial L_i}{\partial U} = 0. \tag{252}$$

The solution of Eq. (252) is given by

$$\begin{aligned}L(U,V) &= L\left(\varepsilon \int_0^V a^6(U,q)\,dq + \int_0^U a^4(p,0)\,dp\right) \\ &= L(r) \\ &= L\left(\varepsilon \int_0^V a^6(U,q)\,dq + \tanh U - \tfrac{1}{3}\tanh^3 U\right),\end{aligned} \tag{253}$$

where the initial pump pulse has been assumed to be given by $\operatorname{sech}(U)$. If the probe pulse shape is also a sech function but of different width, that is, $b(U,0) = \operatorname{sech}(nU)$ and $\beta(U,0) = 0$, then the functions L_1 and L_2 are given by the following parametric representation (where s is the parameter):

$$L_1(x) = (\cosh^4 s)\operatorname{sech}(ns), \tag{254}$$

$$L_2(x) = 2s - 4\varepsilon \tanh s, \tag{255}$$

$$x = \tanh s - \tfrac{1}{3}\tanh^3 s. \tag{256}$$

The frequency extent of the induced supercontinuum centered at the probe frequency (i.e., 2ω) is given by the maximum and minimum of $(1/2K)\partial\beta/\partial U$, where

$$\left(\frac{1}{2K}\frac{\partial \beta}{\partial U}\right) = 1 + 2\varepsilon a^2(U,V) + \frac{1}{2}\left[\frac{\partial L_2(r)}{\partial r}\right] a^4(U,V) \tag{257}$$

and the derivative of L_2 in parametric form is

$$\frac{\partial L_2}{\partial x} = \frac{2 - 4\varepsilon \operatorname{sech}^2 s}{\operatorname{sech}^2 s - (\tanh^2 s)(\operatorname{sech}^2 s)}. \tag{258}$$

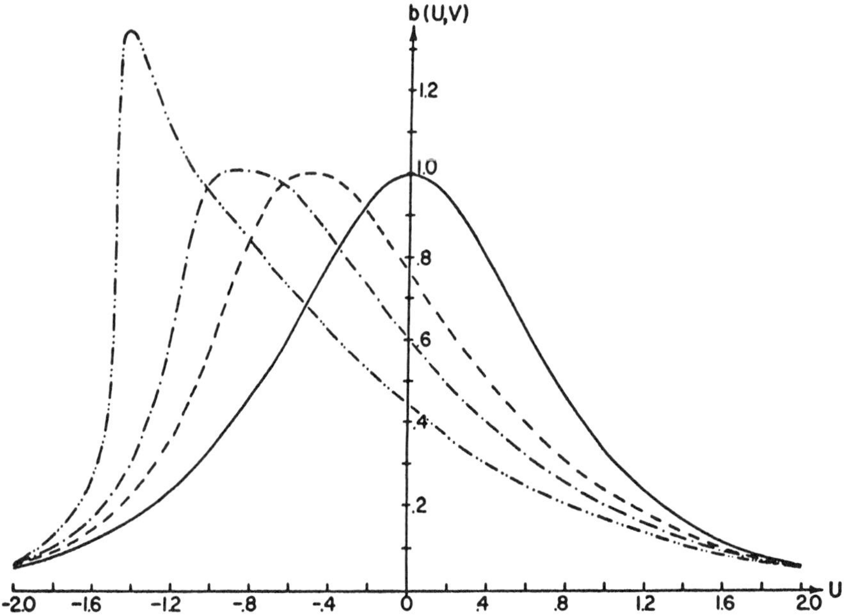

FIGURE 5.58. Amplitude of the induced-phase-modulated steepened second harmonic pulse in a dispersionless $\chi^{(3)}$ medium as a function of U. $\chi^{(2)} = 0$. —— $\varepsilon V = 0.0$; --- $\varepsilon V = 0.3$; -·- $\varepsilon V = 0.5$; -··- $\varepsilon V = 0.8$.

In Figure 5.58, $b(U, V)$ is plotted for different values of εV for $n = 1.7$ and in Figure 5.59 the induced supercontinuum frequency extents are plotted as functions of εV. The induced supercontinuum frequency extents, as functions of εV, grow faster than those corresponding to SPM.

4.6 Induced-Phase Modulation of a Generated Second Harmonic Pulse

In the previous sections we considered the induced-phase modulation on a probe by a pump when both were initially introduced in the system. In this section (Manassah and Cockings, 1987b) we examine the induced-phase modulation of a second harmonic signal generated by the pump; that is, we consider a medium with both $\chi^{(2)}$ and $\chi^{(3)}$ coefficients present (Alfano and Ho, 1986, 1988; Alfano et al., 1987).

To first order in the quadratic susceptibility $[\chi^{(2)}]$ and the Kerr susceptibility, and neglecting all higher-order derivatives of the linear refractive index beyond the first derivative with respect to frequency, the quasi-linear partial differential equations describing the envelopes of the first and second harmonic pulses are given by

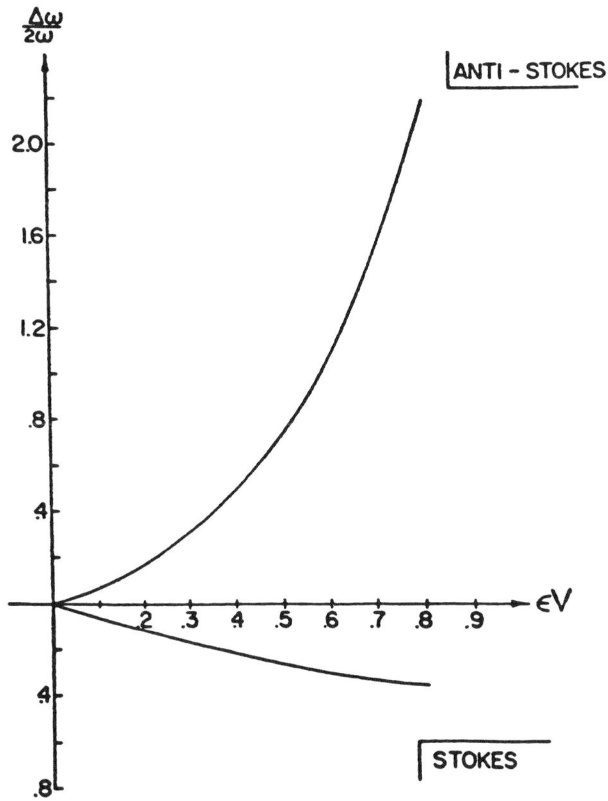

FIGURE 5.59. Induced frequency sweep extents (maxima and minima of the derivative of the induced phase) for a steepened second harmonic pulse in a dispersionless $\chi^{(3)}$ medium as a function of εV. $\chi^{(2)} = 0$.

$$\frac{\partial A}{\partial z} + \frac{1}{v_1}\frac{\partial A}{\partial t} = i\varepsilon A^* B \exp[i(k_2 - 2k_1)z] + i\gamma_a(|A|^2 + 2|B|^2)A, \quad (259)$$

$$\frac{\partial B}{\partial z} + \frac{1}{v_2}\frac{\partial B}{\partial t} = i\varepsilon A^2 \exp[-i(k_2 - 2k_1)z] + i\gamma_b(|B|^2 + 2|A|^2)B - \alpha B, \quad (260)$$

where γ was defined earlier, ε is given by

$$\varepsilon = \mu_0 \omega_1 c d/2, \quad (261)$$

α is the absorption coefficient of the medium at ω_2, v_1 and v_2 refer to the group velocities, respectively, at the primary and second harmonic frequencies, k_1 and k_2 are the corresponding wave vectors, and d is the quadratic nonlinear optical coefficient ($P_{NL} = dE^2$). In the weak second harmonic signal limit (i.e., $|B| \ll |A|$ or no pump depletion), the solutions of the above equations reduce to

$$A = a_0 F\left[\left(t - \frac{z}{v_1}\right)\Big/\tau\right]\exp\{i\gamma_a a_0^2 F^2[(t - z/v_1)/\tau]z\}, \tag{262}$$

$$\begin{aligned} B = i\varepsilon a_0^2 \exp&\left[-\alpha z + i(2\gamma_b a_0^2)\int_0^z F^2(\bar{U} + \eta' z')\,dz'\right] \\ &\times \int_0^z dz'\, F^2(\bar{U} + \eta' z')\exp[i2\gamma_a a_0^2 F^2(\bar{U} + \eta' z')z' - i\zeta' z'] \\ &\times \exp\left[\alpha z' - i2\gamma_b a_0^2 \int_0^{z'} F^2(\bar{U} + \eta' z'')\,dz''\right], \end{aligned} \tag{263}$$

where

$$\bar{U} = (t - z/v_2)/\tau, \tag{264}$$

$$\eta' = (n_2^g - n_1^g)/c\tau, \tag{265}$$

$$\zeta' = (2\omega/c)(n_2^p - n_1^p), \tag{266}$$

n_1 and n_2 are the indices of refraction at ω and 2ω, and the superscripts p and g refer, respectively, to the phase and group quantities. If the pulse form function is a hyperbolic secant, then the expressions for A and B can be written as

$$A = a_0 \operatorname{sech}[(t - z/v_1)/\tau]\exp\{i\gamma_a a_0^2 \operatorname{sech}^2[(t - z/v_1)/\tau]\}, \tag{267}$$

$$\begin{aligned} B = i\varepsilon a_0^2 \exp&\left[-\alpha z + \frac{i2\gamma_b a_0^2}{\eta'}\tanh(\bar{U} + \eta' z')\right] \\ &\times \int_0^z dz' \operatorname{sech}^2(\bar{U} + \eta' z')\exp\Bigg\{(\alpha - i\zeta')z' + i2a_0^2 \\ &\times \left[(\gamma_a \operatorname{sech}^2(U + \eta' z'))z' - \frac{\gamma_b}{\eta'}\tanh(\bar{U} + n' z')\right]\Bigg\}. \end{aligned} \tag{268}$$

We note that A includes self-phase modulation (SPM), and B includes the SPM of A through the $\chi^{(2)}$ term and its induced-phase modulation (IPM) by A through the $\chi^{(3)}$ term. A convenient expression for B suitable for numerical integration is

$$\begin{aligned} B = \frac{i\varepsilon a_0^2}{\eta'}\exp&\Bigg\{-\alpha z + \frac{i2\gamma_b a_0^2}{\eta'}\tanh(\bar{U} + \eta' z) \\ &+ i\frac{\zeta'}{\eta'}\bar{U} - \frac{\alpha}{\eta}U\Bigg\}\int_{\bar{U}}^{\bar{U}+\eta' z} dy \operatorname{sech}^2(y) \\ &\times \exp\left\{\left(\frac{\alpha}{\eta'} - \frac{i\zeta'}{\eta'}\right)y + \frac{i2a_0^2}{\eta'}\left[\gamma_a(y - \bar{U})\operatorname{sech}^2(y) - \gamma_b \tanh y\right]\right\}. \end{aligned} \tag{269}$$

We note that

1. The modulation term inside the integral depends on $\zeta'/\eta' = 2\omega\tau(n_2^p - n_1^p)/(n_2^g - n_1^g)$ and on the phase term due to SPM of A and IPM of B.
2. The dependence on the primary intensity in the integrand is multiplicative through $\chi^{(2)}$ and in the phase through $\chi^{(3)}$.
3. The upper limit of integration in y differs from the lower limit by $\eta' z$, which represents the length of the sample in units of walk-off distance.
4. The above treatment for SPM assumes that we are in the regime where self-steepening is unimportant.

To gain a feeling for the different parameters of the present problem, consider ZnSe (Alfano and Ho, 1986, 1988; Alfano et al., 1987). This material was shown to have large values for both $\chi^{(2)}$ and $\chi^{(3)}$ and is therefore an excellent candidate for observing both the induced-phase modulation of a generated second harmonic signal and its time domain structure. The group indices of refraction for this material at the primary (1.06 μm) and second harmonic frequencies are, respectively $n_1^g = 2.62$ and $n_2^g = 3.44$; therefore, for a sample 2 cm long and a pulse of 10-ps duration, $\eta' z \approx 5$.

In Figure 5.60 the magnitude of $B = |B|$ is plotted. As observed, the time domain structure of $|B|$ exhibits two peaks (Manassah, 1988d), one located near $\bar{U} = -\eta' z$ and the other at $\bar{U} = 0$; these two positions correspond to the time of arrival of two weak pulses, at the primary frequency and at the secondary frequency, going through the linear medium [i.e., $\chi^{(2)} = \chi^{(3)} = 0$]. The appearance of the pump companion (i.e., at $\bar{U} \approx \eta' z$) results from the $\chi^{(2)}$ term; the $\chi^{(3)}$ term shifts slightly this pump companion forward.

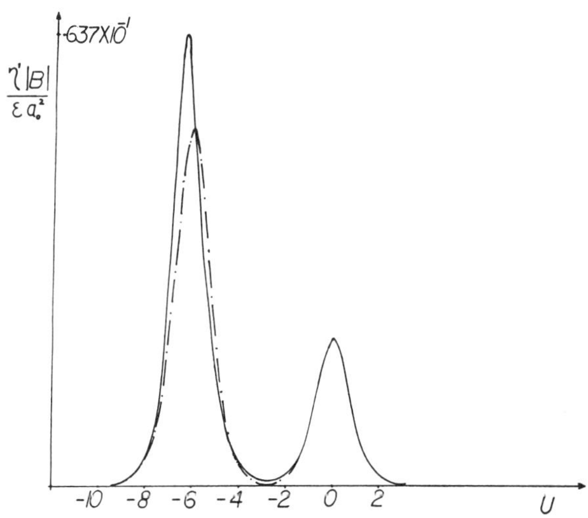

FIGURE 5.60. Magnitude of the generated second harmonic pulse envelope as a function of the normalized time (in units of primary pulse width), in the presence of SPM and IPM: $\eta' z = 6$, $\alpha z = 0.9$, $\zeta'/\eta' = 20$. (Dotted line) $\gamma = 0$; (full line) $\gamma_a = \gamma_b$ and $2\gamma a_0^2/\eta' = 0.5\pi$.

The magnitude of B for weak phase modulation at the points $\bar{U} = \eta' z$ and $\bar{U} = 0$ is given by

$$|B(\bar{U} = -\eta' z)| \approx \varepsilon a_0^2 \frac{1}{(\alpha^2 + \zeta'^2)^{1/2}}, \tag{270}$$

$$|B(\bar{U} = 0)| \approx \varepsilon a_0^2 \exp(-\alpha z) \frac{1}{(\alpha^2 + \zeta'^2)^{1/2}}, \tag{271}$$

These magnitudes are not specific to the detailed pulse shape; for example, they will be the same if, instead of a sech pulse, a Gaussian pulse is used.

The shift and the magnitude value of the pump companion peak are shown in Figure 5.61. As the product of the Kerr constant by the primary intensity is increased (within the range of validity of the above weak phase-modulation equations), the magnitude of the pump companion peak compared with the magnitude of the other peak increases.

Next, we examine the spectral distribution of the second harmonic pulse. We compare the spectral distribution of the second harmonic signal in the absence of $\chi^{(3)}$ (i.e., $\gamma = 0$) with that in its presence (hereafter $\gamma_a = \gamma_b \neq 0$). Furthermore, we compare the phase-modulated primary pulse's spectral distribution with the corresponding incoming pulse's spectral distribution. [It is

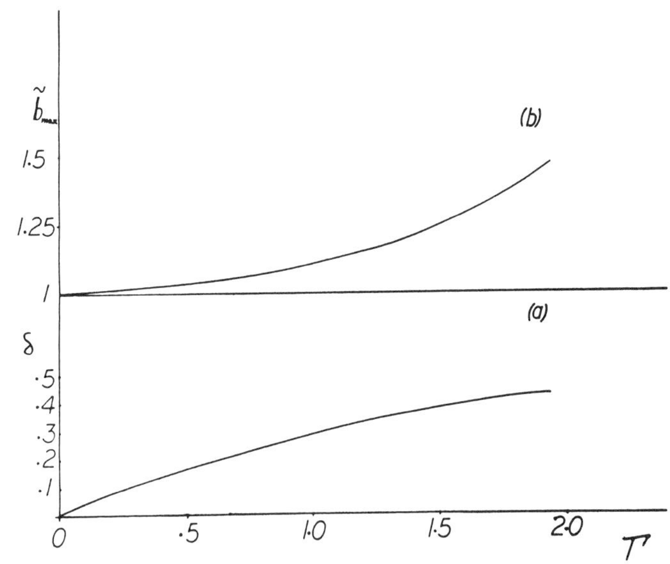

FIGURE 5.61. (a) Position shift, away from the pump, of the pump companion peak as a function of the product of the Kerr constant by the primary pulse intensity. (b) Magnitude of this peak normalized to $\Gamma = 0$ peak: $\zeta'/\eta' = 20$, $\eta' z = 6$, $\alpha z = 0.9$, $\Gamma = 2\gamma a_0^2/\eta'$.

obvious that the spectral distribution of the outgoing primary pulse will have the shape predicted by the conventional self-phase modulation theory.]

In the absence of self-phase modulation, the spectral distribution of the second harmonic signal will exhibit some modulation. Physically, this is a direct result of its two-peak structure. Interference terms will manifest themselves because of the phase difference due to the time separation between the two peaks. The Fourier transform of the second harmonic pulse in the absence of $\chi^{(3)}$ is given by

$$\tilde{B}(\Delta_2) = \int_{-\infty}^{\infty} \exp[i(\omega - \omega_2)t]B(t)\,dt$$

$$= -2\frac{(\varepsilon a_0^2 \tau)}{\eta'} \frac{[\Delta_2 \pi/2]}{\left[\frac{\zeta'}{\eta'} + \Delta_2 + \frac{i\alpha}{\eta'}\right]} \times \frac{\left\{\exp\left[-i\eta' z\left(\frac{\zeta'}{\eta'} + \Delta_2\right)\right] - \exp[-\alpha z]\right\}}{\sinh[\Delta_2 \pi/2]}. \tag{272}$$

Denote the normalized intensities of the primary and second harmonic, respectively, by $\tilde{I}_1$ and $\tilde{I}_2$. In Figures 5.62 and 5.63 these spectral distributions are plotted for specific parameters. The $\tilde{I}$ functions are defined by

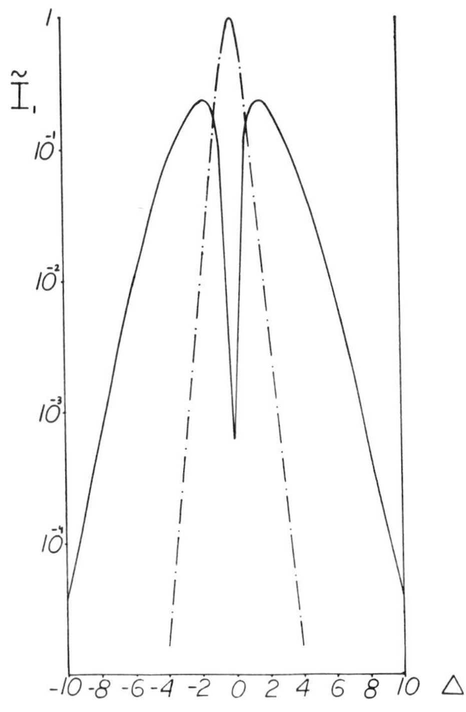

FIGURE 5.62. Normalized spectral distribution of the primary pulse as a function of the normalized frequency difference: $\alpha z = 0.9$, $\eta' z = 6$, $\zeta'/\eta' = 20$. (Dotted line) $\gamma = 0$; (full line) $2\gamma a_0^2/\eta' = 0.5\pi$.

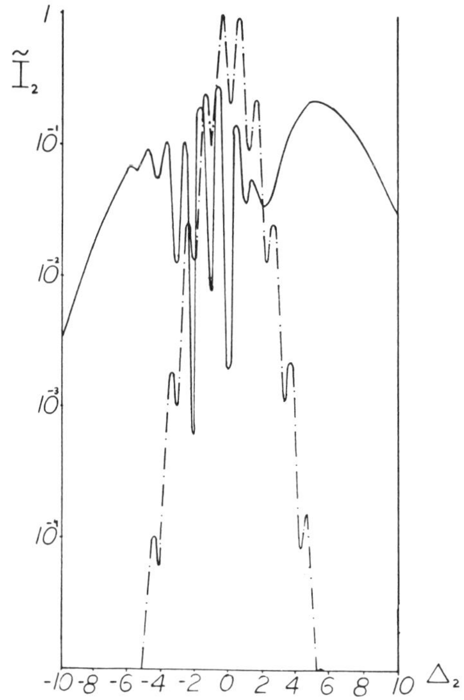

FIGURE 5.63. Normalized spectral distribution of the generated second harmonic signal as a function of the normalized frequency difference: $\eta' z = 6$, $\alpha z = 0.9$, $\zeta'/\eta' = 20$. (Dotted line) $\gamma = 0$; (full line) $2\gamma a_0^2/\eta' = 0.5\pi$.

$$\tilde{I}_1(\Delta_1 \Gamma) = \frac{\left| \int_{-\infty}^{\infty} \exp(i\Delta_1 \bar{U}) A(\bar{U}, \Gamma)\, d\bar{U} \right|^2}{\left| \int_{-\infty}^{\infty} A(\bar{U}, 0)\, d\bar{U} \right|^2}, \tag{273}$$

$$\tilde{I}_2(\Delta_2, \Gamma) = \frac{\left| \int_{-\infty}^{\infty} \exp(i\Delta_2 \bar{U}) B(\bar{U}, \Gamma)\, d\bar{U} \right|^2}{\left| \int_{-\infty}^{\infty} B(\bar{U}, 0) \exp(i\Delta_2 \bar{U})\, d\bar{U} \right|^2_{\max}}, \tag{274}$$

where

$$\Gamma = 2\gamma a_0^2/\eta', \tag{275}$$

$$\Delta_1 = (\omega - \omega_1)\tau, \tag{276}$$

$$\Delta_2 = (\omega - \omega_2)\tau. \tag{277}$$

We note the following:

FIGURE 5.64. The (radiated intensity)$^{1/4}$ $= i$ as a function of $v = (\Delta\omega)\tau$ for $a = n/2$. (*Top*) Full line, $n = 1$; dotted line, $n = 5$. (*Bottom*) Full line, $n = 3$; dashed line, $n = 7$. The intensity curve reaches zero at the minima. ▷

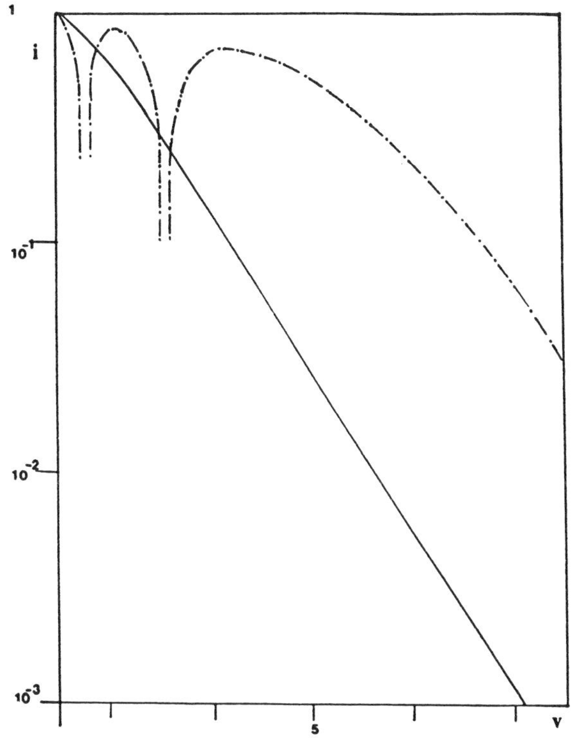
1
i
10^-1
10^-2
10^-3
5
v

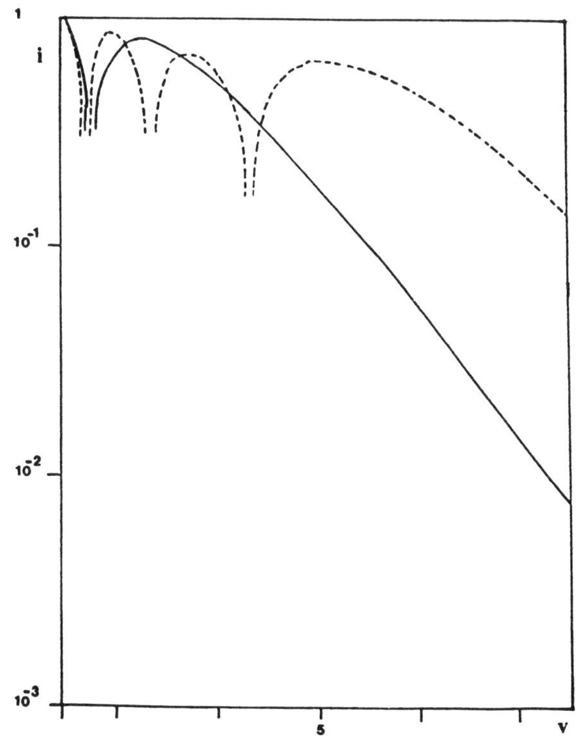
1
i
10^-1
10^-2
10^-3
5
v

1. $\tilde{I}_2$ is not symmetric in Δ_2. This results from the asymmetry in the phase modulation exponents.
2. The frequency extents for the primary and secondary spectral distributions do not scale with γ. The second harmonic spectral distribution is affected by both SPM (through A) and IPM (because of A).

5. Quantum Mechanical Treatment of Supercontinuum Generated in a Thin Sample of Two-Level Resonant Atoms

The phenomenological theories of direct and induced supercontinuum have been developed in previous sections for a Kerr-like medium. In the phenomenological theory, the interaction term in Maxwell's equation is assumed to be instantaneous in the electric field. The microstructure of the nonlinear polarization constant is nowhere incorporated in the phenomenological model. This assumption is reasonable as long as the excitation laser frequency is far off any resonance line of the medium and the medium relaxation constants are short compared to the pulse duration. In this section we investigate the spectrum generated as a result of an incoming resonant ultrashort pulse impinging on a very thin sample of a system of two-level atoms (Manassah, 1986b). We emphasize in our analysis the case of large-amplitude pulses whose width is much shorter than the system inherent relaxation time constant (i.e., natural lifetime).

The model that we adopt for our analysis is Robiscoe's (1978) generalization of the model of Rosen and Zener (1932). In this model, a pulse with a secant hyperbole envelope is incident on a two-level atom system. The natural decay of the atom is incorporated in the Schrödinger equation through the Bethe-Lamb phenomenological prescription (Lamb and Retherford, 1950; Bethe and Salpeter, 1957). The wave function of a two-level atom can be represented by

$$\psi(r,t) = c_1(t)\exp(-i\omega_1 t)u_1 + c_2(t)\exp(-i\omega_1 t)u_2, \tag{278}$$

where u_i and ω_i refer, respectively, to the eigenfunctions and eigenvalues of the unperturbed Hamiltonian. In the presence of an electromagnetic field, central frequency v, and envelope function $f(t)$, the differential equations for c_1 and c_2 are

$$\dot{c}_1 = [V_0 f(t)/i\hbar]e^{+i\omega t}c_2, \qquad \dot{c}_2 = -\tfrac{1}{2}\gamma c_2 + [V_0 f(t)/i\hbar]e^{-i\omega t}c_1, \tag{279}$$

where $\omega = \omega_2 - \omega_1 - v$ and level 1 is assumed to be long lived. If we define the new variables C_1 and C_2 by

$$C_1(t) = c_1(t), \quad \text{and} \quad C_2(t) = c_2(t)e^{\gamma t/2}, \tag{280}$$

the equations of motions reduce to

$$i\hbar\dot{C}_1 = V_0 f(t)C_2 e^{i\Omega t}, \qquad i\hbar\dot{C}_2 = V_0 f(t)C_1 e^{-i\Omega t}, \tag{281}$$

where $\Omega = \omega + \frac{1}{2}i\gamma$. The boundary conditions are assumed to be

$$C_1(t \Rightarrow -\infty) = 1, \qquad C_2(t \Rightarrow -\infty) = 0. \tag{282}$$

The decoupled equations for C_1 and C_2 are given by

$$\ddot{C}_1 - (\dot{f}/f + i\Omega)\dot{C}_1 + |V_0 f(t)/i\hbar|^2 C_1 = 0,$$
$$\ddot{C}_2 - (\dot{f}/f - i\Omega)\dot{C}_2 + |V_0 f(t)/i\hbar|^2 C_2 = 0 \tag{283}$$

for

$$f(t) = \operatorname{sech}(t/\tau,) \quad \text{and} \quad z = \tfrac{1}{2}[\tanh(t/\tau) + 1]. \tag{284}$$

The solutions of (283), satisfying the boundary condition (282), were given by Robiscoe (1978) as

$$C_2(t) = (a/|\phi|)z^{\phi}F(\phi + a, \phi - a; \phi + 1; z),$$
$$C_1(t) = (i\phi/|\phi|)z^{\phi-1/2}(1 - z)^{1/2}xF(\phi + a, \phi - a; \phi; z), \tag{285}$$

where

$$\phi = \tfrac{1}{2} + \tfrac{1}{4}\gamma\tau - \tfrac{1}{2}i\omega\tau \quad \text{and} \quad a = V_0\tau/\hbar.$$

In this section we concentrate henceforth on the physical conditions of on-resonance radiation and pulse duration much shorter than the atomic lifetime (if other relaxation times are present, the pulse duration is also assumed much shorter). The solutions for C_1 and C_2 reduce then to those obtained from Eq. (283) if γ is zero, multiplied by the phenomenological decay function $e^{-\gamma t/2}$. Specifically,

$$C_2(t) = 2az^{1/2}F(\tfrac{1}{2} + a, \tfrac{1}{2} - a; \tfrac{3}{2}; z),$$
$$C_1(t) = i(1 - z)^{1/2}\exp[-(\gamma t/2)]F(\tfrac{1}{2} + a, \tfrac{1}{2} - a; \tfrac{1}{2}; z). \tag{286}$$

The off-diagonal matrix element for the density matrix is given by

$$\rho_{12} = 2iaz^{1/2}(1 - z)^{1/2}F(\tfrac{1}{2} + a, \tfrac{1}{2} - a; \tfrac{3}{2}; z)F(\tfrac{1}{2} + a, \tfrac{1}{2} - a; \tfrac{1}{2}; z)e^{-\gamma t/2}e^{i\omega_0 t}, \tag{287}$$

where ω_0 is the energy level difference.

If $a = n/2$ and the area of the pulse $A = n\pi$, the expression for the off-diagonal matrix element is

$$\rho_{12}^{(a=n/2)} = \tfrac{1}{2}ie^{-\gamma t/2}\sin(2n\arcsin\sqrt{z})e^{i\omega_0 t}. \tag{288}$$

This density matrix element is proportional to the medium complex polarization. For a thin sample, the sheet source approximation, the radiated field is proportional to the polarization (Sargent et al., 1974); specifically:

$$E_R \propto \int P(x, t_r)\exp(i\mathbf{k}_R \cdot \chi)\, d^3x \propto CP(t_r), \tag{289}$$

where t_r refers to the retarded time and C is time independent.

Our next task is to study the radiated electric field in the time and frequency

TABLE 5.3. The function g_n for integers ($x = t/\tau$)

n	$g_{(n)}(x)$
1	$\operatorname{sech} x$
2	$2 \tanh x \operatorname{sech} x$
3	$3 \operatorname{sech} x - 4 \operatorname{sech}^3 x$
4	$\tanh x[4 \operatorname{sech} x - 8 \operatorname{sech}^3 x]$
5	$5 \operatorname{sech} x - 20 \operatorname{sech}^3 x + 16 \operatorname{sech}^5 x$
6	$\tanh x[6 \operatorname{sech} x - 32 \operatorname{sech}^3 x + 32 \operatorname{sech}^5 x]$
7	$7 \operatorname{sech} x - 56 \operatorname{sech}^3 x + 112 \operatorname{sech}^5 x - 64 \operatorname{sech}^7 x$

domains. The zeroes of the polarization in the time domain are located at

$$t = \tau \ln \tan(m\pi/2n), \tag{290}$$

where $m \leq n$. Notice that $t = 0$ is a zero for even n. The specific forms for the polarization, expressed as direct functions of t, are given by

$$\rho_{12}^{(a=n/2)} = \tfrac{1}{2} i[e^{-\gamma t/2}] e^{i\omega_0 t} g_n(t/\tau) \tag{291}$$

where g_n are given in Table 5.3. Notice that g_n determines the asymptotic time dependence given that $\gamma\tau \ll 1$. The values of g_n at $t = 0$ are 1 and 0, respectively, for odd and even n. Observe that as n increases (i.e., the energy of the incoming pulse is increasing) the degree of the polynomial g_n (in sech x) is increasing. The functions $\operatorname{sech}^n x$ are narrower than (sech x) for $n > 1$, which physically implies that as n increases the spectral extent of the radiated field increases. Specifically, the Fourier transform, for odd n, of ρ is given by

$$\rho^{(a=n/2)}(\Delta\omega\,\tau) = \tfrac{1}{2} i\pi\tau \operatorname{sech}(\tfrac{1}{2}\Delta\omega\,\tau\pi) h_n(\Delta\omega\,\tau), \tag{292}$$

where $\Delta\omega$ is the frequency of the radiated field as measured from the center of the resonance line, and $h_n(\Delta\omega\,\tau)$ are polynomials given in Table 5.4. Notice that h_n are even polynomials of degree $n - 1$. The coefficient of the leading power of this polynomial is given by

$$a_{n-1} = (-1)^{(n-1)/2} \frac{2^{n-1}}{(n-1)!}. \tag{293}$$

TABLE 5.4. The function $h_n(v)$ for small odd integers ($v = \Delta\omega\tau$).

n	$h_n(v)$
1	1
3	$1 - 2v^2$
5	$1 - \frac{10}{3}v^2 + \frac{2}{3}v^4$
7	$1 - \frac{196}{45}v^2 + \frac{70}{45}v^4 - \frac{4}{45}v^6$

The behavior at the wings for the Fourier transform of the off-diagonal element of the density matrix is given (for $a = n/2$, n odd) by

$$|\rho^{n/2}(v)| = \tfrac{1}{2}\pi\tau \frac{(2v)^{n-1}}{(n-1)!} e^{-v\pi/2} \tag{294}$$

where $v = \Delta\omega\,\tau$.

Having found ρ on this infinite set of discrete points to be equivalent to the solutions of (279) with $\gamma = 0$ and multiplied by $e^{-\gamma t}$, and given that ρ is an analytical function, then the solution for any arbitrary value of a can similarly be found for the same conditions ($\gamma \ll \tau, u = \omega_0$), by

$$\rho^a(t) = \tfrac{1}{2} i e^{-\gamma t/2} e^{i\omega_0 t} \sin(a\pi + 2a \,\mathrm{gd}\, t/\tau), \tag{295}$$

where gd is the Gudermannian (hyperbolic amplitude), given by

$$\mathrm{gd}\, x = \int_0^x \frac{dt'}{\cosh t'}. \tag{296}$$

For example, ρ^a is given for $a = \frac{1}{4}(2m + 1)$ by

$$\rho^{(a=(2m+1)/4)} = \tfrac{1}{2} i e^{-\gamma t/2} e^{i\omega_0 t} k_{2m+1}(X), \tag{297}$$

where $X = (1 + e^{-2x})^{-1/2}$, $x = t/\tau$, and k_{2m+1} is a polynomial of order $2m + 1$. In Table 5.5, k_{2m+1}, for lowest-order values, is given. Notice that as $t \Rightarrow \infty$, $|\rho^a| \approx e^{-\gamma\tau/2}$. The Fourier component for $\rho^{(a=(2m+1)/4)}$ is given by

$$\tilde{\rho}^{a=(2m+1)/4}(v) = \tfrac{1}{2} i\pi\tau \frac{1}{2\pi} \frac{\Gamma(s)\Gamma(\frac{1}{2} - s)}{\Gamma(\frac{1}{2})} l_m(s), \tag{298}$$

where $v = \Delta\omega\tau$, $s = \frac{1}{4}\gamma\tau + \frac{1}{2}iv$, and l_m is a polynomial of order m. In Table 5.6, $l_m(s)$ is given for the lowest-order values of m. For $\Delta\omega \Rightarrow 0$,

$$|\tilde{\rho}^{a=(2m+1)/4}(0)| \approx \gamma^{-1}. \tag{299}$$

The long tail of $\rho(t)$ for $a = (2m + 1)/4$, at $t \Rightarrow \infty$, is responsible for this large value of $\tilde{\rho}$ at $\Delta\omega = 0$. For $v \gg 1$,

$$|\tilde{\rho}^{a=(2m+1)/4}(v \Rightarrow \infty) \approx \tfrac{1}{2}\pi\tau \left(\frac{1}{v \sinh \pi v}\right)^{1/2} \frac{(4v)^m}{(2m-1)!!}. \tag{300}$$

TABLE 5.5. The function $k_{2m+1}(X)$ for small odd integers $X = (1 + e^{-2x})^{-1/2}$, $x = t/\tau$.

$2m + 1$	$k_{2m+1}(X)$
1	X
3	$3X - 4X^3$
5	$5X - 20X^3 + 16X^5$
7	$7X - 56X^3 + 112X^5 - 64X^7$

TABLE 5.6. The function $l_m(s)$ for small odd integers $s = \frac{1}{2}\tau(i\Delta\omega + \gamma/2)$.

$2m+1$	$l_m(s)$
1	1
3	$-1 + 8s$
5	$1 - \frac{8}{3}s + \frac{64}{3}s^2$
7	$-1 + \frac{144}{15}s - \frac{64}{15}s^2 + \frac{512}{15}s^3$

The steep switch-on in the value of $\rho(t)$, in time of order τ, determines this asymptotic behavior. From Eqs. (294) and (300) it can be observed that the asymptotic form of $\tilde{\rho}(v)$ can be written as

$$\tilde{\rho}(v) \Rightarrow f(a)v^{2a-1}\exp(-\tfrac{1}{2}\pi v) \quad \text{for } v \gg 1, \tag{301}$$

where $f(a)$ is v independent.

In conclusion, using the two-level atom as a model for the interaction of an ultrafast pulse with a resonant system, we exhibited the following features for the generated supercontinuum:

1. At a specific incoming pulse energy, the intensity oscillates as a function of ω, the number of oscillations increasing linearly as a function of the incoming energy.
2. At a specific frequency, the intensity oscillates as a function of the incoming pulse amplitude.
3. On the wings of the line, the intensity is an exponentially decreasing function multiplied by a polynomial whose degree is proportional to the energy of the incoming pulse. Another feature that we found, but that we do not present here, is that if the excitation pulse center frequency is off-resonance, the spectrum of the produced radiation field is asymmetric. Experimentally, the above predictions can easily be tested using 30-fs pulses that have been amplified in a copper vapor laser.

Finally, we reemphasize that in the above analysis we assumed an optically thin sample. This implies that only a very small fraction of the pulse is absorbed by the medium, which in turn implies that the predicted supercontinuum sits on top of a large unaltered component. The portion of the pulse that is absorbed, and thus acts as a source for the supercontinuum, I_s, can be obtained from Beer's law

$$I_s = I(L) - I(0) \quad \text{and} \quad I(L)/I(0) = e^{-\alpha' L}, \tag{302}$$

where $I(0)$ is the incoming pulse intensity, L is the thickness of the sample, α' is the transient effective absorption coefficient, $\alpha' \approx \alpha\tau/T$, α is the CW absorption coefficient, τ is the pulse width, and T is the line width. In the treatment of the optically thick sample case, the full-fledged Maxwell-Bloch equations (Bloch, 1946; Feynman et al., 1957) describing the electromagnetic field–two-level atom system should be used. Under such circumstances, effects such as self-induced transparency (McCall and Hahn, 1969) and superradiant damping (Friedberg and Hartmann, 1971) were shown to be present and are expected to influence the spectral shape of the emitted radiation.

Concluding Remarks

In this chapter, we had the opportunity to review some recent work on simple models of self-phase and induced-phase modulation that can be treated analytically or almost so. What we observe is that:

1. Despite the many simplifying assumptions required to obtain analytical solutions, important qualitatively new features are obtained. Thus, these models serve as all good models should, as theoretical laboratories to seek new effects.
2. The old myth about the existence of "standard supercontinuum characteristics" is as valid as that of any phenomenological categorization, that is, in very limited cases. A rich and diversified variety of supercontinuum structures is shown to exist for different experimental regimes.
3. The partial differential equations derived are amenable to standard numerical computations and approximate methods and would provide the means to the natural extension of the simple analytical cases presented.

This chapter did not treat the following important cases:

1. Instances where the group velocity dispersion effects are important. This precluded any discussion of solitons, an area of intense and extensive research effort.
2. The experimental case where the pump and probe in induced-phase modulation are of the same magnitude. [Our solutions are restricted to weak probes.]
3. The general transverse dependence for large r, that is, for values of r comparable to the value of the pulse radius a. [Our finite beam solutions are restricted to small r/a.]
4. Saturable absorbers, where a full treatment of the Bloch-Maxwell equations is required. Such a treatment will lead to a rigorous inclusion of relaxation times and to the detailed study of frequency-dependent nonlinear indices of refraction.
5. Multimodes of the electromagnetic field and in general the noise problem.

We are actively pursuing work in all of the above areas and will report our results in the literature as they develop.

Finally, I would like to acknowledge the contributions of my coworkers R. Alfano, P. Baldeck, O. Cockings, P. Ho, and M.A. Mustafa, whose participation in the different portions of the work reported in this chapter made it possible. Victoria Okai deserves the credit for typing and producing the manuscript.

References

Alfano, R.R. (1972) GTE Report, TR-72-230-1, April 1972.
Alfano, R.R. and P.P. Ho (1986) Induced phase modulation and induced spectral broadening of propagation laser pulses in condensed matter. *Proc. International Conference on Lasers '86.*
Alfano, R.R. and P.P. Ho (1988) IEEE J. Quantum Electron. **24**, 351.
Alfano, R.R. and S.L. Shapiro (1970) Phys. Rev. Lett. **24**, 592.
Alfano, R.R., Q.X. Li, T. Jimbo, J.T. Manassah, and P.P. Ho (1986) Opt. Lett. **11**, 626.
Alfano, R.R., Q.Z. Wang, T. Jimbo, P.P. Ho, R.N. Bhargava, and B.J. Fitzpatrick (1987) Phys. Rev. A **35**, 459.
Anderson, D. and M. Lisak (1983) Phys. Rev. A **27**, 1393.
Baldeck, P.L., F. Raccah, and R.R. Alfano (1987) Opt. Lett. **12**, 588.
Bethe, H.A. and E.E. Salpeter (1957) *Quantum Mechanics of One- and Two-Electron Atoms.* Academic Press, New York.
Bloch, F. (1946) Phys. Rev. **70**, 460.
Bloembergen, N. (1965) *Nonlinear Optics.* Benjamin, New York.
Born, M. and E. Wolf (1975) *Principles of Optics*, 5th ed. Pergamon, New York.
Dabby, F.W. and J.R. Whinnery (1968) Appl. Phys. Lett. **13**, 284.
De Martini, F., C.H. Townes, T.K. Gustafson, and P.L. Kelley (1967) Phys. Rev. **164**, 312.
Feynman, R.P., F.L. Vernon, Jr., and R.W. Hellwarth (1957) J. Appl. Phys. **28**, 49.
Fork, R.L., C.H. Brito-Cruz, P.C. Becker, and C.V. Shank (1987) Opt. Lett. **12**, 483.
Friedberg, R. and S. Hartmann (1971) Phys. Lett. A **37**, 285.
Gaskill, J.D. (1978) *Linear Systems, Fourier Transforms and Optics.* Wiley, New York.
Goodman, J.W. (1968) *Introduction to Fourier Optics.* McGraw-Hill, New York.
Gordon, J.P. (1986) Opt. Lett. **11**, 662.
Gordon, J.P., R.C.C. Leite, R.S. Moore, S.P.S. Porto, and J.R. Whinnery (1965) J. Appl. Phys. **36**, 3.
Grischkowsky, D. and A. Balant (1986) Appl. Phys. Lett. **41**, 1.
Kelley, P.L. (1965) Phys. Rev. Lett. **15**, 1085.
Kogelnik, H. (1965) Appl. Opt. **4**, 1562.
Kogelnik, H. and T. Li (1966) Proc. IEEE **54**, 1312.
Lamb, W.E. Jr. and R.C. Retherford (1950) Phys. Rev. **79**, 549.
McCall, S.L. and E.L. Hahn (1969) Phys. Rev. **183**, 457.
Manassah, J.T. (1986a) Appl. Opt. **25**, 1737.
Manassah, J.T. (1986b) Phys. Lett. **117A**, 5.
Manassah, J.T. (1987a) Appl. Opt. **26**, 1972.
Manassah, J.T. (1987b) Appl. Opt. **26**, 3747.
Manassah, J.T. (1988a) Opt. Lett. **13**, 755.
Manassah, J.T. (1988b) Appl. Opt. **28**, 206.
Manassah, J.T. (1988c) Opt. Lett. April (1989).
Manassah, J.T. (1988d) Appl. Opt. **27**, 4635.
Manassah, J.T. and O. Cockings (1987a) Appl. Opt. **26**, 3749.
Manassah, J.T. and O. Cockings (1987b) Opt. Lett. **12**, 1005.

Manassah, J.T. and M.A. Mustafa (1988a) Phys. Lett. **A133**, 51.
Manassah, J.T. and M.A. Mustafa (1988b) Opt. Lett. **13**, 752.
Manassah, J.T. and M.A. Mustafa (1988c) Appl. Opt. **27**, 807.
Manassah, J.T. and M.A. Mustafa (1988d) The supercontinuum generated by six-photon mixing. Opt. Lett. **13**, 862.
Manassah, J.T., M.A. Mustafa, R.R. Alfano, and P.P. Ho (1985) Phys. Lett. **113A**, 242.
Manassah, J.T., M.A. Mustafa, R.R. Alfano, and P.P. Ho (1986) IEEE J. Quantum Electron. **QE-22**, 197.
Manassah, J.T., P.L. Baldeck, and R.R. Alfano (1988a) Opt. Lett. **13**, 589.
Manassah, J.T., P.L. Baldeck, and R.R. Alfano (1988b) Opt. Lett. **13**, 1090.
Manassah, J.T., P.L. Baldeck, and R.R. Alfano (1988c) Appl. Opt. **27**, 3586.
Marburger, J.H. (1975) Prog. Quantum Electron. **4**, 35.
Martinez, O.A., J.P. Gordon, and R.L. Fork (1984) J. Opt. Soc. Am. **A-1**, 1003.
Nayfeh, A.H. (1981) *Introduction to Perturbation Techniques*. Wiley, New York.
Robiscoe, R.T. (1978) Phys. Rev. **A-17**, 247.
Rosen, N. and C. Zener (1932) Phys. Rev. **40**, 502.
Rothenberg, J.E. and D. Grischkowsky (1987) Opt. Lett. **12**, 99.
Sargent, M., III, M.O. Scully, and W.E. Lamb, Jr. (1974) *Laser Physics*. Addison-Wesley, Reading, Massachusetts.
Thomas, D.G., L.K. Anderson, M.I. Cohen, E.I. Gordon, and P.K. Runge (1982) Lightwave communication. In *Innovations in Telecommunications*, J.T. Manassah, ed. Academic Press, New York.
Tien, P.K., J.P. Gordon, and J.R. Whinnery (1965) Proc. IEEE **53**, 129.
Tomlinson, W.J., R.H. Stolen, and C.V. Shank (1984) J. Opt. Soc. Am. **B1**, 139.
Treacy, E.B. (1969) IEEE J. Quantum Electron. **OE-5**, 454.
Tzoar, N. and M. Jain (1979) Propagation of nonlinear optical pulses in fibers. In *Fiber Optics*, B. Bendow and S. Mitra, eds. Plenum, New York.
Yang, G. and Y.R. Shen (1984) Opt. Lett. **9**, 510.

Note added in proof: We recently obtained the following new results:

1. An expression for the induced frequency shift of Section 4.1.
2. A scheme to balance Gordon's self-frequency shift by the induced frequency shift.
3. The general solution of the plane wave model for $\chi^{(3)}$ material with relaxation present.

Details of these calculations are given in:

Induced frequency modulation in a nonzero relaxation-time medium. JOSA **B6**, June 1989.

Analytical solution for a self-phase-modulated and self-steepened pulse in a nonlinear medium with material relaxation. JOSA **B6**, June 1989.

Appendix: Analytical Solution for a Self-Phase Modulated and Self-steepened Pulse Propagating in a $\chi^{(5)}$-Medium with Material Relaxation

In this appendix, we derive the analytical solution for a self-phase modulated pulse propagating in a nonlinear $\chi^{(5)}$-medium. The calculation is performed in the plane-wave approximation for dispersionless medium but with self-steepening and material relaxation present.

In the phenomenological model that we consider, the nonlinear source term for Maxwell's equation is taken as:

$$s = \frac{nn_4}{c^2}\frac{\partial^2}{\partial t^2}\left(|E|^4 E - c_1 E \frac{\partial}{\partial t}|E|^4\right). \tag{A1}$$

That is, we are taking the first two terms of the noninstantaneous nonlinear polarization where c_1 is the first moment of the delayed response kernel is essentially equal to the material response time. Variable n_4 is the nonlinear Kerr coefficient, n is the linear index of refraction of the material, and c is the speed of light in vacuum. This phenomenological model is valid for a pulse duration longer than the material response time.

Maxwell's equation reduces to:

$$\left(\frac{\partial^2}{\partial Z^2} - \frac{\partial^2}{\partial T^2}\right)\Phi = \varepsilon' \frac{\partial^2}{\partial T^2}\left(|\Phi|^4\Phi - \gamma'\Phi\frac{\partial}{\partial T}|\Phi|^4\right) \tag{A2}$$

where the dimensional parameters ε', was defined in Section 3, and γ' is given by:

$$\gamma' = \frac{c_1}{\tau} \tag{A3}$$

In the physical coordinates U and V, as defined in Section 3, the amplitude a and the phase α of A, then obey the following quasi-linearized partial differential equations:

$$\frac{\partial a}{\partial V} - \frac{5}{2}\varepsilon' a^4 \frac{\partial a}{\partial U} = 0 \tag{A4}$$

$$\frac{\partial \alpha}{\partial V} - \frac{\varepsilon'}{2}a^4\frac{\partial \alpha}{\partial U} = \frac{K}{2}\varepsilon' a^4 - \frac{K}{8}\varepsilon'^2 a^8 + 2K\gamma'\varepsilon' a^3 \frac{\partial a}{\partial U} \tag{A5}$$

The U-partial derivatives, appearing on the left-hand side of Eqs. (A4) and (A5), are responsible for self-steepening and the term proportional to γ' expresses the finite relaxation time of the medium. We notice that in this model the finite relaxation time does not alter the amplitude equation.

To discuss the impact of self-steepening and finite relaxation time, we shall index a and α by two dummy indices where the first index refers to self-steepening and the second to finite relaxation time. Each index takes the value (0, 1) for the effect being (absent, present). (It is stressed that we can directly solve Eqs. (A4) and (A5) in the most general case, but we are going through these intermediate steps to clarify the role and meaning of the different terms in the equations.) We now write the solutions in the four cases, for an initial sech pulse with zero phase:

(i) $\gamma' = 0,\ \varepsilon' V \ll 1$

$$a_{0,0} = \operatorname{sech}(U) \tag{A6}$$

$$\bar{\alpha}_{0,0} \equiv \frac{\alpha_{0,0}}{K} = \frac{\varepsilon' V}{2} \operatorname{sech}^4(U) \tag{A7}$$

(ii) $\gamma' \neq 0,\ \varepsilon' V \ll 1$

$$a_{0,1} = \operatorname{sech}(U) = a_{0,0} \tag{A8}$$

$$\bar{\alpha}_{0,1} = \frac{\varepsilon' V}{2} \operatorname{sech}^4(U) - 2\gamma'\varepsilon' V \operatorname{sech}^4(U) \tanh(U) \tag{A9}$$

(iii) $\gamma' = 0$

$$a_{1,0}^2 = \operatorname{sech}^2(U + \tfrac{5}{2}\varepsilon' V a_{1,0}^4) \tag{A10}$$

$$\begin{aligned} \bar{\alpha}_{1,0} = {} & -U - \frac{\varepsilon'}{6} \int_0^U a_{1,0}^4(p, V)\, dp - \tfrac{5}{24}\varepsilon'^2 \int_0^V a_{1,0}^8(0, q)\, dq \\ & + \tanh^{-1}(\sin g(U, V)) + \frac{\varepsilon'}{6} \sin g(U, V) - \frac{\varepsilon'}{18} \sin^3(g(U, V)) \end{aligned} \tag{A11}$$

where

$$g(U, V) = \frac{\varepsilon'}{2} \int_0^V a_{1,0}^5(U, q)\, dq + \sin^{-1}(\tanh U) \tag{A12}$$

(iv) general case

$$a_{1,1}^2 = a_{1,0}^2 = \operatorname{sech}^2(U + \tfrac{5}{2}\varepsilon' V a_{1,1}^4) \tag{A13}$$

$$\bar{\alpha}_{1,1} = \bar{\alpha}_{1,0} + \gamma' \ln(a_{1,1}) - \gamma' \ln[\operatorname{sech}[\tanh^{-1}(\sin g(u, v))]] \tag{A14}$$

The foregoing general solution for the amplitude, derived under the condition of no dispersion, is valid for all values of $V < V_{\text{crit}}$, where V_{crit} is the critical value of V.

In Fig. A1, we plot $\bar{\alpha}_{0,1} - \bar{\alpha}_{0,0}$. This quantity, which represents the portion of the phase due to nonzero relaxation time for $\varepsilon' V \ll 1$, is approximately

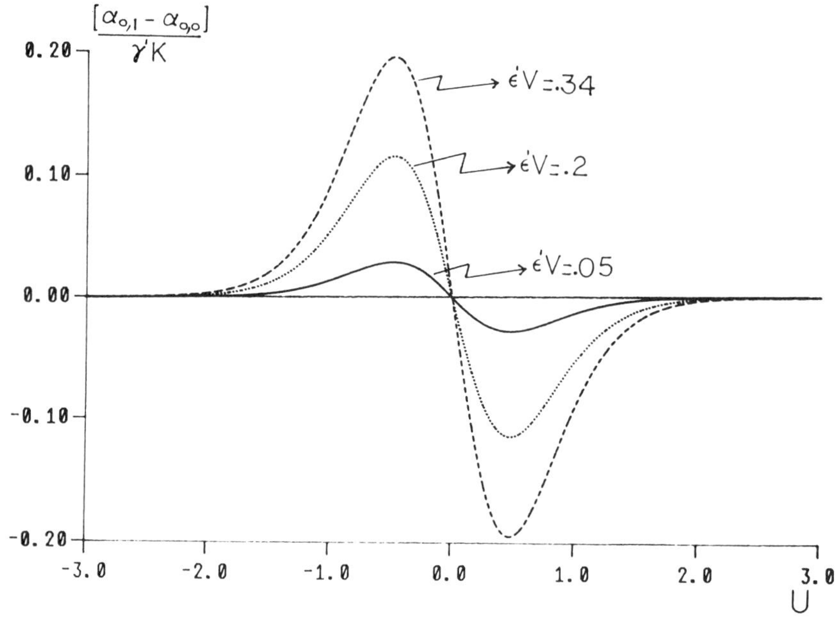

FIGURE A1. The phase portion due to nonzero relaxation time, in the absence of self-steepening, as function of U.

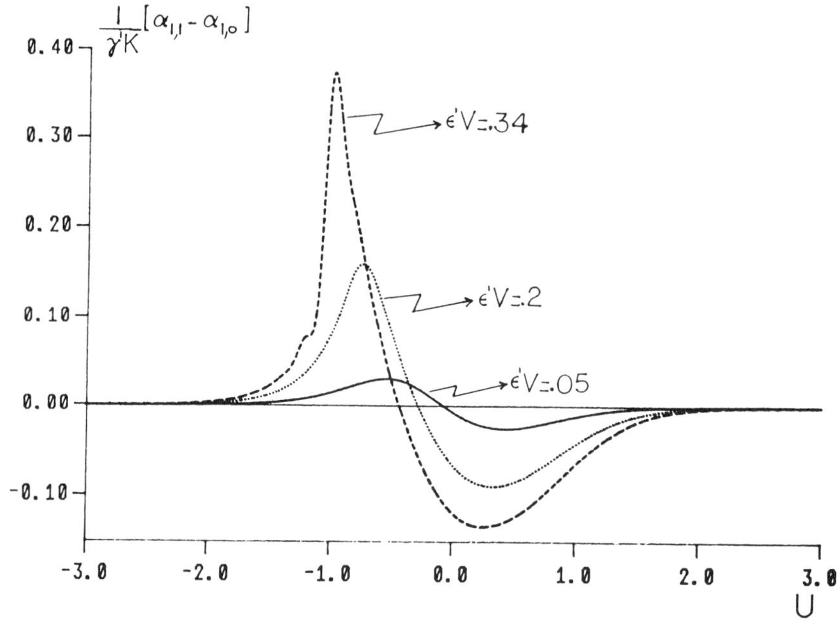

FIGURE A2. The phase portion due to nonzero relaxation time, in the presence of self-steepening, as function of U.

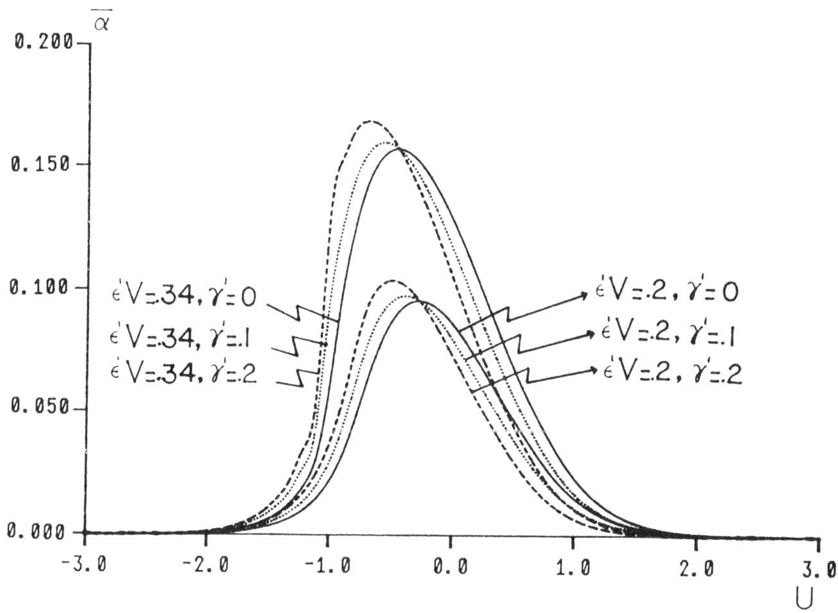

FIGURE A3. The pulse normalized total phase as function of U.

linear in U and has a negative slope, which translates into a Stokes shift in the instantaneous frequency. This shift is similar to Gordon's self-frequency shift, derived for χ^3 material. In Fig. A2, we plot $\overline{\alpha}_{0,1} - \overline{\alpha}_{0,0}$ for different values of $\varepsilon' V$ to qualitatively examine the effects of self-steepening on this Stokes shift.

In Fig. A3, we are plotting $\overline{\alpha}_{1,1}$, the pulse total phase in the general case. (Henceforth, we will omit the subscripts to refer to the general case.) We note that the presence of the γ'-term and the self-steepening term shifts the position of the phase maximum from the $U = 0$ axis. Furthermore, the maximum of the pulse amplitude and that of the phase are shifted with respect to each other, which leads to a shift in the positions of the interference fringes of this pulse, for a Young set-up, from those of *cw* coherent light.

The normalized frequency sweep for this pulse, obtained by taking the U-partial derivative of $\overline{\alpha}$, is given, for the general case, by:

$$\frac{\partial\overline{\alpha}}{\partial U} = -1 - \frac{\varepsilon'}{6}a^4 + \frac{a}{\cos g} + \frac{\varepsilon'}{6}a(\cos g) - \frac{\varepsilon'}{6}a(\sin^2 g)(\cos g) + \gamma'\left[a(\tan g) + \frac{1}{a}\frac{\partial a}{\partial U}\right] \tag{A15}$$

In Fig. A4, we are plotting this quantity. It should be remembered that the maximum of this curve determines in each case the spectral frequency extent on the anti-Stokes side, while its minimum determines the corresponding

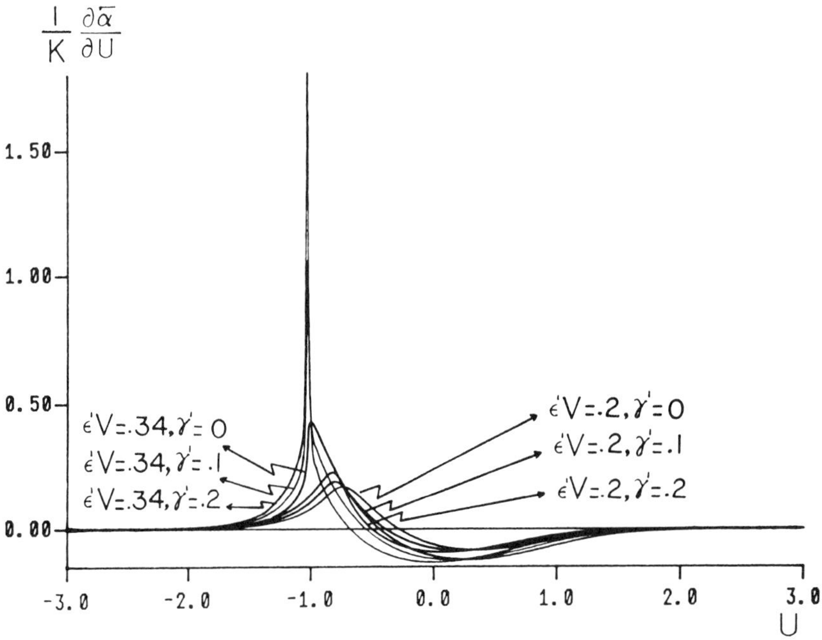

FIGURE A4. The pulse normalized instantaneous frequency sweep as function of U.

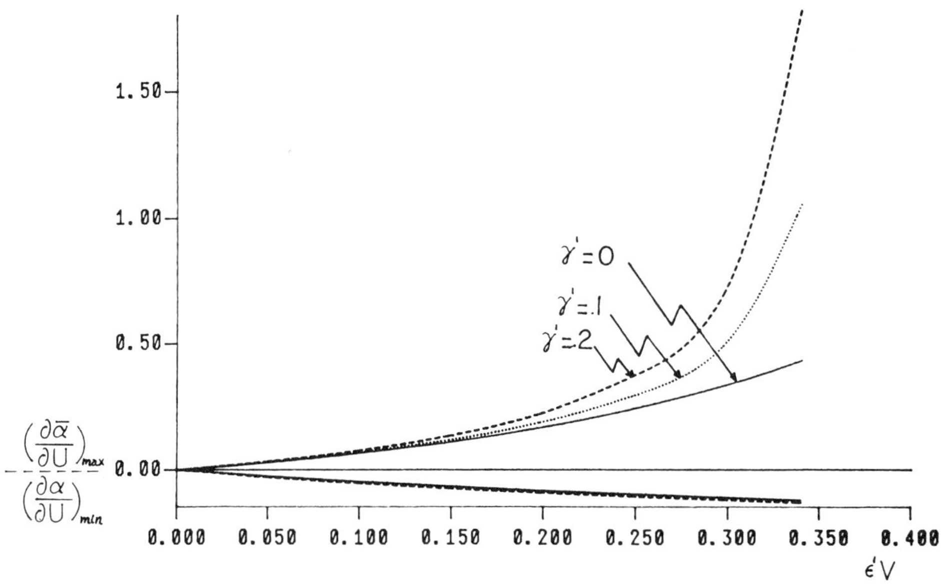

FIGURE A5. The Stokes and anti-Stokes spectral distribution extents as a function of $\varepsilon' V$.

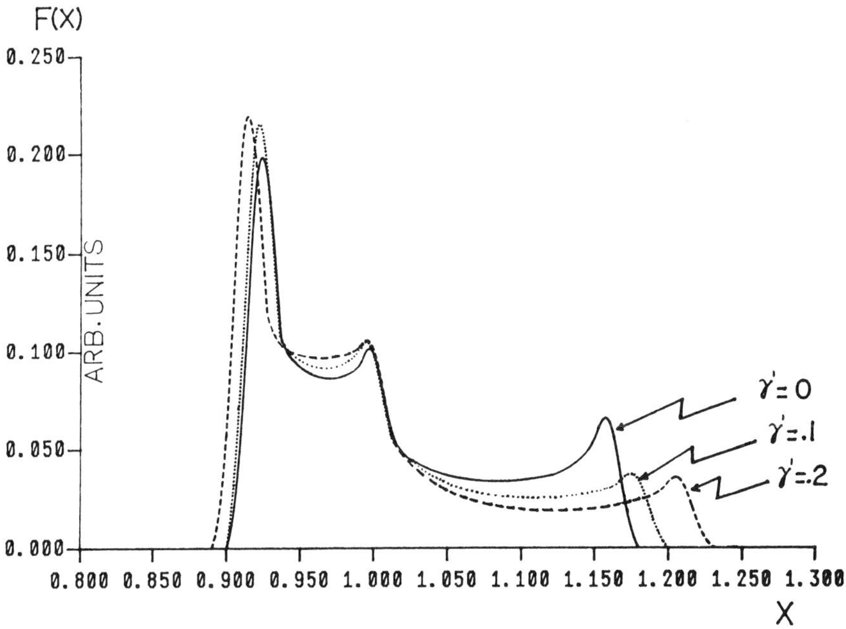

FIGURE A6. The envelope of the Fourier transform of the Young visibility function.

quantity on the Stokes side. As we can observe from the figure, the normalized frequency sweep is asymmetric with respect to the U–0 axis. This results in an asymmetry of the spectral distribution between the Stokes and anti-Stokes portion of the spectrum. Furthermore, the absolute value of the maximum of $\partial\bar{\alpha}/\partial U$ is always bigger than that corresponding to its minimum value, and therefore the anti-Stokes spectral extent is larger than the Stokes extent. In Fig. A5, these extreme quantities are plotted as function of $\varepsilon' V$ for different γ's. These extrema were also shown to determine the domain of $F(x)$, the Fourier transform of the visibility function of the Young intensity distribution with the general solution pulse as input. In Fig. A6, we plot the envelopes of $F(x)$ for different cases. The domain for each case, as computed in Fig. A6, agrees with the results of Fig. A5.

The pulse spectral distribution is obtained by taking the absolute magnitude square of the Fourier transform of the pulse electric field. In Fig. A7, the spectrum is shown for selected values of the parameters. The important features found in our calculations are the following:

1. As $\varepsilon' V$ increases, the spectrum is more asymmetric, and near the Stokes maximum extent the spectrum falls off rapidly.
2. As γ' increases, the spectrum is further shifted to the Stokes side and the maximum frequency extents are consistent with the results of Fig. A5.

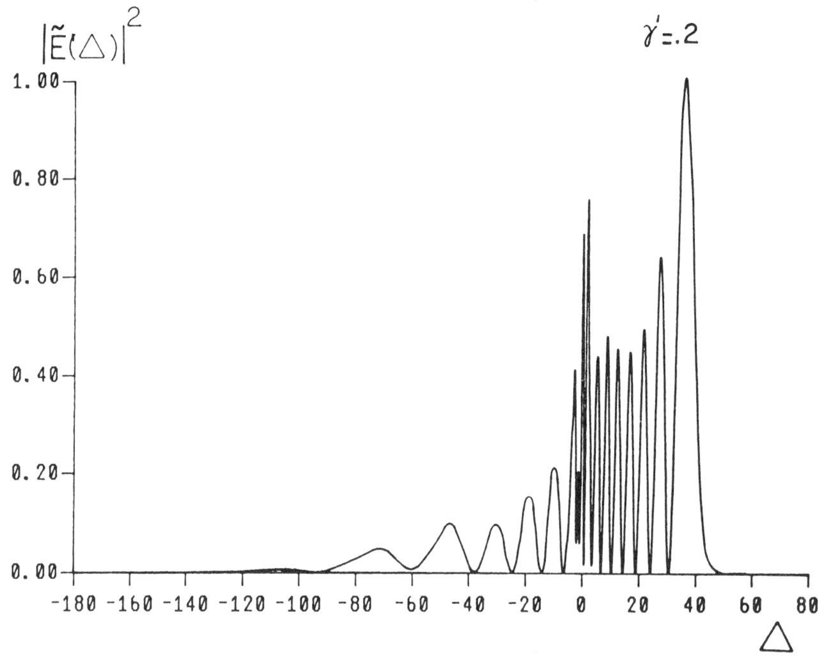

FIGURE A7. The pulse spectral distribution as function of the normalized frequency difference multiplied by the pulse duration. At left is the anti-Stokes side. Normalized center frequency $K = 300$; $\varepsilon' V = 0.2$.

To understand the role of the nonzero relaxation time in the Stokes shift, let us calculate the mean frequency of the pulse normalized to the original center frequency. The normalized first moment of the frequency can be written in the time domain representation as follows:

$$\frac{\bar{K}'}{K} = \frac{\displaystyle\int_{-\infty}^{\infty} \left(1 + \frac{\partial \bar{\alpha}}{\partial U}\right) a^2 \, dU}{\displaystyle\int_{-\infty}^{\infty} a^2(U) \, dU}$$

for small $\varepsilon' V$, this quantity, for the $\chi^{(5)}$ medium, is given by:

$$\lim_{\varepsilon' V \ll 1} \frac{\bar{K}'}{K} \approx 1 - 0.3\gamma'(\varepsilon' V). \tag{A17}$$

The second term on the right-hand side corresponds to the numerically computed Stokes shift. Note that this shift is linear in the thickness of the material, and in the relaxation time and is quadratic in the intensity.

6
Self-Steepening of Optical Pulses

B.R. SUYDAM

Self-steepening of optical pulses has been well described in the literature (De Martini et al., 1967; Marburger, 1975) and has been invoked to explain certain cases of spectral superbroadening (Yang and Shen, 1984). For intense optical pulses the refractive index depends on the intensity. Thus, the peak of the pulse travels at a speed different from that of the leading and trailing edges, so that ultimately the pulse tries to form a shock at the trailing edge if n_2 is positive or on the leading edge if n_2 is negative. A very similar thing occurs in fluid mechanics. However, all fluids exhibit viscosity and heat flow and these effects combine to prevent the formation of a truly discontinuous shock; they give the final shock structure, which has finite thickness and a definite shape. In the optical case, dispersion will play an analogous role; it brings the process of self-steepening to an end before a true discontinuity can form. It is for this last reason that we feel it to be imperative to include the effects of dispersion in our present study of self-steepening.

1. Mathematical Formulation

For mathematical tractability, we restrict our discussion to that of a plane-polarized plane wave traveling in a uniform, isotropic, transparent medium. To arrive at manageable equations we have had to assume also the following:

1. dispersion, absorption, and nonlinearity of the medium are so weak that a light pulse changes shape but slowly as it propagates;
2. higher-order dispersion terms, which would be neglected in a slowly varying envelope theory, are required to take on a certain form; and
3. the nonlinearity develops instantaneously.

Note that we *do not* assume that the envelope varies slowly. Under the foregoing assumptions, if we set

$$E = \tfrac{1}{2}U \exp[i(kz' - \omega t)] + \text{c.c.}$$

into the wave equation for E, we can derive as a good approximation the equation

$$-i\frac{\partial U}{\partial z} = -a\frac{\partial^2 U}{\partial \tau^2} + \gamma|U|^2 U + i\varepsilon\frac{\partial}{\partial \tau}(|U|^2 U). \tag{1}$$

The derivation of this equation is given in Appendix A. In the process we have changed from laboratory coordinates z', t to coordinates z and τ given by

$$\begin{aligned} z &= z', \\ \tau &= t - z/v, \end{aligned} \tag{2}$$

so that $\partial/\partial z$ is calculated in a frame moving with the pulse at the group velocity v.

The coefficients a and γ measure, respectively, the dispersion and the nonlinearity and are given by

$$a = \frac{\lambda^3}{2\pi c^2}\frac{d^2 n}{d\lambda^2} \quad \text{and} \quad \gamma = \frac{3\omega n_2}{4c}$$

where $n(\lambda)$ is the linear refractive index expressed as a function of the free space wavelength $\lambda \equiv 2\pi c/\omega$. Our derivation also yields

$$\varepsilon = \gamma/\omega = 3n_2/4c. \tag{3}$$

We keep the two symbols separate both to facilitate comparison with existing theories and to clarify the physical origin of the various effects we discuss.

In spite of its appearance, Eq. (1) is *not* the result of a slowly varying envelope theory. This is made clear in Appendix A. As a result we are able, without violating the conditions of our approximations, to study pulses of arbitrarily short rise times and of arbitrarily great bandwidths. Assumption (1) is well satisfied for pulses below the damage threshold and frequencies far from an absorption line or band. Assumption (2), spelled out in Appendix A, is automatically met for narrowband signals ($\Delta\omega \ll \omega$); for $\Delta\omega$ of order ω it imposes special, but not nonphysical, conditions on the behavior of $n(\lambda)$ near the band edges. Assumption (3) is not reasonable, but it is at present necessary. Certainly the linear part of the polarization does not respond instantaneously, as otherwise there would be no dispersion, and it would be unrealistic to expect the nonlinear part to respond faster than the linear part. Note also that we have neglected a small amount of absorption, which can be put back in as a perturbation once we have solved Eq. (1).

We shall wish to put numbers into some of our formulas. For this purpose we have chosen from the* *American Institute of Physics Handbook* (1963) and Boling et al., (1978)

$$\begin{aligned} &a = 2.5 \times 10^{-27}(\text{s})^2(\text{cm})^{-1}, \qquad n_2 = 1.27 \times 10^{-13}, \\ &\omega = 3.14 \times 10^{15}(\lambda = 6000\ \text{Å}), \qquad n = 1.67, \\ &\gamma = 10^{-8}, \qquad \varepsilon = 3.2 \times 10^{-24}, \end{aligned} \tag{4}$$

as typical. The number a could vary by a factor 3 or so either way, and as $dn/d\lambda \propto \lambda^{-3}$ over the range considered a scales roughly as λ^{-1}. The dispersion constant a can be either positive for normal dispersion or negative for anomalous dispersion. On the other hand, n_2 can be negative only near to and on the high-frequency side of an absorption line or band, conditions that would strongly violate our assumptions of small dispersion and nearly negligible absorption. Consequently, we shall hereafter assume that γ and ε are positive.

Two special cases of our Eq. (1), both of which include dispersion, have been shown to be completely integrable. The case $\varepsilon \equiv 0$ (no self-steepening), known as the *nonlinear Schrödinger* (NS) equation, was completely solved by Zakharov and Shabat (1972). The case $\gamma \equiv 0$ and $\varepsilon \neq 0$, called by the authors a *derivative nonlinear Schrödinger* (DNS) equation, has been completely solved by Kaup and Newell (1978). In both cases an arbitrary initial pulse of finite energy per unit area will ultimately break up into a set of solitons plus "radiation" (hash that ultimately disperses away). In our case, with general γ and ε, we also find solitary wave solutions, which we call "solitons," which are mathematically scarcely distinguishable from the solitons of the DNS equation and are a direct generalization of the solitons of the NS equation. As do the NS and DNS equations, our equation also admits "kink" or "dark soliton" solutions.

It is convenient for us to rewrite Eq. (1) in terms of a real amplitude A and a real phase ϕ; thus we set

$$U = A\exp(i\phi)$$

and obtain

$$\begin{aligned} A_z &= a[2A_\tau\phi_\tau + A\phi_{\tau\tau}] - 3\varepsilon A^2 A_\tau, \\ A\phi_z &= a[A(\phi_\tau)^2 - A_{\tau\tau}] + A^3(\gamma - \varepsilon\phi_\tau). \end{aligned} \tag{5}$$

The subscripts z and τ denote partial derivatives. These equations describe how the amplitude and the phase evolve as the pulse propagates in z.

2. Early Self-Steepening

There are two ways in which an initially smooth pulse can develop features that sharpen as it propagates, namely by self-steepening and by growth of the modulational instability (longitudinal self-focusing). First let us consider self-steepening. The operator $\partial/\partial\tau$ is of order Ω, which represents the bandwidth of the envelope function U. Thus the dispersion term is of order $a\Omega^2$, whereas the self-steepening term is of order $\varepsilon A^2\Omega$. For very smooth pulses Ω is small; if the power level is not too low we will have

$$\Omega \ll \varepsilon A^2/a \tag{6}$$

and under this condition we can neglect dispersion. In this limit Eqs. (5) reduce to

$$u_z + 3\varepsilon u u_\tau = 0,$$
$$\phi_z + \varepsilon u \phi_\tau = -\gamma u, \tag{7}$$
$$u \equiv A^2.$$

The first is well known (Whitham, 1974), and its general solution can be written in the form

$$u = f(\tau - 3\varepsilon u z). \tag{8}$$

Initially, that is to say at $z = 0$,* $u(\tau, 0) = f(\tau)$, so the form of the arbitrary function f is determined by the pulse shape at $z = 0$. Equation (8) describes shock formation, which we can see as follows. Differentiating Eq. (8) yields

$$u_\tau = \frac{f'}{\Delta}, \qquad u_z = \frac{-3\varepsilon u f'}{\Delta},$$
$$\Delta \equiv 1 + 3\varepsilon z f'. \tag{9}$$

Clearly, these derivatives become infinite and a shock starts to form when Δ first vanishes, and as ε is positive this occurs when f' achieves its most negative value, say f_m'; clearly this occurs on the trailing edge.

The value of z at which the shock first starts to form is clearly given by

$$z_s = -(3\varepsilon f_m')^{-1}. \tag{10}$$

For a Gaussian pulse of width $2T$, we have

$$f(\tau) = A_0^2 \exp[-(\tau/T)^2].$$

Clearly $|f'|$ is maximum when $(\tau/T)^2 = \frac{1}{2}$, so for such a pulse

$$f_m' = -(\sqrt{2/e})A_0^2/T = -0.85776 A_0^2/T,$$

whence, for a Gaussian pulse in a medium of refractive index 1.67,

$$z_s = (0.3886T)/(\varepsilon A_0^2) = (71.42T)/\varepsilon I_0, \tag{11}$$

where I_0 is the peak intensity in watts/cm^2. As an example, for a 100-fs pulse we set $T = 5 \times 10^{-14}$ and with our typical value for ε we would have

$$z_s(100\text{ fs}) = (1.2 \times 10^{11})/I_0 \tag{12}$$

(i.e., a pulse of 1 TW/cm^2 would start to shock after about 12 mm of travel). We have seen that the distance for shock formation is proportional to the pulse width. Thus for a pulse of 10 ps width z_s would be 100 times greater than that for a 100-fs pulse, that is, about 1.2 m for a 1 TW/cm^2 pulse and 1.2 km for a 1 GW/cm^2 pulse.

How a shock actually develops according to Eq. (8) is illustrated in Figure 6.1, in which we have depicted an initially Gaussian pulse, labeled $z = 0$, and

---

* Throughout this chapter we use the expression "initially" or "initial condition" in this sense (i.e., the condition at $z = 0$).

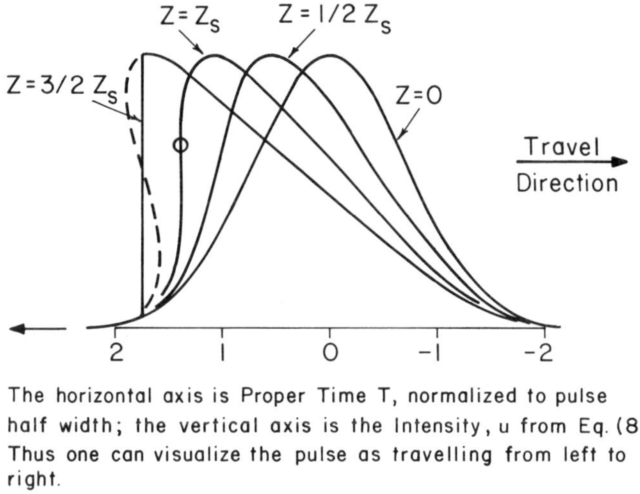

FIGURE 6.1. Various states of self-steepening of an initially Gaussian pulse, according to Eq. (8). Shown are curves for $z = 0$, the initial pulse, $z = \frac{1}{2}z_s$, $z = z_s$, the beginning of shock formation, and $z = \frac{3}{2}z_s$, depicting a well-formed shock. On the $z = z_s$ curve, the circle indicates the point of infinite slope. The dashed curve and vertical shock for $z = \frac{3}{2}z_s$ are discussed in the text.

shown how it changes shape after propagating the distances $z = z_s/2$, $z = z_s$ (the distance for first shock formation), and $z = 3z_s/2$ when the shock is well developed. For $z > z_s$ we must use Eq. (8) with some care. In general, for $z > z_s$, Eq. (8) describes a breaking wave, indicated in our figure by the dashed lines for $z = 3z_s/2$. The intensity cannot, however, be three valued, so we must instead place a shock (i.e., a discontinuous jump) in this region. Equation (8) and, therefore, Eq. (7) also is satisfied on both sides of the jump. The location of the jump is determined by the condition that

$$W = \int_{-\infty}^{\infty} u(\tau, z)\, d\tau = \int_{-\infty}^{\infty} |U|^2\, d\tau = \text{a constant},$$

that is, energy is conserved, as demanded by Eq. (1). We must remember that for $z \geq z_s$ our Figure 6.1 depicts the behavior of the solution of the first of Eqs. (7); in fact, before z reaches z_s, the pulse bandwidth will have become large enough that dispersion can no longer be neglected. We shall now proceed to estimate at what point dispersion must be included.

Once u has been determined, the equation for ϕ is linear and can be solved by the method of characteristics. In the special case of the DNS equation, $\gamma = 0$ and $\phi \equiv 0$ is a solution; in this case phase modulation does not grow as the pulse propagates.

To show more clearly how early self-steepening occurs I think it useful to resort to a simple model problem. To this end we choose the leading edge of

$f(\tau)$ to be anything that rises smoothly to a peak value f_0 at $\tau = 0$. The trailing edge, where the self-steepening occurs, we choose to be of the form

$$f(\tau) = \begin{cases} f_0(1 - \tau/\tau_0) & \text{for } 0 < \tau < \tau_0, \\ 0 & \text{for } \tau > \tau_0. \end{cases} \tag{13}$$

Then by Eq. (8), for $\tau < 0$, u will be some stretched-out version of the original pulse shape $f(\tau)$, and on the trailing edge

$$u = \begin{cases} f_0(1 - \tau/\tau_0)/\Delta & \text{for } 3\varepsilon f_0 z < \tau < \tau_0, \\ 0 & \text{for } \tau > \tau_0, \end{cases} \tag{14}$$

and for this case our previously defined Δ becomes

$$\Delta = 1 - 3\varepsilon z f_0/\tau_0.$$

Setting this value of u into the second of Eqs. (7), we can readily solve by the method of characteristics (in the trailing edge) and find

$$\phi = \begin{cases} \dfrac{\gamma}{\varepsilon}(\tau_0 - \tau)[1 - \Delta^{-1/3}] + \phi_0(\tau_0 - (\tau_0 - \tau)\Delta^{-1/3}) & \text{for } 3\varepsilon f_0 z < \tau < \tau_0, \\ \phi_0(\tau_0) & \text{for } \tau > \tau_0, \end{cases} \tag{15}$$

where $\phi_0(\tau) = \phi(\tau, 0)$ is the initial phase. In the leading edge, $\tau < 3\varepsilon f_0 z$, the phase will build up in a way that depends on the choice of $f(\tau)$ in this region.

Consider now a pulse initially without phase modulation, so $\phi_0 \equiv 0$. We see ϕ reaches its maximum at $\tau = \tau_{\min} = 3\varepsilon f_0 z$ and it is zero at $\tau = \tau_0$. Thus over this period of duration

$$\tau = \tau_0 - 3\varepsilon f_0 z = \tau_0 \Delta$$

the phase swings by

$$\Delta\phi = \omega \tau_0 \Delta[1 - \Delta^{-1/3}]$$

so we can estimate the bandwidth induced solely by this phase modulation, Ω_ϕ, as

$$\Omega_\phi = \Delta\phi/\tau = \omega[1 - \Delta^{-1/3}]. \tag{16}$$

If we insert this into (6) rewritten as an equality we obtain

$$\Delta = [1 + \gamma A_0^2/a\omega^2]^{-3}. \tag{17}$$

When this condition is satisfied dispersion has become just as important as self-steepening and Eqs. (7) have broken down as a decent approximation. For our typical numbers $\gamma A_0^2/a\omega^2 = 1$ implies a peak power level of 3×10^{14} W/cm^2 ($A_0 = 4.7 \times 10^8$ V/cm). For peak power levels significantly lower than this,

$$z_0 = \tau_0/a\omega \tag{18}$$

is the distance at which this breakdown occurs. For our typical numbers and

$\tau_0 = 50$ fs (fall time), this gives $z_0 = 6.4 \times 10^{-3}$ cm, which is far less than the distance for shock formation, z_s. For peak power notably less than 10^{14} W/cm^2 Δ is very nearly unity and the trailing edge has steepened very little. Thus, for such a pulse, self-steepening is of only minor importance during the initial phase; it is the buildup of phase modulation that forces one to include the effects of dispersion.

3. Modulational Instability

Another mechanism for the breakup of an initially smooth pulse is the modulational instability, which has also been called longitudinal self-focusing. This has been well discussed by Marburger (1975) in the case $\varepsilon = 0$. To illustrate how this comes about, let us consider the smoothest possible pulse, namely an unmodulated wave train. Thus we set

$$A = A_0 = \text{a constant} \quad \text{and}$$
$$\phi = \phi_0 \equiv \text{a const} + \gamma A_0^2 z. \tag{19}$$

These functions satisfy Eqs. (5). Now consider a small perturbation to this solution; namely we set

$$A = A_0(1 + w),$$
$$\phi = \phi_0 + \tilde{\phi} = \text{const} + \gamma A_0^2 z + \tilde{\phi}. \tag{20}$$

Setting Eqs. (20) into Eqs. (5) yields equations for w and $\tilde{\phi}$. If the perturbation is small enough, we can neglect squares and products of w, $\tilde{\phi}$, and their derivatives. The resulting linearized equations for the perturbations are

$$w_z + 3\varepsilon A_0^2 w_\tau = a\tilde{\phi}_{\tau\tau},$$
$$\tilde{\phi}_z + \varepsilon A_0^2 \tilde{\phi}_\tau = -a w_{\tau\tau} + 2\gamma A_0^2 w \tag{21}$$

and the question is, what happens to the perturbation as it propagates; does it grow?

We can readily reduce Eqs. (21) to a pair of first-order ordinary differential equations by Fourier analyzing in τ, setting

$$w(z, \tau) = u(z)\exp[i\Omega\tau],$$
$$\tilde{\phi}(z, \tau) = q(z)\exp[i\Omega\tau] \tag{22}$$

and thus eliminate one of the variables, say q, between the two resulting equations. The result is the second-order equation

$$u'' + 4i\varepsilon\Omega A_0^2 u' + [a\Omega^2(a\Omega^2 + 2\gamma A_0^2) - 3\varepsilon^2 A_0^4]u = 0. \tag{23}$$

This has the solutions

$$u_\pm(z) = \text{const.}\exp\{(-2i\varepsilon\Omega A_0^2 \pm \alpha)z\} \tag{24}$$

where

$$\alpha = +[-a\Omega^2(a\Omega^2 + 2\gamma A_0^2) - \varepsilon^2 A_0^4 \Omega^2]^{1/2}. \tag{25}$$

By hypothesis γ is positive; if α is also positive then α is pure imaginary and both functions $u_+(z)$ and $u_-(z)$ simply oscillate as the pulse propagates, that is, as z increases; the perturbation neither grows nor dampens.

Now suppose that the dispersion is anomolous, that is, that a is negative. We set $b = -a =$ positive and obtain

$$\alpha^2 = b\Omega^2(2\gamma A_0^2 - b\Omega^2) - \varepsilon^2 A_0^4 \Omega^2 \tag{26}$$

and this is maximized for

$$\Omega^2 = \Omega_M^2 \equiv (\gamma/b)A_0^2 - (\varepsilon^2/2b^2)A_0^4. \tag{27}$$

If we set this value into Eq. (26) we determine that the maximum value of α^2 is

$$\alpha_M^2 = [\gamma A_0^2 - (\varepsilon^2/2b)A_0^4]^2. \tag{28}$$

Now if

$$A_0^2 < 2b\gamma/\varepsilon^2 \tag{29}$$

(which for our typical numbers means peak power less than 10^{15} W/cm^2), Ω_M^2 and α_M^2 are positive and $u_+(z)$ oscillates with an exponentially growing amplitude as the beam propagates. Thus, no matter how small the initial disturbance, the modulation of the beam will grow exponentially until it interacts nonlinearly with itself. This is the modulational instability. The reader may notice that had we set $\varepsilon = 0$ our analysis would have been identical to that of the self-focusing instability (Bespalov and Talanov, 1966; Suydam, 1973). One might therefore call the phenomenon longitudinal self-focusing rather than modulational instability; the two terms mean the same thing.

It is interesting to compare the effects of the modulational instability and of self-steepening. To this end we will consider a beam of a few terawatts power or less in a medium of anomalous dispersion. The inequality (29) is strongly satisfied, so the maximum growth rate is

$$\alpha_M = \gamma A_0^2$$

very nearly. Now multiply this by the shock formation distance z_s, given by Eq. (11); we get

$$\alpha_M z_s = 0.3886 T\omega = 2.44N$$

where N is the half-width of the pulse measured in cycles of the carrier frequency ω. In propagating over the shock-forming distance z_s the modulational instability grows by a factor $\exp[2.44N]$, which amounts to a factor 10^5 for $N = 4.7$ cycles half-width. Clearly, unless we are dealing with pulses of less than 10 cycles full width, the modulational instability, if it occurs (i.e., if a is negative), far outstrips self-steepening. Thus, for the exceptional case of anomalous dispersion, the breakup of a smooth pulse into nascent solitons does not involve self-steepening. However, as we shall see in the next section, the ultimate soliton (or solitons) possesses strong phase modulation, which would have been absent were $\varepsilon = 0$. Thus even here the self-steepening term is very important in establishing the ultimate bandwidth of the pulse.

4. Solitonlike Solutions

When dispersion is normal, the only way a pulse can develop sharp features and undergo spectral broadening is through the processes of self-steepening and self-phase modulation. As we saw in Section 2 [see Eqs. (15) and (16)], self-steepening enhances the self-phase modulation. For the DNS equation self-steepening interacting with dispersion ultimately breaks the pulse up into a set of solitons plus, perhaps, some radiation. We therefore expect something similar to happen in our case, so we turn now to the study of the solitary wave solutions of Eq. (1).

For a traveling wave solution, the amplitude will be of the form

$$A = A(\tau - \eta z/v) \equiv A(\xi) \tag{30}$$

where η is an as yet undetermined constant. With this ansatz the first of Eqs. (5) becomes

$$-(\eta/v)A' = a[2A'\phi_\tau + A\phi_{\tau\tau}] - 3\varepsilon A^2 A'. \tag{31}$$

This equation is solved identically in A by the choice

$$\phi_\tau = -(\eta/2av) + (3\varepsilon/4a)A^2. \tag{32}$$

If we define a new function

$$H(\xi) = \int^{\xi} [A(\xi')]^2 \, d\xi' \tag{33}$$

we can integrate Eq. (32), obtaining

$$\phi(z, \tau) = \phi_0(z) - (\eta\tau/2av) + (3\varepsilon/4a)H(\xi) \tag{34}$$

whence

$$\phi_z = \phi_0' - (3\varepsilon\eta/4av)A^2. \tag{35}$$

Setting Eqs. (32) and (35) into the second of Eqs. (5) yields

$$A'' - \left(\frac{\eta^2}{4a^2v^2} - \frac{b}{a}\right)A - \frac{1}{a}\left(\gamma + \frac{\varepsilon\eta}{2av}\right)A^3 + \frac{3\varepsilon^2}{16a^2}A^5 = 0. \tag{36}$$

We have set $\phi_0'(z) = b = \text{constant}$, as is required by our insistence that A depend only on $\xi \equiv \tau - \eta z/v$. Equation (36) has the first integral

$$(A')^2 = C + PA^2 - 2QA^4 - R^2A^6 \tag{37}$$

where C is an arbitrary integration constant and we have set

$$P = \left(\frac{\eta}{2av}\right)^2 - \left(\frac{b}{a}\right), \qquad Q = -\frac{1}{4a}\left(\gamma + \frac{\varepsilon\eta}{2av}\right), \qquad R^2 = \frac{\varepsilon^2}{16a^2} \tag{38}$$

for short. If we set

$$u = A^2$$

into Eq. (37) we can write its solution as

$$\pm 2\xi = \int \frac{du}{\sqrt{Cu + Pu^2 - 2Qu^3 - R^2u^4}}. \tag{39}$$

For general values of C, this is an elliptic integral and the solution represents the anharmonic oscillations of a particle of mass 2 and total energy C in the potential well

$$V = -PA^2 + 2QA^4 + R^2A^6.$$

In the special case $R = 0$, these are the so-called cniodal waves.

For two special choices of C the integration becomes elementary. First, if we set $C = 0$, u^2 factors out of the radical so that we can carry out the integration and then solve for u, obtaining

$$u \equiv A^2 = \frac{A_0^2}{\cosh^2 v\xi + y^2 \sinh^2 v\xi} \tag{40}$$

where

$$\begin{aligned} A_0^2 &= \frac{1}{R^2}\left[-Q + \sqrt{Q^2 + PR^2}\right], \\ v^2 &= P, \\ y^2 &= R^2A_0^4/v^2. \end{aligned} \tag{41}$$

This solution requires that P be positive. The material constants a, γ, and ε together with the two integration constants b and η determine A_0 and v. Alternatively, we could assign A_0 and v and solve for the integration constants, obtaining

$$\begin{aligned} \eta &= \frac{4a^2v}{\varepsilon A_0^2}(R^2A_0^4 - v^2) - \frac{2av\gamma}{\varepsilon}, \\ b &= \frac{1}{4av^2}\left[\frac{4a^2v}{\varepsilon A_0^2}(R^2A_0^4 - v^2) - \frac{2av\gamma}{\varepsilon}\right]^2 - av^2. \end{aligned} \tag{42}$$

The quantities η and b can be viewed as nonlinear modifications of the group and phase velocities, respectively.

To complete our solution, we set Eq. (40) into Eq. (33) and obtain

$$H(\xi) = \frac{1}{R}\left\{\arctan y - \arctan\left[\frac{2y}{(1 + y^2)\exp(v\xi) + (1 - y^2)}\right]\right\}. \tag{43}$$

With this value of H the phase function is

$$\phi(\tau, z) = bz - (\eta\tau/2av) + (3\varepsilon/4a)H(\tau - \eta z/v). \tag{44}$$

Using the addition formula for the tangent, one sees that H is odd, that is, $H(-\xi) = -H(\xi)$. Clearly

$$W \equiv \int_{-\infty}^{\infty} A^2\,d\tau = (2/R)\arctan y. \tag{45}$$

The solutions we have found above we shall call solitons for short. They are the only traveling wave solutions for which W is finite. Clearly W is proportional to the total soliton energy per unit area; we shall simply call it the total energy for short.

Our solution goes to a perfectly regular limit as $P \to 0$, namely

$$A = \pm A_0[1 + R^2 A_0^4 \xi^2]^{-1/2} \qquad (v = 0), \tag{46}$$

and

$$H(\xi) = \frac{1}{R}\arctan(RA_0^2\xi) \qquad (v = 0). \tag{47}$$

Both our general soliton and our limiting algebraic soliton are formally identical to those of the DNS equation (Kaup and Newell (1978)); our general solitons go over to the solitons of the NS equation in the limit $\varepsilon \to 0$.

Another traveling wave solution, known as a "kink" or as a "dark soliton," can be found as a special case of Eq. (39). As it contains infinite total energy in the sense that the integral of Eq. (45) diverges, it is not germane to our discussion. Nevertheless, the reader might find it interesting, so we give this solution in Appendix B.

As is seen, our solitons differ from those of Kaup and Newell (1978) only in the presence of the constant γ in our Eqs. (42). Soliton solutions exist quite independently of the algebraic signs of the material constants a, γ, and ε. In both cases the phase ϕ is given by Eqs. (43) and (44), so with identical values of ε the two show identical phase modulation. The solitons of the NS equation, for which $\varepsilon = 0$, are quite different. In this case Eqs. (40) and (44) become

$$A = \pm A_0 \operatorname{sech}(v\xi),$$
$$\phi = bz - (\eta\tau/2av),$$

so for this case there is no phase modulation. Equations (41) become

$$v^2 = P,$$
$$A_0^2 = P/2Q = -2(a/\gamma)v^2.$$

Thus such solitons can exist only if the dispersion is anomalous. Moreover, A_0^2 and v^2 are not independent so we cannot determine η or b but only the relationship between them:

$$v^2 = (\eta/2av)^2 - (b/a).$$

Note also that Eq. (45) goes into

$$\int_{-\infty}^{\infty} A^2\, d\tau = 2A_0\sqrt{-2a/\gamma}$$

so there is no upper limit to the total energy that an NS soliton may carry. The two constants that uniquely specify an NS soliton ($\varepsilon \equiv 0$) are its peak amplitude and its velocity; that is, A_0 and η may be arbitrarily chosen.

It is useful to make estimates of the total bandwidth of one of our solitons. We can define the effective width of our soliton, T_W, as that of a square pulse with the same total energy and the same peak intensity as our soliton. Then clearly, from Eq. (45)

$$T_W = (2/RA_0^2)\arctan y. \tag{48}$$

We know the spectrum and can compute T_W for a Gaussian pulse. If we define the spectral bandwidth Ω to be that of the interval $(-\Omega/2, \Omega/2)$ that contains half the pulse energy, we find that a Gaussian pulse yields

$$\Omega_A = 0.8455RA_0^2/\arctan y. \tag{49}$$

The subscript A means that this is the bandwidth arising solely from the temporal shape of the real amplitude.

In addition to Ω_A, there is frequency modulation on the pulse. If we define the instantaneous frequency to be the time derivative of the total phase, we have seen that this frequency swings over a range $(\omega, \omega + 3RA_0^2)$. Thus it seems reasonable to estimate the total bandwidth owing to phase modulation to be

$$\Omega_\phi = 3RA_0^2. \tag{50}$$

For our typical numbers $\Omega_\phi = \omega$ at a peak power of 6.5×10^{14} W/cm^2. If instead we took $\Delta\phi = \phi(\infty) - \phi(-\infty)$ as given by Eq. (44) and estimated this bandwidth as $\Delta\phi/T_W$, we would exactly recover Eq. (50) above. The ratio of the phase-to-amplitude modulational bandwidths is

$$\Omega_\phi/\Omega_A = 3.55\arctan y. \tag{51}$$

If we define W for the actual soliton by Eq. (45) and call W_M its maximum possible energy, that is, that when $v \to 0$, $y \to \infty$, then Eq. (51) can also be written as

$$\Omega_\phi/\Omega_A = 5.57W/W_M. \tag{52}$$

When W approaches W_M, the bandwidth of the soliton is dominated by phase modulation, whereas for $W \ll W_M$ the quantity v becomes large, the soliton is very narrow, and this effect dominates the bandwidth. For our typical numbers W_M corresponds to about 2 J/cm^2. In general,

$$W_M = \frac{\pi}{R} = \frac{16\pi|a|c}{3n_2} \tag{53}$$

so it can readily vary a factor 3 or so either way from the figure quoted above.

Note that the frequency modulation disappears if we set $\varepsilon = 0$ into Eq. (44), so although self-phase modulation is a dominant contributor to the bandwidth of very energetic solitons (W approximately W_M), it arises from the interaction of dispersion with the self-steepening term of Eq. (1) and is, in this sense, a self-steepening effect.

5. An Attempted Synthesis

We have seen that the trailing edge of an initially very smooth pulse will steepen and will develop phase modulation as it propagates. In the absence of dispersion, this process would continue until a discontinuous shock formed on the trailing edge and the phase modulational bandwidth would become infinite at the shock. We have further seen that when dispersion is taken into account there are traveling wave solutions, which we have called solitons, for which the self-steepening and the dispersion just balance. It seems very reasonable, then, to postulate that the self-steepening process is limited by soliton formation. In fact, for the DNS equation, our Eq. (1) with $\gamma \equiv 0$, this postulate is a rigorous result (Kaup and Newell, 1978).

According to the above ideas, I now describe my view of the development of an initially smooth pulse as it propagates. At first, dispersion is unimportant and the pulse gradually sharpens up on its trailing edge, and this self-steepening process is accompanied by the growth of self-phase modulation, all as described in Section 2. As we have seen, when the phase modulation during the self-steepening phase reaches a value

$$(\Omega_\phi)_{ss} = \gamma A_0^2/a\omega, \tag{54}$$

Equations (7) break down as a model for the propagation because they ignored dispersion, which has a very important influence on the further development. Actually, dispersion starts to play an important part somewhat earlier, and by the time Eq. (54) is satisfied it will already have modified the trailing edge shape to somewhat resemble that of a soliton. In Section 4 we saw that the phase modulational contribution to the bandwidth of a soliton is

$$(\Omega_\phi)_{\text{sol}} = 3\gamma A_0^2/4a\omega. \tag{55}$$

The similarity of Eqs. (54) and (55) reinforces our belief that at this transitional stage the trailing edge of our steepened pulse somewhat resembles that of a soliton.

During the self-steepening phase the velocity of the trailing edge (actually of its peak) is given by

$$v_{ss} = v/(1 + 3\varepsilon u_0 v) \tag{56}$$

where u_0 is the peak value of A^2 for the initial pulse. A soliton propagates with velocity

$$v_{\text{sol}} = v/(1 + \eta),$$
$$\eta = \tfrac{1}{4}\varepsilon A_0^2 v - 2av(\omega + 2av^2/\varepsilon A_0^2) \tag{57}$$

so that $v_{\text{sol}} > v_{ss}$. In fact, for our typical numbers v_{sol} exceeds group velocity unless the peak power exceeds 4×10^{15} W/cm^2. The solitonlike trailing edge will therefore advance into the smooth leading section of the pulse. What subsequently happens depends on the total energy W of the initial pulse.

If the initial pulse energy is

$$W \leq W_M,$$

the nascent soliton can engulf the whole pulse and form a genuine soliton. Our parameter $y = RA_0^2/v$ will be determined by

$$y = \tan(\pi W/2W_M). \tag{58}$$

To estimate separately A_0^2 and v, we note that the trailing edge need not steepen beyond what it had reached at the end of the self-steepening stage, Eq. (17). Thus, this trailing edge advances until the pulse is symmetrical, and from this stage onward there are only detailed shape adjustments to arrive finally at the exact soliton form. Using as our model an initial triangular pulse with rise time T_1 and decay time T_0, we have clearly

$$W = \tfrac{1}{2}f_0(T_0 + T_1) = \tfrac{1}{2}A_i^2(T_0 + T_1), \tag{59}$$

where A_i denotes the peak value of A in the initial pulse. As the pulse advances its decay time becomes $T = T_0\Delta$ and this continues until Δ reaches the value given by Eq. (17). From this time on, the decay time remains $T_0\Delta$ and the pulse compresses until the rise time is the same. Thus at the time of formation of the quasi-soliton

$$W = A_0^2 T_0 \Delta, \tag{60}$$

whence for the peak amplitude of the quasi-soliton we have

$$A_0^2 = (A_i)^2(T_0 + T_1)/2\Delta T_0, \tag{61}$$

with Δ given by Eq. (17). As we saw in Section 2, unless we are dealing with initial pulses of peak power in excess of 10^{14} W/cm^2, Δ differs little from unity, so an initially symmetrical pulse would undergo little compression unless its peak intensity was very high.

If $W > W_M$ the situation is more complicated and our present understanding more conjectural. Again the trailing edge starts to eat up the smooth part of the pulse until it can swallow no more. Somewhere along this process the advancing "snowplow" must wrinkle the part of the pulse immediately ahead of itself, which can then proceed to self-steepen on its own and ultimately develop a new soliton. If W exceeds W_M by only a modest amount, an alternative to the formation of a new soliton is simply to throw the remaining energy away as radiation. Something analogous to this is known to happen in self-focusing.

6. Conclusion

We conclude this chapter by examining the virtues and the weaknesses of what we have presented. The virtues are easily disposed of; there are two, namely:

1. we do not employ a slowly varying envelope approximation, and
2. we include linear dispersion in an approximation that is reasonably good for many common transparent media.

Our mathematical model contains three weaknesses. First, dispersion is always accompanied by some small absorption and this we have neglected. This is not a serious fault, as the effects of a small absorption can readily be estimated. The method is to start with Eqs. (40) and (44) for the amplitude and the phase. In these equations one now allows A_0 and v to be slowly varying functions of z; they are determined by substituting the resulting expressions into Eqs. (5), to which a small absorption term has been added. One then solves for $A_0(z)$, $v(z)$ by a "slowly varying A_0 and v approximation." Normally such corrections are important only for fiber transmission lines.

The second fault of our mathematical model is that it is limited to plane waves, for which nothing varies in the x or y direction. We could extend our results to a nearly plane wave for which the beam diameter is so large that the $\nabla_\perp^2$ operator (missing from our Eq. (1)) is negligible compared to all the other operators in our equation. This seems to rule out any really practical application of our treatment to transmission in fibers.

The third fault is that we do not include nonlinear dispersion; that is, we postulate that the nonlinear response develops instantaneously. The linear part of the polarization does not respond instantaneously, as otherwise there would be no linear dispersion. Surely the nonlinear polarization cannot respond faster than does the linear. Thus, one really should write the nonlinear polarization in the form

$$\Pi_{NL} = \int_{-\infty}^{t} dt' \int_{-\infty}^{t'} dt'' \int_{-\infty}^{t''} G(t - t', t' - t'', t'' - t''')E(t')E(t'')E(t''')\,dt''$$

in the cubic approximation. If we set

$$E = \tfrac{1}{2}U \exp[i(kz - \omega t)] + \text{cc}$$

in this, it becomes a three-dimensional generalization of our expression for the linear polarization in Appendix A. Assuming a very quick (but not instantaneous) response, one could expand as in Appendix A, obtaining something like

$$4\pi P_{NL} = \gamma |U|^2 U + i[\alpha |U|^2 U_\tau + \beta(U)^2 U_\tau^*],$$

and now γ, α, β, etc. would all be complex. If their imaginary parts can be neglected, the first-order effect would seem to be to modify the constant we called ε. A guess as to what this might do is that, at least in some cases, we would have something very like Eq. (1) for which ε is no longer γ/ω but instead rather larger, thus shortening the distance over which self-steepening effects develop.

A fourth fault, not with our model but with our mathematics, is the handwaving in Section 5 that replaces sound mathematics. This fault I do not know how to cure unless someone can show our Eq. (1), or an appropriate improvement, to be completely integrable. However, our estimate of the final bandwidth owing to phase modulation, Eq. (50), does not depend on this handwaving, but is based on balancing the self-steepening term against the dispersion term. Similarly, the estimate of the propagation distance at which this broadening has essentially developed, given by Eq. (17), depends on rather modest handwaving.

References

American Institute of Physics Handbook, 2nd ed. (1963) McGraw-Hill, New York. See p. 6–13 for dispersion data.

Bespalov, V.I. and V.I. Talanov (1966) Zh. Eksp. Teor. Fiz. Pis'ma **3**, 307–310 (June 15).

Boling, N.J., A.J. Glass, and A. Owyoung (1978) IEEE Jour. Quant. Electronics **QE-14**, 601 (August).

DeMartini, F., C.H. Townes, T.K. Gustafson, and P.L. Kelley (1967) Phys. Rev. **164**, 312.

Hasegawa, A. and F. Tappert (1973) Appl. Phys. Lett. **23** (No. 4), 171 (August 15).

Kaup, D.J. and A.C. Newell (1978) J. Math. Phys. **19** (No. 4), 798 (April).

Marburger, J.H. (1975) In *Progress in Quantum Electronics*, J.H. Sanders and S. Stenholm, eds., vol. 4, part 1, p. 35. Pergamon, New York.

Suydam, B.R. (1973) Self focusing of very powerful laser beams. In *Laser Induced Damage in Optical Materials*: *1973*, A.J. Glass and A.H. Guenther, eds. NBS Special Publication 387.

Whitham, G.B. (1974) *Linear and Nonlinear Waves*, chapters 1 and 3. Wiley, New York.

Yang, G. and Y.R. Shen (1984) Opt. Lett. **9**, 510.

Zakharov, V.E. and A.B. Shabat (1972) Sov. Phys. JETP **34** (No. 1), 62.

Appendix A: Approximations Leading to the Extended Nonlinear Schrödinger Equation

Equation (1) of the main text forms the foundation of all of our discussion of self-steepening and of spectral broadening of short intense pulses. It is our purpose here to derive this as an approximation to the true wave equation without assuming the envelope to be slowly varying in comparison with the oscillating factor, $\exp[i(kz - \omega t)]$.

For strictly plane waves traveling in a dielectric medium Maxwell's equations reduce to the simple wave equation

$$\frac{\partial^2 E}{c^2\,\partial t^2} - \frac{\partial^2 E}{\partial z^2} = -\frac{\partial^2}{c^2\,\partial t^2}(4\pi\Pi), \tag{A.1}$$

where E is the electric vector and Π is the polarization vector. We assume our medium to be uniform and isotropic and the electric field to be plane polarized in the direction of the unit vector e; then Π is also parallel to e and we can factor this vector out. In the resulting scalar equation we set

$$\begin{aligned} E &= \tfrac{1}{2}U\exp[i(kz-\omega t)] + \text{cc}, \\ \Pi &= \tfrac{1}{2}P\exp[i(kz-\omega t)] + \text{cc}, \end{aligned} \tag{A.2}$$

and the result is the equation

$$\begin{aligned} &(k^2-\omega^2/c^2)U - \left(\frac{2i\omega}{c^2}\frac{\partial U}{\partial t} + 2ik\frac{\partial U}{\partial z}\right) + \frac{\partial^2 U}{c^2\,\partial t^2} - \frac{\partial^2 U}{\partial z^2} \\ &= \frac{4\pi\omega^2}{c^2}\left[P + \frac{2i}{\omega}\frac{\partial P}{\partial t} - \frac{1}{\omega^2}\frac{\partial^2 P}{\partial t^2}\right]. \end{aligned} \tag{A.3}$$

In general, we can split the polarization into a linear and a nonlinear part

$$\Pi = \Pi_L + \Pi_{NL}; \qquad P = P_L + P_{NL}. \tag{A.4}$$

In Eq. (A.2) we have ignored the higher harmonic terms in Π_{NL} because we are only calculating the propagation of the fundamental.

In general, in our uniform isotropic medium, the linear part of the polarization can be represented as

$$4\pi\Pi_L(t) = \int_{-\infty}^{t} K(t-t')E(t')\,dt',$$

which, using Eq. (A.2), translates as

$$4\pi P_L = \int_{-\infty}^{t} K(t-t')\exp[i\omega(t-t')]U(t')\,dt'. \tag{A.5}$$

As Π_L and E are real, so is K. We have suppressed the z dependence of P and of U; K is independent of z.

Our first assumption is that the dispersion is small. If $K(t)$ were equal to $\kappa_0\delta(t)$ (κ_0 = a constant), we would have $4\pi P = \kappa_0 U$ and there would be no dispersion. Weak dispersion then must mean that $K(t)$ is very sharply spiked near $t = 0$; this will soon be made precise. The nature of K described above suggests expansion of Eq. (A.5) by partial integration. To this end, we define a sequence of functions

$$\begin{aligned} H_0(t) &= K(t)\exp(i\omega t), \\ H_{n+1}(t) &= \int_{t}^{\infty} H_n(t')\,dt'; \qquad n = 0, 1, 2, \ldots. \end{aligned} \tag{A.6}$$

With the aid of these functions, successive partial integration of Eq. (A.5) yields

$$4\pi P_L(t) = H_1(0)U - H_2(0)\frac{\partial H}{\partial t} + H_3(0)\frac{\partial^2 U}{\partial t^2} - H_4(0)\frac{\partial^3 U}{\partial t^3}$$
$$+ \int_{-\infty}^{t} H_4(t-t')\left[\frac{d^4 U(t')}{(dt')^4}\right] dt'. \tag{A.7}$$

This is still exact; our assumption of weak dispersion means that the series converges quickly, because the $H_n(0)$ rapidly decrease with increasing n, and that the integral remainder is small.

From the recursion relations, Eqs. (A.6), it readily follows that

$$H_{n+1}(t) = \frac{1}{n!}\int_{t}^{\infty} (t'-t)^n K(t')\exp(i\omega t')\,dt', \tag{A.8}$$

whence, if we define

$$\kappa(\omega) = \int_{0}^{\infty} K(t)\exp(i\omega t)\,dt, \tag{A.9}$$

then

$$H_{n+1}(0) = \frac{1}{n!}\left(\frac{-id}{d\omega}\right)^n \kappa(\omega) \tag{A.10}$$

and Eq. (A.7) is just the Fourier transform of the Tayler series expansion of $\kappa(\omega)$. If we now differentiate Eq. (A.7) an appropriate number of times we obtain

$$4\pi\left(P_L + \frac{2i}{\omega}\frac{\partial P_L}{\partial t} - \frac{1}{\omega^2}\frac{\partial^2 P_L}{\partial t^2}\right) = \kappa U + i[\kappa' + 2\kappa/\omega]\frac{\partial U}{\partial t}$$
$$- [\tfrac{1}{2}\kappa'' + 2\kappa'/\omega + \kappa/\omega^2]\frac{\partial^2 U}{\partial t^2}$$
$$- i[\tfrac{1}{6}\kappa''' + \kappa''/\omega + \kappa'/\omega^2]\frac{\partial^3 U}{\partial t^3} + R, \tag{A.11}$$

where R is the remainder, given by

$$R = \int_{-\infty}^{t}\left[H_4(t-t') - \frac{2i}{\omega}H_3(t-t') - \frac{1}{\omega^2}H_2(t-t')\right]\left[\frac{d^4 U(t')}{(dt')^4}\right] dt', \tag{A.12}$$

κ is given by Eq. (A.9), and primes on κ mean $d/d\omega$. From Eq. (A.9), as K is real it follows that $\kappa(\omega)$ *cannot* be pure real unless K is a δ-function, and in fact the Kramers-Kronig relations follow from Eq. (A.9). Small dispersion implies very small absorption, which we ignore. Thus, from now on κ is assumed to be real.

Our next step is to set Eq. (A.11) into the right-hand side of Eq. (A.3) and to combine the terms in U, $\partial U/\partial t$, and $\partial^2 U/\partial t^2$. The choice

$$k^2 = ((\omega^2)/c^2)(1 + \kappa) \equiv \omega^2 n^2/c^2 \tag{A.13}$$

makes the term in U vanish; the rest of the terms yield

$$-2ik\left[\frac{\partial U}{v\,\partial t} + \frac{\partial U}{\partial z}\right] + \frac{\partial^2 U}{v^2\,\partial t^2} - \frac{\partial^2 U}{\partial z^2} = -\alpha\frac{\partial^2 U}{\partial t^2} - i\beta\frac{\partial^3 U}{\partial t^3} + \frac{\omega^2}{c^2}R$$
$$+ \frac{4\pi\omega^2}{c^2}\left[P_{NL} + \frac{2i}{\omega}\frac{\partial P_{NL}}{\partial t} - \frac{1}{\omega^2}\frac{\partial^2 P_{NL}}{\partial t^2}\right], \tag{A.14}$$

where we have defined new constants

$$1/v = (1 + \kappa + \tfrac{1}{2}\omega\kappa')/(c\sqrt{1 + \kappa}) = \frac{\partial k}{\partial \omega},$$
$$\alpha = [(1 + \kappa)(\omega\kappa' + \tfrac{1}{2}\omega^2\kappa'') - \tfrac{1}{4}(\omega\kappa')^2]/c^2(1 + \kappa) = k\frac{\partial^2 k}{\partial \omega^2}, \tag{A.15}$$
$$\beta = [\kappa' + \omega\kappa'' + \tfrac{1}{6}\omega^2\kappa''']/c^2,$$

and R is as defined by Eq. (A.12). Owing to our assumption of weak dispersion, the expansion of Eq. (A.7) is a rapidly converging one and $\alpha\,\partial^2 U/\partial t^2$, $\beta\,\partial^3 U/\partial t^3$, and R are all small.

We are now ready for the main step in our approximation. Defining the two operators

$$D_{\pm} = \frac{\partial}{v\,\partial t} \pm \frac{\partial}{\partial z}, \tag{A.16}$$

we can write Eq. (A.14) as

$$\left(1 + \frac{i}{2k}D_-\right)D_+ U = \frac{i}{2k}X, \tag{A.17}$$

where X represents the right-hand side of Eq. (A.14). We assume both the dispersion and the nonlinearity to be weak, so X is small. If X were zero, a solution would be $U = f(t - z/v)$, which yields $D_+ U = 0$. With X small but not quite zero $D_+ U$ must be very small; that is, the operator D_+ is small. Now

$$D_- = \frac{2\partial}{v\,\partial t} - D_+ \tag{A.18}$$

or, as D_+ is small,

$$D_- = \frac{2\partial}{v\,\partial t} \tag{A.19}$$

very nearly. Again

$$\frac{i}{2k}D_- = \frac{i}{kv}\frac{\partial}{\partial t} = \frac{i}{\omega}\left(1 + \frac{\omega\kappa'}{2(1 + \kappa)}\right)\frac{\partial}{\partial t}. \tag{A.20}$$

For weak dispersion $\omega\kappa'/2(1+\kappa) \ll 1$, so finally

$$\frac{iD_-}{2k} = p \equiv \frac{i\partial}{\omega\,\partial t} \tag{A.21}$$

is a good approximation. Thus Eq. (A.17) can be written as

$$-iD_+U = \frac{1}{2k}(1+p)^{-1}X. \tag{A.22}$$

First note that the nonlinear part of X can be written as

$$X_{NL} = \frac{4\pi\omega^2}{c^2}(1+p)^2 P_{NL}. \tag{A.23}$$

whence we immediately get

$$(1+p)^{-1}X_{NL} = \frac{4\pi\omega^2}{c^2}(1+p)P_{NL} = \frac{4\pi\omega^2}{c^2}\left[P_{NL} + \frac{i}{\omega}\frac{\partial P_{NL}}{\partial t}\right]. \tag{A.24}$$

To carry out a similar trick on the linear part of X, which arises from dispersion, we must assume something special about the behavior of $\kappa(\omega)$ at frequencies far removed from the carrier. Thus we write

$$X_L = -(1+p)\left(\alpha\frac{\partial^2 U}{\partial t^2}\right) - i(\beta - \alpha/\omega)\frac{\partial^3 U}{\partial t^3} + \frac{\omega^2}{c^2}R \tag{A.25}$$

and we assume that $\kappa(\omega)$ or $K(t)$ takes a special form such that the terms $(\beta - \alpha/\omega)\,\partial^3 U/\partial t^3$ and $(\omega^2/c^2)R$ either vanish identically or are at least negligibly small even for U that are not slowly varying in time. Under this special assumption

$$(1+p)^{-1}X_L = -\alpha\frac{\partial^2 U}{\partial t^2}$$

is a good approximation and Eq. (A.22) becomes

$$-iD_+U = -a\frac{\partial^2 U}{\partial t^2} + \frac{2\pi\omega}{nc}\left[P_{NL} + \frac{i}{\omega}\frac{\partial P_{NL}}{\partial t}\right] \tag{A.26}$$

where

$$a = \alpha/2k = \frac{1}{2}\frac{\partial^2 k}{\partial\omega^2}, \qquad n = \sqrt{1+\kappa}. \tag{A.27}$$

It is convenient to express things in terms of the retarded time $\tau = t - z/v$ rather than the true time t. Thus we make the transformation of independent variables

$$\begin{aligned} z' &= z, \\ \tau &= t - z/v \end{aligned} \tag{A.28}$$

and Eq. (A.26) becomes

$$-i\frac{\partial U}{\partial z'} = -a\frac{\partial^2 U}{\partial \tau^2} + \frac{2\pi\omega}{nc}\left[P_{NL} + \frac{i}{\omega}\frac{\partial P_{NL}}{\partial \tau}\right]. \tag{A.29}$$

Our derivation of Eq. (A.29) assumes nothing about P_{NL} except that it is small. Our assumption about the dispersion, however, makes this equation really a model rather than a description of a real material. The function $\kappa(\omega)$ can be expanded about a frequency ω_0 as

$$\kappa(\omega) \equiv \kappa(\omega_0 + \Omega) = \kappa_0 + \kappa_0'\Omega + \tfrac{1}{2}\kappa_0''\Omega^2 + \tfrac{1}{6}\kappa_0'''\Omega^3 + \text{etc.}$$

where κ_0' means κ' at $\omega = \omega_0$ etc. Our dropping all the terms on the right-hand side of Eq. (A.25) except the first means that there are special relations for determining all the κ_0''', κ_0'''', etc. from the first three κ_0, κ_0', and κ_0'', or at least these relations must be satisfied to a good approximation. For narrowband signals Ω remains small and the values of κ_0''', etc. do not matter. If, however, the bandwidth becomes wide enough that these terms do matter, they must be such as to make $\beta - \alpha/\omega \approx 0$ and $R \approx 0$.

Another thing to note about our derivation is that it works only for a plane wave. For a general wave, Eq. (A.1) acquires two new terms on the left-hand side, namely

$$-\nabla_\perp^2 E = -\left(\frac{\partial^2 E}{\partial x^2} + \frac{\partial^2 E}{\partial y^2}\right)$$

and

$$\nabla\nabla\cdot E = -4\pi\nabla\nabla\cdot\Pi$$

and this latter, owing to the nonlinear part of Π, cannot vanish in general. Our above derivation would throw these two extra terms into the X of Eq. (A.22) and the operator $(1 + p)^{-1}$ would make a mess of them. An additional idea is required to include diffraction in the model.

For our study we make the very simply choice

$$4\pi\Pi_{NL} = 2nn_2E^3,$$

which using Eq. (A.2) becomes

$$4\pi P_{NL} = \tfrac{3}{2}nn_2|U|^2U. \tag{A.30}$$

Thus we assume a nonlinearity that responds instantaneously. Equation (1) of the main text, with $\varepsilon = \gamma/\omega$, follows.

Appendix B: The "Kink" or "Dark Soliton"

An interesting solitary wave solution known as a "kink" or as a "dark soliton" can also be found as a special case of Eq. (39). To find it we must postulate that P and Q are both negative and moreover that

$$4Q^2 > -3PR^2. \tag{B.1}$$

Under these conditions if one defines

$$u_0 \equiv \frac{1}{3R^2}[2Q + \sqrt{4Q^2 + 3PR^2}],$$
$$u_1 \equiv \frac{2}{3u_0R^2}(Qu_0 - P) \tag{B.2}$$

and chooses

$$C = R^2 u_1 u_0^2 = \tfrac{2}{3}(Qu_0 - P)u_0, \tag{B.3}$$

one finds that

$$Cu + Pu^2 - 2Qu^3 - R^2u^4 = R^2u(u - u_0)^2(u_1 - u) \tag{B.4}$$

and the integration of Eq. (39) becomes elementary. Carrying it out and solving the result for u yields

$$A = \frac{A_0\sqrt{1 + y^2}\sinh v\xi}{\sqrt{\cosh^2 v\xi + y^2 \sinh^2 v\xi}}, \tag{B.5}$$

where

$$A_0^2 \equiv u_0^2 = -\frac{1}{3R^2}[2Q + \sqrt{4Q^2 + 3PR^2}],$$
$$v^2 = -\frac{1}{3R^2}[4R^2 + 3PR^2 + 2Q\sqrt{4Q^2 + 3PR^2}], \tag{B.6}$$
$$y^2 = R^2A_0^4/v^2.$$

Note that condition (B.1) guarantees that both A_0^2 and v^2 are positive.

As before, we can express η and b in terms of A_0^2 and v^2, namely

$$\eta = \frac{2av}{\varepsilon A_0^2}[2a(v^2 + 3R^2A_0^4) - \gamma A_0^2],$$
$$b = a\left\{\frac{1}{\varepsilon^2 A_0^4}[2a(v^2 + 3R^2A_0^4) - \gamma A_0^2]^2 + 2v^2 + 3R^2A_0^4\right\}. \tag{B.7}$$

It is interesting to note that this kink solution is, in a sense, complementary to our soliton. Writing A_{sol} for the soliton and A_k for the kink and choosing A_0^2 and v^2 identical for the two, we have

$$[A_{\text{sol}}(x)]^2 + [A_k(x)]^2 = A_0^2; \qquad x \equiv v\xi. \tag{B.8}$$

For this reason this kink solution has, in the case $\varepsilon = 0$, been called a dark soliton by Hasegawa and Tappert (1973). As the kink and the soliton have different velocities (different values of η), Eq. (B.8) can be true only instantaneously, e.g., at $z = 0$. However, it immediately follows that for the kink

$$H(\xi) = A_0^2\xi + \frac{1}{R}\left\{\arctan y - \arctan\left[\frac{2y}{(1 + y^2)\exp(2v\xi) + (1 - y^2)}\right]\right\}. \tag{B.9}$$

As does the soliton, the kink goes to a perfectly regular solution in the limit $v \to 0$, namely the algebraic kink

$$\left.\begin{cases} A = A_0(RA_0^2\xi)[1 + R^2A_0^4\xi^2]^{-1/2} \\ H(\xi) = A_0^2\xi + \dfrac{1}{R}\arctan(RA_0^2\xi) \end{cases}\right\} \quad \text{for } v = 0. \tag{B.10}$$

If one looks at the amplitude A rather than the intensity A^2, these kinks are somewhat reminiscent of a plane hydrodynamic shock in which viscosity and heat conduction are taken into account. This kink solution is also a solution to the DNS equation and is mentioned, but not discussed, by Kaup and Newell (1978).

7 Self-Focusing and Continuum Generation in Gases

PAUL B. CORKUM and CLAUDE ROLLAND

1. Introduction

This book attests to the fact that continuum generation has become both technically and conceptually important. Discovered in 1970 (Alfano and Shapiro, 1970a, 1970b), continuum generation is a ubiquitous response of transparent materials (liquids, solids, and gases) to high-power, ultrashort-pulse radiation. This chapter highlights some of these aspects while presenting the sometimes unique characteristics of continuum generation in gases. In addition, we introduce some related results that reflect on the light-atom interaction at high intensities.

Gases are ideal media in which to study nonlinear phenomena, such as continuum generation. The choice of low-density rare gases makes the nonlinearity simple since the susceptibility will be purely electronic in nature. Experimentally, the strength of the nonlinearity can be precisely controlled by varying the gas pressure. Gases are ideal in another way. There is a strong conceptual link between the susceptibility and the transition probability. Since there is a lot of emphasis, at present, on understanding multiphoton ionization in rare gases,* concepts being developed in this area can provide a framework for further advances of nonlinear optics in general and continuum generation in particular.

In gases, the lowest-order contribution to the nonlinear susceptibility is $\chi^{(3)}$. The magnitude of the nonresonant $\chi^{(3)}$ for the rare gases (Lehmeier et al., 1985) and for many molecular gases is well known. For xenon $\eta_2 = (3\chi^{(3)}_{1111}/\eta_0 = 2.4 \times 10^{-25}$ m^2/V^2 atm, where the refractive index η is given by $\eta = \eta_0 + \eta_2 E^2 + \cdots$, E being the rms electric field. $\chi^{(3)}$ is proportional to the gas pressure.

This chapter is organized around the pressure-dependent strength of the nonlinearity. Much of the content originates from six experimental papers

* See for example, papers in *Multiphoton Ionization of Atoms* (S.L. Chin and P. Lambropoulos, eds.), Academic Press, New York (1984), and special issue on Multielectron Excitation of Atoms, J. Opt. Soc. Am. **B4**, no. 5 (1987).

(Corkum et al., 1986a, 1986b; Corkum and Rolland, 1987, 1988a, 1988b; Chin et al., 1988) describing related work at the National Research Council of Canada.

Section 2 discusses the aspects of the experiment that are common to all parts of the chapter.

Section 3 describes the interaction of ultrashort pulses with very low-pressure gases. Low pressure ensures that nonlinear optics plays no role in the interaction (Corkum and Rolland, 1988a; Chin et al., 1988). This allows the ionization properties of xenon to be established. We will see that relatively high intensities are required to ionize gases with ultrashort pulses ($\sim$100 fs). In this way, we establish an upper intensity limit for the nonlinear interaction in a purely atomic system.

Section 3 also introduces the concept of transient resonances. Although transient resonances are a characteristic of the interaction of ultrashort pulses with matter in the intensity and wavelength range discussed in this chapter, their role in multiphoton ionization depends on the pulse duration.

As the gas pressure is increased, we enter the traditional realm of nonlinear optics. If the intensity for the production of significant plasma is not exceeded, changes to the spectrum of the pulse can be investigated under conditions where self-phase modulation is the dominant mechanism. We will see in Section 4 that high-order nonlinear terms must contribute to the spectral bandwidth if the laser intensity reaches 10^{13} W/cm^2 or higher (Corkum and Rolland, 1988b).

A qualitative explanation of why high-order terms must contribute to self-phase modulation is given in Section 5.

At still higher pressures the region of continuum generation (Corkum et al., 1986a, 1986b) and self-focusing (Corkum and Rolland, 1988b) is reached. Section 6.1 describes the spectral aspects of continua in gases. In particular, it shows that the spectra are similar for condensed media and for gases.

The spatial characteristics of continuum generation are particularly striking (Corkum and Rolland, 1988b). These are described in Section 6.2 with special emphasis on the role of self-focusing in continuum generation. There is a wide range of conditions over which continua are produced with virtually the same beam divergence as the incident diffraction-limited beam (Corkum and Rolland, 1987, 1988b). As the intensity or the gas pressure is increased, conical emission is observed.

2. Experimental Aspects

Pulses of three different durations (22, $\sim$90, and $\sim$900 fs full width at half-maximum (FWHM)) were used in various parts of the experiment. This section discusses the experimental aspects that are common to all parts of the chapter. Each subsection includes experimental details of specific interest.

Laser pulses were produced by amplifying the output of either a spectrally filtered synchronously pumped dye laser (900 fs) or a colliding-pulse mode-locked dye laser (90 fs). The temporal, spatial, and spectral characteristics of the pulses have been fully described (Corkum and Rolland, 1988a; Rolland and Corkum, 1986). The wavelength of the 900-fs pulse was centered at 616 nm and its bandwidth ($\Delta\nu$) was slightly greater than the transform limit ($\Delta t\,\Delta\nu = 0.52$; Δt is the FWHM pulse duration). The 90-fs pulse was centered at 625 nm and had $\Delta t\,\Delta\nu = 0.5$. The pulse durations were measured by autocorrelation and fit by a sech^2 (90 and 22 fs) or to a Gaussian (900 fs) pulse shape. After amplification the 90- and 900-fs pulses were spatially filtered to ensure diffraction-limited beam profiles.

The 22-fs pulses were created from the 90-fs pulses using large-aperture pulse compression techniques (Rolland and Corkum, 1988). The resulting 100-μJ pulses were diffraction limited with a signal-to-background power contrast ratio of approximately 30:1 (5:1 in energy). Compensation for the dispersion in all optical elements (lenses, windows, beam splitters, etc.) was accomplished by predispersing the pulse. Thus, the pulse measured 22 fs only in the target chamber and at the autocorrelation crystal. Since the 350-Å bandwidth of the 22-fs pulse gives rise to serious chromatic aberration in a single-element lens, an achromatic lens ($f = 14.3$ cm) was used to focus the pulses into the vacuum chamber (and onto the autocorrelation crystal).

All focal spot measurements were made by either scanning a pinhole (900 and 90 fs) through the focus or observing the portion of the energy transmitted through a pinhole (22 fs) of known diameter. Within the accuracy of the scans, the beam profiles were Gaussian.

3. Multiphoton Ionization

Some time ago it was proposed (Bloembergen, 1973) that ionization could play a major role in continuum generation though a time-dependent change in the plasma density. Plasma density changes impress a frequency chirp on a transmitted pulse. However, to influence continuum generation (Corkum et al., 1986a, 1986b) even to a small degree by plasma production there is a price to pay in energy absorption and in the distortion of the spatial beam profile (Corkum and Rolland, 1988b). Since we will show that these signatures of plasma production are not observed, we can conclude that ionization plays no role in gaseous continuum generation. The absence of ionization can be used to establish a maximum intensity in the laser focus where the continuum is being generated and hence the maximum value of $\eta_2 E^2$.

Continuum experiments were the first to indicate that it is difficult to ionize xenon and krypton with ultrashort pulses (Corkum et al., 1986a, 1986b) relative to extrapolations of 0.53 μm, 1.06 μm experiments (l'Huillier et al., 1983) (25 ps). Since the ionization threshold is a major uncertainty in continuum generation, we have performed two experiments (Corkum and Rolland, 1988a; Chin et al., 1988) whose specific aim was to study multiphoton ionization. The more recent and more quantitative of these is described in this section (Chin et al., 1988).

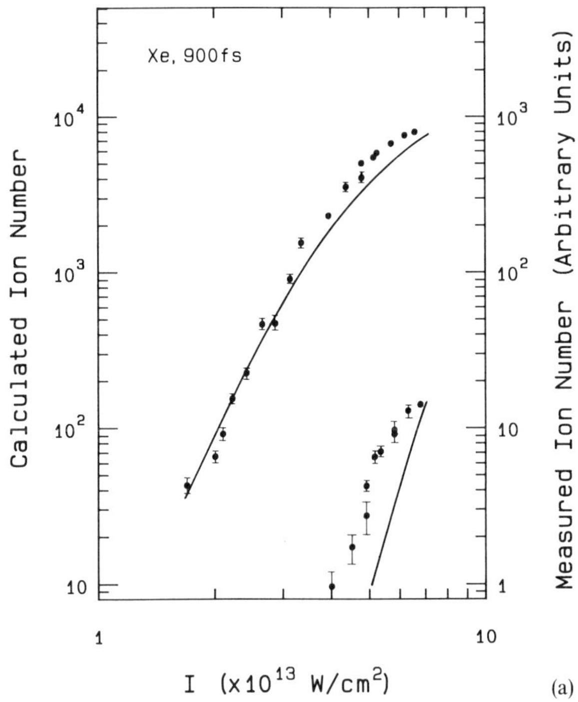

(a)

FIGURE 7.1. Ion yield of Xe for 900-, 90-, and 22-fs pulses. The solid curves are calculated from a modified Keldysh theory (Szöke, 1988). The calculations give an absolute number of ions for the measured focal geometry and the neutral gas density. The experimental number of ions is plotted in relative units. The data have been positioned on the graph so as to emphasize the agreement between experiment and theory. The error bars show the standard error of the experimental data.

Femtosecond pulses were focused into a vacuum cell filled with $\sim 4 \times 10^{-6}$ torr of xenon. Ions were extracted with an ~ 80 V/cm static field into a time-of-flight mass spectrometer. Data were obtained using a microcomputer, coupled to a boxcar integrator that was programmed to accept only laser pulses within a narrow energy range ($\pm 2.5\%$). The computer recorded and averaged the associated ion signals. The intensity in the vacuum chamber was varied by rotating a $\lambda/2$ plate placed in front of a polarizer (reflection from a Brewster's angle germanium plate was used as a dispersion-free polarization selector for the 22-fs pulse).

Figure 7.1 is a graph of the number of ions as a function of the peak laser intensity for both rare gases and all three laser pulse durations. (Higher ionization states were observed but not plotted since they were too weak.) The solid curves were obtained from a modified Keldysh theory (Szöke, 1988). Although we did not measure the absolute number of ions, we estimate the threshold sensitivity (the lowest ion signals in Figure 7.1) of our ion collector to be approximately 10 ions. The relative scaling between experi-

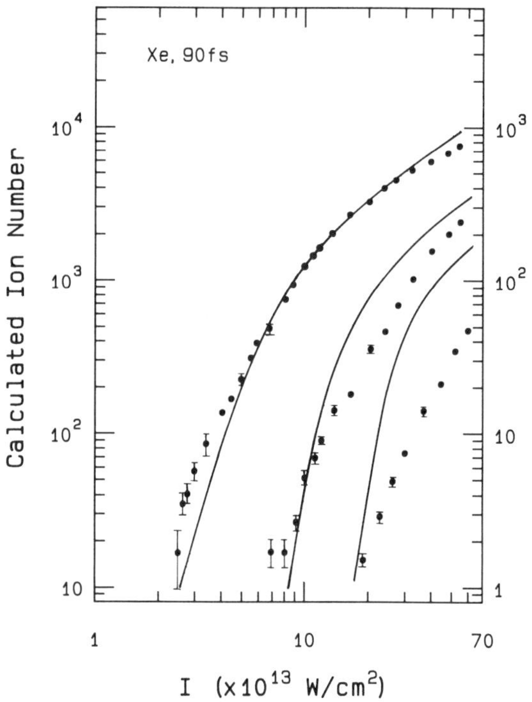

(b)

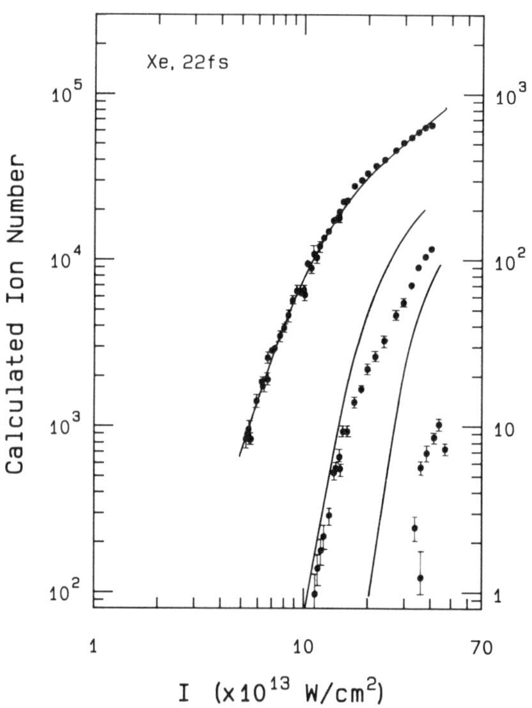

(c)

FIGURE 7.1 (*continued*)

mental and calculated ion signals for Xe (Figure 7.1) is consistent with this estimate.

We performed this experiment to find the intensity at which ionization would need to be considered in nonlinear optics experiments. Not only does Figure 7.1 answer this question qualitatively, but the agreement between theory and experiment allows us to make quantitative predictions. However, the agreement raises an important issue. How can a Keldysh theory, which assumes that resonances are unimportant, be consistent with electron spectral measurements (Freeman et al., 1987) which indicate that resonances play a major role in ionization? Because of the importance of this issue for nonlinear optics, we discuss it below with respect to ionization and, in Section 5, with respect to high-order nonlinear optics.

An important feature of the high-power light-atom interactions is the ac Stark shift. At $I = 10^{13}$ W/cm^2, the laser field exceeds the atomic field (of hydrogen) for all radii greater than $R = 4$ Å. At this radius the atomic potential is 3.8 eV below the ionization potential. For $R > 4$ Å, it is appropriate to consider the electron oscillating in the laser field as the lowest-order solution and the atomic field as a perturbation. Nearly all excited states, therefore, have an energy of oscillations (ac Stark shift) approximately equal to the ponderomotive potential ($U_{osc} = (qE)^2/2m\omega^2$ where q is the electronic charge, ω is the laser angular frequency, and m is the electron mass). At 10^{13} W/cm^2 and 620 nm the ponderomotive potential is approximately 0.4 eV. Thus resonances are transiently produced and resonant enhancement of high-order terms in the susceptibility will occur.

In view of the transient resonance induced in the medium, we might expect resonances always to be important. However, small deviations from Keldysh models appear only at relatively low intensities. Transient resonances appear to play a significant role in the overall ionization rate over, at most, a limited intensity/time range. (Note that detailed electron spectral measurements have so far been performed only in the $1\text{–}3 \times 10^{13}$ W/cm^2 intensity range with ~500-fs pulses (Freeman et al., 1987).)

To understand why the contributions of transient resonances to ultrashort pulse ionization should be so small, consider just how transient these resonances can be. Assuming that all high-lying states move with the ponderomotive potential, we can write the maximum rate of change of the ponderomotive shift as $dU/dt)_{max} = 2\sqrt{2}\, U_0(\ln(0.5)^{1/2}/\Delta t \exp(0.5))$ where U_0 is the maximum value of the ponderomotive shift during the pulse and a Gaussian pulse shape has been assumed. In the case of the 90-fs pulse with a characteristic peak intensity of 10^{14} W/cm^2, $dU/dt)_{max} = 0.1$ eV/optical cycle. In the even more extreme case of the 22-fs pulse, the same peak intensity gives $dU/dt)_{max} =$ 0.4 eV/optical cycle.

The significance of such large ponderomotive shifts can be seen by considering a two-level system. For a two-level system both the pulse duration dependence and the intensity dependence of the dephasing between the transition (transition frequency $= \omega_{ab}$) and the near-resonant harmonic of the laser frequency can be estimated. For a constantly shifting transition, $\omega_{ab} + (dU/dt)t/\hbar$, the dephasing time ($T$) is given by the condition that $\delta\phi \sim 2\pi$. That is, $T \sim (2h/dU/dt)^{1/2}$ where dU/dt is assumed constant. For a 90-fs pulse at 10^{14} W/cm^2, $T \sim 13$ fs. At the same intensity $T \sim 6$ fs for a 22-fs pulse. Resonances that last only a few cycles are hardly resonances at all and can be expected to have only minor effects on the overall ionization rate. Only for relatively small dU/dt can transient resonances play an important role. They may account for the deviations of the experimental and calculated curves observed in the 900-fs and low-power 90-fs xenon results.

The above discussion does not imply that transient resonances cannot lead to observable nonlinear optical consequences. In fact, nonlinear optics may provide one of the best methods of observing transient resonances.

In summary, these experiments show that ionization will be barely significant for 90-fs pulses at intensities of 10^{13} W/cm^2. In addition, the slopes of the ion curves in Figure 7.1 indicate that a lowest-order perturbation expansion for the transition rate (and, therefore, the susceptibility) will be incomplete for intensities greater than $\sim 3 \times 10^{13}$ W/cm^2 for 0.6-μm light. This intensity can be used to estimate the maximum value of $\eta_2 E^2$ that is experimentally accessible with 90-fs pulses.

4. Self-Phase Modulation

One of the most studied nonlinear processes with ultrashort pulses is self-phase modulation. It is the basis of optical pulse compression, which is widely used in femtosecond technology. In many cases continuum generation is believed to be an extreme version of self-phase modulation. Thus, it seems natural to adjust the strength of the nonlinearity by varying the gas pressure so that only modest self-phase modulation occurs. We can then follow the magnitude of the spectral broadening as the intensity or the nonlinearity is increased. Analogous experiments can be performed in fibers by increasing the length of the fiber.

Self-phase modulation is more complex in unbounded media than in fibers because, in unbounded media, self-phase modulation is inescapably related to self-focusing. (This relationship ensured that pulse compression based on self-phase modulation remained a curiosity until fiber compression became available.) It is possible to minimize the effects of self-focusing by keeping the medium shorter (Rolland and Corkum, 1988; Fork et al., 1983) than the self-focusing length. High-power pulse compression experiments use precisely this technique to control self-focusing (Rolland and Corkum, 1988). However, long before the self-phase modulation has become strong enough to generate continua, the beam propagation can no longer be controlled (Rolland and Corkum, 1988). In spite of this complexity, most continua are produced in long, unbounded media. Much of the remainder of the chapter addresses some of the physics issues associated with continuum production in this kind of medium.

The self-phase modulation experiment (Corkum and Rolland, 1988b) was performed with the 90-fs, 625-nm pulse with a maximum energy of $\sim 500\ \mu$J. A vacuum spatial filter with aperture diameter less than the diffraction limit of the incident beam produced an Airy pattern from which the central maximum was selected with an iris. The resulting diffraction-limited beam was focused into a gas cell that was filled to a maximum pressure of 40 atm. We report here mainly on the results obtained with xenon. However, where other gases have been investigated, we have found similar behavior.

As the gas pressure or laser power is increased, spectral broadening due to self-phase modulation is observed. In the η_2 limit (i.e., terms of higher order than $\eta_2 E^2$ are negligible) and neglecting dispersion, the spectral width depends only on the laser power

$$\delta\lambda)_{\max} \sim \frac{8\sqrt{2}\,\eta_2 P_0}{c^2 \varepsilon_0 \tau \exp(0.5)} + \delta\lambda)_{\text{init}} \tag{1}$$

where the power in the pulse is given by $P = P_0 e^{-(t/\tau)^2}$ and $\delta\lambda)_{\max}$ and $\delta\lambda)_{\text{init}}$ are the maximum and initial bandwidth of the pulse, respectively. All other symbols have their conventional meaning. The factor exp(0.5) arises because the maximum broadening for a Gaussian pulse occurs at $t = \tau/\sqrt{2}$. It will be present in Eq. (2) for the same reason. We can evaluate* $\delta\lambda)_{\max}$ for $\eta_2 = 2.4 \times 10^{-25}$ m^2/V^2 atm and obtain $\delta\lambda)_{\max} = 3.9 \times 10^{-7}$ Å/W atm. Equation (1) is valid only below the self-focusing threshold.

Equation (1) shows that modifications of the nonlinearity can be observed through an intensity dependence of the spectral broadening. The results obtained with two different focusing lenses ($F/70$ and $F/30$) and a selection of pressures are presented in Figure 7.2. In all cases, the power was maintained below the self-focusing threshold. The solid lines are a fit to the experimental data using Eq. (2) and a saturation intensity of $I_{\text{sat}} = 10^{13}$ W/cm^2.

$$\delta\lambda)_{\max} = \frac{8\sqrt{2}\,P_0}{(2.1)c^2 \varepsilon_0 \tau \exp(0.5)} \left[\frac{\eta_2}{1 + I/(I_{\text{sat}} \exp(0.5))}\right] + \delta\lambda_{\text{init}}. \tag{2}$$

In Figure 7.1, the saturation intensity of 10^{13} W/cm^2 corresponds to a power of 1.5×10^8 W for the $F/30$ and 8×10^8 W for the $F/70$ lens. The dashed lines are obtained using Eq. (2) and $I_{\text{sat}} = \infty$. Compared to Eq. (1), a factor of 2.1 is required in the denomimator of Eq. (2) to fit the data. This factor is attributed to uncertainties in the input parameters such as the value of η_2, the pulse shape and duration (measured by autocorrelation), and approximations made in deriving Eq. (1).

---

* η_2 of xenon at atmospheric pressure was erroneously reported to be $\eta_2 = 4 \times 10^{-26}$ m_2/V_2 in Corkum et al. (1986a, 1986b).

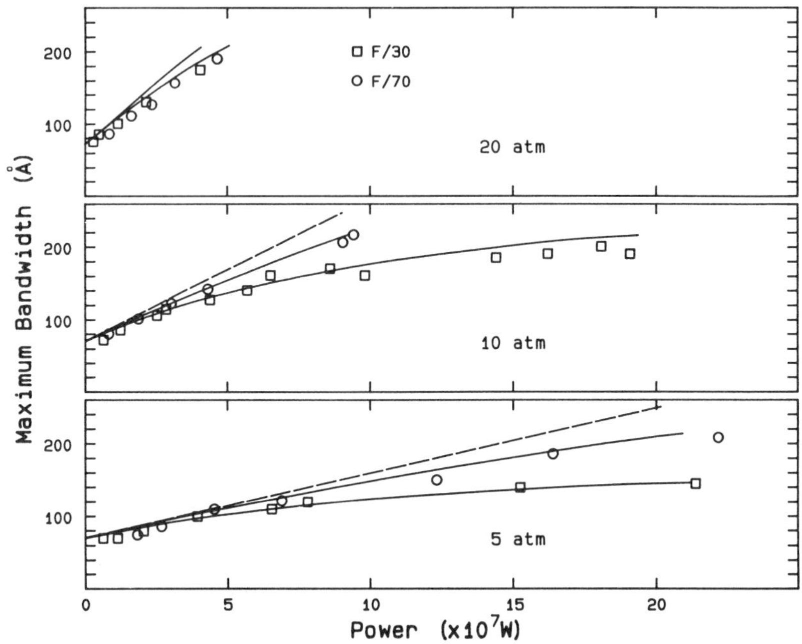

FIGURE 7.2. Spectral width of the radiation transmitted through a cell filled with 5, 10, and 20 atm of xenon as a function of the peak laser power. The circles and squares represent data obtained with an $F/70$ and an $F/30$ lens, respectively. The solid curves are plots of Eq. (2) with a saturation intensity of 10^{13} W/cm^2 corresponding to 1.5×10^8 W for the $F/30$ lens ($\omega_0 = 21$ μm).

Saturationlike behavior of the nonlinearity could be caused by plasma produced by ionization (Corkum and Rolland, 1988b). There are four experimental reasons to believe that saturation is a fundamental phenomenon, not directly related to ionization:

1. At 10^{13} W/cm^2, sufficient ionization to modify the beam propagation by one diffraction-limited beam divergence, or to modify the spectrum measurably, would require $\sim$25% of the beam energy. We measure an absorption of less than 3%.
2. Ionization would produce asymmetric self-phase modulation since the plasma would most affect the trailing region of the pulse. We observe a nearly symmetrical spectrum.
3. Figure 7.1 shows that insignificant plasma density is produced by $I < 3 \times 10^{13}$ W/cm^2.
4. Ionization would produce irreversible distortion of the transmitted beam profile. In fact, beam distortion is frequently used as a diagnostic of ionization (Corkum and Rolland, 1988a; Guha et al., 1985). We see little beam distortion.

5. Saturation of the Nonlinear Response in Gases

Since the nonlinear response is not modified by ionization, we must consider other explanations. For xenon and 0.6-μm light, the first excited state is 4.2 photon energies above the ground state. Any pulse duration or intensity-dependent changes must come from higher-order terms. In the absence of resonances, high-order terms should contribute to the nonlinear response approximately in the ratio $\chi^{(3)}E^2/\chi^{(1)}$. At 10^{13} W/cm^2 the ratio is $\sim$0.04. To explain the observations in Figure 7.2, resonant enhancement is required and, as we have already indicated in Section 3, resonant enhancements are inevitable.

In discussing transient resonances, we have already pointed out that bound carriers in a high-lying resonant level respond as free electrons. It was just this fact that required that the ac Stark shift be equal to the ponderomotive potential. These bound electrons *must* reduce the refractive index as would truly free electrons. Since the high-lying states are only transiently resonant, they are only virtually occupied. Thus, aside from resonantly enhanced ionization, which is discussed below, the reduced refractive index need not be associated with net absorption from the beam. The change in refractive index due to bound electrons in high-lying levels is equivalent to $\eta_2 E^2$ when only 10^{-3} of the ground state population is in these levels. (It is interesting to note that the connection between the ac Stark shift and the susceptibility, implicit in this description, can also be shown for a weakly driven two-level system (Delone and Krainov, 1985).)

It is essential to consider whether a transiently resonant population of 10^{-3} is consistent with low ionization levels, since transiently resonant states in xenon lie within one or two photon energies of the continuum. Resonantly enhanced ionization of xenon has been observed in multiphoton ionization experiments with $\sim$500-fs pulses (Freeman et al., 1987). If we assign a cross section (Mainfray and Manus, 1980) of $\sigma = 10^{-19}$ (10^{-20}) cm^2 to the single-photon ionization from a near-resonant state, we can calculate the ratio of resonantly excited electrons (N_e) to free electrons (N_i) at an intensity of 10^{13} W/cm^2 ($N_e/N_i = \hbar\omega/\sigma I \tau$). The ratio for a 90-fs pulse is $\sim$3 (30). Note the pulse duration scaling. Thus the resonant population can exceed the free-electron population for ultrashort pulses.

It may seem that the small cross section used above is in contrast to what would be calculated from Keldysh-type theories (Szöke, 1988; Keldysh, 1965) for 10^{13} W/cm^2, assuming an ionization potential of <1 eV. This apparent discrepancy is explained by the fact that a transiently resonant electron is only weakly bound. Since an unbound electron cannot absorb photons from a plane electromagnetic wave, as we cross the boundary between an unbound and a weakly bound electron we should not expect the electron to absorb photons readily. (In the long-wavelength limit this is no longer valid because of the Lorentz force contribution to ionization.)

Values for σ are very poorly known experimentally, especially for levels near the continuum. Recent UV measurements (Landen et al., 1987) for the krypton $4p5d$ and $4p4d$ levels (1 and 1.7 eV below the continuum) yielded values of $\sigma = 3 \times 10^{-18}$ and 8×10^{-18} cm^{-2}, respectively. These results satisfy the trend of decreasing σ as the continuum is approached.

It is useful to reexpress the above discussion in more general terms. Many high-order nonlinear terms will be enhanced by transient resonance due to the dense packing of levels at high energies. Our qualitative description of the plasmalike response of the electron is equivalent to summing a series of nonlinear terms. It should be emphasized that transient resonances will influence all nonlinear processes in this intensity range. Their effects could well exceed the nonresonant contributions to the susceptibility. Note that enhancement of the nonlinear response is also observed in partially ionized plasma due to excited state population (Gladkov et al., 1987).

6. Self-Focusing—$\chi^{(3)}$ Becomes Large

The modification to the nonlinear response of the medium that we have described has important consequences for self-focusing. In Figure 7.2 the highest-intensity data points (for a given F-number and gas pressure) give approximately the threshold above which the nature of the spectral broadening changes nearly discontinuously. For $F/70$ optics this value is approximately a factor of 2 above the calculated self-focusing threshold. A factor of 2 discrepancy is consistent with the correction factor of 2.1 that we required to make Eq. (2) agree with experimental data. The critical power is clearly not a useful parameter if the intensity at the geometric focus exceeds 10^{13} W/cm^2.

To ensure that self-focusing will be initiated all remaining results are taken with large F-number optics ($F/200$).

6.1 Spectral Characteristics of Gaseous Continua

When the beam intensity is increased above that plotted in Figure 7.2, the wavelength scale of the spectral broadening increases dramatically. In Figure 7.3 typical multishot spectra are plotted for the 70-fs and 2-ps pulses (with characteristics similar to the 90- and 900-fs pulses described previously) transmitted through a gas cell filled with various gases. Shown in Figure 7.3a are spectra for 30 atm and 15 atm of xenon illuminated with 70-fs and 2-ps pulses, respectively. Figure 7.3b shows spectra for 40 atm of N_2 (2 ps) and 38 atm of H_2 (70 fs). The spectra in Figure 7.3 are typical of spectra obtained with all gases that we have investigated, provided only that the laser intensity was sufficient to exceed the critical power for self-focusing.

The similarity in the blue spectral component for all the curves in Figure

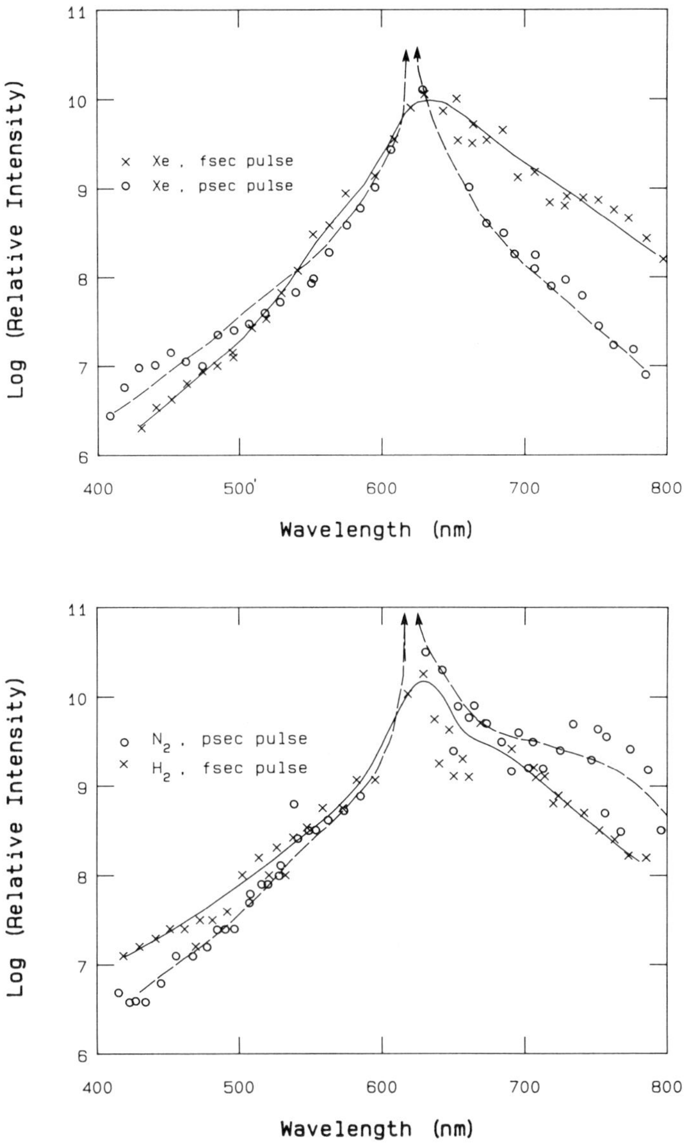

FIGURE 7.3. Continuum spectra. (a) In xenon: $P = 30$ atm, $\Delta t = 70$ fs (crosses); $P = 15$ atm, $\Delta t = 2$ ps (circles). (b) In N_2: $P = 40$ atm, $\Delta t = 2$ ps (circles) in H_2: $P = 38$ atm, $\Delta t = 70$ fs (crosses).

7.3 should be noted. In fact, the blue spectral component is nearly universal for all gases that produce continua regardless of the (above-threshold) intensity or pressure. (It is also typical of a chaotic spectrum (Ackerhalt et al., 1985).) The red component, however, varies with the laser and gas parameters. We have investigated the red cutoff only with CO_2 using femtosecond pulses. The maximum wavelength for 30 atm of CO_2 exceeded the 1.3-μm limit of our S1 photocathode. (Because of an orientational contribution to the nonlinearity, continua can be produced at a particularly low threshold intensity with picosecond pulses in CO_2.)

Figure 7.4 shows that spectral modulation is another characteristic of the spectrum of gaseous continua. Modulation has been noted previously on the single-shot spectra of gases (Glownia et al., 1986). Spectral modulation is characteristic of continua from condensed media as well (Smith et al., 1977). Figure 7.4 illustrates the intensity and η_2 scaling of the spectral modulation of a xenon continuum as measured in the region of 450 nm. Figures 7.4a and 7.4b show that the modulation frequency is reproducible from shot to shot. The modulation depth is not always as great as shown in Figure 7.4. Figures 7.4c and 7.4d demonstrate that the modulation frequency varies with $\eta_2 E^2$ near the continuum threshold. However, in Figure 7.4e to 7.4g we see that the simple $\eta_2 E^2$ scaling is eventually lost at higher pressure-power products. In all cases, the modulation frequency increases further from the laser frequency. This behavior is in contrast with that expected for self-phase modulation (Smith et al., 1977).

The characteristics of gaseous continua described so far are similar to those of condensed-medium continua. However, the extra flexibility provided by pressure dependence of the nonlinearity allows issues like the $\eta_2 E^2$ dependence of the spectral modulation to be addressed. We will see that it also allows us to correlate self-focusing with continuum generation much more precisely than previously possible.

As already mentioned, continuum generation showed a sharp threshold, below which spectral broadening is described by Eq. (2) and above which full continua are produced. The threshold power for continuum generation equals the self-focusing threshold power to the accuracy to which the self-focusing threshold is known. The functional dependence of the continuum threshold on laser power, gas pressure, and the hyperpolarizability is also the same as that for self-focusing. This dependence is shown in Figure 7.5, where the product of the gas pressure and the laser power at threshold for all gases investigated with the femtosecond pulse is plotted as a function of the laser power. Comparing the pressure-power products for each gas, we find that they are inversely proportional to the hyperpolarizabilities. Similar data were obtained (but are not plotted) with the picosecond pulse. For the rare gases and H_2, the picosecond data would fit on their respective lines in Figure 7.5. Both N_2 and CO_2, however, have lower thresholds than would be indicated from their purely electronic nonlinearities. This is due to orientational effects that are important in both gases.

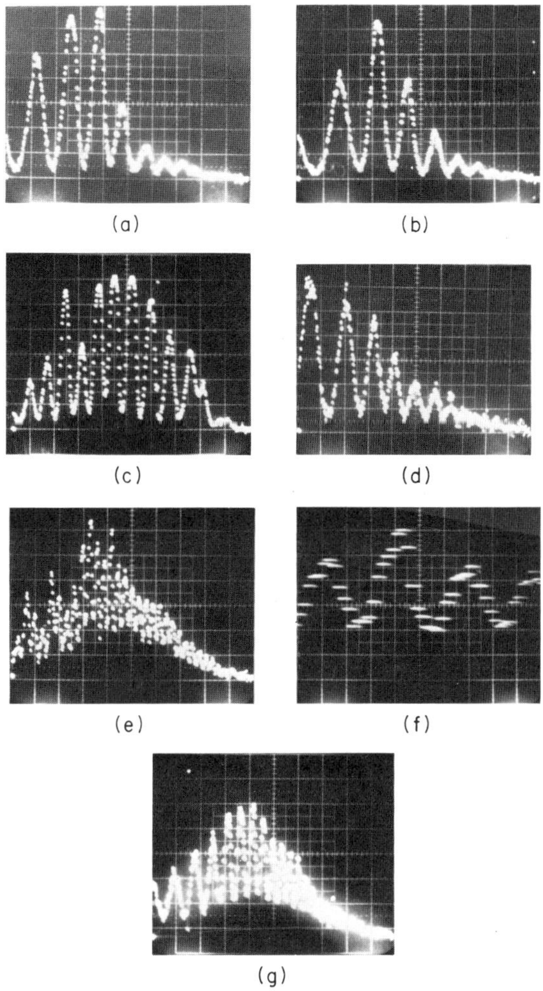

FIGURE 7.4. Details of the single-shot continuum spectrum centered near $\lambda = 450$ nm illustrating the spectral modulation on the continuum. (The horizontal scale for all traces except (f) is 137 Å/div.) (a) and (b) (7 atm pressure of xenon) show the reproducibility of the spectral modulation. (c) and (d) (14 atm pressure of xenon) show the $\eta_2 E^2$ scaling of the spectral modulation. (c) was taken with the same laser power as (a) and (b). (d) was taken with one-half the laser power of (a)–(c). (e) to (g) (21 atm pressure of xenon) show that the $\eta_2 E^2$ scaling is not valid well above the continuum threshold. (e) and (f) were taken with the same laser power as (a)–(c). The wavelength scale in (f) has been expanded by a factor of 10. (g) was taken with one-third of the laser power of (a)–(c).

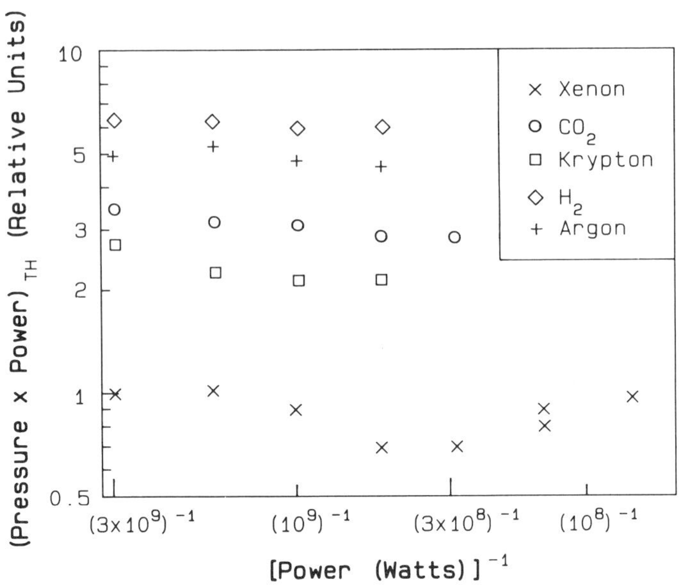

FIGURE 7.5. Laser power multiplied by the gas pressure at the continuum threshold, plotted as a function of the inverse of the laser power for different gases.

6.2 Spatial Characteristics of Gaseous Continua

One might expect that spatial changes in a beam that has experienced at least the onset of self-focusing would be severe. Considering that the spectrum of the beam is catastrophically modified, can we expect anything but a severely distorted transmitted beam?

Figure 7.6 shows the near-field and far-field distributions of the beam after passing through the gas cell. The first row is composed of reproductions of Polaroid photos of the near-field spatial distribution as viewed through an ~0.5-mm-wide slit and recorded on an optical multichannel analyzer (OMA). The second row shows far-field distributions recorded in a similar manner. From left to right are distributions taken through a filter that blocks all wavelengths $\lambda < 650$ nm (left column), with the gas cell evacuated (middle column), and through a filter that blocks $\lambda > 525$ nm (right column). The left- and right-hand columns were obtained with the gas cell filled with sufficient pressure to ensure that the laser peak power exceeded the continuum threshold. On the basis of the spatial profile alone, it is virtually impossible (with large F-number optics) to distinguish between the presence and absence of self-focusing and continuum generation for powers near the self-focusing threshold.

As the laser power is increased to approximately four times the continuum threshold, conical emission is observed. At first the ring structure is simple, but it becomes increasingly complex at higher powers. Figure 7.7 shows

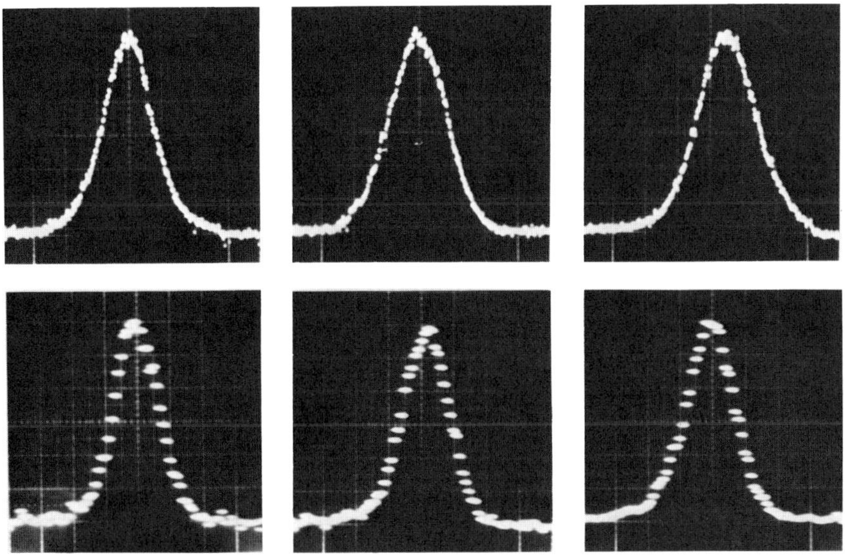

FIGURE 7.6. Near-field (top row) and far-field (bottom row) distributions of the beam after passing through the gas cell. From left to right are shown the red spectral component ($\lambda > 650$ nm), the beam with the gas cell evacuated, and the blue spectral component ($\lambda < 525$ nm).

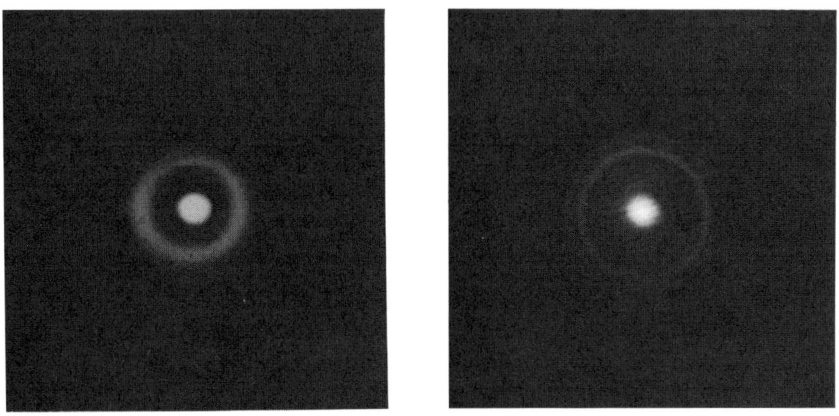

FIGURE 7.7. Conical emission as observed in the near field through the red ($\lambda >$ 650 nm, left image) and blue ($\lambda < 525$ nm, right image) filters.

the ring structure only slightly above the threshold for conical emission. Conical emission was previously observed in condensed-medium continuum generation (Smith et al., 1977; Alfano and Shapiro, 1970a).

Figure 7.6 indicates that the transverse beam distribution is almost totally reconstructed after self-focusing. The spatial reconstruction of the beam stands in stark contrast to the catastrophic change in the initial spectrum. With such beam reconstruction, it is natural to reexamine whether self-focusing was ever initiated. Several reasons to associate continuum generation with self-focusing in gases are listed below:

1. Conceptually, self-focusing is just the free-space spatial manifestation of self-phase modulation. Since Eq. (1) is valid for large F-number optics, η_2 must be the dominant nonlinearity, rendering self-focusing inevitable. Of course, when the intensity approaches 10^{13} W/cm^2, η_2 is no longer dominant.
2. The continuum threshold has approximately the same magnitude as the calculated self-focusing threshold. It also has the same functional dependence on the gas pressure, laser power, and hyperpolarizability.
3. Conical emission has been predicted by the moving-focus model of self-focusing (Shen, 1975). It is also a characteristic of a saturating nonlinearity (Marburger, 1975) and high-order nonlinear mixing processes. With large F-number optics, all of these potential explanations of conical emission require self-focusing to increase the peak intensity.
4. We have projected conical emission to its source and find that it originates from the prefocal region. This origin can be graphically illustrated by placing a 3-mm-diameter opaque disk at the geometric focus. Significant conical emission escapes around its side.
5. The $\eta_2 E^2$ dependence of the spectral modulation implies that η_2 plays at least a limited role in continuum generation.
6. In no case have we been able to observe continuum generation without the laser power exceeding the calculated self-focusing power.

6.3 Discussion

It is not possible to consider gaseous continuum generation as if it were produced by self-phase modulation in the η_2 limit alone. The conceptual link between self-phase modulation and self-focusing makes this approach unrealistic. The very small value of $\eta_2 E^2$, even at the ionization threshold (Corkum et al., 1986a, 1986b), gives additional evidence that continuum generation is not only an η_2 process.

There is a second conceptual problem. If plasma is not created (as we have shown experimentally), then high-order nonlinearities are required to stabilize self-focusing: consequently, η_2 is no longer the dominant nonlinear term. (It is interesting to note that nonlinear optics will be very different in the long-wavelength limit, where the Lorentz force severely limits the lifetime of most high-lying states.)

Out of this apparent complexity, however, very simple and near-universal behavior emerges. This simple behavior will have to be explained by continuum theories. In particular, theory will have to explain the periodicity of the modulation, the spatial properties of the beam, conical emission, and the universality of the blue spectral component of the continuum.

7. Conclusions

With the recent development of ultrashort pulses, it is now possible to perform nonlinear optics experiments in new limits of intensity and pulse duration. Due to the sweep of the focus, earlier self-focusing experiments may already have explored this region, although unwittingly.

This chapter has described experiments explicitly performed to investigate subpicosecond nonlinear optics. It described the first high-intensity experiments performed with pulses as short as 22 fs and showed that ionization cannot be described by perturbation theory for pulse durations shorter than 1 ps. It also discussed the role of transient resonances in multiphoton ionization and in high-intensity self-phase modulation experiments. For ultrashort pulses, these transient resonances dominate the nonlinear optical response of gases in much the same way that high-lying resonances dominate in partially ionized plasmas (Gladkov et al., 1987). (Presumably, the same is true in condensed media near the multiphoton ionization threshold.)

Continuum generation in gases (and indeed all nonlinear optical phenomena in this intensity and pulse duration range) will be understood only in the context of transient resonances and limited convergence of perturbation theory.

In conclusion, it should be emphasized that the observations in this chapter are very much in keeping with the condensed-media results. The difference is only that gases show the properties of continua in such a dramatic form as to strongly challenge conventional ideas of continuum generation.

Acknowledgments. The author acknowledges important contributions to this work by Dr. S.L. Chin, during a five-month sabbatical at NRC. His experience with multiphoton ionization experiments was invaluable, as was his contribution to other concepts expressed in this chapter. Dr. T. Srinivasan-Rao's contributions are also acknowledged. Without her short visit to NRC we would never have begun this set of experiments. D.A. Joines has cheerfully provided technical support throughout all of the experiments. Our rapid progress would not have been possible without him. Discussion with many colleagues at NRC are gratefully acknowledged.

References

Ackerhalt, J.R., P.W. Milonni, and M.-L. Shih (1985). Phys. Rep. **128**, 207.

Alfano, R.R. and S.L. Shapiro (1970a) Phys. Rev. Lett. **24**, 584.

Alfano, R.R. and S.L. Shapiro (1970b) Phys. Rev. Lett. **24**, 592.

Bloembergen, N. (1973) Opt. Commun. **8**, 285.

Chin, S.L., C. Rolland, P.B. Corkum, and P. Kelly (1988). Phys. Rev. Lett. **61**, 153.

Corkum, P.B. and C. Rolland (1987) Summary of postdeadline papers, XV International Quantum Electron. Conference, April, Baltimore, Maryland, Paper PD-21.

Corkum, P.B. and C. Rolland (1988a) In *NATO ASI Series—Physics B*, vol. 171, A. Bandrauk, ed., p. 157. Plenum, New York.

Corkum, P.B. and C. Rolland (1988b) unpublished.

Corkum, P.B., C. Rolland, and T. Srinivasan-Rao (1986a) Phys. Rev. Lett. **57**, 2268 (1986a).

Corkum, P.B., C. Rolland, and T. Srinivasan-Rao (1986b) In *Ultrafast Phenomena V*, G.R. Fleming and A.E. Siegman, eds., p. 149. Springer-Verlag, New York.

Delone, N.B. and V.P. Krainov (1985) *Atoms in Strong Light Fields*, p. 174, Springer-Verlag, Berlin.

Fork, R.L., C.V. Shank, C. Hirliman, R. Yen, and W.J. Tomlinson (1983) Opt. Lett. **8**, 1.

Freeman, R.R., P.H. Bucksbaum, H. Milchberg, S. Darack, D. Schumacher, and M.E. Geusic (1987) Phys. Rev. Lett. **59**, 1092.

Gladkov, S.M., N.I. Koroteev, M.V. Ruchev, and A.B. Fedorov (1987) Sov. J. Quantum Electron. **17**, 687.

Glownia, J.H., J. Miswich, and P.P. Sorokin (1986) J. Opt. Soc. Am. **B3**, 1573.

Guha, S., E.W. Van Stryland, and M.J. Soileau (1985) Opt. Lett. **10**, 285.

l'Huillier, A., L.A. Lompre, G. Mainfray, and C. Manus (1983) Phys. Rev. A **27**, 2503.

Keldysh, L.V. (1965) JETP **20**, 1307.

Landen, O.L., M.D. Perry, and E.M. Campbell (1987) Phys. Rev. Lett. **59**, 2558.

Lehmeier, H.J., W. Leupacher, and A. Penzkofer (1985) Opt. Commun. **56**, 67.

Mainfray, G. and C. Manus (1980) Appl. Opt. **19**, 3934.

Marburger, J.H. (1975) Prog. Quantum Electron **4**, 35.

Rolland, C. and P.B. Corkum (1986) Opt. Commun. **59**, 64.

Rolland, C. and P.B. Corkum (1988) J. Opt. Soc. Am. B **5**, 641.

Shen, Y.R. (1975) Prog. Quantum Electron **4**, 1.

Smith, W.L., P. Liu, and N. Bloembergen (1977) Phys. Rev. A **15**, 2396.

Szöke, A. (1988) *NATO ASI Series—Physics B*, vol. 171, A. Bandrauk, ed., p. 207. Plenum, New York.

8
Utilization of UV and IR Supercontinua in Gas-Phase Subpicosecond Kinetic Spectroscopy

J.H. GLOWNIA, J. MISEWICH, and P.P. SOROKIN

1. Introduction

Through the work of photochemists extending over many decades, there now exists a wealth of information on the various reactions that photoexcited gas phase molecules undergo. Most of this information relates to the product molecules that are formed, either as the direct result of a primary photochemical act, such as photodissociation, or through subsequent secondary reactions, involving collisions with other molecules in the gas. Recently, there has been an extensive effort directed at determining the exact energy distributions of the primary products formed in photodissociation. With the use of nanosecond tunable-laser techniques, such as laser-induced fluorescence (LIF) and coherent anti-Stokes Raman spectroscopy (CARS), scientists have successfully determined the nascent electronic, vibrational, and rotational energy distributions of various diatomic fragments such as CN, OH, NO, and O_2 that are directly formed in the photodissociation of many kinds of molecules. The ready availability of high-quality, tunable, nanosecond lasers has made determination of the above-mentioned collisionless energy distributions a relatively straightforward process. The determination of product translational energies has long effectively been handled by angularly resolved time-of-flight (TOF) spectroscopy, or by sub-Doppler resolution spectroscopy, including a recently improved version of the latter, velocity-aligned Doppler spectroscopy (Xu et al., 1986).

Of great interest, but until recently unobtainable, is detailed knowledge of the time sequences of the various internal conversions, rearrangements, dissociations, etc. that molecules typically undergo upon photoexcitation. To illustrate this point, let us consider Figure 8.1, which depicts an alkyl azide analog of the Curtius rearrangement for an acyl azide molecule. It is known that the weakest bond in the covalently bonded azide group lies between the two nitrogen atoms closest to the carbon atom. This fact explains the finding that molecular nitrogen is invariably produced in the photolysis of covalently bonded azides. The existence of the stable methylenimine product in the case of Figure 8.1 implies a breaking of the C–H bond and forming of an

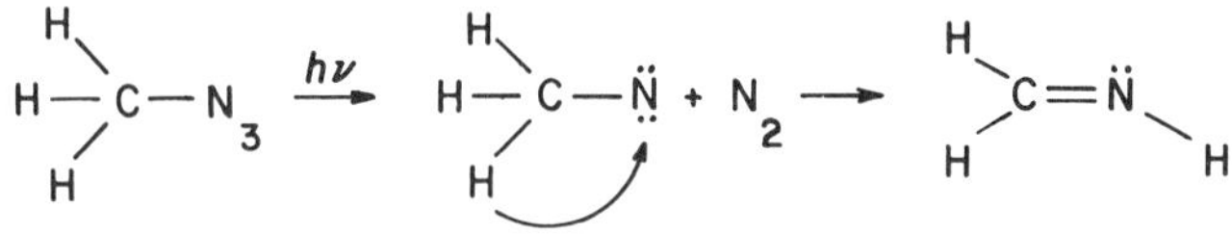

FIGURE 8.1. Rearrangement mechanism proposed for the photolysis of methyl azide.

N–H bond. However, as to whether the above reactions actually happen sequentially in the order described, or whether the whole sequence occurs in a simultaneous, concerted manner, the relevant literature is quite contentious. It would appear that an advanced technique of *kinetic spectroscopy*,* having a spectral range that includes portions of the infrared and having a subpicosecond time resolution, could provide direct answers in the above example, provided that the time to form the final methylenimine product is no less than several hundred femtoseconds. Experimentally, one would monitor the times for disappearance of the azide symmetric or antisymmetric stretches and compare these with the appearance time for the N–H stretch.

While time-resolved broadband infrared probing of the vibrational modes of photoexcited molecules should thus give easily interpretable results, it appears, unfortunately, to be a rather difficult technique to develop and to apply. In general, the optical absorbances associated with purely vibrational transitions are a few orders of magnitude weaker than those associated with electronic transitions. Thus, signal-to-noise considerations become of paramount importance. Aside from questions of sensitivity, there is also the problem that only a very limited spectral range has thus far been demonstrated for subpicosecond kinetic spectroscopy in the infrared. For many photoexcited molecules one can, of course, determine exact times of dissociation by observing the appearance times of products of the photodissociation. Here one can fortunately utilize as monitors electronic transitions with their intrinsically high molar absorbances. Examples are given further on in this chapter. Electronic transitions between excited states of a molecule can also be utilized to monitor the motion of a molecule along an excited state potential surface, tracking the times when the molecule internally converts or fragments. This is illustrated by another example, discussed further on, in which an infrared subpicosecond kinetic spectroscopy probe is actually utilized.

Of course, the problem of spectroscopically monitoring in real time the unimolecular reactions of a photoexcited molecule can also be attacked in ways other than through kinetic spectroscopy. A.W. Zewail's group uses a different time-resolved approach, which is illustrated by their recent study of

* The term "kinetic spectroscopy" was used by G. Porter, R. Norrish, and others, who pioneered the field of flash photolysis. Since the technique we describe in this chapter also relies on the use of a time-resolved, broadband, absorption spectroscopy probe, we choose to describe our work with the same term, with the addition, however, of the prefix "subpicosecond."

ICN photodissociation (Dantus et al., 1987). After a subpicosecond UV pump pulse has initiated ICN fragmentation, a tunable LIF subpicosecond probe pulse induces the CN fragments to fluoresce. They recorded the CN* excitation spectrum as a function of pump-probe delay, providing new information about the photodissociation dynamics.

We have tried briefly in the preceding paragraphs to stress the potential utility of a subpicosecond kinetic spectroscopy approach to the real-time study of intramolecular photoinitiated reactions of gas-phase molecules. As is well known, subpicosecond kinetic spectroscopy has already been employed successfully in the condensed phase by several research groups to elucidate the dynamics of biological processes, to follow the approach to equilibrium in photoexcited dyes, to study ultrafast processes in semiconductors, etc. However, such studies, almost without exception, have relied on the use of broadband time-resolved spectroscopic probes in the visible. Our group has recently started to make advances toward developing equipment capable of providing a wider subpicosecond spectral probing range. In recent papers (Glownia et al., 1986a, 1987a, 1987b) we have described an apparatus capable of simultaneously generating both intense subpicosecond UV (308, 248.5 nm) excitation pulses and subpicosecond continua for probing photoexcited molecules via broadband absorption spectroscopy. Both UV (230 to 450 nm) and IR (2.2 to 2.7 μm) continua have thus far been produced. A method of upconverting the latter to the visible for ease of detection has also been demonstrated. In the present chapter this apparatus is described in detail. Also given is an account of some of the first experiments performed using this equipment.

The organization of this chapter is as follows. Since the pulses in both excitation and probe channels are derived from subpicosecond UV pulses amplified in XeCl gain modules, we begin, in Section 2, with a description of the apparatus we have built for producing intense subpicosecond 308-nm pulses. In Section 3 we show how these 308-nm pulses can be used to generate ultrafast UV supercontinuum pulses and also how the latter can be used to seed an amplification process in a KrF excimer gain module in order to produce intense subpicosecond pulses at 248 nm. Section 4 describes our technique for producing an ultrafast IR continuum, as well as the method we use for upconverting it to the visible for ease of detection. Exactly how the UV and IR continua are utilized in subpicosecond kinetic spectroscopy experiments is shown in the examples discussed in the balance of the chapter. Section 5 describes an IR experiment we have performed, the measurement of the $\tilde{B} \rightarrow \tilde{A}$ internal conversion rate in 1,4-diazabicyclo[2.2.2]octane (DABCO) vapor, while Section 6 describes some preliminary results obtained in the case of two subpicosecond kinetic spectroscopy experiments recently attempted, photolysis of thallium halide vapors at 248 and 308 nm and photolysis of chlorine dioxide vapor at 308 nm. Section 7 concludes with a brief description of two promising directions that can now be taken in our approach to subpicosecond kinetic spectroscopy.

2. 160-fs XeCl Excimer Amplifier System

The suitability of commercially available excimer gain modules for amplification of ultrafast UV pulses has been apparent for many years. Around 1982, various groups (Corkum and Taylor, 1982; Egger et al., 1982; Bucksbaum et al., 1982; Szatmári and Schäfer, 1984a) successfully utilized discharge-pumped excimer gain modules for amplification of UV pulses having durations of a few picoseconds. However, it was well known (e.g., see Corkum and Taylor, 1982) that the gain bandwidth of these systems is such that amplification of much shorter pulses can also be accomplished. The first published accounts of subpicosecond pulse amplification in excimers were published some four years later (Glownia et al., 1986b; Schwarzenbach et al., 1986). Glownia et al. (1986b) used a pair of XeCl gain modules to amplify 350-fs, 308-nm pulses to $\sim$10-mJ energies with $<$1-mJ amplified spontaneous emission (ASE) content. Seed pulses for the excimer amplifier were formed through the combined use of a synch-pumped mode-locked dye laser tuned to 616 nm, a single-mode fiber pulse compressor (Nakatsuka et al., 1981; Nikolaus and Grischkowsky, 1983), a four-stage Nd^{3+}:YAG-laser-pumped dye amplifier, and, finally a KDP frequency-doubling crystal. Schwarzenbach et al. (1986) used generally similar methods to produce subpicosecond seed pulses suitable for amplification at 248.5 nm in KrF gain modules.

In 1987 further significant advances in excimer-based UV subpicosecond amplification were made. Szatmári et al. (1987a) reported the generation of 220-fs pulses at 308 nm from an XeCl amplifier. Shortly afterwards, the same Göttingen group (Szatmári et al., 1987b) reported having obtained 15-mJ, 80-fs, 248.5-nm pulses from a KrF amplifier and having then amplified these pulses to 900 GW peak power in a wide-aperture KrF discharge amplifier. A remarkable feature of the above work is that the seed pulses in each case were formed directly with nanosecond excimer-pumped dye laser sources. A novel method discovered earlier (Szatmári and Schäfer, 1983, 1984b) of generating single, picosecond pulses through the combined use of an excimer-pumped, quenched, dye laser and a distributed feedback dye laser (DFDL) was improved on by Szatmári et al. (1987a, 1987b) to the degree that subpicosecond pulses were produced. These pulses were amplified and then frequency doubled prior to final amplification in an excimer gain module. One of the many advantages of the technique employed by the Göttingen group is that, due to the wide wavelength range accessible with DFDLs, all of the known rare-gas halide wavelengths can be reached through frequency doubling or mixing. Possible disadvantages of the Göttingen technique center around the inherent difficulties in adjusting and stabilizing the DFDL. It will be interesting to follow the development of this unique approach and to see also if it eventually benefits from commercial product engineering.

In 1987 a full description was published (Glownia et al., 1987b) of the XeCl excimer-based system our group currently employs, which generates bandwidth-limited, 160-fs, 308-nm pulses at a 10-Hz rate. Subpicosecond pulses at ~616 nm are formed in a colliding-pulse mode-locked (CPM) laser (Fork et al., 1981; Valdmanis et al., 1985), amplified in a four-stage Nd^{3+}:YAG laser-pumped amplifier chain, and then frequency doubled in a 1-mm-long KDP crystal, forming seed pulses at 308 nm for further amplification in the XeCl excimer gain module. Since this sytem is the heart of our subpicosecond pump-probe apparatus, we summarize its main features here.

The design of the CPM laser is generally similar to that of Valdmanis et al. (1985), incorporating four dispersion-compensating prisms in the seven-mirror ring cavity arrangement of Fork et al. (1981). The pulse repetition rate is 116 MHz. An 80-mm focal-length lens is used to focus the 514.5-nm CW Ar^+-ion laser pump beam into the Rhodamine 6G gain jet. Surprisingly, optimum mode locking at 616 mm, described below, requires only 1.4 W of pump power. Both the gain and DODCI absorber jets are standard Coherent dye laser nozzles (dye stream thickness ~100 μm). A relatively dilute concentration of DODCI (50 mg in 2l ethylene glycol) was found to be optimum for operation at 616 nm. The CPM output power is about 20 mW in each arm when the laser is optimized at this frequency.

We generally tune the intraprism path length for minimum amplified laser-pulse duration (see below) while maintaining the peak of the output spectrum near 616 nm. With this adjustment, the CPM pulse width is about 240 fs. However, when measured after the beam has propagated through an additional 5 cm of H_2O, the pulse width is 200 fs, showing that the laser operates with excess negative dispersion in the cavity. It should be noted that this CPM laser can readily produce much shorter pulse widths (~70 fs) at longer wavelengths. However, the spectrum of the amplified and frequency-doubled pulses cannot then properly match the XeCl gain profile. The first three stages of the dye amplifier are excited transversely; the fourth, longitudinally. Kiton Red 620 dye is used in the first stage; Sulforhodamine 640 in the last three stages. The solvent used in all stages is H_2O + 4% Ammonyx LO. Malachite Green bleachable absorber dye jets are used between the first three amplifier stages to control amplified spontaneous emission. The small-signal attenuations of the absorber jets are roughly 10× and 200×. Pumping of the dye amplifier chain is accomplished with ~125 mJ of 532-nm light from a Quanta-Ray DCR-2A Nd^{3+}:YAG laser operating in the short pulse (2-ns central peak) mode. We typically measure a total amplified pulse energy of ~0.6 mJ. The autocorrelation trace of the amplified CPM dye laser pulses at full power indicates a pulse width of ~200 fs.

The ~200-fs amplified 616-nm pulses are spatially compressed to a beam diameter of ~2 mm, then frequency doubled in 1-mm-long KDP crystals to form seed pulses for amplification in the XeCl excimer gain module. In Glownia et al. (1987b) extensive frequency broadening of the second harmonic was noted when the size of the input beam in the KDP crystal was allowed to be less than 2 mm in diameter. This was attributed to self-phase modulation (SPM) occurring in the KDP crystals. Since the UV spectral width of the seed pulses was observed to be typically 10× the XeCl gain bandwidth, it was argued by Glownia et al. (1987b) that the positive frequency sweep associated with the most intense part of the pulse would result in a nearly in-phase excitation of all the frequencies lying under the XeCl gain curve. This is an exact prescription for forming bandwidth-limited pulses. This argument was used to explain the observation (see below) that the UV pulses amplified by the XeCl gain module are bandwidth limited.

The UV seed pulses are amplified in a single pass through a Lambda-Physik EMG101-MSC excimer gain module (45-cm-long discharge). Provided that the seed pulse spectrum is relatively flat over the XeCl gain profile, the spectrum of the amplified 308-nm pulse appears as shown in Figure 8.2. Figure 8.3 displays a typical amplified 308-nm pulse autocorrelation trace. (Two-photon ionization in DABCO vapor is used for the 308-nm pulse autocorrelation measurements.) Measurements made on this system over a period of more

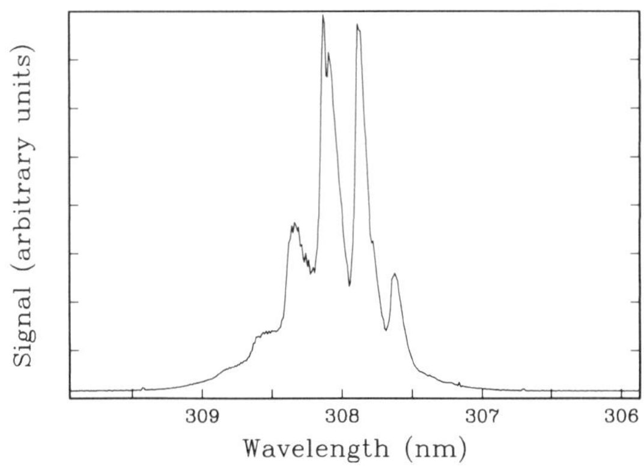

FIGURE 8.2. Spectrum of amplified 308-nm pulses.

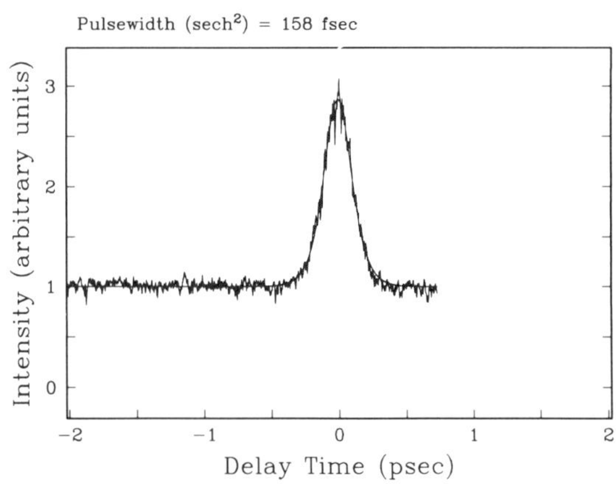

FIGURE 8.3. Autocorrelation trace of amplified 308-nm pulses.

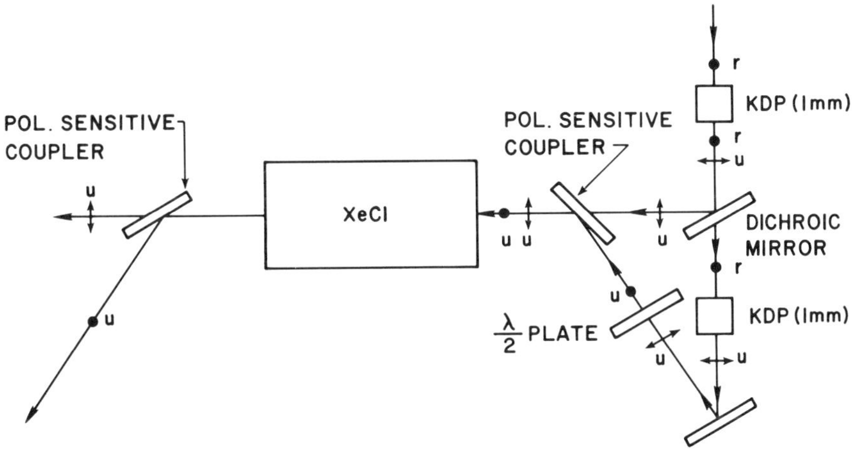

FIGURE 8.4. Scheme employed for multiplexing the 308-nm pulses (u, ultraviolet; r, red).

than one year have consistently shown the amplified 308-nm pulse width to be near 160 fs, a number very close to the bandwidth limit calculated (Glownia et al., 1987b) for the spectrum of Figure 8.2. This observed facile generation of bandwidth-limited UV pulses is rationalized by the argument mentioned in the preceding paragraph.

With the use of 1-mm-long KDP doubling crystals, pulse energies of 4 to 5 mJ are achieved in a single pass through the XeCl excimer gain module. With the scheme shown in Figure 8.4, however, a second UV pulse, having roughly the same energy, can be obtained during the same excimer discharge. Because the efficiency of second harmonic generation in the first 1-mm-long KDP crystal is only ~10%, enough 616-nm light remains to generate a second UV seed pulse having almost the same energy as the first. If the two UV seed pulses are spaced apart by 2 or 3 ns, there is sufficient time for repumping the XeCl B state, according to Corkum and Taylor (1982). Thus amplification in the XeCl gain module occurs in the form of pairs of orthogonally polarized pulses, each pulse 4 to 5 mJ in energy. The 160-fs amplified UV pulse pairs are then separated by a Brewster polarizer into pump and probe channels (see Figure 8.14).

3. Ultraviolet Supercontinuum Generation

It was noted by Glownia et al. (1986b) that gentle focussing in air of the XeCl-amplified subpicosecond pulses resulted in a spectral broadening of the pulses by roughly a factor ten. Figure 8.5 shows a typical single-shot spectrum of an amplified 308-nm, 160-fs pulse after it has been focused in the laboratory atmosphere with a 1-m lens. The spectrum is almost 1000 cm^{-1} wide, with the

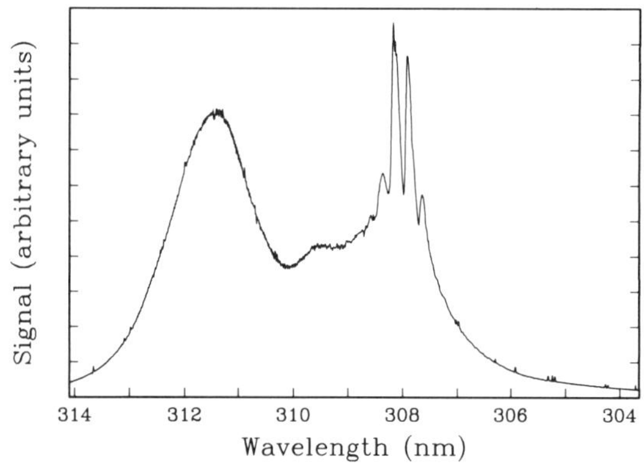

FIGURE 8.5. Single-shot spectrum of amplified 308-nm pulse, recorded after the latter was focused in air with a 1-m lens.

spectral enhancement being predominantly on the Stokes side. At the same time it was observed that the far-field pattern of the beam beyond the focal point of the 1-m lens usually contained bright spots in which the light was concentrated. These observations qualitatively suggested to us that SPM and self-focusing were involved in the above phenomena. However, due to the multimode spatial character of the beam, no direct steps were taken to verify these speculations. Instead, it was decided to experiment with a variety of gases (Ar, H_2, N_2, and CO_2) under high pressure to see if any differences in spectral broadening could be discerned between the various gases and also whether spectral continua with widths in excess of $\sim 1000\ cm^{-1}$ could be produced by this method. The results of our measurements were reported in Glownia et al. (1986a, 1986c). The main features are summarized below. Independently of us, P. Corkum's group at National Research Council, Canada, observed the same basic phenomenon of supercontinuum generation from high-pressure gases (Corkum et al., 1986a, 1986b). In their case, both subpicosecond and picosecond amplified red pulses were successfully utilized as pump pulses. The basic physics of this newly discovered phenomenon is discussed by Corkum et al. (1986b). More recent observations and deductions about gas-phase supercontinuum generation are contained in Corkum and Rolland (Chapter 7 in the present volume).

Figure 8.6 displays the spectrum of the energetic UV supercontinuum beam that emerges from a high-pressure Ar cell when high-power subpicosecond UV (308 nm) pulses are focused into the cell. Consistent with the finding of Corkum et al. (1986b), we observe nearly full transmission ($>80\%$) of energy through the pressurized cell, with no significant degradation of the beam profile. There is thus adequate probe energy to pass through a photoexcited sample on to a high-dispersion visible-UV spectrograph, to be then recorded on an unintensified silicon diode array. With the supercontinuum source and

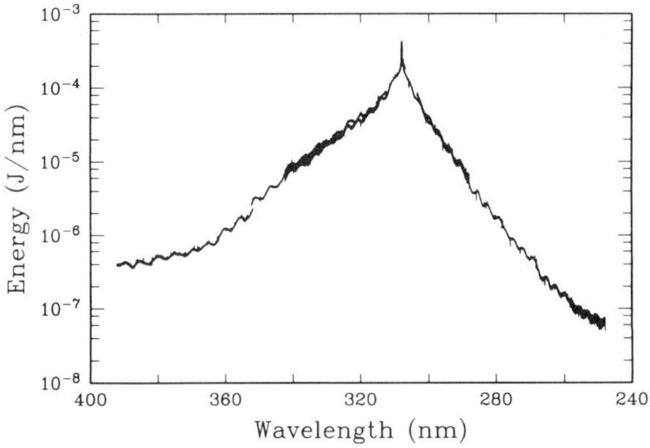

FIGURE 8.6. Spectrum of UV supercontinuum beam emerging from Ar cell, $p = 40$ atm. Pump pulses (4 mJ, 308 nm, 160 fs) were focused into the middle of the 60-cm-long Ar cell with a 50-cm lens. Average of 64 shots.

spectrograph/detector system we normally employ, a spectral resolution of better than 0.3 Å is achieved. For probe continua at longer wavelengths, one must continue to rely on the various condensed-matter supercontinua (Alfano and Shapiro, 1970a, 1970b; Fork et al. 1982) or else on gas-phase supercontinua pumped by a red laser (Corkum et al. 1986a, 1986b), since all supercontinua peak at the pump wavelengths employed. It should be pointed out, however, that condensed-matter supercontinua peaking in the UV evidently cannot easily be generated. Our attempts to achieve this effect with the use of amplified subpicosecond 308-nm pulses as pump pulses were unsuccessful, possibly due to nonlinear absorption in the various liquids tried.

Because of the large spectral extent of supercontinuum pulses, they are broadened in time by group velocity dispersion (GVD). For condensed-matter visible supercontinua, Li et al. (1986) have measured spectral delays with the use of a streak camera and filters. Utilizing a cross-correlation technique, Fork et al. (1983) have measured the sweep of a supercontinuum generated in an ethylene glycol jet. We have ultilized a novel method, based on time-resolved absorption spectroscopy, to measure with subpicosecond resolution the frequency sweep of the supercontinuum displayed in Figure 8.6. This method, in the form in which it was originally demonstrated (Misewich et al., 1988a), is now briefly described.

Thallium chloride molecules, contained in a vapor cell at 450°C, were irradiated by 250-fs, 248-nm pump pulses derived (by a method to be described below) from 160-fs, 308-nm pulses. Supercontinuum probe pulses were simultaneously obtained from the same apparatus by focusing 4-mJ, 160-fs, 308-nm pulses into 40-atm Ar with a 50-cm lens. The probe pulses were directed into the TlCl vapor colinearly with the pump pulses, then dispersed

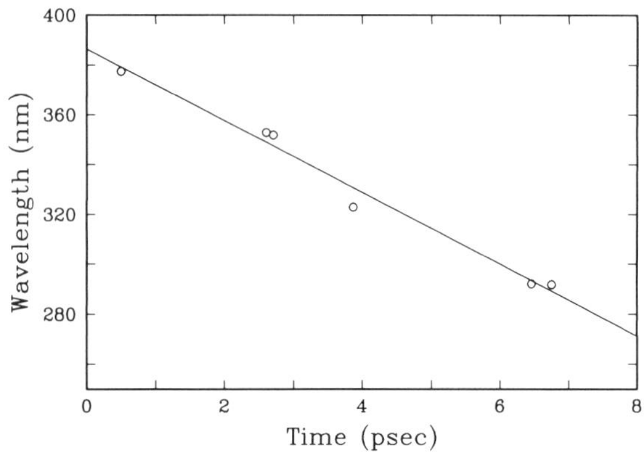

FIGURE 8.7. Frequency sweep of 40-atm Ar supercontinuum beam whose spectrum is shown in Figure 8.6. Data points correspond to Tl absorptive transitions whose onsets were measured.

in a spectrograph and recorded on an optical multichannel analyzer (OMA). Absorbances were computed by comparison of supercontinuum intensities recorded with and without the UV pump blocked.

Thallium chloride molecules irradiated at 248 nm undergo prompt dissociation into Tl and Cl atoms, with the former being distributed into ground $^2P_{1/2}$ and first-excited $^2P_{3/2}$ (7793 cm^{-1}) states in roughly a 30:70% ratio (van Veen et al., 1981). Several allowed transitions, spanning a wide range of frequencies, connect the two states with various higher excited states. By plotting the individual rise times of the above atomic transitions as a function of pump-probe delay, one obtains the plot in Figure 8.7, which shows that the 40-atm Ar supercontinuum is characterized by a positive chirp of approximately 1340 cm^{-1}/ps.

Our technique utilizing TlCl photodissociation to measure the UV supercontinuum frequency sweep has also revealed some interesting features of the photodissociation itself. We defer a discussion of the transient absorption spectra of photodissociating TlCl and TlI molecules until Section 6. From the results presented there, however, it can be stated that while the total duration of the gas-phase supercontinuum pulse emerging from the high-pressure cell is on the order of 10 ps, the effective time resolution is much better. As shown above, the probe continuum pulse has a fast red-to-blue linear sweep. Our spectral results (Section 6) indicate that the cross-correlation between the 160-fs, 308-nm pump pulse and a given wavelength interval of the swept probe pulse is stable to at least ± 50 fs from shot to shot. It is this observed stability that gives the UV gas-phase supercontinuum pulse its good time resolution.

The UV supercontinuum pulses of Figure 8.6 can be directly used as seed pulses suitable for further amplification in KrF gain modules. From

Figure 8.6 one sees that there is roughly a microjoule of energy in the supercontinuum between 248 and 249 nm, the wavelength range over which KrF amplification occurs. With this amount of input energy one easily obtains 6-mJ output pulses through single-pass amplification in a KrF gain module with a 45-cm-long discharge region (Glownia et al., 1986a, 1986c, 1987c). In Glownia et al. (1986a), the high-pressure cell used to form the KrF seed pulses contained H_2 gas. We originally used hydrogen because of a coincidence between an anti-Stokes Raman wavelength and the wavelength at which maximum KrF gain occurs. Stimulated Raman scattering (SRS) occurs in high-pressure H_2 gas when 160-fs, 308-nm pump pulses are applied, but not without the simultaneous occurrence of UV continuum generation. In several other molecular gases we failed to observe SRS with our 308-nm, 160-fs pump source, but we do observe UV continuum generation. Since SRS is known to be a ubiquitous phenomenon when high-power pulses of a few picoseconds duration are applied to high-pressure molecular gases (Mack et al., 1970), one must conclude that the conditions for its occurrence are made far less favorable as one proceeds to the subpicosecond domain, while the occurrence of supercontinuum generation becomes much more likely.

4. Subpicosecond Time-Resolved Infrared Spectral Photography

Time-resolved infrared spectral photography (TRISP) (Avouris et al., 1981; Bethune et al., 1981, 1983; Glownia et al., 1985) is a nonlinear optical technique by which a broadband ($\Delta\nu \sim 1000$ cm^{-1}) infrared absorption spectrum can be recorded in a single shot of a few nanoseconds duration. The IR spectral range that has thus far been convered with this technique is 2 to 11 μm (Bethune et al., 1983). Recently, we reported a successful extension of the TRISP technique to the subpicosecond time domain (Glownia et al., 1987a). The IR spectral region that can be probed with our present ultrafast apparatus is only 2.2 to 2.7 μm, but extension of subpicosecond capability to other IR ranges seems possible.

In this section we present details of the 2.2–2.7-μm subpicosecond TRISP apparatus. In the following section we describe the actual measurement of a subpicosecond photophysical event with the use of this apparatus.

In a TRISP apparatus, means for generating a broadband infrared sample-probing pulse are combined with a method for upconverting and detecting this signal in the visible. Our ultrafast TRISP apparatus combines a new subpicosecond IR continuum generator with a standard TRISP upconverter. We first discuss formation of an ultrashort IR continuum. Powerful subpicosecond IR pulses at $\sim$2.4 μm are produced by stimulated electronic Raman scattering (SERS) in Ba vapor (Figure 8.8). The latter is contained at $\sim$10 torr pressure inside an Inconel pipe heated to $\sim$1050°C. The length of the heated

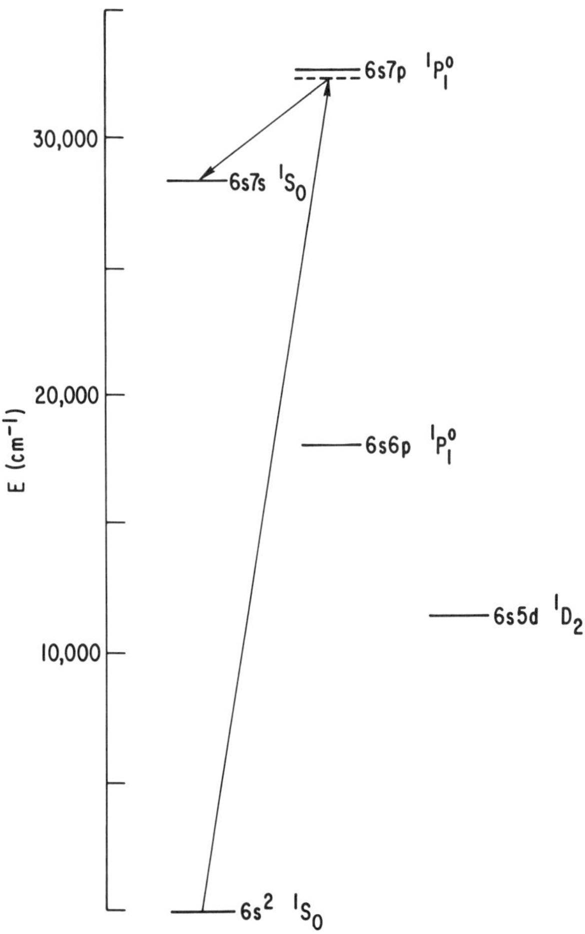

FIGURE 8.8. Diagram of the Ba SERS process.

region is ~0.5 m. As pump pulses we directly utilize amplified 308-nm, 160-fs pulses. Although application of ~20-ns XeCl laser pulses to Ba vapor is known (Burnham and Djeu, 1978; Cotter and Zapka, 1978) to produce SERS only on the $6s^2\,{}^1S_0 \rightarrow 6s5d\,{}^1D_2$ transition, with a Stokes output near 475 nm, we find, by contrast, that with ultrashort 308-nm excitation SERS occurs only on the $6s^2\,{}^1S_0 \rightarrow 6s7s\,{}^1S_0$ transition, with a Stokes output peaked near 2.4 μm. The 2.4-μm SERS output is highly photon efficient, with measured IR output pulse energies of ~0.4 mJ for ~5-mJ UV input pulses. The SERS threshold is lower than 1 mJ, even for an unfocused pump beam. Spectrally, the IR output is found to be a continuum, extending from 2.2 to 2.7 μm (Figure 8.9). Occurrence of the 2.4-μm SERS process is accompanied by the presence of a

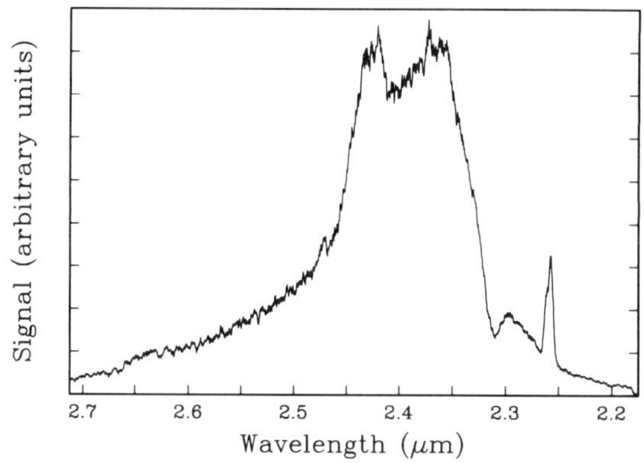

FIGURE 8.9. Spectrum of Raman Stokes light, recorded with the use of a scanning monochromator and PbS detector.

parametrically generated beam of light in the vicinity of the 535-nm Ba resonance line.

We have also measured the $\sim$2.4-μm IR continuum pulse width by non-background-free autocorrelation, with the second harmonic being generated in a thin $LiIO_3$ crystal. For the 250-fs, 308-nm pump pulses that were applied (Glownia et al., 1986d), the IR pulse width was determined to be $\sim$160 fs.

In Glownia et al. (1987a), a plausible argument was given to account for the switch of Stokes wavelength from 475 nm to 2.4 μm that occurs with ultrashort excitation. It was suggested that this switch could be attributed to the change in Raman gain regime (from stationary to transient) that occurs in going from $\sim$20 ns to subpicosecond 308-nm excitation pulses. In the stationary regime ($t_p > T_2$), the intensity of the Stokes wave increases in accordance with the law

$$I_S(z) = K_0 \exp(\Gamma_0 z), \tag{1}$$

where Γ_0 is the static gain, inversely proportional to the Raman linewidth. In the transient regime, the intensity of the Stokes wave assumes (for a square input pulse) the value (Akhmanov et al., 1972; Carman et al., 1970)

$$I_S(z) \sim K_1 \exp[2(2\Gamma_0 t_p T_2^{-1} z)^{1/2}]. \tag{2}$$

Since Γ_0 is directly proportional to T_2, one sees that there is no dependence of Stokes gain on Raman linewidth in the transient regime. A possible explanation for the Stokes wavelength switch would thus be that the collisional linewidth of the $6s5d\,^1D_2$ state is sufficiently narrow compared with that of the $6s7s\,^1S_0$ state to favor Raman Stokes generation of 475 nm in the stationary case, even though the remaining factors in Γ_0 favor Stokes generation at

2.4 μm. In the transient regime the gain is independent of T_2, and the above-mentioned remaining factors entirely determine the Stokes wavelength. No Ba linewidth data are available to support this contention. However, a crude calculation of the van der Waals interaction between a ground state Ba atom and a Ba atom in either the $6s7s\,^1S_0$ state or the $6s5d\,^1D_2$ state, with use of London's general formula (Margenau, 1939), indicates a larger width for the $6s7s$ state.

We now discuss upconversion of the IR. With the use of a polished Si wafer, the horizontally polarized ultrashort IR pulses (ν_{IR}) are colinearly combined with the vertically polarized $\sim$15-ns pulses (ν_L) from a tunable narrowband furan 1 dye laser. With the timing between the two sources adjusted so that the subpicosecond IR pulses occur within the 15-ns-long dye laser pulses, both beams are sent into an Rb upconverter (Glownia et al., 1985) where the dye laser beam induces SERS on the Rb $5s \rightarrow 6s$ Stokes transition, producing a narrowband vertically polarized Stokes wave ν_S. Horizontally polarized, visible continuum pulses at $\nu_L - \nu_S \pm \nu_{IR}$ are then observed to emerge from the Rb cell when ν_L is tuned to phase match either upconversion process.

Figure 8.10 shows a recording of the (lower-sideband) upconverted spectrum. In Figure 8.11 portions of two upconverted spectra are superimposed. In one case, the IR pulse was passed through an empty 20-cm cell; in the other case it was passed through the same cell filled with 200 torr of CO. The deduced absorbance is shown in Figure 8.12. A surprising finding is the observed increase in upconverted signal at the peaks of the CO 2–0 bands. This is explained as follows (Glownia et al., 1987a). Under the conditions of Figure 8.11 and 8.12, the upconverted signal was heavily saturated by the subpicosecond IR pulse; that is, too few photons at ν_L and ν_S were available

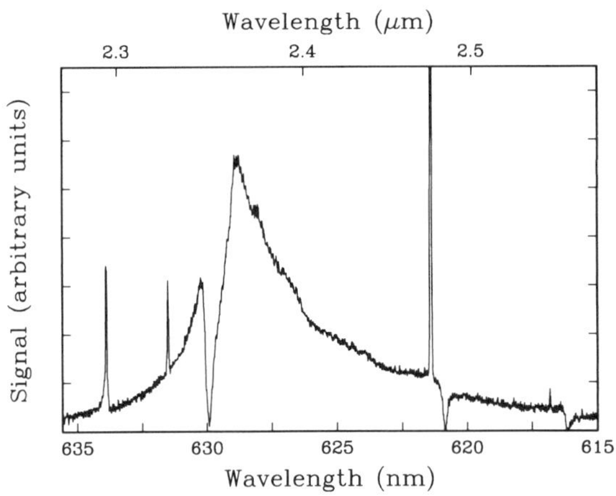

FIGURE 8.10. Upconverted TRISP spectrum (128-shot average, lower sideband). The spectrum is saturated (see text). The three absorptions are Rb excited state ($5p$) absorptions occurring at the visible wavelengths shown.

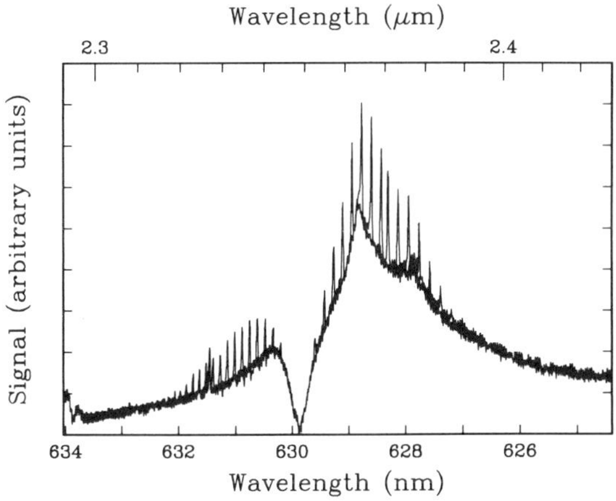

FIGURE 8.11. Superimposed upconverted spectra (cell empty and cell filled with 200-torr CO gas). Each spectrum is the average of 128 shots.

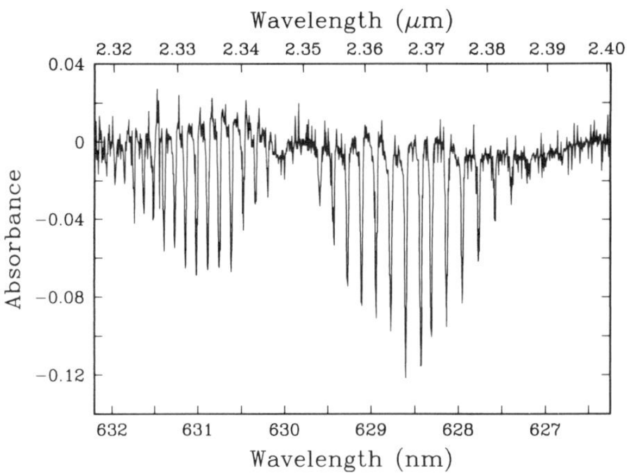

FIGURE 8.12. Absorbance formally deduced from Figure 8.11.

during the actual IR pulse to allow efficient upconversion of the latter. Therefore a decrease in transmitted light due to molecular resonance absorption during the IR pulse did not result in a measurable decrease in upconverted signal. However, the coherently reemitted light of the molecules (Hartmann and Laubereau, 1984), occurring for a time the order of T_2 after the IR pulse, when the upconverter is no longer saturated, was able to be efficiently upconverted, resulting in the observed peaks. Positive IR absorption is observed in the upconverted spectra when the subpicosecond IR probe beam is sufficiently attenuated.

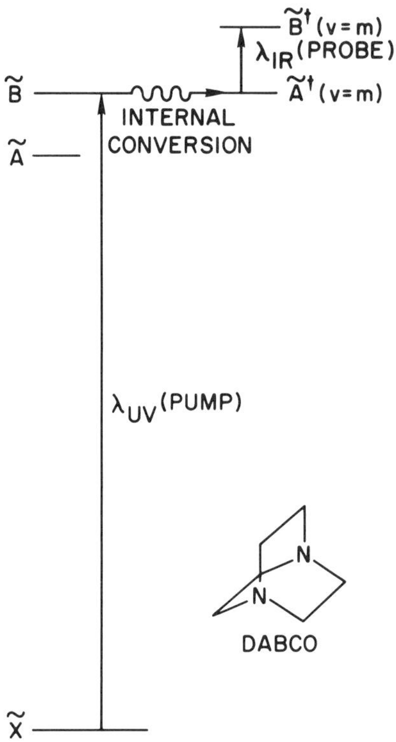

FIGURE 8.13. Diagram of photophysical processes involved in the subpicosecond DABCO experiment.

5. Application of Subpicosecond TRISP: Measurement of Internal Conversion Rates in DABCO Vapor

We recently reported (Glownia et al., 1987c) the first application of subpicosecond time-resolved IR absorption spectroscopy to measure an ultrafast molecular process. This experiment combined subpicosecond 248-nm excitation with subpicosecond IR continuum probing to measure the $\tilde{B} \rightarrow \tilde{A}$ internal conversion rate in DABCO vapor. A diagram of the photophysical processes involved is shown in Figure 8.13. The idea that the $\tilde{B} \rightarrow \tilde{A}$ internal conversion rate in DABCO might be high enough to require ultrafast techniques for its measurement is contained in an earlier study (Glownia et al., 1985), in which the population of the $\tilde{A}$ state was monitored following the application of a 30-ns, 248.5-nm KrF laser excitation pulse. A high $\tilde{B} \rightarrow \tilde{A}$ internal conversion rate for DABCO was also implied in a recent two-color laser photoionization spectroscopy study (Smith et al., 1984). In the above-mentioned earlier study of DABCO by our group, nanosecond TRISP was used to monitor the $\tilde{A}$ state population. This was because the $\tilde{B} \leftarrow \tilde{A}$ transition (occurring at $\sim$2.5 μm) was found to have a much higher oscillator strength than all other transitions connecting the $\tilde{A}$ state with higher electronic states (Glownia et al., 1985). Since the subpicosecond TRISP apparatus described in the last section monitors the region 2.2 to 2.7 μm, we decided to measure the DABCO $\tilde{B} \rightarrow \tilde{A}$ internal conversion with the greatly improved time resolution this apparatus offers.

Collimated 2-mJ, 250-fs, 248.5-nm pulses were sent unfocused (beam dimensions: 2 cm × 1 cm) into a 60-cm-long cell containing DABCO at its ambient vapor pressure ($\sim$0.3 torr) together with 100 torr of H_2. The linear absorption of the DABCO at 248.5 nm (40,229 cm^{-1}) was more than 50%, even though this wavelength lies near the point of minimum absorbance between the $v' = 0 \leftarrow v'' = 0$ (39,807 cm^{-1}) and next highest vibronic peaks of the lowest-energy, dipole-allowed, band (Halpern et al., 1968; Hamada et al., 1973). This band system has been assigned (Parker and Avouris, 1978, 1979) as $\tilde{B}^1E'[3p_{x,y}(+)] \leftarrow \tilde{X}^1A_1'[n(+)]$. Optical transitions from the ground state $\tilde{X}$ to the first excited state, the $\tilde{A}^1A_1'[3s(+)]$ (origin at 35,785 cm^{-1}), are one-photon forbidden, two-photon allowed (Parker and Avouris, 1978, 1979).

The 160-fs IR continuum pulses that probe the $\tilde{A}$ state population were directed through the vapor collinearly with the UV photoexcitation pulses, upconverted to the visible, and then dispersed in a spectrograph equipped with an unintensified OMA detection system. The pump-probe delay could be varied up to ± 1 ns by means of an optical delay arm. Absorbances were computed by comparison of upconverted intensities recorded with and without the UV pump blocked.

A block diagram of the experiment is shown in Figure 8.14. As described in Section 2, subpicosecond pulses at $\sim$616 nm formed in a CPM laser are amplified, then frequency doubled, forming seed pulses at $\sim$308 nm for further amplification in the XeCl excimer gain module. Amplification of the UV pulses

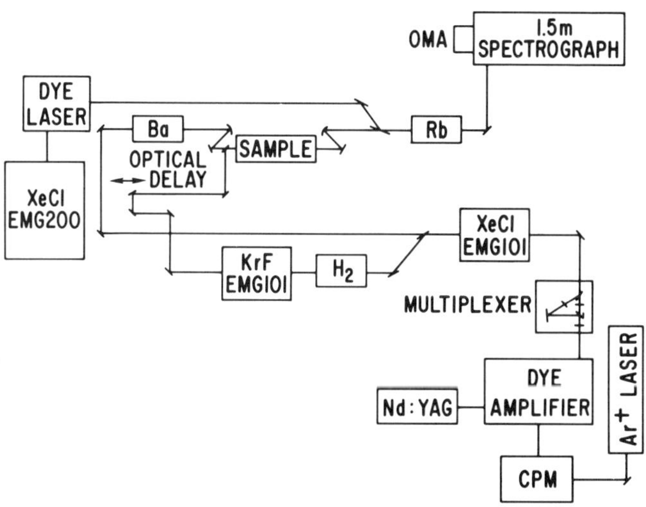

FIGURE 8.14. Diagram of experimental apparatus for the DABCO experiment.

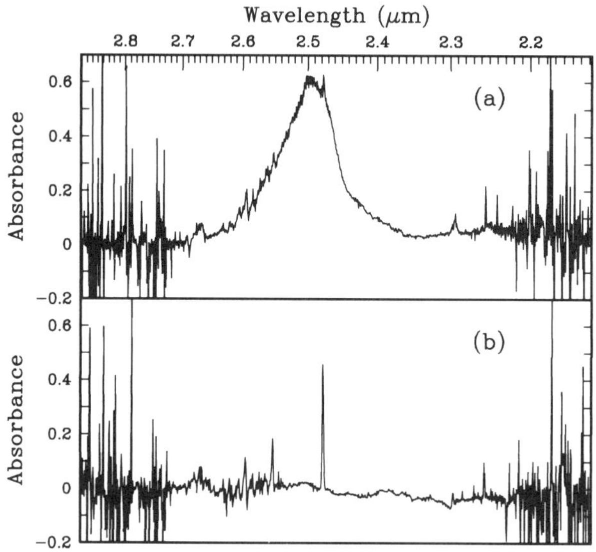

FIGURE 8.15. (a) Absorbance (base 10) with probe delayed ≈4 ps with respect to pump. (b) Absorbance with probe pulse preceding pump pulse.

in the latter occurs in the form of pairs of orthogonally polarized pulses, spaced 2 to 3 ns apart, formed in the multiplexer described also in Section 2. The 160-fs amplified UV pulse pairs are separated by a polarization-sensitive coupler into pump and probe channels. The pump channel 308-nm pulses are focused into high-pressure gas to form seed pulses for amplification at 248.5 nm in a KrF module (see Section 3). The probe channel pulses are Raman shifted in Ba vapor to form IR probe continuum pulses (see Section 4). The narrowband pulsed dye laser drives the Rb upconverter (see Section 4).

Figure 8.15a shows the absorbance recorded when the probe is delayed ~4 ps with respect to the pump (point (a) in Figure 8.16), while Figure 8.15b displays the absorbance with the probe arriving just before the pump (point (b) in Figure 8.16). The absorbance recorded at 2.494 μm, as a function of probe delay, is shown in Figure 8.16. The large absorption band that develops represents transitions $\tilde{B}^{\dagger} \leftarrow \tilde{A}^{\dagger}$ of vibrationally excited $\tilde{A}$ state molecules, containing up to 4400 cm^{-1} of vibrational energy. Since the $\tilde{B} \leftarrow \tilde{A}$ transition is one that occurs between Rydberg states, vertical ($\Delta v = 0$) transitions are expected. Thus it is not surprising that the peak of the band in Figure 8.15a appears very close to the $\tilde{B} \leftarrow \tilde{A}$ peak for vibrationally equilibrated $\tilde{A}$ state molecules, shown here in Figure 8.17 and described in detail in Glownia et al. (1985). However, the width of the $\tilde{B} \leftarrow \tilde{A}$ band is obviously greater in the vibrationally excited case.

The computer-generated curve in Figure 8.16 is a nonlinear least squares fit to the data. The fit indicates a rise time of ~500 fs. Although the infrared

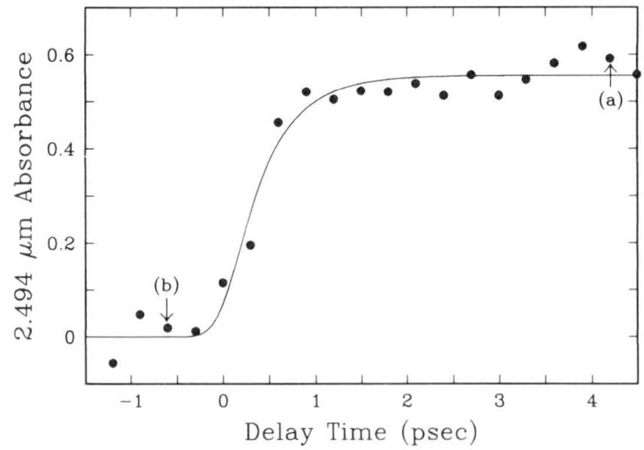

FIGURE 8.16. Peak $\tilde{B}^{\dagger} \leftarrow \tilde{A}^{\dagger}$ absorbance as a function of probe pulse delay time with respect to pump pulse.

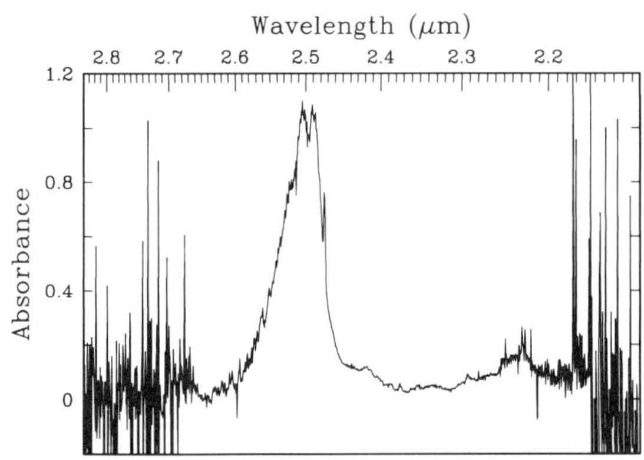

FIGURE 8.17. Absorbance measured with probe pulse delayed 75 ns with respect to pump pulse. The latter was an 8-mJ, 25-ns pulse obtained by operating the KrF gain module as a conventional laser. 60-cm DABCO cell, 3-atm H_2 pressure.

and ultraviolet pulse widths were determined by autocorrelation to be ~160 and ~250 fs, respectively, the cross-correlation between these pulses has not been measured. Thus, the rise time in our experiment could be limited by the laser system cross-correlation. In any case, the process converting DABCO states accessed by the subpicosecond 248.5-nm pump beam into vibrationally excited $\tilde{A}$ states is observed to occur on a time scale that is at least as fast as ~500 fs. That internal conversion to vibrationally excited $\tilde{A}$ state molecules is the dominant process for photoexcited DABCO molecules, even at UV pump intensities of ~4 GW/cm^2, is also underscored by the fact that there is no apparent decrease in the integrated intensity of the 2.5-μm absorption band induced at these pump intensities, as compared with the 2.5-μm integrated intensity induced by 25-ns UV pulses at comparable fluence levels (compare Figures 8.15a and 8.17).

The DABCO experiment shows how subpicosecond TRISP can be used to monitor the internal conversion of photoexcited molecules in real time. In the sequence of spectra corresponding to the data points in Figure 8.16, it is clearly seen that the wavelength at which the peak absorbance occurs undergoes a definite blue-to-red shift as the magnitude of the absorbance grows in. It is tempting to attribute this to an intramolecular vibrational redistribution (IVR) process. However, heavy caution must be applied here, since the same apparent phenomenon could easily be induced by a red-to-blue sweep of the IR probe continuum. We have no information at present as to whether or not the IR continuum is swept.

6. Preliminary Results on the Application of the UV Supercontinuum Probe

In Section 3 we described a convenient method, based on photofragmentation of thallium halides, by which the sweep of the UV supercontinuum can be measured. In the process of measuring the rise times of the Tl absorption lines, we have consistently noted that the latter assume unusual line shapes, with enhanced integrated intensities, for a period lasting roughly a picosecond, beginning the moment the atomic absorption is first discerned and ending when the asymptotic, normal appearing, absorption line profile is finally attained. We present some of these preliminary spectral results in this section. These results are qualitatively discussed in terms of a model based on the transient behavior of the polarization induced by the subpicosecond swept UV continuum as the latter interacts with the time-varying population of two-level atoms produced by the photolysis pulse. A detailed description of our model will be presented elsewhere (Misewich et al., 1988b).

There has been broad interest for some time in the spectroscopy of the thallium halides. The ultraviolet absorption cross sections have been measured by Davidovits and Bellisio (1969). The UV absorption spectra comprise a number of well-defined bands whose conformity between the various halides is striking. In the most recent thallium halide photofragmentation study (van Veen et al., 1981), these bands are simply labeled A, B, C, and D. In that work, the time-of-flight spectra and angular distributions of photofragments were measured for the thallium halides at a variety of UV wavelengths, including 308 and 248 nm. We have now utilized subpicosecond pulses, at both 308 and 248 nm, to separately excite TlI and TlCl molecules. In the case of the former, 308 nm is very close to the peak of the C band, while 248 nm lies on the high-frequency side of the D band. For the latter, 308 nm is near the peak of the B band, while 248 nm excites the C band.

Figures 8.18, 8.19, and 8.20 show the appearance of time-resolved absorption spectra recorded in the vicinity of the 377.6-nm Tl $7S_{1/2} \leftarrow 6P_{1/2}$ resonance

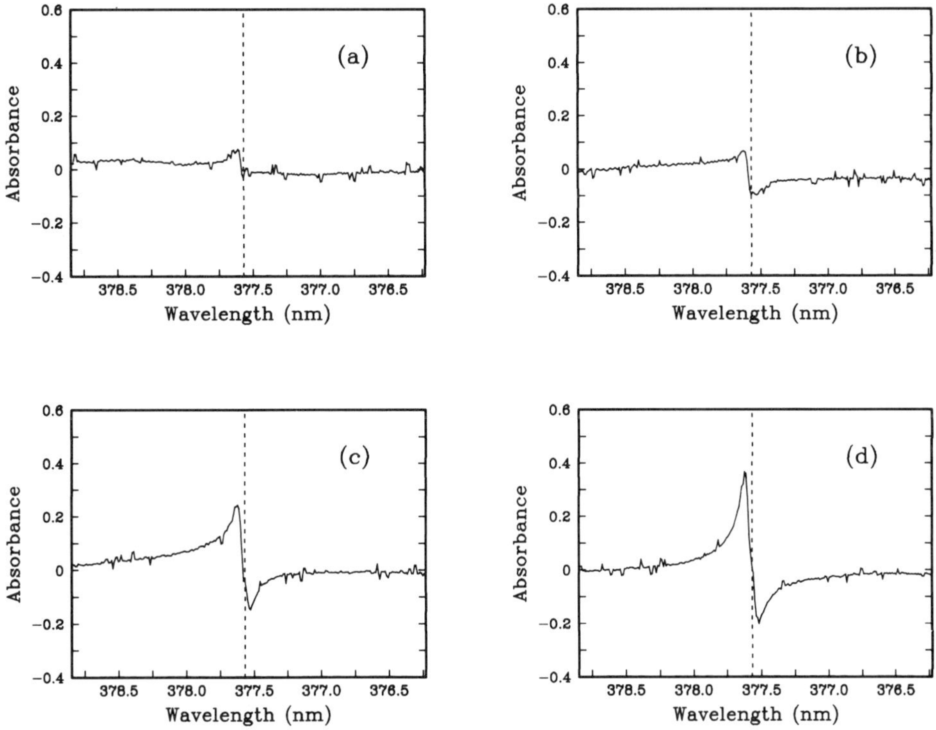

FIGURE 8.18. Time-resolved absorption spectra, recorded in the vicinity of the 377.6-nm Tl $7S_{1/2} \leftarrow 6P_{1/2}$ resonance line, following the application of 160-fs, 308-nm pulses to TlI vapor. Pump-probe separation increased by 100-fs between each spectrum shown. Each spectrum represents the average absorbance (base 10) deduced from dual 64-shot continuum accumulations, one with and one without the 308-nm pump blocked. The dashed line marks the position of the asymptotic resonance peak.

line following the application of 160-fs, 308-nm pump pulses to TlI vapor. It is seen that roughly 1 ps elapses from the moment the atomic transition appears to the point at which no further changes in the appearance of the atomic resonance line occur. From the TOF data presented in van Veen et al. (1981), iodine atoms produced by 308-nm photolysis of TlI have a relatively broad distribution of translational velocities peaking at $\sim 4.4 \times 10^4$ cm/s. Hence the average Tl–I separation at large distances must increase as $\sim 7.1 \times 10^4$ cm/s. However, at smaller distances the average rate of increase of separation is much smaller, because the atoms are accelerated from rest. We have been unable to mark the exact time of occurrence of the 308-nm pump pulse with respect to the times shown in Figures 8.18 to 8.20. However, if one assumes the pump pulse occurs somewhere in the vicinity of Figures 8.18a and b, one deduces that the asymptotic line shape must be attained well before the Tl-I separation has increased by 7.1 Å.

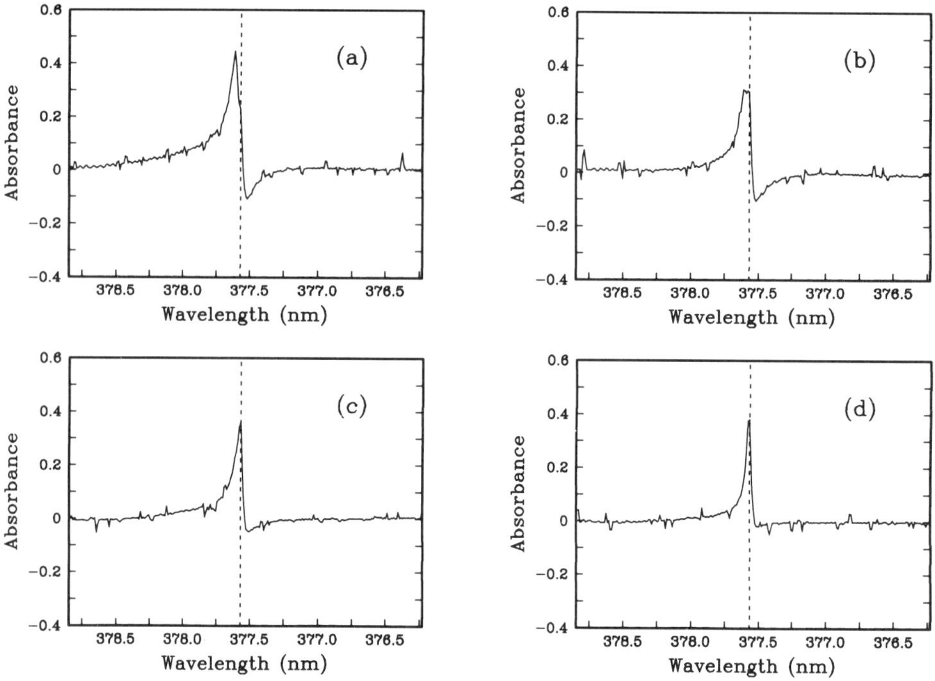

FIGURE 8.19. Continuation of Figure 8.18, with pump-probe separation in (a) increased by 100 fs over that in Figure 8.18 (d).

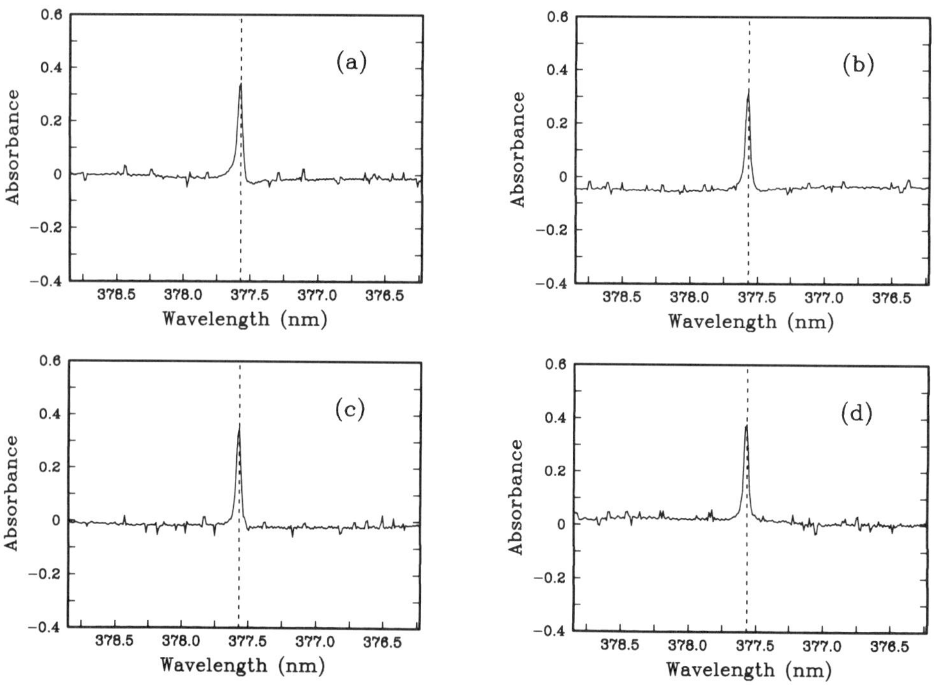

FIGURE 8.20. Continuation of Figure 8.19, with pump-probe separation in (a) increased by 100 fs over that in Figure 8.19 (d).

The most striking feature of Figures 8.18 to 8.20 is, of course, the dispersion-like appearance of the atomic absorption feature for the first 600 or 700 fs, with evidence of a spectral region in which apparent gain prevails. A superficially similar phenomenon was recently observed by Fluegel et al. (1987) in their femtosecond studies of coherent transients in semiconductors. However, in that work the dispersive structure observed in the normalized differential transmission spectra in the region of the exciton resonance, when pump and probe pulses overlapped in time, was attributed to the frequency shift of the exciton resonance, i.e., the optical Stark shift. In the case of Figures 8.18 to 8.20 there is no preexisting absorption line to be shifted when the pump is applied.

The appearance of the absorption spectra when the 308-nm pump intensity is reduced by roughly a factor 3 (0.5 ND filter inserted in the pump arm) is shown in Figures 8.21 and 8.22. From the entire sequence of spectra constituting this particular experiment, we have selected the eight consecutive spectra that most closely correspond with Figures 8.18 and 8.19. (Exact correspon-

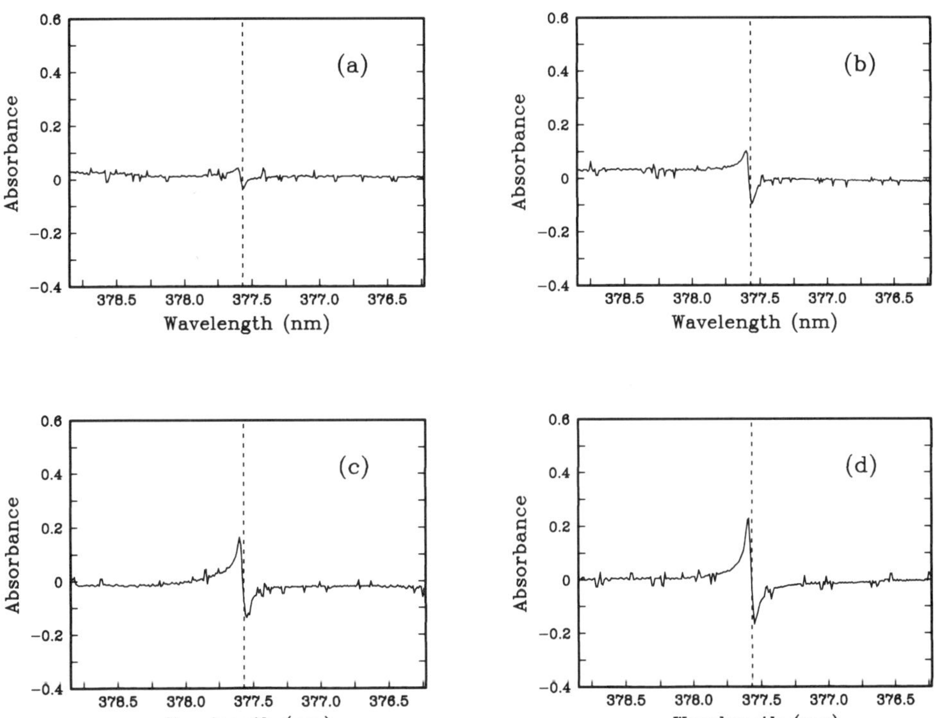

FIGURE 8.21. Time-resolved spectra taken from a sequence with conditions generally similar to those in Figures 8.18 to 8.20, except that the 308-nm pump intensity was reduced by a factor 3. Pump-probe separation increased by 100 fs between spectram.

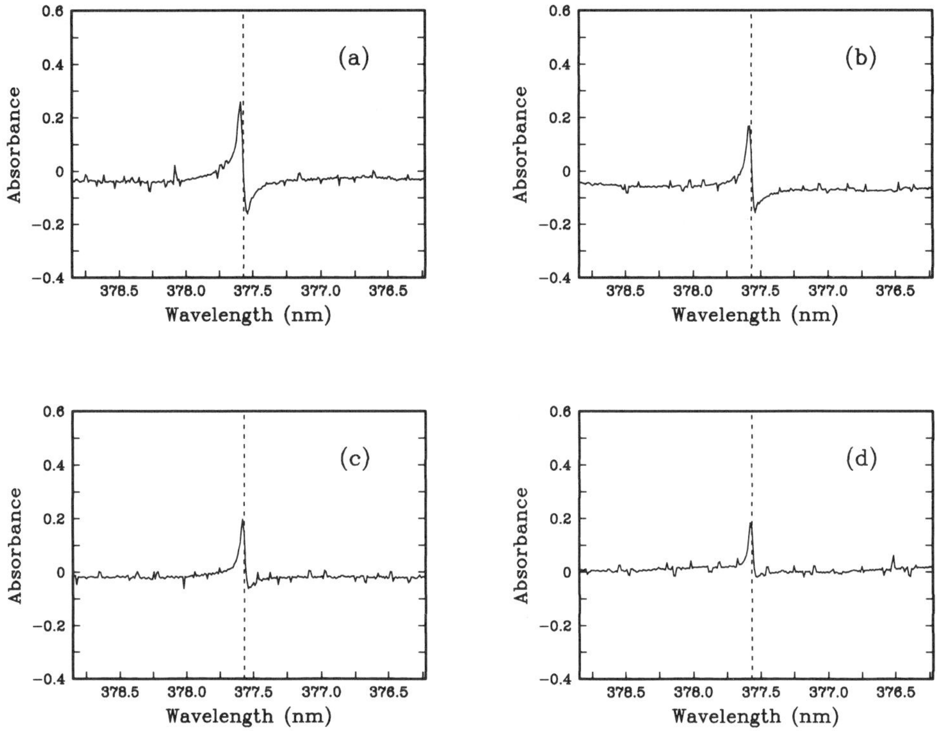

FIGURE 8.22. Continuation of Figure 8.21, with pump-probe separation in (a) increased by 100 fs over that in Figure 8.21 (d).

dence between the pump-probe delays of Figures 8.18 to 8.20 was not possible to maintain because of the insertion of the 0.5 ND filter in the pump arm.) The peak absorbances are clearly saturated. However, the integrated absorbances may not be so, since the spectra in Figures 8.21 and 8.22 are clearly narrower. This line broadening is a feature we have observed in all our Tl-halide spectra, with both 308- and 248-nm pumping and in both TlI and TlCl. Widths of all Tl absorption lines, even those measured at very long pump-probe separations, are dependent on the UV pump intensity applied. A reasonable explanation of this phenomenon is Stark broadening due to creation of ions or electrons in the vapor by the UV pump pulse.

Figures 8.23 to 8.26 show the appearance of the absorption spectra as a function of time in the vicinity of the 351.9-nm $6D_{5/2} \leftarrow 6P_{3/2}$ and 352.9-nm $6D_{3/2} \leftarrow 6P_{3/2}$ absorption lines, following application of a 308-nm, 160-fs excitation pulse to TlI vapor. These transitions thus monitor thallium atoms in the excited $6P_{3/2}$ state (7793 cm^{-1}). Note the absence of any evident spectral region with apparent gain during any part of the sequence. However, there are again strong transiently appearing asymmetries in the two line shapes.

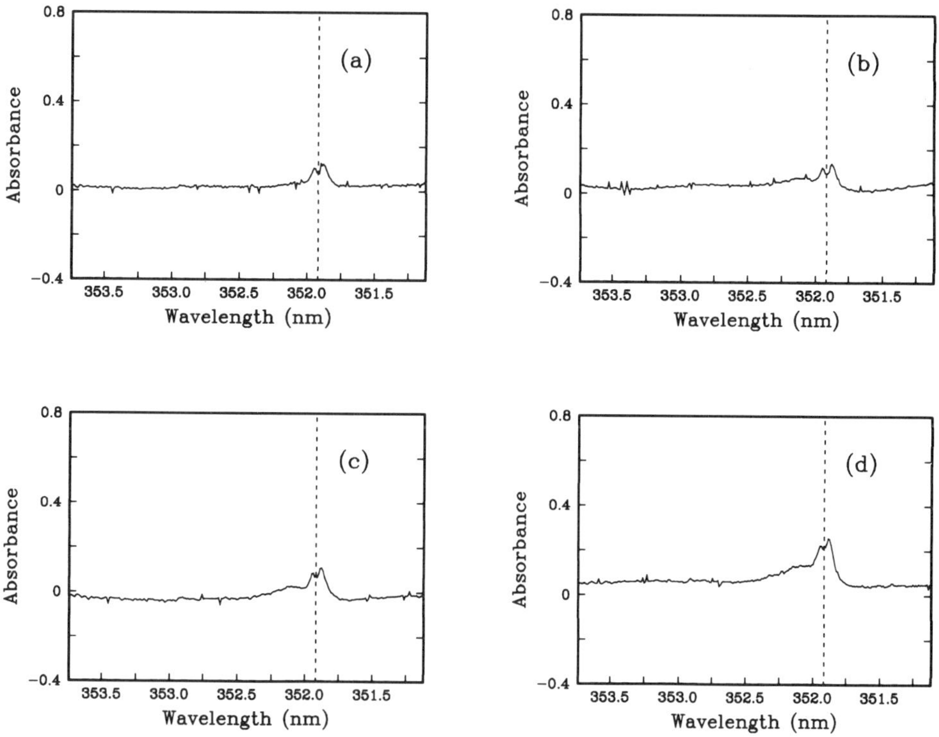

FIGURE 8.23. Time-resolved spectra, recorded in the vicinity of the 351.9-nm $6D_{5/2} \leftarrow 6P_{3/2}$ and 352.9-nm $6D_{3/2} \leftarrow 6P_{3/2}$ Tl absorption lines, following the application of 160-fs, 308-nm pulses to TlI vapor. Pump-probe separation increased by 100 fs between spectra.

Generally similar results were obtained in the case of 248-nm pump excitation of TlI, and with both 248- and 308-nm excitation of TlCl. In the case of TlI excited by a 248-nm pump, the time evolution of the line shapes of the two resonances near 352 and 353-nm is similar to that shown in Figures 8.23 to 8.26, with pronounced red wings during a period of again approximately a picosecond. The 377.6-nm line is much less intense and as a result comparatively difficult to minotor. There appears again to be a transiently appearing negative absorption on the high-frequency side of the line, but its magnitude is much less than the magnitude of the differential positive absorption appearing on the low-frequency side of the line. The latter absorbance monotonically grows to a final value of ~ 0.1 for the same UV pump powers for which the 353-nm absorbance (the weak line in Figures 8.23 to 8.26) almost attains the value 0.4. Clearly, a large inversion on the $6P_{3/2}$–$6P_{1/2}$ transition is produced by 248-nm photolysis of TlI.

For the TlCl, with 248-nm excitation, the 352-nm absorption develops a pronounced *blue* wing, in contrast with the cases discussed above. This blue

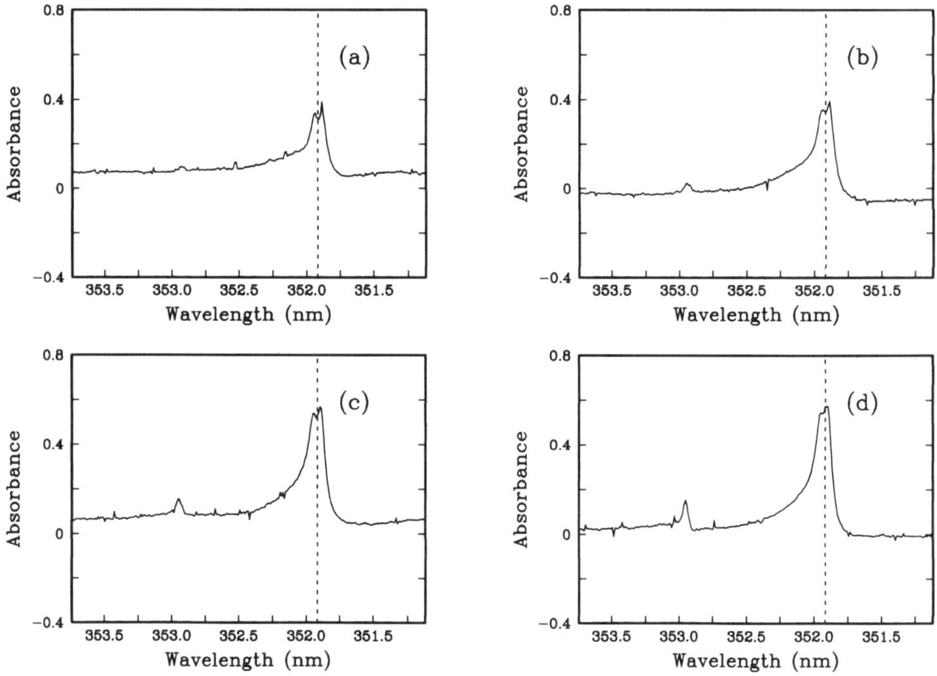

FIGURE 8.24. Continuation of Figure 8.23, with pump-probe separation in (a) increased by 100 fs over that in Figure 8.23 (d).

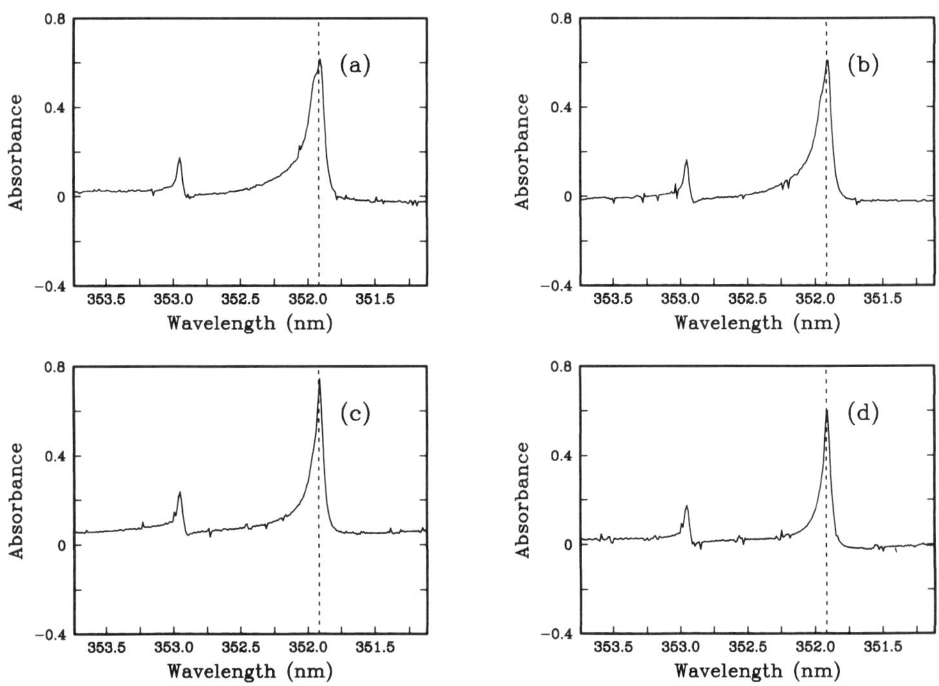

FIGURE 8.25. Continuation of Figure 8.24, with pump-probe separation in (a) increased by 100 fs over that in Figure 8.24 (d).

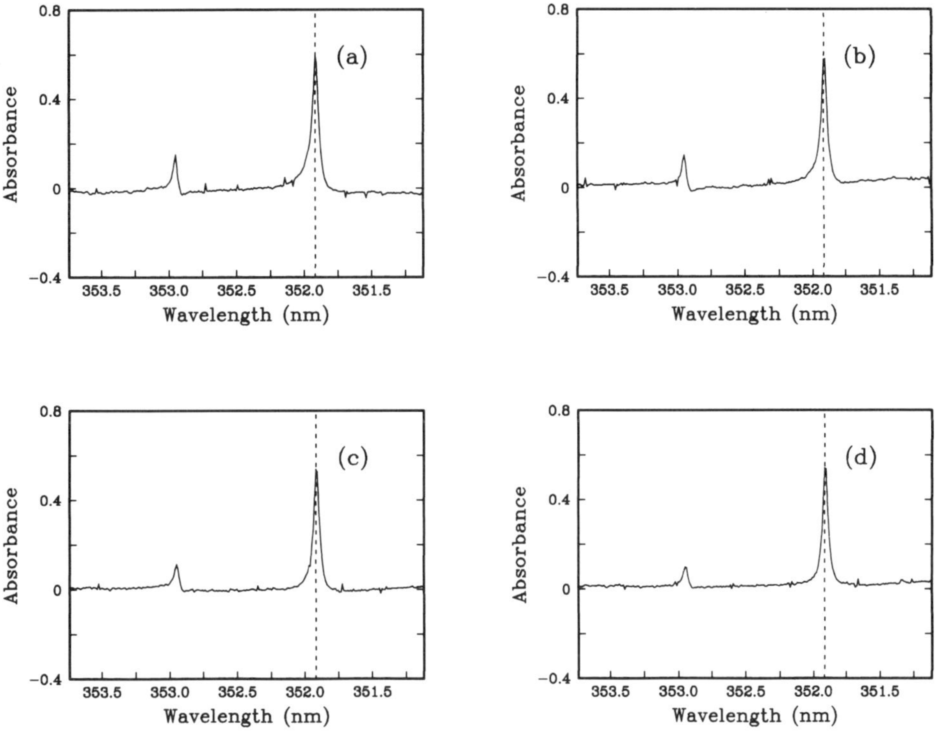

FIGURE 8.26. Continuation of Figure 8.25, with pump-probe separation in (a) increased by 100 fs over that in Figure 8.25 (d).

wing is evident for roughly 1.2 ps, then quickly disappears as the final line shape is assumed. During the 1.2-ps interval, the magnitude of the integrated absorption of the 352-nm line is enhanced by at least a factor 2, relative to that for the asymptotically attained line shape. The 377.6-nm line appears to grow monotonically to its asymptotic value, without noticeable line shape distortions other than a slight negative differential absorption on the high-frequency side of the line. For TlCl with 308-nm pumping, there is *no* observed 352- or 353-nm absorption feature, in agreement with the specific finding of van Veen et al. (1981) that only one dissociative channel (either Tl + Cl or Tl + Cl*) is active when TlCl is pumped at 308 nm. Since this should be the simplest situation to analyze, we present in Figures 8.27 to 8.29 some of the observed spectra for this specific case. A region of negative differential absorption is again clearly seen in some of the spectra (Figures 8.28a–d, Figure 8.29a).

We now present a qualitative explanation for the unusual spectral line shapes observed for the first picosecond following the photolysis pulse. Our model is based on the transient behavior of the polarization induced by the subpicosecond swept UV continuum pulse as the latter interacts with the time-

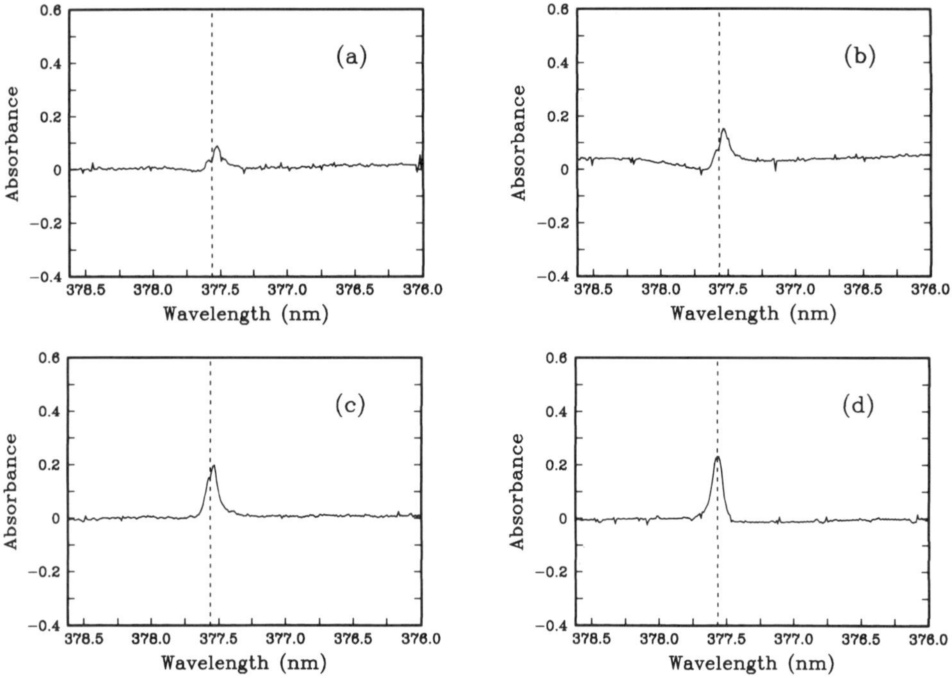

FIGURE 8.27. Time-resolved spectra, recorded near the 377.6-nm Tl line, following application of 160-fs, 308-nm pulses to TlCl vapor. 100-fs steps between successive spectra.

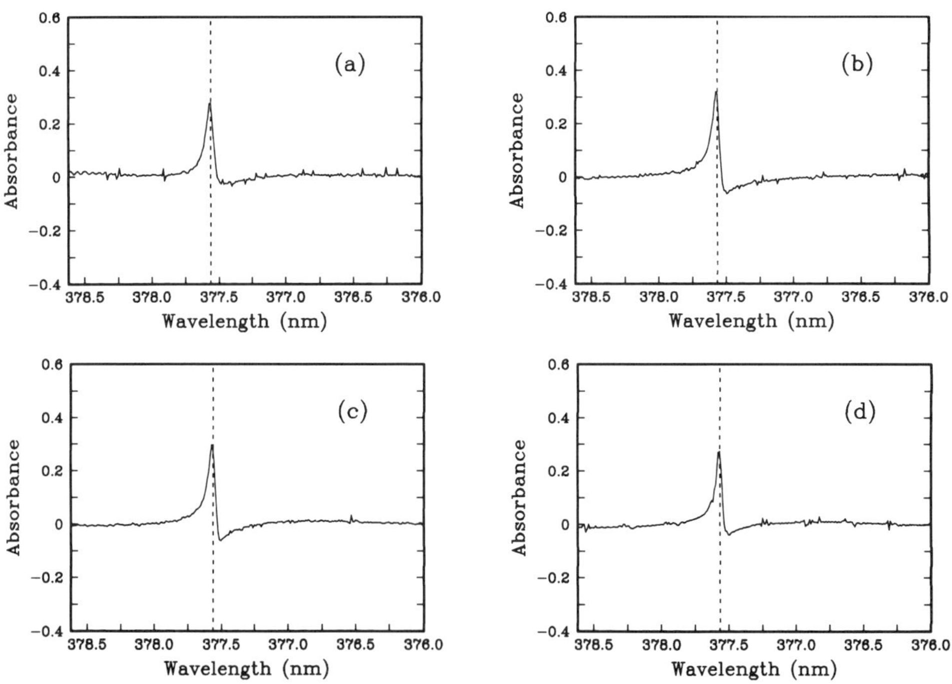

FIGURE 8.28. Continuation of Figure 8.27, with pump-probe separation in (a) increased by 100 fs over that in Figure 8.27 (d).

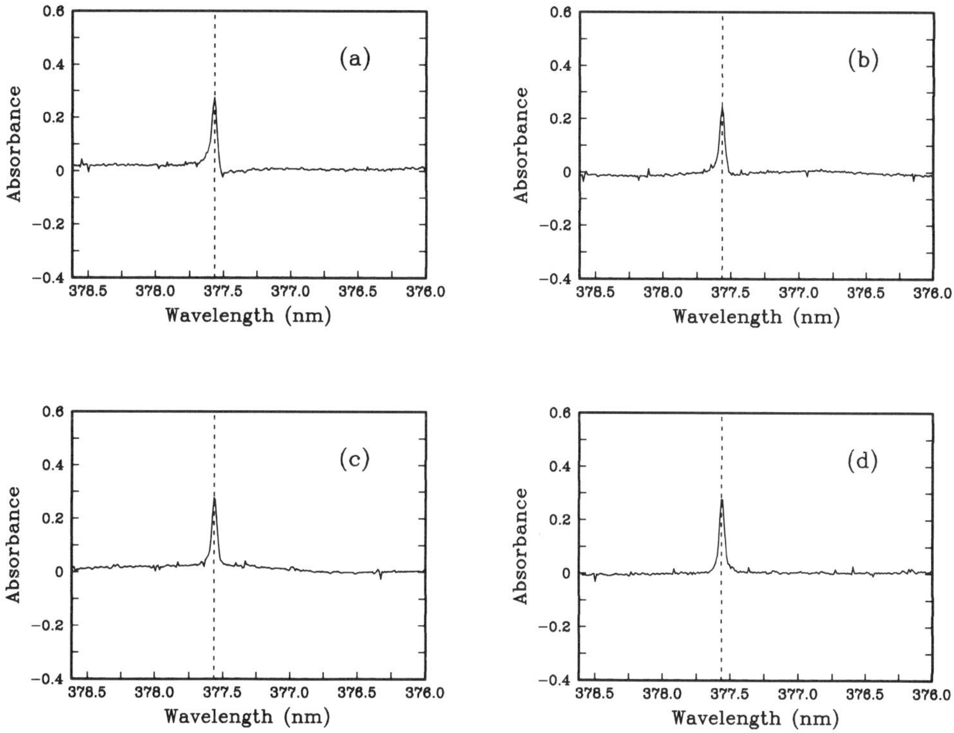

FIGURE 8.29. Continuation of Figure 8.28, with pump-probe separation in (a) increased by 100 fs over that in Figure 8.28 (d).

varying (growing) population of two-level atoms produced by the photolysis pulse. As we will show below, in order to get a reasonable correspondence of the line shapes calculated in this manner with the observed line shapes, especially with those possessing the most unusual features (e.g., Figures 8.18 to 8.20), one has to allow the created atoms to undergo continuous frequency shifts in time for a period after the photolysis pulse.

The numerical calculations we have performed are basically straightforward. An optically thin sample is assumed. There are two contributions to the output field from the resonant vapor, i.e., $E_{\text{tot}}(t, z) = E_{\text{in}}(t, z) + E_{\text{rad}}(t, z)$, where $E_{\text{in}}(t, z)$ is the swept UV continuum and $E_{\text{rad}}(t, z)$ is the field radiated by the polarization it induces in the medium. Various contributions to the latter are numerically calculated, based on solutions of Schrödinger's equation for a two-level atom interacting with a swept continuum pulse

$$\dot{a}_1 = \left\{ \frac{ier_{10}}{2\hbar} \mathscr{E}(t) \exp[i\omega_c z/c] \exp[-i(\omega_c - \omega_a)t] \right\} a_0 - \frac{\gamma}{2} a_1, \tag{3}$$

where a_1 and a_0 are the time-dependent amplitudes of the upper and lower atomic states, and ω_a is the atomic resonance frequency, equal to $\omega_1 - \omega_0$. The rotating-wave approximation has been made in Eq. (3). We assume the swept continuum pulse to have the form

$$E_{\text{in}}(t,z) = \text{Re}\left\{\mathscr{E}(t)\exp\left[-i\omega_c\left(t - \frac{z}{c}\right)\right]\right\}, \tag{4}$$

with

$$\mathscr{E}(t) = \mathscr{E}_0(t)\exp(-ibt^2/2), \tag{5}$$

and

$$\mathscr{E}_0(t) = \mathscr{E}_0[\exp(-t^2/\Delta^2)]^{1/2}. \tag{6}$$

Equations (4) to (6) imply that the instantaneous frequency of the UV continuum sweeps across the frequency ω_c at time $t = 0$. The value of b was taken to correspond with the value we measured for the supercontinuum sweep rate, 1340 cm^{-1}/ps. Equation (6) shows the form of the swept continuum pulse amplitude that we assumed in our numerical integrations. We generally specified a width Δ on the order of a picosecond. The wave function for the atom is

$$\psi = a_0 e^{-i\omega_0 t} u_0 + a_1 e^{-i\omega_1 t} u_1, \tag{7}$$

and the polarization P is generally expressed as

$$P = N\langle\psi| -er|\psi\rangle, \tag{8}$$

where N is the atomic density.

We now outline the general procedure that was followed in obtaining numerical solutions. Let $a_0 = R(t - t_i)$ be the amplitude of the lower state for an atom created at time t_i. The solution to Eq. (3) can be formally expressed as

$$a_1(t) = \frac{ier_{10}}{2\hbar} e^{-\gamma t/2} \int_{-\infty}^{t} \mathscr{E}(t')\exp[i\omega_c z/c]\exp[-i(\omega_c - \omega_a)t']e^{\gamma t'/2} R(t' - t_i)\, dt'. \tag{9}$$

The contribution to the polarization (per atom) is

$$p(t, t_i) = -2e\,\text{Re}\{r_{01} a_0^* a_1 e^{-i\omega_a t}\}. \tag{10}$$

Equation (9) was numerically integrated with the use of a specific rise function

$$R(t' - t_i) = \frac{1}{2}\left[1 + \tanh\left(\frac{t' - t_i}{\text{WTANH}}\right)\right]. \tag{11}$$

The numerical integration in Eq. (9) was combined with an additional integration over another distribution function: $D(t) = dN(t)/dt$, where $N(t)$ represents the atomic population. We specified $D(t)$ to be proportional to the quantity $\text{sech}^2[(t - \text{TSECH2})/\text{WSECH2}]$. Here the quantity TSECH2 marks the time at which the growth rate of $N(t)$ achieves its maximum, and WSECH2 characterizes the width of the growth period. Thus the total polarization P is given by

$$P(z,t) = \int_{-\infty}^{\infty} p(t,t_i)D(t_i)\,dt_i. \tag{12}$$

We are modeling the dissociation of a diatomic molecule to create two atoms. At early times in the dissociation, when the two atoms are close to one another, the atomic transition frequencies are perturbed by the bending of the potential surfaces. We allowed for the existence of a continuous red (or blue) shift of the atomic transition frequency by making ω_a in the above equations a function of time relative to the creation of the atom:

$$\omega_a(t,t_i) = \omega_a^0 - \mathrm{RMAX}\left[\exp\left(-\frac{t-t_i}{\mathrm{RTAU}}\right)\right], \tag{13}$$

where ω_a^0 is the unperturbed atomic transition frequency.

In the slowly varying envelope approximation (SVEA), it is assumed that the total field $E_{\mathrm{tot}}(t,z)$ and polarization $P(t,z)$ can be written in the following forms:

$$E_{\mathrm{tot}}(t,z) = \mathrm{Re}\left\{\mathscr{E}(t,z)\exp\left[-i\omega_c\left(t-\frac{z}{c}\right)\right]\right\} \tag{14}$$

and

$$P(t,z) = \mathrm{Re}\left\{\mathscr{P}(t,z)\exp\left[-i\omega_c\left(t-\frac{z}{c}\right)\right]\right\}, \tag{15}$$

where $\mathscr{E}(t,z)$ and $\mathscr{P}(z,t)$ are complex functions of z and t that vary little in an optical period or wavelength. Following the usual procedure of neglecting second derivatives of the slowly varying quantities $\mathscr{E}(t,z)$ and $\mathscr{P}(t,z)$, one obtains on substitution of (14) and (15) into the wave equation the well-known complex field self-consistency equation

$$\frac{\partial\mathscr{E}(z,t)}{\partial z} + \frac{1}{c}\frac{\partial\mathscr{E}(z,t)}{\partial t} = \frac{i\omega_c}{2c\varepsilon_0}\mathscr{P}(z,t). \tag{16}$$

If one defines a retarded time $\tau = t - z/c$, one can rewrite Eq. (16) as

$$\frac{\partial\mathscr{E}(z,\tau)}{\partial z} = \frac{i\omega_c}{2c\varepsilon_0}\mathscr{P}(z,\tau). \tag{17}$$

From Eqs. (9), (10), (12), and (15), $\mathscr{P}(z,\tau)$ is seen to have no explicit dependence on z: $\mathscr{P}(z,\tau) = \mathscr{P}(\tau)$. Thus we can integrate Eq. (17) to obtain

$$\mathscr{E}(z,\tau) = \mathscr{E}(0,\tau) + \frac{i\omega_c}{2c\varepsilon_0}\mathscr{P}(\tau)z. \tag{18}$$

Hence, in general,

$$E_{\mathrm{tot}}(z,t) = E_{\mathrm{in}}(z,t) + \frac{i\omega_c z}{2c\varepsilon_0}P(z,t), \tag{19}$$

with the last term of Eq. (19) representing the field $E_{\mathrm{rad}}(z,t)$ radiated by the polarization induced in the medium by $E_{in}(z, t)$. The spectral dependence of the total field $E_{tot}(z, t)$ is given by the sum of the Fourier transforms $\tilde{\mathscr{E}}_{in}(\omega)$ and $\tilde{\mathscr{E}}_{rad}(\omega)$, and the spectral dependence of the collected intensity at the end of the vapor cell is given by

$$I_{out}(\omega) = |\tilde{\mathscr{E}}_{in}(\omega)|^2 + |\tilde{\mathscr{E}}_{rad}(\omega)|^2 + 2\mathrm{Re}\{\tilde{\mathscr{E}}^*_{in}(\omega)\tilde{\mathscr{E}}_{rad}(\omega)\}. \tag{20}$$

With the use of fast Fourier transform numerical computation techniques, and with the quantity ω_c in Eq. (9) set equal to ω_a^0, we obtained computer plots of the quantity $\log_{10}[I_{out}(\omega)/|\tilde{\mathscr{E}}_{in}(\omega)|^2]$ for various choices of the parameters introduced above.

An example of a sequence of calculated spectra is shown in Figures 8.30 to 8.32. The parameters used here were chosen in an attempt to approximate the shape of the observed spectra shown in Figures 8.18 to 8.20. It is seen that a qualitative agreement exists between theory and experiment, with the particular observed feature of a transient spectral region of negative absorbance clearly captured in the calculated spectra. In Figures 8.30 to 8.32, the quantity

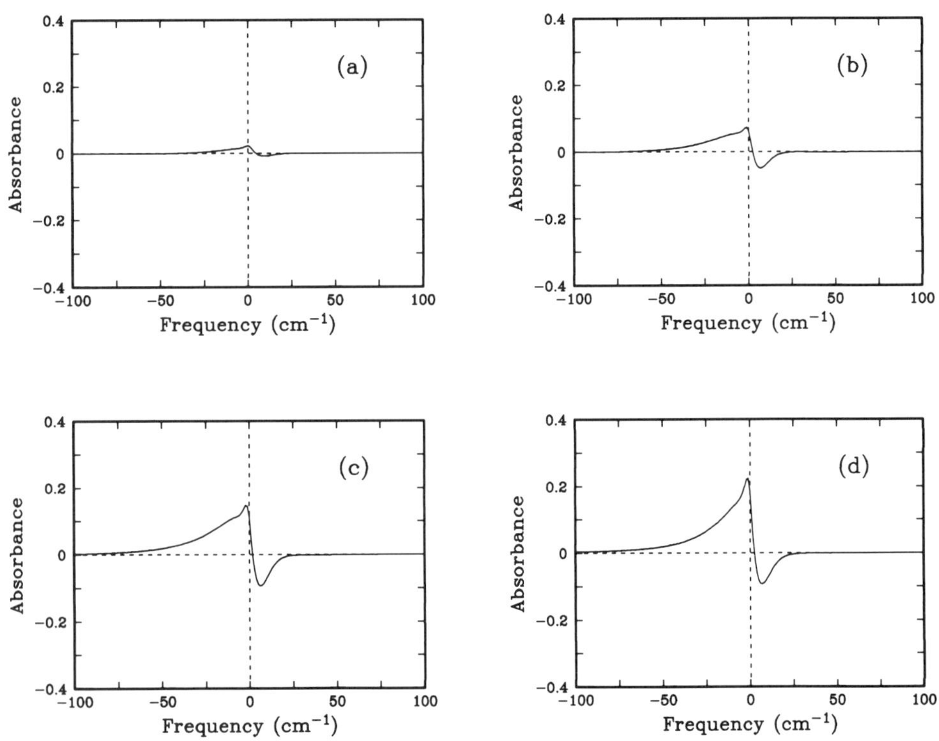

FIGURE 8.30. Calculated transient absorption spectra for a sequence of pump-probe separations increasing by 100 fs between successive spectra. The following parameter values were used: $\gamma = 0.0015$ fs^{-1}, WSECH2 = 150 fs, WTANH = 50 fs, RMAX = 40 cm^{-1}, RTAU = 600 fs.

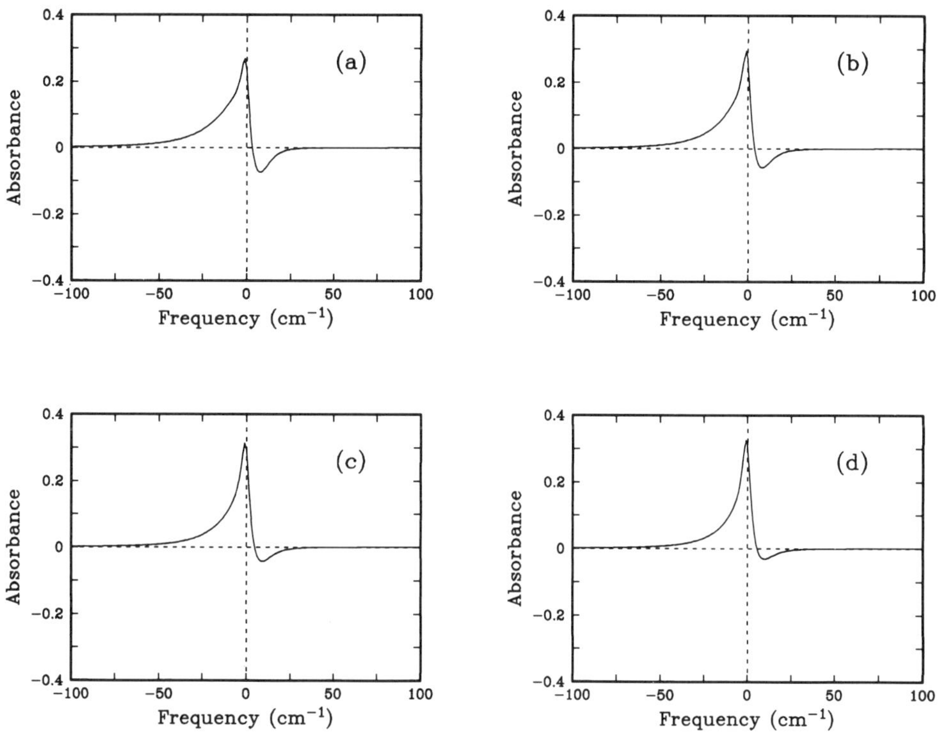

FIGURE 8.31. Continuation of Figure 8.30, with pump-probe separation in (a) increased by 100 fs over that in Figure 8.30 (d).

γ was chosen to correspond to a polarization dephasing time of $T_2 = 1.3$ ps. This value results in an asymptotic atomic linewidth (~ 6 cm^{-1} FWHM) that closely approximates the measured value.

Several general conclusions can be drawn by examining the shapes of the various calculated spectra. Most important, *without the inclusion of a red shift RMAX very little asymmetry appears in the spectra, and there is no significant negative absorption.*

The rise time of the absorption seems to be simply related to the quantity WSECH2, for all values of RMAX.

For a given T_2, even for a relatively large RMAX (e.g., 60 cm^{-1}), decreasing RTAU below $T_2/10$ has the effect of reducing the height of the transiently appearing wings, so that the spectra are dominated at all times by a symmetric peak centered at ω_a^0. To get more pronounced wings and a greater asymmetry for a given RMAX, one has to increase RTAU. However, if RTAU becomes comparable to T_2, additional oscillatory half-cycles of absorption and gain begin to appear in the absorption spectrum. For RTAU $\gg T_2$, a simple growth of the absorption line at a frequency ω_a^0 − RMAX is observed.

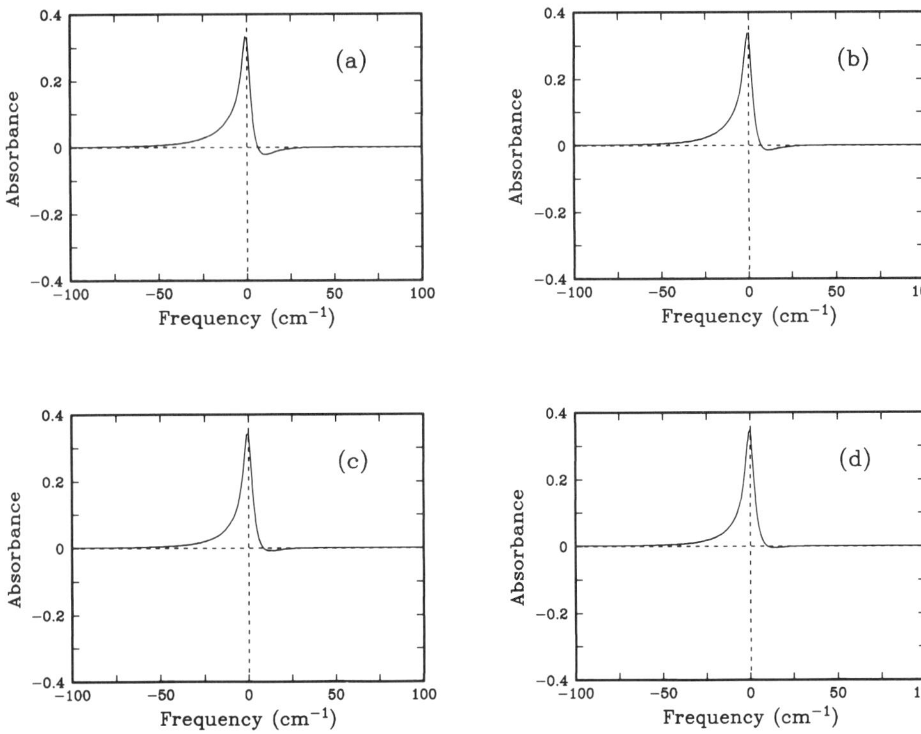

FIGURE 8.32. Continuation of Figure 8.31, with pump-probe separation in (a) increased by 100 fs over that in Figure 8.31 (d).

A blue atomic resonance frequency shift reverses the asymmetry, producing transient negative absorption in the region $\omega < \omega_a^0$. Changing the direction of the probe continuum sweep, however, does not appear to affect the appearance of the absorption spectrum, at least with the use of the b value appropriate for our case.

We generally used a single value (50 fs) for WTANH. The spectra were seen to be generally insensitive to the choice of this parameter, provided it was taken to be short enough.

To summarize very briefly, it appears that the transient absorption spectra we have obtained of photolytically produced atoms contain qualitative information regarding the "transition state" that occurs between the time a molecule has absorbed a UV photon and the time its constituent atoms have fully separated. Specifically, information can be obtained about the rise time of the atomic population and the frequency shift that the atom undergoes during dissociation. However, information about the latter tends to be diluted by the polarization dephasing time T_2 of the separating atoms.

We now describe a brief result obtained by applying the UV super-

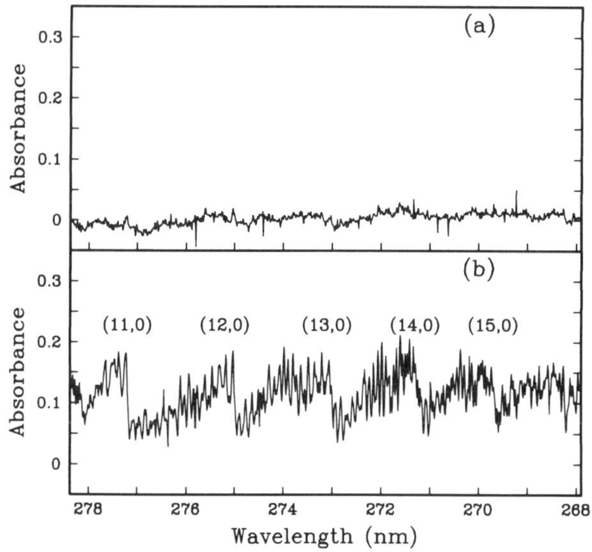

FIGURE 8.33. (a) Single-pass absorbance spectrum recorded through a 10-cm cell filled with 5-torr OClO with probe pulse occurring at the end of a 25-ns, 308-nm photolysis pulse. Pump energy of 3 mJ was sent through a 3-mm aperture. (b) Same as (a) but recorded 900 ns after the photolysis pulse.

continuum to the gas-phase molecule chlorine dioxide, OClO. For several decades it has been assumed that the sole result of applying light to the structured OClO ($\tilde{A}^2A_2 \leftarrow \tilde{X}^2B_1$) band system, which extends from ~460 to 280 nm, is the photolytic production of ClO + O (Bethune et al., 1983). From OClO absorption spectra taken at very high spectral resolution, in which the widths of individual vibronic lines could be measured, it was inferred that OClO predissociates at a rate strongly dependent on how far above the origin one photoexcites the molecule (Michielson et al., 1981).

As an initial test and possible calibration of the UV gas-phase supercontinuum, we attempted an experiment to photolyze OClO with a 160-fs, 308-nm UV pulse and then probe the appearance of the ClO radical with the UV supercontinuum pulse. The ClO radical strongly absorbs from ~310 to 250 nm, a region largely nonoverlapping with the OClO absorption band. With 308-nm excitation, subpicosecond predissociation rates for OClO were expected.

When the above spectral region was probed, no ClO was detected, with pump-probe delays of up to a nanosecond tried. The 160-fs, 308-nm excitation pulse was then replaced by a standard 25-ns, 308-nm pulse with the same 0.04 J/cm^2 fluence. (We sent 3 mJ through a 3-mm aperture into a 10-cm cell containing 5 torr OClO.) Again, no significant absorption was detected at relatively short delays (Figure 8.33a). However, if the probe was delayed by several hundred nanoseconds, the fully developed ClO spectrum was easily observed (Figure 8.33b). We thus conclude that 308-nm photolysis of OClO produces Cl + O_2^*, with O_2^* most likely being in high vibrational levels of the $^1\Sigma_g^+$ state. The observed slow appearance of ClO results from the combination of Cl atoms with unphotolyzed parent OClO molecules

$$\mathrm{Cl} + \mathrm{OClO} \rightarrow 2\mathrm{ClO}. \tag{21}$$

For 5 torr of unphotolyzed OClO present in the cell and with use of the known 300 K rate constant for (21), 6×10^{-11} cm^3 molec^{-1} s^{-1}, one predicts the ClO would appear in $\sim$100 ns, approximately what was observed. This OClO result should have a large impact on further understanding the interesting sequence of chemical reactions initiated by photolysis of OClO (Bethune et al., 1983).

7. Promising Directions for Subpicosecond Kinetic Spectroscopy

In the Introduction we presented an example of a unimolecular photochemical reaction, the rearrangement of methyl azide (Figure 8.1). The example was primarily meant to be illustrative. The IR subpicosecond continua required to probe the most relevant vibrational transitions have not yet been developed. However, through the use of the UV subpicosecond gas-phase continuum discussed in the preceding sections, an alternative method of attacking this problem, and many similar problems, now appears to be possible. Specifically, it should now be possible to observe momentarily in the UV the singlet spectrum of methylnitrene (CH_3N) and then to watch it disappear as the molecule isomerizes to singlet methylenimine (CH_2NH). Demuynck et al. (1980) have predicted little or no barrier for isomerization from singlet CH_3N to singlet CH_2NH, but they also predict a sizable barrier (53 kcal/mol) for isomerization of triplet CH_3N. In recent years, searches for optical spectra of CH_3N were made in several photolysis and pyrolysis studies, but no CH_3N was ever observed in any of these studies, most likely because of rapid isomerization of singlet CH_3N to CH_2NH. (By the spin conservation selection rule, photolysis of methyl azide has to produce singlet nitrene, since the ground state of N_2 is a singlet.) Recently, the $\tilde{A}^3E$–$\tilde{X}^3A_2$ ultraviolet emission spectrum of triplet CH_3N was observed (Carrick and Engelking, 1984; Franken et al., 1970) by reacting methyl azide with metastable ($A^3\Sigma_u^+$) N_2 in a flowing afterglow. This result qualitatively confirms the prediction of a high triplet isomerization barrier height made by Demuynck et al. (1980). The triplet CH_3N (0, 0) band occurs at 314.3 nm, not too distant from the 336-nm origin of the $A^3\Pi - X^3\Sigma^-$ band system of the isoelectronic radical NH. Since NH has also an allowed singlet system ($c^1\Pi$–$a^1\Delta$), with origin at 324 nm, one should expect an analogous singlet CH_3N system to exist, with an origin somewhere in the region of 300 nm. Methyl azide can be photolyzed at either

248 or 308 nm. With the effective good time resolution of the swept UV supercontinuum demonstrated in Section 6, one should be able to observe the transient singlet band system of CH_3N, even if the latter isomerizes in a time of half a picosecond.

We conclude by mentioning one improvement in the subpicosecond kinetic spectroscopy technique we employ that is scheduled to be tried soon. This is the incorporation of a reference arm for improved sensitivity. With the use of two matched spectrograph-OMA systems, it should be possible to cancel in every shot the effect of random spectral variations in the supercontinuum intensity. At present, we require two independent accumulations of several tens of shots, one with the pump blocked and one with the pump unblocked, from which the spectral variation of absorbance is electronically calculated. Incorporation of such a reference arm is, of course, simpler with the UV supercontinuum than in the case of the IR continua, due to the need to upconvert the latter. It will be interesting to see how much subpicosecond kinetic spectroscopy can benefit from such an improvement once it is implemented.

Acknowledgments. We are indebted to J.E. Rothenberg for suggesting to us an initial conceptual framework for the calculations. We also thank M.M.T. Loy and D. Grischkowsky for helpful discussions. L. Manganaro assisted us with some of the instrumentation. This research was sponsored in part by the U.S. Army Research Office.

References

Akhmanov, S.A., K.N. Drabovich, A.P. Sukhorukov, and A.K. Shchednova (1972) Combined effects of molecular relaxation and medium dispersion in stimulated Raman scattering of ultrashort light pulses. Sov. Phys. JETP **35**, 279–286.

Alfano, R.R. and S.L. Shapiro (1970a) Emission in the region 4000 to 7000 Å via four-photon coupling in glass. Phys. Rev. Lett. **24**, 584–587.

Alfano, R.R. and S.L. Shapiro (1970b) Observation of self-phase modulation and small scale filaments in crystals and glasses. Phys. Rev. Lett. **24**, 592–594.

Avouris, Ph., D.S. Bethune, J.R. Lankard, J.A. Ors, and P.P. Sorokin (1981) Time-resolved infrared spectral photography: study of laser-initiated explosions in HN_3. J. Chem. Phys. **74**, 2304–2312.

Bethune, D.S., J.R. Lankard, P.P. Sorokin, R.M. Plecenik, and Ph. Avouris (1981) Time-resolved infrared study of bimolecular reactions between *tert*-butyl radicals. J. Chem. Phys. **75**, 2231–2236.

Bethune, D.S., A.J. Schell-Sorokin, J.R. Lankard, M.M.T. Loy, and P.P. Sorokin (1983) Time-resolved study of photo-induced reactions of chlorine dioxide. In B.A. Garetz and J.R. Lombardi (eds.), *Advances in Laser Spectroscopy*, vol. 2, pp. 1–43. Wiley, New York.

Bucksbaum, P.A., J. Bokor, R.H. Storz, and J.C. White (1982) Amplification of ultra-short pulses in krypton fluoride at 248 nm. Opt. Lett. **7**, 399–401.

Burnham, R. and N. Djeu (1978) Efficient Raman conversion of XeCl-laser radiation in metal vapors. Opt. Lett. **3**, 215–217.

Carman, R.L., F. Shimizu, C.S. Wang, and N. Bloembergen (1970) Theory of Stokes pulse shapes in transient stimulated Raman scattering. Phys. Rev. A **2**, 60–72.

Carrick, P.G. and P.C. Engelking (1984) The electronic emission spectrum of methylnitrene. J. Chem. Phys. **81**, 1661–1665.

Corkum, P.B. and R.S. Taylor (1982) Picosecond amplification and kinetic studies of XeCl. IEEE J. Quantum Electron. **QE-18**, 1962–1975.

Corkum, P.B., C. Rolland, and T. Srinivasan-Rao (1986a) Supercontinuum generation in gases: a high order nonlinear optics phenomenon. In G.R. Fleming and A.E. Siegman (eds.), *Ultrafast Phenomena V*, pp. 149–152. Springer-Verlag, New York.

Corkum, P.B., C. Rolland, and T. Srinivasan-Rao (1986b) Supercontinuum generation in gases. Phys. Rev. Lett. **57**, 2268–2271.

Cotter, D. and W. Zapka (1978) Efficient Raman conversion of XeCl excimer laser radiation in Ba vapour. Opt. Commun. **26**, 251–255.

Dantus, M., M.J. Rosker, and A.H. Zewail (1987) Real-time femtosecond probing of "transition states" in chemical reactions. J. Chem. Phys. **87**, 2395–2397.

Davidovits, P. and J.A. Bellisio (1969) Ultraviolet absorption cross sections for the thallium halide and silver halide vapors. J. Chem. Phys. **50**, 3560–3567.

Demuynck, J., D.J. Fox, Y. Yamaguchi, and H.F. Schaefer III (1980) Triplet methyl nitrene: an indefinitely stable species in the absence of collisions. J. Am. Chem. Soc. **102**, 6204–6207.

Egger, H., T.S. Luk, K. Boyer, D.F. Muller, H. Pummer, T. Srinivasan, and C.K. Rhodes (1982) Picosecond, tunable ArF* excimer laser source. Appl. Phys. Lett. **41**, 1032–1034.

Fluegel, B., N. Peyghambarian, G. Olbright, M. Lindberg, S.W. Koch, M. Joffre, D. Hulin, A. Migus, and A. Antonetti (1987) Femtosecond studies of coherent transients in semiconductors. Phys. Rev. Lett. **59**, 2588–2591.

Fork, R.L., B.I. Greene, and C.V. Shank (1981) Generation of optical pulses shorter than 0.1 psec by colliding pulse mode locking. Appl. Phys. Lett. **38**, 671–672.

Fork, R.L., C.V. Shank, R.T. Yen, and C. Hirlimann (1982) Femtosecond continuum generation. In K.B. Eisenthal, R.M. Hochstrasser, W. Kaiser, and A. Laubereau (eds.), *Picosecond Phenomena III*, pp. 10–13. Springer-Verlag, New York.

Fork, R.L., C.V. Shank, C. Hirlimann, R. Yen, and W.J. Tomlinson (1983) Femtosecond white-light continuum pulses. Opt. Lett. **8**, 1–3.

Franken, Th., D. Perner, and M.W. Bosnali (1970) UV-absorptionsspektren von methyl- und äthylnitren mittels pulsradiolyse in der gasphase. Z. Naturforsch. A **25**, 151–152.

Glownia, J.H., G. Arjavalingam, and P.P. Sorokin (1985) The potential of DABCO for two-photon amplification. J. Chem. Phys. **82**, 4086–4101.

Glownia, J.H., J. Misewich, and P.P. Sorokin (1986a) Ultrafast ultraviolet pump-probe apparatus. J. Opt. Soc. Am. B **3**, 1573–1579.

Glownia, J.H., G. Arjavalingam, P.P. Sorokin, and J.E. Rothenberg (1986b) Amplification of 350-fsec pulses in XeCl excimer gain modules. Opt. Lett. **11**, 79–81.

Glownia, J.H., J. Misewich, and P.P. Sorokin (1986c) New excitation and probe continuum sources for subpicosecond absorption spectroscopy. In G.R. Fleming and A.E. Siegman (eds.), *Ultrafast Phenomena V*, pp. 153–156. Springer-Verlag, New York.

Glownia, J.H., J. Misewich, and P.P. Sorokin (1986d) Amplification in a XeCl excimer gain module of 200-fsec UV pulses derived from a colliding pulse mode locked (CPM) laser system. Proc. Soc. Photo-Opt. Instrum. Eng. **710**, 92–98.

Glownia, J.H., J. Misewich, and P.P. Sorokin (1987a) Subpicosecond time-resolved infrared spectral photography. Opt. Lett. **12**, 19–21.

Glownia, J.H., J. Misewich, and P.P. Sorokin (1987b) 160-fsec XeCl excimer amplification system. J. Opt. Soc. Am. B **4**, 1061–1065.

Glownia, J.H., J. Misewich, and P.P. Sorokin (1987c) Subpicosecond IR transient absorption spectroscopy: measurement of internal conversion rates in DABCO vapor. Chem. Phys. Lett. **139**, 491–495.

Halpern, A.M., J.L. Roebber, and K. Weiss (1968) Electronic structure of cage amines: absorption spectra of triethylenediamine and quinuclidine. J. Chem. Phys. **49**, 1348–1357.

Hamada, Y., A.Y. Hirikawa, and M. Tsuboi (1973) The structure of the triethylenediamine molecule in an excited electronic state. J. Mol. Spectrosc. **47**, 440–456.

Hartmann, H.-J. and A. Laubereau (1984) Transient infrared spectroscopy on the picosecond time-scale by coherent pulse propagation. J. Chem. Phys. **80**, 4663–4670.

Li, Q.X., T. Jimbo, P.P. Ho, and R.R. Alfano (1986) Temporal distribution of picosecond super-continuum generated in a liquid measured by a streak camera. Appl. Opt. **25**, 1869–1871.

Mack, M.E., R.L. Carman, J. Reintjes, and N. Bloembergen (1970) Transient stimulated rotational and vibrational Raman scattering in gases. Appl. Phys. Lett. **16**, 209–211.

Margenau, H. (1939) Van der Waals forces. Rev. Mod. Phys. **11**, 1–35.

Michielson, S., A.J. Merer, S.A. Rice, F.A. Novak, K.F. Freed, and Y. Hamada (1981) A study of the rotational state dependence of predissociation of a polyatomic molecule: the case of ClO_2. J. Chem. Phys. **74**, 3089–3101.

Misewich, J., J.H. Glownia, and P.P. Sorokin (1988a) Measurement with subpicosecond resolution of the frequency sweep of an ultrashort supercontinuum. In *Conference on Lasers and Electro-Optics Technical Digest Series 1988*, vol. 7, pp. 420–421. Optical Society of America, Washington, D.C.

Misewich, J., J.H. Glownia, J.E. Rothenberg, and P.P. Sorokin (1988b) Subpicosecond UV kinetic spectroscopy; Photolysis of thallium halide vapors. Chem. Phys. Lett. **150**, 374–379.

Nakatsuka, H., D. Grischkowsky, and A.C. Balant (1981) Nonlinear picosecond-pulse propagation through optical fibers with positive group velocity dispersion. Phys. Rev. Lett. **47**, 910–913.

Nikolaus, B. and D. Grischkowsky (1983) 90-fsec tunable optical pulses obtained by two-stage pulse compression. Appl. Phys. Lett. **43**, 228–230.

Parker, D.H. and Ph. Avouris (1978) Multiphoton ionization spectra of two caged amines. Chem. Phys. Lett. **53**, 515–520.

Parker, D.H. and Ph. Avouris (1979) Multiphoton ionization and two-photon fluorescence excitation spectroscopy of triethylenediamine. J. Chem. Phys. **71**, 1241–1246.

Schwarzenbach, A.P., T.S. Luk, I.A. McIntyre, V. Johann, A. McPherson, K. Boyer, and C.K. Rhodes (1986) Subpicosecond KrF* excimer-laser source. Opt. Lett. **11**, 499–501.

Smith, M.A., J.W. Hager, and S.C. Wallace (1984) Two-color laser photoionization spectroscopy in a collisionless free-jet expansion: spectroscopy and excited-state dynamics of diazabicyclooctane. J. Phys. Chem. **88**, 2250–2255.

Szatmári, S. and F.P. Schäfer (1983) Simple generation of high-power, picosecond, tunable excimer laser pulses. Opt. Commun. **48**, 279–283.

Szatmári, S. and F.P. Schäfer (1984a) Generation of intense, tunable ultrashort pulses in the ultraviolet using a single excimer pump laser. In D.H. Auston and K.B. Eisenthal (eds.), *Ultrafast Phenomena IV*, pp. 56–59. Springer-Verlag, New York.

Szatmári, S. and F.P. Schäfer (1984b) Excimer-laser-pumped psec-dye laser. Appl. Phys. B **33**, 95–98.

Szatmári, S., B. Racz, and F.P. Schäfer (1987a) Bandwidth limited amplification of 220 fs pulses in XeCl. Opt. Commun. **62**, 271–276.

Szatmári, S., F.P. Schäfer, E. Müller-Horsche, and W. Mückenheim (1987b) Hybrid dye-excimer laser system for the generation of 80 fs, 900 GW pulses at 248 nm. Opt. Commun. **63**, 305–309.

Valdmanis, J.A., R.L. Fork, and J.P. Gordon (1985) Generation of optical pulses as short as 27 femtoseconds directly from a laser balancing self-phase modulation, group-velocity dispersion, saturable absorption, and saturable gain. Opt. Lett. **10**, 131–133.

van Veen, N.J.A., M.S. deVries, T. Baller, and A.E. deVries (1981) Photofragmentation of thallium halides. Chem. Phys. **55**, 371–384.

Xu, Z., B. Koplitz, S. Buelow, D. Bauch, and C. Wittig (1986) High-resolution kinetic energy distributions via Doppler shift measurements. Chem. Phys. Lett. **127**, 534–540.

9
Applications of Supercontinuum: Present and Future

R. DORSINVILLE, P.P. HO, J.T. MANASSAH, and R.R. ALFANO

1. Introduction

Over the past two decades the supercontinuum light source has been extensively used in laser spectroscopy research such as inverse Raman scattering (Alfano and Shapiro, 1971), time-resolved induced absorption in liquids and solids (Alfano and Shapiro, 1971; Greene et al., 1978), primary vision processes (Doukas et al., 1980), energy transfer mechanisms in photosynthesis (Searle et al., 1978; hot carrier and exciton relaxation processes in semiconductors (von der Linde and Lambrich, 1979; Shank et al., 1982), and time-resolved multiplex coherent anti-Stokes Raman scattering (Goldberg, 1982), to name a few.

In this chapter, we review some of the applications in time-resolved absorption and excitation spectroscopy and pulse compression and propose several new possible applications using supercontinuum light for ranging, imaging, remote sensing, and computation. It is impossible to review all the research involved using the supercontinuum. Only a small sample was selected, reflecting the interests of the authors. We apologize in advance to those whose work is not mentioned.

2. Time-Resolved Absorption Spectroscopy

The ultrafast supercontinuum pump-and-probe absorption technique has been used to study ultrafast relaxation processes in solid-state physics, chemistry, and biological systems. An intense pulse photoexcites the sample into an excited state. The dynamics of the excited state is followed using the white light supercontinuum at subsequent delay times. Optically thin samples are usually pumped by a pulse $I(t)$ at wavelength λ, and the induced transient optical density is observed at various wavelengths at time τ (delay time between pump and probe pulses). The OD can be written (Greene et al., 1978; Gayen et al., 1987) as

$$\mathrm{OD}(\tau, \lambda) = -\ln\left\{\int I(t-\tau)\exp[-\sigma L N_1(t)]\,dt \Big/ \int I(t)\,dt\right\}, \tag{1}$$

where σ is the excited state absorption cross section at λ, L is the sample thickness, and $N_1(t)$ is the instantaneous population density in the probing lower-lying energy level. Polarization selection in time can be investigated using polarized pump and probe pulses.

A typical experimental setup for the pump and probe is shown in Figure 9.1. The arrival time of the pump pulse to the sample can be adjusted with an optical delay line. Transient absorption change in the sample at various wavelengths are measured and processed with a spectrograph and an optical

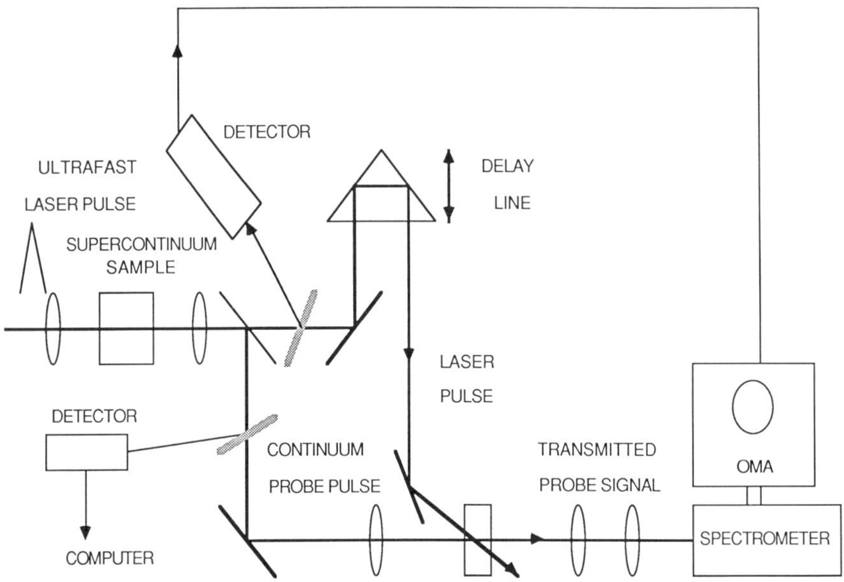

FIGURE 9.1. Schematic of an experimental setup for measuring ultrafast relaxation kinetics using the supercontinuum.

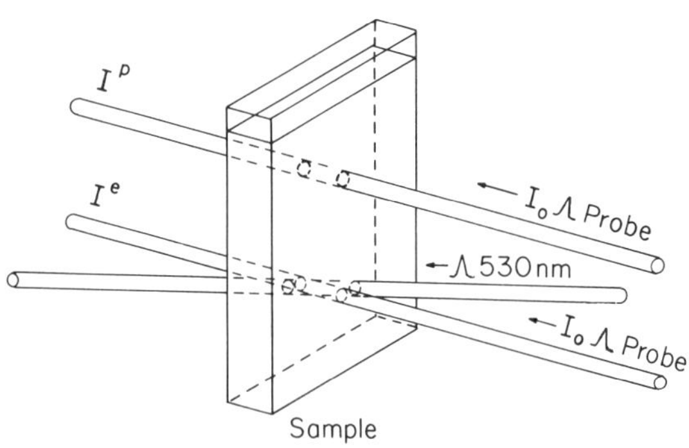

FIGURE 9.2. Geometry of the optical excitation and probing beams at the sample site.

multichannel analyzer. The supercontinuum probing pulse is usually separated into two beams at the sample site as shown in Figure 9.2. Accuracy of the measured induced OD change in the order of OD 0.005 can be achieved. By coupling a dispersion-corrected spectrometer to a streak camera (Masuhara, 1983), it is possible to measure simultaneously the temporal and spectral information of the sample. To give the reader a feeling for what is going on in the field, several experiments are reviewed using the supercontinuum induced absorption technique. The next section gives typical examples in the fields of solid-state physics, chemistry, and biology.

2.1 Solid-State Physics

2.1.1 Supercontinuum Spectroscopy of Semiconductor Microstructures

Carrier and elementary excitation relaxation, intervalley scattering, and phonon-carrier dynamics in very thin (50 to 200 Å) GaAs-AlGaAs semiconductor microstructures and bulk GaAs have been characterized using supercontinuum picosecond and femtosecond probe pulses. The optical density change of a 205-Å-thick GaAs layer sample is plotted in Figure 9.3 at three different probing times (Fork et al., 1980). Information on the induced bleaching of exciton absorption peaks at each of the 2-dimensional ($n = 1, 2, 3$) band edges can be clearly identified. Ionization dynamics of excitons were also followed using femtosecond interactions. These techniques have effectively probed the carrier and excitation dynamics in quantum wells. A similar experiment has been applied in the understanding of subpicosecond optical nonlinearities in GaAs multiple-quantum-well structures (Hulin et al., 1986).

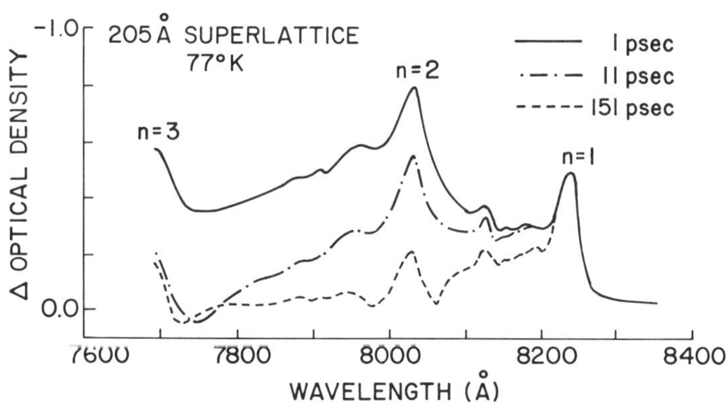

Figure 9.3. Optical density change in a superlattice optically pumped by a 625-nm femtosecond beam and probed by the supercontinuum at different delay times (1, 11, and 151 ps).

Nonlinearities based on the optical Stark effect can be a potential application for the ultrafast optical logic gates and switches.

2.1.2 Nonequilibrium Electron Processes in Metals

Nonequilibrium electron heating in gold films has been measured using an amplified 65-fs colliding-pulse mode-locked (CPM) laser pulse pump at 625 nm and supercontinuum probe technique (Schoenlein et al., 1986). Time-resolved reflectivity allows a characterization of both the excited electron thermal distribution and its cooling dynamics. Differential measurements of transient reflectivity can be obtained by monitoring the sample-reflected signal along with a reference signal from the continuum. Transient reflectivity was found to depend on the conduction band energy being probed and on the incident fluence. This technique allows the separation of electronic from phonon processes.

2.1.3 Transient Lattice Defects in Alkali Halide Crystals

Dynamics of self-trapped exciton, F, and F-like species in semiconducting or insulating crystals by two 266-nm photons (Willians et al., 1984) were studied using supercontinuum absorption spectroscopy with 3-ps pulses. Kinetics from the NaCl measurements pertain to thermally activated defect formation from relaxed self-trapped excitons. Photochemical defect production follows promotion to a potential sheet on which the barriers to halogen diffusion are small.

2.2 Chemistry

2.2.1 Iodine Photodissociation in Solution

Supercontinuum absorption spectroscopy (Berg et al., 1984) has identified the excited states of I_2 molecules using a 2-ps, 590-nm pump laser pulse. There is a fast predissociation in 5 to 10 ps with rapid partitioning between separated species excited state molecules and vibrationally excited ground state atoms. This is followed by a slower relaxation process that involves the transition from the excited electronic state to ground state vibrational relaxation. This study can help to understand the early partitioning among various intermediate states following photodissociation.

2.2.2 Photoinduced Electron Transfer in Porphyrin-Quinone

A way to avoid the ultrafast deactivation of the porphyrin-quinone photoinduced electron transfer is to construct intramolecular exciplex-type compounds where porphyrin and quinone are combined by methylene chains. These chain will weaken the chromophore interactions and prevent the strong solvation by intervening chains. The transient absorption spectra (Hirata et al., 1983) indicate a considerable contribution from the S_1 state at a short decay time. The long decay component of a few hundred picoseconds has demonstrated the characteristic of the electron transfer state. This information can be used to design a biomimetic photosynthetic system.

2.3 Biology

2.3.1 Photosynthesis

The present understanding of oxygen-evolving photosynthesis is that the process is driven by two light reactions (Wong, 1982). Associated with each photoreaction is a reaction center that utilizes the energy of a captured photon to promote an electron to a state of higher reduction potential.

Using supercontinuum induced-absorption spectroscopy, green plant photosystem I reaction centers, isolated from pea chloroplasts, have been measured (Gore et al., 1986). Transient absorption spectra of photosystem I reaction centers spanning supercontinuum wavelengths from 625 to 765 nm are displayed in Figure 9.4 at different delay times. In the samples containing chemically reduced P700, there is a significant delay between the grow-in of the excited antenna chlorophyll signal and the signal due to the excitation of P700. The 15- to 20-ps energy transfer time is consistent with the antenna chlorophyll lifetime.

Matveetz et al. (1985) have applied 300-fs pulses to study the transient differential absorption spectra of reaction centers of *Rhodopseudomonas viridis*. Kobayashi and Iwai (1984) have investigated the primary process in the photoconversion of protochlorophyllide to chlorophyllide a. This reveals the reduction of the precursor protochlorophyllide into chlorophyllide in the development of chloroplasts in higher plants.

2.3.2 Vision

Energy of absorbed photons in visual pigments is converted to a change in electrical potential across the photoreceptor cells, which is transmitted to the nervous system through standard synaptic processes. The first intermediate process in the photoreaction of rhodopsin is formed in subpicoseconds. Lippitsch et al. (1982a) have applied the supercontinuum from 340 to 460 nm in linear dichroism spectroscopy to study the primary processes in retinal. Conformational changes of retinal following excitation have been studied. Doukas et al. (1980) and Monger et al. (1979) studied laser induced absorbance changes as a function of time in squid and bovine rhodopsin using a supercontinuum generated by the 7-ps, 530-nm second harmonic pulse from an Nd:glass laser. Their data show that bathorhodopsin formation is complete within 3 ps, supporting the cis-trans isomerization model of the primary event in vision. Kobayashi et al. (1984) used the 500 to 700-nm supercon-

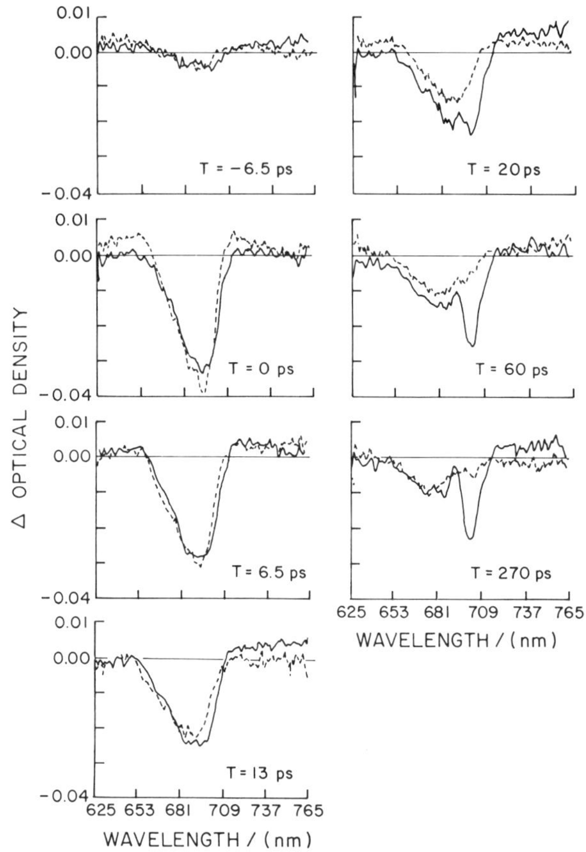

FIGURE 9.4. Transient absorption spectra of PS I reaction centers (——) P700 chemically reduced; (·······) P700 chemically oxidized.

tinuum to probe the primary process in the photocycles of the purple membrane bacteriorhodopsin. They claimed a pH-dependent fast relaxation process for the change in structures of chromophore and near residues. These and similar studies should be able to establish the sequence and dynamics of the events initiating the visual process and the structural information.

2.3.3 Hemoproteins

Hemoproteins contain one or more heme groups embedded as active centers in the folded polypeptide chain. They occur in all aerobic and many anaerobic cells and perform oxygen storage and transport, electron transfer, and catallysis (Eisenstein and Frauenfelder, 1982). Martin et al. (1984) have used a 100-fs supercontinuum from 390 to 500 nm to probe a photodissociation process of hemoproteins and protoheme with a lifetime of 350 fs. Information or the primary events of local motions inside the heme pocket following ligand detachment can be obtained. Cornelius and Hochstrasser (1982) have examined the supercontinuum difference spectra from 400 to 480 nm of myoglobins with ligands of O_2 and CO pumped by 350 and 530 nm. No subnanosecond geminate recombination occurs in either case. Both species are efficiently photolyzed with the 350-nm pump. Details of spectral shape dynamics are useful for understanding some of the energy transfer mechanisms.

In addition, Lippitsch et al. (1982b) have found that biliverdin plays an important role in photobiological processes. Supercontinuum absorption spectroscopy from 400 to 650 nm suggested that the predominant relaxation mechanism is single-bond rotation after internal conversion.

3. Time-Resolved Excitation Spectroscopy

Several ultrafast excitation spectroscopy techniques using the continuum have been developed and successfully used. These are near-IR excitation and probing techniques, coherent anti-Stokes Raman scattering, and Raman induced-phase conjugation. In this section we present a brief description of these techniques and a short review of the use of the supercontinuum for optical pulse compression. A detailed discussion of pulse compression is given in Chapter 10 by Johnson and Shank.

3.1 Generation of Subpicosecond IR Laser Pulses

Picosecond IR pulses can be generated by difference frequency mixing a strong monochromatic picosecond pulse with a white light continuum, as recently shown by Jedju and Rothberg (1987). The apparatus is depicted in Figure 9.5. A CW mode-locked Nd^{3+}:YAG laser is used to pump synchronously a Rhodamine-6G-based dye laser to obtain 5-ps nearly transform-limited pulses of 1 nJ each at a repetition rate of 82 MHz. After amplification, 1-mJ pulses at 580 nm are obtained. Half of this energy is focused by a 15-cm lens into a 3-cm cell containing a liquid to generate a white continuum extending from 300 to 900 nm. The generated continuum and the other half of the laser pulse are loosely focused into an $LiIO_3$ crystal to generate the IR pulses. The wavelength of the output pulse is determined by the frequency within the continuum that is phase matched at the given crystal angle. The authors report a photon conversion efficiency of about 5×10^{-3}, 40 nJ maximum energy per pulse, 4 to 5 ps pulse duration, and tunability from 2 to 5 μm. The generated pulses were used to study transient absorption of photogenerated charged solitons in polyacetylene (Rothberg et al., 1986). In a similar arrangement, Moore and Schmidt (1987) have used the 25-μJ output of an amplified CPM laser to generate 2-nJ, 0.2-ps pulses tunable from 1.7 to 4.0 μm.

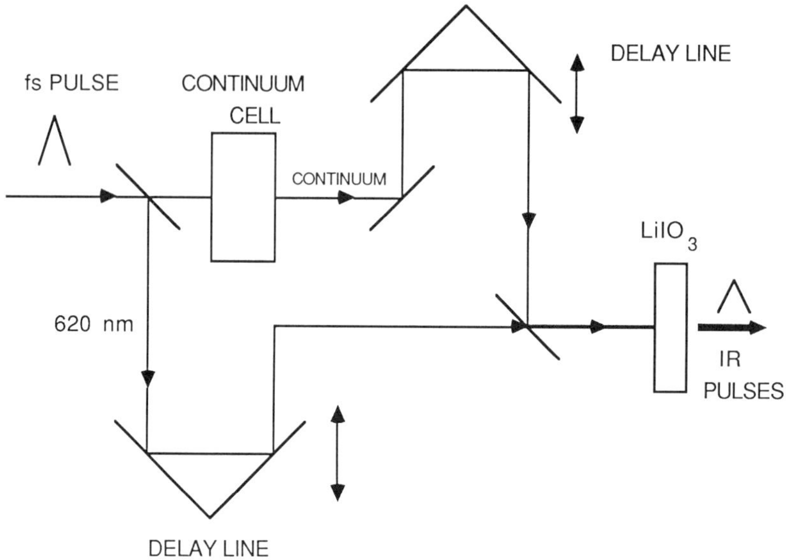

FIGURE 9.5. Experimental arrangement for generating subpicosecond near-IR pulses by difference mixing a strong monochromatic picosecond pulse with the supercontinuum.

3.2 Coherent Anti-Stokes Raman Scattering (CARS)

Coherent anti-Stokes Raman scattering (CARS) spectroscopy uses two electromagnetic fields, E_1 and E_s, with wave vectors k_1 and k_s and photon frequencies ω_1 and ω_s, where $\omega_p = \omega_1 - \omega_s$ is the frequency of a Raman-active vibrational mode in the system under consideration. The interaction through the vibrational resonance components of the third-order nonlinear coefficient χ^3 generates strong anti-Stokes signals at the frequency $\omega_{as} = 2\omega_1 - \omega_s$. Picosecond white light continuum can be used as the Stokes pulse E_s and enables an extensive anti-Stokes spectrum to be obtained in a single ultrafast laser pulse. The method was first used by Goldberg (1982) to obtain transient CARS spectra of benzene and toluene vapors. The picosecond pulse continuum, extending throughout the visible and near-infrared spectrum, was produced by focusing the 25-mJ, 5-ps, 1054-nm fundamental from a mode-locked Nd:phosphate glass laser system into a 5-cm liquid D_2O cell. The continuum beam was then collimated and filtered to pass wavelengths > 530 nm, providing a broad band of light at Stokes frequencies. The second harmonic at 530 nm was used as the second beam, E_1. For photolysis purposes a third beam, the fourth harmonic at 264 nm, was sent through an independent delay path, recombined with the other two beams, and focused into the sample cell. The CARS spectrum was recorded using a grating spectrograph and a vidicon system. The sample was probed about 200 ps after arrival of the UV pulse.

The author observed a complex spectrum, which was attributed to the formation of a C_2 diradical.

3.3 Raman-Induced Phase Conjugation (RIPC)

Phase conjugation allows the generation of a time-reversed replica of an optical wave front using nonlinear optical effects. Saha and Hellwarth (1983) have used the phase conjugation geometry in conjunction with coherent Raman spectroscopy techniques to obtain vibrational spectra in liquids. In this Raman induced phase conjugation technique, two nanosecond single-pulse laser beams at ω and $\omega - \Omega$ or $(\omega + \Omega)$ (where Ω corresponds to a vibrational frequency in a nonlinear medium) mix with a third laser beam to generate a fourth beam at $\omega - \Omega$ or $(\omega + \Omega)$, nearly phase conjugate to one of the beams at ω. The resonance enhancement of the χ^3 nonlinear coefficient generates the signals at the Stokes and anti-Stokes frequencies. The main characteristics of this technique are a wide frequency range (thousands of cm^{-1}) and a broad acceptance angle for phase matching (40 mrad). For example, under identical conditions, the phase-matching limited frequency range for RIPC is one to two orders of magnitude larger than the frequency range in CARS spectroscopy. Dorsinville et al. (1987) later extended the RIPC technique to the picosecond regime by using a mode-locked YAG laser as the laser source at ω and a picosecond continuum as the $\omega \pm \Omega$ beam. Picosecond RIPC spectra covering a 2000-cm^{-1} range were obtained for different liquids and solids. By delaying one of the interacting beams relative to the other two beams, vibrational lifetimes could be determined. The kinetic information on a picosecond time scale is obtained using a slow detector from the convolution of the probe pulse shape with the response function of the material. Delfyett et al. (1987) have shown that by combining the RIPC geometry with streak camera techniques, phonon and vibrational lifetimes can be measured in *real time*. In this method, the generated phase conjugate pulse using RIPC is passed into a streak camera and a video computer system to record its time profile. The rise and decay times of the generated pulse are directly related to the phonon formation and decay times. The technique was used to determine phonon and vibrational lifetimes of solids and liquids in different organic liquids and solids.

4. Optical Pulse Compression

Pulse compression techniques have been used to generate optical pulses of few femtoseconds (Knox et al., 1985) from CPM dye lasers and to generate 2-ps pulses from a 35-ps Nd:YAG pulse (Kafka et al., 1984). Details of pulse compression are given in Chapter 10 by Johnson and Shank.

The fundamental limit on the shortness of the temporal duration of an ultra-

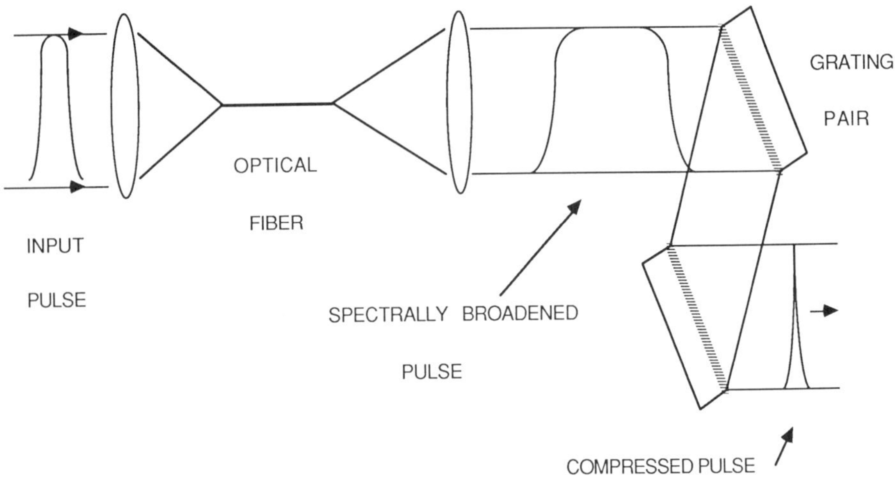

FIGURE 9.6. Experimental arrangement for compressing a laser pulse using an optical fiber and a grating pair.

fast pulse can be determined by the uncertainty relation ($\Delta\nu\,\Delta t = K; K = 0.44$ for a Gaussian pulse). This suggests that the first condition for obtaining a short pulse is a broad bandwidth. This can be achieved by self-phase modulation (SPM). The SPM-generated continuum pulse is chirped (has a time-dependent frequency shift). The lower frequencies are produced at the leading edge of the pulse, the higher frequencies at the trailing edge. If this positive chirp is linear and the chirped pulse is allowed to traverse a medium with negative linear dispersion, the pulse can be compressed to a temporal duration close to that allowed by the uncertainty principle.

Figure 9.6 shows a typical pulse compression setup consisting of an optical fiber and a grating pair. An ultrashort laser pulse is focused into the optical fiber. The pulse is spectrally broadened by SPM and acquires a positive linear chirp, i.e., $\phi = \omega_0 t + Bt^2$ and $\omega(t) = \omega_0 + 2Bt$. A positively chirped frequency increases in time from the leading edge to the trailing edge. That is, the pulse is blue in the rear and red in the front. Group velocity dispersion (GVD) temporally broadens the pulse. The grating pair acts as the negative dispersive element by forcing the lower frequencies (red) to travel a longer optical path than the higher frequencies (blue) (Treacy, 1969). The higher frequencies of the trailing edge of the pulse catch up with the leading edge, resulting in a compressed pulse (Treacy, 1969; Shank et al., 1982).

Single-mode fibers have several advantages over bulk materials for pulse compression applications:

- There are no complications from other nonlinear processes, such as self-focusing. This is due to the relatively low light intensity inside the fiber.
- The chirp is independent of transverse position on the output beam.

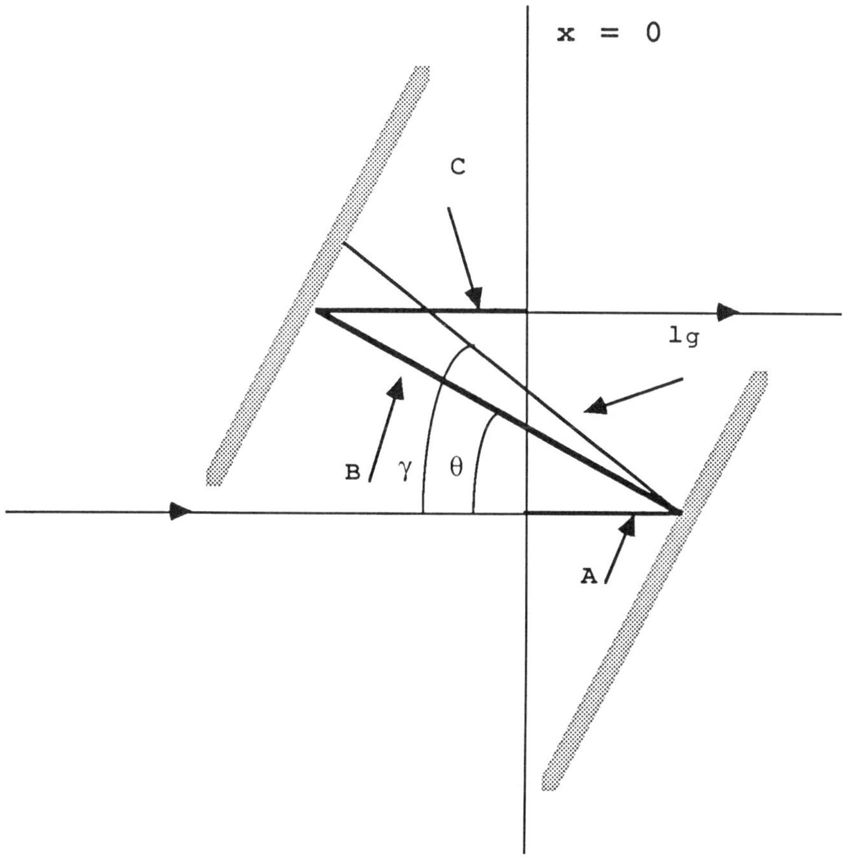

FIGURE 9.7. Ray path of a monochromatic wave through a pair of gratings used for pulse compression.

- The path length in fibers is usually longer than in bulk media and group velocity dispersion plays a more important role. GVD acts to linearize the chirp over most of the length of the pulse.

All these characteristics act to produce stable, reproducible, self-phase-modulated and positively linearly chirped pulses (Grischkowsky and Balant, 1982).

Figure 9.7 illustrates the ray path of a monochromatic wave through a pair of gratings used for pulse compression. Here θ is the angle between incident and diffracted rays, γ is the angle of incidence, l_g is the distance between the two gratings, and the relation between γ and θ for first-order diffraction is given by $\sin(\gamma - \theta) = (\lambda/d) - \sin\gamma$, with d being the groove spacing. The line $x = 0$ represents a reference plane between the two gratings.

The ray path length $p(\omega)$ is the sum of the paths A, B, and C and is given by (see Figure 9.7)

$$p(\omega) = A + B + C = l_g \frac{1 + \cos\theta}{\cos(\gamma - \theta)}. \tag{2}$$

The resulting frequency-dependent phase is given by

$$\Phi(\omega) = \omega \frac{p(\omega)}{c} + R(\omega), \tag{3}$$

where R is a complex geometric function specific to the grating pair (Treacy, 1969) and c is the velocity of light.

For pulse compression purposes, the important characteristic of a grating pair is the variation of group delay time $\tau(\omega)$ with frequency. This characteristic determines whether the grating pair can compensate for the linear chirp introduced by SPM and the fiber GVD. $\tau(\omega)$ is defined as follows:

$$\tau = \frac{\delta\Phi}{\delta\omega} = \frac{p(\omega)}{c} + \frac{\omega\, dp(\omega)}{c\, d\omega} + \frac{dR(\omega)}{d\omega}. \tag{4}$$

For gratings the last two terms cancel and $\tau(\omega) = p(\omega)/c$.

The group delay time can also be expressed as

$$\tau = \tau_0 + \frac{\delta\tau(d\omega)}{\delta\omega} + \frac{\delta^2\tau}{\delta\omega^2}(d\omega)^2 + \cdots. \tag{5}$$

Keeping only the linear term and taking into account that θ is frequency dependent (formula 2), an analytical approximation of the variation of the group delay with frequency can be derived by differentiating $p(\omega)/c$ with respect to ω:

$$d\tau = \frac{\delta\tau}{\delta\omega} d\omega = \frac{\delta^2\Phi}{\delta\omega^2} d\omega = -4p^2 c l_g\, d\omega / (\cos(\gamma - \theta)\omega^3 d^2 \{1 - [(2pc/\omega d) - \sin\gamma]^2\}). \tag{6}$$

This formula gives the time separation $\Delta\tau$ introduced by a grating pair for a spectral separation $\Delta\omega$. For example, for $d^{-1} = 1200$ lines/mm, $\lambda = 1.06\ \mu$m, and $\gamma = 60°$, formula (6) gives a relative time separation of 0.6 ps per centimeter of grating spacing l_g for two wavelength components differing by 100 Å.

Grating pairs also introduce nonlinear terms such as the quadratic term in Eq. (5). This term is positive and proportional to the second derivative of the group delay $(\delta^2\tau/\delta\omega^2)$ or the third derivative of the phase $(\delta^3\Phi/\delta\omega^3)$. At large bandwidth and pulse durations in the order of 10 fs, this introduces undesirable phase distortions that prevent further compression.

A prism-grating combination substantially reduces the phase distortion (Brito Cruz et al., 1988). Figure 9.8 shows the propagation of an optical pulse through the prism sequence. The optical path variation with frequency is caused by the refracted rays leaving the first prism at different angles for different wavelengths. In the case of normal dispersion, the low frequencies will be less refracted and reach the second prism earlier than the high fre-

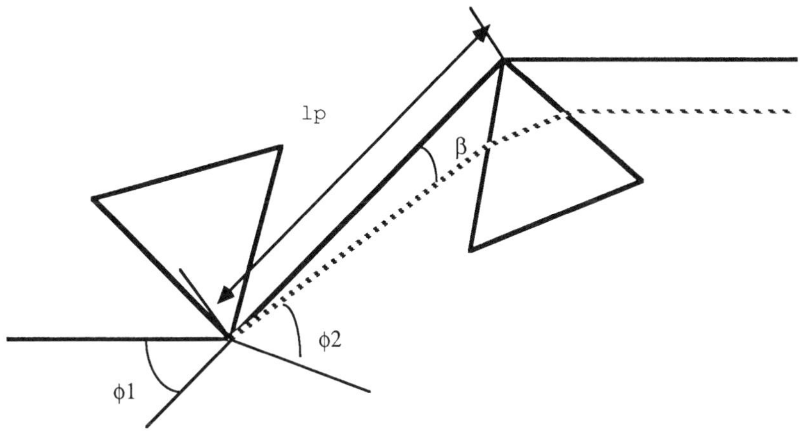

FIGURE 9.8. Propagation of the optical pulse through a prism pair. ϕ_1 is the angle of incidence at the entrance face of the first prism, ϕ_2 is the angle with respect to the normal to the exit face, β is the angle between a line drawn between prism apices and the direction of a ray leaving the prism at ϕ_2; and l_p is the distance between prism apices.

quencies, causing a path difference in air. An additional path variation is caused by the difference in travel inside the two prisms. By changing the distance l_p between the two prisms, the group delay time can be widely adjusted. The phase shift is given by (Fork et al., 1984):

$$\Phi_p = \frac{\omega l_p}{c} \cos \beta,$$

where β is the angle between a line drawn between prism apices and the direction of a ray leaving the first prism. Prism pairs provide adjustable second- and third-order phase corrections (Fork et al., 1984) that can be either positive or negative. The prisms can be adjusted to cancel the third derivative of the phase introduced by the grating pair.

In general, for pulse compression, group delay times should satisfy the condition

$$\Delta\tau_g + \Delta\tau_p = -\Delta\tau, \tag{7}$$

where $\Delta\tau$ is the group delay time difference before the prism-grating compressor and the subscripts g and p refer to prism and grating group delay times, respectively.

Using the Taylor expansion for the group delay time (5), condition (7) gives

$$\begin{aligned} \frac{\delta^2\Phi_g}{\delta\omega^2} + \frac{\delta^2\Phi_p}{\delta\omega^2} &= \frac{\delta^2\Phi}{\delta\omega^2}, \\ \frac{\delta^3\Phi_g}{\delta\omega^3} + \frac{\delta^3\Phi_p}{\delta\omega^3} &= \frac{\delta^3\Phi}{\delta\omega^3}, \end{aligned} \tag{8}$$

TABLE 9.1. Second and third derivatives of the phase with respect to frequency for a double prism pair and double grating pair.[a]

Derivative	Prisms	Gratings	Material
$\frac{d^2\Phi}{d\omega^2}$ (fs^2)	$+648 - 32(l_p)$	$-3640(l_g)$	$+2900(l_m)$
$\frac{d^3\Phi}{d\omega^3}$ (fs^3)	$+277 - 49(l_p)$	$+3120(l_g)$	$+1620(l_m)$

[a] l_g is the grating spacing, l_p the prism spacing, and l_m the material length.

where Φ is the phase before the prism-grating compressor and Φ_g, Φ_p are the contributions to the phase due to the gratings and prisms, respectively.

A short transform-limited pulse is obtained if the different phase derivatives can be adjusted to satisfy (8), by choosing the gratings and prisms with the correct parameters and adjusting the angle and the separation between the two gratings and the two prisms forming the compressor pairs (Brito Cruz et al., 1988). This is illustrated in Table 9.1, which gives the second and third derivatives of the phase with respect to frequency for a double prism pair and double grating pair described by Fork et al. (1988). The expressions for the second and third derivatives of the phase for the prism pairs can be positive or negative depending on the prism separation l_p. The lengths l_p, l_g (the grating separation), and l_m (the length of material other than the minimum material path contributed by the prisms) can be adjusted to satisfy Eq. (8).

For example, using a value of $+700$ fs^2 for the quadratic phase distortion (linear chirp) for the pulse emerging from the optical fiber and using the experimentally determined value for the grating spacing, the values of the prism spacing and material length can be calculated from the expressions in Table 9.1 to obtain a quadratic phase distortion of -700 fs and a net cubic phase distortion of zero. For this particular case, $l_g = 0.5$ cm, $l_p = 68$ cm, and $l_m = 1.0$ cm, in good agreement with the experimental values (Fork et al., 1987).

The compression techniques we have described were used to obtain picosecond and femtosecond pulses from different laser systems. Table 9.2 reviews the pulse compression characteristics obtained from different laser systems.

Substantial pulse compression can also be obtained without gratings or prisms. As a result of the SPM process, the Stokes and anti-Stokes shifts are proportional to the intensity gradients on the sides of the pulse. Therefore, a pulse obtained by passing the SPM beam through a narrow spectral window could be shorter than the original beam (Masuhara et al., 1983). Gomes et al. (1986) demonstrated a pulse compression technique where a spectral window within the broadened spectral profile eliminates the wings of the SPM-generated pulse where the high- and low-frequency components are located.

TABLE 9.2. Pulse compression performances of different laser systems.

References	Pulse duration Before	Pulse duration After	Repetition rate	Wavelength (nm)
Gomes et al., 1985	85 ps	2.9 ps	500 Hz	1064
Zysset et al., 1986	90 ps	200 fs	82 MHz	1064
Johnson and Simpson, 1986	33 ps	410 fs	100 MHz	532
Shank et al., 1982	90 fs	30 fs	20 Hz	620
Knox et al., 1985	40 fs	8 fs	5 kHz	620
Palfrey and Grischkowsky, 1985	5.4 ps	16 fs	200 Hz	587
Fork et al., 1987	30 fs	7 fs	8 kHz	620

They obtained a 3-fold shortening of 80-ps pulses from an Nd:YAG laser. The pulses were broadened from 0.3 to 4 Å after propagation through 125 m of optical fiber. A monochromator was used as a spectral window. Dorsinville et al. (1988) generated pulses down to 3 ps using a continuum generated in a 5-cm D_2O cell by an intense 25-ps Nd:YAG second harmonic laser pulse and narrowband filters for spectral selection. The pulse duration was measured by a 2-ps time resolution streak camera. More details on pulse compression are given in Chapter 10.

5. Exploration of Future Applications

The supercontinuum can have a number of new applications in instances where knowledge is needed on specified phase and intensity for well-separated wavelengths (Manassah et al., 1984). Specifically, time delay and relative intensity measurements for different wavelengths become simple.

5.1 Ranging

The main limitation on accurate determination of optical lengths through the uncontrolled atmosphere is the uncertainty in the average refractive index over the optical path due to the nonuniformity and turbulence of the atmosphere. Simultaneous measurements over the same path using two or more different wavelengths of light could be used to provide the base values. The dispersive delays between any two wavelengths can be calculated using the expressions of Owens (1967) and Topp and Orner (1975) for the optical refractive index of air as a function of the ambient pressure, temperature, and composition. For path lengths of a few tens of meters, Erickson (1962) proposed the use of direct interferometry for the dispersive delay measurements. In effect, his system consists of an automatic fringe-counting Michelson interferometer. For lengths greater than 100 m, the method of direct interferometry is impractical. A system using a pulsed subpicosecond single-wavelength laser source and a synchroscan streak camera with a time resolution of 3 ps as a detector can measure distances of a few kilometers to an accuracy of 1 mm if the index of refraction of ambient air is known to an accuracy of 0.1 ppm; however, the index of refraction of air varies by as much as 100 ppm over the operational range of such range finders and the measurement accuracy is thus reduced to ± 1 m. (The time delay between two signals in different media is $\Delta\tau = (L/c\delta n)$.) On the other hand, if a system consists of a multiwavelength source, such as the supercontinuum, with a multichannel synchroscan streak camera (Tsuchiya, 1983) in the receiver, the arrival time data for the different wavelengths determine the parameters in the Owens index of refraction formula (wavelength, temperature, pressure, and relative humidity dependence) to an accuracy of 0.4 ppm, and consequently the accuracy of distance measurements is restored to the ± 0.4-cm range. Simultaneously, the pressure can be determined to an accuracy of 1.5 mbar and temperature to $\pm 1/2$ K. Furthermore, the wide frequency band of the continuum source allows for selective tuning and encoding (for example, the addition of a constant phase) of the different emitted lines, thus providing the added feature of system integrity under adverse field conditions (jamming, interference, etc.).

5.2 3-D Imaging

Direct nondestructive in situ measurement of the contour of surface, as well as the internal structure of an object, with an accuracy of 30 μm can be accomplished using 100-fs laser pulses derived from the supercontinuum.

The reflection of an incoming signal from surface imperfections will be delayed in time, with respect to other points on the surface, arrival at the detector. Experimentally, these measurements are carried out by modulating the arrival times to the target using an oscillating delay prism and passing the reflected signal together with a reference signal into a second harmonic correlation crystal. The convoluted harmonic signal is then detected by a video system in synchronism with the modulation frequency. Using three different optical frequencies from the supercontinuum and collimating the corresponding beams in a three-dimensional orthogonal configuration, a 3-D image of the imperfection can be directly deduced. The S/N ratios for each frequency channel and interchannels are excellent. This technique can, for example, measure the defects and wear on the ball bearings in delicate equipment, semiconductor surfaces, and tissue topography.

For the diagnostics of an object inside a material, the different reflection and absorption coefficients for the different optical frequencies at the interfaces and in the materials can be used to measure the location and contour of cracks or impurities inside a sample. It is to be noted that the accuracy of the measurement can be further enhanced from ± 30 μm by selecting one of the frequencies to be close to a resonance line, thus endowing it with a large optical path in the medium and resulting in longer time delays and more accurate spatial measurements. These techniques can, for example, measure locations of impurities inside Si and GaAs. Measurements in biological and medical samples are possible.

5.3 Induced-Phase Modulation Based Optical Computational Switches

The development of optical systems for ultrafast signal processing, communication, and computation has been a very active area of research. The use of nonlinear properties of materials for implementing this program has received considerable attention. Using induced-phase modulation, it is possible to configure canonical logical elements (Lattes et al., 1983; Manassah and Cockings, 1987) for computation. The configurations of XOR (exclusive OR) and NOT gates presented here are only for illustrative purposes. We refer the reader to Section 4.2 of the Manassah (Chapter 5) for the underlying models and to Section 9 of Baldeck et al. (Chapter 4) for additional applications using cross-phase modulation.

In Figure 9.9, possible gates are displayed. The input signals and seed center frequencies are ω_2 and the pump center frequency is ω_1. The Kerr phase shifter(s) (KPS) is in the path of the seed signal. The seed, following its passage through the KPS, is multiplexed with the pump and both are fed into a fiber (or waveguide) with molecular vibration (phonon) energy equal to $\hbar(\omega_1 - \omega_2)$; that is, under proper time delays between the seed and pump the seed can be Raman amplified or left unchanged. The amplification factor can easily change by two orders of magnitude over a few pulse widths (Chapter 4). Denote this phase shift by Φ. At the end of the fiber a filter allows only the seed to pass through to the next space.

The NOT gate pump arm and seed arm are chosen such that in the absence of an input signal the phase between the seed and pump corresponds to the maximum of the amplification curve, while in the presence of an input signal the KPS introduces in the seed phase a shift equal to Φ_0.

The XOR gate seed arm and pump arm differ in the absence of any input by a phase $-\Phi_0$ from the maximum and amplification position. In the presence of either input the seed is amplified, while in the presence of both inputs the phase is $+\Phi_0$ from the maximum position and the seed is not amplified.

Other configurations that are more energy efficient and have fewer ports are possible but are not covered here.

5.4 Atmospheric Remote Sensing

The use of differential absorption lidar for remote sensing of atmospheric species is a viable experimental technique for the detection and identification of a wide range of molecular constituents. The lidar equation is given by

$$\frac{P_r}{P_t} = \frac{K\rho A}{\pi R^2} \exp[-2(\sigma_a(\lambda)N_a + \alpha(\lambda))R], \tag{9}$$

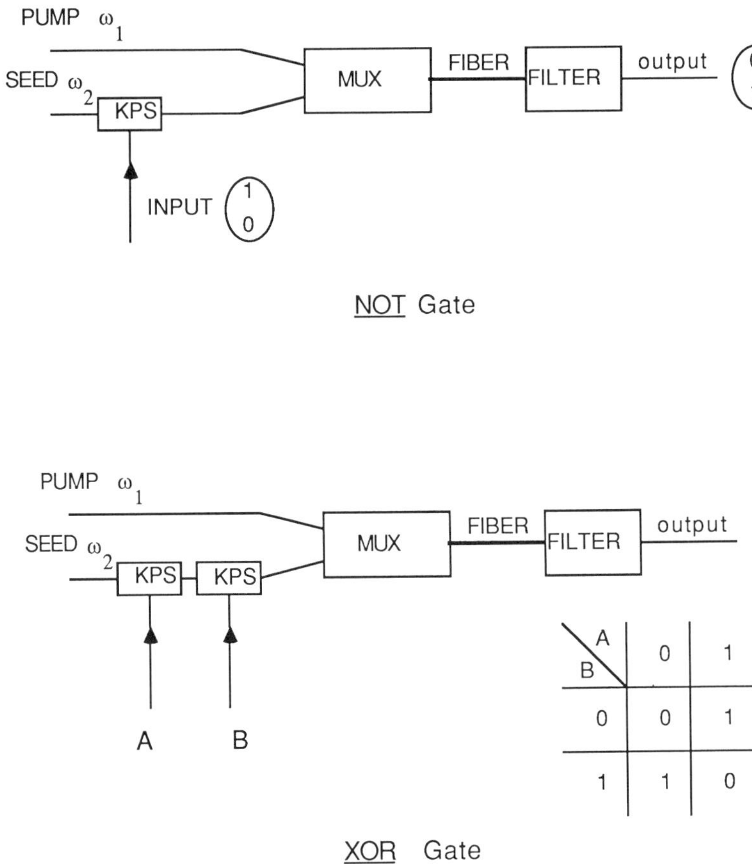

FIGURE 9.9. (Top) NOT gate and (bottom) XOR gate.

where P_r and P_t are respectively the received and transmitted powers, K is the overall system efficiency, ρ is the target reflectivity, A is the receiving telescope area, σ_a is the absorption cross section of the absorbing molecule a, R is the range, and α is the background extinction coefficient of the atmosphere. A measurement of N_a based on a single wavelength requires an accurate knowledge of K, ρ, α, and σ_a. However, K, ρ, and α are generally known with poor accuracy. The differential absorption lidar (Tsuchiya, 1983) approach attempts to overcome some of these difficulties by performing the lidar experiment with two or more frequencies. For a dual-wavelength source (with frequencies ν^1 and ν^2),

$$N_a = \frac{1}{2(\sigma_a^2 - \sigma_a^1)R}\left[\ln\left(\frac{P_a^1}{P_a^2}\right) - \ln\left(\frac{\rho^2}{\rho^1}\right) - 2(\alpha^1 - \alpha^2)R\right], \tag{10}$$

where $P_a = P_r/P_t$, and the superscript refers to the wavelength. A multiwavelength source such as the supercontinuum, with an inbuilt calibrated power spectrum, provides the possibility for a more accurate determination of N_a. The differential spectral reflectance (DSR) error can be reduced significantly by using three or more frequencies because the DSR Taylor's expansion terms can be directly determined from the ratios of $P_a^{(i)}$ and $P_a^{(j)}$ for i and j off the molecular absorption line. Furthermore, the background extinction coefficient can be factored out and the water vapor absorption tail contribution can be Taylor expanded around the molecular absorption line. Another advantage of the supercontinuum is the virtual elimination of time interval (1 ns) between the different probing wavelengths. This interval should be less than 10^{-2} s to neglect temporal fluctuations in the atmospheric parameters.

5.5 Optical Fiber Measurements

Signal attenuation and pulse spreading are the two characteristics that determine the suitability of a particular optical fiber for communication use. Signal attenuation determines the distance between repeaters and pulse spreading determines the maximum bit rate over a channel (channel capacity). In an optical fiber, three sources of attenuation are present: Mie scattering (wavelength independent), Rayleigh scattering (λ^{-4} law), and molecular absorption. The fiber loss is written as

$$\alpha = \frac{A}{\lambda^4} + B + C(\lambda), \tag{11}$$

where $C(\lambda)$ includes impurity absorption and other wavelength-dependent phenomena and B includes Mie scattering and losses introduced by microbending or coating defects. The molecular absorption is a function of the impurities present in the raw material or introduced in the manufacturing process. Quality control and component specifications are achieved through a measurement of attenuation as a function of wavelength. Present techniques use near-IR fiber Raman laser to perform these measurements for discret points in the wavelength region 1.1 to 1.6 μm (pulse duration $\sim$100 ps).

The supercontinuum in the IR band allows the measurement of a normalized attenuation profile for all points in the band 0.7 to 2 μm and with the option of reducing the pulse duration to 100 fs. This data can further provide the needed parameters for the design of wavelength division multiplexing (WDM) systems. Furthermore, the SPM source can actually be used in situ to calibrate existing network losses for the different wavelengths and assess their suitability for WDM upgrading.

Pulse spreading in fibers is dominated by the material dispersion term $(dn/d\omega)$. The supercontinuum provides an experimental tool for measuring pulse spreading at different wavelengths and meaningfully comparing the results for WDM applications. Furthermore, this experimental technique is useful in exploring the inherent lower limits on the duration of a pulse that is suitable for optical communication.

5.6 Kinetics of Optical Nonlinearities

Spectral and temporal information on nonlinear optical susceptibilities is important for the design and application of future optoelectronic devices. Using the pump-and-probe technique to determine ultrafast temporal responses of nonlinear processes requires a great number of successive shots to obtain all the information. Based on the techniques of single-shot Kerr gate (Ho, 1984) and real-time picosecond oscilloscopes (Valdmanis, 1986), a single-shot supercontinuum Kerr method can be developed to probe for complete information on nonlinearities with single-shot excitation (Saux et al., 1988). A long chirped probe pulse from a supercontinuum is used to transform a temporal modulation into a spectral modulation. A spectrograph then converts the wavelength-encoded temporal information to the spatial domain for readout. The chirp of the supercontinuum is created mainly by the group velocity dispersion in the generation sample. Subpicosecond temporal resolution of optical nonlinearity measurements of glasses has been demonstrated.

6. Conclusion

We have reviewed some of the applications of the continuum in ultrafast spectroscopy and nonlinear optics and suggested some future applications including ranging, imaging, and remote sensing. The supercontinuum source can be directly applied in many other areas of solid-state physics, biology, chemistry, and medicine. Exciting and far-reaching future applications are expected to be related to the possibility of producing femtosecond pulses using SPM. Ultrashort pulses are expected to lead to breakthroughs in communication and optical computing. Data transmission rates may be increased three or more orders of magnitude. Optical computing based on ultrafast logic units has the potential for revolutionizing the field of computers.

References

Ahmed, S. (1973) Appl. Opt. **12**, 901.

Alfano, R.R. and S.L. Shapiro (1971) Chem. Phys. Lett. **8**, 631.

Berg, M., A.L. Harris, J.K. Brown, and C.B. Harris (1984) Opt. Lett. **9**, 50.

Brito Cruz, C.H., P.C. Becker, R.L. Fork, and C.V. Chank (1988) Opt. Lett. **13**, 123.

Cornelius, P.A. and R.M. Hochstrasser (1982) In *Picosecond Phenomena III*, Eisenthal et al., eds., p. 288. Springer-Verlag, New York.

Delfyett, P.J., R. Dorsinville, and R.R. Alfano (1987) Opt. Lett. **12**, 1002.

Dorsinville, R., P.J. Delfyett, and R.R. Alfano (1987) Appl. Opt. **26**, 3655.

Dorsinville, R., P.J. Delfyett, and R.R. Alfano (1988) Appl. Opt. **27**, 16.

Doukas, A.G., V. Stefancic, T. Suzuki, R.H. Callender, and R.R. Alfano (1980) Photobiochemistry and Photobiophysics **1**, 305.

Eisenstein, L. and Frauenfelder (1982) In *Biological Events Probed by Ultrafast Laser Spectroscopy*, R.R. Alfano, ed., chapter 14. Academic Press, New York.

Erickson, K.E. (1962) J. Opt. Soc. Am. **52**, 777.
Fork, R.L., C.V. Shank, and B.I. Greene (1980) In *Picosecond Phenomena II*, R.M. Hochrasser et al., eds. Springer-Verlag, New York.
Fork, R.L., O.E. Martinez, and J.P. Gordon (1984) Opt. Lett. **9**, 150.
Fork, R.L., C.H. Brito Cruz, P.C. Becker, and C.V. Shank (1987) Opt. Lett. **12**, 483.
Gayen, S.K., W.B. Wang, V. Petricevic, K.M. Yoo, and R.R. Alfano (1987) Appl. Phys. Lett **50**, 1494.
Goldberg, L.S. (1982) In Proceedings of the Third International Conference on Picosecond Phenomena, *Picosecond Phenomena III*, p. 94. Springer-Verlag, New York.
Gomes, A.S.L., W. Sibbet, and J.R. Taylor (1985) IEEE J. Quantum Electron. **QE-21**, 1157.
Gomes, A.S.L., A.S. Gouveia-Neto, J.R. Taylor, H. Avramopoulos, and G.H.C. New (1986) Opt. Commun. **59**, 399.
Gore, B.L., L.B. Giorgi, and G. Porter (1986) In *Picosecond Phenomena V*, G.R. Fleming and A.E. Seigman, eds., 298. Springer-Verlag, New York.
Greene, B.I., R.M. Hochstrasser, and R.B. Weisman (1978) In *Picosecond Phenomena I*, p. 12. Springer-Verlag, New York.
Grischkowsky, D. and A.C. Balant (1982) Appl. Phys. Lett. **41**, 1.
Hirata, Y., Y. Kanda, and N. Mataga (1983) J. Phys. Chem. **87**, 1659.
Ho, P.P. (1984) In *Semiconductors Probed by Ultrafast Spectroscopy*" R.R. Alfano, ed., Chapter 25. Academic Press, New York.
Hulin, D., A. Mysyrowicz, A. Antonetti, A. Migus, W.T. Masselink, H. Morkoc, H.M. Gibbs, and N. Peyghambarian (1986) Phys. Rev. B **33**, 4389.
Inada, K. (1976) Opt. Commun. **19**, 436.
Jedju, T.M. and L. Rothberg (1987) Appl. Opt. **26**, 2877.
Johnson, A.M. and M. Simpson (1986) IEEE J. Quantum Electron. **QE-22**, 133.
Kafka, J.D., B.H. Kolner, T. Baer, and D.M. Bloom (1984) Opt. Lett. **9**, 505.
Knox, W.H., R.L. Fork, M.C. Downer, R.H. Stolen, and C.V. Shank (1985) Appl. Phys. Lett. **46**, 1120.
Kobayashi, T. and J. Iwai (1984) In *Picosecond Phenomena IV*, D.H. Auston and K.B. Eisenthal, eds., p. 484. Springer-Verlag, New York.
Kobayashi, T., H. Ohtani, and J. Iwai (1984) In *Picosecond Phenomena IV*, D.H. Auston and K.B. Eisenthal, eds., P. 481. Springer-Verlag, New York.
Lattes, A., H.A. Haus, F.J. Leonberger, and E.P. Ippen (1983) IEEE J. Quantum Electron. **QE-19**, 1718.
Lippitsch, M.E., M. Rieger, F.R. Aussenegg, L. Margulies, and Y. Mazur (1982a) In *Picosecond Phenomena III*, K.B. Eisenthal et al., eds., p. 319. Springer-Verlag, New York.
Lippitsch, M.E., M. Riegler, A. Leitner, and F.R. Aussengg (1982b) In *Picosecond Phenomena III*, K.B. Eisenthal et al., eds., p. 323. Springer-Verlag, New York.
Manassah, J.T. (1986) Appl. Opt. **25**, 3979.
Manassah, J.T. and O. Cockings (1987) Appl. Opt. **26**, 3749.
Manassah, J.T., P.P. Ho, A. Katz, and R.R. Alfano (1984) Photonics Spectra, November issue.
Martin, J.L., A. Migus, C. Poyart, Y. Lecarpentier, A. Astier, and A. Antonetti (1984) In *Picosecond Phenomena IV*, D.H. Auston and K.B. Eisenthal, eds., p. 447. Springer-Verlag, New York.
Masuhara, H., H. Miyasaka, A. Karen, T. Uemiya, N. Mataga, and M. Koishi (1983) Opt. Commun. **44**, 426.

Matveętz, Yu. A., S.V. Chekalin, and A.V. Sharkov (1985) J. Opt. Soc. Am. B **2**, 634.
Menyuk, N. and D.K. Killinger (1983) Appl. Opt. **22**, 2690.
Monger, T.G., R.R. Alfano, and R.H. Callender (1979) Biophys. J. **27**, 105.
Moore, D.S. and S.C. Schmidt (1987) Opt. Lett. **12**, 480.
Owens, J.C. (1967) Appl. Opt. **6**, 51; and Topp, M. and G. Orner (1975) Opt. Commun. **13**, 276.
Palfrey, S.L. and D. Grischkowsky (1985) Opt. Lett. **10**, 562.
Rothberg, L., T.M. Jedju, S. Etemad, and G.L. Baker (1986) Phys Rev. Lett. **57**, 3229.
Saha, S.K. and R.W. Hellwarth (1983) Phys. Rev. A **27**, 919.
Saux, G. Le, F. Salin, P. Georges, G. Roger, and A. Brun (1988) Appl. Opt. **27**, 777.
Schoenlein, R.W., W.Z. Lin, J.G. Fujimoto, and G.L. Eesley (1986) In *Ultrafast Phenomena V*, G.R. Fleming and A.E. Siegman, eds., p. 260. Springer-Verlag, New York.
Shank, C.V., R.L. Fork, R.T. Yen, R.J. Stolen, and W.J. Tomlinson (1982) Appl. Phys. Lett. **40**, 761.
Searle, G.F.W., J. Barbet, G. Porter, and C.J. Tredwell (1978) Biochim. Biophys. Acta **501**, 246–256.
Singer, S.S. (1969) Appl. Opt. **7**, 1125.
Thomas, D.G., L.K. Anderson, M.I. Cohen, E.I. Gordon, and P.K. Runge (1982) In *Innovations in Telecommunications*, J.T. Manassah. ed., Academic Press, New York.
Tomlinson, W.J., R.H. Stolen, and C.V. Shank (1984) J. Opt. Soc. Am. B **1**, 139.
Treacy, E.B. (1969) IEEE J. Quantum Electron. **QE-5**, 454.
Tsuchiya, Y. (1983) "Picosecond streak camera and its applications," Hamamatsu Tech. Bull. **14**, June; C.S. Gardner, B.M. Tsai, and K.F. Im (1983) Appl. Opt. **22**, 2571; Abshire, J.B. and G.E. Kalshoven (1983) Appl. Opt. **22**, 2578.
Valdmanis, J.A. (1986) In *Ultrafast Phenomena V*, Fleming and Siegman, eds., p. 82.
von der Linde, D., and R. Lambrich (1979) Phys. Lett. **42**, 1090.
Williams, R.T., B.B. Craig, and W.L. Faust (1984) Phys. Rev. Lett. **52**, 1709.
Wong, D. (1982) In *Biological Events Probed by Ultrafast Laser Spectroscopy*, R.R. Alfano, ed., chapter 1. Academic Press, New York.
Zysset, B., W. Hodel, P. Beaud, and H.P. Weber (1986) Opt. Lett. **11**, 156.

10 Pulse Compression in Single-Mode Fibers—Picoseconds to Femtoseconds

A.M. JOHNSON and C.V. SHANK

1. Introduction

The compression of frequency swept (in time) or "chirped" optical pulses was independently proposed by Gires and Tournois (1964) and Giordmaine et al. (1968). Optical pulse compression is the optical analog of microwave pulse compression or chirp radar developed by Klauder et al. (1960). The compression is accomplished in two steps. First, an optical frequency sweep is impressed on the pulse. The next step is the compensation of this frequency sweep by using a dispersive delay line, where the group velocity or group delay varies with optical frequency. Ideally, the dispersive delay line would impress the opposite chirp on the pulse, resulting in the compression of the pulse to its minimum width, $\sim 1/\Delta\omega$, where $\Delta\omega$ is the frequency sweep. Treacy (1968, 1969) was the first to recognize that a pair of diffraction gratings was a suitable dispersive delay line for a linearly chirped pulse; he used gratings to compress the inherently chirped output of a mode-locked Nd : glass laser. Similar experiments were later performed by Bradley et al. (1970). Duguay and Hansen (1969) used an $LiNbO_3$ phase modulator and Gire-Tournois interferometer to compress pulses from a mode-locked He-Ne laser.

A chirp can be impressed on a intense optical pulse as it passes through a medium with an intensity-dependent refractive index, i.e., an optical Kerr medium. The phase of the intense optical pulse is modulated by the nonlinear refractive index. Extreme small spectral broadening of optical pulses in optical Kerr liquids was first observed in self-focused filaments by Bloembergen and Lallemand (1966), Brewer (1967), and Ueda and Shimoda (1967). The weak spectral broadening was first explained by Shimizu (1967) as due to a rapid time-varying phase shift arising from the nonlinear refractive index. Gustafson et al. (1969) further elaborated on Shimizu's explanation with detailed numerical calculations of the spectra of self-phase-modulated pulses, including the effects of dispersion and relaxation of the nonlinearity. Alfano and Shapiro (1970) made the first measurements of self-phase modulation (SPM) in crystals, liquids, and glasses (see Chapter 2). Spectral broadening data in glasses were also obtained by Bondarenko et al. (1970).

Fisher et al. (1969) suggested that optical pulses in the range 10^{-13} to 10^{-14} s could be achieved as a result of the SPM obtained by passing a short pulse through an optical Kerr liquid followed by a dispersive delay line. Laubereau (1969) used several cells of the optical Kerr liquid CS_2 and a pair of diffraction gratings to compress 20-ps-duration pulses from a mode-locked Nd:glass laser by 10×. Zel'dovich and Sobel'man (1971) proposed the possibility of using alkali metal vapors to both spectrally broaden optical pulses by SPM and compress the pulses by the strong dispersion of the group velocity near the atomic resonance. Lehmberg and McMahon (1976) compressed 100-ps-duration pulses from a mode-locked and amplified Nd:YAG laser by 14×, using a series of liquid CS_2 cells and diffraction gratings separated by 23 m. Spectral broadening of picosecond pulses from a flashlamp-pumped, passively mode-locked Rhodamine 6G dye laser was reported by Arthurs et al. (1971) and was attributed to SPM. Ippen and Shank (1975b) compressed 1-ps-duration pulses from a CW pumped, passively mode-locked Rhodamine 6G dye laser by 3× to a duration of 0.3 ps using diffraction gratings separated by 10 cm.

The early measurements of SPM (Bloembergen and Lallemand, 1966); Brewer, 1967; Ueda and Shimoda, 1967; Shimizu, 1967; Gustafson et al., 1969; Alfano and Shapiro, 1970; Bondarenko et al., 1970) occurred in self-focused filaments, where the intensity was high and there were problems with competing nonlinear effects and uncertainties concerning the filament size (see Chapter 2). Ippen et al. (1974) reported the first measurement of SPM in the absence of self-trapping or self-focusing with the use of a guiding multimode optical fiber filled with liquid CS_2. Stolen and Lin (1978) reported measurements of SPM is single-mode silica core fibers. In fibers, any additional confinement caused by self-focusing is negligible. An additional advantage of this guiding structure over bulk crystals or liquid cells is that the modulation can be imposed over the entire transverse spatial extent of the beam, and the problem of unmodulated light in the wings of the beam is eliminated (Ippen et al., 1974). Perhaps the most important feature of SPM in optical fibers is that significant spectral broadening can be achieved at power levels much lower than those required in bulk media.

The first fiber pulse compression experiments utilized the fiber as a dispersive delay line to compress chirped optical pulses. Suzuki and Fukumoto (1976) used an $LiNbO_3$ phase modulator to chirp 1-μm laser pulses, which were subsequently compressed by the normal or positive group velocity dispersion (GVD) (red frequencies lead blue) of a silica optical fiber. Wright and Nelson (1977) compressed the chirped output of a GaAs semiconductor laser operating at 0.894 μm using a positive GVD optical fiber delay line. Iwashita et al. (1982) demonstrated 5× compression of 1.7-ns, 1.54-μm pulses from a chirped InGaAsP injection laser using a 104-km negative GVD fiber delay line. Mollenauer et al. (1980) performed the first pulse compression experiments using optical fibers as a Kerr medium, in their work on soliton compression of pulses from a color center laser. In these experiments, the laser wavelength ($\lambda = 1.55\ \mu m$) was in the anomalous or negative GVD (blue frequencies lead red) region for silica and did not require a separate dispersive delay line. In this instance, the fiber material forms an integrated dispersive delay line and self-compresses the pulse. Using soliton compression, Mollenauer et al. (1983) compressed 7-ps-duration pulses by $27\times$ to a duration of 0.26 ps with a 100-m length of single-mode fiber. This compression was achieved with only 200 W of peak power at the fiber input, thus further attesting the low power requirements of nonlinear effects in optical fibers. It is beyond the scope of this chapter to consider soliton compression in optical fibers. Further information on soliton compression and its applications can be found, for example, in excellent discussions by Mollenauer and co-workers (Mollenauer and Stolen, 1982; Mollenauer, 1985; Mollenauer et al., 1986). This chapter is limited to pulse compression, in silica core fibers, in the normal or positive GVD region ($\lambda \leq 1.3\ \mu m$), where a separate dispersive delay line is necessary. The compression of positively chirped optical pulses passing through a dispersive medium possessing negative GVD is schematically illustrated in Figure 10.1.

Nakatsuka and Grischkowsky (1981) demonstrated distortion-free pulse propagation of synchronously mode-locked dye laser pulses by using the positive GVD of fibers to chirp the pulses. In this experiment, low-power (to avoid SPM) 3.3-ps dye laser pulses were chirped and temporally broadened to 13 ps and recompressed back to 3.3 ps by the negative GVD of a near-resonant atomic Na-vapor delay line. Subsequently, Nakatsuka et al. (1981) performed the first pulse compression experiment using fibers as a Kerr medium in the positive GVD region. This experiment utilized both the positive

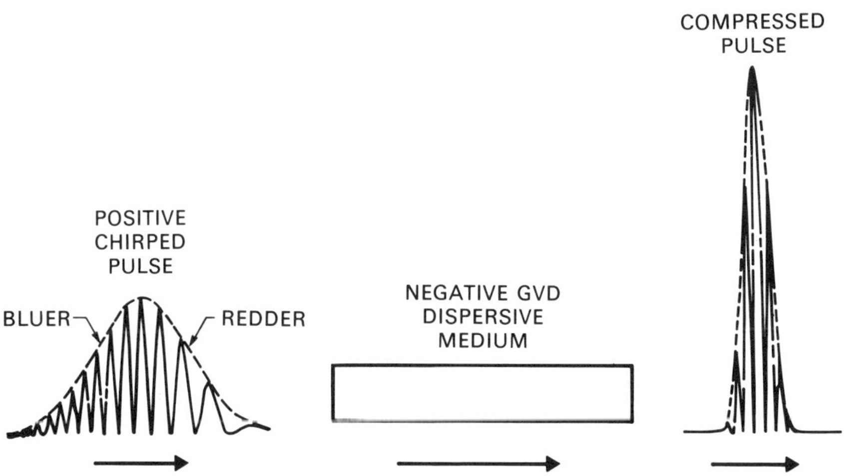

FIGURE 10.1. Compression of positively chirped optical pulses (red frequencies leading blue) using a dispersive medium possessing negative group velocity dispersion (GVD). Negative GVD: red frequencies delayed with respect to blue.

GVD and SPM to temporally and spectrally broaden 5.5-ps dye laser pulses with subsequent compression by $>3\times$ to 1.5 ps by passage through a near-resonant atomic Na-vapor delay link. Shank et al. (1982) replaced the atomic vapor delay line with a Treacy (1968; 1969) grating pair to compress the 90-fs amplified output of a colliding-pulse mode-locked (CPM) dye laser by $3\times$ to a duration of 30 fs, using a 15-cm fiber. Subsequently, Nikolaus and Grischkowsky (1983a) compressed the 5.4-ps output of a synchronously mode-locked and cavity-dumped dye laser by $12\times$ to a duration of 450 fs, using a grating-based dispersive delay line and a 30-m fiber. Using two stages of fiber-grating compression, Nikolaus and Grischkowsky (1983b) compressed 5.9-ps pulses from the aforementioned dye laser by $65\times$ to a duration of 90 fs. In the technological push to generate optical pulses of less than 10-fs duration, amplified CPM dye laser pulses were next compressed to 16 fs by Fujimoto et al. (1984). Compression to 12 fs by Halbout and Grischkowsky (1984) was soon followed by compression to 8 fs by Knox et al. (1985). Each of these compression achievements occurred with an important concomitant increase in repetition rate. Recently, Fork et al. (1987) achieved compression to 6 fs, the shortest to date, by using a grating pair followed by a prism sequence in order to compensate the cubic phase distortion of these large-bandwidth pulses by the grating pair.

Optical fiber compression of "long" duration picosecond pulses from CW mode-locked (CWML) Nd:YAG-based systems has also been achieved. Subpicosecond pulses can be generated with these sources without the use of a mode-locked dye laser. In addition, compressed CWML Nd:YAG-based systems can be used as pump sources for synchronously mode-locked dye lasers. Johnson et al. (1984a, 1984b) performed the first "long" pulse fiber compression experiments in a system other than a dye laser. In these experiments, 33-ps pulses at 0.532 μm from a CWML and frequency-doubled Nd:YAG laser were compressed $80\times$ to a duration of 410 fs, using a 105-m fiber and a grating pair. Shortly thereafter, Dianov et al. (1984a) compressed 60-ps pulses, at 1.064 μm from a CWML and Q-switched Nd:YAG laser (1 kHz repetition rate), by $15\times$ to a duration of 4 ps using a 10-m fiber and a grating-based delay line. The compression of 1.064-μm pulses from a CWML Nd:YAG laser was later performed independently by Kafka et al. (1984) and Heritage et al. (1984). Kafka et al. (1984) demonstrated the compression of 80-ps pulses by $45\times$ to a duration of 1.8 ps, while Heritage et al. (1984) compressed 90-ps pulses by $30\times$ to a duration of 3 ps. Dupuy and Bado (1984) reported the compression of 110-ps pulses from a CWML argon-ion laser by $5\times$. Further studies of the compression of 1.064-μm pulses from CWML Nd:YAG lasers were reported by Heritage et al. (1985a), Kafka and Baer (1985), and Gomes et al. (1985a). Using two stages of fiber-grating compression, Gomes et al. (1985b) compressed 85-ps pulses, at 1.064 μm, from a CWML Nd:YAG laser by $113\times$ to a duration of 750 fs. Damm et al. (1985) reported on the use of large-core (50 μm) graded-index fiber to compress 5-ps pulses at 1.054 μm from a mode-locked Nd:phosphate glass laser by $7\times$ to a duration of 700 fs. CWML and Q-switched Nd:YAG laser pulses at 1.064 μm were compressed 29 × to a duration of 2.9 ps by Gomes et al. (1985c) in a manner similar to that reported by Dianov et al. (1984). Blow et al. (1985) reported on all-fiber compression of a CWML Nd:YAG laser at 1.32 μm by adjusting the waveguide dispersion of two lengths of fiber. In this experiment, 130-ps pulses were compressed to a photodiode limit of 70 ps by using a dispersion-shifted positive GVD fiber followed by a negative GVD fiber. Kai and Tomita (1986a) reported the compression of 100-ps pulses at 1.32 μm from a CWML Nd:YAG laser by 50 × to a duration of 2 ps using 2 km of dispersion-shifted fiber and a grating pair. Using two stages of fiber-grating compression, Zysset et al. (1986) compressed 90-ps pulses, at 1.064 μm, from a CWML Nd:YAG laser by 450 × to a duration of 200 fs. Kai and Tomita (1986b) demonstrated the compression of 100-ps, 1.32-μm pulses from a CWML Nd:YAG laser by 1100 × to a duration of 90 fs by using one stage of fiber-grating compression (dispersion-shifted fiber) followed by soliton compression in a length of negative GVD fiber.

In Section 2 we present results for picosecond fiber-grating compression in a normalized form, from which one can calculate the optimum fiber length, the achievable compression, and the proper grating separation for a given input pulse and fiber. Section 3 deals with the subtleties and nuances of femtosecond fiber pulse compression, that is, higher order dispersion compensation of very large bandwidth pulses.

2. Picosecond Pulse Compression

2.1 Optical Kerr Medium

Optical fibers are usually considered to be linear media; that is, as the input power is increased, one expects only a proportional increase in output power (Stolen, 1979b). However, dramatic nonlinear effects can occur that can cause strong frequency conversion, optical gain, and many other effects generally associated with very intense optical pulses and highly nonlinear optical materials. These nonlinear processes depend on the interaction length as well as the optical intensity. In small-core fibers high intensities can be maintained over kilometer lengths. If this length is compared with the focal region of a Gaussian beam of comparable spot size, enhancements of 10^5 to 10^8 are possible using fibers. This enhancement lowers the threshold power for nonlinear processes—in some cases to less than 100 mW (Stolen, 1979b). For example, single-mode fibers with core diameters less than 10 μm possess core areas of $<10^{-6}$ cm^2, which serves to translate powers in watts to intensities of MW/cm^2. An intensity-dependent refractive index leads to SPM and self-focusing within a single optical pulse. In fibers, however, any additional confinement caused by self-focusing is negligible (Stolen and Lin, 1978). Recently, Baldeck et al. (1987) reported on the observation of self-focusing in optical fibers with 25-ps pulses from an active-passive mode-locked and frequency-doubled Nd:YAG laser. (see Chapter 4) There are several caveats to this observation of self-focusing: (1) self-focussing occurred with pulse energies greater than 10 nJ, in a multimode fiber with a core diameter of 100 μm; (2) self-focusing appeared primarily at Stokes-shifted stimulated Raman frequencies, for which the effectof the nonlinear refractive index is enhanced by cross-phase modulation; (3) self-focusing occurred at stimulated Raman conversion efficiencies of approximately 50%. The experimental conditions under which Baldeck et al. (1987) were able to observe self-focusing lend further substantial support to the claim that self-focusing is negligible under the standard experimental conditions for pulse compression in single-mode fibers. Hence single-mode fibers represent a nearly ideal nonlinear Kerr medium for the generation of the SPM necessary for pulse compression.

Fisher et al. (1969) suggested that picosecond pulses could be compressed to femtosecond durations by employing the large positive chirp obtainable near the center of a short pulse as a result of SPM in optical Kerr liquids. SPM results from the passage of an intense pulse through a medium with an intensity-dependent refractive index. When the relaxation time of the nonlinearity is much less than the input pulse duration, the region where the positive chirp is largest and least dependent on time occurs at about the peak of the pulse and large compression ratios are possible. For longer relaxation times, this region is delayed with respect to the peak of the pulse. Compression is diminished by the influence of relaxation, which not only delays the maximum chirp but also decreases the linear chirp in magnitude and extent. In the limit of the input pulse duration being much shorter than the relaxation time, the resultant chirp would be nonzero only on the wings of the pulse. In fact, if such a pulse were passed through the dispersive delay line, the most intense portion would remain uncompressed. Consequently, Fisher et al. (1969) limited their discussion to picosecond (>5 ps) pulses incident on Kerr liquid CS_2. The dominant contribution to the optical Kerr effect in CS_2 is molecular orientation, with a relaxation time of ~ 2 ps (Shapiro and Broida, 1967; Ippen and Shank, 1975a). In the optical Kerr gate experiments of Ippen and Shank (1975a) utilizing subpicosecond pulses incident on CS_2, the asymmetries in the transmission of the optical gate have been attributed to the relatively long relaxation time. In the case of fused silica, the dominant contribution to the Kerr coefficient is the optically induced distortion of the electronic charge distribution and is expected to have a relaxation time of $\sim 10^{-15}$ s (Alfano and Shapiro, 1970; Owyoung et al., 1972; Duguay, 1976). Thus, relaxation time effects should be negligible, even for the case of femtosecond duration input pulses, for compression in silica fibers.

In general, when an intense optical pulse passes through a nonlinear medium, the refractive index n is modified by the electric field E,

$$n = n_0 + n_2 \langle E^2(t) \rangle + \cdots , \tag{1}$$

where n_0 is the refractive index at arbitrarily low intensity and n_2 is the optical

Kerr coefficient (see Chapters 1 and 2). The time-dependent portion of the refractive index modulates the phase of the pulse as it propagates through the medium. A phase change $\delta\phi(t)$ is therefore impressed on the propagating pulse:

$$\delta\phi(t) = n_2 \langle E^2(t) \rangle \frac{\omega z}{c}, \tag{2}$$

where ω is the optical frequency, z is the distance traveled in the Kerr medium, and c is the velocity of light. When relaxation time effects can be neglected, according to Shimizu (1967) and DeMartini et al. (1967), the approximate frequency shift at any retarded time (t) is given by the time derivative of the phase perturbation, which is therefore proportional to the time derivative of the pulse intensity

$$\delta\omega(t) = -\frac{d}{dt}\delta\phi(t) = -\frac{\omega z}{c} n_2 \frac{d}{dt}\langle E^2(t) \rangle. \tag{3}$$

The instantaneous frequency $\omega(t)$ will shift from the input optical frequency ω_0 by an amount that depends on the intensity profile

$$\omega(t) = \omega_0 \left[1 - \frac{z}{c} n_2 \frac{d}{dt} \langle E^2(t) \rangle \right]. \tag{4}$$

A positive value of n_2 for silica implies that the increasing intensity in the leading edge of a short pulse results in an increasing refractive index, or a decreasing wave velocity. As a result of the negative sign in Eq. (4), the instantaneous frequency of the leading edge of the pulse will decrease with respect to ω_0. This time-dependent slowing of the wave reduces the rate at which the wave fronts pass a given point in the fiber, thus reducing the optical frequency. The leading edge of the pulse is therefore red-shifted. On the trailing edge of the pulse there is a corresponding frequency increase, or blue shift, resulting in an increase in the spectral bandwidth of the pulse.

For the propagation of low-intensity optical pulses the input and output frequency spectra would be the same (Ippen et al., 1974). As the intensity increases, the transmitted spectrum is broadened and spectral interference maxima and minima appear as the peak phase shift passes through multiples of π. As discussed in the next section, the peaks in the self-phase-modulated spectrum can be washed out or filled in by the presence of GVD.

The spectral broadening of an optical pulse is much easier to treat in the time domain than in the frequency domain (Stolen and Lin, 1978). In the time domain the intensity-dependent refractive index causes a phase shift of the center of the pulse with respect to the wings. A phase modulation introduces sidebands on the frequency spectrum. In the absence of GVD, the shape of the pulse does not change with distance along the fiber and the instantaneous phase depends on the pulse intensity. A phase-modulated or chirped pulse is illustrated in Figure 10.2a, indicating that the instantaneous frequency is lower

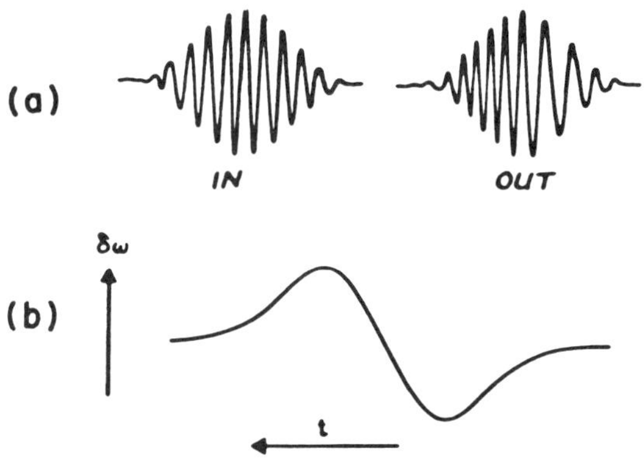

FIGURE 10.2. Effect of an intensity-dependent refractive index on the phase and instantaneous frequency of an optical pulse after propagation down a single-mode fiber.

in the leading half of the pulse and higher, than the carrier frequency, in the trailing half of the pulse. The magnitude of this frequency chirp, in the absence of pulse-shaping effects, builds up in direct proportion to the length of fiber traversed. As illustrated in Figure 10.2b, there is a nearly linear chirp through the central part of the pulse. This region of linear chirp (positive) can be compressed by the linear dispersion (negative) of a grating-pair delay line, by reassembling its frequency components. Treacy (1978, 1979) showed that when two wavelength components, λ and λ' are incident on a grating pair, the longer wavelength experiences a greater group delay. This group delay is determined by the optical path length traversed. With reference to Figure 10.3a, the relationship between the first-order diffraction angles is

$$\sin(\gamma - \theta) = \frac{\lambda}{d} - \sin\gamma$$

where d is the grating ruling spacing, θ is the acute angle between incident and diffracted rays, and γ is the angle of incidence measured with respect to the grating normal. The slant distance AB between the gratings is b, which equals G sec $(\gamma - \theta)$ where G is defined as the perpendicular distance between the gratings. The ray path length $PABQ$ (see Figure 10.3a) is given by

$$p = b(1 + \cos\theta) = c\tau$$

where τ is the group delay. After considerable algebraic manipulation, the variation of the group delay with wavelength, for various ray path lengths, is found by differentiating p/c with respect to λ leading to

$$\frac{\partial\tau}{\partial\lambda} = \left(\frac{b}{d}\right)\left(\frac{\lambda}{d}\right)\frac{1}{c[1 - (\lambda/d - \sin\gamma)^2]}$$

(see also Chapter 9).

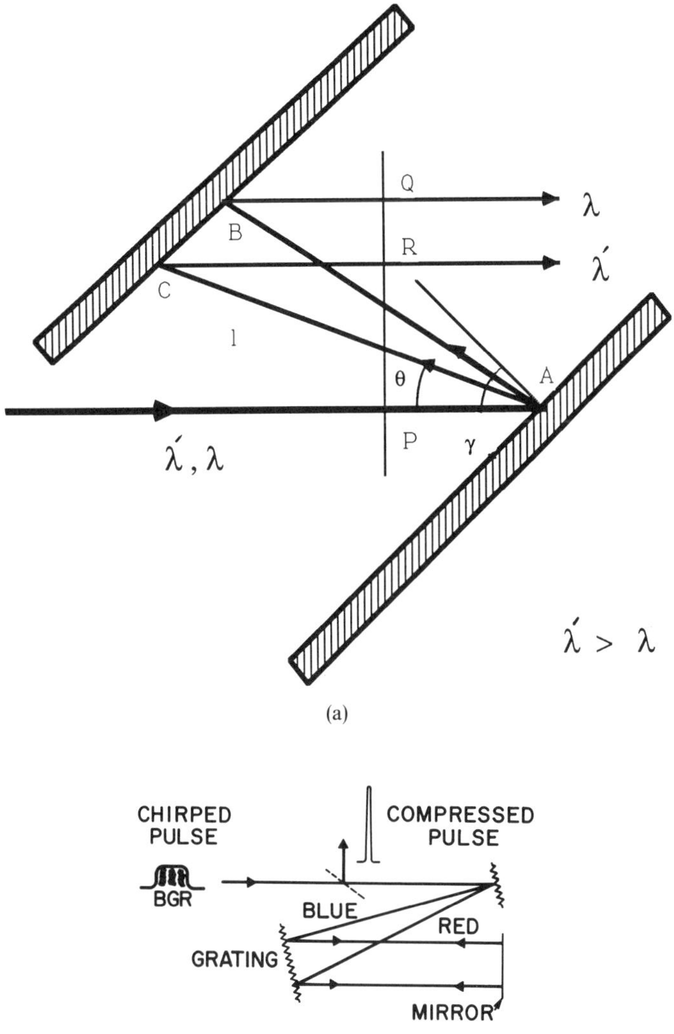

FIGURE 10.3. (a) Treacy (1978, 1979) "single-pass" geometrical arrangement of diffraction gratings used for pulse compression. The angle of incidence with respect to the grating normal is γ, and θ is the acute angle between the incident and diffracted rays. The ray paths are shown for two wavelength components with $\lambda' > \lambda$. Since the path length for λ' is greater than that for λ, longer wavelength components experience a greater group delay. (b) A positively chirped optical pulse with red (R) frequencies leading the blue (B) incident on a typical "double-pass" grating-pair compressor.

A typical "double-pass" (Johnson et al., 1984a, 1984b) grating-pair dispersive delay line is illustrated in Figure 10.3. Here the first grating disperses the beam and the second grating makes the spectral components parallel. The path for the red-shifted light is longer than that for the blue, so if the spacing is chosen correctly all the spectral components will be lined up after the second grating. However, the spectral components will not be together spatially and the output beam looks like an ellipse with red on one side and blue on the other. This is corrected by reflecting the beam back through the grating pair, and hence "double-pass," which puts the spectral components back together and doubles the dispersive delay; that is, the rays (blue and red) undergo double delay or retrace. Johnson et al. (1984a, 1984b) revived the use of the "double-pass" grating-pair delay line, introduced by Desbois et al. (1973) and Agostinelli et al. (1979), for the temporal expansion (picoseconds to nanoseconds) and shaping of mode-locked Nd:glass and Nd:YAG laser pulses. In large-compression-ratio experiments, the double-pass delay line cancels the large transverse displacement of the spatially dispersed spectral components of the output beam evident in Treacy's (1968, 1969) "single-pass" grating-pair delay line. The first compressor application of the double-pass delay line was in the $80\times$ compression of mode-locked and frequency-doubled Nd:YAG laser pulses by Johnson et al. (1984a, 1984b).

In the next section the parameters necessary for constructing an optical pulse compressor based on a single-mode fiber and grating pair are described.

2.2 Nonlinear Pulse Propagation and Grating Compression

Tomlinson et al. (1984) and Stolen et al. (1984c) have shown, over a fairly broad range of experimental parameters, that the propagation of short, high-intensity pulses in a single-mode fiber can be accurately described by a model that includes only the lowest-order terms in GVD (positive) and SPM. The pulse propagation is modeled by the dimensionless nonlinear wave (Schrödinger) equation:

$$\frac{\partial \mathscr{E}}{\partial(z/z_0)} = -i\frac{\pi}{4}\left[\frac{\partial^2 \mathscr{E}}{\partial(t/t_0)^2} - 2|\mathscr{E}|^2\mathscr{E}\right], \tag{5}$$

where $\mathscr{E}$ is the (complex) amplitude envelope of the pulse. (For a derivation of the nonlinear wave equation see Chapter 3). The time variable t is a retarded time and is defined such that for any distance z along the fiber, the center of the pulse is at $t = 0$, and we assume an input pulse envelope of the form

$$\mathscr{E}(z = 0, t) = A\,\mathrm{sech}(t/t_0). \tag{6}$$

The normalized length z_0 and the peak amplitude A are defined by

$$z_0 = 0.322\frac{\pi^2 c^2 \tau_0^2}{|D(\lambda)|\lambda} = 0.322\frac{\pi^2 c\tau_0^2}{|D|\lambda^2} \tag{7}$$

and

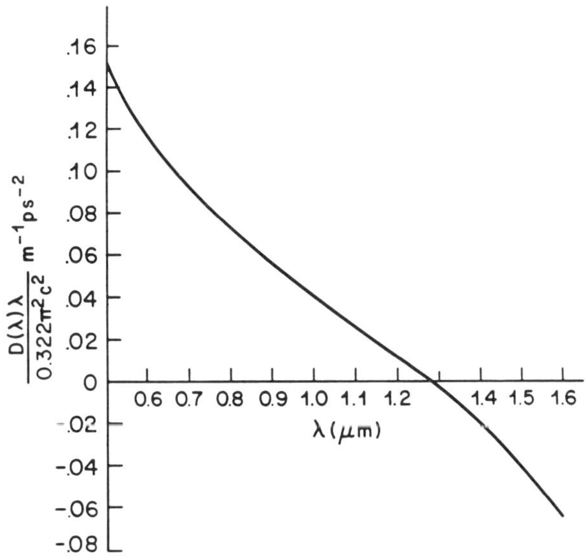

FIGURE 10.4. Plot of the normalized group velocity dispersion for a silica-core fiber. The normalized length z_0 is obtained in meters by dividing the square of the input pulse width in picoseconds by the value from the plot.

$$A = \sqrt{P/P_1}, \tag{8}$$

where

$$P_1 = \frac{nc\lambda A_{\text{eff}}}{16\pi z_0 n_2} \times 10^{-7}\ \text{W}, \tag{9}$$

τ_0 is the pulse width (full width at half-maximum) of the input pulse ($\tau_0 = 1.76t_0$), $D(\lambda)$ or D is the GVD, $D(\lambda)$ in dimensionless units or D in ps-nm/km, n is the refractive index of the core material and n_2 is its nonlinear Kerr coefficient in electrostatic units (1.1×10^{-13} esu for silica), c is the velocity of light (cm/s), and λ is the vacuum wavelength (cm). Figure 10.4 is a plot of the normalized GVD for a silica-core fiber, based on values of $D(\lambda)$ derived from Gloge (1971) and Payne and Gambling (1975). The peak power of the input pulse is given by P, and the quantity A_{eff} is an effective core area (cm^2), which for typical fiber parameters is fairly close to the actual core area. These normalized parameters come out of the theory for optical solitons in fibers (*negative* GVD), where z_0 is the soliton period and P_1 is the peak power of the fundamental soliton (Mollenauer et al., 1980). In the present regime of *positive* GVD there are no solitons, but these parameters are still useful because z_0 is actually the length of fiber required for GVD to approximately double the width of the input pulse (in the *absence* of SPM), and P_1 is the peak power required for SPM to approximately double the spectral width of the input pulse in a fiber of length z_0 (in the *absence* of GVD).

As discussed earlier, an intense optical pulse will be spectrally broadened and frequency chirped on exiting the fiber. The next step of the pulse compression process is to reassemble the chirped pulse with a compressor. The action of the compressor is most easily described in the frequency domain since it is simply a frequency-dependent time delay. The Fourier transform of the pulse can be expressed in the form

$$\mathscr{E}(z,\omega) = A(\omega)e^{i\phi(\omega)}, \tag{10}$$

where $A(\omega)$ and $\phi(\omega)$ are the amplitude and phase (for simplicity we do not indicate their z dependence explicitly). The effect of the compressor can be described by a phase function $\phi_c(\omega)$, so that the Fourier transform of the compressed pulse is given by

$$\mathscr{E}_c(z,\omega) = A(\omega)\exp\{i[\phi(\omega) + \phi_c(\omega)]\}. \tag{11}$$

If $\phi_c(\omega) = -\phi(\omega)$, then at $t = 0$ all the frequency components of the pulse will be in phase and will thus create the pulse with the maximum peak amplitude. We assume that it is also the shortest possible compressed pulse or close to it. We define this compressor as *ideal.*

One of the most useful types of compressors is the Treacy (1968, 1969) grating-pair compressor. A grating-pair compressor has a delay function that is approximately of the form

$$\phi_c(\omega) = \phi_0 - a_0\omega^2. \tag{12}$$

The compressor constant a_0 can easily be adjusted by varying the grating separation and is thus a directly accessible experimental parameter. We use the term *quadratic* compressor to refer to a compressor with a response function of the form of Eq. (12). It can be shown that the Fourier transform of a pulse with a linear frequency chirp, which means that the temporal phase is proportional to t^2, has a phase that is proportional to ω^2, so that, for a linearly chirped pulse, a quadratic compressor is the *ideal* compressor. To the extent that the frequency chirp on a pulse is nonlinear, a quadratic compressor is not the ideal compressor for that pulse, but if the departure is not too large, a quadratic compressor can still give reasonably good compression. The expression for the grating constant is

$$a_0 = \frac{b\lambda^3}{4\pi c^2 d^2 \cos^2\gamma'}, \tag{13}$$

where b is the center-to-center distance between the two gratings, d is their groove spacing, λ is the center wavelength of the pulse, and γ' is the angle between the normal to the input grating and the diffracted beam at λ. From the numerical solutions of Eq. (5), which includes GVD, Tomlinson et al. (1984) and Stolen et al. (1984c) were able to derive an expression for the grating constant in terms of the input pulse duration and peak amplitude,

$$a_0/t_0^2 \approx \tau/\tau_0 \approx 1.6/A \tag{14}$$

or

$$a_0 \approx 0.52\tau_0^2/A. \tag{15}$$

An expression for the grating separation can also be generated in terms of these experimental parameters by combining Eqs. (13) and (15):

$$b \approx \frac{2.08\pi c^2 d^2 \cos^2 \gamma'}{\lambda^3}\left(\frac{\tau_0^2}{A}\right). \tag{16}$$

For the limiting case of zero GVD, the numerical results for the optimum quadratic compressor lead to

$$a_0/t_0^2 \approx 0.25(A^2 z/z_0)^{-1}. \tag{17}$$

Using Eqs. (13) and (17) as well as the pulse length normalization ($\tau_0 = 1.76t_0$), the grating separation in the absence of GVD is given by

$$b \approx \frac{0.323\pi c^2 d^2 \cos^2 \gamma'}{\lambda^3}\left(\frac{\tau_0^2}{A^2 z/z_0}\right). \tag{18}$$

It is interesting to discuss the role GVD plays in the pulse compression process. In simulating the compression of picosecond pulses chirped by propagating in the nonlinear Kerr liquid CS_2, Fisher and Bischel (1975) concluded that GVD would have the influence of expanding the temporal region over which the chirp was relatively linear, resulting in optimum compression. Grischkowsky and Balant (1982a, 1982b) were the first to realize the significance of GVD in fiber-grating compression. During passage through the fiber, both the pulse shape and the frequency bandwidth are broadened by the combined action of SPM and positive GVD. Thus, the red-shifted light generated at the leading edge of the pulse travels faster than the blue-shifted light generated at the trailing edge, and this leads to pulse spreading and rectangular pulse shapes. Because the new frequencies are generated primarily at the leading and trailing edges, which gradually move apart in time, the pulse develops a linear frequency chirp over most of the pulse length. These "enhanced frequency chirped" (Grischkowsky and Balant, 1984a, 1984b) pulses can lead to almost ideal compression by a grating pair. Negligible GVD, on the other hand, can lead to large deviations from linearity of the chirp, which can result in substantial wings or sidelobes on the compressed pulse.

Optimum fiber-grating compression requires the appropriate choice of the fiber length and grating spacing for optimum chirp and chirp compensation, respectively. These length scales can vary over an enormous range. For example, Knox et al. (1985) compressed 40-fs pulses to a duration of 8 fs using 7 mm of fiber and a grating separation of approximately 1 cm. On the other extreme, Johnson et al. (1984a, 1984b) compressed 33-ps pulses to a duration of 410 fs using 105 m of fiber and a grating separation of 7.2 m. For a particular input pulse width, peak power, wavelength, and fiber core area, the optimal chirp occurs for a single fiber length. The optimal fiber length varies as

$$z_{\text{opt}} \propto \tau_0^2/\sqrt{P}. \tag{19}$$

Not unexpectedly, the grating separation has the same dependence on input pulse width and peak power (see Eqs. (8) and (16)).

Two limiting fiber length regimes of practical interest can be identified (Tomlinson et al., 1984; Stolen et al., 1984). The first is that of a fiber of optimum length to provide the best linear chirp (Grischkowsky and Balant, 1982a, 1982b). The second regime is that for which the length is much less than optimal and the effects of GVD can be neglected. Some of the properties of the chirp and compression in these two limiting regimes are illustrated in Figure 10.5, where the compression factor is about $12.5\times$ in each case. Displayed in Figure 10.5 are the pulse shape exiting the fiber, the frequency spectrum, the chirp, and the compressed pulse for both optimum quadratic and ideal compressors. If GVD is negligible, the fiber output pulse will have the same shape and intensity as the input pulse, while in a fiber of optimum length GVD will broaden the output pulse by about a factor of $3\times$. As pointed out by Grischkowsky and Balant (1982a, 1982b), a "squared" or "rectangular" fiber output pulse will have a linear frequency chirp over most of the length of that pulse. The overall width of the frequency spectrum is about the same

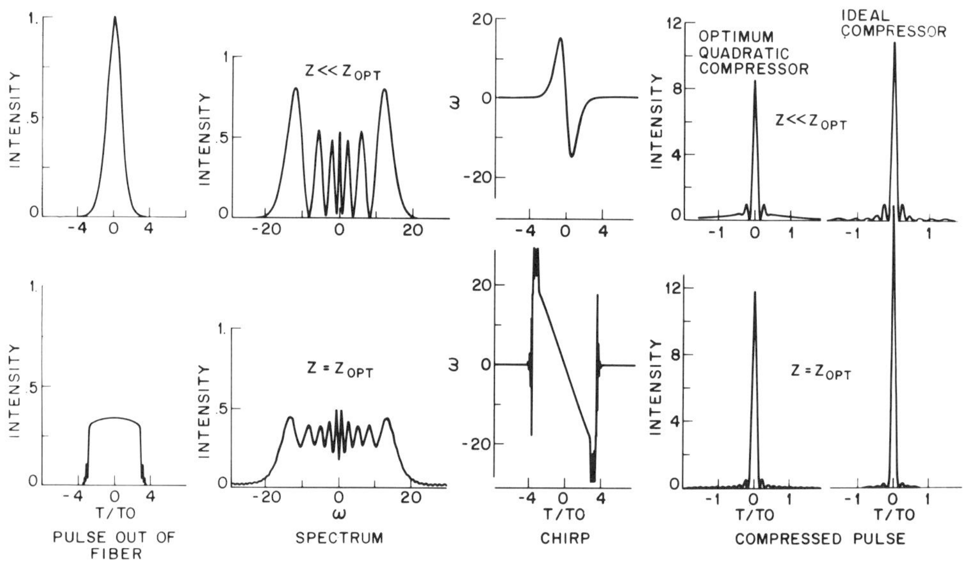

FIGURE 10.5. Pulse shapes before and after grating compression, frequency spectra, and chirp for the limiting regimes of optimal fiber length and of negligible group velocity dispersion. The upper curves are for the case of negligible GVD and an intensity-length product $A^2z/z_0 = 12.5$. The lower curves are for $A = 20$ and the corresponding optimum fiber length of $z_{\text{opt}} = 0.075z_0$. To compare the quality of compression in the two regimes, a common compression factor of approximately $12.5\times$ was chosen in each case.

TABLE 10.1. Fiber-grating compressor parameters.[a]

$$C_1 = \frac{D(\lambda)\lambda}{0.322\pi^2 c^2} = \begin{cases} 0.144 & (\lambda = 0.5145\ \mu\text{m}) \\ 0.138\ \text{m}^{-1}\ \text{ps}^2 & (\lambda = 0.532\ \mu\text{m}) \\ 0.117 & (\lambda = 0.600\ \mu\text{m}) \\ 0.031 & (\lambda = 1.064\ \mu\text{m}) \end{cases}$$

$$P_1 = \frac{nc\lambda A_{\text{eff}}}{16\pi z_0 n_2} = 7.92\left[\frac{\lambda(\text{cm})A_{\text{eff}}(\text{cm}^2)}{z_0(\text{cm})}\right] \times 10^{14}\ \text{W}$$

$z = z_{\text{opt}}$	$z \ll z_{\text{opt}}$
$\tau_0/\tau \approx 0.63A$	$\tau_0/\tau \approx 1 + 0.9[A^2 z/z_0]$
$z_{\text{opt}}/z_0 \approx 1.6/A$	—
$b \approx 84C_2 \dfrac{\text{cm}}{\text{ps}^2}\left[\dfrac{\tau_0^2}{A}\right]$	$b \approx 13C_2 \dfrac{\text{cm}}{\text{ps}^2}\left[\dfrac{\tau_0^2}{A^2 z/z_0}\right]$

$$C_2 = \left[\frac{d(\text{cm})}{5.56 \times 10^{-5}}\right]^2 \left[\frac{600}{\lambda(\text{nm})}\right]^3 \cos^2\gamma'$$

[a] $z_0 = \tau_0^2/C_1$; $A = \sqrt{P/P_1}$; $\tau_0 = 1.76t_0$.

in the two cases, but GVD acts to fill in the spectrum. For each length, the grating separation was optimized to give the maximum peak intensity. The optimum fiber length was chosen to maximize the energy in the compressed pulse. In Figure 10.5 it is interesting to note that when the fiber length and grating separation are optimized, the quality of the compressed pulse (optimum quadratic compressor) is better than that with the ideal compressor in the absence of GVD.

The procedure of Tomlinson et al. (1984) and Stolen et al. (1984c) for calculating the compression, the optimal fiber length, and the grating separation is presented in Table 10.1. There are two important normalized parameters. The first is the normalizing length z_0 defined in Eq. (7), and the second is the normalized amplitude A defined in Eq. (8). For a silica-core fiber the GVD parameter C_1 has been given for several common laser wavelengths but can also be read directly from Figure 10.4 for an arbitrary wavelength. Table 10.1 also gives approximate expressions for the compression factor τ_0/τ, the optimum fiber length z_{opt}, and the grating separation b in both limits of fiber length (see Eqs. (16) and (18)). These expressions are supported by a recent approximate analytical theory of the compression process reported by Meinel (1983). The angle γ' is between the normal to the grating and the diffracted beam. For $A \leq 3$, the compression and pulse quality are not strong functions of fiber length, so there is no clear optimum length. This fact is consistent with the results of Shank et al. (1982) in which 90-fs pulses were compressed with fiber lengths between 4 and 20 cm and produced the same factor of 3 × compression.

Johnson et al. (1984a, 1984b) used these numerical calculations to design the first "long" pulse fiber-grating compressor for 33-ps pulses (0.532 μm) from

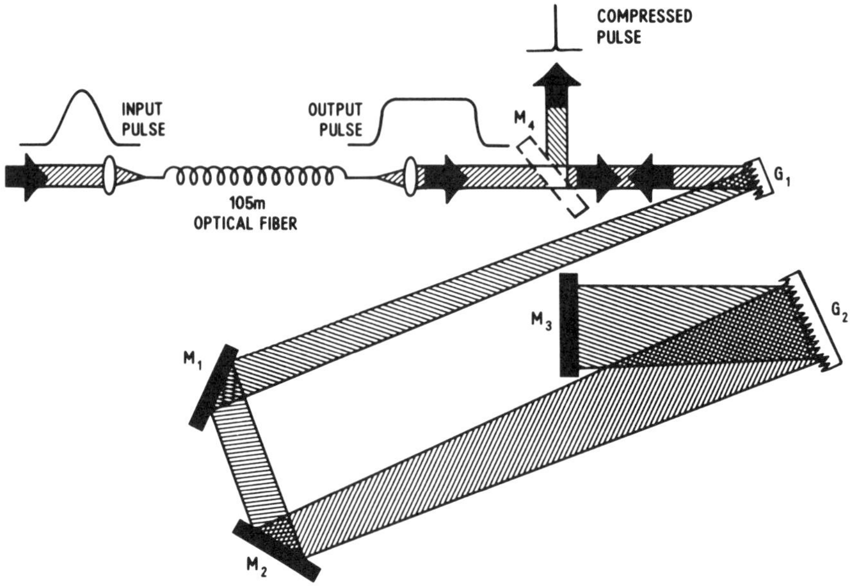

FIGURE 10.6. Schematic drawing of the "double-pass" fiber-grating pulse compressor used for the $80\times$ compression of 0.532-μm pulses. The dispersive delay line consists of gratings G_1 and G_1 (1800 grooves/mm) and mirrors M_1, M_2, and M_3. Mirror M_4 is cut in half to allow the fiber output pulse to pass over it. M_3 is slightly tilted downward to allow the output beam to be reflected by M_4 out of the compressor. The round-trip distance between the gratings was 724 cm.

a frequency-doubled Nd:YAG laser. The calculation deduced an optimum fiber length $z_{opt} = 83$ m, a grating separation $b = 606$ cm (1800 grooves/mm), and a compressed pulse width $\tau = 350$ fs. The experiment consisted of coupling 240-W pulses into a 105-m single-mode polarization-preserving fiber (Stolen et al., 1978) with a core diameter of 3.8 μm. A schematic drawing of the fiber-grating compressor is displayed in Figure 10.6. Compressed pulses as short as 410 fs or a compression of $80\times$ was obtained with a grating separation of 724 cm. The fiber input pulse and the compressed pulse are displayed together in Figure 10.7. The agreement between calculation and experiment was quite remarkable in light of the fact that the calculations were for normalized amplitudes of $A \leq 20$. The compression experiments had normalized amplitudes of $A > 150$.

As a result of the limited availability of high-reflectivity gratings at 0.532 μm with greater groove densities than 1800 grooves/mm it was important to keep the input pulse width as short as possible to avoid an unreasonable grating separation. The grating separation varies as the square of the input pulse width (see Eq. (16)). This is not as serious a problem at 1.064 μm, where the cubic wavelength dependence is quite helpful in keeping

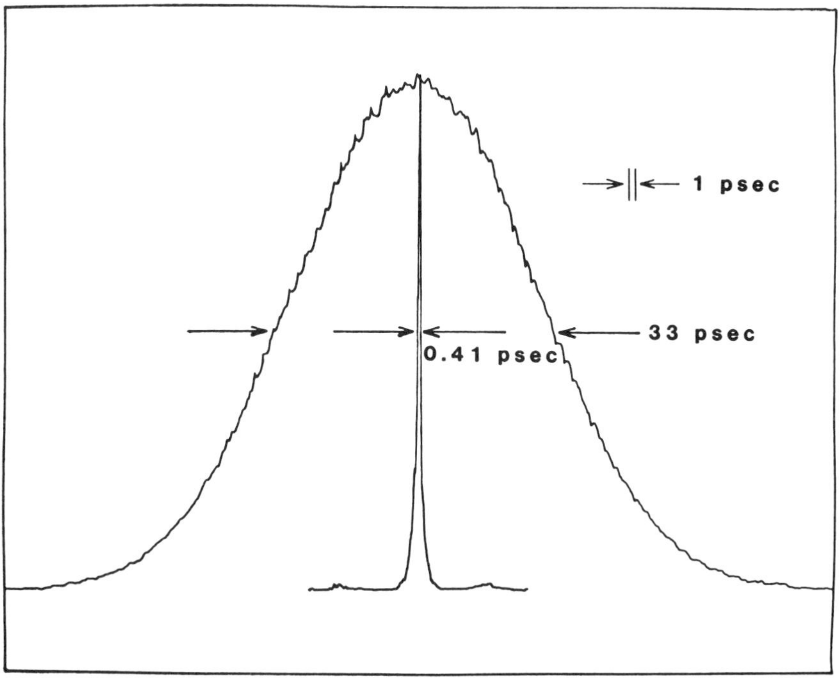

FIGURE 10.7. 80 × compression of 33-ps, 0.532-μm pulses. Standard background-free autocorrelation of the input and compressed pulses displayed on the same scale. (Input: Gaussian pulse shape. Compressed pulse: sech^2 pulse shape.)

the grating separation reasonable. The 33-ps fiber input pulses were obtained by frequency doubling a harmonically mode-locked CW Nd:YAG laser (Johnson and Simpson, 1983, 1985a; Keller et al., 1988). As opposed to fundamental mode locking, typical harmonic mode-locked pulse widths are 50 ps at 1.064 μm (Johnson and Simpson, 1985a) (see Figure 10.8). If 50- to 60-ps fiber input pulses (0.532 μm) from a standard fundamentally mode-locked laser were used, a grating separation of greater than 14 m would have been needed.

At this point it is useful to give a pulse compression example replete with experimental parameters, numerical calculation parameters, and the resultant numerical simulation of the compression process. This next example of 0.532 μm pulse compression was performed using a separate, larger-core fiber distinct from that discussed earlier in Johnson et al. (1984a, 1984b). A "flat" polarization-preserving fiber made by preform deformation (Stolen et al., 1984b) with a 4.1-μm silica-core diameter. A micrograph of the fiber is shown in Figure 10.9. The birefringence resulting from the stress cladding lifts the degeneracy of the two orthogonal modes of propagation. Thus linearly polarized light propagating along the well-defined principal fiber axis will be

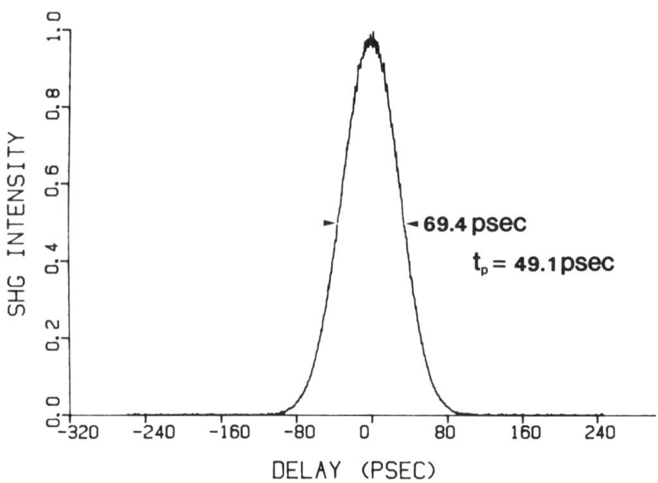

FIGURE 10.8. Background-free autocorrelation of the 1.064-μm output of a harmonically mode-locked CW Nd:YAG laser (Gaussian pulse shape). These pulses are frequency doubled in $KTiOPO_4$ (KTP) to yield pulses of 33 to 35 ps duration at 0.532 μm.

FIGURE 10.9. "Flat" polarization-preserving fiber made by preform deformation. The 4.1-μm pure silica core is surrounded by a $B:Ge:SiO_2$ stress cladding, an $F:SiO_2$ outer cladding, followed by a pure silica support cladding. The "rectangular" fiber has overall dimensions of 100×200 μm.

preserved. The diffraction efficiency of the gratings is a very sensitive function of the polarization of the fiber output. Thus, polarization-preserving fiber is extremely useful in reducing amplitude fluctuations in the compressed output due to polarization "scrambling" effects in fibers. One disadvantage of using polarization-preserving fibers is that the threshold for stimulated Raman scattering (SRS) is reduced by a factor of 2 in fibers maintaining linear polarization (Stolen, 1979a). Experimentally, it was found that the reduced SRS threshold was a small price to pay for the increased amplitude stability afforded by polarization-preserving fiber. The effects of SRS on compression are discussed in the next section.

For comparable compression, this larger core diameter fiber resulted in the reduction of SRS by about a factor of 2. The fiber length was 93.5 m and had a loss of 16 db/km at 0.532 μm. At the input lens ($10\times$ objective) the peak power was 235 W (820 mW average, 100 MHz repetition rate). However, this is not the best estimate of the peak power actually coupled into the fiber. The best approach is to measure the light coupled out of the fiber and correct for the transmission of the output lens and the known loss in the given length of fiber. This approach corrects for the loss in the input lens, mode-matching effects, light coupled into the cladding, and the reflection loss on the the fiber input face. In this instance the peak power actually coupled into the fiber is closer to 172 W. The fiber-grating compressor parameters given in Table 10.1 can be calculated with the following information: $P = 172$ W, $\tau_0 = 35$ ps, $n = 1.46$, $A_{\text{eff}} = 0.92 A_{\text{core}} = 1.21 \times 10^{-7}$ cm^2, $n_2 = 1.1 \times 10^{-13}$ esu, $P_1 =$ 5.74 mW, $A = 173$, $z_0 = 8.88$ km, $d = 5.56 \times 10^{-5}$ cm, and $\gamma' = 32.5°$. The calculations indicate an optimum fiber length $z_{\text{opt}} = 82$ m, a grating separation of 603 cm, and a compressed pulse width of 320 fs. With this fiber, pulses have been compressed to durations as short as 430 fs ($80\times$ compression). The typical day-to-day duration of the compressed pulses falls in the range 460 to 470 fs (Johnson and Simpson, (1985a, 1986). The actual grating separation used to generate the 460-fs pulse displayed in Figure 10.10 was 698 cm.

What actually happened to the 35-ps, 235-W pulse as it propagated through the fiber-grating compressor to produce the clean 460-fs pulses at the output? The spectral width of the fiber input pulses was measured to be 0.27 Å and was limited by the slits on the spectrometer. The self-phase-modulated spectrum of the output pulse was broadened to 17.2 Å and is displayed in Figure 10.11. GVD acted to fill in the self-phase-modulated spectrum, resulting in a flattened spectrum. The fiber output pulse was substantially broadened to a duration of 142 ps and is displayed in Figure 10.12. The triangular autocorrelation function is indicative of a rectangular intensity profile. As pointed out by Grischkowsky and Balant (1982a, 1982b), a rectangular fiber output pulse will have a linear frequency chirp over most of its length and result in optimum compression by a grating pair.

The two symmetrically located sidelobes on the fiber output frequency spectrum in Figure 10.11 are not expected from pure SPM. To determine the origin of these sidelobes Tomlinson et al. (1985) performed numerical

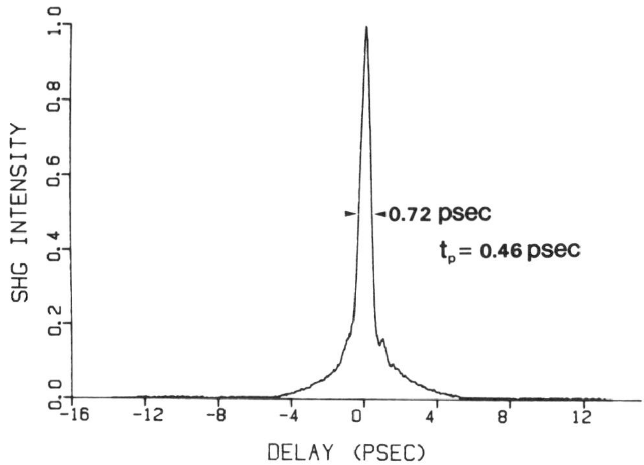

FIGURE 10.10. Typical autocorrelation of the compressed 0.532-μm pulses using a 93.5-m length of the "flat" polarization-preserving fiber displayed in Figure 10.9. Typical day-to-day pulses fall into the range of 460 to 470 fs.

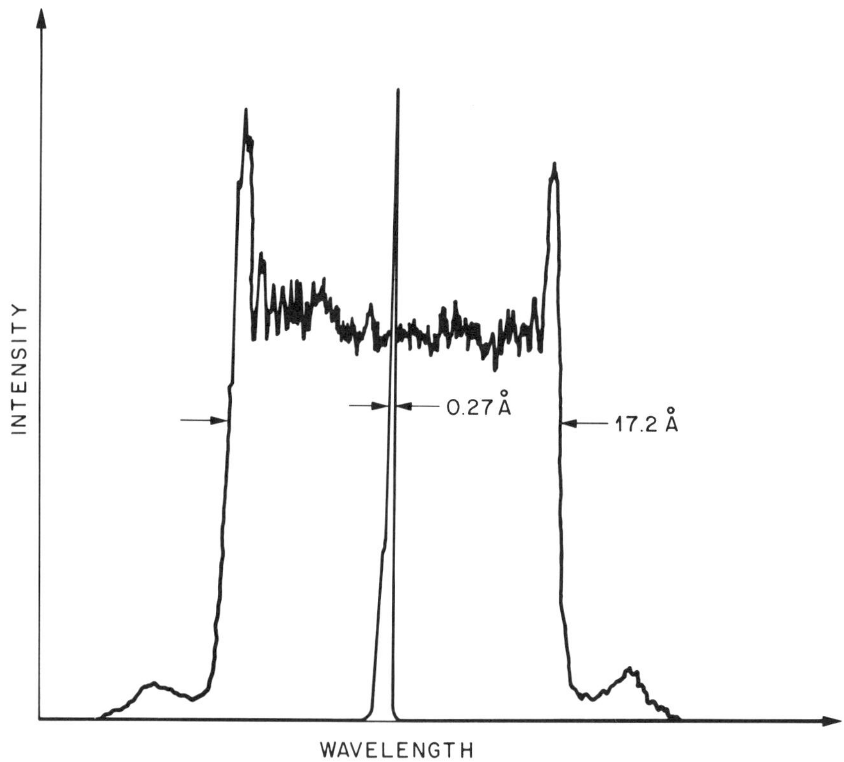

FIGURE 10.11. Spectral width of the 35-ps, 235-W, 0.532-μm fiber input pulses and the spectrally broadened (by SPM) output pulses after propagation down a 93.5-m length of the 4.1-μm core diameter polarization-preserving fiber.

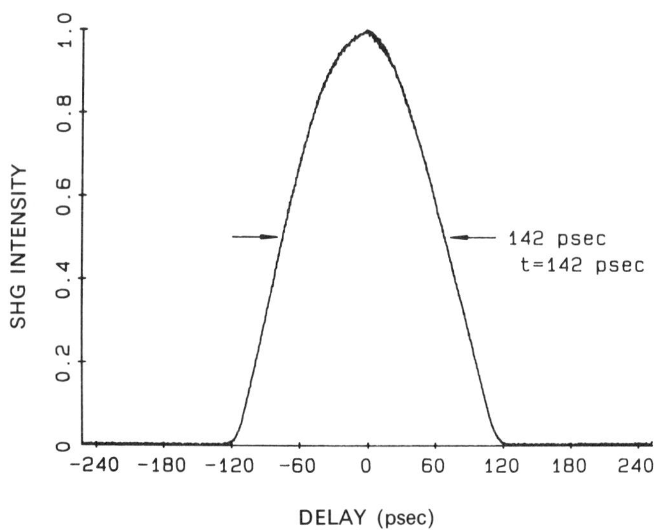

FIGURE 10.12. Autocorrelation of the fiber output pulse broadened by GVD. A triangular autocorrelation function is indicative of a rectangular intensity profile that has a deconvolution factor of unity.

simulations of the nonlinear pulse propagation using the experimental and calculated numerical parameters given previously. The results of several of the numerical calculations are given in Figure 10.13. This figure presents the nonlinear pulse propagation as a function of fiber length z/z_0. In Figure 10.13a, at $z/z_0 = 0.0020$ ($z = 18$ m), the temporal shape of the pulse is only slightly broadened (the input pulse width is $1.76t_0$), and the instantaneous frequency function and spectrum are characteristic of pure SPM. Recall that the calculated optimum fiber length is $z_{\text{opt}} = 82$ m. In Figure 10.13b, at $z/z_0 = 0.0054$ ($z = 48$ m), the temporal shape has become more "rectangular" as a result of the influence of GVD. The instantaneous frequency function indicates a nearly linear frequency chirp over a significant portion of the pulse width. In Figure 10.13c, at $z/z_0 = 0.0060$ ($z = 53$ m), as the fiber length approaches z_{opt} the chirp "linearization" proceeds as a result of the concomitant pulse broadening.

Figure 10.14 displays the instantaneous frequency function, the temporal pulse shape, and its frequency spectrum, for a length $z/z_0 = 0.01$ ($z = 89$ m), for a lossless fiber and for a fiber with a loss of 16 db/km (normalized loss parameter $\alpha = 16.36$). The nonlinear Schrödinger equation (Eq. (5)) with the inclusion of a normalized linear loss parameter α is given by

$$\frac{\partial \mathscr{E}}{\partial(z/z_0)} = -i\frac{\pi}{4}\left[\frac{\partial^2 \mathscr{E}}{\partial(t/t_0^2)} - 2|\mathscr{E}|^2\mathscr{E}\right] - \alpha\mathscr{E}. \tag{20}$$

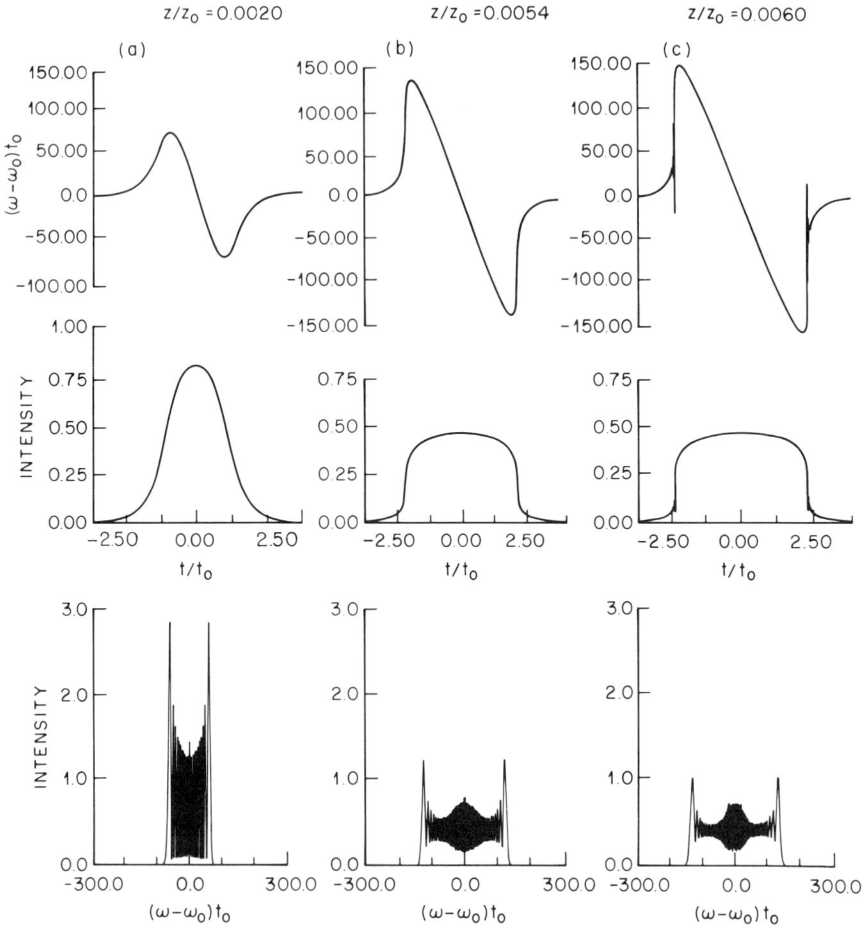

FIGURE 10.13. Numerical simulations of the nonlinear pulse propagation of 34.4-ps pulses ($t_0 = 19.5$ ps) at 0.532 μm with a normalized input amplitude of $A = 173$ and normalizing fiber length of $z_0 = 8.88$ km displayed as a function of fiber length z/z_0. The upper curves show the instantaneous frequency as a function of time, the middle curves show the intensity as a function of time, and the lower curves show the frequency spectra of the pulses. For the 93.5-m fiber used the normalized length was $z/z_0 = 0.0105$.

In Figure 10.14a and b the resulting pulse shape shows well-developed interference fringes on the leading and trailing edges. The frequency spectra clearly display the symmetrically located sidelobes. The spectrum for the fiber with loss (Figure 10.14b) is in excellent agreement with the experimental spectrum of Figure 10.11. (Since the experimental spectrum is an average over many pulses, we do not expect to see the fine structure displayed in the calculated spectrum.) Simulations of the effect of a grating-pair compressor on these fiber output pulses gave a compression of 98 × for the lossless fiber and of 84 ×

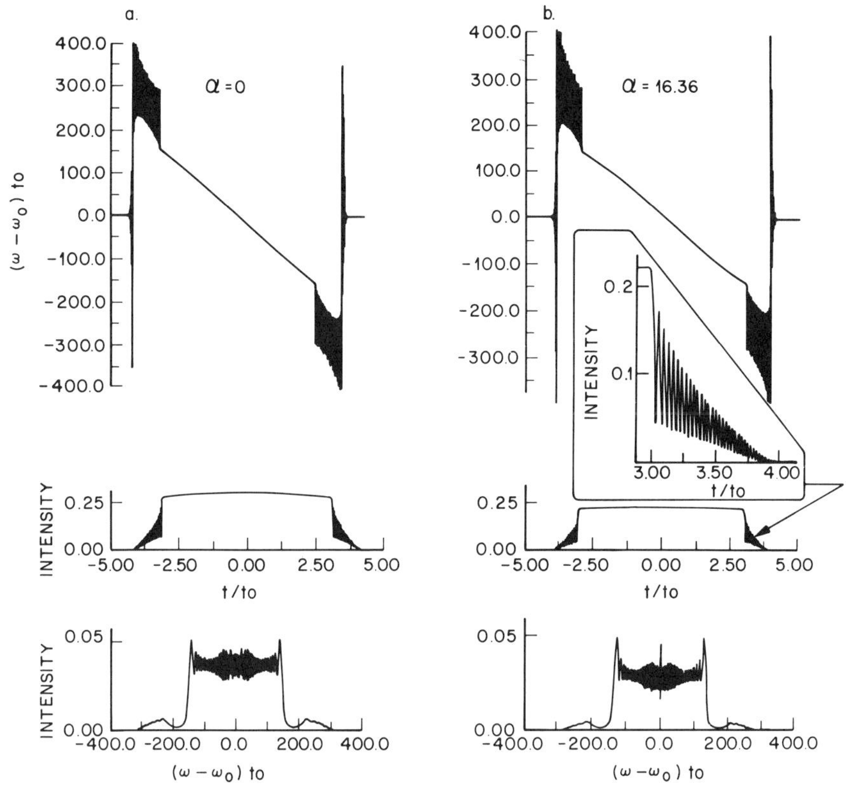

FIGURE 10.14. Numerical simulations of nonlinear pulse propagation for $A = 173$ and a fiber length of $z/z_0 = 0.01$ for (a) a lossless fiber and (b) a fiber with a normalized loss parameter $\alpha = 16.36$ (16 db/km). The upper curves are the instantaneous frequency, the middle curves are the temporal shape of the output pulse, and the lower curves show the frequency spectra of the output pulses. The inset in (b) displays a detail of the interference region (optical wave-breaking) on the edge of the pulse.

for the fiber with loss. Thus by including the fiber loss in the nonlinear Schrödinger equation, numerical solutions of Eq. (20) are in excellent agreement with the experimentally observed 80 × compression.

The origin of the sidelobes in the frequency spectrum of the fiber output pulses on Figures 10.11 and 10.14 has been attributed to a phenomenon that Tomlinson et al. (1985) dubbed "optical wave-breaking." Briefly, when an intense pulse propagates down an optical fiber, the leading edge of the pulse experiences a frequency decrease or red shift, while the trailing edge experiences a blue shift. For large-compression experiments (large values of the normalized amplitude A) in the presence of GVD, the red-shifted light near the leading edge of the pulse travels faster than, and overtakes, the unshifted light in the forward tails of the pulse (and vice versa on the trailing edge). Therefore, the leading and trailing regions of the pulse will contain light at two different frequencies, which will interfere and generate new frequencies. These new frequencies appear as the sidelobes on the fiber output spectrum and result in a small increase in the background on the compressed pulse. This phenomenon is somewhat analogous to the "breaking" of water waves and has been described as optical wave-breaking. Optical wave-breaking can also occur in small-compression-ratio experiments (small values of A) if the fiber is longer than the optimum fiber length, so that there is sufficient GVD to mediate the interference process. The interference fringes resulting from the optical wave-breaking are prominent in the calculated temporal pulse shapes (see inset of Figure 10.14b). Since the spectral bandwidth that contributes to the compressed pulse is approximately twice the frequency difference involved in the wave-breaking interference, the period of the interference fringes is approximately twice the width of the compressed pulse. Additional evidence for optical wave-breaking has appeared in the numerical studies of nonlinear pulse propagation by Lassen et al. (1985).*

These experiments and numerical simulations demonstrate the enormous range of applicability of the nonlinear Schrödinger equation (Eq. (20)) for describing nonlinear pulse propagation in single-mode fibers. Each of the various terms in Eq. (20) represents the lowest-order approximation to the phenomenon that it is describing, and it is assumed that the higher-order terms will be significant for very high compression ratios and/or very short input pulses. The present results indicate that large compression ratios of 80 × can accurately be described by Eq. (20) without invoking any higher-order terms. The limits of this fiber-grating compression approach have recently been studied by Bourkoff et al. (1987a, 1987b), Tomlinson and Knox (1987), and Golovchenko et al. (1988).

2.3 Stimulated Raman Scattering and Pulse Compression

The interplay between optical fiber pulse compression ($\lambda < 1.3\ \mu$m) and stimulated Raman scattering, or more appropriately the interplay between SPM, SRS, and GVD, could easily fill a book chapter. It is much beyond the scope of this chapter to discuss the role of SRS in great detail. Instead, the reader is referred to a number of excellent articles on SRS and nonlinear pulse propagation in fibers in the region of positive GVD: Auyeung and Yariv, 1978; Butylkin et al., 1979; Dianov et al., 1984b, 1985, 1986a, 1987; Gomes et al., 1986a, 1986b, 1988a; Heritage et al., 1988; Hian-Hua et al., 1985; Johnson et al., 1986; Kuckartz et al., 1987, 1988; Lin et al., 1977; Nakashima et al., 1987; Ohmori et al., 1983; Roskos et al., 1987; Schadt et al., 1986, 1987; Smith, 1972; Stolen and Ippen, 1973; Stolen and Johnson, 1986; Stolen et al., 1984a, 1972; Stolz et al., 1986; Valk et al., 1984, 1985; Weiner et al., 1988. Several of the salient features of SRS are briefly discussed in this section.

---

* Note added in proof: Optical wave-breaking was recently observed (temporally) by Rothenberg and Grischkowsky (1989).

The maximum power of an optical pulse in a fiber is usually limited by SRS. In the region of positive GVD, a Raman Stokes pulse will travel faster than the pump pulse. Thus the role of GVD is important in determining the limitations of SRS on the self-phase-modulated pump pulse. Stolen and Johnson (1986) discussed a simple picture of the SRS process that assumes that the Stokes power builds up from a weak injected signal rather than from spontaneous scattering. This follows the approach of Smith (1972) for CW Raman generation, in which the integrated spontaneous Raman scattering along the fiber can be replaced by a weak effective Stokes input power. At the top of Figure 10.15, a portion of the injected CW Raman signal enters the fiber along with the leading edge of the pump pulse. Because of GVD, this part of the signal will travel faster than the pump and never experience Raman amplification. In the second line of Figure 10.15, a portion of the signal enters the fiber along with the peak of the pump pulse. As the pump pulse travels along the fiber with velocity V_p, the faster-traveling Stokes signal (V_s) is amplified by extracting energy from the pump. Amplification ceases when the signal has passed through the pump pulse. Maximum amplification will occur for a Stokes signal that passes through the entire pump pulse, and this is the portion of the CW signal that enters the fiber along with the trailing edge of the pump pulse (third line of Figure 10.15). If a significant part of the pump energy has been shifted to the Stokes frequency (pump depletion), subsequent portions of the CW signal (fourth and fifth lines of Figure 10.15) will see a much reduced amplification.

The net result is that the amplification of the injected CW signal by the pump pulse has produced a Stokes pulse with a peak that is ahead of the pump pulse by about one pump pulse length (line 7 of Figure 10.15). If we define a walk-off length l_w,

$$l_w = \left(\frac{V_s V_p}{V_s - V_p}\right)\tau_0, \tag{21}$$

as the distance in which the Stokes signal passes through one pump pulse width τ_0 the Stokes maximum will be produced about two walk-off lengths into the fiber. All of the Stokes conversion will occur within about four walk-off lengths.

The signal gain depends on distance (L) along the fiber, and the net amplification involves an integral over the region where the Stokes and pump pulses interact. In the limit where the Stokes signal sees the entire pump pulse the gain becomes

$$P(L) = P(0)e^G, \tag{22a}$$

$$G \to \frac{g_0 P_0}{A_{\text{eff}}} \int dz \exp\left[\frac{-1.67z}{l_w}\right]^2 = 1.06 \frac{g_0 P_0 l_w}{A_{\text{eff}}}, \tag{22b}$$

where g_0 is the peak Raman gain coefficient, A_{eff} is the effective core area, and P_0 is the peak pump power. Thus, the maximum gain is approximately the peak power times the walk-off length. Typical values of G will be around 16

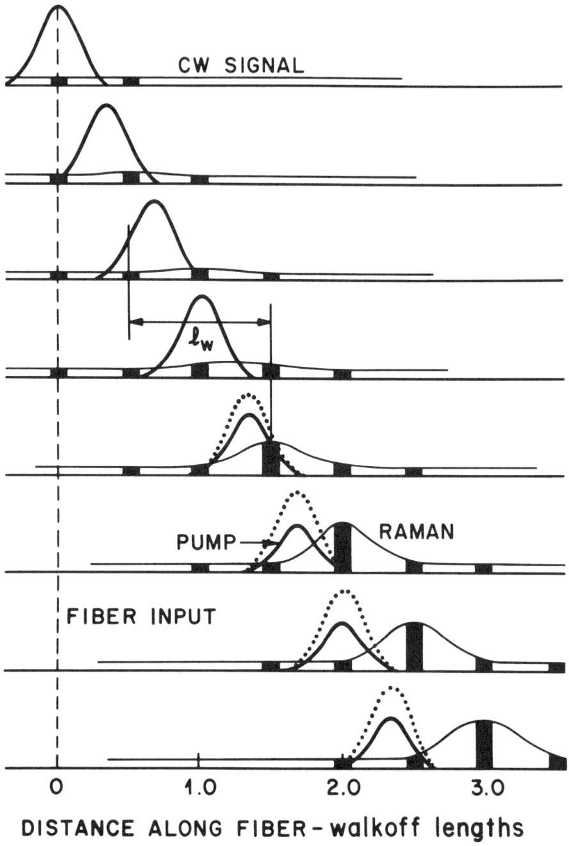

FIGURE 10.15. Schematic representation of a CW Stokes signal amplified as it passes through a pump pulse because of GVD. Portions of the CW signal are separately identified to illustrate that maximum amplification occurs for the part of the CW signal that enters the fiber with the trailing edge of the pump pulse. Earlier portions of the CW signal see reduced gain because they do not pass through the entire pump pulse, and later portions of the signal see reduced gain because the pump pulse has been depleted by earlier Raman conversion. The dotted curve represents the propagation of the pump pulse in the absence of Raman conversion.

but can go as high as 20 for significant Raman conversion and small walk-off or interaction lengths. For example, Stolen and Johnson (1986) estimated a value of $G = 19.7$, for 20% Raman conversion of 35-ps, 0.532-μm pulses, and a walk-off length of 6.2 m.

An estimate of the critical pump power $P_p = P_c$ entering the fiber, for which the intensities of the first Stokes component of SRS and of the pump were equal at the fiber output, was derived for the case of CW SRS by Smith (1972).

This approach has proved to be fairly reliable even in the pulsed case. For the case of *polarization-preserving* fiber, the critical pump power is estimated to be

$$P_c \approx \frac{GA_{\text{eff}}}{g_0 l_w}, \tag{23}$$

and for *non-polarization-preserving* fiber the critical pump power is

$$P_c \approx 2\frac{GA_{\text{eff}}}{g_0 l_w}. \tag{24}$$

The Raman gain in fibers is a factor of 2 higher if linear polarization is maintained (Stolen, 1979) and accounts for the factor of 2 in Eqs. (23) and (24). The peak Raman gain coefficient g_0 at a pump wavelength of 0.526 μm is 1.86×10^{-11} cm/W (Stolen and Ippen, 1973). The Raman gain varies linearly with pump frequency and the peak coefficient for a 1.064-μm pump is $g_0 = 0.92 \times 10^{-11}$ cm/W (Lin et al., 1977).

Fiber-grating compression of optical pulses can be strongly affected by SRS, which at high intensities will distort the pulse profile and consequently the chirp on the pulse. SRS limits the power available in the compressed pulse. Above the Raman threshold, further increases in pump power result only in increased Raman conversion. SRS does not seem to have a major effect on the compression of femtosecond pulses, and this can be attributed to the very short walk-off lengths involved (see Eqs. (21), (23), and (24)). Longer pulses translate into longer walk-off lengths and lower critical powers for the onset of SRS. The compression of the fundamental and the second harmonic of mode-locked Nd:YAG lasers falls squarely into this region of competition between SRS, SPM, and GVD. Under conditions of walk-off of the generated Stokes pulse, intense SRS will preferentially deplete the leading edge of the pump pulse and steepen its rising edge. SPM of the reshaped pump pulse causes nonsymmetric spectral broadening and a nonlinear chirp. An example of the distortion of the self-phase-modulated pump spectrum by the presence of 20% Raman conversion (Stolen and Johnson, 1986) is displayed in Figure 10.16. This figure shows the Stokes and pump spectra for 35-ps, 0.532-μm pump pulses after propagation down 101 m of single-mode polarization-preserving fiber (walk-off length = 6.2 m). The long-wavelength component, or red-shifted frequency component, shows signs of depletion. The resultant nonlinear chirp, of course, leads to very poor fiber-grating compression.

In general, compression in the presence of strong SRS using fiber lengths less than z_{opt} (i.e., negligible GVD) results in compressed pulses accompanied by broad wings. In addition, the spectral fluctuations lead to severe fluctuations in compressed pulse amplitude and shape. Recently, Weiner et al. (1988) demonstrated that high-quality stabilized compression could be achieved, under these circumstances, only by utilizing an asymmetric spectral window (Heritage et al., 1985a) to select out a linearly chirped portion of the broadened spectrum.

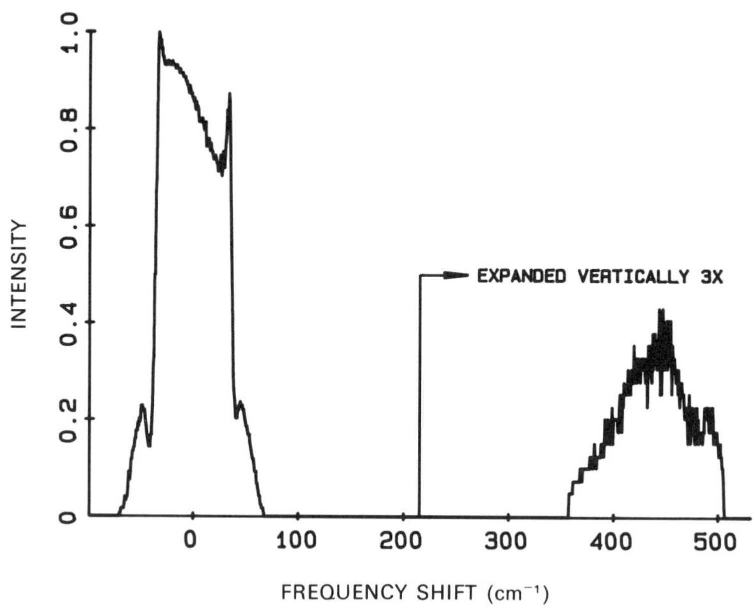

FIGURE 10.16. Raman and pump spectra of 35-ps, 0.532-μm pulses as measured from a 101-m fiber at about 20% Raman conversion. The self-phase-modulated pump spectrum shows signs of depletion of the long-wavelength or red-shifted frequency components by SRS leading to a nonlinear chirp. (Walk-off length = 6.2 m.)

Another approach to obtaining high-quality stable pulse compression is to avoid or severely limit SRS and use fibers of length z_{opt} to obtain the necessary linear chirp. Johnson et al. (1984a, 1984b) generated high-quality 410-fs pulses (80 × compression) at 0.532 μm with a fiber length greater than z_{opt} and less than 5% Raman conversion. Using fibers of nearly optimum length and operating below the Raman threshold, Roskos et al. (1987) and Dianov et al. (1987) generated high-quality pulses as short as 550 fs (110 × compression) at 1.064 μm.

Earlier pulse compression calculations for shorter fiber lengths clearly indicated that the chirp would be severely distorted and asymmetric as a result of strong SRS (Schadt et al., 1986; Schadt and Jaskorzynska, 1987; Kuckartz et al., 1987). However, Kuckartz et al. (1988) recently demonstrated that in sufficiently long fibers the combined action of GVD and SPM could cause a further reshaping and linearization of the chirp, which then could be efficiently compressed by a grating pair. High-quality pulses with comparatively low substructure as short as 540 fs (130 × compression) at 1.064 μm using 120 m of polarization-preserving fiber were obtained in the presence of strong SRS (Kuckartz et al., 1988). Heritage et al. (1988) recently demonstrated that with a 400-m length of polarization-preserving fiber, significant *third* Raman

Stokes generation, and an asymmetric spectral window (Heritage et al., 1985a), high-quality ultrastable compressed pulses as short as 550 fs (130 × compression) at 1.064 μm could be obtained. They found that most of the pump spectrum was linearly chirped by the strong reshaping due to strong SRS, SPM, and GVD.

Several review and extended-length articles on nonlinear pulse propagation, pulse shaping, and compression in fibers have recently been published: Alfano and Ho, 1988; Dianov et al., 1988; Golovchenko et al., 1988; Gomes et al., 1988b; Kafka and Baer, 1988; Thurston et al., 1986; Zhao and Bourkoff, 1988. (See Chapter 3.) This list of articles is by no means complete and furthermore is limited to discussions in the region of positive GVD.

Thus far, this chapter has dealt with negative dispersive delay lines consisting of Treacy, (1968, 1969) grating pairs in reflection mode. Several alternatives to this approach deserve mentioning. Yang et al. (1985) demonstrated femtosecond optical fiber pulse compression using a holographic volume phase transmission grating pair. Prisms were used as negative dispersive delay lines by Fork et al. (1984), Martinez et al. (1984), and Bor and Racz (1985) and were used in femtosecond optical fiber pulse compression by Kafka

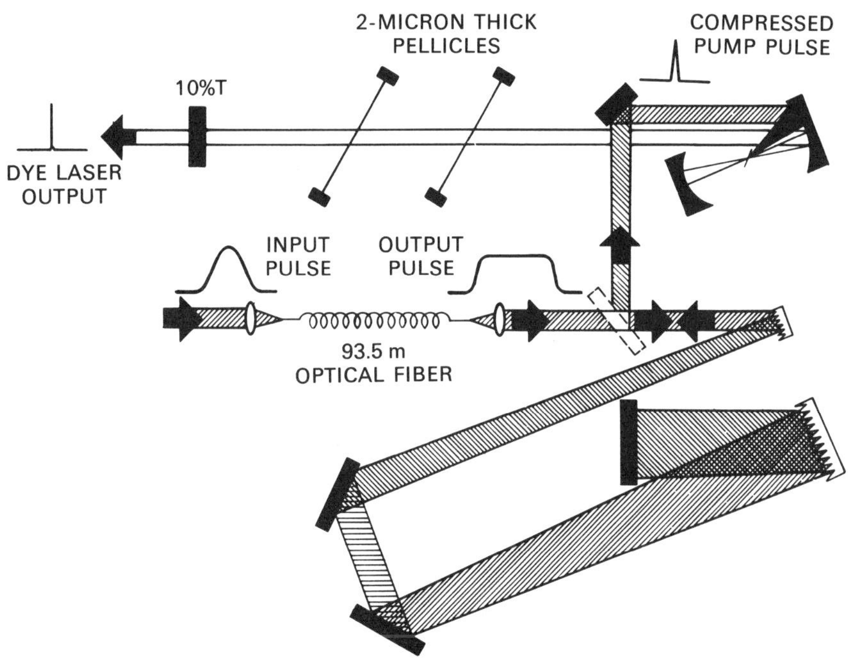

FIGURE 10.17. Schematic of a synchronously mode-locked Rhodamine 6G dye laser pumped by fiber-grating compressed 0.532-μm pulses. A pair of 2-μm-thick pellicles or a single-plate birefringent filter was used for wavelength tuning and bandwidth control.

and Baer (1987). In another variation, Nakazawa et al. (1988) used a TeO_2 acousto-optic light deflector and corner cube combination as a negative dispersive delay to demonstrate femtosecond optical fiber pulse compression.

One of the first applications of "long" fiber-grating compressed pulses was their use by Johnson et al. (1984b), Johnson and Simpson (1985a, 1985b, 1986), Kafka and Baer (1985, 1986), and Beaud et al. (1986) as a source of ultrashort pump pulses for wavelength-tunable femtosecond dye lasers. A schematic of a Rhodamine 6G synchronously mode-locked dye laser pumped by compressed 0.532-μm pulses is displayed in Figure 10.17. Wavelength-tunable pulses as short as 180 fs (Johnson and Simpson, 1986) were obtained from the dye laser synchronously pumped with 470-fs-duration 0.532-μm pulses (see Figure 10.18).

Johnson et al. (1984a, 1984b) reported that the duration and functional form of the compressed 0.532-μm pulses were extremely sensitive to the grating separation (for constant fiber input power). As the grating separation was decreased from its optimum, the compressed pulse would broaden smoothly. Compressed pulses of 460 to 470 fs (Johnson and Simpson, 1985a) duration were obtained with a grating separation of 698 cm (see Figure 10.10). By decreasing the grating separation by 4.2, 15.6, and 27% the compressed pulses were broadened to 920 fs, 12.3 ps, and 22 ps, respectively (see Figure 10.19).

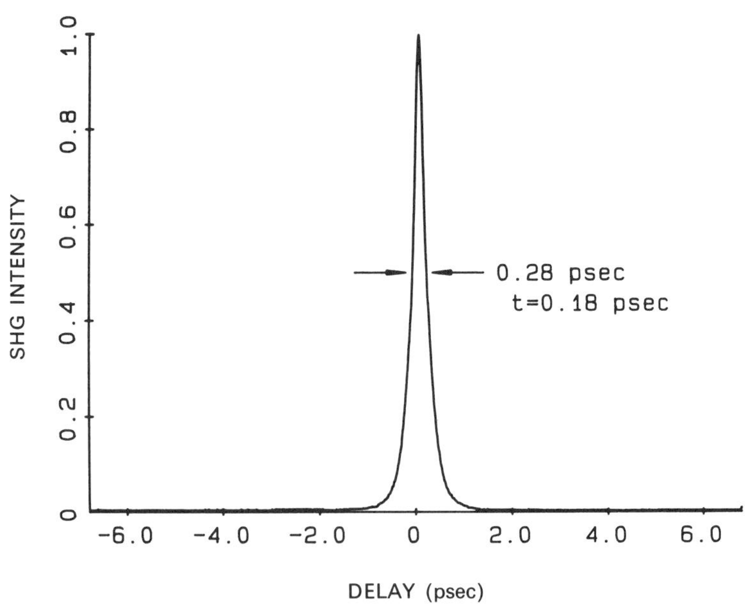

FIGURE 10.18. Autocorrelation function of the pellicle-tuned dye laser synchronously pumped by 470-fs-duration 0.532-μm pulses, tuned to a wavelength of 0.592 μm ($sech^2$ pulse shape assumed).

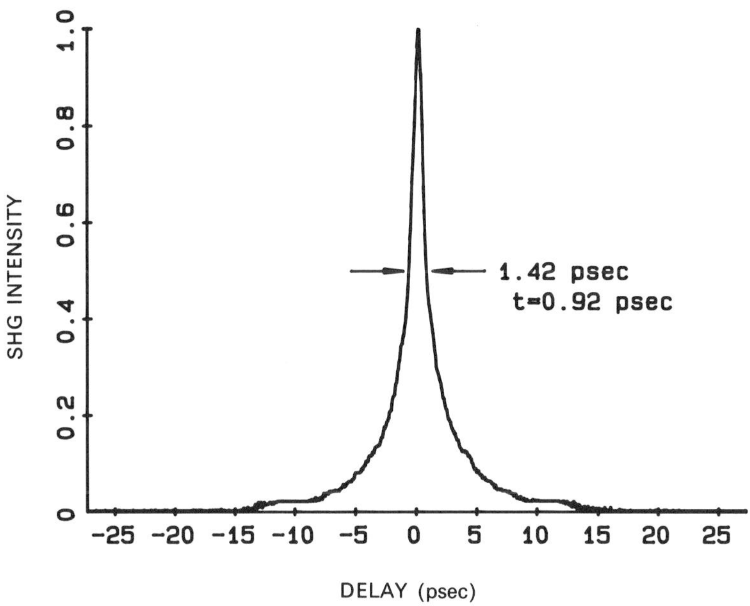

(a)

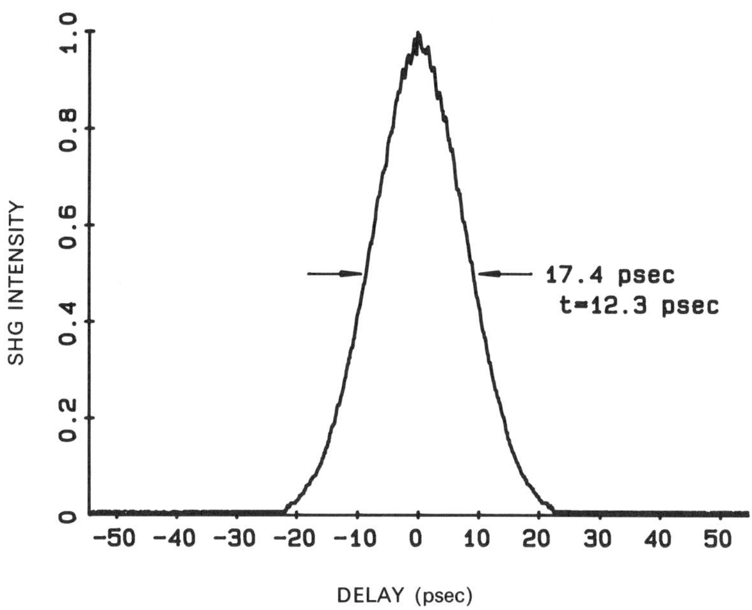

(b)

FIGURE 10.19. *Temporal* tuning of compressed 0.532-μm pulses with decreasing grating separation from the optimum (see Figure 10.10). The grating separation was reduced from the optimum of 698 cm by (a) 4.2% (sech^2 pulse shape), (b) 15.6% (Gaussian pulse shape, and (c) 27% (Gaussian pulse shape).

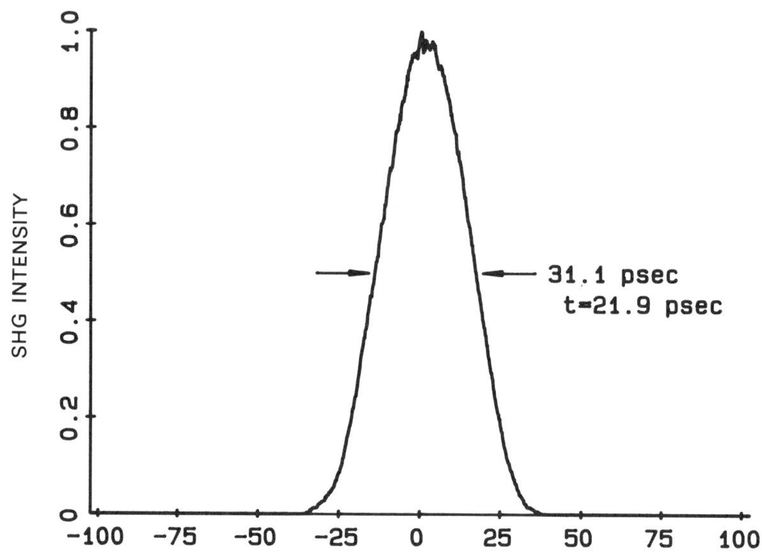

(c)

FIGURE 10.19 (*continued*)

Thus a *temporally* tunable source of ultrashort pulses was demonstrated. With this source of temporally tunable pump pulses, the first reported investigation of the dynamics of synchronous mode locking as a function of pump pulse duration was made by Johnson and Simpson (1985a). The experimental variation of the dye laser pulse width as a function of the pump pulse width (see Figure 10.20) was

$$t_{\text{dye}} \sim t_{\text{pump}}^{0.52}, \tag{25}$$

in excellent agreement with the square root dependence predicted by Ausschnitt and Jain (1978) and Ausschnitt et al. (1979).

Palfrey and Grischkowsky (1985) generated 16-fs frequency-tunable pulses by using a two-stage fiber pulse compressor together with an optical amplifier. Ishida and Yajima (1986) generated pulses of less than 100 fs tunable over 0.597 to 0.615 μm by taking the output from a cavity-dumped, hybridly mode-locked CW dye laser and coupling it to a single-stage fiber compressor. Damen and Shah (1988) reported on femtosecond luminescence spectroscopy of III-V semiconductors with 60-fs compressed pulses. The 60-fs pulses were derived from a compressed pulse-pumped synchronously mode-locked dye laser that was further compressed by a fiber-prism pulse compressor.

Applications of fiber-grating compressed pulses include picosecond photoconductive sampling characterization of semiconductor epitaxial films de-

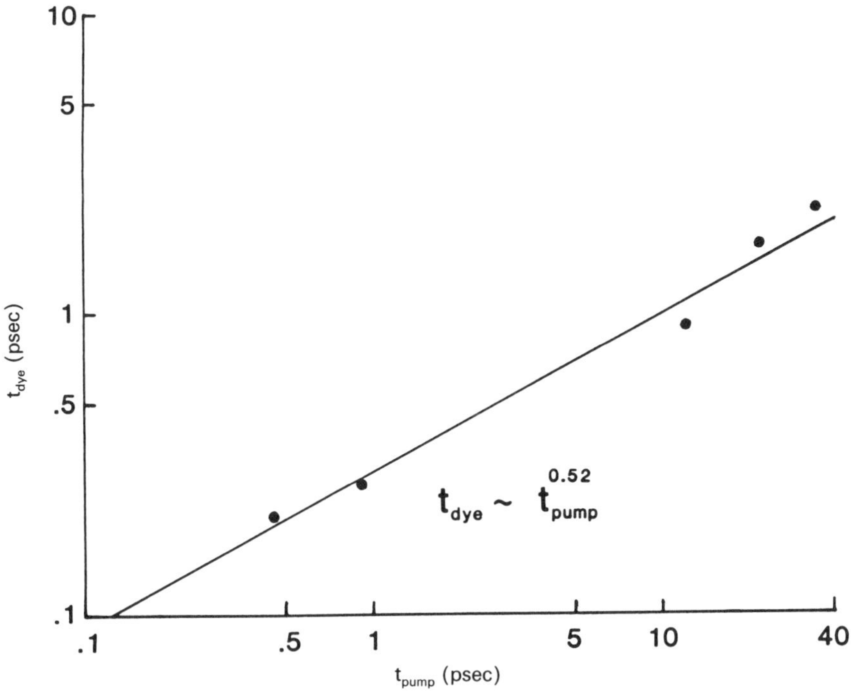

FIGURE 10.20. Temporal dynamics of the synchronous mode locking of the pellicle-tuned Rhodamine 6G dye laser as a function of the pump pulse duration, with the dye laser tuned to 0.595 μm.

posited on lattice-mismatched substrates (Johnson et al., 1985, 1987; Feldman et al., 1988), picosecond electro-optic sampling of GaAs integrated circuits (Kolner and Bloom, 1984, 1986; Weingarten et al., 1988), picosecond photoemission sampling of integrated circuits (Bokor et al., 1986; May et al., 1987, 1988) picosecond vacuum photodiode (Bokor et al., 1988), picosecond optical pulse shaping and spectral filtering (Dianov et al., 1985, 1986b; Haner and Warren, 1987; Heritage et al., 1985a, 1985b; Weiner et al., 1986), and the demonstration of an ultrafast light-controlled optical fiber modulator (Halas et al., 1987) and its use in the first experimental investigation of dark-soliton propagation in optical fibers (Krokel et al., 1988).

Recently, pulse compression techniques have been applied to the amplification of high-energy 1.06-μm pulses. The onset of self-focusing limits the amplification of ultrashort optical pulses. Fisher and Bischel (1974) proposed avoiding self-focusing in Nd:glass amplifiers by temporally broadening the input pulse to lower the pulse intensity. They noted that under certain circumstances, the glass nonlinearity would impress a chirp on the pulse that could subsequently be compressed by a dispersive delay line. Strickland and

Mourou (1985) and Maine et al. (1988) used an optical fiber to stretch a short optical pulse, amplify, and then recompress using a grating pair. Since the stretched pulse is amplified, the energy density can be increased, thereby more efficiently extracting the stored energy in the amplifier.

3. Femtosecond Pulse Compression

3.1 Theory

Remarkable progress has taken place in the compression of optical pulses. The theoretical limit in the visible spectrum is just a few femtoseconds. Already optical pulses as short as 6 fs have been generated and used in experiments (Fork et al., 1987; Brito Cruz et al., 1986). Such a pulse contains spectral components covering nearly the entire visible and near-infrared region of the spectrum. The short pulse itself is nearly an ideal continuum source.

It is useful to explore the limits of pulse compression in order to understand and appreciate the processes involved in compressing optical pulses in the femtosecond time regime. Attacking the limits provides a pathway for both utilizing and generating ever shorter optical pulses.

When an optical pulse propagates through any dielectric medium, group velocity dispersion broadens the pulse. For example, an 8-fs pulse will have its width doubled by passage through ~1 mm of glass or ~3 m of air. These linear propagation effects are not fundamental and can in principle be corrected by a linear compensation scheme.

One of the most useful pulse compressors, the grating-pair compressor devised by Treacy (1969) has been discussed earlier in this chapter. In his original paper, Treacy pointed out some of the limitations of this compressor for very short optical pulses having a large bandwidth. A grating pair induces a phase distortion on an optical pulse that becomes more severe as the ratio of the pulse bandwidth to the carrier frequency begins to approach unity.

The problem of generating ultrashort optical pulses reduces to minimizing the phase distortion. A useful way to discuss this problem is in terms of the Taylor series expansion of the phase:

$$\phi(\omega) = \phi(\omega_0) + \left(\frac{d\phi}{d\omega}\right)_{\omega_0}(\omega - \omega_0) + \frac{1}{2}\left(\frac{d^2\phi}{d\omega^2}\right)_{\omega_0}(\omega - \omega_0)^2 + \frac{1}{6}\left(\frac{d^3\phi}{d\omega^3}\right)_{\omega_0}(\omega - \omega_0)^3, \tag{26}$$

which is made around the central frequency, ω_0, of the pulse spectrum. Treacy has shown that a pair of diffraction gratings can be used to compensate for the quadratic phase distortion, $(d^2\phi/d\omega^2)_{\omega_0}$, of a frequency-broadened optical pulse. He pointed out in the same paper that the principal remaining problem in pulse compression of large-bandwidth signals using gratings is

TABLE 10.2. Second and third derivatives of phase with respect to frequency for a double prism pair, a double grating pair, and material.

Prism	Grating	Material
$\dfrac{d^2\phi_p}{d\omega^2} = \dfrac{\lambda^3}{2\pi^2c^2}\dfrac{d^2P}{d\lambda^2}$	$\dfrac{d^2\phi_g}{d\omega^2} = \dfrac{\lambda^3 l_g}{\pi c^2 d^2}\left[1-\left(\dfrac{\lambda}{d}-\sin\gamma\right)^2\right]^{-3/2}$	$\dfrac{d^2\phi_m}{d\omega^2} = \dfrac{\lambda^3 l_m}{2\pi c^2}\dfrac{d^2n_m}{d\lambda^2}$
$\dfrac{d^3\phi_p}{d\omega^3} = \dfrac{-\lambda^4}{4\pi^2c^3}\left(3\dfrac{d^2P}{d\lambda^2}+\lambda\dfrac{d^3P}{d\lambda^3}\right)$	$\dfrac{d^3\phi_g}{d\omega^3} = \dfrac{-d^2\phi_g}{d\omega^2}\dfrac{6\pi\lambda}{c} \times \dfrac{\left(1+\dfrac{\lambda}{d}\sin\gamma-\sin^2\gamma\right)}{\left[1-\left(\dfrac{\lambda}{d}-\sin\gamma\right)^2\right]}$	$\dfrac{d^3\phi_m}{d\omega^3} = \dfrac{-\lambda^4 l_m}{4\pi^2c^3}\left(\dfrac{3d^2n_m}{d\lambda^2}+\dfrac{\lambda d^3n_m}{d\lambda^3}\right)$

Derivatives of the path P in the prism sequence with respect to wavelength

$$\frac{d^2P}{d\lambda^2} = 4\left[\frac{d^2n}{d\lambda^2}+(2n-n^{-3})\left(\frac{dn}{d\lambda}\right)^2\right]l_p\sin\beta - 8\left(\frac{dn}{d\lambda}\right)^2 l_p\cos\beta$$

$$\frac{d^3P}{d\lambda^3} = 4\frac{d^3n}{d\lambda^3}l_p\sin\beta - 24\frac{dn}{d\lambda}\frac{d^2n}{d\lambda^2}l_p\cos\beta$$

TABLE 10.3. Second and third derivatives of phase with respect to frequency for the double prism pair and double grating pair described in the text.[a]

Derivative	Prisms	Gratings	Material
$\dfrac{d^2\phi}{d\omega^2}$ (fs^2)	$+648-32l_p$	$-3640l_g$	$+2900l_m$
$\dfrac{d^3\phi}{d\omega^3}$ (fs^3)	$+277-49l_p$	$+3120l_g$	$+1620l_m$

[a] Lengths are in centimeters.

uncompensated cubic phase distortion, $(d^3\phi/d\omega^3)_{\omega_0}$. Christov and Tomov (1986) also recognized this problem in a recent publication on optical fiber-grating compressors Tables 10.2 and 10.3 show phase derivatives for prisms and gratings.

To overcome the problem of unwanted cubic phase distortion an elegant solution has been devised. Both a grating pair and a prism pair induce a cubic phase distortion. We can take advantage of the fact that the cubic phase distortion for gratings and prisms is of the opposite sign by using a configuration where the compressed pulse is passed sequentially through a pair of gratings and a pair of prisms. In this manner it is possible to cancel the cubic phase distortion (Treacy, 1969).

The effect of a combination of prisms, gratings, and material on a pulse is described by a total phase shift

$$\phi_T(\omega) = \phi_p(\omega) + \phi_g(\omega) + \phi_m(\omega), \tag{27}$$

where the subscripts p, g, and m refer to prisms, gratings, and material, respectively. The material of length l_m contributes a phase shift

$$\phi_m(\omega) = \frac{\omega l_m}{c} n_m(\omega), \tag{28}$$

where c is the speed of light and $n_m(\omega)$ the refractive index. For the prism-and-grating pair we follow the method described by Martinez et al. (1984). A grating pair in a double-pass configuration causes a phase shift

$$\phi_g(\omega) = \frac{2\omega l_g}{c}\left[1 - \left(\frac{2\pi c}{\omega d} - \sin\gamma\right)^2\right]^{1/2}, \tag{29}$$

where l_g is the grating spacing, d is the groove spacing, and γ is the angle of incidence.

For a double prism pair the phase shift is

$$\phi_p(\omega) = \frac{2\omega l_p}{c} \cos[\beta(\omega)], \tag{30}$$

where l_p is the distance between prism apices and $\beta(\omega)$ is the angle between the refracted ray at frequency ω and the line joining the two apices (Figure 10.21). For prisms with apex angle α and refractive index $n_p(\omega)$ the angle $\psi 2(\omega)$ at which the refracted ray leaves the first prism can be calculated by a straightforward application of Snell's law as a function of the angle of incidence ψ_1.

We define $\psi_{2\max}$ as the maximum angle at which a ray can leave the first prism and still intersect the apex of the second prism. Equation (30) can then be rewritten as

$$\phi_p(\omega) = \frac{2\omega l_p}{c} \cos[\psi_{2\max} - \psi_2(\omega)]. \tag{31}$$

Typical experimental values are $\alpha = 60°$, $\psi_1 = 47°$ (minimum deviation), $n_p(\omega_0) = 1.457$ (quartz prisms) at $\omega_0 = 3.1$ rad/fs ($\lambda_0 = 615$ nm), and $\psi_{2\max} = 49°$.

The total phase shift $\phi_T(\omega)$ can be calculated numerically for conditions typical of recent ultrashort-pulse-compression experiments by using Eqs. (28), (29), and (31) to provide a group delay dispersion $d^2\phi_T/d\omega^2|_{\omega_0}$ of -700 fs^2 and to cause the derivative of the group delay dispersion, $d^3\phi_T/d\omega^3$, to be zero at the center frequency of the pulse. The value of $d^2\phi_T/d\omega^2$ is such as to compensate the linear frequency sweep generated on a 60-fs, 200-kW pulse propagated through a 0.9-cm quartz fiber with a 4-μm core diameter. The prism spacing, grating spacing, and material length used in the numerical calculations are $l_p = 74$ cm, $l_g = 0.7$ cm, and $l_m = 0.5$ cm of quartz, respectively. The angle of incidence at the first grating was 45°, and the number of grooves per millimeter was 600.

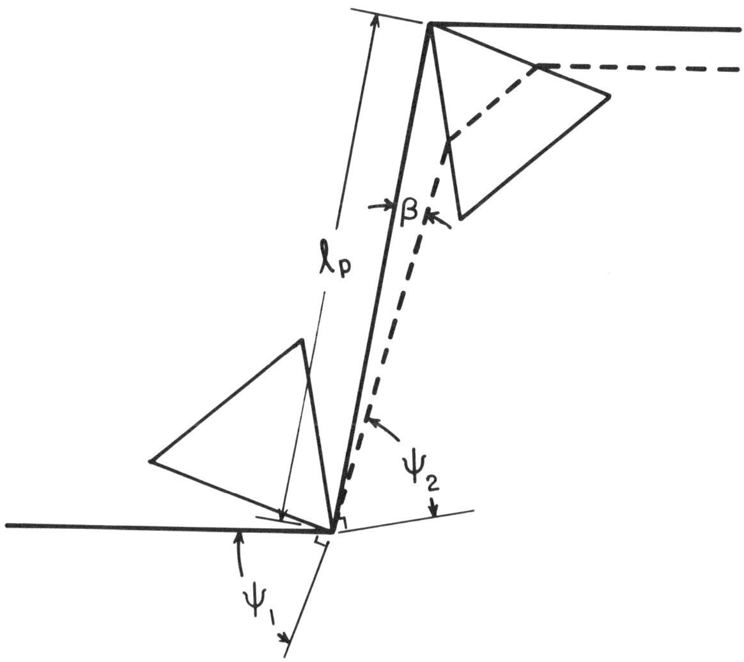

FIGURE 10.21. Parameters used in describing propagation of the optical pulse through the prism sequence. The angle of incidence at the face of the first prism is ψ_1 and the angle with respect to the normal to the exit face is ψ_2. The angle between a line drawn between prism apices and the direction of a ray exiting the first prism at ψ_2 is denoted by β. The distance between prism apices is l_p.

The departure of this compressor based on prisms and gratings from an ideal quadratic compressor can be evaluated by examining the variation of the group delay dispersion, $d^2\phi_T(\omega)/d\omega^2$, with frequency (Figure 10.22). One sees that this combination of prisms, gratings, and material provides the value of group delay dispersion required to compensate for the linear chirp in the pulse. At the same time, this combination of prisms, gratings, and material makes the derivative of the group delay dispersion zero at the center frequency of the pulse. This minimizes the cubic distortion and leaves as the main contribution to the phase that which is due to the curvature of the group delay dispersion ($d^4\phi/d\omega^4 \neq 0$) across the spectral range of the pulse.

The consequence of the departure of these actual compressors from an ideal quadratic compressor can be examined by calculating the temporal profile of the compressed pulse given a hypothetical incident pulse with an ideal quadratic phase distortion. In particular, we compare a compressor using prisms and gratings with a compressor using the gratings alone. Figure 10.23 shows the calculated intensity profile for the case when prisms, gratings, and material are used with the same parameters as in Figure 10.22. The bandwidth

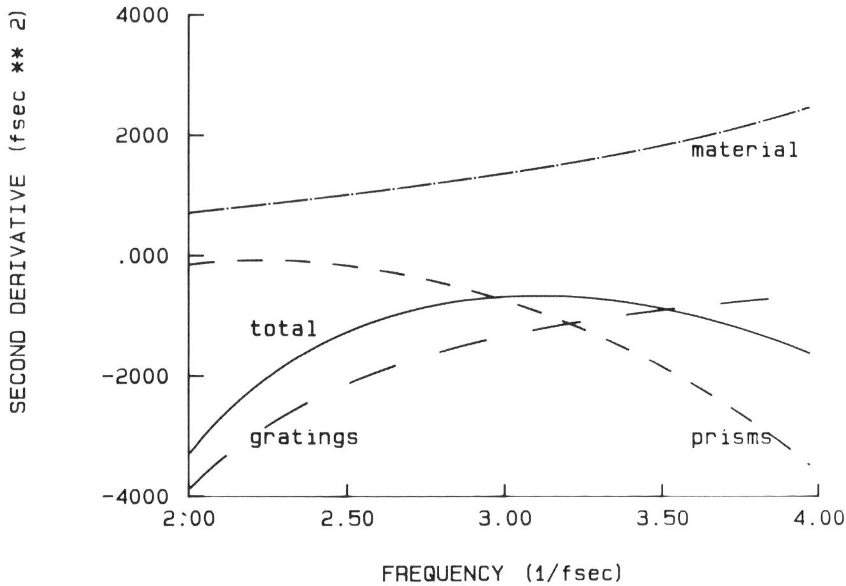

FIGURE 10.22. Plot of the second derivative of phase with respect to frequency for the prisms (short dash), gratings (long dash), and material (dash-dot) and for the total phase shift (solid).

of the incoming pulse was chosen to be 0.5 rad/fs, which corresponds to a transform-limited pulse duration of 4 fs. In both Figures 10.23a and 10.23b the linear frequency sweep has been compensated for, but only in the case of the compressor with prism pair, grating pair, and dispersive material (Figure 10.23b) was it possible to compensate for the parabolic frequency sweep by setting the cubic phase distortion to zero.

The oscillatory trailing edge on the pulse shown in Figure 10.23a is due to the uncompensated cubic distortion, which causes the high- and low-frequency edges of the pulse spectrum to lag with respect to the center frequency. These delayed frequency components beat with each other to create an oscillatory trailing edge on the pulse. If prisms are used alone, the compressed pulse is similar to that shown in Figure 10.23a but with the time axis reversed; that is, the oscillatory trailing edge becomes an oscillatory leading edge.

The dominant residual distortion of the phase-corrected pulse is that which is due to the uncorrected negative curvature of the group velocity dispersion $d^4\phi/d\omega^4 < 0$. The effect is to leave small oscillatory wings on the leading and trailing edges of the pulse, as is evident from Figure 10.23b, and to broaden the main peak slightly. The lower limit on the duration of pulses compressed in this manner depends on the specific shape of the input-pulse spectrum and its precise distortions. For the ideal secant hyperbolic shape input pulse

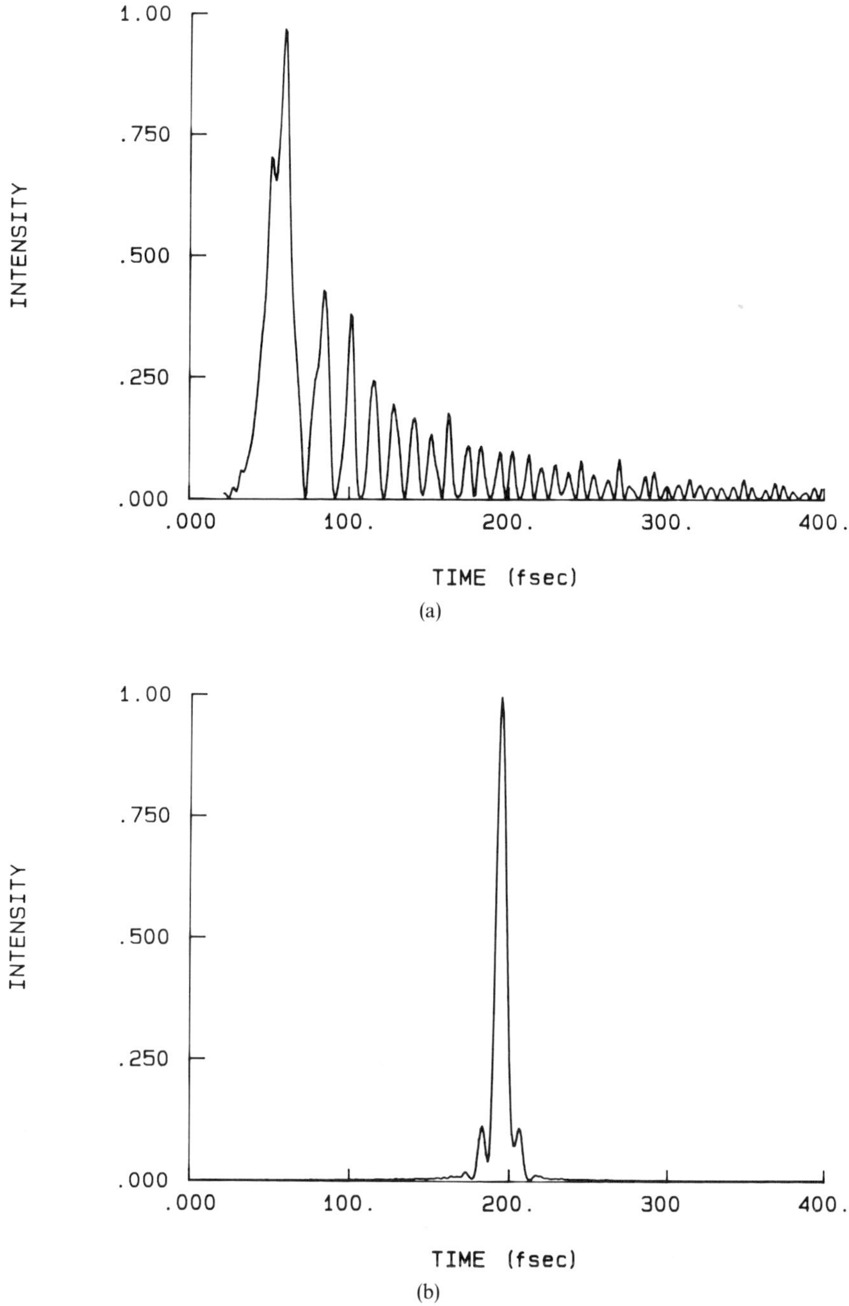

FIGURE 10.23. Calculated pulse intensity vs. time for the case of compression using only gratings and material dispersion (a) and for the case of compression using a combination of prisms, gratings, and material dispersion (b).

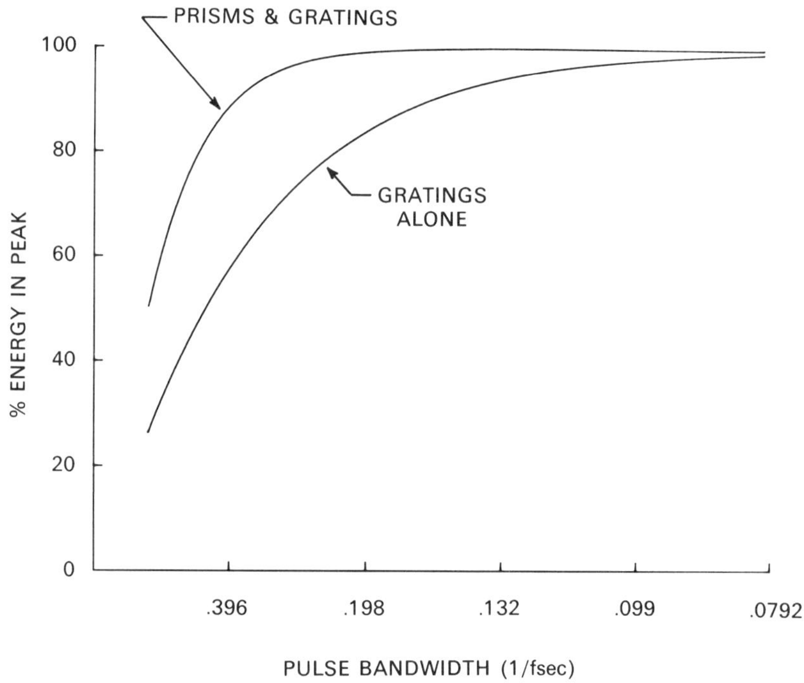

FIGURE 10.24. Plot of energy in the main peak of the compressed pulse for the case of combined prisms and gratings and for the case of gratings alone.

assumed above, the minimum compressed pulse width is between 6 and 7 fs, which is in approximate agreement with recently observed pulses compressed with grating and prism pairs (Fork et al., 1987). In Figure 10.24 the energy in the pulse peak is plotted versus pulse bandwidth for the case of gratings alone and the grating-prism pair combination.

3.2 Experiment

The arrangement of gratings and prisms for pulse compression is illustrated in Figure 10.25. The experimental study was carried out using optical pulses generated in a colliding-pulse mode-locked laser that contained an intracavity prism sequence identical to the four-prism set shown in Figure 10.25. These pulses were amplified at a repetition rate of 8 kHz in a copper-vapor laser-pumped amplifier to energies of $\sim 1\ \mu$J. The amplified pulses had durations of 50 fs and a spectrum centered at 620 nm. A fraction of the amplified pulse energy was coupled into a polarization-preserving quartz fiber with core dimensions of $\sim 4\ \mu$m and a length of 0.9 cm. The optical intensity in the fiber was $1\text{–}2 \times 10^{12}$ W/cm.

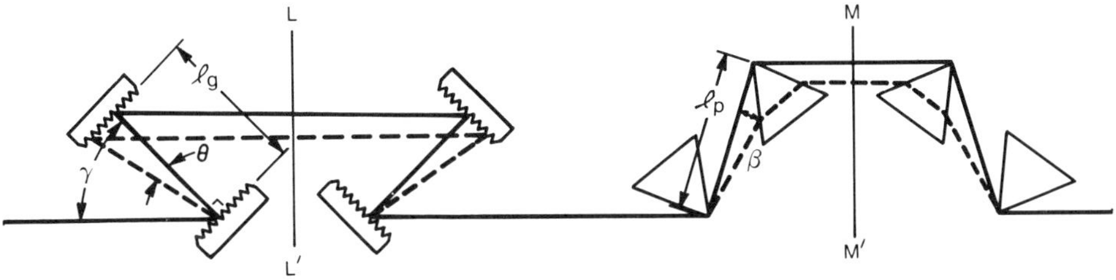

FIGURE 10.25. Combined grating and prism sequence used to remove both quadratic and cubic phase distortion. The solid line is a reference line. The dashed line is the path of a plane wave that propagates between the gratings at an angle θ with respect to the normal to the grating faces and between the prism pairs at an angle β with respect to a line drawn between the prism apices. The plane LL' is a plane of symmetry for the grating sequence, and the plane MM' is a plane of symmetry for the prism sequence.

A four-prism sequence was then introduced, so the combined prism and grating sequence was equivalent to that shown in Figure 10.25. It was then possible to adjust the spacing of the prism and grating pairs so the maxima of the six different upconverted intensity traces all occurred at the same phase delay. Subsequent optimization was done by monitoring the interferometric autocorrelation (Diels et al., 1978, 1985) trace of the compressed pulse while adjusting the prism spacing l_p and the grating spacing l_g. It is not possible to use the more conventional background-free autocorrelation technique for pulses this short since even a small relative angle between wave vectors of the interacting beams introduces measurable error. It was also necessary to use an extremely thin (32-μm) KDP crystal to double the compressed pulse so as to minimize distortion by group velocity dispersion within the doubling crystal.

The interferometric autocorrelation trace obtained on optimizing l_p and l_g is shown in Figure 10.26. The prism spacing for this trace was $l_p = 71$ cm, and the grating spacing was $l_g = 0.5$ cm. For purposes of comparison we have used crosses to indicate the calculated maxima and minima for an interferometric autocorrelation trace of a hyperbolic-secant-squared pulse having zero phase distortion and a full width at half-maximum of 6 fs. The close fit between the calculated and experimental interferometric autocorrelation functions indicates an absence of significant phase distortion over the bandwidth of the pulse. The well-resolved interference maxima also provide a rigorous calibration of the relative delay.

3.3 Applications

The success in generating optical pulses as short as 6 fs opens up the domain of physical processes that take place in a few femtoseconds to study. A pulse

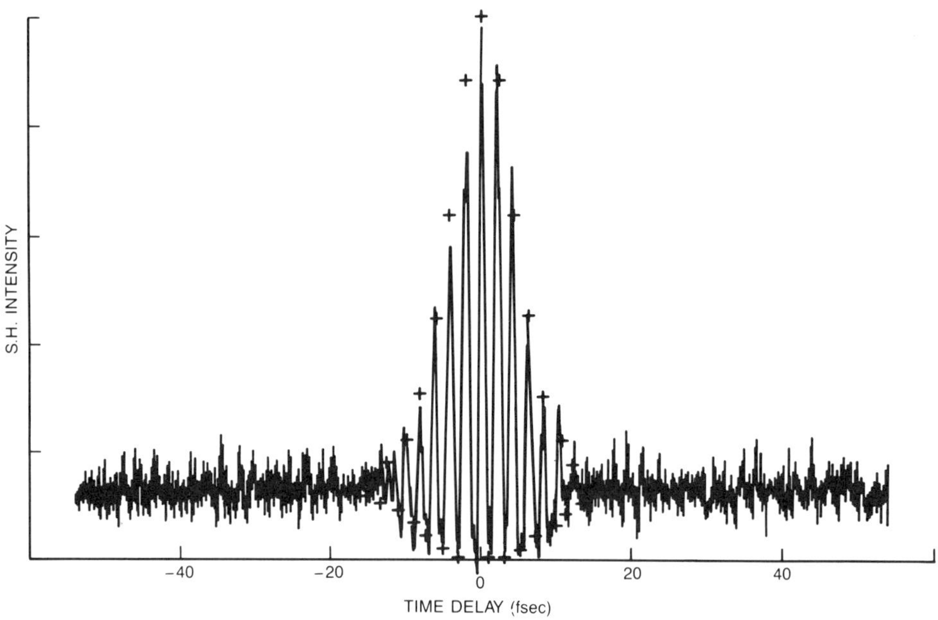

FIGURE 10.26. Experimental interferometric autocorrelation trace for a compressed pulse for $l_p = 71$ cm and $l_g = 0.5$ cm. The interference maxima and minima calculated for the interferometric autocorrelation trace for a hyperbolic-secant-squared pulse of 6-fs duration are indicated by crosses. The close agreement between experiment and theory demonstrates absence of significant phase distortion in the compressed pulse.

that is so short contains frequency components from almost the entire visible region of the spectrum. Such a pulse is a nearly ideal continuum source. The well-defined temporal and spectral character of the pulse makes it quite useful for time-resolved spectroscopic problems.

Ultrashort optical pulse techniques provide a unique means for investigating nonequilibrium energy redistribution among vibronic levels in large organic molecules in solution. Previously, the dynamics of induced absorbance changes have been measured using pump and probe pulses having the same frequency spectrum. In the experiments described here induced absorption changes of optically excited molecules over a broad spectral range of 2400 cm^{-1} centered at the energy of the excitation pulse were measured while maintaining a 10-fs time resolution. These experiments permit the observation of time-resolved hole burning and the process of equilibration to a thermalized population distribution on a femtosecond time scale (Brito Cruz et al., 1986, 1988b).

The absorption spectrum of a large dye molecule is dominated by vibronic transitions from a thermalized group state. Typically, these large molecules, which have a molecular weight of 400 or more, have a large number of degrees of freedom. The optical absorption coefficient may be written as a sum over transitions from occupied vibrational levels in the ground state to vibrational levels in the excited state. The absorption coefficient is given by

$$\alpha(\nu) = C \sum_{if} P_i M^2 \chi_{if} \nu g(\nu - \nu_{if}), \tag{7}$$

where C is a constant, P_i is the thermal occupation probability of the initial state, M is the dipole moment of the electronic transition, χ_{if} is the Franck-Condon factor, and g is the line shape profile for each transition. The above expression describes the molecular system in thermal equilibrium. With a short optical pulse it is possible to excite a band of states that are resonant with the pumping energy. Before the molecular system comes into equilibrium, bleaching is observed in a spectral range determined by the convolution of the pump spectrum with the line shape profile of the individual transitions. As time progresses, the system relaxes to thermal equilibrium due to interaction with the thermal bath. The thermal bath couples to the vibronic levels by both intramolecular and intermolecular processes. The large number of degrees of freedom in the molecular backbone can form a thermal bath within the molecule itself. It is also possible for intermolecular energy transfer to take place on a somewhat longer time scale by collisions, dipole-dipole interaction, etc.

The experimental apparatus is arranged to perform a pump-probe type of experiment with one important modification over previous experiments. The probe pulse is approximately 10 fs in duration and has a significantly broader bandwidth than the 60-fs pump pulse. The pumping and probing pulses are derived from the same initial amplified 60-fs optical pulse having an energy of 1 μJ with a center frequency at 618 nm. The probe pulse is formed by passing a portion of the initial pulse through a 12-nm length of optical fiber followed by a grating-pair compressor. The shorter pulse is then used to probe the absorption spectrum by passing through the excited sample into a spectrometer and diode array. Care is taken to compensate for group velocity dispersion in the probe optical path. The experiments are performed at a repetition rate of 8 kHz.

The dyes are dissolved in ethylene glycol at concentrations that yield optical attenuations of less than $1/e$ when the dye solution is flowed through a jet with a thickness of 100 to 300 μm. The pump pulses are attenuated to levels that induce absorption changes of a few percent or less.

The data are collected by a differential measuring technique. The pump beam is periodically blocked by a shutter at a frequency of 10 Hz and the transmitted spectra are recorded in the computer memory in phase with the chopped pumping beam. Spectra are recorded at different time delays as determined by the optical path delay, which is controlled by a stepping motor translation stage. Integration time for a single spectrum is typically 30 s.

In Figure 10.27 the absorption spectrum for cresyl violet is plotted before and after excitation with a 60-fs optical pulse at 618 nm for zero relative time delay. A decrease in absorption is clearly observed in the spectral region close

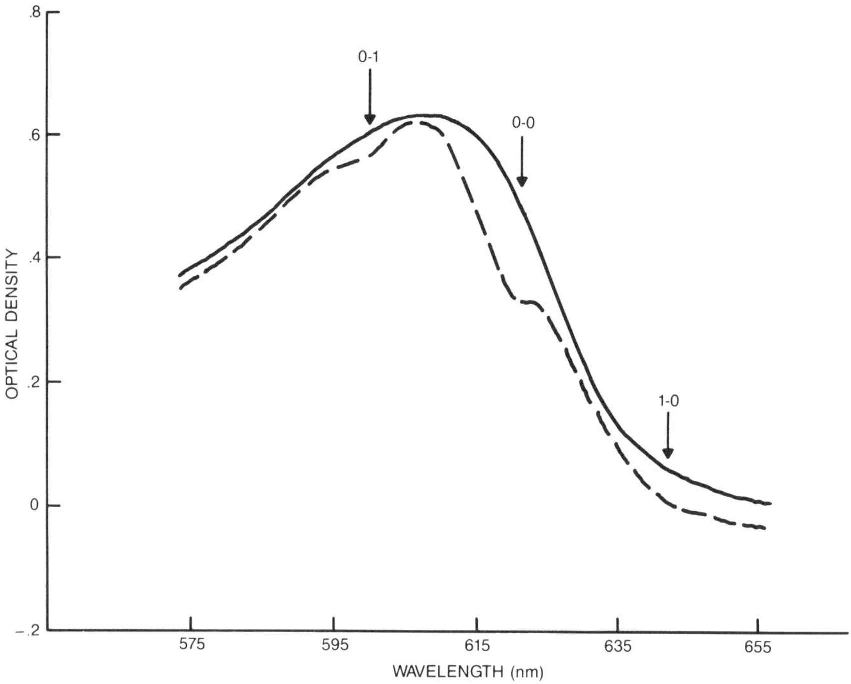

FIGURE 10.27. Plot of the absorbance spectrum of the molecule cresyl violet near zero time delay before (solid line) and after (dashed line) excitation with a 60-fs optical pulse.

to the pumping wavelength. In addition, replica holes are seen approximately 600 cm^{-1} above and below the excitation energy. In Figure 10.28 the time-resolved differential spectra are plotted for cresyl violet. The time delay between spectra is 25 fs. The central hole and the two adjacent replica holes are seen to broaden and form a thermalized spectrum in the first few hundred femtoseconds following excitation.

The mechanism for the formation of the replica holes is readily understood. Measurements of the Raman spectra of cresyl violet reveal the presence of a strong mode at 590 cm^{-1}. In a large molecule with a large number of degrees of freedom a correspondingly large number of modes can contribute to the absorption spectrum, as illustrated with Figure 10.27. Usually only a few modes with energies larger than kT change their occupation number during the optical transition to the excited state. These modes are called active or system modes and have large Franck-Condon factors. The strength of the absorption is determined by the Franck-Condon factor χ_{if}. The 598-cm^{-1} mode appears to be the dominant mode in the absorption spectrum as evidenced by bleaching both at the 0–0 transition, which is at the excitation energy, and at the 0–1 and 1–0 positions of the Franck-Condon progression.

The relative strengths of the bleaching at the central hole and at the replica

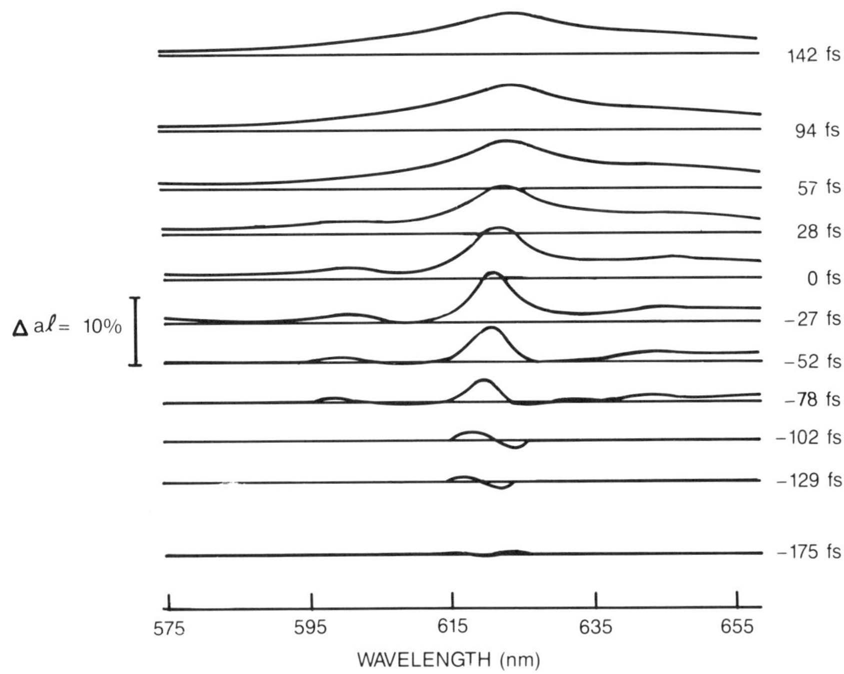

FIGURE 10.28. Differential absorbance spectra plotted as a function of relative time delay following excitation with a 60-fs optical pulse at 618 nm for the molecule cresyl violet.

holes can be determined by estimating the Franck-Condon factors in the harmonic approximation. We calculate $\chi_{00}/\chi_{01,10} = 0.26$, which is consistent with the experimental observation value.

The hole observed in Figure 10.28 broadens and relaxes to the quasi-equilibrium spectrum within the first few hundred femtoseconds. If we assume that the inhomogeneous linewidth is much larger than the homogeneous linewidth, we can estimate the polarization dephasing time T_2 from the width of the hole burned in the spectrum. For the case of a Lorentzian profile, where $\Delta\lambda$ is the half-width of the hole, the expression for T_2 is given by $T_2 = 2\lambda^2/\pi c\Delta\lambda$. Using the above expression we determine T_2 to be 75 fs for cresyl violet.

Some insight into the energy relaxation of the excited molecules can be obtained by looking at the time evolution of the differential absorption at different spectral regions within our range of observation. In Figure 10.29 we plot this evolution for a region 7 nm wide around 587, 625, and 654 nm. The curve for 625 nm shows the evolution of the population in levels that are very close in energy to the levels excited by the pump. In this spectral region an overshoot in the bleaching occurs as a consequence of the spectral hole

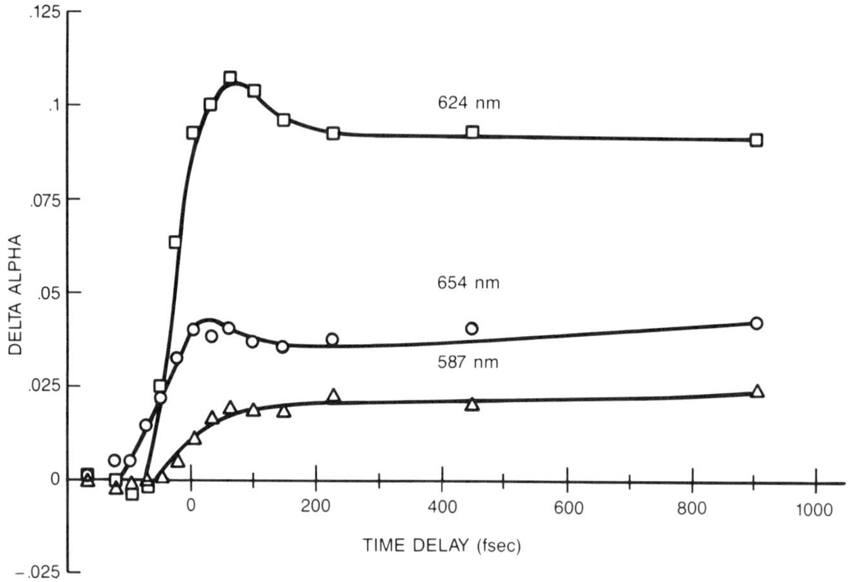

FIGURE 10.29. Plot of the differential absorption spectrum in time for three selected wavelengths for the molecule cresyl violet. (The lines are to guide the eye.)

burning and a rapid recovery on the order of 50 fs is observed as the ground and excited states become thermalized. The bleaching at 654 nm also shows a small overshoot, which recovers as the nonequilibrium distribution transits to a thermalized distribution in the first few hundred picoseconds. On the high-energy side at 587 nm a slower rise is observed as the molecular distribution thermalizes.

References

Agostinelli, J., G. Harvey, T. Stone, and C. Gabel (1979) Appl. Opt. **18**, 2500.

Alfano, R.R. and P.P. Ho (1988) IEEE J. Quantum Electron.**QE-24**, 351.

Alfano, R.R. and S.L. Shapiro (1970) Phys. Rev. Lett. **24**, 592.

Arthurs, E.G., D.J. Bradley, and A.G. Roddie (1971) Appl. Phys. Lett. **19**, 480.

Ausschnitt, C.P. and R.K. Jain (1978) Appl. Phys. Lett. **32**, 727.

Ausschnitt, C.P., R.K. Jain, and J.P. Heritage (1979) IEEE J. Quantum Electron. **QE-15**, 912.

Auyeung, J. and A. Yariv (1978) IEEE J. Quantum Electron. **QE-14**, 347.

Baldeck, P.L., F. Raccah, and R.R. Alfano (1987) Opt. Lett. **12**, 588.

Beaud, P., B. Zysset, A.P. Schwarzenbach, and H.P. Weber (1986) Opt. Lett. **11**, 24.

Bloembergen, N. and P. Lallemand (1966) Phys. Rev. Lett. **16**, 81.

Blow, K.J., N.J. Doran, and B.P. Nelson (1985) Opt. Lett. **10**, 393.

Bokor, J., A.M. Johnson, R.H. Storz, and W.M. Simpson (1986) Appl. Phys. Lett. **49**, 226.

Bokor, J., A.M. Johnson, W.M. Simpson, R.H. Storz, and P.R. Smith (1988) Appl. Phys. Lett. **53**, 2599.

Bondarenko, N.G., I.V. Eremina, and V.I. Talanov (1970) JETP Lett. **12**, 85.

Bor, Z. and B. Racz (1985) Opt. Commun. **54**, 165.

Bourkoff, E., W. Zhao, R.I. Joseph, and D.N. Christodoulides (1987a) Opt. Lett. **12**, 272.

Bourkoff, E., W. Zhao, R.I. Joseph, and D.N. Christodoulides (1987b) Opt. Commun. **62**, 284.

Bradley, D.J., G.H.C. New, and S.J. Caughey (1970) Phys. Lett. **32A**, 313.

Brewer, R.G. (1967) Phys. Rev. Lett. **19**, 8.

Brito Cruz, C.H., R.L. Fork, W.H. Knox, and C.V. Shank (1986) Chem. Phys. Lett. **132**, 341.

Brito Cruz, C.H., P.C. Becker, R.L. Fork, and C.V. Shank (1988a) Opt. Lett. **13**, 123.

Brito Cruz, C.H., J.P. Gordon, P.C. Becker, R.L. Fork, C.V. Shank (1988b) IEEE J. Quantum Electron. **24**, 261.

Butylkin, V.S., V.V. Grigoryants, and V.I. Smirnov (1979) Opt. Quantum Electron. **11**, 141.

Christov, I.P. and I.V. Tomov (1986) Opt. Commun. **58**, 338.

Damen, T.C., and J. Shah (1988) Appl. Phys. Lett. **52**, 1291.

Damm, T., M. Kaschke, F. Noack, and B. Wilhelmi (1985) Opt. Lett. **10**, 176.

DeMartini, F., C.H. Townes, T.K. Gustafson, and P.L. Kelley (1967) Phys. Rev. **164**, 312.

Desbois, J., F. Gires, and P. Tournois (1973) IEEE J. Quantum Electron. **QE-9**, 213.

Dianov, E.M., A. Ya. Karasik, P.V. Mamyshev, G.I. Onishchukov, A.M. Prokhorov, M.F. Stel'makh, and A.A. Fomichev (1984a) Sov. J. Quantum Electron. **14**, 726.

Dianov, E.M., A.Y. Karasik, P.V. Mamyshev, G.I. Onishchukov, A.M. Prokhorov, M.F. Stel'makh, and A.A. Fomichev (1984b). JETP Lett. **39**, 691.

Dianov, E.M., A.Y. Karasik, P.G. Mamyshev, A.M. Prokhorov, and V.N. Serkin (1985) Sov. Phys. JETP **62**, 448.

Dianov, E.M., L.M. Ivanov, A.Y. Karasik, P.V. Mamyshev, and A.M. Prokhorov (1986a) Sov. Phys. JETP **64**, 1205.

Dianov, E.M., L.M. Ivanov, A.Y. Karasik, P.V. Mamyshev, and A.M. Prokhorov (1986b) JETP Lett. **44**, 156.

Dianov, E.M., A.Y. Karasik, P.V. Mamyshev, A.M. Prokhorov, and D.G. Fursa (1987) Sov. J. Quantum Electron. **17**, 415.

Dianov, E.M., P.V. Mamyshev, and A.M. Prokhorov (1988) Sov. J. Quantum Electron. **18**, 1.

Diels, J.-C., E.W. Stryland and D. Gold (1978) In *Proceedings of the First International Conference on Picosecond Phenomena*, p. 117. Springer-Verlag, Berlin.

Diels, J.-C., J.J. Fontaine, I.C. McMichael, and F. Simoni (1985) Appl. Opt. **24**, 1270.

Duguay, M.A. (1976) The ultrafast optical Kerr shutter. In *Progress in Optics XIV*, E. Wolf, ed., p. 161. North-Holland, Amsterdam.

Duguay, M.A. and J.W. Hansen (1969) Appl. Phys. Lett. **14**, 14.

Dupuy, C.G. and P. Bado (1984) Dig. Conf. Lasers and Electro-Opt., Anaheim, Calif., paper TUE2, 58.

Feldman, R.D., R.F. Austin, P.M. Bridenbaugh, A.M. Johnson, W.M. Simpson, B.A. Wilson, and C.E. Bonner (1988) J. Appl. Phys. **64**, 1191.

Fisher, R.A. and W.K. Bischel (1974) Appl. Phys. Lett. **24**, 468.

Fisher, R.A. and W.K. Bischel (1975) J. Appl. Phys. **46**, 4921.

Fisher, R.A., P.L. Kelley, and T.K. Gustafson (1969) Appl. Phys. Lett. **14**, 140.
Fork, R.L., O.E. Martinez, and J.P. Gordon (1984) Opt. Lett. **9**, 150.
Fork, R.L., C.H. Brito Cruz, P.C. Becker, and C.V. Shank (1987) Opt. Lett. **12**, 483.
Fujimoto, J.G., A.M. Weiner, and E.P. Ippen (1984) Appl. Phys. Lett. **44**, 832.
Giordmaine, J.A., M.A. Duguay, and J.W. Hansen (1968) IEEE J. Quantum Electron. **QE-4**, 252.
Gires, F. and P. Tournois (1964) C.R. Acad. Sci. (Paris) **258**, 6112.
Gloge, D. (1971) Appl. Opt. **10**, 2252.
Golovchenko, E.A., E.M. Dianov, P.V. Mamyshev, and A.M. Prokhorov (1988) Opt. Quantum Electron. **20**, 343.
Gomes, A.S., U. Osterberg, W. Sibbett, and J.R. Taylor (1985a) Opt. Commun. **54**, 377.
Gomes, A.S., W. Sibbett, and J.R. Taylor (1985b) Opt. Lett. **10**, 338.
Gomes, A.S., W. Sibbett, and J.R. Taylor (1985c) IEEE J. Quantum Electron. **QE-21**, 1157.
Gomes, A.S.L., W. Sibbett, and J.R. Taylor (1986a) Appl. Phys. B **39**, 43.
Gomes, A.S.L., W.E. Sleat, W. Sibbett, and J.R. Taylor (1986b) Opt. Commun. **57**, 257.
Gomes, A.S.L., V.L. da Silva, and J.R. Taylor (1988a) J. Opt. Soc. Am. B **5**, 373.
Gomes, A.S.L., A.S. Gouveia-Neto, and J.R. Taylor (1988b) Opt. Quantum Electron. **20**, 95.
Grischkowsky, D. and A.C. Balant (1982a) Appl. Phys. Lett. **41**, 1.
Grischkowsky, D. and Balant, A.C. (1982b) Optical pulse compression with reduced wings. In *Picosecond Phenomena III*, K.B. Eisenthal, R.M. Hochstrasser, W. Kaiser, and A. Laubereau, eds., p. 123. Springer-Verlag, Berlin.
Gustafson, T.K., J.P. Taran, H.A. Haus, J.R. Lifsitz, and P.L. Kelley (1969) Phys. Rev. **177**, 306.
Halas, N.J., D. Krokel, and D. Grischkowsky (1987) Appl. Phys. Lett. **50**, 886.
Halbout, J.-M. and D. Grischkowsky (1984) Appl. Phys. Lett. **45**, 1281.
Haner, M. and W.S. Warren (1987) Opt. Lett. **12**, 398.
Heritage, J.P., R.N. Thurston, W.J. Tomlinson, A.M. Weiner, and R.H. Stolen (1984) J. Opt. Soc. Am. A **1**, 1288A.
Heritage, J.P., R.N. Thurston, W.J. Tomlinson, A.M. Weiner, and R.H. Stolen (1985a) Appl. Phys. Lett. **47**, 87.
Heritage, J.P., A.M. Weiner, and R.N. Thurston (1985b) Opt. Lett. **10**, 609.
Heritage, J.P., A.M. Weiner, R.J. Hawkins, and O.E. Martinez (1988) Opt. Commun. **67**, 367.
Hian-Hua, L., L. Yu-Lin, and J. Jia-Lin (1985) Opt. Quantum Electron. **17**, 187.
Ippen, E.P. and C.V. Shank (1975a) Appl. Phys. Lett. **26**, 92.
Ippen, E.P. and C.V. Shank (1975b) Appl. Phys. Lett. **27**, 488.
Ippen, E.P., C.V. Shank, and T.K. Gustafson (1974) Appl. Phys. Lett. **24**, 190.
Ishida, Y. and T. Yajima (1986) Opt. Commun. **58**, 355.
Iwashita, K., K. Nakagawa, Y. Nakano, and Y. Suzuki (1982) Electron. Lett. **18**, 873.
Johnson, A.M. and W.M. Simpson (1983) Opt. Lett. **8**, 554.
Johnson, A.M. and W.M. Simpson (1985a) J. Opt. Soc. Am. B **2**, 619.
Johnson, A.M. and W.M. Simpson (1985b) Proc. SPIE **533**, 52.
Johnson, A.M. and W.M. Simpson (1986) IEEE J. Quantum Electron. **QE-22**, 133.
Johnson, A.M., R.H. Stolen, and W.M. Simpson (1984a) Appl. Phys. Lett. **44**, 729.
Johnson, A.M., R.H. Stolen, and W.M. Simpson (1984b) Generation of 0.41-picosecond pulses by the single-state compression of frequency doubled Nd:YAG laser pulses.

In *Ultrafast Phenomena IV*, D.H. Auston and K.B. Eisenthal, eds., p. 16. Springer-Verlag, Berlin.

Johnson, A.M., D.W. Kisker, W.M. Simpson, and R.D. Feldman (1985) Picosecond photoconductivity in polycrystalline CdTe films prepared by UV-enhanced OMCVD. In *Picosecond Electronics and Optoelectronics*, G.A. Mourou, D.M. Bloom, and C.-H. Lee, eds., p. 188. Springer-Verlag, Berlin.

Johnson, A.M., R.H. Stolen, and W.M. Simpson (1986) The observation of chirped stimulated Raman scattered light in fibers. In *Ultrafast Phenomena V*, G.R. Fleming and A.E. Siegman, eds., p. 160. Springer-Verlag, Berlin.

Johnson, A.M., R.M. Lum, W.M. Simpson, and J. Klingert (1987) IEEE J. Quantum Electron. **QE-23**, 1180.

Kafka, J.D. and T. Baer (1985) Proc. SPIE **533**, 38.

Kafka, J.D. and T. Baer (1986) Proc. SPIE **610**, 2.

Kafka, J.D. and T. Baer (1987) Opt. Lett. **12**, 401.

Kafka, J.D. and T.M. Baer (1988) IEEE J. Quantum Electron. **QE-24**, 341.

Kafka, J.D., B.H. Kolner, T. Baer, and D.M. Bloom (1984) Opt. Lett. **9**, 505.

Keller, U., J.A. Valdmanis, M.C. Nuss, and A.M. Johnson (1988) IEEE J. Quantum Electron. **QE-24**, 427.

Klauder, J.R., A.C. Price, S. Darlington, and W.J. Albersheim (1960) Bell System Tech. J. **39**, 745.

Knox, W.H., R.L. Fork, M.C. Downer, R.H. Stolen, C.V. Shank, and J.A. Valdmanis (1985) Appl. Phys. Lett. **46**, 1120.

Kolner, B.H. and D.M. Bloom (1984) Electron. Lett. **20**, 818.

Kolner, B.H. and D.M. Bloom (1986) IEEE J. Quantum Electron. **QE-22**, 79.

Krokel, D., N.J. Halas, G. Giulinai, and D. Grishckowsky (1988) Phys. Rev. Lett. **60**, 29.

Kuckartz, M., R. Schulz, and H. Harde (1987) Opt. Quantum Electron. **19**, 237.

Kuckartz, M., R. Schulz, and H. Harde (1988) J. Opt. Soc. Am. B **5**, 1353.

Lassen, H.E., F. Mengel, B. Tromborg, N.C. Albertsen, and P.L. Christiansen (1985) Opt. Lett. **10**, 34.

Laubereau, A. (1969) Phys. Lett. **29A**, 539.

Lehmberg, R.H. and J.M. McMahon (1976) Appl. Phys. Lett. **28**, 204.

Lin, C., L.G. Cohen, R.H. Stolen, G.W. Tasker, and W.G. French (1977) Opt. Commun. **20**, 426.

Maine, P., D. Strickland, P. Bado, M. Pessot, and G. Mourou (1988) IEEE J. Quantum Electron. **24**, 398.

Martinez, O.E., J.P. Gordon, and R.L. Fork (1984) J. Opt. Soc. Am. A **1**, 1003.

May, P., J.-M. Halbout, and G. Chiu (1987) Appl. Phys. Lett. **51**, 145.

May, P., J.-M. Halbout, and G. Chiu (1988) IEEE J. Quantum Electron. **QE-24**, 234.

Meinel, R. (1983) Opt. Commun. **47**, 343.

Mollenauer, L.F. (1985) Philos. Trans. R. Soc. London A **315**, 437.

Mollenauer, L.F. and R.H. Stolen (1982) Laser Focus **18(4)**, 193.

Mollenauer, L.F., R.H. Stolen, and J.P. Gordon (1980) Phys. Rev. Lett. **45**, 1095.

Mollenauer, L.F., R.H. Stolen, J.P. Gordon, and W.J. Tomlinson (1983) Opt. Lett. **8**, 289.

Mollenauer, L.F., J.P. Gordon, and M.N. Islam (1986) IEEE J. Quantum Electron. **QE-22**, 157.

Nakashima, T., M. Nakazawa, K. Nishi, and H. Kubota (1987) Opt. Lett. **12**, 404.

Nakatsuka, H. and D. Grischkowsky (1981) Opt. Lett. **6**, 13.

Nakatsuka, H., D. Grischkowsky, and A.C. Balant (1981) Phys. Rev. Lett. **47**, 910.
Nakazawa, M., T. Nakashima, and H. Kubota (1988) Opt. Lett. **13**, 120.
Nikolaus, B. and D. Grischkowsky (1983a) Appl. Phys. Lett. **42**, 1.
Nikolaus, B. and D. Grischkowsky (1983b) Appl. Phys. Lett. **43**, 228.
Ohmori, Y., Y. Sasaki, and T. Edahiro (1983) Trans. IECE Jpn. **E-66**, 146.
Owyoung, A., R.W. Hellwarth, and N. George (1972) Phys. Rev. B **5**, 628.
Palfrey, S.L. and D. Grischkowsky (1985) Opt. Lett. **10**, 562.
Payne, D.N. and W.A. Gambling (1975) Electron. Lett. **11**, 176.
Roskos, H., A. Seilmeier, W. Kaiser, J.D. Harvey (1987) Opt. Commun. **61**, 81.
Rothenberg, J.E. and D. Grischkowsky (1989) Phys. Rev. Lett. **62**, 531.
Schadt, D. and B. Jaskorzynska (1987) J. Opt. Soc. Am. B **4**, 856.
Schadt, D., B. Jaskorzynska, and U. Osterberg (1986) J. Opt. Soc. Am. B **3**, 1257.
Shank, C.V., R.L. Fork, R. Yen, R.H. Stolen, and W.J. Tomlinson (1982) Appl. Phys. Lett. **40**, 761.
Shapiro, S.L. and H.P. Broida (1967) Phys. Rev. **154**, 129.
Shimizu, F. (1967) Phys. Rev. Lett. **19**, 1097.
Smith, R.G. (1972) Appl. Opt. **11**, 2489.
Stolen, R.H. (1979a) IEEE J. Quantum Electron. **QE-15**, 1157.
Stolen, R.H. (1979b) Nonlinear properties of optical fibers. In *Optical Fiber Telecommunications*, S.E. Miller and A.G. Chynoweth eds., chapter 5. Academic Press, New York.
Stolen, R.H. and E.P. Ippen (1973) Appl. Phys. Lett. **22**, 276.
Stolen, R.H. and A.M. Johnson (1986) IEEE J. Quantum Electron. **QE-22**, 2154.
Stolen, R.H. and C. Lin (1978) Phys. Rev. A **17**, 1448.
Stolen, R.H., E.P. Ippen, and A.R. Tynes (1972) Appl. Phys. Lett. **20**, 62.
Stolen, R.H., V. Ramaswamy, P. Kaiser, and W. Pleibel (1978) Appl. Phys. Lett. **33**, 699.
Stolen, R.H., C. Lee, and R.K. Jain (1984a) J. Opt. Soc. Am. B **1**, 652.
Stolen, R.H., W. Pleibel, and J.R. Simpson (1984b) IEEE J. Lightwave Technol. **LT-2**, 639.
Stolen, R.H., C.V. Shank, and W.J. Tomlinson (1984c) Procedure for calculating optical pulse compression from fiber-grating combinations. In *Ultrafast Phenomena IV*, D.H. Auston and K.B. Eisenthal, eds., p. 46. Springer-Verlag, Berlin.
Stolz, B., U. Osterberg, A.S.L. Gomes, W. Sibbett, and J.R. Taylor (1986) IEEE J. Lightwave Technol. (1986) **LT-4**, 55.
Strickland, D. and G. Mourou (1985) Opt. Commun. **56**, 219.
Suzuki, T. and T. Fukumoto (1976) Electron. Commun. Jpn. **59-C(3)**, 117.
Tai, K. and A. Tomita (1986a) Appl. Phys. Lett. **48**, 309.
Tai, K. and A. Tomita (1986b) Appl. Phys. Lett. **48**, 1033.
Thurston, R.N., J.P. Heritage, A.M. Weiner, and W.J. Tomlinson (1986) IEEE J. Quantum Electron. **QE-22**, 682.
Tomlinson, W.J. and W.H. Knox (1987) J. Opt. Soc. Am. B **4**, 1404.
Tomlinson, W.J., R.H. Stolen, and C.V. Shank (1984) J. Opt. Soc. Am. B **1**, 139.
Tomlinson, W.J., R.H. Stolen, and A.M. Johnson (1985) Opt. Lett. **10**, 457.
Treacy, E.B. (1968) Phys. Lett. **28A**, 34.
Treacy, E.B. (1969) IEEE J. Quantum Electron. **QE-5**, 454.
Ueda, Y. and K. Shimoda (1967) Japan. J. Appl. Phys. **6**, 628.
Valk, B., W. Hodel, and H.P. Weber (1984) Opt. Commun. **50**, 63.
Valk, B., W. Hodel, and H.P. Weber (1985) Opt. Commun. **54**, 363.
Weiner, A.M., J.P. Heritage, and R.N. Thurston (1986) Opt. Lett. **11**, 153.

Weiner, A.M., J.P. Heritage, and R.H. Stolen (1988) J. Opt. Soc. Am. B **5**, 364.
Weingarten, K.J., M.J.W. Rodwell, and D.M. Bloom (1988) IEEE J. Quantum Electron. **QE-24**, 198.
Wright, J.V. and B.P. Nelson (1977) Electron. Lett. **13**, 361.
Yang, T.Y., P.P. Ho, A. Katz, R.R. Alfano, and R.A. Ferrante (1985) Appl. Opt. **24**, 2021.
Zel'dovich, B.Y. and I.I. Sobel'man (1971) JETP Lett. **13**, 129.
Zhao, W. and E. Bourkoff (1988) IEEE J. Quantum Electron. **QE-24**, 365.
Zysset, B., W. Hodel, P. Beaud, and H.P. Weber (1986) Opt. Lett. **11**, 156.

Index

ABOUT THE AUTHOR

Gale Craig is retired from General Motors Corporation research and engineering where he was involved in battery research, gas turbine engines, electronics and automotive lighting. He holds a Masters degree in Physics and a Bachelors degree in Education with a license to teach high school math and physics. He is named as sole inventor in sixteen U.S. patents plus numerous foreign filings. His creative accomplishments include design, construction and testing of the world's first transposed open Brayton cycle gas turbine engine (Patent 3,369,361), for possible incinerator application, and the design, construction and testing of a high temperature battery, as a research project for missile application, which produced a world record power density of 43 watts per square centimeter of electrode area (Patent 3,560,265). His electronic stall-sensing motor control (Patent 4,558,259) has been installed on all General Motors cars with hidden headlamps through a period of seventeen years to date.

His lifelong interest in how things work, and dissatisfaction with traditional aerodynamics explanations led to this effort to provide a better explanation for others. He resides with his Okinawan wife, Kiku, in Anderson, Indiana.

To my Wife, Kiku, who provided moral support and tolerated without complaint my withdrawals into study and writing.

Introduction to Aerodynamics

Preface

This effort was preceded by *Stop Abusing Bernoulli!-- How Airplanes Really Fly* (ISBN 09646806-2-9). Although printed in limited number and little advertised, that book has reached many countries around the world, and has been translated into Korean for use in Pusan National University (ISBN 89-89951-22-4).

The newer effort describes principles of flight in essentially the same terms as the previous work, but with added content, beginning with basic Newtonian physics principles of action-reaction presented in aerodynamic context. It then proceeds to further develop aerodynamics in Newtonian context, wherein aerodynamic forces are associated with accelerations of air mass in interaction with aircraft surfaces. As examples, airplane propellers gain forward thrust in reaction to accelerating air mass rearward, rotating wings of helicopters gain upward lift in reaction to accelerating air mass downward, and similarly, airplane wings produce upward lift in reaction to accelerating air mass downward.

Strangely, the action-reaction principle is hardly mentioned in popular theories of flight. Instead, lift explanation is usually centered on the inverse pressure to velocity relationship derived by Daniel Bernoulli (1700-1782), and known variously as *Bernoulli's law*, *Bernoulli effect*, *Bernoulli theorem* and *Bernoulli principle*. Bernoulli's law does not specify that pressure change is caused by velocity change, or that velocity change is caused by pressure change. It only specifies the changes occur simultaneously.

The most popular explanation of wings says air passing over a more convex wing upper surface must transit the surface in the same time as that of air passing below, and hence must travel faster and therefore produce pressure difference and lift according to Bernoulli's law. Although Bernoulli's law is quite valid, the equal transit time claim is easily proven wrong. Yet this erroneous explanation has been very widely taught through the past century.

False equal transit time explanation of lift is not taught to college level students of aerodynamics who must calculate real lift forces. Unfortunately, the texts involved in these teachings are generally too complex and abstract for concise learning of basic principles, while the lower level simplistic and false teachings of equal transit time are useless, or worse. Thus there is a need for basic aerodynamics description in Newtonian terms. This effort is intended for application to that need.

Gale M. Craig
August, 2002

Introduction to Aerodynamics

CONTENTS

Entrance to United States Air Force Museum Dayton, Ohio.

Introduction to Aerodynamics

I. Basic Physics of Flight

Successful heavier-than-air manned flight has been achieved with airplanes, sailplanes, autogyros, helicopters and rockets. These various aircraft depend upon acceleration of mass to produce forces for lift, propulsion and control. Although rockets are used in special applications, the only practical force available for sustained passenger-carrying flight in the atmosphere is that derived from acceleration of air mass in interactions with aircraft surfaces. The science of these interactions, known as *aerodynamics,* is explainable in terms of Newton's laws of physics. Understanding of aerodynamics therefore requires some understanding of physics and Newton's laws.

Although Sir Isaac Newton (1642-1727) is not often mentioned in connection with flight, his contributions to physics and mathematics are of the greatest importance to flight theory and the design, construction and operation of aircraft. It was only after Newton introduced his laws of physics that aerodynamics could progress as a true science.

Speed, Velocity and Acceleration

We and all objects around us are in motion. We live on a rotating planet which orbits a Sun which moves through the universe. Thus motion, or lack of it, is only as we perceive it in reference to a position or body we choose to regard as stationary. Choosing a reference to regard as stationary can be a matter of convenience. In popular explanations of how wings work, they are usually regarded as stationary in passing air flow despite the fact that they exist on airplanes and birds which we regard as flying through relatively stationary air. Since the choice of stationary reference is an arbitrary one, either perspective, of a moving wing in still air or a still wing in moving air, is acceptable, and in the situations to follow we shall vary the choice as a matter of convenience for explaining different situations. At times we shall consider a wing as stationary in passing flow, as perceived in a wind tunnel, and at other times we shall consider a wing as moving forward in relatively still air, as we would perceive it on a moving airplane.

Speed, Velocity and Acceleration

Speed is the ratio of distance traveled to time of travel, and is expressed in various ways, some of which are knots, miles per hour, kilometers per hour and feet per second. If speed and direction of motion are both specified, as for example, fifty miles per hour in an East direction, then that is a specification of *velocity*. Change of velocity, whether in change of speed or direction, or both, involves *acceleration*, defined as rate of change of velocity. As indicated in Figure 1.1, a car turning

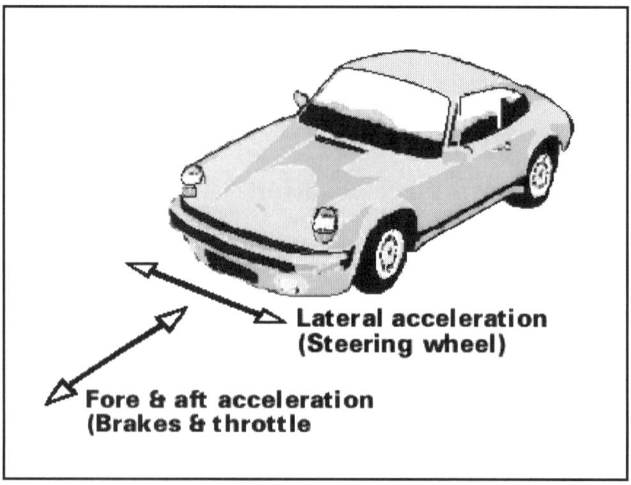

Figure 1.1 **Accelerations of an Automobile**

left or right is accelerating laterally. If brakes are applied to a car in forward motion, the velocity change, or acceleration, is rearward. If the throttle is opened further, acceleration is forward. Thus we drive our automobiles by operating a system of forward, rearward and lateral accelerators which we refer to as throttle, brakes and steering wheel.

Mass, Inertia, Force and Momentum

*Inerti*a is the property of a body to oppose acceleration, the change in velocity. This property is quantified as *mass* of the body. In absence of accelerating force, a body will continue in straight-line motion or remain stationary. That is considered to be Newton's first law, although it was described earlier by Galileo, who died in 1642, the year Newton was born. Newton's second law says a body subjected to applied force will accelerate at a rate equal to force divided by mass of the body. This is mathematically expressed as $a = F/m$, where a is acceleration, or rate of velocity change, F is force and m is mass of the body. The third law says an accelerating body will resist acceleration with force equal and opposite to that applied.

Mass in motion has *momentum,* equal to mass times velocity. Momentum changes at a rate equal to applied force, and mass reacts to the momentum change with equal opposing force. Forward thrust of a propeller is due to rearward momentum production in the acceleration of air. Similarly, upward lift of an airplane wing or helicopter rotor is a reaction to downward momentum produced in air. Leonardo Da Vinci recognized, even before Newton's time, that lift could be generated by downward acceleration of air mass. He envisioned a sort of man powered helicopter, as shown in the photograph of Figure 1.2, having a rotating "airscrew" to propel air downward.

Figure 1.2 **DaVinci's Helicopter Concept**
Model in the Dayton, Ohio USAF Museum.

We shall never know whether or not Davinci realized that torque involves action and reaction, for his version of a helicopter lacked a tail rotor, common on most helicopters today. Thrust of a tail rotor is needed to counter torque of the lift rotor and prevent rotation of the airframe, or in this case to prevent rotation of the airframe as the men run on the revolving platform.

An airplane propeller, which is a sort of rotating wing, develops forward thrust equal to the rate of momentum produced in accelerating air mass rearward. If during pre-takeoff run-up the propeller accelerates initially stationary air rearward with velocity of 100 feet per second at a rate of air mass throughput of five slugs per second, then the rearward momentum change rate, equal to velocity change times mass flow rate, would be 500 foot slugs per second squared, equating to 500 pounds thrust. Equivalently in metric units, the mass flow rate of 72.6 kilograms per second accelerated to velocity of 30.5 meters per second would produce thrust of 2214 Newtons.

Upward thrust of wings, or lift, is equal to the downward rate of momentum change produced as the wings accelerate air mass downward. We shall refer to downwardly departing air behind a wing as *downwash.* Pilots, especially sailplane pilots, are well aware that upward-moving air in a thermal updraft produces lift when wings intercept the upward movement. Interception of upward air movement also constitutes downward acceleration of air mass. In addition to thermal updrafts, pressures produced in the lift process cause *upwash*, a rising air movement ahead of a wing, which when intercepted also contributes to lift. This will be covered in a later section. Total upward lift of a wing is thus equal to the sum of the rates of upward air mass momentum interception plus downward air mass momentum production.

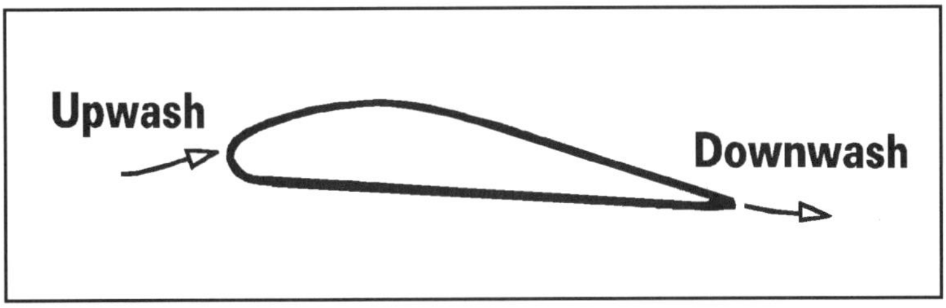

Figure 1.3 **Upwash and Downwash**
In perspective of a stationary wing in passing flow, air rises ahead of a lifting wing and departs in downward direction pointed by the wing rear portion. Rising air ahead will be referred to as "upwash" while the departing air behind will be termed "downwash."

Conservation of Momentum

Momentum is said to be *conserved* in mass interaction. This occurs in mixing masses of differing velocity, where total momentum remains constant. As an example, consider the illustration of Figure 1.4, where a jet of air issues from a nozzle at high velocity. The jet mixes turbulently with surrounding air, carrying an increasing volume of slowing air in the direction of jet travel. Momentum gained by surrounding air is equal to momentum loss of air issued from the nozzle, so that total momentum remains equal to that issued from the nozzle.

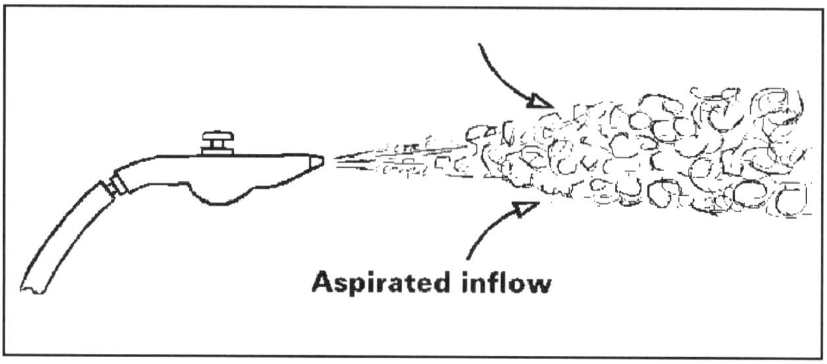

Figure 1.4 **Conservation of Momentum**
Air issuing from a high pressure source mixes turbulently into surrounding air and carries it downstream, where the volume of movement is much greater than that at the nozzle, but the momentum rate carried in slower movement of greater mass remains equal to that issuing from the nozzle.

Moving of air or other fluid by mixing with a higher velocity discharge is widely practiced in pumping applications. A nozzle located within a larger conduit, as in Figure 1.5, produces a jet which mixes into the fluid being pumped so that fluid drawn into the upstream entry is carried to the downstream exit for discharge.

Among applications are "jet pumps," for pumping water, and steam driven "ejectors" to provide vacuum for industrial processes. Fluid pumped may be different from that of the jet. For example, water jets can pump air and air jets can pump water. Although energy efficiency of such pumps is poor because of turbulence losses in flow mixing, they are attractive for many applications because of their simplicity, low cost and reliability.

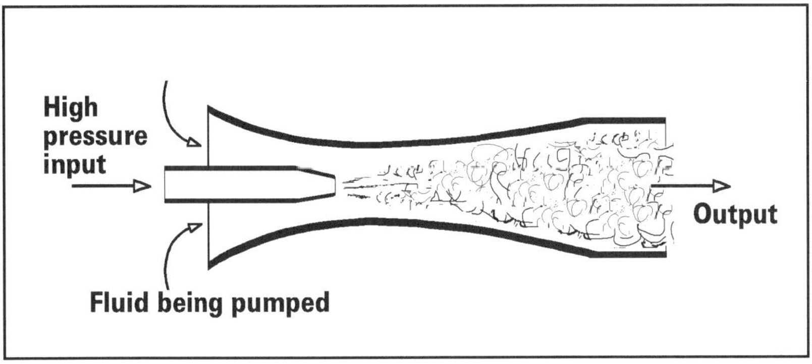

Figure 1.5 **Jet Pump or Ejector**
A jet of high velocity flow can pump a greater mass flow rate through mixing of momentum issuing from the jet into flow surrounding the jet. This is not a highly efficient process because energy is lost into turbulence in the mixing. However, simplicity and low cost makes this a practical pump for many applications.

Carrying away of slower air through mixing with faster flow is important in wing operation. Although momentum, the mass times velocity product, is conserved in the mixing, kinetic energy, the product of one-half mass times velocity squared, decreases due to losses in generation of turbulence and heat.

Pressure Gradient

Where pressure varies in relation to distance, we say there is a *pressure gradient.* Direction of the gradient is toward higher pressure. For example, pressure of the Earth's atmosphere increases at lower altitude. Thus the pressure gradient is downward, toward the Earth's surface, as illustrated in Figure 1.6. A pressure gradient is also produced in the acceleration of air mass. In Figure 1.7, where a tube is accelerating to the right, the closed end of the tube pushes on the column of air contained within. Force on the closed end, in accordance with Newton's laws, is equal to air mass in the tube times acceleration, or rate of change of velocity of the tube. Pressure rise at

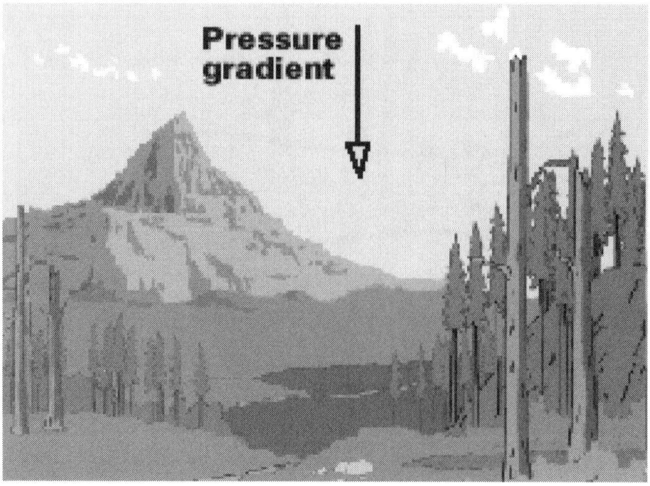

Figure 1.6 **Atmospheric Pressure Gradient**
The Earth's surface pushes upward against atmospheric pressure with force equal to the weight of atmosphere pushing down. At higher altitudes pressure is less. Since pressure decreases at higher altitude the *gradient* of pressure is downward.

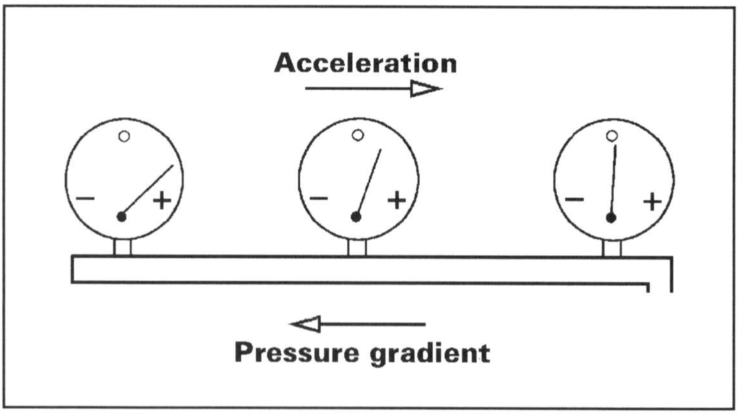

Figure 1.7 **Pressure Gradient in Acceleration**
When the tube is accelerated to the right, the opposing air mass reaction force is to the left. This creates a gradient of pressure variation along the tube, as indicated by the gauges. Near the open end, pressure is near that of outside air, while pressure rise at the closed end is greatest.

the closed end is equal to that force divided by end area. At positions nearer the open end, where there is less air mass between the measuring position and open end, pressure rise is proportionately less. Distribution of pressure difference along the tube constitutes a pressure gradient having direction (of increasing pressure) opposite to the direction of acceleration. Pressure gradients are always opposite in direction to that of acceleration.

Centrifugal Pressure Gradient in Curving Flow

Momentum, the product of mass times velocity, has the direction of velocity. Changing the speed or direction of momentum involves acceleration and force. In the case of circular motion, force applied to curving the mass movement is center-directed, or *centripetal*, and the opposing outward reaction force is known as *centrifugal*, or center-fleeing. Outward centrifugal force created when air or other fluid is accelerated toward a center of curving motion creates an outward pressure gradient, as illustrated in Figure 1.8. Man-made devices

Figure 1.8 **Centrifugal Pressure Gradient**
In rotational movement of air or other fluid, acceleration is inward toward the center. Opposing centrifugal force, outward from the center, creates a centrifugal pressure gradient which opposes surrounding pressure, In the gradient, pressure is least at the center and increases outwardly.

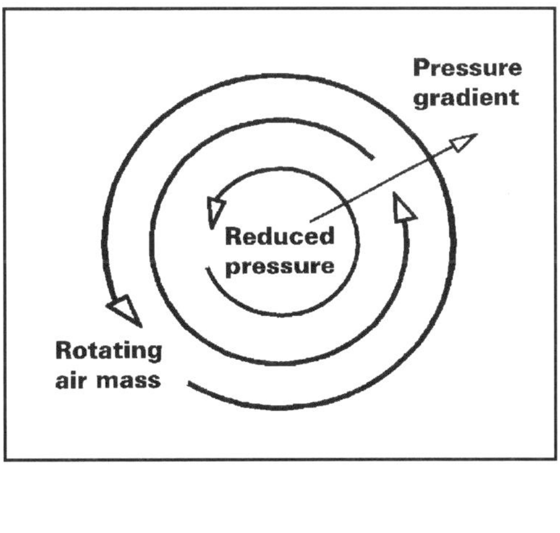

like water pumps, vacuum cleaners and hair dryers are dependent on centrifugal pressure gradients in rotating fluid. A vacuum cleaner uses reduced center pressure for suction, while a hair dryer uses outwardly increased pressure for blowing warm air. We shall refer to the outward pressure gradient in rotating or curving flow as a *centrifugal pressure gradient.*

Centrifugal pressure gradients occur naturally in tornados and hurricanes. Outward pressure gradient in a tornado opposes pressure of surrounding atmosphere so that pressure is reduced within the rotating funnel. Buildings sometimes explode from retained internal pressure as a tornado passes.

Motion of air or other fluid need not be in complete circles to produce pressure gradients. Where wind passes over the convex surface of a hill or mountain peak, as indicated in Figure 1.9, centrifugal gradients in curving flow increase pressure at the base and reduce pressure at the summit. In accelerating toward reduced pressure at the summit, wind velocity increases. From summit rearward, velocity decreases due to opposition of rising pressure. Departing wind resumes original direction,

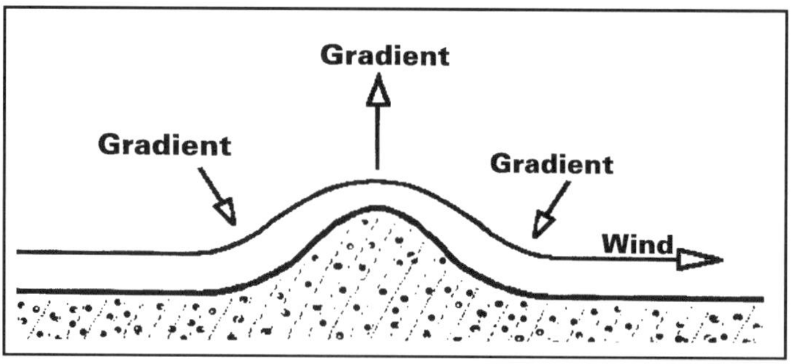

Figure 1.9 **Hilltop Pressure Reduction**
At the upwind base of the hill pressure is increased by the downward gradient in upward curving flow. Above the summit the curvature is downward, creating an upward pressure gradient which reduces summit pressure. At the downwind base pressure is again increased in upward curvature of flow.

parallel to the horizontal surface, with no net downward momentum production and hence no net lift on the surface.

Paper Lift Demonstration

A demonstration often mistakenly presented as showing pressure reduction due to "Bernoulli principle" is illustrated in Figure 1.10 The mistake is in declaring that air velocity causes pressure to be reduced above the paper, making it rise. But if the paper rests flat on a horizontal screen without curvature, there is no tendency to rise when air is blown over it.

The real cause of pressure reduction is curvature of flow following the surface, creating an upward pressure gradient opposing atmospheric pressure so that surface pressure is reduced. Tendency for flow to follow a surface, explained in a later section, is known as *Coanda effect* or *surface attachment*.

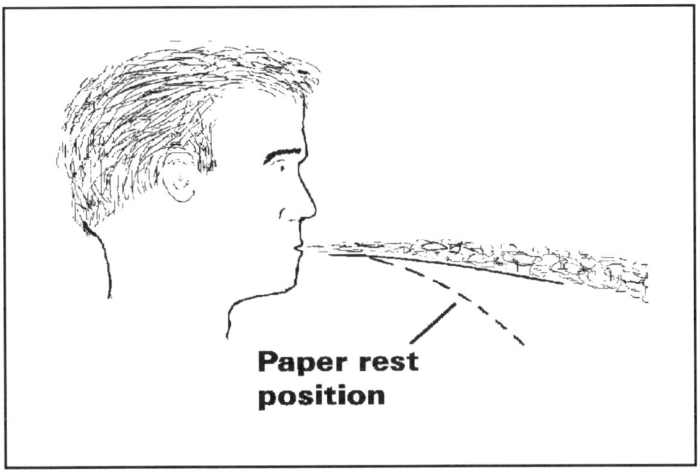

Figure 1.10 **Lift Demonstration**
Air flow following the downward-curving convex surface creates an upward centrifugal pressure gradient partially offsetting atmospheric pressure at the surface. Less pressure at the upper surface as compared to the lower causes the paper to rise.

In a counter-demonstration, illustrated in Figure 1.11, pressure is shown to increase in passing the paper lower surface. In this case downward curvature of flow again produces an upward centrifugal pressure gradient, but here the gradient adds to atmospheric pressure, again causing the paper to rise.

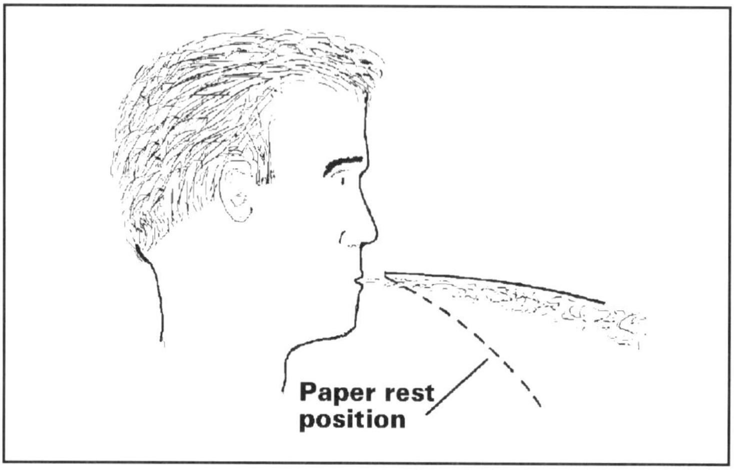

Figure 1.11 **Alternate Lift Demonstration**
When air is blown beneath the drooping paper it rises as the upward centrifugal pressure gradient adds to atmospheric pressure below.

Like lift of the paper, lift of a wing can be explained in terms of centrifugal pressure gradients. As explained in a later section, air rises ahead of a lifting wing (upwash) and descends behind (downwash). In perspective of a stationary wing in passing flow, the flow curves from upward ahead of the wing to downward behind, as in Figure 1.12. In curvature from upward to downward, upward pressure gradients are created. The upward gradient adds to atmospheric pressure at the lower surface, while at the upper surface the upward gradient opposes atmospheric pressure, causing surface pressure decrease. Because flow curves from upward to downward, wing profiles are usually curved to some degree so as to match curvature of flow. This curvature, called *camber*, is effected by making the upper surface more convex than the lower.

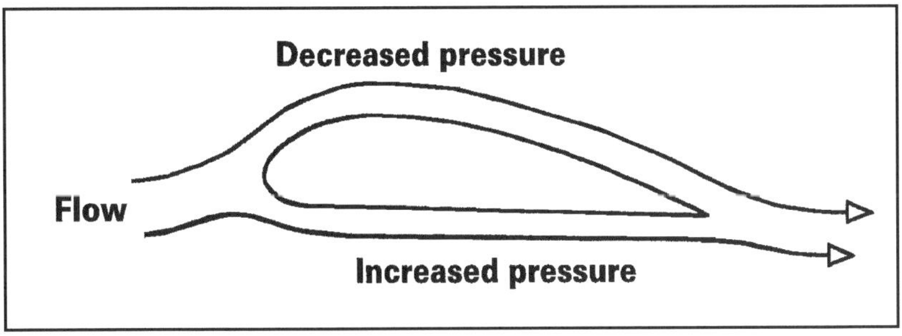

Figure 1.12 **Wing Pressures**
Upward pressure gradients in curvature of flow, from upward arrival at the front toward downward departure at the rear, produce decreased pressure above and increased pressure below.

Conservation of Angular Momentum

Circular motion of mass around a central position involves *angular momentum*, equal to mass times velocity times length of radius from the center of curvature. Angular momentum changes at a rate equal to applied torque. Torque requires at least a *couple* of two noncolinear forces, such as the offset string tensions at opposite sides of the cylinder illustrated in Figure 1.13.

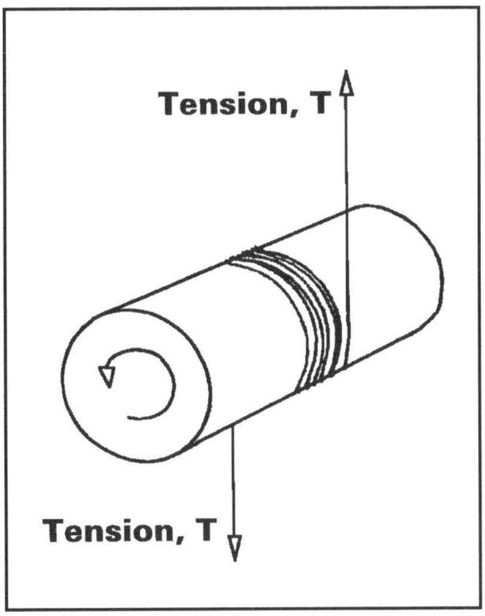

Figure 1.13 **A Couple**
String wound around a cylinder applies a couple equal to string tensions of magnitude, T, times cylinder diameter, the distance between strings at opposite sides. Angular momentum of the cylinder increases at a rate equal to magnitude of the torque moment.

In a fluid vortex, total angular momentum is equal to the integral, or addition over the entire vortex, of mass times velocity times radius. In absence of torque, angular momentum remains constant.

Angular momentum of a closed system having no torque input is said to be *conserved*, meaning unchanging. However, with radially directed force, energy in rotation can be changed without change of angular momentum. As illustrated in figure 1.14, inward movement of rotating mass toward the center of rotation requires energy in overcoming centrifugal force. That energy goes into increasing speed of the rotating mass, but the inward force of string tension conveys no torque, hence speed and kinetic energy increase without change of angular momentum. For angular momentum to be constant, rotational speed of the mass must vary in inverse proportion to radius from the center. An ice skater uses angular momentum conservation when pulling arms inward to increase body rotation rate.

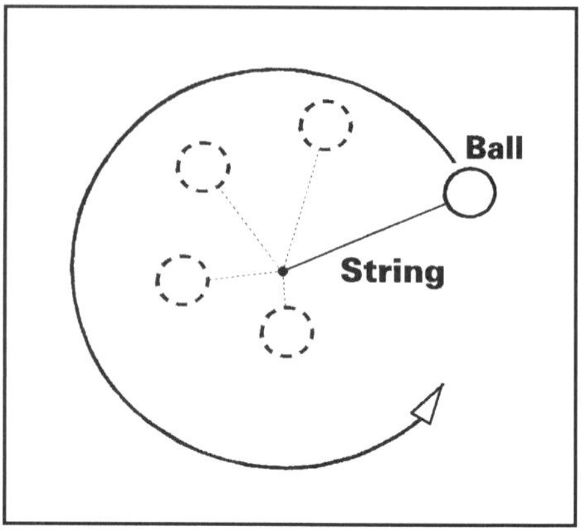

Figure 1.14 **Constant Angular Momentum**
The round mass rotating about a center hole as it is restrained by a string, tends to rotate in a circle, represented by the solid line. As the withdrawing string pulls the mass inward, as represented in dashed outline, the mass speeds up with increased kinetic energy, but without change of angular momentum.

Rotational movements of vortices trailing behind airplane wings, indicated in Figure 1.15, contain angular momentum.

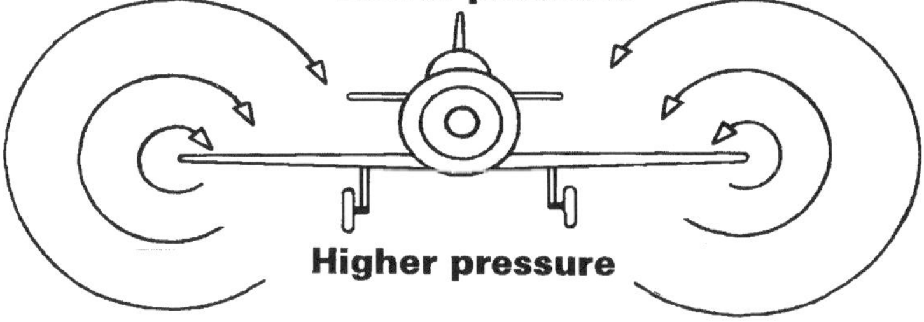

Figure 1.15 **Trailing Vortices**
In stable flight, airplane wings leave continuous trails of oppositely rotating vortices behind, produced by pressure force couples which accelerate air away from higher pressure below and toward lower pressure above.

The vortices, which are of opposite direction of rotation, can be seen at air shows where planes are equipped with smoke generators. In the case of a sailplane in stable flight, opposite vortices are equal, adding algebraically to zero total rate of angular momentum production. In the case of a propeller driven plane, where lift of opposite wings must be unequal in order to compensate for engine torque, opposite vortices differ in angular momentum production in degree equal to angular momentum produced by the rotating propeller. The total rate of angular momentum production, when total angular momenta produced by wings and propeller are algebraically added, is zero in stable normal flight.

In rolling into a turn, airplane mass has angular momentum created through momentary wing lift difference. In reaction, equal and opposite angular momentum change is caused in the air. In this case the system in which angular momentum is conserved includes airplane and surrounding air. In stopping the roll, opposite angular momenta changes of airplane and air would again be produced in association with torque required to stop the airplane roll.

Conservation of Energy

Energy is the capacity to do work. According to the *law of conservation of energy*, energy cannot be created or destroyed, but may be converted from one form to another. Airplane engines convert chemical energy of fuel and oxygen into heat, and then into mechanical energy for thrust. Thrust moves the plane forward against drag losses of friction and turbulence, which convert the energy finally back into heat which is dissipated in the atmosphere.

Mechanical energy expended is often in the form of applied force times displacement, or distance of movement in direction of the force. In fluid displacement, force is that of pressure, and displacement is that of volume moved. Energy involved in fluid movement is equal to pressure times displaced volume, expressed mathematically as PV. For energy stored as weight of mass held at some height above a reference level, the force is weight and the displacement is height. These expressions of energy apply to the hydropower system of Figure 1.16, where succeeding positions of a small volume of water are numbered. In exiting the nozzle, the mass has kinetic energy equal to one-half the mass times velocity squared.

Altitude energy of a volume depicted is equal to its weight times height above water discharge nozzle. As the volume drifts downward, this height times weight energy is converted into pressure times volume, or PV energy, and then is converted into kinetic energy at the nozzle outlet for driving the Pelton wheel. If we dismiss as negligible the frictional effects which would convert a small amount of the energy into heat, we can say that the law of conservation of energy dictates that in transition the sum of the three different forms, altitude, pressure and kinetic energy, remains constant in the volume considered.

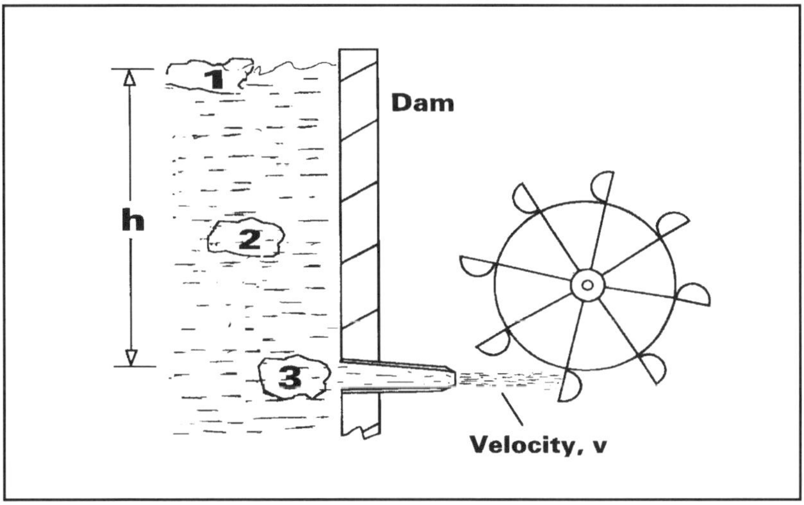

Figure 1.16 **Water Energy Conversion**
The outlined water volume in position "1" originally arrived at the surface as rain, where in relation to the nozzle discharge it has altitude energy equal to height, h, times its weight. At position "2" it has lost some height, but has gained pressure. Arriving at the nozzle entry position "3" its height energy in relation to the nozzle is zero, as it has been converted to PV energy. Pressure pushing the water through the nozzle to position "4" converts the PV energy into kinetic energy. Except for frictional or viscous effects the total energy is unchanged.

The small water volume, V, has weight equal to its mass, m, times gravity, G. Mass, m, is equal to density, ρ, times volume, V. Thus potential energy of the volume in relation to height, h, above the nozzle, can be expressed as:

(1-1) $$\text{Altitude energy} = \rho VGh$$

As the water descends toward the nozzle entry port, height energy is converted to PV (pressure times volume) energy. But total energy is unchanged. Thus at position 2, total energy is:

(1-2) $$\text{Pressure plus height energy} = \rho VGh_2 + PV$$

Leaving the nozzle, PV energy content of the water has been converted to kinetic energy, equal to one-half the water mass, m, times velocity, v, squared. Since the mass is equal to density, ρ, times volume, V, the kinetic energy of the volume of water becomes:

$$\text{(1-3)} \qquad \text{Kinetic energy} = 1/2\ \rho V v^2$$

These conversions between forms of energy occur progressively as the water moves from the reservoir surface toward the Pelton wheel. Altitude, pressure and velocity all vary as the water moves, but the total energy of volume, V, remains constant. We can express this mathematically by stating that the unchanging total of the three energy forms of a small volume at any position is equal to a constant times the volume, V, or:

$$\text{(1–4)} \qquad \rho V G h + PV + 1/2\ \rho V v^2 = \text{constant x V}$$

Dividing through by volume, V, we get:

$$\text{(1-5)} \qquad \rho G h + P + 1/2\ \rho v^2 = \text{constant}$$

This is Bernoulli's law as it is often stated in engineering texts. It is just a statement of energy conservation. For those taught the Bernoulli equation in an airplane context, the first term, ρGh, may not appear familiar because it is of little importance to airplanes, but it is important if you are a balloonist. Lift of a balloon is due to greater atmospheric pressure below the balloon than above it. Buoyancy of less importance exists even for more dense objects. Buoyancy of a commercial airplane amounts to several hundred pounds, but that is considered insignificant in relation to airplane weight, and is generally disregarded. Thus for airplane aerodynamics the Bernoulli principle, or Bernoulli's law, is customarily shortened to:

(1-6) $$P + 1/2\ \rho v^2 = \text{constant}$$

The P term, sometimes called *static pressure* represents pressure as would be measured by a barometer. For the sum of $P + 1/2\ \rho v^2$ to be constant, the $1/2\ \rho v^2$ term must represent a pressure equivalent, and is termed *dynamic pressure*. Together the total of pressures, constant according to Bernoulli's law, is known as *total pressure* or *stagnation pressure*. In the case of the hydropower plant, if we discount atmospheric pressure, total pressure of unmoving water behind the dam at nozzle level would be only static, while that of water issuing from the nozzle would be only dynamic. At any arbitrary point in the nozzle where flow is accelerating, energy is shared between static and dynamic forms.

PV Energy in Other Context

It is worth noting that the expression of PV energy, as common in hydrodynamics or aerodynamics, where the fluid is considered incompressible, differs from PV energy expression in the study of gases in thermodynamics. In relation to the hydropower plant of Figure 1.16, PV energy entering the nozzle is that which is exerted by pressure, P, of the reservoir in pushing the volume, V, into the nozzle. In the case of a venturi tube, PV at the entry is that of atmospheric pressure, P, pushing volume, V, into the tube. In such cases PV energy of the fluid displacement is exerted by the environment upon the fluid. In the study of gases in physics, chemistry and thermodynamics, however, PV energy is defined as internal thermal energy of the quantity considered, and varies with temperature.

Bernoulli's law is based on consideration of incompressible fluid, and although air is compressible, we can usually ignore the small compressibility effects in subsonic flight where most airplanes operate. There are, however, some effects of compressibility of which a pilot should be aware. Anyone who has ever used a hand pump to inflate a tire knows the pump becomes warmed by the effort. This is because work energy converts to thermal PV energy in the air being compressed, causing temperature to rise. Oppositely, when air expands it cools, as thermal energy is converted into work of expansion. Cooling of a pneumatic hand grinder when operated is evidence of this effect.

Air expands in passing over an airplane wing as pressure is reduced, and although associated cooling may be of small degree, if that degree is near the water freezing point it can cause moisture in clouds to freeze onto the wing. Another matter to consider is rapid formation of fog which can occur when temperature is near the dew point. In that case, expansion cooling due to a small decrease in atmospheric pressure can trigger fog in a very short time. Thus a pilot should be knowledgable of the local spread between temperature and dew point. Above a wing in flight, especially military aircraft wings during maneuvers involving higher G forces at air shows, one can occasionally see fog form in the cooling of reduced pressure .

The Pitot Tube

French physicist, Henri Pitot (1695-1771), devised the *Pitot tube*, based on Bernoulli's law, for measuring fluid velocities. The device, sketched in Figure 1.17, has an opening at the front where fluid impact produces stagnation pressure, while openings at the sides admit static pressure. The difference between these pressures, equal to dynamic pressure, can be interpreted as a measure of flow velocity. In an airplane cockpit the difference is read on a pressure differential gauge calibrated in units of airspeed.

In the figure, air can be regarded as stationary while the Pitot tube moves with the airplane. Static pressure then becomes that of unmoving air, and stagnation pressure rise occurs where the forward-moving tube impacts the air it encounters. The alternate perspective, of moving air passing a stationary tube yields the same result, for either way it is acceleration of air in encounter with the tube which produces the pressure rise.

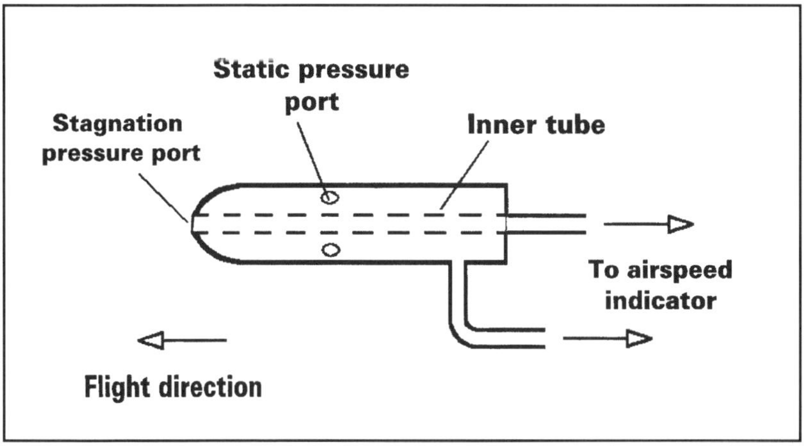

Figure 1.17 **Pitot Tube**
This Pitot tube consists of two concentric tubes having an annular space between which is closed at both ends. The inner tube is open to stagnation pressure at the front, and the outer tube is open to static pressure at the sides. The two pressures are piped to a pressure differential indicator calibrated to read *indicated airspeed*.

On an airplane, static pressure and stagnation pressure ports are usually located separately, with static pressure port or ports located somewhere along the fuselage, where air undergoes little acceleration and pressure change during passage of the plane, as shown in Figure 1.18. Fuselage static ports, in addition to connecting to airspeed indicators, also connect to altimeters, rate-of-climb indicators, autopilots and altitude reporting transponders.

Figure 1.18 **Static Port**
The static port, shown larger in the inset, has an orifice about one millimeter in diameter. An identical port is located on the other side of the plane

Often planes have dual static ports, one on each side, connected together in the plumbing. This reduces error caused in sideslipping of the plane, and provides redundancy in case of a plug, such as from ice or polishing wax.

The stagnation pressure port for airspeed indication is located where it normally projects into oncoming flow, often below a wing leading edge, as shown in Figure 1.19. Airplane pitch attitude affects orientation of the Pitot stagnation port in relation to flow, and location of the static port affects the reference pressure. These factors cause *indicated airspeed* to differ from that which would result solely from Bernoulli's law considerations. For a pilot wanting to know the *calibrated airspeed*, as would be indicated precisely in accordance with Bernoulli's law and relative speed of the plane through the air, a correction chart may be included in the pilot's operating handbook.

Figure 1.19 **Pitot Stagnation Pressure Port**
This stagnation pressure port is located below the wing leading edge at the end of a tube which projects into oncoming flow. The outside diameter of the tube is made large enough to accommodate an electrical heating element for deicing.

In addition to being functions of *true airspeed,* the actual speed of a plane in relation to the air, indicated airspeed and calibrated airspeed vary in relation to air density, which is affected by altitude and temperature. True airspeed is of importance in estimating time of travel between airports or landmarks, but indicated or calibrated airspeed is more closely related to aerodynamic forces on the airplane, and hence is of more importance to airplane operation.

Never exceed speeds and maximum airspeeds for extension of flaps and landing gear are marked on indicated airspeed instruments. Also minimum speeds for landing approach and takeoff are determined in reference to indicated airspeed. In less dense air of a high altitude airport, true airspeeds required may be much greater for takeoff and landing, requiring longer runways, but indicated airspeed requirements, which more closely relate to airplane flying performance, would be unchanged.

The Venturi Tube

Figure 1.20 represents a venturi tube for supplying vacuum to airplane instruments in some older airplanes. In the venturi tube, it is clear that velocity must be greater in the narrowed portion in order to pass the same volume rate as passed in the larger portions. Pressure is reduced in the faster flow. Does this mean that pressure is decreased because of the velocity increase? That would satisfy Bernoulli's law, but if velocity increases because pressure decreases, that would also satisfy Bernoulli's law. The law only states an inverse relationship between pressure and velocity. It does not assign cause and effect.

Velocity change occurs in response to pressure difference. We may ponder what process might cause mid-tube pressure reduction in a venturi so as to cause air to accelerate toward it.

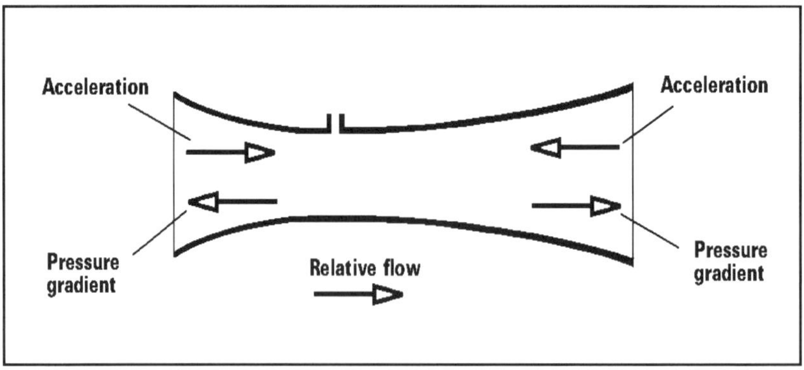

Figure 1.20 **Venturi Tube**
Air entering the inlet accelerates toward mid-tube reduced pressure with increasing velocity. From the minimum pressure position onward, reversed acceleration toward the minimum pressure position reduces air velocity as pressure rises.

Let's suppose we have a 1947 Luscombe as in Figure 1.21, which has a venturi for instrument vacuum. Before flight we cap venturi inlet and outlet. In flight we find pressure rise in front of the inlet cap and pressure reduction behind the outlet cap. If we quickly remove the caps, we find that as flow accelerates through the tube, pressure difference between inlet and outlet ends disappears. Initial accelerative force is supplied by pressure difference between inlet and outlet, and PV energy is converted to kinetic energy of tube flow. Stabilized flow still accelerates toward mid-tube, but instead of being driven by pressure difference between inlet and outlet it is driven by pressure drop at mid-tube. From mid-tube to the exit end, acceleration is reversed by opposition of rising pressure, so that exit velocity decreases toward that of air passing outside.

Figure 1.21 **Venturi on a 1940s Luscombe**
This airplane was designed without an electrical system. The only instruments available were vacuum operated. Vacuum was obtained in flight from air passage through the side-mounted venturi.

Energy conservation occurs in the Venturi tube as PV energy expended in work of atmospheric pressure pushing air volume toward the reduced pressure is converted to kinetic energy at mid-tube. From mid-tube rearward, kinetic energy is converted back into work against rising pressure. This is in accordance with Bernoulli's law which expresses conversion between PV energy and kinetic energy. Pressure change associated with acceleration of air passing through the Venturi tube of Figure 1.22 can be calculated through application of Newton's laws and a bit of first semester calculus, as follows:

Each element of mass, dm, which has volume equal to area, s, times thickness, dx, opposes acceleration through the tube with force of minus mass times acceleration, or -(dm a). The force is distributed over surface area, s, so that the opposing increment of pressure, dp, becomes:

$$dP = -\frac{dm\ a}{s} \qquad (1\text{-}10)$$

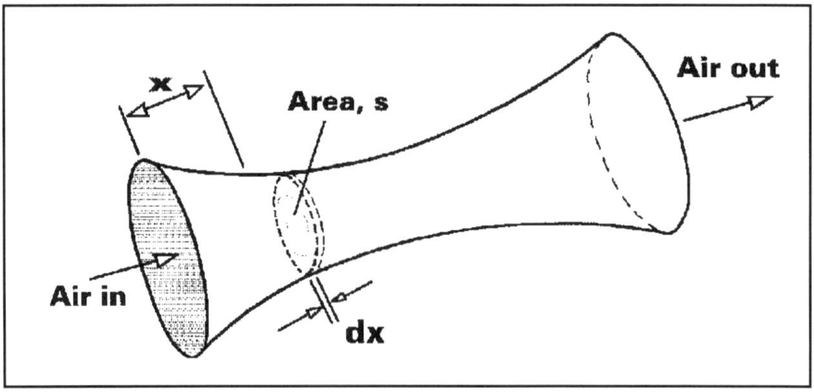

Figure 1.22 **Incremental Mass Elements**
Each imaginary disk element of air volume in the Venturi tube is of mass, dm, which is equal to the product of its cross section area, s, its length, dx, and density, ρ.

Since mass element, dm, is equal to the product, $\rho s\ dx$, and a is equal to dv/dt:

(1-11)
$$dP = -\rho\ dx\ \frac{dv}{dt}$$

And with $v = dx/dt$:

(1-12)
$$dP = -\rho\ v\ dv$$

We can obtain total pressure change by adding up, or integrating, the pressure increments due to accelerating all the air mass elements filling space between the tube entry and any point of interest downstream. When integrated from the tube entrance to a downstream distance, x, the integral becomes:

(1-13)
$$\Delta P = -\rho \int_{v_0}^{v_x} v\ dv$$

And after integrating:

(1-14)
$$\Delta P = \frac{\rho(v_0^2 - v_x^2)}{2}$$

Here ΔP is the pressure change, v_0 is entry velocity, and v_x is velocity at the point where the pressure change is taken.

Equation (1-14) provides pressure change accompanying velocity change. If flow of initial velocity, v_0, is stopped in impact with a fixed object, so that v_x becomes zero, then the pressure rise becomes:

(1-15)
$$\Delta P = \frac{\rho v_0^2}{2}$$

Total pressure, or stagnation pressure, equal to static plus dynamic pressure, would be:

$$\text{(1-16)} \qquad \text{Total pressure} = P + \frac{\rho v^2}{2}$$

This represents energy density associated with volume, v, which remains constant as the work of moving air through difference in pressure converts to kinetic energy of flow, so that:

$$\text{(1-17)} \qquad P + \frac{\rho v^2}{2} = \text{constant}$$

Again, this relationship expresses Bernoulli's law, as derived from consideration of incompressible frictionless flow passing through a channel of varying cross section, as Daniel Bernoulli (1700-1782) derived it long ago.

Bernoulli was a physician with a good education in mathematics and physics who developed a means of measuring blood pressure by seeing how high blood would rise in a glass tube. He noticed an apparent inverse relation between vein size and pressure. When he applied Newton's calculus and physics to find a mathematical expression of the relationship, he was amazed at how simple it was and that no one before him had found it.

Bernoulli's equation does not consider effects of friction or compressibility, but nevertheless gives good results in subsonic flight applications where pressure variation and frictional effects are relatively small.

Vortex Flow

We can apply Bernoulli's law to uniform flow through a circular bend in a pipe or conduit, as illustrated in Figure 1.23. As flow curves around the bend it develops an outward pressure gradient due to centrifugal force so that pressure at the outer wall is greater than pressure at the inner wall. Then in accordance with Bernoulli's law, velocity in lowest pressure near the inner wall must be greater than that farther out.

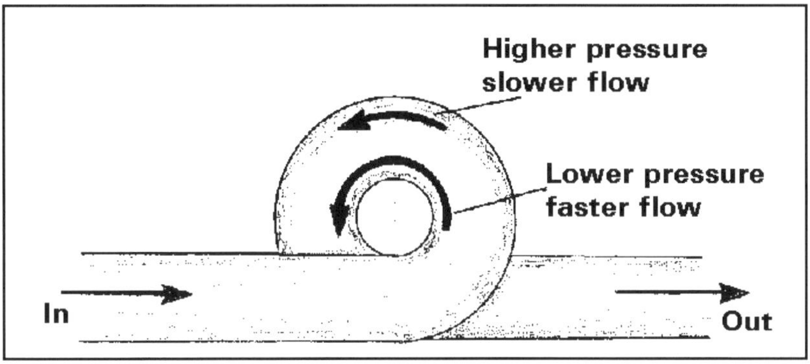

Figure 1.23 **Bernoullian Rotational Flow**
As uniform bernoullian flow passes through a bend it develops a centrifugal pressure gradient toward the outside wall, and in accordance with Bernoulli's law, simultaneously develops an inward velocity gradient such that velocity is greatest at the inside wall.

If we differentiate equation (1-6) we get:

$$dP + (\rho v)\, dv = 0 \qquad (1\text{-}7)$$

This says that effects of a small increment of pressure change, dp, and accompanying opposite small change in velocity, dv, in Bernoullian flow produce zero energy change. Another expression, from basic physics, says centripetal acceleration is equal to velocity squared divided by the radius of curvature.

Multiplying this acceleration times fluid density, ρ, and a small increment of radius, dr, we get an outward pressure increase increment, dp, as:

(1-8) $$dp = [(\rho v^2)/r]dr$$

Substituting this into the previous equation, simplifying and separating variables we get:

(1-9) $$dv/v = - dr/r$$

Upon integrating, this leads to:

(1-10) $$v = c/r$$

This says velocity in the curving flow is equal to a constant, c, divided by radius of flow curvature, r. Although we have found velocity to be inversely proportional to radius in the confinement of a pipe, it also applies in open air, as in a tornado, where an outward centrifugal pressure gradient is exerted against atmospheric pressure. The inverse velocity to radius characteristic defines a *free vortex*. In a free vortex, if velocity is known at some arbitrary radius, then it can be found at any other radius from the preceding equation. This will be important in finding an equation for lift in a later section where we consider *circulation*, which has free vortex properties. Since velocity is inversely proportional to radius, and angular momentum is a function of velocity times radius, angular momentum content is uniform throughout a free vortex.

As we proceed, we shall consider a vortex to be a rotation of air mass about a center. This differs in some aerodynamics publications which regard a vortex as a center around which rotation is induced.

Pressure Recovery

In the Venturi tube, pressure decrease in rearward acceleration of flow between inlet and throat is regained, or recovered, in deceleration of flow bctwccn throat and outlet. Pressure recovery is very important in aerodynamics because it relates to drag. In flight, an object such as that of Figure 1.24 develops impact pressure rise at the front, pressure reduction farther aft, where curving flow produces an outward gradicnt, and pressure recovery at the rear.

If pressure recovery is without loss, then opposition to forward flight from pressure rise at the front is equaled by assistance to forward flight from pressure rise in flow convergence at the rear, resulting in no net drag due to pressure difference.

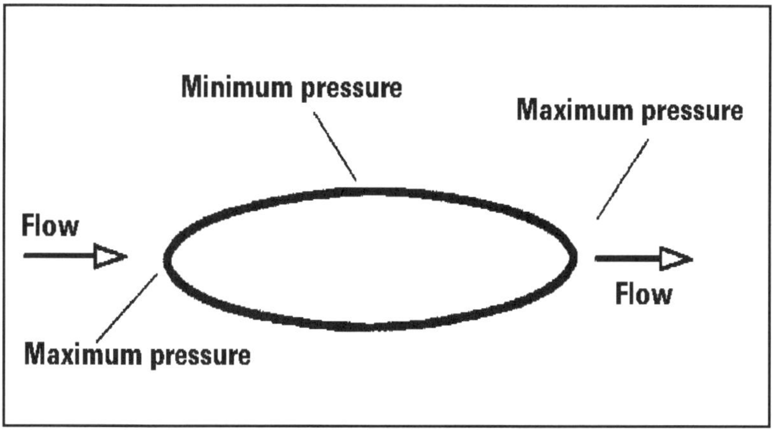

Figure 1.24 **Pressure Recovery In Lossless Flow**
From the front, where air impact produces pressure rise, flows accelerate rearward toward mid-surface positions above and below, where pressures are reduced due to centrifuging action in curving flows. From mid-positions rearward, flows decelerate in the opposition of aft convergence pressure increase equaling that at the front. Converged flow accelerates rearward with velocity and pressure returning to that of undisturbed flow.

Pressure recovery is affected by losses due to viscosity, the property of a fluid to exert frictional drag, which we shall be considering. In the real world where all fluids, including air, have viscosity, pressure recovery is never complete.

Newton's Lift Concept

Sir Isaac Newton reasoned that upward lift of a wing would be due to downward acceleration of air mass, and the magnitude of lift would equal the rate of downward momentum production. To evaluate his concept he considered a flat plate angled in relation to the direction of flight, or in relation to the direction of flow passage, as in Figure 1.25.

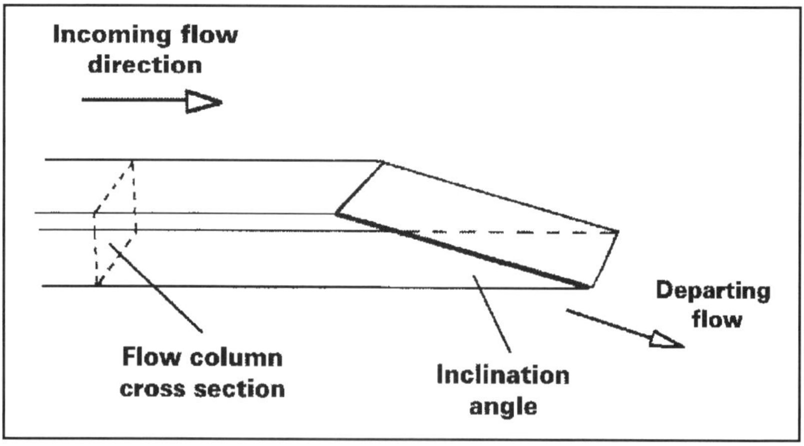

Figure 1.25 **Newton's Flat Plate**
Newton imagined an inclined flat plate held in horizontally flowing air would intercept and deflect the column of air directly ahead of the plate. The column vertical cross section area would equal plate area times sine of the angle of inclination. Deflecting the stream of air downward would produce lift force equal to the downward momentum rate produced, and would produce drag equal to the rate of rearward momentum reduction. (This description ignores frictional and edge losses which could be expected.)

Newton considered the rate of air mass involvement as equal to air mass density, ρ, times cross section area of the column of air ahead of the plate times relative velocity, v, of flow. Cross section of the column would equal plate area, A, times sine of the inclination angle, α, or:

(1.18) $$\text{Mass rate} = \rho v A \sin \alpha$$

The vertical component of velocity in departing flow would equal airspeed, v, times sine of the deflection angle, α, or:

(1.19) $$\text{Downwash vertical velocity} = v \sin \alpha$$

The vertical momentum rate in downwash then becomes the product of mass rate times vertical velocity, so that lift would become equal to the product of the two preceding expressions:

(1-20) $$\text{Lift} = \rho v^2 A \sin^2 \alpha$$

This result is far short of what we now know to be reality. Newton was unaware that flow would also follow the plate upper surface, and that the depth of air affected would be much greater than that of the column ahead of the plate. In addition, Newton did not realize that downward displacement of flow would produce upward recirculation. As we shall see, this is a very important factor in lift.

Recirculation

Air accelerates away from higher pressure and toward lower pressure. Thus air is accelerated away from higher pressure beneath the wing, and toward lower pressure above. Resulting air movements are three dimensional, including vertical, longitudinal (forward and rearward) and lateral, or spanwise.

Wherever air is displaced by a moving wing, recirculation occurs away from the region displaced by the wing and toward the region the wing vacates. To prevent air movements around wing test section ends, laterally uniform wing sections are tested in two-dimensional flow, where a section spans between tunnel walls, as in Figure 1.26.

By virtue of lateral symmetry, flow develops in only two dimensions. Such a test is for evaluation of a design profile, as in Figure 1.3, called an *airfoil*. A wing may be of uniform profile across the span, based on a single airfoil design, but often includes two or more airfoil profiles blended together somewhere along the span for aerodynamic advantage.

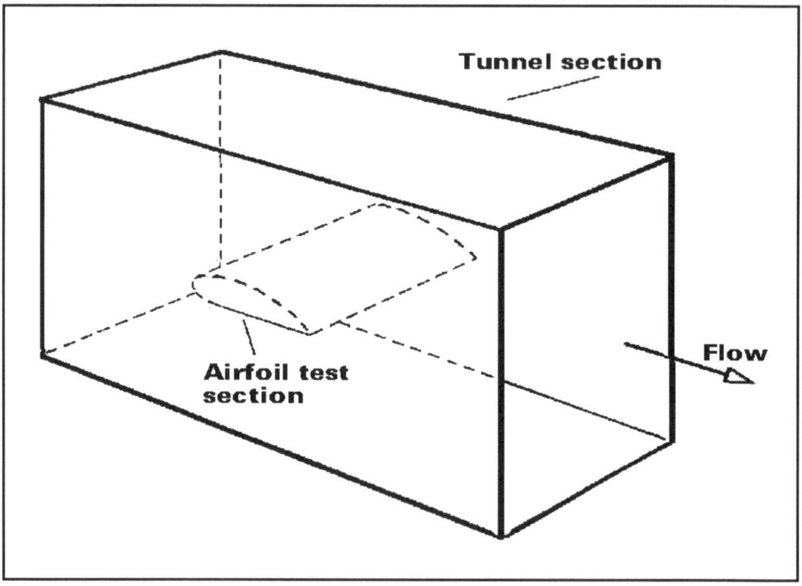

Figure 1.26 **Wall-to-Wall Wind Tunnel Airfoil Section Test**
The above illustration illustrates the mounting of a uniform airfoil section in a wind tunnel with airfoil ends in contact with the tunnel walls. This prevents lateral air movements around the ends from higher pressure below to lower pressure above. An actual test would have tunnel dimensions much larger as compared to the airfoil in order to minimize interference effects.

Perspectives

Popular teachings from throughout the 21st century usually regard wings as stationary in passing flow, despite the fact that we regard airplanes as moving in relatively still air. The apparent reason for this is that the Bernoulli based lift explanation works only in the perspective of a fixed wing in passing flow. In the sections to follow we shall find use for both perspectives, of a fixed wing in passing air, and a moving wing in stationary air, with the choice being a matter of convenience for the situation being considered.

In developing a concept of wing operation we shall begin with as much simplification as possible, and add detailed factors as we proceed. As is customary in beginning discussion of flight fundamentals we shall omit consideration of lateral air movements, which are associated with losses of efficiency, and consider only the remaining two dimensions in which movements are vertical and longitudinal along the flight path. A two-dimensional perspective is sufficient for initial lift concepts, although inadequate for evaluation of a real wing, which develops lateral recirculatory air movements around the ends from higher pressure below to lower pressure above. The more complex effects of three dimensional movements are to be considered later.

We shall also limit consideration to flight speeds less than about half that of sound, in which compressibility can be disregarded as having little effect, and air movements can be considered as circulatory in complete closed paths.

Whirlpool
This whirlpool occurs where water from a creek passes underground to an adjacent lake when the creek level rises. Angular momentum is conserved as water drawn into the whirlpool maintains angular momentum of creek flow. Velocity of rotation increases as gravity accelerates flow inward toward the lower level at the center.

Introduction to Aerodynamics

II. Differing Theories

Airfoil Nomenclature

Discussion of wings and airfoils requires defining a few terms, beginning with those illustrated in Figure 2.1. More terms will be introduced as needed in later sections.

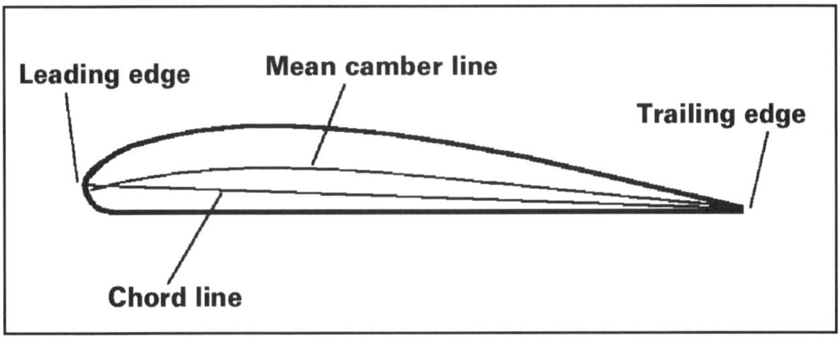

Figure 2.1 **Nomenclature**

Leading edge: Foremost point of a wing section or airfoil.
Trailing edge: Rearmost point of a wing section or airfoil.
Chord: Straight line between leading and trailing edges.
Mean camber line: Line midway between upper and lower surfaces, extending from leading to trailing edge.
Camber: Curvature of the mean camber line.
Angle of attack: Angle between chord line and flight path through the air.

Plausible Falsehoods of Popular Teaching

William Reese Sears wrote of his career in aerodynamics in an interesting book titled *Stories From a 20th Century Life*. In it he describes being puzzled in his earlier years when his mentor, famous aerodynamicist Dr. Theodore Von Karman (1881-1963), gave visitors a false explanation of why a model airplane in a wind tunnel was inverted. Having noticed Sears' puzzled expression, Von Karman later explained, "I must teach you an important rule. When you are speaking to technically illiterate people you must resort to the Plausible Falsehood instead of the Difficult Truth."

Plausible falsehood is involved in a quite significant portion of aerodynamics teaching worldwide. The lift explanation below, which typifies popular lay teaching, has at the time of this writing been posted for several years near the beginning of a historical sequence of exhibits in the United States Air Force Museum at Dayton, Ohio.

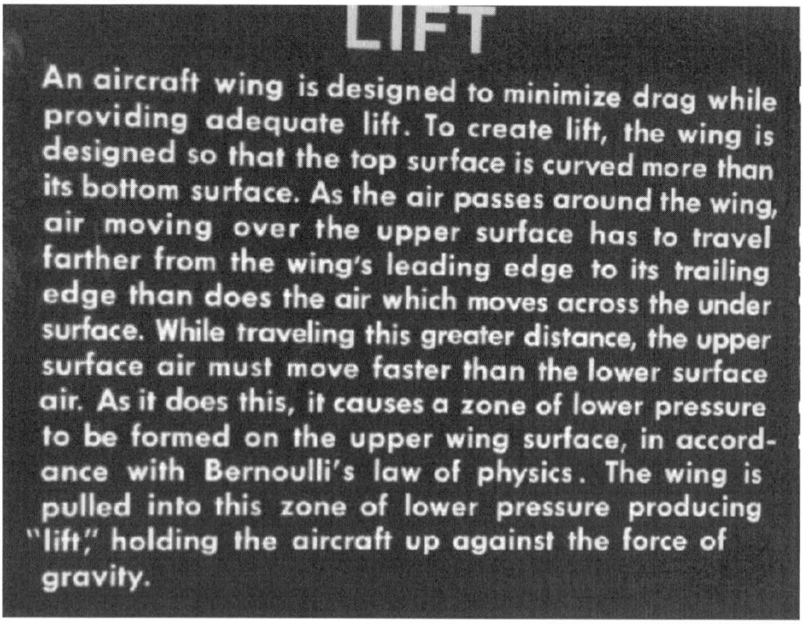

An official of the AF Museum, when queried about this plausible but wrong explanation of lift, replied that it had to be "suitable for an average visitor of eighth grade education." It seems this reasoning is similar to that of Dr. Von Karman. Unfortunately, false teaching can impede real learning. As we shall see, examples of inappropriate aerodynamics teaching exist at both layman level and professional level.

The popular explanation reflected in the museum posting is based on an erroneous myth which declares transit time for air passing a wing upper surface must equal that of air passing below. Some books purporting to explain wing operation with equal transit time contain illustrations such as that of Figure 2.3, wrongly indicating that air particles separated ahead of a wing are reunited at the rear.

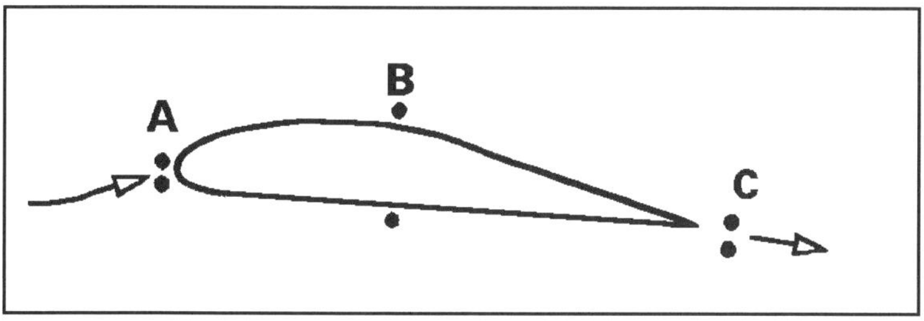

Figure 2.3 **Principle of equal transit times**
One widely taught but false theory says air particles separating at point A in front of a lifting airfoil travel together as in B, and are rejoined at point C in the rear.

According to this theory, a wing upper surface is made more convex than the lower in order to create a longer path. If transit times must be equal, then upper surface flow must be faster in the longer path. Hence in accordance with Bernoulli's law, it is said that pressure in the faster upper surface flow must be less than that at the lower surface, resulting in lift.

Equal transit time lift theory does not survive reasonable thought. First, it would indicate that a wing having equal leading edge to trailing edge path lengths across upper and lower surfaces, should not produce lift. Thus aerobatic planes with equal upper and lower surface curvatures should not fly, and balsa toy planes with flat wings should not fly. Furthermore, there is no known principle in physics on which to base the equal transit time conclusion.

Although Bernoulli's law is cited in the false equal transit time explanation of lift, the law can better be used to prove transit times unequal. Consider a flat sheet, which will indeed produce lift, as anyone who has ever handled plywood sheets in a wind can testify. A flat sheet develops lift when pressure difference between upper and lower surfaces is produced in passing flow. Bernoulli's law states that flow velocities will differ between surfaces in inverse relation to pressure difference. Therefore we can reason, as illustrated in Figure 2.4, that air transit time in lower pressure of the upper surface must be less than in the greater pressure of the lower surface.

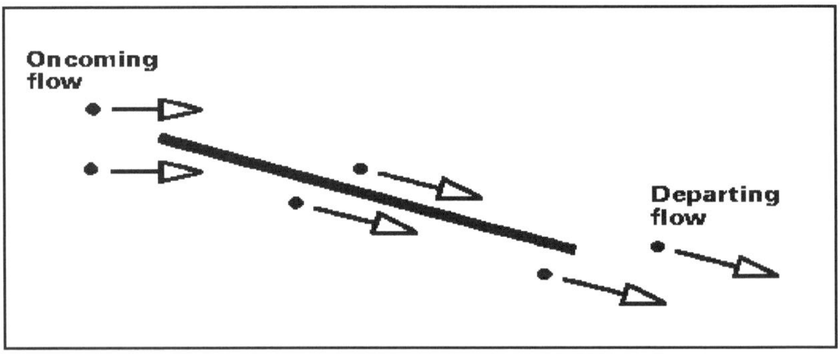

Figure 2.4 **Velocities of Flow Passing a Flat Sheet**
Equal flow velocities approaching a flat sheet, which is inclined to produce lift, are increased above the sheet where pressure is decreased, and decreased below where pressure is increased, in accordance with Bernoulli's law, which specifies inverse relation between velocities and pressures. Upper surface transit time must then be less than lower surface transit time

We may consider equal transit time theory applied to a common light Cessna 182 airplane. For a first approximation let's assume constant velocities across upper and lower surfaces which differ as necessary to produce the required lift of 16.1 pounds per square foot of wing area. The owner's manual rates cruise speed at 158 miles per hour, or about 232 feet per second in air density near sea level of about .0024 slug per cubic foot. If velocity difference between surfaces is achieved with velocity increase across the upper surface equal to velocity decrease at the lower surface, we find that at cruise upper surface velocity must become 246.5 feet per second, and that of the lower surface 217.5 feet per second, a difference of about 13 percent, in order to achieve the 16.1 pounds per square foot pressure difference. Plugging these numbers into Bernoulli's equation, (1-14) from the preceding chapter, yields:

$$(2\text{-}1)\quad \Delta p = (.0024/2)\,(246.5^2 - 217.5^2) = 16.1 \text{ lb./ft}^2$$

At minimum no-flaps speed of 64 miles per hour the upper surface path would have to be more than **twice** that of the lower to achieve the required lift. Actual measurements of upper and lower distances, shown in Figure 2.5, differ by less than two percent. Although velocities are not constant across the surfaces, and airplane pitch attitude is a small factor, it seems unlikely that upper and lower surface transit times could be equal. In fact, experimental results published by Martin Ingleman-Sundberg, now retired former head of low speed aerodynamic research of the Aeronautical Research Institute of Sweden, conclusively show upper surface transit time of a lifting wing to be less than that of the lower surface.

A popular textbook used in several universities is *Foundations of Aerodynamics*, by Arnold Kuethe and Chen-Yen Chow. It has an illustration on page 96 like that of Figure 2.6, specifically indicating shorter upper surface transit time.

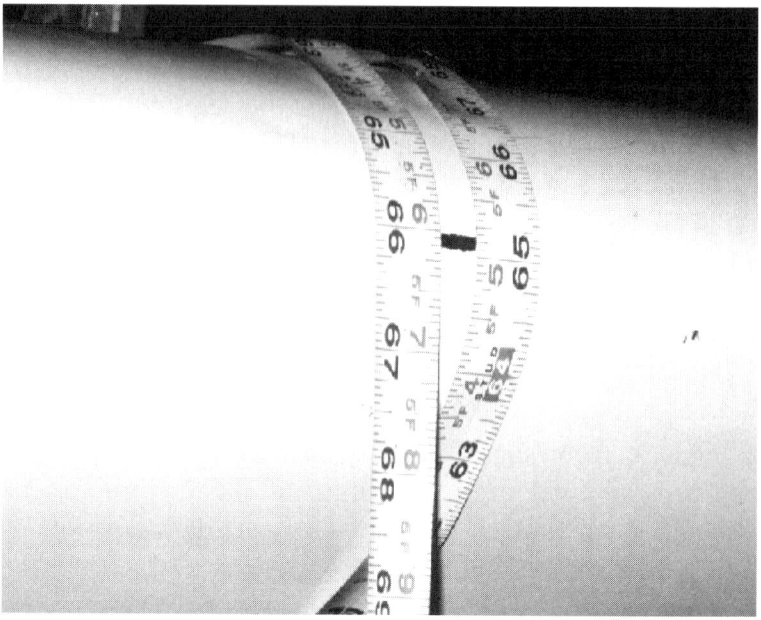

Figure 2.5 **Cessna 182 Flow Path Measurements**
The upper and lower surface path lengths of the author's Cessna 182 wing. Distances as measured in the photograph average about 65.5 inches. Difference between distances is only about one inch.

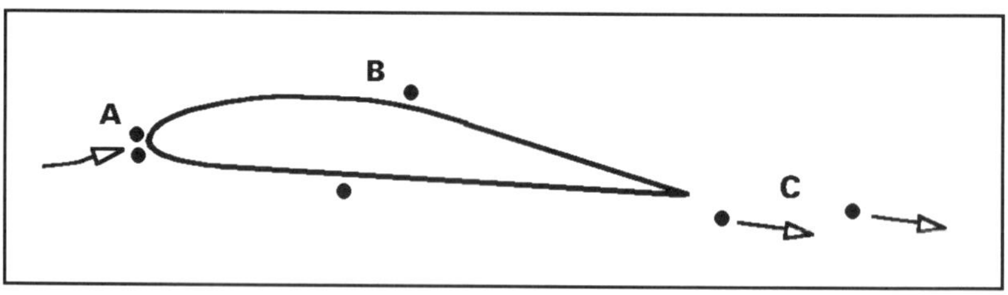

Figure 2. 6 **More Realistic Transient Positions**
Higher level aerodynamics teaching does not include equal transit time theory, for it is well known that upper surface time is generally less than that at the lower surface.

Half-Venturi Theory

Another simplistic and false but less popular concept of lift visualizes a wing profile as like half of a Venturi tube, as in Figure 2.7. It is imagined that a wing upper surface restricts the flow path in the manner of a Venturi, and thus, like a Venturi, produces velocity increase. Like theory of equal transit times, lift is then described as due to pressure reduction in the faster flow, in accordance with Bernoulli's law. This theory also does not allow for lift of a flat or inverted wing.

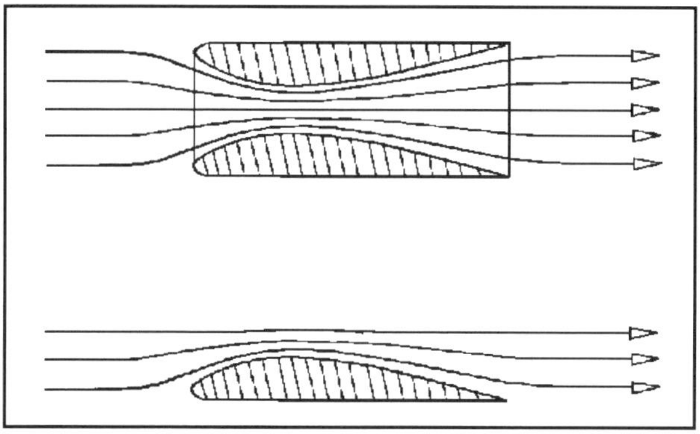

Figure 2.7 **Half Venturi Theory**
A wing section (lower illustration) is visualized as like half of a Venturi (upper illustration).

Figure 2.8 is a photograph of a Cessna 150 instrumented to measure relative air flow velocities above and below the wing along a line six inches ahead of the leading edge. Measured relative velocities, as plotted in Figure 2.9, are greater above the wing as compared to below. Near the wing, difference between velocities is about thirty percent, This is clearly much greater than relative distance between upper and lower surface path lengths.

Figure 2.8 **Cessna 150 Velocities Instrumentation**
Pitot tubes mounted on a vertical pipe connect to dyed water manometers in the plane window. The Pitot tubes were moveable vertically along a line six inches ahead of the leading edge, and the pipe position could be changed from above-wing to below-wing so as to also determine relative velocities at intervals below the wing.

Why false aerodynamic theories are still taught in our schools remains a mystery. We can assume a great number of students taught equal transit time theory in lower level schools have found it to be wrong in higher level teaching. Why have the better educated not passed better understanding to the lower levels? Could it be contempt for the less learned, as it seems in Von Karman tale? Or is it a mistaken belief that basic concepts of classical teaching cannot be presented without a great amount of complex mathematics? Or could it be that higher level theory when stripped of mathematical complexity seems no more credible than equal transit time theory?

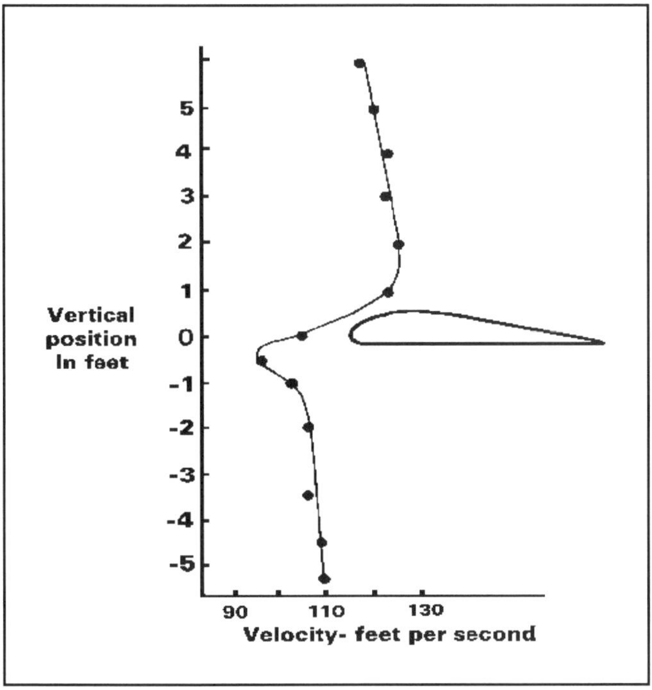

Figure 2.9 **Velocities plot**
Airplane airspeed is represented by wing leading edge position. Indicated above-wing velocities are greater, while those below are less. This is because rearward air movement above adds to Pitot pressure while forward movement below detracts from it.

Individuals protesting the state of aerodynamics teaching have, at the time of this writing, received little attention from the academic world. However, with more formal effort, the situation may improve. In the 39th Aerospace Sciences Meeting of the *American Institute of Aeronautics and Astronautics* in year 2001, Jaako Hoffren, of the Helsinki University of Technology, presented a paper titled *QUEST FOR AN IMPROVED EXPLANATION OF LIFT*. The paper characterizes classical explanations of lift, including higher level teaching, as "abstract, nonphysical and difficult to comprehend," and proposes that improved explanations of lift be based on Newtonian principles.

Circulation Theory

According to classical theory, a rotational motion, known as *circulation*, is superimposed on passing flow to cause oncoming air flow to slope upward in *upwash*, flow above a wing to have increased velocity, flow below the wing to have reduced velocity, and departing flow to slope downward in *downwash*. In Figure 2.10, relative flows are represented by dark arrows, while circulation velocity is represented by phantom arrows.

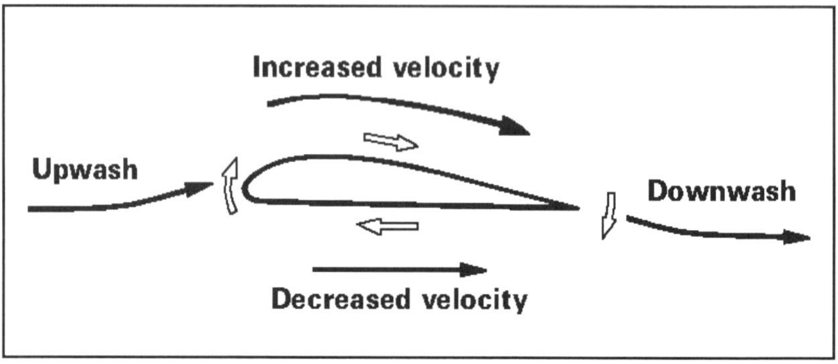

Figure 2.10 **Circulation Superimposed on Flow**
Dark arrows represent relative flow lines and phantom arrows represent the circulation component.

With relative upper and lower flow velocities differing due to inclusion of circulation, pressure difference and lift can be explained in connection with Bernoulli's law, but to be complete, explanation should include why circulation occurs.

Magnus Effect

In one case, known as *Magnus effect*, circulation around a rotating cylinder is attributed to air being dragged around the cylinder by viscosity. Lift of a rotating cylinder is easy to demonstrate using a few commonly available items, including a sheet of ordinary typing paper, about twelve feet of light string or thread, a ruler and a few inches of cellophane tape.

For the cylinder lift demonstration, the paper is rolled into a cylinder of about two inches diameter with the longer dimension of the paper establishing the cylinder axial length. The cylinder is taped in two or three places so it will not unroll. The thread is cut into two equal pieces and one end of each piece is taped near opposite ends of the cylinder. On a flat surface, the threads are aligned perpendicular to the cylinder axis, and the loose ends taped near opposite ends of the ruler. Then the threads are wound onto the cylinder by rolling it toward the ruler. When completed and ready to run, the apparatus should appear as shown in Figure 2.11.

When cylinder and ruler are held horizontally at head height above the floor, and the cylinder is released, so that the threads unwind as the cylinder falls, it will be seen that the cylinder does not fall vertically, but follows a curved path, as in Figure 2.12. Path curvature indicates lift is being developed in direction perpendicular to the path of cylinder travel.

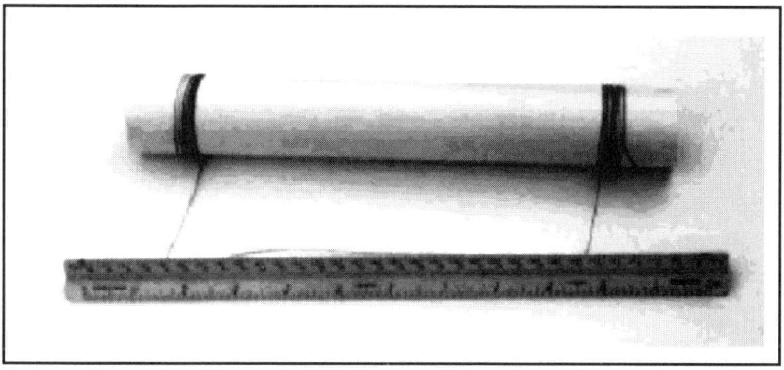

Figure 2.11 **Paper Cylinder Ready to Run**

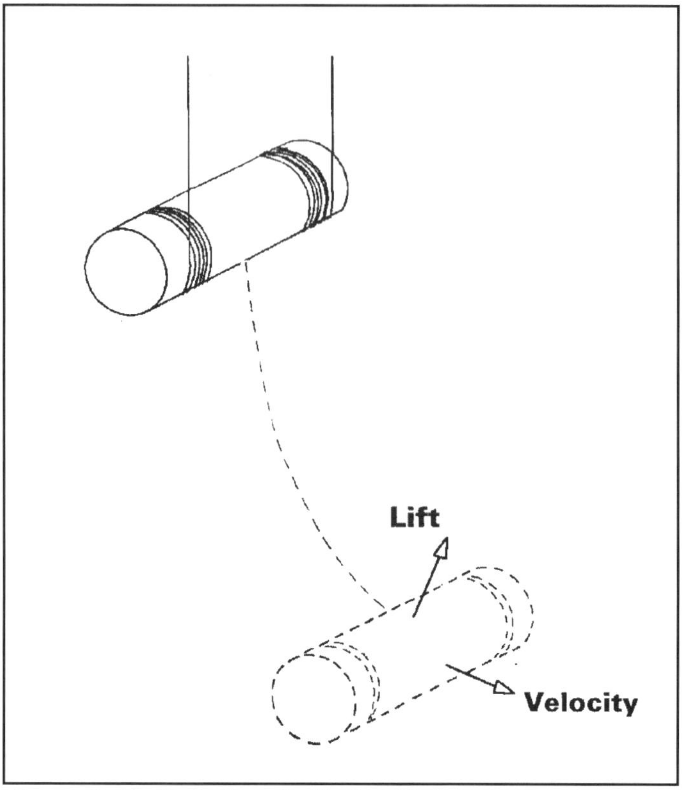

Figure 2.12 **Rotating Cylinder Fall**
As the cylinder falls in counterclockwise rotation with strings unwinding, viscous drag opposition to upward air recirculation is greater on the downgoing side. Lift, perpendicular to the cylinder path, is toward the side of lesser viscous drag.

Magnus effect is named in honor of a German engineer credited with explaining it. If presented in perspective of stationary cylinder in upward-moving air, the passing flow explanation would describe upward air movement as impeded more by viscous drag at the downgoing side, as compared to the side from which the strings unwind. Upward velocity at the side from which strings unwind would be greater, and lift could be explained as due to greater pressure reduction on the side of faster upward flow, in accordance with Bernoulli's law.

There has been much experimentation with Magnus effect. It was applied to ship propulsion by Anton Flettner, another German engineer, in 1924. Sails of a ship were replaced with two vertical rotors, each nine feet in diameter and fifty feet high. Operationally, Flettner's ship was a success. He could even turn it in place by rotating the cylinders in opposite directions so as to produce opposite direction lifts. Unfortunately, the boat was an economic failure due to advances in steam propulsion and expense of maintaining the cylinder rotating machinery.

A Circulation-Electromagnetism Analogue

Similarity of the mathematical equations of electricity and magnetism to the mathematical equations of mechanical processes has long been recognized. For example, equations relating force and mechanical energy storage in springs are essentially the same as the equations relating voltage and electrical energy storage in capacitors, and mathematical expression of kinetic energy associated with moving mass is equivalent to mathematical expression of magnetic energy associated with electrical current in a coiled wire. These relationships have enabled engineers of the past, before sophisticated digital computers became available, to use electrical circuits for simulating dynamics of mechanical operations.

Similarities of electromagnetism concepts to those of aerodynamics led to vortices in air being described as "induced" from central vortex cores in the manner that magnetic fields are regarded as induced by electrical currents in wires. This analogy has produced aerodynamics terms of "induced downwash," "induced angle of attack" and "induced drag." Like current in a wire is said to induce magnetic field rotation around the center of a wire, circulation around a wing, which adds to passing flow velocities, is sometimes described as induced from an imaginary central vortex core residing in the wing. This electromagnetic analogue of aerodynamics is mathematically useful for engineering purposes, but is too often regarded as reality rather than analogy.

Vortex laws, applicable to electromagnetism and aerodynamics, were formulated by German physicist and mathematician, Hermann Helmholtz (1821-1894). One of the laws says creation of one vortex is accompanied by creation of another equal vortex of opposite rotation. As an example, electrical currents flow in complete, closed circuits wherein flows occur in opposite directions. Because current is equal in opposite directions, magnetic field vortices of equal magnitude and opposite direction are created. The concept of simultaneous creation of equal and opposite vortices has been adopted in explaining generation of circulation around a wing or airfoil.

In experiments of a moving airfoil shape in water, Ludwig Prandtl (1875-1953) photographed streaks made by movement of suspended light reflecting particles. The streaks showed that when the airfoil shape was started in forward movement, circulation appeared around it, while simultaneously a vortex of opposite rotation was left behind as in Figure 2.13.

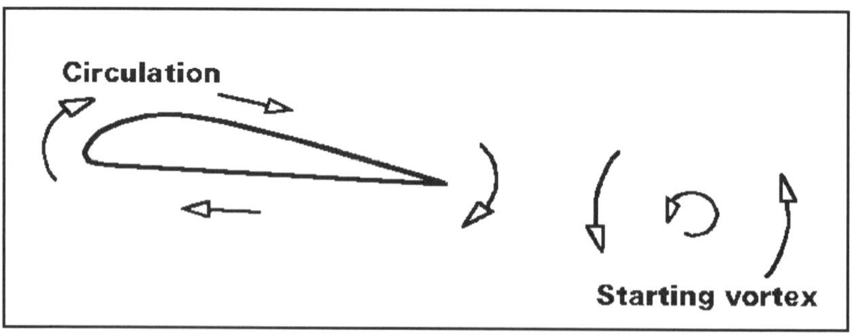

Figure 2.13 **Starting Vortex and Circulation**
When an airfoil begins forward motion and lift it leaves a growing vortex behind while circulation also grows around the airfoil.

Classical circulation theory says the aft vortex, known as a *starting vortex,* left behind a wing or airfoil, is caused by viscous drag of flow passing the sharp trailing edge. Since Helmholtz' vortex theory says one vortex cannot be created in a fluid without simultaneous creation of an equal vortex of opposite rotation, it is asserted that circulation around a wing or airfoil is an unavoidable vortex of opposite direction, caused in response to the starting vortex. Photographs of the starting movements can be found in the Prandtl-Tietjens book, *Applied Hydro-and Aeromechanics*.

Growths of starting vortex and circulation are said to continue until surface flows merge smoothly at the sharp trailing edge. This stabilized state is known as the "*Kutta Condition.*" The starting vortex remains centered near its starting position while circulation around the airfoil, sometimes referred to as a "bound vortex" is carried with the airfoil in its travel.

The classical view, of circulation being a response to the starting vortex caused by a sharp trailing edge, is convenient for explanation based on induction principles of electromagnetism, but in the real world a sharp trailing edge is not needed for generation of lift. A non-rotating cylinder of round cross section spontaneously develops lift over a wide range of Reynolds number. Although the spontaneous lift oscillates in reversing direction, it is nonetheless as real as lift which occurs when a round cylinder is rotated. Classical Magnus effect reasoning says rotating cylinder lift develops as the rotating movement drags air around with it. However, since lift of alternating direction can develop without rotation, we may speculate that the main effect of rotation is to stabilize lift direction.

Another questionable point in explaining a starting vortex as due to viscous drag at the trailing edge concerns conservation of angular momentum. If the airfoil exerts viscous force to cause air mass rotation in a starting vortex, then it must be coupling angular momentum into passing flow. For a non-rotating airfoil this would not be possible. Another important consideration is that this explanation ignores forces of pressure difference which unavoidably must produce air movement in accordance with Newton's laws.

A Logical Explanation of Circulation

Let's consider the following:

- Air accelerates away from higher pressure and toward lower pressure.
- Pressure disturbances in air propagate outward at the speed of sound.
- Air movements in a finite atmosphere are circulatory, so that where air is displaced downward there is equal recirculation upward.
- Pressure above a normal lifting wing is less than that of surrounding atmosphere.
- Pressure below a lifting wing is normally greater than that of surrounding atmosphere.
- An upward lifting wing accelerates air downward, as can be seen when observing operation of crop-dusters, helicopters, and aerobatic planes with smoke generators.

Since air accelerates away from higher pressure and toward lower pressure, and pressure effects in air travel with sonic velocity, we must assume that air surrounding a lifting subsonic airfoil is affected by the lower pressure above and higher pressure below. This means air approaching the subsonic airfoil must unavoidably be accelerated toward reduced pressure above and away from higher pressure below, with the result that oncoming flow curves upward ahead of the leading edge, as indicated in Figure 2.14.

In accordance with Bernoulli's law, velocity of flow accelerating into reduced pressure above the airfoil is increased while velocity of flow into increased pressure below is decreased. At the airfoil trailing edge, convergent flow following the mean camber line departs in a downward direction.

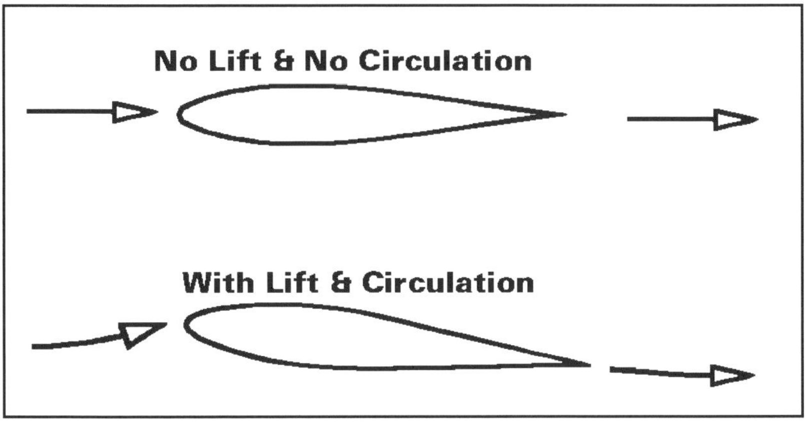

Figure 2.14 **Flows With and Without Lift**
At zero angle of attack, the symmetric airfoil in the upper illustration develops no downwash or lift. In the lower illustration, positive angle of attack produces downward acceleration of flow following the mean camber line. In response to pressure difference, air rises ahead of the leading edge

Rise of air ahead of the airfoil, descent behind, velocity increase above and velocity decrease below account for variations considered in classical theory as circulation addition to passing flow. Thus circulation can readily be attributed to pressure effects, without resort to invocation of induction. Circulatory movement begins when flow following the mean camber line is first accelerated downward, initiating pressure difference which causes upward movement around the airfoil leading edge, and an initially upward movement around the trailing edge in beginning a starting vortex.

Circulation In Moving Wing Perspective

It seems strange that although we think of airplanes as flying through relatively still air, popular explanations of how wings work describe them as stationary in passing flow. In principle, this is acceptable since we live on a moving planet, and can only describe motion in relation to some object or position we choose to regard as stationary. Bernoulli's law, used in popular explanations, properly states an inverse relation between pressure and velocity in flow passing fixed positions, but it is useless where the wing and its pressure differences are regarded as in motion. However, a moving wing in otherwise stationary air perspective, as in Figure 2.15, is advantageous in permitting circulation to be more easily explained and visualized.

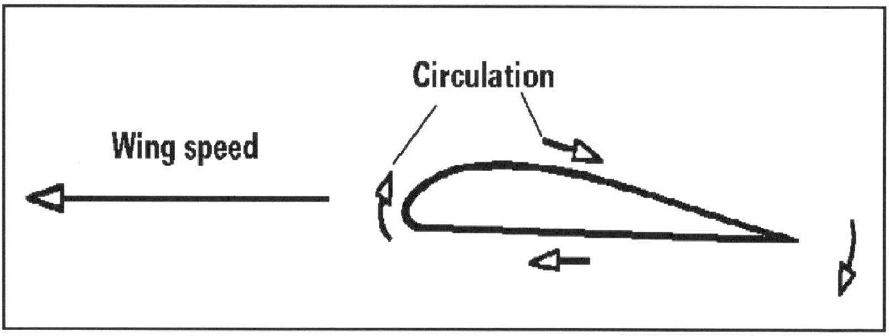

Figure 2.15 **Circulation in Moving Wing Perspective**
As the wing moves forward with velocity labeled "Wing speed," air ahead of the leading edge rises away from higher pressure below and toward lower pressure above. Behind the wing air continues downward after following the descent of departing wing surface contours. Above the wing air has accelerated rearward into decreased pressure while below the wing air has accelerated forward due to increased pressure. As indicated by lengths of the arrows, circulation velocity is, with little exception, less than forward velocity of the wing.

In moving subsonic wing perspective, disturbances of greater pressure below and lesser pressure above extend ahead, causing air to move upward around the leading edge, away from higher pressure below and toward lower pressure above. Above the leading edge, air accelerates rearward into lower pressure, while air below accelerates forward away from higher pressure. Thus as the wing passes there is rearward movement of air above and forward movement of air below. The rearward and forward movements decelerate in the recovering pressures as the trailing edge approaches. Behind the wing, downward air movement developed in following the mean camber line continues. If a stationary observer could photograph in slow motion the passage of a wing and movements of dust particles carried in the air, the result would show circulation movements typically much slower than forward wing movement.

Circulation increases with pressure difference required for lift and decreases with wing speed, which affects time available for air acceleration. Greater circulation velocity, which occurs when airspeed is low in relation to lift, can operate a switch, as in Figure 2.16, to power an electrical stall warning horn.

Figure 2.16
Cessna 182 Stall Warning Tab
When airspeed is low in relation to lift, intense upward air movement around the leading edge lifts the switch tab to actuate an electrical stall warning horn.

Alternatively, though less popular, stall warning can operate from suction produced in centrifuging of upward flow around a leading edge port, as in Figure 2.17. The warning alerts a pilot to approaching stall condition when air speed is low in relation to lift.

Figure 2.17 **Cessna Stall Warning Port**
Centrifuging circulation around the leading edge produces pressure reduction, or suction, which is piped to a stall warning horn in the cabin.

The Kutta Condition

Classical theory describes circulation as a reaction to the starting vortex. It is supposed the starting vortex is caused when circulation downwash is not adequate to supply the downward velocity component required for flow to depart the trailing edge in the pointed direction. According to the theory, growth of a starting vortex causes circulation of opposite direction around the airfoil to grow equally until the downwash demand is met. Then upward flow across the trailing edge is said to cease, whereupon growth of starting vortex and circulation also cease. The mature condition is illustrated in Figure 2.18, where circulation magnitude is just that required for flow to conform to the departure angle pointed by the airfoil.

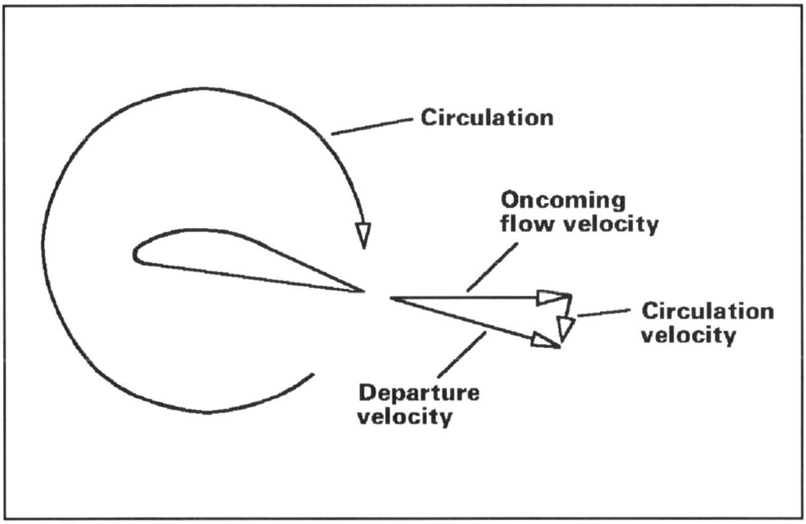

Figure 2.18 **Classical Kutta Condition**
With the downward movement of circulation at the trailing edge added to passing flow velocity, flow departs smoothly in the direction pointed by the airfoil.

The classical explanation can be expressed mathematically to give numerical results which seem to conform to reality, but it ignores factors of flow attachment, otherwise known as "Coanda effect," and accelerations due to pressure difference. It also leaves the physical connection between starting vortex and circulation not well defined.

Let's consider another possibility for relating a starting vortex to circulation. We can begin with analyzing the stable Kutta condition in moving wing perspective. In a stabilized condition of lift, above-airfoil pressure is less than that of surrounding atmosphere, while pressure below is greater than that of surrounding atmosphere. This creates two equal pressure couples of opposite direction, as indicated in Figure 2.19.

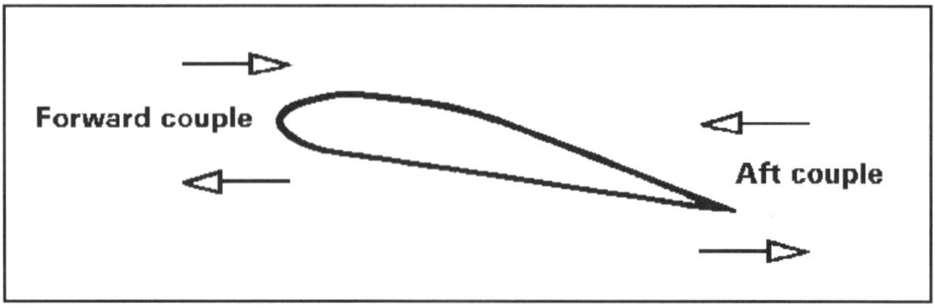

Figure 2.19 **Pressure Couples in Stable Kutta Condition**
Lower pressure above the airfoil and higher pressure below create equal and opposite couples fore and aft which, in perspective of moving airfoil in still air, recover circulation kinetic energy from air being left behind and transfer it into air being approached ahead. The two couples have equal and opposite effects on circulation angular momentum, which remains stable.

Pressure forces of the forward couple tend to accelerate air upward around the leading edge so as to increase circulation velocity and kinetic energy in air ahead, while pressure forces of the rear couple equally tend to decrease circulation velocity and kinetic energy in air being left behind. Thus, while kinetic energy is being coupled into air being approached by the advancing airfoil, kinetic energy is being removed from air being left behind. In this process kinetic energy is recovered from air behind and recycled into air ahead. In passing flow perspective above-airfoil flow could be compared to that of a venturi where kinetic energy is recovered aft of the throat and recycled through Bernoullian process into acceleration of flow ahead of the throat. In passing flow perspective below the airfoil, energy conservation would occur in rise and fall of pressure, in reversed sequence as compared to a venturi, but still in accordance with Bernoulli's law.

The equal opposing couples in Kutta condition have equal and opposite effects on airfoil circulation angular momentum, which remains constant as the forward moment for angular momentum increase is equaled by the aft moment for decrease.

If angle of attack increases from that of the stable Kutta condition, the airfoil pressure gradients are increased. Forward and rearward couples both increase in magnitude, but remain balanced so as to produce no net change in angular momentum. However, as the increased forward couple increases circulation kinetic energy in air being approached, the rear couple temporarily exceeds the requirement for aft recovery of circulation kinetic energy.

As depicted in Figure 2.20, the temporary excess of aft couple magnitude is spent in producing a new aft vortex of rotation opposite to that of circulation. Growth rate of the new vortex diminishes as increasing circulation magnitude approaches capacity of the aft couple for kinetic energy recovery. When the aft couple again becomes fully spent in recovery of circulation kinetic energy, growths of aft vortex and airfoil circulation cease, the Kutta condition is again met, and the aft vortex is left behind. Kinetic energy content of stabilized airfoil circulation and of the aft vortex is in each case would be equal to reduction of PV energy in its centrifugal pressure gradient.

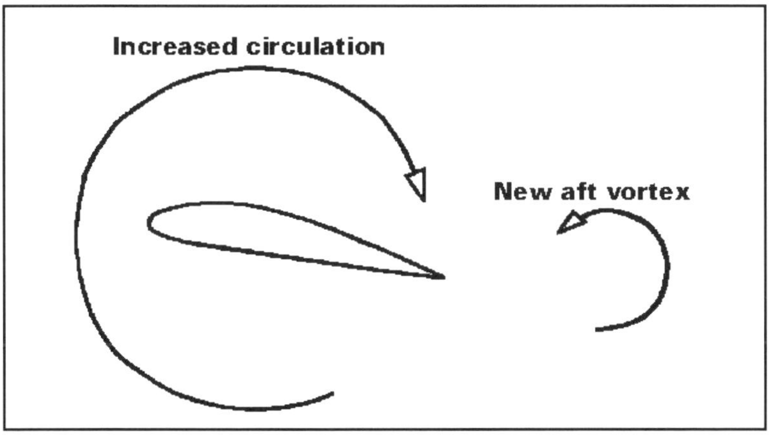

Figure 2.20 **Circulation Increase Adjustment**
During circulation increase to a new level created by increased downwash demand, the downwash rate is temporarily supplemented in production of an aft vortex.

If angle of attack is lessened from that of stable Kutta condition, reduction in pressure difference makes the aft couple momentarily less than adequate for recovering the departing circulation energy. Kinetic energy is then left behind in a new aft vortex having the rotational direction of circulation, as in Figure 2.21. In the adjustments, angular momentum is conserved as change in angular momentum of circulation is equaled by opposite change in angular momentum left behind.

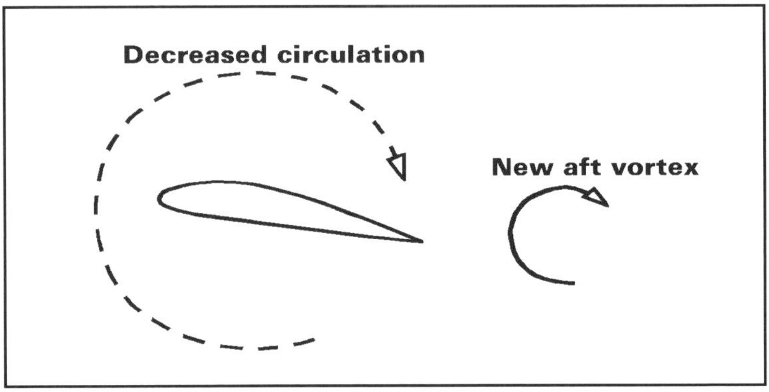

Figure 2.21 **Circulation Decrease Adjustment**
During angle of attack decrease, circulation kinetic energy not recovered by the diminished aft couple is shed into a new vortex of the direction of circulation.

Simultaneous creation of starting vortex and circulation observed by Prandtl can be explained as initial division of starting downwash into opposite patterns of recirculation in which angular momentum is conserved. As illustrated in Figure 2.22, inertia of air mass opposes downward acceleration, producing pressure increase below the airfoil and pressure decrease above. Fore and aft pressure couples of direction opposite to the pressure gradients produce rotational movements traditionally identified as circulation and a starting vortex. The starting vortex center remains behind while circulation is carried with the airfoil.

Opposite rotations of airfoil circulation and aft vortex grow equally in angular momenta, slowing in growth rate as circulation contribution to downwash approaches the downwash rate demanded by the combination of airfoil chord, downwash angle and airspeed. When the downwash demand is matched by circulation the need for aft vortex involvement ceases.

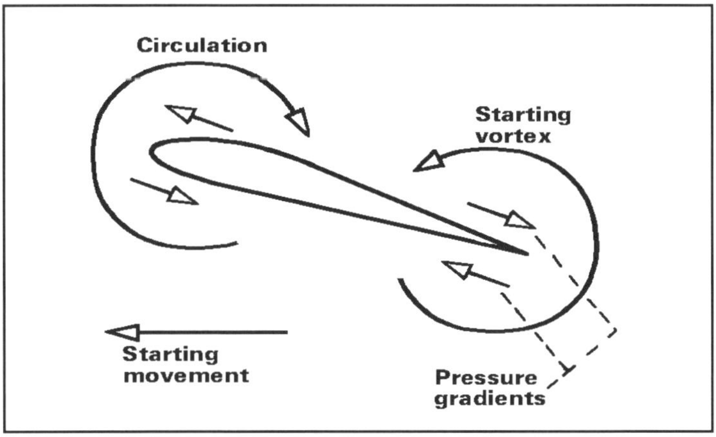

Figure 2.22 **Starting Vortex and Circulation**
In initial forward airfoil movement, pressure gradients created by air displacement constitute fore and aft couples, producing air accelerations in direction opposite to the gradients. The rear pressure gradients produce an aft starting vortex, while forward gradients initiate circulation ahead.

Ideal Lift Calculation

Circulation has characteristics of a free vortex, and except for distortion where airfoil shape interferes, movement is circular in nature, with circulation velocity inversely proportional to radius from the rotation center. Circulation strength, customarily represented by the Greek letter, Γ, is equal to circulation velocity, v_C, times length of circular path, $2\pi r$, where in circular motion r is circulation radius:

(2-2) $$\Gamma = 2\pi r\ v_C$$

Because circulation velocity in a free vortex varies inversely in relation to radius, the above expression holds for any radius.

For convenience of illustration in Figure 2.23, we choose an airfoil of uniform curvature having its center of circulation at mid-chord. Assuming departing velocity equal in magnitude to that of arriving flow, the vertical velocity component at the trailing edge is equal to incoming velocity times sine of the angle of departure. The circulation radius, r, is one-half the chord length, or C/2, and circulation velocity at the trailing edge is $v_0 \sin\theta$. Substituting these for v_C and r in (2-2), circulation strength, Γ, is established as:

(2-3) $$\Gamma = \pi C\ v_0 \sin\theta$$

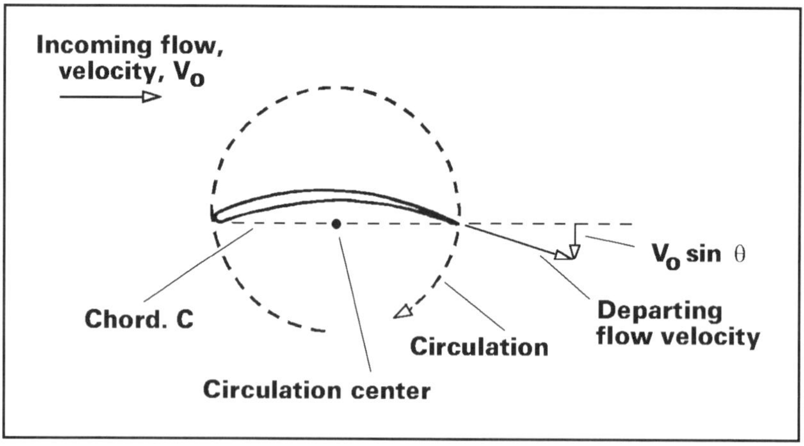

Figure 2.23 **Trailing Edge Circulation Velocity**
Circulation velocity at the trailing edge is equal to magnitude of flow velocity times sine of the departing downwash angle.

The Kutta-Joukowski Theorem

Circulation velocity at any radius, as derived from (2-2), the expression for circulation strength, is:

(2-4) $$v_C = \Gamma/(2\pi r)$$

From consideration of this expression we can derive circulation velocity components at any position in relation to the circulation center, as represented in Figure 2.24.

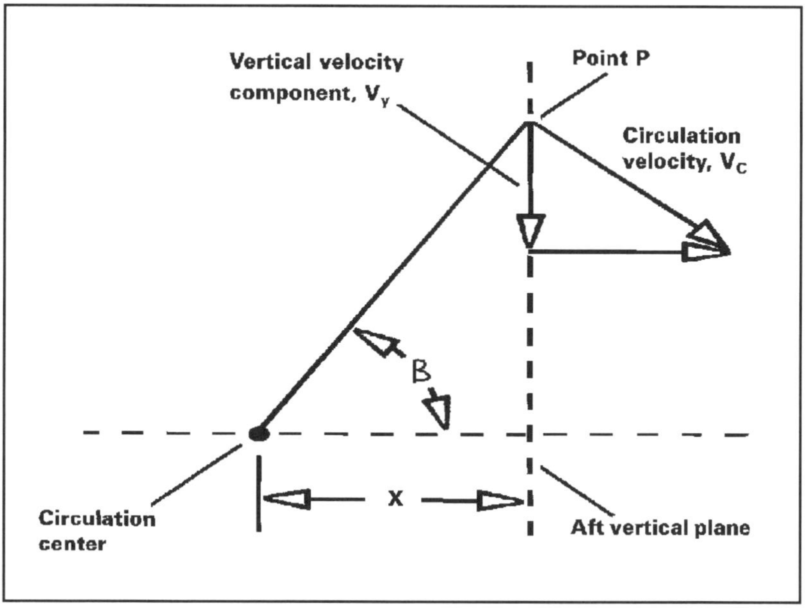

Figure 2-24 **Circulation Components**
From consideration of the above parameters and circulation strength, circulation velocities can be derived for varying position (point P) in the trailing edge vertical plane.

Circulation velocity components are depicted at a position on a vertical plane, at distance, x, behind the airfoil circulation center, and at radius, R. β represents the angle between flight path and radius to the position, P.

At position, P the variables for velocities calculation are as follow:

$$r = x \sec \beta$$

$$y = x \tan \beta$$

$$dy = x \sec^2 \beta \, d\beta$$

$$v_C = \Gamma/(2\pi r) = \Gamma/(2\pi x \sec \beta)$$

$$v_y = \Gamma \cos \beta/(2\pi x \sec \beta) = \Gamma \cos^2 \beta/(2\pi x)$$

ρv_y = vertical momentum density, or:

$$\rho v_y = \rho\Gamma \cos^2 \beta/(2\pi x)$$

To obtain lift due to downwash momentum production, we can integrate vertical momentum passage rates through aft plane segments of height, dy, and span width, S, between limits of plus and minus infinity vertically, as carried by flow of airspeed, v_0, passing through. From this, the integral of lift due to downwash vertical momentum rate becomes:

$$\text{(2-5)} \qquad \text{Lift}_{dw} = v_0 \, S \int_{-\infty}^{\infty} \rho v_y \, dy$$

Substituting from preceding expressions and simplifying:

(2-6) $$\text{Lift}_{dw} - \rho v_0 S\Gamma/(2\pi)\int_{-\pi/2}^{\pi/2} d\beta$$

Which integrates to indicate lift due to downwash as:

(2-7) $$\text{Lift}_{dw} = \rho v_0 S\Gamma/2$$

Doubling this to include upwash interception lift in no-loss confined condition where the rate of upwash momentum interception is equal to the rate of downwash momentum production:

(2-8) $$\text{Total Lift} = \rho v_0 S\Gamma$$

This equation is known as the *Kutta-Joukowski theorem* in honor of W.M. Kutta (1867-1944) and N.E. Joukowski (1847-1921), jointly credited with deriving it long ago.

For the equation to be useful we need to specify circulation strength, Γ. If we substitute the value of Γ from equation 2-3, we get:

(2-9) $$\text{Lift} = \pi\rho v_0^2 SC \sin\theta$$

Since the product, SC, is airfoil span times chord, or airfoil area, the lift equation can be written as:

(2-10) $$\text{Lift} = \pi\rho v_0^2 A \sin\theta$$

Here A is airfoil area, and θ is the angle of downwash departure from the trailing edge. This equation applies generally to different airfoils, but it is customary to specify lift in relation to angle of attack, the angle between airfoil chord and flight path, rather than the angle of flow departure. Since air follows the mean camber line toward departure direction, an uncambered airfoil at low angle of attack has a departure angle equal to the angle of attack, and thus, where the angle of attack is expressed as α, has lift:

$$\text{Lift} = \pi\rho v_0^2 A \sin\alpha \tag{2-11}$$

This equation is valid only for an uncambered airfoil. For airfoils in general, whether or not cambered:

$$\text{Lift} = \pi\rho v_0^2 A \sin(\alpha - \alpha_0) \tag{2-12}$$

Here α_0 is the angle of attack at zero lift. The angle difference, $\alpha - \alpha_0$, known as *absolute angle of attack*, is equivalent to downwash departure angle, θ.

Performance Coefficients

Aerodynamic forces, including dynamic pressure, are proportional to density times velocity squared. It has become customary to eliminate these factors, as well as airfoil area, from performance data plots by dividing aerodynamic forces by dynamic pressure and airfoil area. The results are known as *performance coefficients*. For example, dividing lift magnitude of (2-11) by dynamic pressure, $\rho v_0^2/2$, and area, A, yields the theoretical lift coefficient, C_L, of an uncambered airfoil as:

$$C_L = 2\pi \sin\alpha \tag{2-13}$$

Lift coefficient of an uncambered airfoil closely follows the theoretical lift line, as illustrated in Figure 2-25, until effects of viscosity begin to interfere. When lift peaks and begins to decline as angle of attack increases, the airfoil is said to be *stalled.* Reynolds number affects deviation of lift coefficient slope from the theoretical line. Thus published airfoil data includes Reynolds number at which data was plotted.

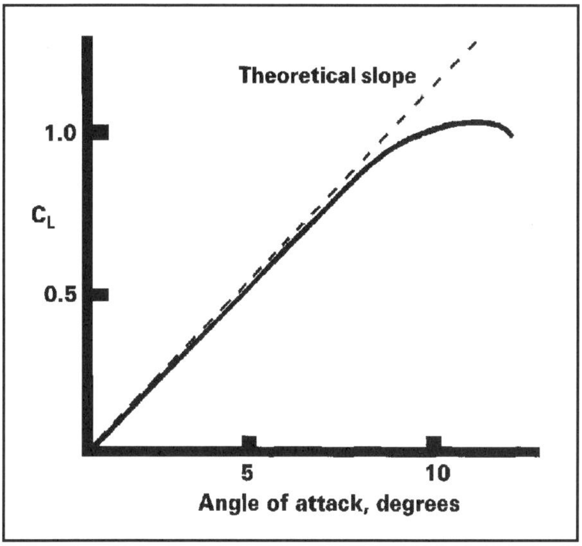

Figure 2.25 **Lift Coefficient vs Angle of attack**
Lift coefficient follows the theoretical line drawn to correspond with equation (2-13) except for effects of viscosity, which causes boundary layer accumulation to interfere with flow attachment. Greater Reynolds number would allow for closer conformance to higher lift coefficient.

It is not uncommon to publish lift coefficients in terms of angles in radians, rather than degrees. At low angles where airfoils are efficient, a radian angle and its sine are very nearly equal. Thus for angles in radians the theoretical lift of an uncambered airfoil could be expressed as either $2\pi \sin \alpha$ or $2\pi\alpha$, since the expressions would be essentially equivalent.

To find lift from published lift coefficient data, one multiplies lift coefficient by airfoil area, A, and the expression for dynamic pressure, $\rho v_0^2/2$. Lift of an airfoil thus becomes:

$$\text{Lift} = C_L A\rho v_0^2/2 \quad (2\text{-}14)$$

Similarly, drag based on drag coefficient, $C_{D,}$ becomes:

$$\text{Drag} = C_D A\rho v_0^2/2 \quad (2\text{-}15)$$

Differing Lift Calculation Methods

Our preceding lift calculation was based on momentum effects occurring in interaction between airfoil and air, in accordance with Newton's laws of physics. Classical aerodynamics more often bases calculation of lift on velocity and Bernoulli's law. The result is the same because Bernoulli's law is based on Newton's laws. Thus either method results from application of Newtonian principles.

In the Prandtl-Tietjens book, *Applied Hydro- And Aerodynamics* it is shown that lift can be calculated through applying Bernoulli's law to passing velocities. This is the common method in higher level teaching. The book also demonstrates that lift can be calculated partially through Bernoulli's law and partially through calculation of momentum effects. If the airfoil is surrounded by an imaginary square box, as in Figure 2.26, and lift is calculated as a sum of upward momentum rate entering the box plus the downward momentum rate leaving, the resulting lift is exactly half of that obtained when integrating to plus and minus infinity.

If lift is calculated as due to pressure difference between top and bottom of the box in accordance with Bernoulli's law, then that result is also equal to the half the proper lift, and the two results can be added to get the proper result. This seems to have led to belief among some that lift is half due to Bernoullian considerations and half due to Newtonian considerations. However, pressure difference between upper and lower box surfaces can also be attributed to Newtonian momentum effects, so that all lift is nevertheless in accordance with Newtonian principles.

Apportioning of lift in the box scenario depends on the ratio of box width to height. For example, suppose we assume an imaginary box having vertical sides tangent to airfoil leading

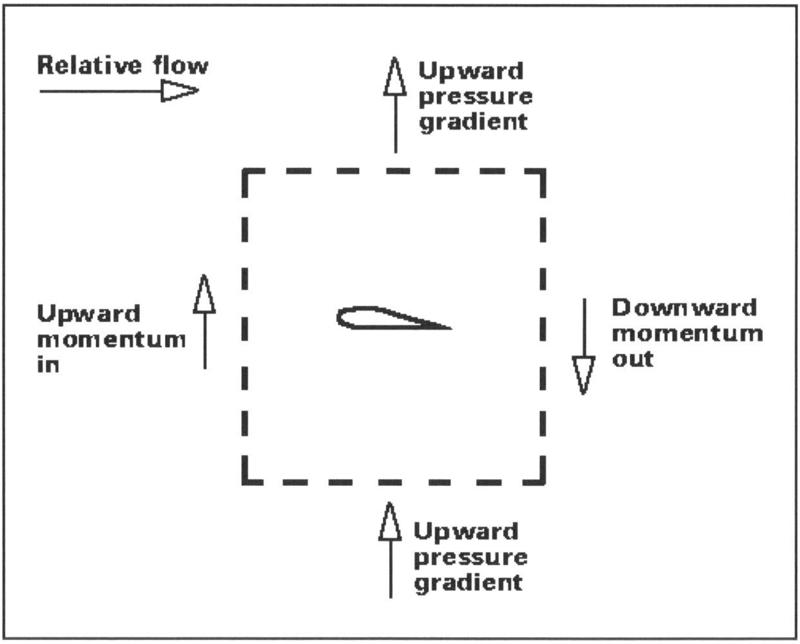

Figure 2.26 **Lift Apportioning**
Lift can be calculated as the sum of the upward momentum passage rate through the forward side of the square box, downward momentum rate passage through the aft side, pressure reduction lift at the top and pressure increase lift at the bottom.

and trailing edges, and extending vertically to plus and minus three chord lengths, as in Figure 2.27. We can put these limits into integral (2-6) and double the result to get lift associated with vertical momentum change in flow through the elongated box. Angles associated with a radius line from mid-chord to vertical distances in an aft plane of plus and minus three chord lengths will be plus and minus arc tangent 6.0, or plus and minus 1.41 radians.

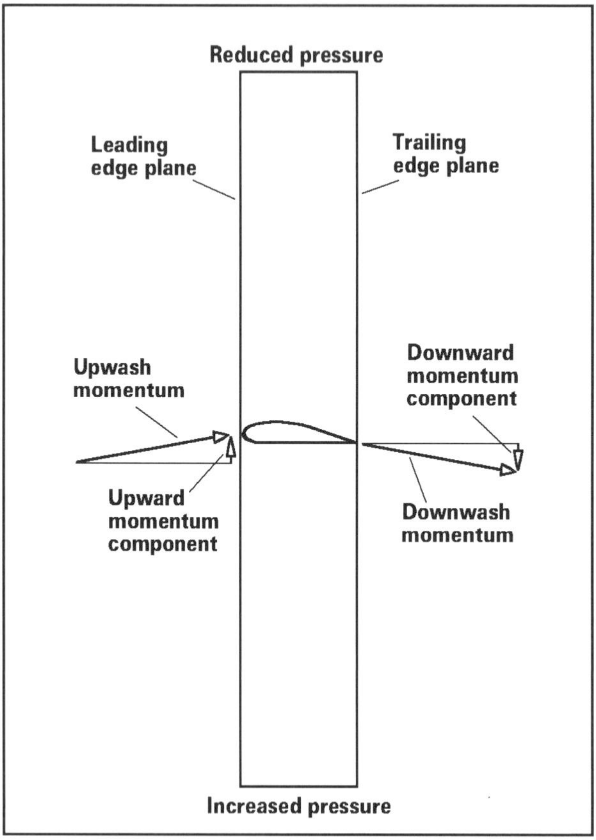

Figure 2.27 **Momentum Passage Box**

The half-lift integral becomes:

$$(2\text{-}16) \qquad \text{Half-lift} = \rho v_0 S\Gamma/(2\pi)\int_{-1.41}^{1.41} d\beta \;= 0.45\, \rho v_0 S\Gamma$$

When doubled to include lift due to both upwash and downwash momenta, the result is ninety percent of lift indicated by the Kutta-Joukowski theorem, even without considering pressure changes at the box upper and lower surfaces. This suggests that most of lift generation is associated with air mass passage within a few chord lengths of the airfoil or wing.

Octave Chanute Glider Model
Chanute Air Museum, Rantoul Illinois

Introduction to Aerodynamics

III. Viscosity, Flow Attachment and Stall

Viscosity

The word "viscosity" brings to mind the apparent flow resistance of liquids such as motor oil or honey. In liquids, viscosity is due to the tendency of molecules to adhere together. In rising temperature, viscosity decreases as thermal agitation loosens the intermolecular bonds. Oppositely, in gases such as air, where molecules are separated and intermolecular attraction is insignificant, viscosity increases in rising temperature as greater random thermal motion increasingly couples momentum between differing velocities of mass movement.

Air consists mostly of empty space. A volume of air at normal conditions is only about 0.1 percent occupied by molecules of nitrogen and oxygen plus other minor gases. The molecules are constantly in random colliding motion of all directions with average speed, related to temperature, of about 1,500 feet per second, but the average molecule travels only about two millionths of an inch before colliding with another one. For a volume of air we regard as not being in motion, average molecular velocities are equal in all directions so that the average velocity in any direction is zero. Air we regard as being in motion, as in a ten miles per hour wind for example, would have average molecular *velocity* of about 15 feet per second in the wind direction, although average molecular *speed* would still be in the neighborhood of 1,500 feet per second.

In a solid, molecules are anchored in an elastic manner, able to vibrate in differing directions. Heat content determines vibration amplitudes. The solid surface, even if highly polished, is still rough at the molecular level. Vibrating surface molecules interact with impacting air molecules. A surface molecule may absorb energy from an impacting air molecule, or it may bat an incoming air molecule away with increased departing energy. Coupling of random molecular motion between a solid surface and adjacent air is the mechanism of heat transfer.

If air molecules adjacent to a surface have nonzero average velocity, then the air is considered to be flowing, and shear force develops between surface and air as momentum is coupled in molecular collisions. The capacity of a fluid to develop shear force drag between flow and a surface or between flows of differing velocity is known as *viscosity*. It is quite important in aerodynamics.

The Boundary Layer

Flow near a stationary surface loses relative velocity and dynamic pressure through molecular collisions, so that a layer of retarded flow develops. Depth of the retarded layer, known as the *boundary layer*, becomes deeper as flow proceeds aft, as illustrated in Figure 3.1.

Flow velocity at the surface is virtually stopped by intermolecular collisions. At increasing distance from the surface, the boundary layer is dragged more rearward by passing flow, until at some distance, velocity becomes essentially that of unretarded flow. Thus a velocity gradient perpendicular to the surface is established, as represented in Figure 3.2.

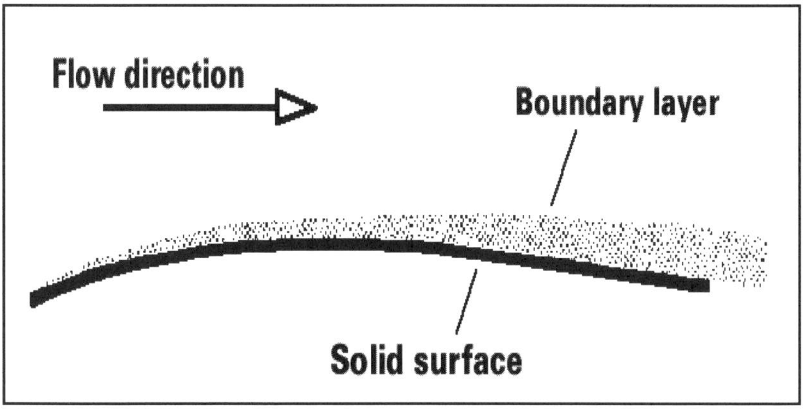

Figure 3.1 **Boundary Layer**
Viscous drag between surface and flow produces a *boundary layer* of retarded flow which increases in depth rearward. While surface drag opposes rearward movement of the boundary layer, adjacent passing flow drags the boundary layer rearward.

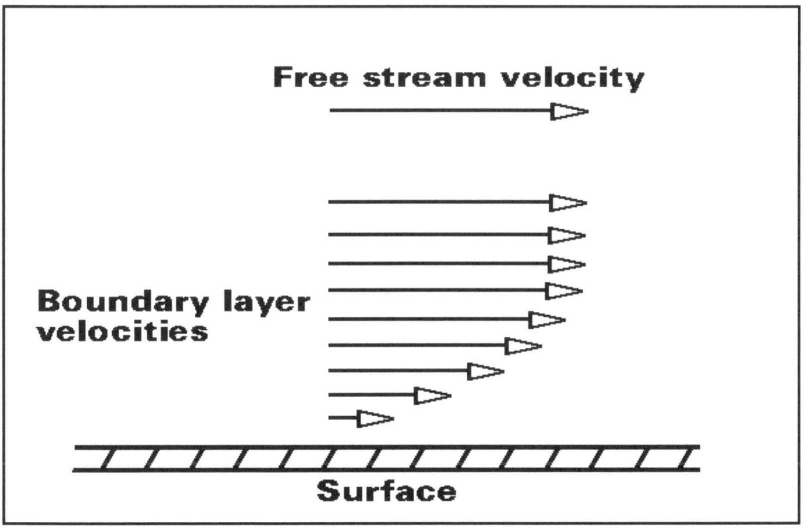

Figure 3.2 **Boundary Layer Velocity Gradient**
In the boundary layer, velocity is virtually zero at the surface and increases toward that of the free stream at some distance out. In this illustration flow is indicated as *laminar*, in which velocity change is gradual and there is no turbulence.

Coanda Effect, or Surface Attachment

Viscous drag of passing air tends to drag lower velocity flow with it. This is very important to wing operation. When passing flow drags slower or relatively stagnant air away from a wing surface it is replaced by flowing air which follows the surface, even if the surface curves away from the direction of flow, as illustrated in Figure 3.3. This *surface attachment* of flow is popularly known as *Coanda effect*, in honor of Henri Coanda (1885-1972), a Romanian. Flow attachment is essential to winged flight.

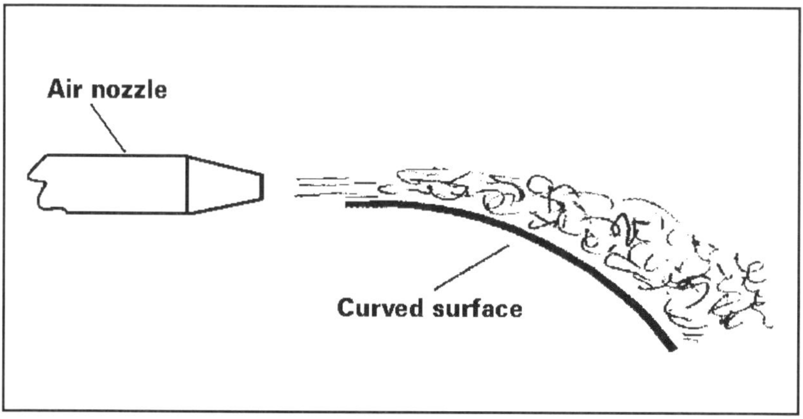

Figure 3.3 **Momentum Coupling in Coanda Effect**
As flow passes over a curved surface, stagnant air between flow and the surface is dragged away by viscous and perhaps turbulent coupling between flow and relatively stationary air. Atmospheric pressure then forces flow to curve so as to fill the space vacated by stagnant air removal.

Although surface attachment is said to have been recognized before Coanda's name became associated with it, perhaps it is so named because it was a factor in an accident which nearly cost Coanda's life. Coanda, like many other creative persons, was willing to take risks in pursuit of accomplishment. In his earlier years he conducted air flow experiments while riding the cowcatcher of a railroad locomotive at night. This risky perch gave him the air flow he needed, and he was able to control illumination of his experiments with oil lamps. Presumably he did not encounter any cows, for he lived to be credited as the first to fly a jet propelled airplane, although not very far, and later became Chief Engineer of Bristol Aircraft.

Coanda's jet was not a turbojet, for that technology had not yet developed. He coupled a forward-mounted fifty horsepower piston engine to a centrifugal compressor which forced air into two annular combustion chambers, each somewhat like a modern afterburner. From the combustion chambers, hot gases passed through jets aimed rearward on opposite sides of the airplane. In static tests the system is said to have produced 480 pounds of thrust, nearly equal to half the airplane weight.

Coanda had no experience as a pilot when he sat in his airplane without a seat belt to demonstrate his propulsion system in a ground run-up at a Paris aeronautical exhibition in 1912. The plane unexpectedly accelerated and took to the air with terrified Coanda seeing the jet exhausts attach to the airplane surface and set the fabric afire. The unintentional test pilot quickly crashed, but fortunately was thrown clear of the burning wreck. Although injured, he survived and lived sixty more years as an important contributor to manned flight.

Flow Separation

Even with Coanda effect, a boundary layer remains at a surface. Although rearward drag of passing flow tends to carry boundary layer accumulation aft, the movement can be slowed or even reversed by opposition of increasing pressure. If there were no viscous loss to reduce stagnation pressure of flow passing a surface where pressure is reduced, conversions between PV energy and kinetic energy would be reversible, and pressure recovery at the trailing edge would be complete. However, some energy is lost in viscous drag, reducing boundary layer pressure rise capacity so that it cannot regain pressure necessary to reach the trailing edge without assistance.

Assistance for rearward boundary layer flow comes from rearward drag of passing adjacent flow. If rearward movement is sufficiently opposed by force of rising pressure, boundary layer buildup can displace passing flow, as indicated in Figure 3.4, and seriously degrade wing performance. Flow displacement due to boundary layer accumulation is referred to as *flow separation.* Rising aft pressure can even reverse boundary layer flow, causing interaction with passing flow to produce turbulence which can further increase flow separation.

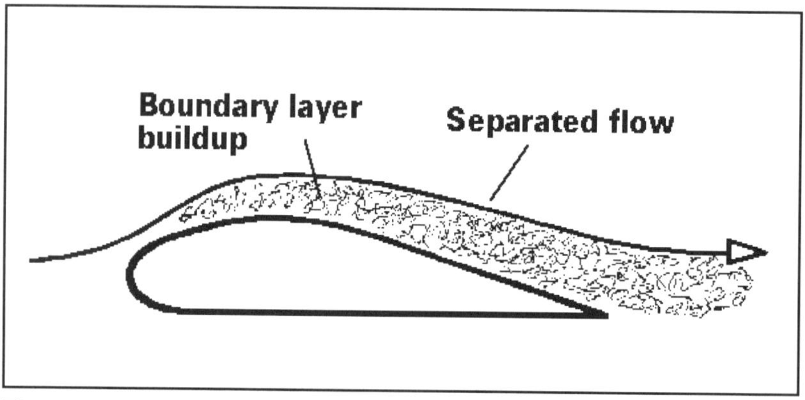

Figure 3.4 **Flow Separation**
Boundary layer accumulation, especially where pressure is rising in the rearward direction, can displace passing unretarded flow so that lift is seriously impaired, leading to wing stall. The retarded flow accumulation may be laminar or turbulent.

Flow separation can lead to *stall* of a wing or airfoil. A condition in which further increase of angle of attack produces decreasing lift. Separation also decreases pressure recovery, thereby increasing drag. Drag due to flow separation can be a problem on other objects as well. Drag forces of wind passing fixed objects such as power lines, smokestacks and other structures create stresses. Drag affects flight of balls used in sports. A smooth ball develops a laminar boundary layer in the front half, and turbulent separation in the rear half. By dimpling the surface of a golf ball, as shown in Figure 3.5, a small degree of turbulence is generated, mixing momentum of passing flow into the boundary layer to assist its rearward movement so that flow separation occurs farther aft. This results in greater pressure recovery, reduced drag and increased travel distance. Surfaces are roughened for better aerodynamics also on tennis balls by giving the balls "fuzzy" exteriors of light fibrous material.

Figure 3.5 **Golf Ball Dimples**
To aid boundary layer flow, a golf ball is dimpled so as to produce a bit of turbulence in passing air.

Reynolds Number

The tendency to develop turbulence in flow passing a smooth surface was investigated by Osborne Reynolds (1842-1912) while studying flow in pipes. He found the tendency was related to velocity, length of flow path, fluid density and viscosity, which he related in the mathematical expression known as Reynolds number:

$$(3\text{-}1) \qquad Re = \rho v l / \mu$$

In this expression Re is Reynolds number, ρ is fluid density, v is flow velocity, l is length of path and μ is fluid viscosity. Greater Reynolds number means greater tendency to develop turbulence.

As compared to golf balls, airplane wings have much longer flow paths and usually faster speeds so that Reynolds numbers are greater. Upper surface turbulence begins naturally in a forward region where the rearward pressure gradient causes localized boundary layer flow reversal in a *separation bubble*, as indicated in Figure 3.6. From the separation bubble rearward, turbulent mixing with passing flow assists movement of the boundary layer, which reattaches as a turbulent layer. Higher Reynolds number provides greater boundary layer assistance so that greater maximum lift can be achieved before stall.

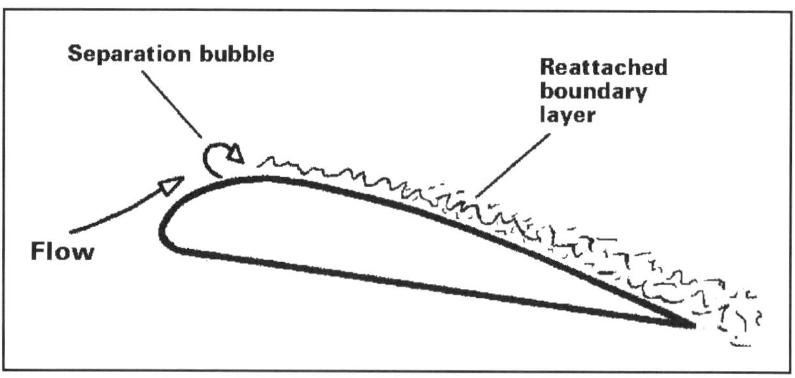

Figure 3.6 **Separation Bubble**
Where flow separation begins, reversed boundary layer flow produces a local "separation bubble." Turbulence beginning at the separation bubble mixes rearward momentum of higher velocity flow into the boundary layer, assisting its rearward movement as a turbulent boundary layer.

Because surface drag is least in laminar flow, a goal in wing design is to adjust the wing contour for farthest aft position of the separation bubble so as to minimize drag. A turbulent boundary layer can still separate and degrade performance if coupling of rearward momentum from passing flow is insufficient to overcome opposition of rising aft pressure.

Stall Delay in Rapidly Increasing Lift

In testing of airfoils for helicopter rotor application, the angle-of-attack may be varied rapidly to simulate "cyclic pitch," the variation in blade angle during each cycle of rotation. Tests with varying rates of angle-of-attack increase produce differences in peak lift achieved before stall occurs. As represented in Figure 3.7, maximum lift may be about fifty percent greater with rapid increase as compared to slow increase.

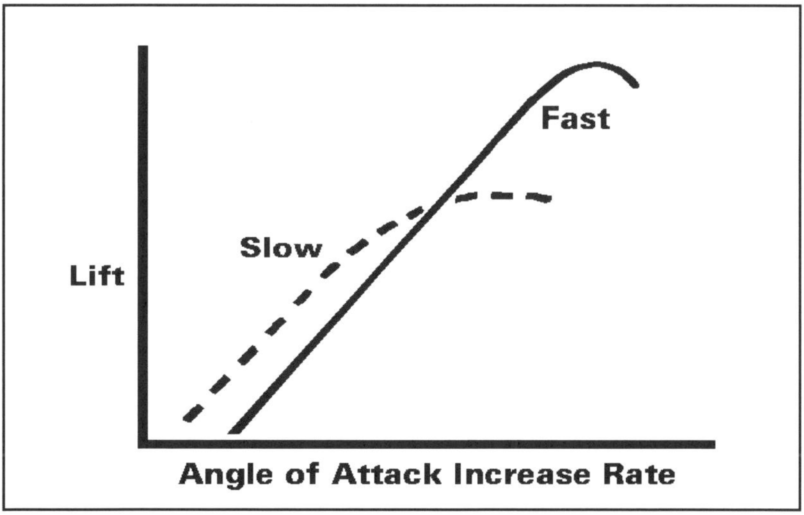

Figure 3.7 **Peak Lift vs Rate of Increase of Angle of Attack**
The dashed line indicates lift versus angle of attack for a helicopter airfoil during slow rate of increase of angle of attack. The solid line indicates lift at a fast rate of increase. Note that the fast rate of lift increase produces peak lift nearly fifty percent greater than that of slow increase.

Difference in peak lift as related to rate of increase is not just a peculiarity of helicopter airfoils. It can also occur on airplane wings, even though airplane pitch attitude may not be changed quickly enough to obtain comparably great effect. Normally, stall serves to protect an airplane from excessive stress, but if an airplane operating in turbulent weather encounters a downdraft such that lift is greatly diminished, or even goes negative, and then quickly encounters an updraft such that lift increases at a very fast rate, the maximum lift achieved may overstress the plane sufficiently to cause structural failure. Because of this, pilots are taught to slow to *maneuvering speed* to reduce peak forces when in turbulence. Maneuvering speed is specified as the maximum at which abrupt control input will not overstress the aircraft structure. Greater peak lift is achieved with rapid rate of increase because the level of boundary layer buildup lags behind the forces which produce it. With lift increasing rapidly, a higher level of lift is achieved before boundary layer accumulation becomes sufficient to produce stall.

Regeneration

In wing operation, pressure difference produces lift and promotes circulation upwash. Interception of upwash momentum adds to pressure difference. Thus pressure difference promotes more pressure difference in a regenerative feedback effect. Many physical processes involve regenerative feedback. Such processes tend to increase in magnitude until a limit is reached. The process may remain at its limit, or if there is delayed opposition, it may reverse periodically to produce oscillations. In the case of a wing in normal stable operation, lift rises to a limit related to the factors in equation (2-12), and remains at that level until change occurs in angle of attack or other lift-related parameter.

Round form lift can have both regenerative and degenerative effects. If unbalance of opposite surface flows occurs, as indicated in Figure 3.8, the longer and faster flow path has more pressure decrease over a longer path while pressure change in the shorter path is less. The pressure difference tends to regeneratively promote more upwash and more circulation.

While pressure difference is regenerative in promoting increased circulation, greater viscous drag in the longer path opposes difference of flows, and thus is degenerative. At low Reynolds number, where flow is laminar, the degenerative effect of drag dominates over the regenerative effect of pressure difference, so that opposite flows remain equal and stable.

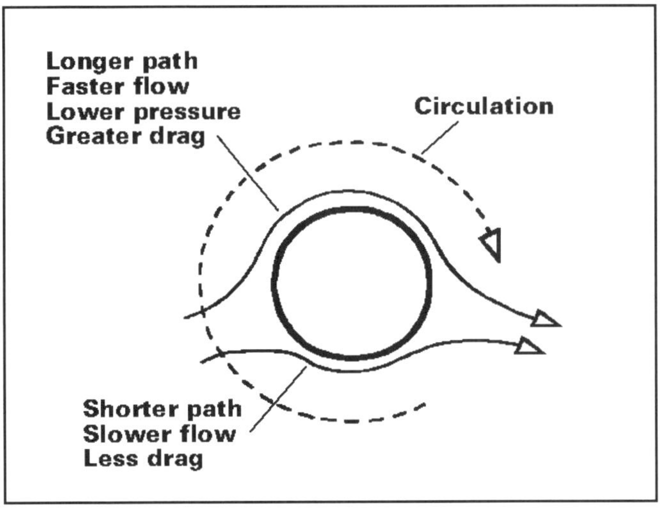

Figure 3.8 **Regenerative and Degenerative Effects** Pressure difference promotes positive feedback favoring more circulation and lift, while greater viscous drag in the longer and faster flow path opposes circulation. At low Reynolds number viscous drag stabilization dominates so that circulation and lift indicated here do not develop.

The stabilizing effect of viscous surface drag decreases at higher Reynolds number as passing flow is separated more by increasing boundary layer buildup. When pressure rise in convergence of boundary layers is sufficiently less than pressure rise in farther aft convergence of separated flows, the resulting aft pressure gradient causes boundary layer reversal and formation of attached vortices, as indicated in Figure 3.9, which add to the flow separation.

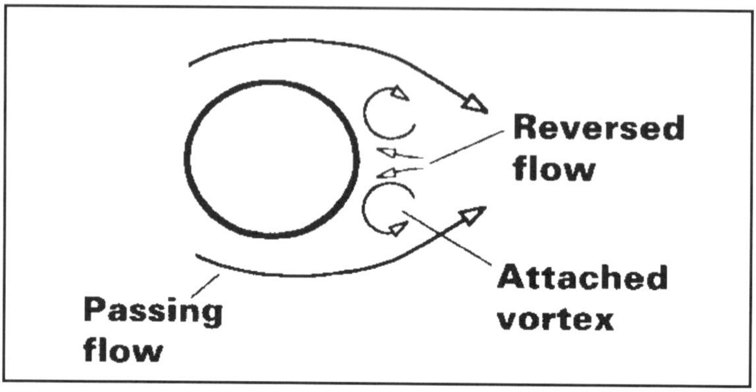

Figure 3.9 **Attached Vortices**
When boundary layer stagnation pressure is sufficiently less than that of passing flow, pressure rise in convergence of flows at the rear produces pressure rise sufficient to reverse low energy boundary layer flow and create vortices in the shear between reversed and passing flows.

With further Reynolds number increase, flow separation increases further, reducing the stabilizing effect of surface drag sufficiently to allow the regenerative effect of circulation to dominate over the degenerative effect of surface drag. Any random unbalance can then initiate circulation which increases regeneratively as increased velocity aids boundary layer flow on the side where circulation is in the direction of flow, and causes further flow separation at the other side where boundary layer assistance is diminished.

Where boundary layer flow is assisted, the attached vortex shrinks as the other grows, as indicated in Figure 3.10. Difference of angular momenta of the vortices is equaled by opposite angular momentum of circulation around the cylinder.

Figure 3.10 **Regenerative Circulation**
Clockwise circulation adds boundary layer assistance where velocity is increased above the cylinder, and decreases assistance where velocity is decreased below. Reversed flow, encountering increasing circulation downwash, diverts toward growth of the lower vortex, causing the other to shrink.

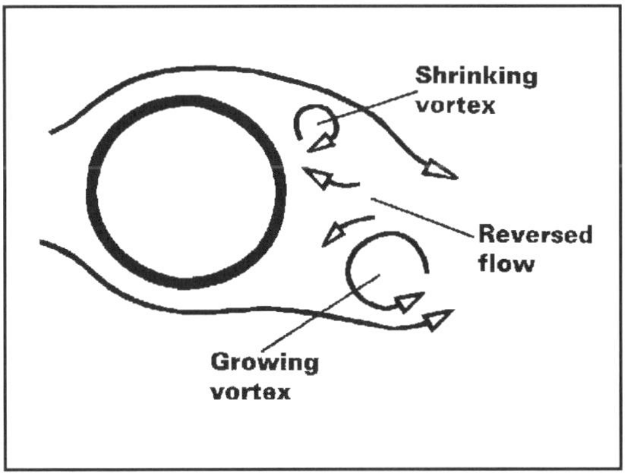

As circulation increases regeneratively toward maturity, the smaller vortex disappears while the other reaches maximum growth, as in figure 3.11. This limit occurs when boundary layer buildup in the increased velocity flow and increasing adverse pressure gradient prevent further circulation growth. This is not a stable condition. Boundary layer buildup, which lags behind circulation, increases further to allow the adverse

Figure 3.11 **Mature Lift**
When lift approaches maximum, circulation reinforcement of the upper boundary layer has caused the smaller vortex to disappear while the other has grown and become detached.

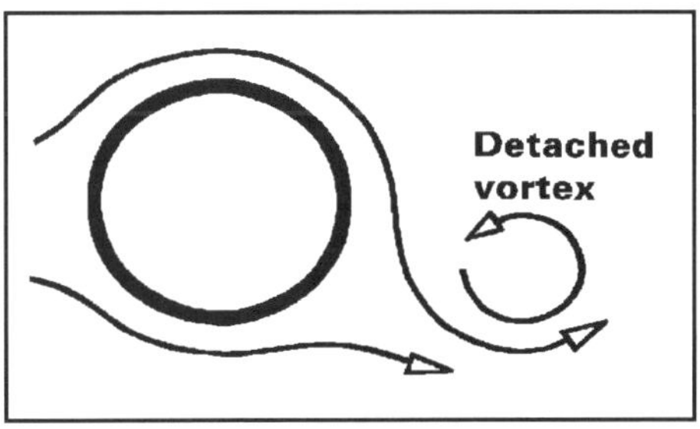

pressure gradient to initiate lift decay followed by reversal of downwash and lift. Reversed lift increases until stall and reversal again occurs in a repeating process which produces a downstream trail of starting vortices, as in Figure 3.12, known as the "Karman vortex street" in honor of Dr. Von Karman.

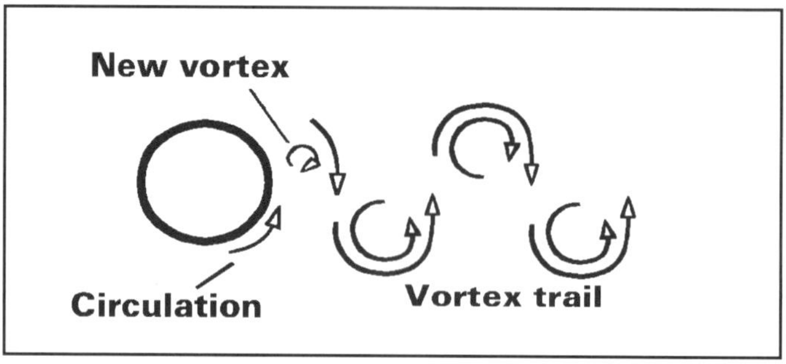

Figure 3.12 **Karman Vortex Street**
Continued stall and reversal produces a trail of vortices downstream of the cylinder. Each vortex has angular momentum equal and opposite to that of simultaneously created circulation around the cylinder.

Stabilization of Cylinder Lift

Figure 3.13 shows oscillating deflection of downstream yarn tufts in flow passing a cylinder which has narrow surface slots for boundary layer removal by suction.

Figure 3.13 **Cylinder in Down-flow**
Boundary layer suction is not applied, and the stationary black strip indicates that the cylinder is not rotating, Alternating lift direction causes the yarn tufts to deflect left and right, producing blurred fanlike images.

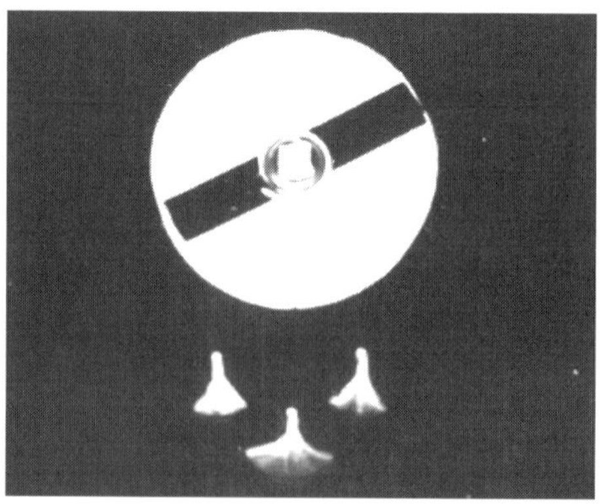

When the cylinder is rotated, the tufts deflect in direction of circulation favored by rotation, as seen in Figure 3.14. When rotation direction is reversed, lift direction also reverses, as indicated in Figure 3.15

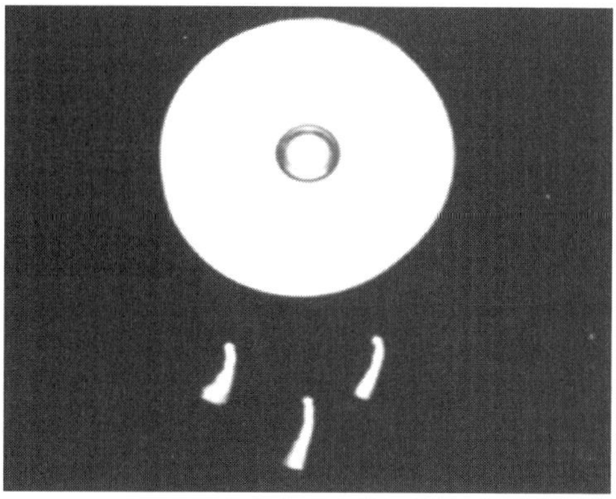

Figure 3.14 **Cylinder in Clockwise Rotation**
With clockwise rotation, downward surface movement on the right side carries the boundary layer through the aft pressure gradient. Lift toward the right is sustained without stall, and tuft deflection is sustained toward the left.

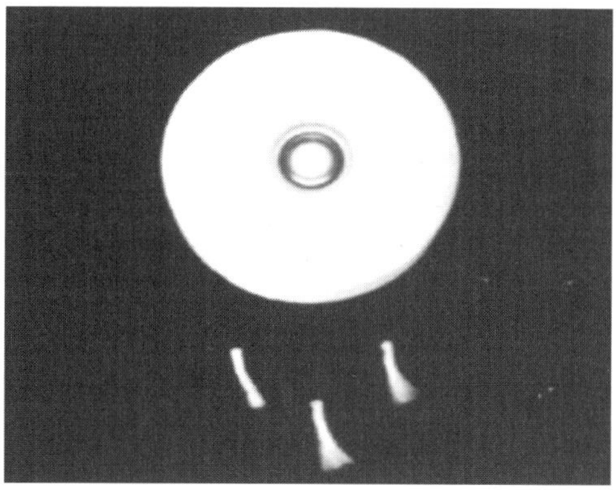

Figure 3.15 **Cylinder in Counterclockwise Rotation**
With counterclockwise rotation, stall is again prevented. Sustained lift is toward the left, and tuft deflection is toward the right.

As shown in Figure 3.16, where there is no rotation, withdrawal of boundary layer accumulation by suction through narrow surface slits produces lift comparable to that obtained with counterclockwise rotation. The direction of lift can be reversed, as shown in Figure 3.17, by inserting an object into the flow to upset the stabilized lift direction.

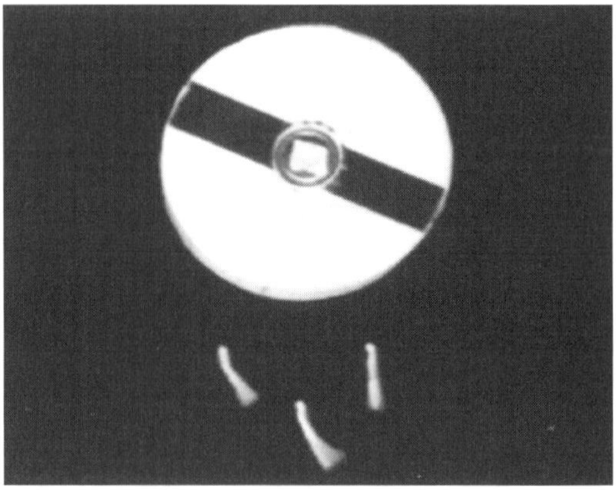

Figure 3.16 **Cylinder With Boundary Layer Removal**
Without rotation, but with boundary layer suction removal through narrow surface slots, stable lift develops spontaneously. In this photo, lift is toward the left with tuft deflection to the right.

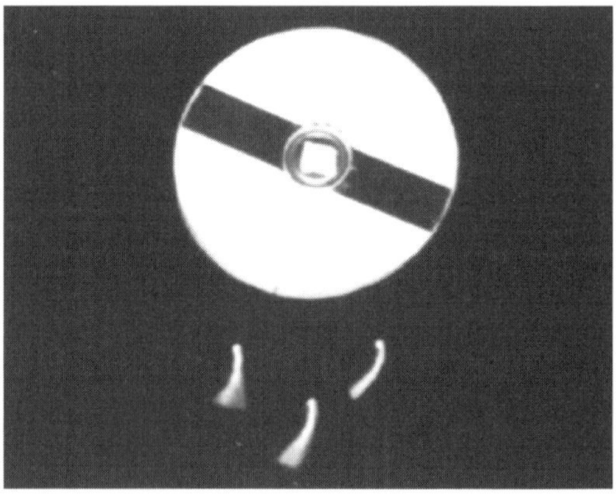

Figure 3.17 **Reversal**
From conditions of the preceding photo, the direction of lift has been tripped into reversal by momentarily inserting an object near the cylinder to disturb the flow.

In the four cases of stabilized lift direction, a degree of oscillation still exists because the round cylinder has no precise control of the angle of departing flow. This result relates to airplane wing operation near stall, in which boundary layer accumulation masks the wing trailing edge so that departing flow direction is not well defined, and a degree of lift oscillation, known as *buffeting,* may occur. This buffeting may be felt by passengers as a shuddering of the airplane.

Aerodynamics of the Environment

We must deal with aerodynamics of the natural environment in which we fly. An important aerodynamic effect of nature is that, like a boundary layer which develops in flow passing a wing surface, a deep layer of retarded wind exists near the ground, with a gradient of velocity. In some publications for pilots, near-surface wind is regarded as affecting only distance of takeoff or landing run, and speed relative to the ground in flight. More subtle but quite important effects of wind gradient on airplane performance during climbout after takeoff, and during descent for landing are often overlooked. Figure 3.18 illustrates a popular misconception of the effect of wind on climb performance after takeoff.

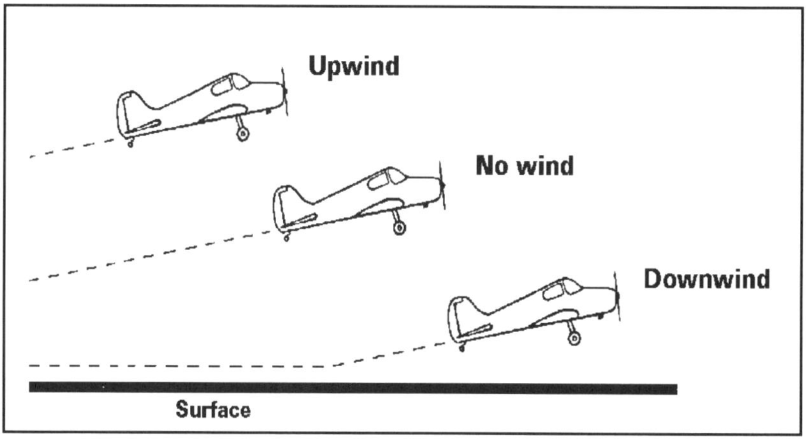

Figure 3.18 **Popular Concept of the Effect of Wind on Climbout**
If wind were of constant speed without regard to distance from the surface, the effect of wind on climb after takeoff would be only on groundspeed. Upwind and downwind takeoffs would differ only in length of ground run and in ground speed at lift-off. There would be no difference in rate of climb or pitch attitude for holding constant airspeed.

Wind gradient near ground level, like that of a boundary layer, is caused by surface resistance to air flow. Turbulence and gradient depth are affected by trees, buildings and other objects.

Figure 3.19 better illustrates how a strong wind can affect climbout after takeoff. The difference in pitch attitude for holding constant airspeed during climb into a strong wind can be impressive. Significant surface-related wind gradients may exist to altitudes of two hundred to over one-thousand feet.

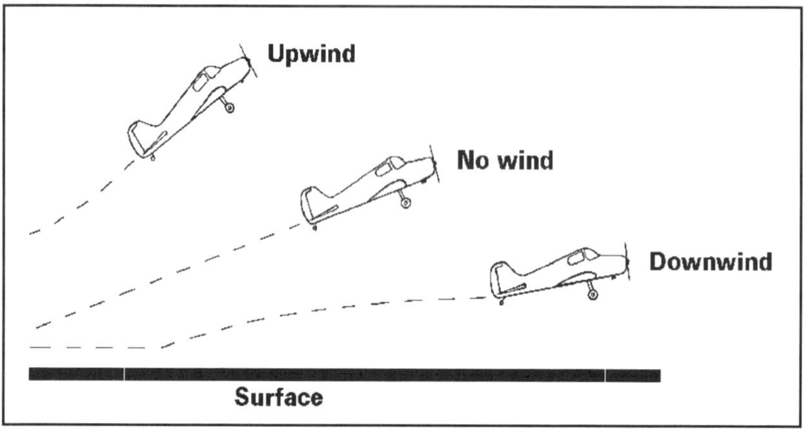

Figure 3.19 **True Effect of Wind Gradient on Climbout**
As an airplane climbs into a gradient of increasing headwind, the increasing wind adds to airspeed. If constant airspeed is held by pitch attitude, then ground speed continuously decreases and the climb rate is made greater as kinetic energy of the slowing airplane is converted to potential energy associated with altitude. In downwind takeoff the result is opposite. The increasing tailwind during climb subtracts from airspeed, and pitch attitude may have to be lowered to maintain airspeed and avoid stall.

During a windy, sunny day, thermal stirring from surface heating energizes near-surface wind just as separation bubble turbulence energizes a wing boundary layer. Then at night, surface cooling by radiation into a clear night sky produces a cool blanket of relatively stagnant surface air, and a strong wind gradient near the top of the cool layer. In downwind takeoff into the apparent calm, the effect of entering rapid tail wind increase at perhaps fifty to one-hundred feet is much like partial loss of engine power. Climb rate can decrease dramatically, and clearance of obstructions may become cause for concern. Wind gradient also affects landing. In final approach through the gradient of a strong headwind, an inexperienced pilot may be surprised to find airspeed decreasing quickly, and sink rate increasing as the headwind tapers off, so that additional power may be required to maintain desired descent rate.

The Downwind Turn Controversy

In early days of powered flight a fear of "downwind turns" developed. Many have claimed that fear of the downwind turn is without any reasonable basis, because an airplane flies in the air environment alone and is not affected by the relation of air movement with respect to the ground. This seems reasonable if one does not consider the effects of wind gradient.

In climbout after takeoff into a headwind with a strong wind gradient, the angle of ascent may safely be made unusually steep. Then if a turn to crosswind is made, the increasing headwind in climb ceases, and the climb attitude may have to be reduced to prevent stall. A perhaps more important consideration however, is that in a turn one wingtip is higher than the other.

In a steep low altitude turn from upwind to crosswind in a strong gradient, as indicated in Figure 3.20, the higher wingtip of a typical lightplane may be as much as 25 feet higher than the left where difference in wind speed will tend to roll the plane into a greater bank. If the pilot is hesitant in applying hard aileron to counter the roll, or if aileron effectiveness of the plane is not sufficient, the roll may continue to an inverted condition followed by a dive to the ground. To reduce potential for such accidents, students are usually taught to avoid turns below 400 feet.

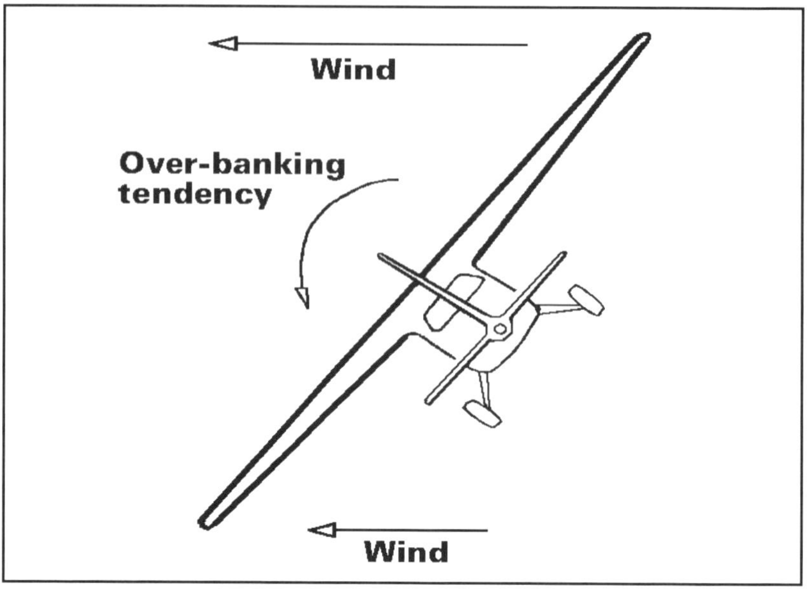

Figure 3.20 **Turn From Upwind to Crosswind**
In a low altitude turn from upwind in a strong wind gradient, a tendency to over-bank begins when the bank is initiated, due to greater airspeed at the higher wing. The situation worsens as the turn approaches crosswind, as indicated above, where the wind gradient makes the angle of attack of the higher wing greater than that of the lower.

During training of pilots in World War One the danger of downwind turns was emphasized to pilots. Perhaps the danger was great because some airplanes of the time, such as the Jenny, had long wing spans and flew slowly so that the effect of wind gradient in a low altitude turn was more pronounced than it would be on more modern, faster airplanes with shorter wingspan.

Clouds, Fog, Icing and Tornado Funnels

In considering aerodynamics of flight at less than about half the speed of sound, we ignore small air temperature changes which we deem insignificant. Temperature change can be caused by pressure change. When air is reduced in pressure, it expands to occupy more space. In this expansion, the work of expansion is converted from thermal energy of air molecules. Since temperature is a measure of molecular energy, the temperature drops. If temperature drops below the dew point, visible moisture appears.

As warm moist air rises, it similarly expands and cools in the decreasing pressure to produce visible condensation in clouds. Cloud bases develop in rising air where temperature and dew point coincide, and if upward movement continues from there, the growing moisture particles may coalesce into rain drops.

Fog can form when temperature drops below the dew point, as from radiation of ground heat into a clear sky. It can also form if temperature falls below the dew point due to atmospheric pressure decrease, producing air expansion in which the work of expansion is at the expense of thermal energy. Conversely, fog can dissipate if pressure rises. These effects can be seen where slow moving air crosses a valley. As air descends into rising pressure of the valley, fog and clouds may disappear and then reappear on the other side where air rises on the up-slope. A pilot needs to understand that, when temperature is just a bit above the dew point, visibility can be quickly lost if there is atmospheric pressure drop. This is especially important in flying at night. In the darkness, fog formation below may not be noticed until it is time to land, when visibility for landing may not exist.

Except when air is very dry, moisture condensation makes tornado funnels visible. As air is drawn laterally into reduced pressure of a tornado, there is no torque exerted on it. There is, however, increase in kinetic energy as work of air movement into reduced pressure converts into rotational velocity. Because no torque is exerted, angular momentum content remains constant. This means that velocity of rotation is inversely proportional to radius from the tornado center. There is also conversion of thermal energy into kinetic energy of rotation as air expands in moving into decreasing pressure. When temperature drops below the dew point, the tornado becomes visible, and at higher altitude, where pressure is less, more cooling and increasing condensation widens the apparent diameter of the tornado so that it appears as a funnel shape.

Wingtip vortices and propeller tip vortices sometimes become visible in condensation due to air expansion and temperature drop. In high "G" maneuvers of military aircraft, as can be seen at airshows, pressure drop above a wing can be substantial, and small clouds of moisture condensation may appear to be carried above the wings.

In earlier times of flying, when instrument vacuum was normally obtained from a venturi, the possibility of venturi icing existed. Vacuum operated instruments require a vacuum of about three inches of mercury. This would be about ten percent of sea-level atmospheric pressure, much more than the pressure drop above a typical airplane wing. When flying at or near freezing temperatures in moist air, moisture could condense and freeze in the venturi, causing it to fail, and the vacuum operated instruments would fail with it. More modern airplanes have engine driven pumps for instrument vacuum, and have part of the basic flight instruments operated by electricity. Electrical systems can fail, and vacuum pumps can fail, but failure of both in a flight is quite unlikely. In event of failure of either system, flight can continue with partial instrument operation.

Introduction to Aerodynamics

IV. Airfoil Applications

Camber

In Sir Isaac Newton's time it was known a flat object would produce lift in a wind. Thus a flat plate was the logical choice for attempted lift calculation. At low angle of attack, flat plate lift will indeed follow the theoretical $2\pi \sin \alpha$ slope. Later experimenters, noting curvature between leading and trailing edges of bird wings, included curvature, now known as camber, in their experiments. One of the first to determine the importance of camber was Otto Lilienthal (1848-1896), who made thousands of downhill glider flights, finally dying in an accident after a wing failed. Camber adapts the airfoil or wing to curvature of flow from upwash to downwash, and thus allows greater level of lift coefficient before stall occurs.

The Wright brothers experimented with camber in lift and drag experiments conducted on a moving bicycle, and later in a wind tunnel built for the purpose. Their successful airplane wings had curved wooden ribs, such as pictured in Figure 4.1, over which canvas was attached to form curved surfaces.

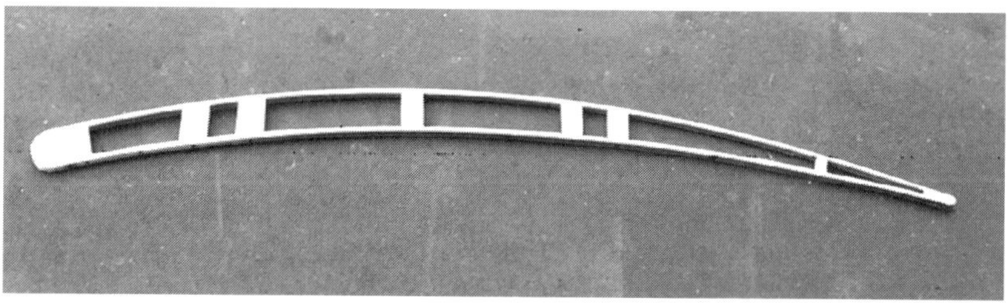

Figure 4.1 **Early Cambered Wing Rib**
Early airplane designers including the Wrights, found that, like curvature of bird wings, airfoils worked best if curved, or cambered.

Camber is common to most modern wings, but there are exceptions. Where lift is required to alternate in direction, as in the case of a horizontal or vertical stabilizer, or wings of an aerobatic plane, camber is not desired.

Pitching Moment

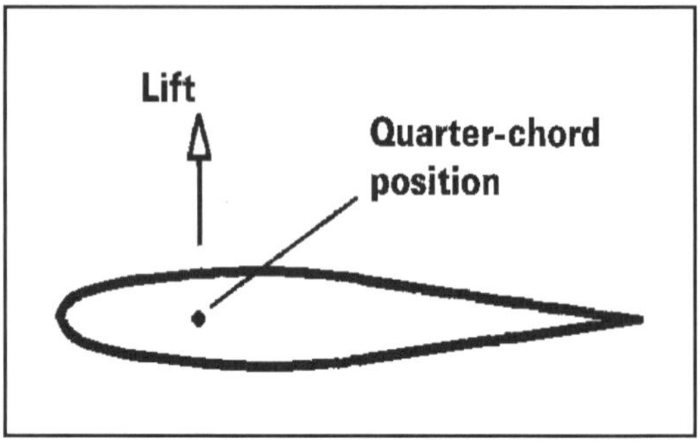

Figure 4.2 **Center of Lift Without Camber** With the center of lift coinciding with the quarter-chord position over a wide range of angle of attack there is no pitching moment between upward lift force and downward load located at that position.

The center of lift of an uncambered airfoil is located at the quarter-chord position, as illustrated in Figure 4.2, and remains there through a wide range of angles of attack and lift variation. Thus in relation to this position, lift develops no torque, or *pitching moment*. A cambered airfoil, on the other hand, develops a pitching moment, even at zero lift angle, defined as negative because it is in negative direction of angle of attack. At zero lift angle, as illustrated in Figure 4.3, upward acceleration of arriving flow produces downward force, while equal upward force is produced by interception of the upward movement before flow departs. A pitching moment is produced between the opposing upward and downward forces.

As angle of attack is increased from that of zero lift, downward force due to upward deflection decreases, and flow is made to depart with downward momentum, so that upward force is greater than downward force, producing net lift.

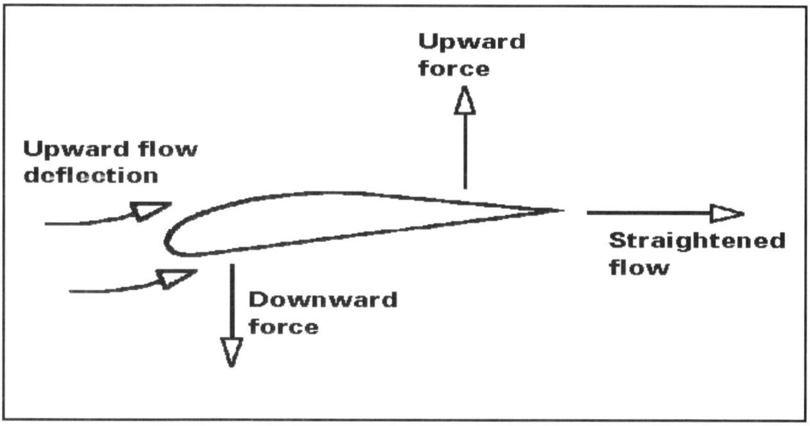

Figure 4.3 **Pitching Moment at Zero Lift**
A cambered airfoil at a negative angle of attack which produces zero lift nevertheless produces a negative pitching moment due to upward air acceleration in the forward region, and interception of the upward movement farther aft.

In Figure 4.4, an airfoil of uniform curvature operating at zero angle of attack produces lift centered at mid-chord, where the weight load is located. If the angle of attack is increased, re-direction of flow from upwash to downwash centers farther forward, so that the center of lift moves forward, which would create an upward pitching moment between lift and load.

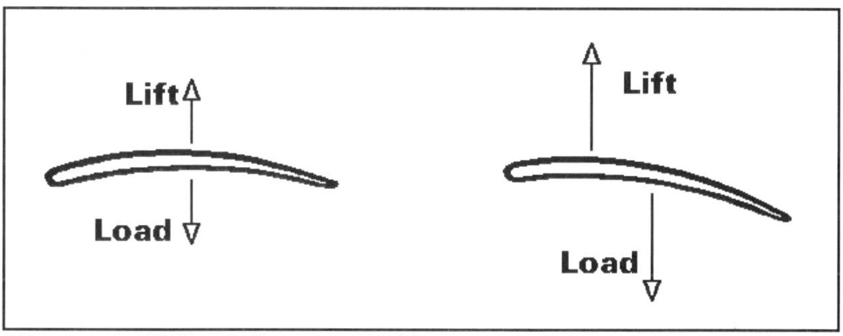

Figure 4.4 **Shift of Center of Lift**
The center of lift of a cambered airfoil moves forward with increasing angle of attack. If lift coincides with load at a particular angle of attack, an increase will produce forward movement of the center of lift, producing a couple for further pitch increase.

If pitching moment is evaluated, not about the mid-chord position but about the *aerodynamic center*, located at or near the quarter-chord position, as in Figure 4.5, the pitching moment remains essentially constant and equal to that at zero lift over a range of angle of attack. Varying angle of attack produces varying lift, but simultaneous chordwise movement of the center of lift inversely varies length of the moment arm between lift and aerodynamic center such that, in relation to the aerodynamic center, the product of lift times moment arm remains essentially constant.

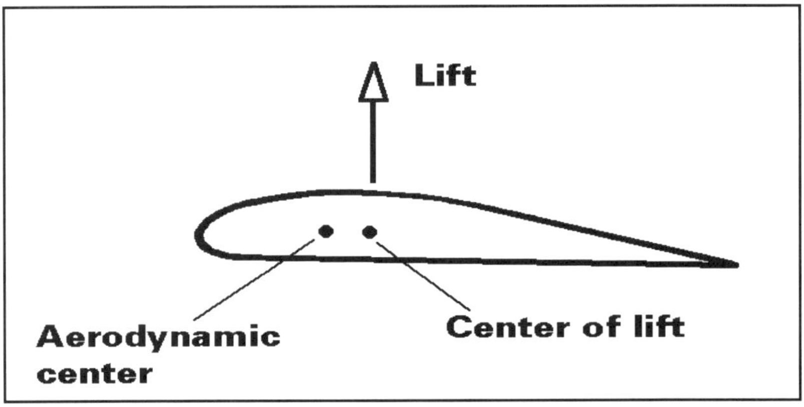

Figure 4.5 **Moment Referenced to Aerodynamic Center**
In relation to the aerodynamic center, as if it were a fixed pivot point, lift produces a counterclockwise moment equal to lift times distance between lift center and the aerodynamic center located approximately at the quarter-chord position. Lift and moment arm distance vary inversely as angle of attack changes, so that pitching moment remains essentially constant and equal to the zero lift moment.

Pitching moment is of importance to airplane design and operation. Wings must accommodate the twisting force without failure or impairment of operation, and the moment must be resolved in connection with other forces to achieve flight equilibrium.

Although pitching moment relative to the aerodynamic center is essentially constant over a range of angle of attack, it does vary as other factors vary. Pitching moment data is plotted and published with other airfoil performance data. In recording airfoil performance data, forces of lift and drag are divided by dynamic pressure of flow and by airfoil area to obtain lift and drag coefficients, which can then be multiplied by planned dynamic pressure and area to find forces for a particular application. Pitching moment performance is, in addition, divided by chord length to obtain the *moment coefficient*, C_M. This is because distance between upward and downward forces of a cambered wing is related to chord length. Thus in terms of moment coefficient, pitching moment of an airfoil becomes:

(4-1) $$\text{Moment} = (1/2)\,\rho\, v^2 C\, C_M$$

In this expression, ρ is air density, v is airspeed, C is chord length and C_M is the moment coefficient.

Wing Thickness

Early designers, like the Wright Brothers, made wings thin with both upper and lower surfaces upwardly convex. Popular belief was that wings should be thin in order to minimize drag. Because they were thin, and strong exotic materials of today did not exist early in the 20th century, it was convenient to brace thin wings with complex drag-producing systems of wires and struts. Successful monoplanes were built, having wings braced with wires and vertical gin poles. A notable early monoplane example was that of Louis Bleriot, first to fly across the English Channel in July 1909, but biplanes with inter-wing bracing, as illustrated in Figure 4.6, became more popular until later, and are still being built by individuals as homebuilts under provisions of FAA rules for experimental aircraft.

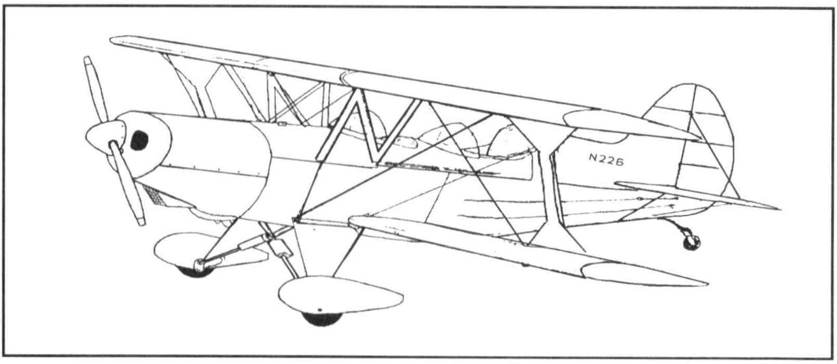

Figure 4.6 **Biplane Structure**
Biplanes achieve strength through inter-wing bracing. Bracing was an important structural feature of early airplanes which had flimsy thin wings.

In WW1, German aerodynamicist Ludwig Prandtl determined that thicker wings with more rounded leading edges were better adapted to changing lift conditions. A more rounded, and therefore less directional, leading edge can accommodate a wider range of upwash angles associated with varying angle of attack. Thicker wings also allow for stronger structure so that less bracing is required.

When it was learned that lower wing surface contour was not as important as that of the upper surface, where adverse pressure gradient is a problem to be dealt with, wings were made more flat on the bottom, thereby gaining thickness. This allowed greater strength needed for cantilevered wings so that complex wing bracing systems were no longer necessary, and storage of fuel in the wings became practical. Wing location of fuel tanks improved wing load distribution and reduced fire hazard. Later, some lower surfaces were also made convex, though usually less convex than upper surfaces. By making the upper surface more convex than the lower, the mean camber line, as was defined in Figure 2.1, is curved to effect wing camber needed to efficiently redirect flow curvature from upwash to downwash.

Camber is desirable for high lift coefficients, as when operating heavily loaded in low density air, or at low speeds during takeoff and landing, when flow curvature from upwash to downwash is greater. But camber designed for high lift coefficients can be a drag handicap when not needed. As the plot of lift versus drag coefficients of Figure 4.7 shows, drag of a cambered airfoil operating at low lift coefficient, where flow curvature from upwash to downwash is minimal, can be greater than that of an uncambered airfoil.

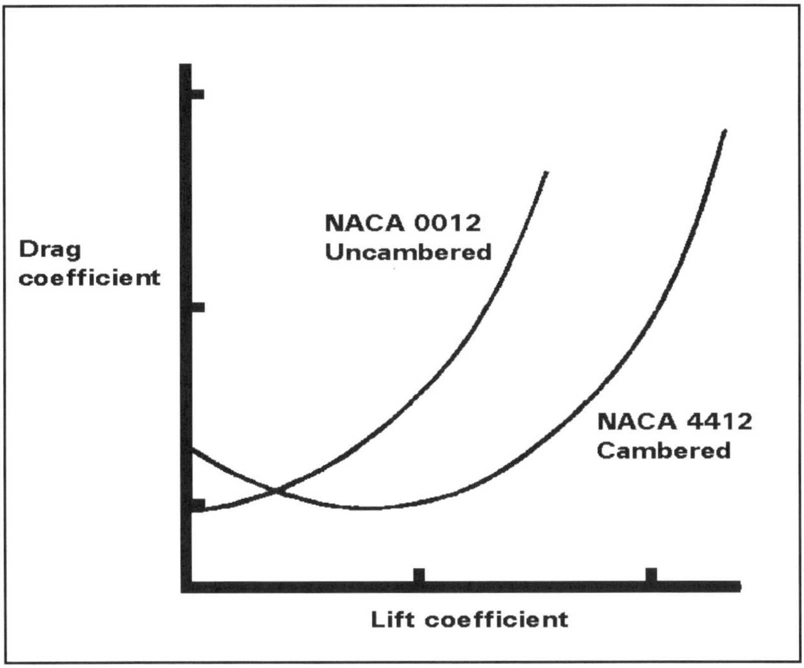

Figure 4.7 **Camber Effect on Drag vs Lift**
As angle of attack of a cambered airfoil (NACA 4412) is first increased from that of zero lift, there is a decrease in drag due to better match of flow curvature to camber. The uncambered airfoil (NACA 0012) produces less drag at zero lift, but is less suited to flow curvature so that drag immediately begins increasing as lift begins with increasing angle of attack.

Lift and Drag of Thin Airfoils

Drag is customarily classified into categories of *parasitic drag* and *induced drag*. Parasitic drag is due to frictional or viscous effects, while induced drag is often referred to as "drag due to lift." Induced drag is alternatively considered as due to wing end losses. It does not occur in airfoil testing where air movements around wing ends are blocked by tunnel walls.

The division of drag forces into "due to lift" and parasitic is not a clear one. Parasitic drag is also lift-related, and can even vary inversely in relation to lift, as can be seen in Figure 4.7, where at low lift coefficient drag decreases as lift increases due to effects of mismatch between airfoil curvature and the curvature of flow from upwash to downwash.

Figure 4.8 shows a cambered airfoil and a thin flat plate airfoil. Lift in the forward portion of the cambered airfoil, where upwash is redirected rearward, is tilted forward, while in the aft portion, where rearward air movement is redirected downward, lift is tilted rearward. Thus it can be seen that with camber there can be at least partial cancellation of forward and rearward forces related to lift. In the case of a flat plate, pressure forces which produce lift are, except for viscous drag, normal to the plate and thus have rearward but no forward tilt. Thus camber in the case of a thin plate provides a drag advantage. In experiments, Ludwig Prandtl found that by adding a small degree of camber to a thin flat plate operating at around four degrees angle of attack, the lift to drag ratio could be doubled.

Flat plate drag loss begins at the leading edge where flow cannot smoothly round the sharp corner to be abruptly redirected. A resulting vortex at the leading edge introduces turbulence and energy loss. Although vortex loss adds drag, it is useful on small flat wings of insects where it aids flow attachment. If airfoil thickness permits having a rounded leading edge, energy which would otherwise be lost into vortex turbulence can be recovered in thrust of leading edge suction.

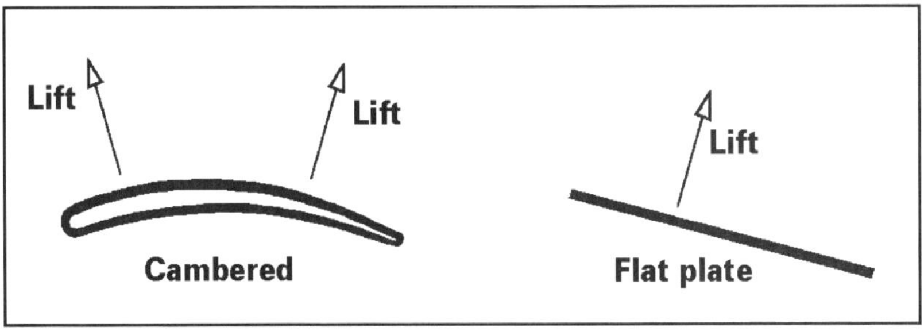

Figure 4.8 **Parasitic Drag Related to Lift**
The cambered airfoil has cancelling forward and rearward lift components, while the thin flat plat has only rearward lift. Thus the flat plate has greater drag in proportion to lift.

Circulation Energy Cycle

Pressure reduction in flow rounding a leading edge, otherwise known as "leading edge suction," was described in an earlier section in connection with centrifuging circulation and stall warning. The pressure reduction is very important to flight, for without it engine power would have to be much greater. In WW1, large bombers sometimes carried protective cables, stretched in "V" shape, across the front from a wingtip to the aircraft nose and on to the other wingtip. The purpose was to deflect barrage balloon anchoring cables intended to destroy low flying airplanes. In pullout from dives the protective cables slackened as the wings pulled forward in thrust exerted against drag of other parts of the aircraft structure.

Although circulation centrifuging around a wing leading edge toward rearward direction produces forward thrust, it does not provide free propulsion. Circulation also centrifuges around the rear of a wing, curving from rearward to forward direction, producing rearward thrust. Because of wing end losses, rearward thrust associated with circulation is normally greater than the forward thrust. The difference appears as drag, known, for historical reasons as "induced drag" because the classical explanation of drag associated with wing end losses invokes induction concepts from electromagnetism. We shall consider drag due to end loss in a later section.

As rearward circulation above a wing curves from upward to downward, it produces an upward gradient and pressure drop at the upper surface. Since circulation produces forward, upward and rearward pressure gradients around a wing, it would be logical to expect circulation to also produce a downward gradient below a wing so as to also reduce pressure there. That does not happen.

To understand the centrifuging actions and resulting pressure gradients, let's consider air movements during wing passage, as illustrated in Figure 4.9. As illustrated, air movements both above and below during wing passage are convex upward, involving upward centrifugal pressure gradients which produce upper surface pressure reduction and lower surface pressure increase. Ahead of the wing circulation is convex in forward direction, producing leading edge pressure drop aiding propulsion, while circulation behind is convex rearward, producing a rearward pressure gradient associated with the rearward component of lift at the downward sloping upper surface.

If there were no losses due to parasitic drag or wing end effects, the forward and rearward thrusts of centrifuging circulation would be equal, and continuous recycling between rearward drag expenditure and forward thrust gain would be without energy loss so that there would be no net drag associated with circulation.

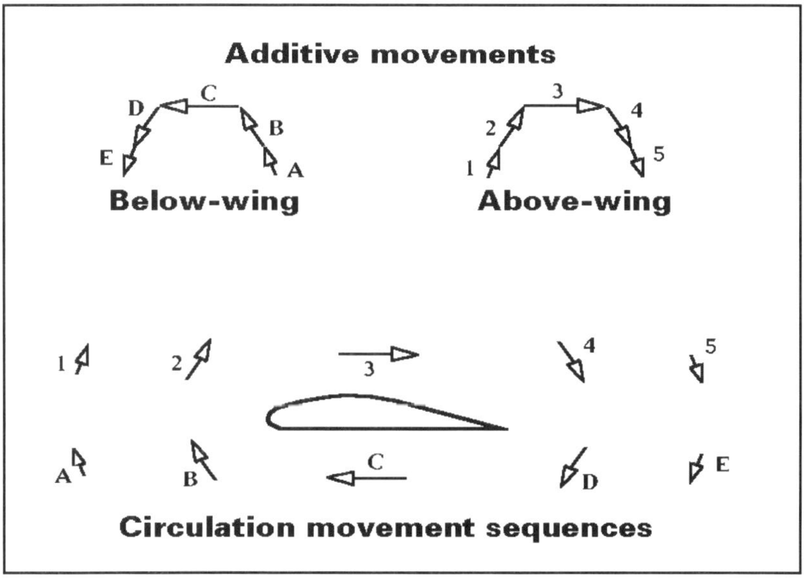

Figure 4.9 **Circulation Movements**
Circulation movements during wing passage are indicated by the numbered and lettered arrows. Numbers 1 through 5 indicate progress of above wing movement. Letters A through E indicate movement progression below. The added sequences of movements indicate that motion during wing passage is convex upward in both cases, meaning that momentum change is downward in wing passage. Note that the summation of air movement below is actually counterclockwise. Ahead of and behind the wing, circulation movements are convex outward from the wing, meaning centrifugal pressure gradients produce opposing forward and rearward thrusts.

Surface pressure forces associated with circulation are depicted in Figure 4.10, where longer arrows indicate greater pressure drop around the leading edge, producing lift and leading edge suction which aids forward propulsion. Net lift is perpendicular to relative flow direction, or in the case of moving wing in stationary air perspective, perpendicular to the flight path relative to the air. In level airplane flight, lift perpendicular to the flight path is upward, while in the descending path of gliding flight, lift is in a somewhat forward direction, as illustrated in Figure 4.11. Downward movement of a wing carries more air

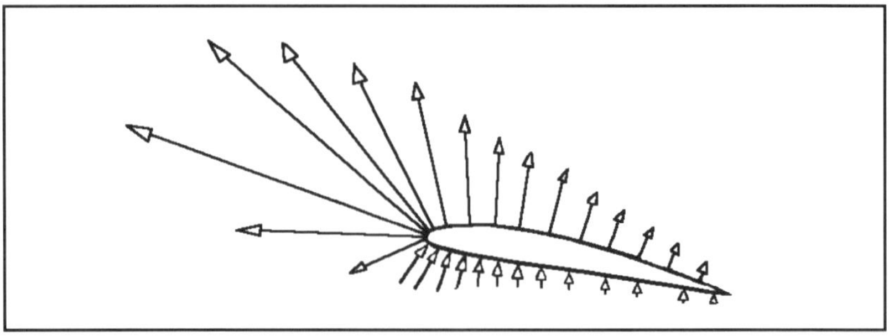

Figure 4.10 **Aerodynamic Force Concentration**
The arrows, normal to the surface, indicate relative surface pressure change as compared to undisturbed atmosphere and the directions of resulting force.

downward, thus increasing recirculation upwash and leading edge thrust. In this process, gravitational work of downward movement in opposition to upward lift is converted into work of forward thrust, making forward thrust of circulation greater than rearward thrust, and propelling the plane forward against drag forces.

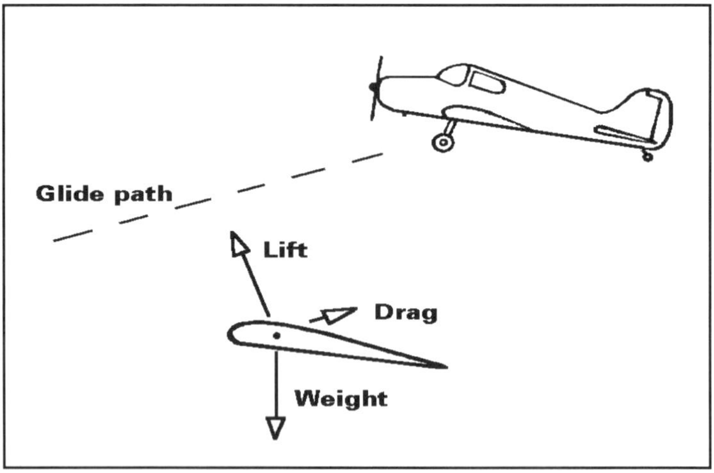

Figure 4.11 **Gliding Descent**
Wing lift is perpendicular to the flight path. In stable gliding, the flight path slopes downward, and wing lift is tilted forward. Added lift and weight forces act in balance against drag.

"Induced" Drag

Pressure differences cause upward air recirculation, away from increased below-wing pressure and toward decreased pressure above. Confined airfoil recirculation develops in patterns of circulation around the airfoil and around the starting vortex left behind. In addition, a wing operating in open air also develops lateral upward recirculation into large *trailing vortices* which are left behind, as depicted in Figure 4.12.

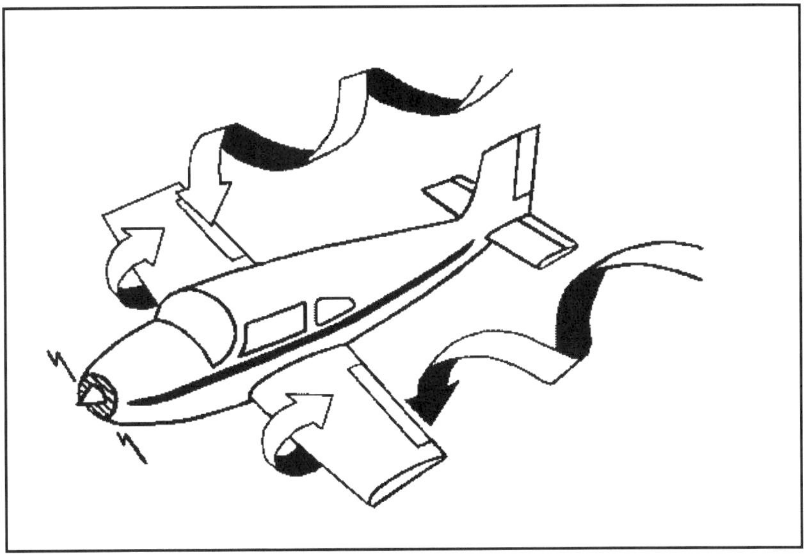

Figure 4.12 **Overall Recirculation in Moving Wing Perspective** Downward acceleration of air mass, which produces pressure difference and lift in accordance with Newton's laws, results in upward recirculation as air mass responds to force of the difference between increased below-wing pressure and decreased above-wing pressure. Upward recirculation rising ahead of the wing is intercepted by wing forward movement, and re-curves rearward around the leading edge to produce lift and forward thrust. Upward recirculation occurring laterally around wing ends combines with downwash to produce twin trailing vortices, represented as helixes above.

Because part of upward recirculation is lost laterally, energy recovered in forward thrust associated with circulation ahead of the wing is less than energy expended in opposition to rearward thrust of centrifuging circulation behind. The difference appears as drag, popularly referred to as "induced drag." Several wing design factors affect lateral loss. A primary factor is wing span because it affects the length of lateral recirculation path and thereby affects air mass opposition to acceleration of lateral recirculation.

In a wind tunnel situation where lateral loss is prevented, lift magnitude coincides with the theoretical equation, lift = $\rho v S \Gamma$. Rearranging: Γ= lift/$\rho v S$, where ρ is air density, v is airspeed, S is span and Γ represents circulation magnitude or velocity content. Circulation around a wing and around the trailing vortices is produced by the same pressure factors. Thus strength of the trailing vortices is proportional to wing circulation, Γ, and at any point in a plane perpendicular to axes of the trailing vortices, vortex velocity varies in proportion to Γ. Vortex kinetic energy volume density at any aft position is thus proportional to $\rho\Gamma^2$. Squaring the above equation for Γ, multiplying by density, ρ, and expressing the result as a proportion gives vortices kinetic energy volume density at any aft position in proportion to the factors of lift, air density, airspeed and wingspan:

(4–2) $$\rho\Gamma^2 \propto \text{lift}^2/\rho v^2 S^2$$

Integrated throughout a plane perpendicular to the path of flight, this expression would yield the same proportion to represent vortices kinetic energy linear density in the direction of the flight path. Multiplying linear energy density by airspeed, v, would yield power loss into trailing vortices, so that:

(4-3) $$\text{"Induced" drag power loss} \propto \text{lift}^2/\rho v S^2$$

And drag force is equal to power lost divided by airspeed, v, so that:

(4-4) "Induced" drag $\propto \text{lift}^2/\rho v^2 S^2$

Power loss into vortices is greater in flight at higher temperatures, and higher altitudes, both of which decrease air density. Another important fact is that engine power decreases as air density decreases. Further compounding the considerations of flight in less dense air is that ground speeds for takeoff and landing are increased, resulting in longer runway requirements, especially in takeoff.

A pilot whose experience is mainly in operating from low level airports in moderate temperatures may be surprised by conditions of mountain flying in the summer time. Too often, planes are lost in the mountains by flying into rising terrain where climb rate or room for a turn around of a poorly performing airplane may not be adequate to avoid ground impact.

For operating in lower density air, span can be quite important. Since circulation lateral loss drag decreases with wingspan squared, greater span can offset other factors to some degree. The single seat U-2 spyplane, which flew at altitudes over 70,000 feet which no fighter plane could approach, had an impressive slender wing of 80 feet span. The plane served well until the Soviets managed to shoot one down with a new rocket which they would have preferred to keep secret.

Since vortices loss drag increases with lift squared, drag can become important in turns, where lift increase causes airspeed loss. Accidents too frequently occur in final turns to landing where low airspeed is made even lower by tight turning in which lift must increase. Students are usually taught to make no turns below 400 feet after takeoff, and only gentle turns near the surface in landing so as to reduce the risk.

Total Drag

The parasitic and lateral circulation loss drags are affected differently by airspeed. While parasitic drag varies directly in proportion to airspeed squared, lateral vortices loss drag, or "induced" drag, varies inversely in relation to airspeed squared. In Figure 4.13, opposite trends of the two types of drag are plotted, along with total drag, the sum of the two.

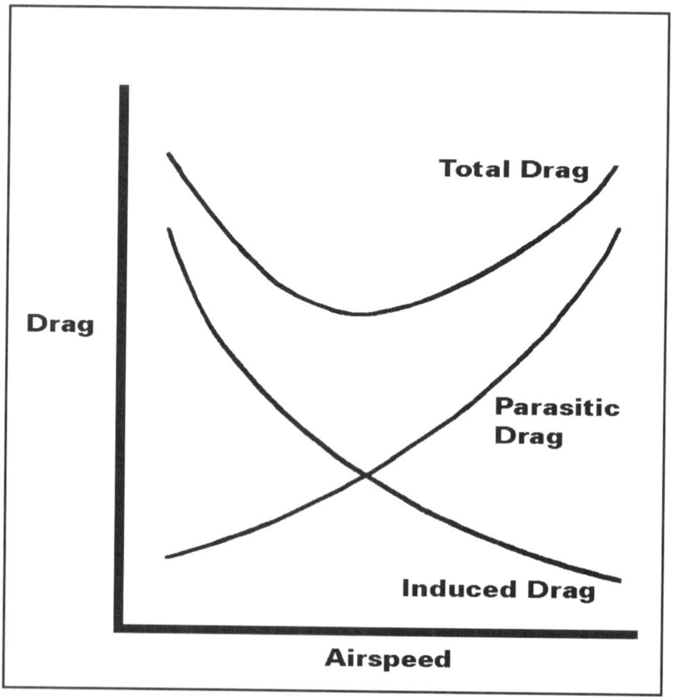

Figure 4.13 **Drag Components in Relation to Airspeed** Total drag, represented by the uppermost plot, consists of parasitic and induced drags combined. Induced drag decreases as airspeed increases, while parasitic drag increases with airspeed.

If other factors would remain unchanged as airspeed varies, best fuel economy could be gained by flying at minimum total drag airspeed. However, since engine and propeller efficiencies tend to be better at higher speeds, best fuel economy would occur at slightly greater than minimum drag speed.

An airplane is not capable of flying level at such low speed that total drag exceeds maximum engine thrust. Flying in this condition results in sinking descent, and is known as flying "behind the power curve." To recover to level flight it is necessary to increase the rate of descent so as to gain airspeed for drag reduction. This condition is of less concern with propeller driven planes than with jets because propeller thrust increases at lower speeds, while jet thrust is relatively constant.

Aspect Ratio

Aspect ratio, important in wing performance, is the ratio of wing span squared to wing area. For a rectangular wing this reduces to the ratio of span to chord. Span, the distance between wing tips, includes fuselage width. Figure 4.14 is a plot of lift coefficients for differing aspect ratios. It can be seen that aspect ratio has considerable effect on the slope of lift versus angle of attack, but has only a small effect on maximum lift coefficient at stall.

Figure 4.15 shows the lift to drag relationship as affected by aspect ratio. Although aspect ratio has little effect on maximum lift coefficient, it has great effect on lift to drag ratio when operating with high lift coefficient. At low lift coefficient parasitic drag of excess camber is the primary cause of drag, and aspect ratio has little effect. From zero lift coefficient, drag decreases as lift begins and curvature of flow begins to match the camber.

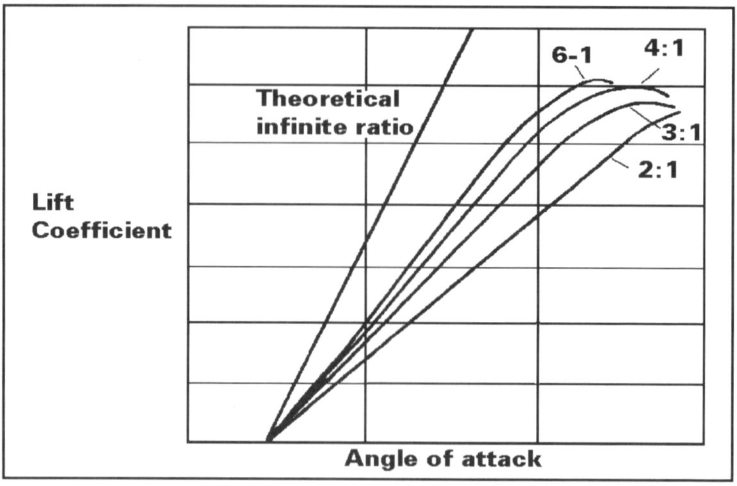

Figure 4.14 **Lift coefficients with Differing Aspect Ratios**
Lift coefficients are better with greater aspect ratios. The theoretical maximum line is from equation 2-10 for lossless lift.

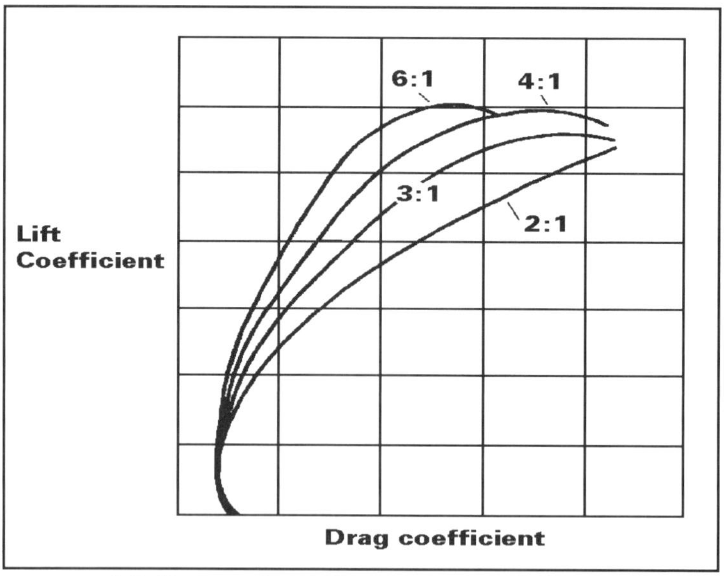

Figure 4.15 **Lift Versus Drag Coefficients of a Cambered Wing**
Because lateral circulation loss increases with span loading, drag coefficient in relation to lift coefficient is greatly affected by aspect ratio when lift coefficient is high, as when operating at low airspeeds or in low density air of high altitude.

Aspect ratio is important to lift and drag. Vortices loss drag is related to span loading, while stall speed is related more to wing area. If wing area is chosen to meet stall speed requirements, and span is chosen for induced drag requirements, the aspect ratio, or ratio of span to chord is determined.

Ideally, span should be maximized for minimizing drag. However, there are practical limits for reasons of space needs in ground handling, and for bending moment stresses of long span, which can require exotic and expensive materials and construction techniques. *Voyager*, the Rutan-built, carbon fibre reinforced composite plane, which flew around the world unrefueled in December 1986, had a wing aspect ratio of 29:1, and span of 111 feet. Tips of the slender wings, flexing because of the distributed fuel load, dragged on the runway during takeoff.

Ground Effect

Early in flight training a student learns that in landing, gliding efficiency of an airplane increases as the surface is neared. An airplane will tend to continue gliding just above the runway without settling onto its landing gear. The apparent greater efficiency of flight near the ground is said to be due to *ground effect.*

Ground effect occurs when operating in close proximity to the surface, for reasons illustrated in Figures 4.16 and 4.17. As vertical distance between wings and surface becomes less during landing, the lateral recirculation path becomes more vertically narrow and restricted so that lateral loss decreases.

The forward recirculation path below the wing likewise becomes narrowed, but as illustrated in Figure 4.17, there is a compensating effect as downwash behind the wing impacting on the surface acts

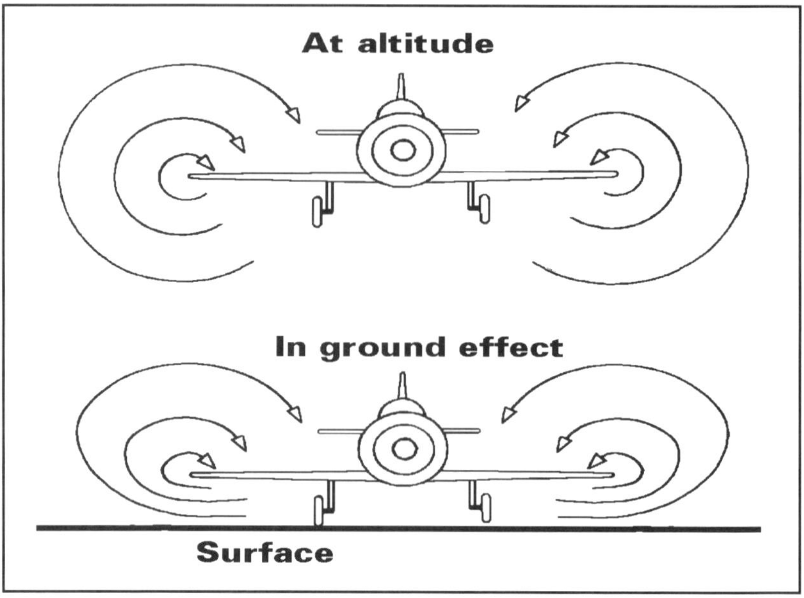

Figure 4.16 **Lateral Loss Reduction in Ground Effect**
When operating near the surface, lateral recirculation loss is impeded in the restrictive, shallow lateral path between wing and surface. The decrease of lateral loss improves aerodynamic performance.

as a sort of *momentum curtain*, tending to push air forward and increase pressure under the wing. Increased lift and reduced drag then tend to make a landing airplane seem to "float" above the runway surface with reduced rates of descent and deceleration.

Ground effect helps in making smooth landings but adds to landing distance. In takeoff, ground effect diminishes as distance from the surface increases. If lift-off is accomplished before airspeed is sufficient for flight out of ground effect, an airplane may sink back onto the runway. A lightly loaded Piper Warrior, for example, operating "behind the power curve" in takeoff with full up elevator, will repeatedly lift off and resettle to the ground so that it becomes necessary to reduce up elevator in order to gain flying speed to avoid running off the end of the runway.

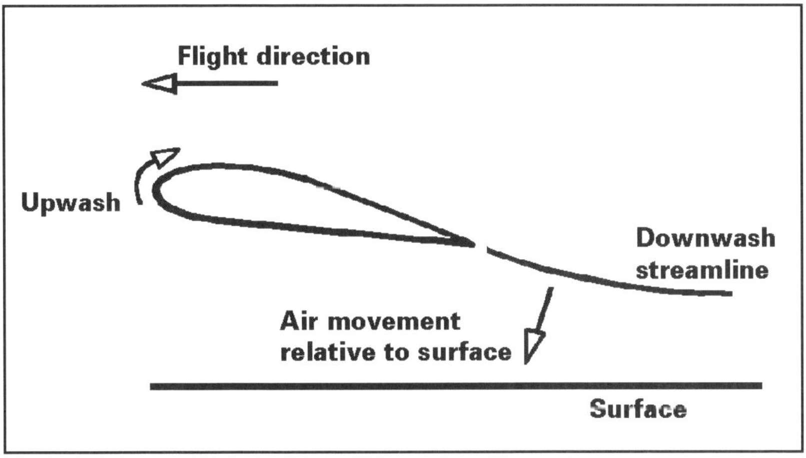

Figure 4.17 **Near-surface Momentum Curtain Effect.**
In ground effect, although the forward recirculation path is vertically narrowed, downwash behind the wing impacts upon the surface and acts as a forward-moving *momentum curtain,* sweeping air before it to increase below-wing pressure and maintain forward circulation.

Propellers

A propeller is basically a set of two or more wings, usually called "blades," which rotate in a plane about a central hub. Because the propeller advances axially through the air as it rotates, it is sometimes referred to as an "airscrew." Compared to a fixed airplane wing it has much different operating requirements. Air velocity in relation to a wing is basically that of airplane airspeed, with some difference associated with circulatory air movements. Air velocity in relation to a propeller is more complex.

The angle and relative airspeed of air encounter with any position along a propeller blade varies with airplane airspeed, propeller rotation rate and radial distance from the hub. To accommodate this variation a propeller blade must be twisted, as can be seen in Figure 4.18, in order to maintain efficient angle of attack along the blade.

Figure 4.18 Blade Twist
In feathered condition, twist of the four blades on this turbine driven propeller is evident.

Like wings, pressure difference exists between opposite surfaces of propellers. Thus, like wings, they have end loss and leave trailing vortices behind the tips. A tractor propeller (front mounted) is also affected by upwash ahead of the airplane wing, which makes relative airspeed of a downgoing blade greater than that of an upgoing blade. A pusher propeller is similarly affected by wing downwash, which makes relative airspeed of an upgoing blade greater than that of a downgoing blade. The resulting thrust impulses produce a degree of vibration.

A fixed pitch propeller designed for cruise will not allow engine speed for producing rated power at lower airspeeds. If designed for good performance in takeoff and climb, a fixed pitch propeller is operated with reduced throttle in cruise to prevent engine overspeeding. Some propellers can be switched between takeoff and cruise blade angles.

Planes with engines of greater than about 160 horsepower are usually equipped with constant speed variable pitch propellers. These serve a function comparablc to that of an automatic transmission in a car. A propeller with variable pitch will allow rated engine speed for maximum power in takeoff and climb, and can then be adjusted to hold efficient speed for cruise or good economy as desired. Constant speed is usually controlled automatically by oil pressure delivered from a speed sensing control system. After a pilot adjusts the propeller speed as desired, the propeller governing pitch control system maintains that speed, even as throttle is varied.

Figure 4.19 Constant Speed Propeller on Cessna 182
A variable pitch propeller such as this can be identified by round shanks which pass through bearings and seals into the pitch change mechanism.

As in the case of wings, induced drag of propeller blades is inversely related to span loading. Thus propeller blades need to be long, but the ends must not approach sonic velocity where parasitic drag and noise level would greatly increase.

A small, fast airplane may have a small diameter propeller driven directly by a high speed engine crankshaft, as in the case of a homebuilt Sonerai having a Volkswagen engine, but a large airplane requires a large propeller, and consequently must have a lower rotation rate. Thus large engines generally have gearing to reduce propeller shaft speed.

Modern propellers are usually made of aluminum, but reinforced plastic is growing in blade use. Some older factory-built airplanes and homebuilt airplanes have wooden propellers. In general, plastic and wooden propellers are lighter in weight, which is advantageous in terms of aircraft weight, but for piston engines a disadvantage is that lighter propellers have less moment of inertia to carry them in rotation between power impulses of the engine, so that they require higher engine idling speeds. There may also be an efficiency disadvantage if power impulse spacing in relation to moment of inertia allows substantial rotational speed variation between impulses. Such efficiency loss would be greatest with a light propeller on a single cylinder four stroke engine, as on a few ultralights. Some light airplane operators have reported improved performance in switching from a wooden propeller to a metal one. However, such finding could possibly be more related to better blade design than to blade material change.

Windmilling

When the engine of a typical propeller driven lightplane quits producing power in flight, the propeller may continue to turn as a "windmill," delivering power to turn the dead engine at the expense of added drag to the airplane. If the plane has an adjustable speed propeller, the drag can be minimized by setting rotation speed to the lowest level. If the plane has a featherable propeller which can be pitched for zero rotational speed, that is even better. A single engine plane with engine out will glide farther if propeller rotation can be stopped. Multi-engine planes are usually equipped with featherable propellers, and will fly more efficiently if the propeller on a dead engine is "feathered," to zero torque blade position.

Windmills

Windmill (more properly wind turbine) blades are designed for power production, and hence are cambered oppositely as compared to propellers, with the down wind side more convex. Thrust for torque and power production is similar to a wing in gliding descent. Lift for blade propulsion comes from wind crossing the path of blade movement, as in Figure 4.20.

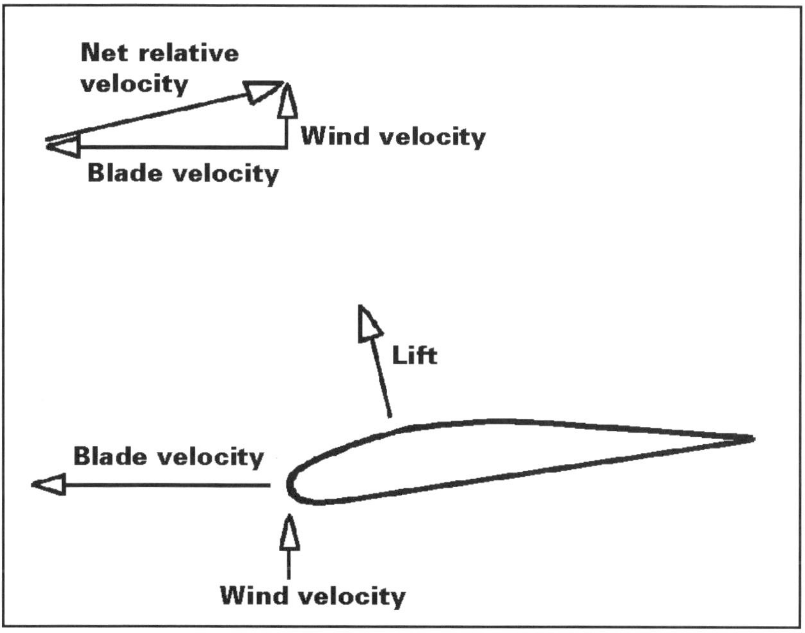

Figure 4.20 **Windmill Blade Thrust**
Wind movement across the path of blade travel produces lift perpendicular to velocity of air in relation to the blade, Thus lift is tilted in forward direction so as to produce rotational thrust.

Old farm windmills, used for pumping water in times before rural electrification, as shown in the photograph of Figure 4.21, rotate about horizontal axes. Designed for slow rotation, they are multi-bladed with radial blades close together for sharing lift. This blade arrangement, known as a *cascade*, is stall resistant, providing good torque at low speed so that little speed reduction gearing is required for an eccentric journal which drives a slow vertically oscillating pump rod.

Figure 4.21 **Farm Windmill**

Other wind turbines rotate about vertical axes, as in Figure 4.22. Those rotating about vertical axes have vertical blades, or at least substantially vertical blades, parallel to the rotation axes. Blades of a vertical axis windmill change direction of movement relative to the wind as they rotate. At different positions of each cycle of rotation they move respectively upwind, downwind and crosswind, and in combinations thereof.

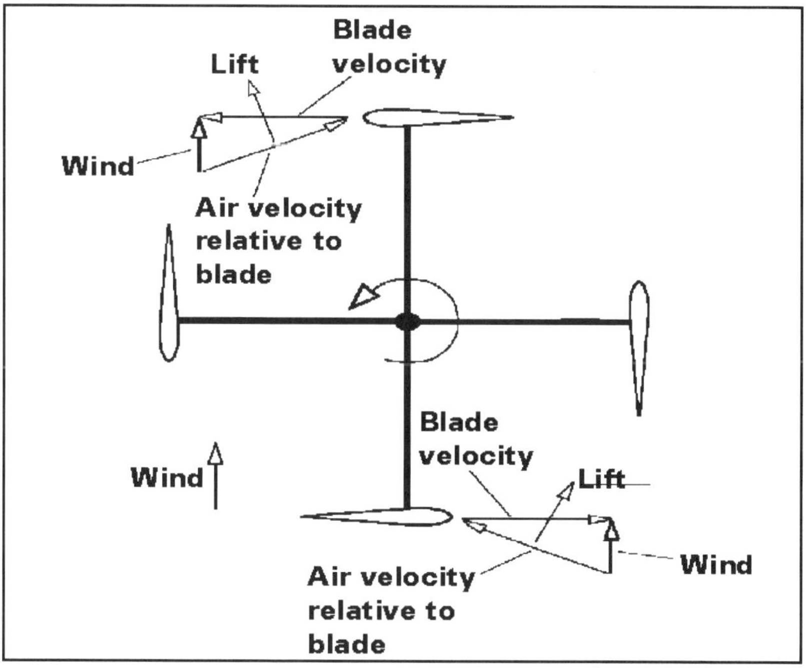

Figure 4.22 **Four Blade Vertical Axis Windmill (top view)**
Each of these four blades operating around a vertical axis would produce maximum lift and contribution to torque when in crosswind movement. Lift, perpendicular to flow in relation to the blade, drives the blades forward.

Because vertical axis windmill blades operate in alternating orientation with respect to the wind as they rotate, they are made without camber, and have chord lines tangent to the path of rotation. Rotational speed of such a windmill must be sufficient to maintain angle of attack below that of stall. This means that for an angle of attack of 10 degrees, for example, when in crosswind movement, blade forward speed would be about six times that of the wind. To allow for avoiding excess angle of attack in partially downwind movement, rotation speed could be greater, perhaps ten times that of the wind.

Vertical axis *Darrieus* windmills driving electrical generators are seen in some areas of Western United States. A Darrieus windmill has a distinct appearance. Outwardly bowed blades attached at top and bottom of a rotating vertical support shaft make it resemble a giant eggbeater. The advantages of this design are a non-directional characteristic, so that orientation with wind direction is unnecessary, simplicity of construction, and location of power absorbing machinery below the rotor for easy access. A disadvantage of the symmetric blades is lack of starting torque because of being stalled. In generator applications Darrieus windmills can be provided with needed initial operating speed by energizing the coupled generator as a motor. Darrieus windmills usually have three blades. This provides thrust pulsation peaks of six times per revolution as each blade reaches peak thrust twice in each cycle. Pulsation rate in relation to torque affects smoothness of power transfer.

The *Savonius* windmill is another type which usually has a vertical axis, but may operate in any orientation. As viewed from above, a Savonius rotor is shaped in approximate form of an "S," as indicated in Figure 4.23.

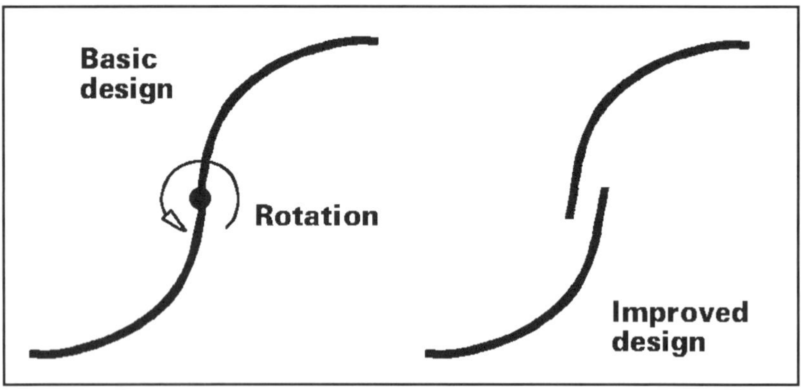

Figure 4.23 **Savonius Rotor Vertical View**
The channel between halves of the improved rotor design allows air passage from the convex side of one rotor half to the concave side of the other.

Torque is produced as average pressure during rotation is reduced in air flow centrifuging over the convex surfaces and increased at the concave surfaces. Simple in construction and self-starting, low cost Savonius windmills may be found on primitive or remote farms having no electrical power, where they may be used to pump water or aerate animal waste ponds.

Figure 4.24 shows a Savonius rotor, built of tube and fabric, and mounted on an automotive rear axle which provides an initial speed step-up of about five-to-one. The reversed auto transmission has a second ratio of about six-to-one for driving a generator to maintain charge on a battery-powered sump pump.

Figure 4.24 **Savonius Windmill**
Built to maintain charge of a battery powering a sump pump, the rotor is mounted on an automotive axle which increases rotational speed belt-coupled to a reversed auto transmission driving a generator.

In past decades, before proliferation of attractive and inexpensive electrical signs, small Savonius windmills were often seen as advertising signs outside business establishments. Savonius windmill signs usually had different paintings or printing on opposite sides. As they rotated in the wind, the messages might alternate, or images on opposite sides could appear to blend at higher speed.

Savonius windmill rotors have been made from cut sections of 55 gallon drums for use in remote areas of the world where wind power is more economical and available as compared to other power possibilities. Because rotor torque varies during the cycle of rotation, dual rotors are sometimes stacked with rotation angle offset of 90 degrees between them so as to reduce torque variation.

Introduction to Aerodynamics

V. Wing Variable Elements

Airfoils vary as needed for different applications, and wings which incorporate them have further variations, including aspect ratio and span, as previously considered, but also sweep, washout, dihedral and taper. In addition, wings have variable elements for purposes of stability and control. These include flaps, ailerons, spoilers, slats and slots. Other aerodynamic components, not referred to as wings but having wing characteristics, include vertical and horizontal stabilizers, with attached rudders and elevators. All of these elements are aerodynamically involved in stable and controllable flight.

An airplane rotates around vertical, horizontal and longitudinal axes which pass through the airplane center of mass, as illustrated in Figure 5.1.

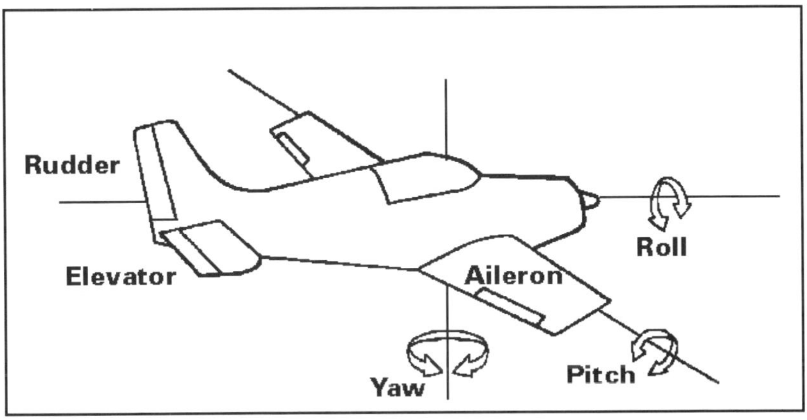

Figure 5.1 **Axes of Control**
An airplane is guided by rotation about three axes which pass through the center of airplane mass. Rotation about the longitudinal (fore-aft) axis is known as roll, rotation about the vertical axis is known as yaw, and rotation about the lateral axis is known as pitch.

Motion around the longitudinal axis is known as *roll*, about the lateral axis as *pitch* and about the vertical axis as *yaw*. As pictured in Figure 5.2, a vertical stabilizer for yaw stabilization has a hinged rudder attached for yaw control. The horizontal stabilizer for pitch stabilization has a hinged elevator attached for pitch control.

Figure 5.2 **Cessna 182 Tail, or Empennage**

For roll control the Wright brothers used a system of wing warping operated by lateral hip movements of a prone pilot. The movements were transmitted to the wings by cables. A simpler mechanism of roll control was invented by Glenn H. Curtiss (1878-1930). Curtiss, in the biplane "June Bug," made the first public flight of an airplane in July 1908. The Wrights were greatly distressed to learn that Curtiss had achieved roll control by means of small, triangular shaped hinged tips on his wings, and claimed the method infringed on their wing warping patents. In later airplanes, roll control evolved into the now widely used hinged *ailerons* on wing trailing edges. A common undesirable characteristic of both aileron and wing warping methods of roll control is *adverse yaw*.

Adverse yaw is the tendency for an airplane to yaw in direction opposite to that desired as the plane is rolling into a bank for turning. For example, to turn to the left the plane is banked to the left, but adverse yaw tends to point the nose to the right. This is inconvenient, but not a great problem, for it is easily overcome by application of rudder.

As illustrated in Figure 5.3, when rolling into a turn, the wing moving downward develops a forward component of lift because downward movement makes relative flow slope upward. While upward slope of relative flow tilts lift direction of the

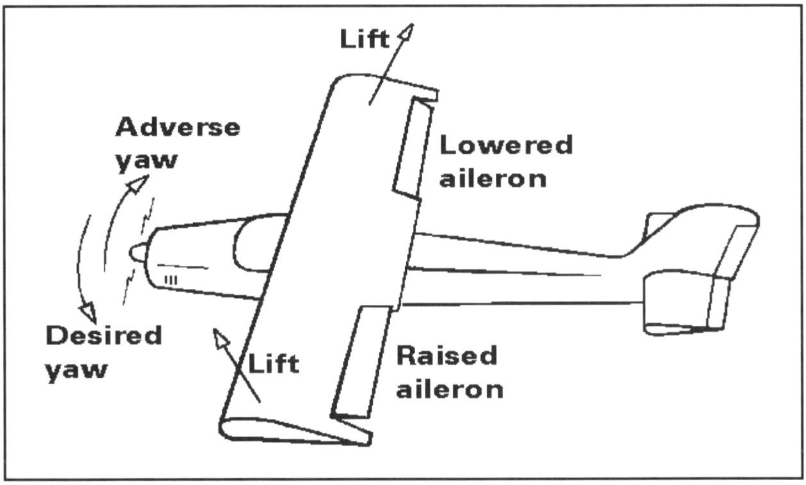

Figure 5.3 **Ailerons and Adverse Yaw**
As an airplane is rolled into a bank for turning, the downgoing wing develops a forward lift component due to the relative direction of wing movement in relation to the air. Upward movement of the opposite wing produces a rearward component.

downgoing wing forward, lift of the opposite upgoing wing is tilted rearward. The forward and rearward tilts of lifts of opposite wings produces tendency to turn in direction opposite to the desired turn.

Spoilers

Spoilers, an alternative to ailerons for roll control, operate by interfering with circulation. Spoilers are drag devices, and because they are located on upper wing surfaces, they oppose circulation when deployed. Unlike ailerons, which are located in wing trailing edges, spoilers are normally located farther forward, as illustrated in Figure 5.4.

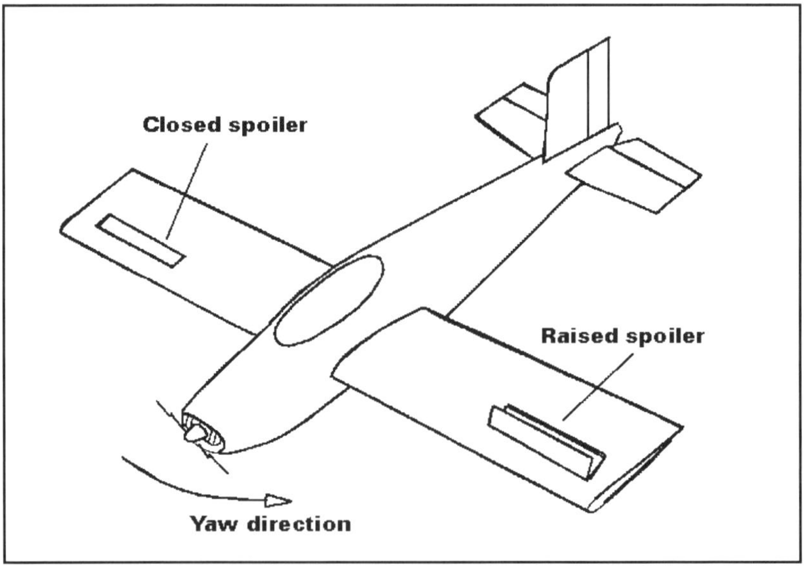

Figure 5.4 **Spoilers**
Spoilers oppose circulation by producing above-wing drag, causing reduction of lift when raised. They are operated independently as roll control devices which also produce desired yaw. Operated together they produce drag and reduce lift. Sailplanes use spoilers to control speed in descent, and to spoil lift for good braking at touchdown.

Spoilers have a definite yaw advantage in turn coordination. For example, raising a left spoiler will produce left bank due to reduction of left wing lift, and simultaneous left yaw due to spoiler drag.

If located in the aft portion of wings, spoilers must be raised higher than would be required if they were forward-mounted, in order to extend above the boundary layer and separated flow to be effective, especially at higher angle of attack when separated flow is deeper. Thus a small degree of spoiler movement for minor roll correction might not be effective, making it seem to a pilot there is loss of authority in the system.

When mounted in the forward regions, where there is little flow separation, spoilers are more effective, but should seal reasonably well when retracted to prevent air leakage which would add to boundary layer retardation. In sealing they must also be resistant to freezing in closed position with consequent loss of function.

While above-wing obstructions oppose circulation, obstructions below aid circulation. In moving wing and stationary air perspective, below wing items, such as landing gear or floats, drag air forward, and hence add to forward movement of below-wing circulation and to lift. A retractable geared airplane can fly slower with gear extended because of the drag-produced increase of circulation. Because circulation energy is partially recovered in leading edge thrust, below-wing drag which adds to circulation is not totally lost to airplane efficiency. Were it not for this fact, fixed gear airplanes would be undoubtedly be less popular.

Although pilots may be taught to not fly when upper wing surfaces are covered with frost or snow, impressively high drag, flow-disturbing items are sometimes carried below wings, as pictured in Figure 5.5 on the Cessna T37 "Tweetiebird" jet in the AF Museum.

Figure 5.5 **Cessna T37 Armament Load (WPAFB Museum)**

Farmers have been known to lash bales of hay to lightplane wing struts in order to feed cattle stranded in heavy snow. Bicycles have been carried on wing support struts of planes as small as a Piper Cub. This is illegal and potentially dangerous. If a load is lost from one side, so that drag is unbalanced, the resulting yaw force may be greater than rudder application can overcome.

Flaps

Flaps are somewhat like ailerons, but operate in unison to increase camber of both wings in the inboard regions. In addition, flaps increase downwash angle and lift in the wing midsection without increasing aircraft pitch attitude. By keeping the airframe pitch low, over-the-nose visibility is maintained for the pilot, and by keeping the angle of attack lower in outboard regions where ailerons are located, flow separation there is limited so that ailerons remain effective at lower speeds.

There are many types of flaps, some of which are illustrated in Figure 5.6. All flap designs increase wing midsection downwash angle, recirculation and lift when deployed.

The plain flap and split flap are simplest, with the plain flap being more popular on light airplanes. The split flap is essentially a flat section which is held against the underside of the wing during cruise, and is pivoted downward to extend operating speed capability downward, mainly for landing. Split flaps of a Cessna 195, located well ahead of the trailing edge, drag below-wing air forward. The forward movement adds to circulation

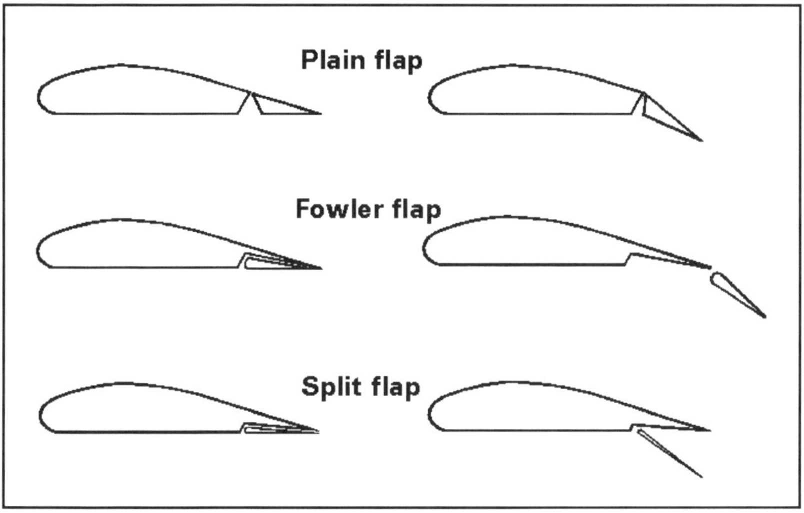

Figure 5.6 **Flaps**
Three basic types of flaps are shown. On the left the flaps are shown in cruise mode while on the right they are in deployed mode as for takeoff or landing. Of these types, the *Fowler flap* provides best performance but is also the most costly because it requires a track mechanism for support and travel. On large airplanes, flaps are usually more complex, often with multiple extensions in tandem.

Fowler flaps, pictured in Figure 5.7, extend trailing edges rearward, as well as downward, to increase local chord length for a gain in wing area.

Figure 5.7 **Cessna Wing Flaps**
Fowler flaps at the rear of wings on this Cessna 182 are positioned downward for operation with higher lift coefficients, as when airspeeds are low during takeoff or landing. They also extend rearward to increase wing area.

Flaps increase drag as well as lift. Although flaps lower takeoff speed, the drag increase adversely affects takeoff acceleration, and light planes often take off without flaps. Drag is desirable however, when gravitational work energy of descent must be dissipated in order to limit speed in descent to landing.

Flaperons

Flaps are normally located only on inboard portions of wings, while ailerons are located farther out. The functions of ailerons and flaps can be combined into flaperons, which may extend fully across wing trailing edges. Flaperon control involves a mechanical mixing arrangement in which flap operation is accomplished by lowering the flaperons in unison, while aileron function is accomplished by flaperon movement differentially.

Slats and Slots

Lower operating speeds can be achieved with leading edge slots, as in Figure 5.8. A slot supplies higher velocity flow from greater pressure below the wing to reinforce rearward movement of the upper surface boundary layer. A slot may be permanently open, or may be closed for drag reduction in cruise flight. Opening and closing is accomplished by having a moveable leading edge, called a slat. Slat operation can be passively automatic. At high lift coefficient the stagnation line, or line of division between upper and lower flows, and location of greatest pressure, moves downward, resulting in lower leading edge pressure force for moving the slat forward to open position. In cruise, the stagnation line moves upward to force the slat to closed position.

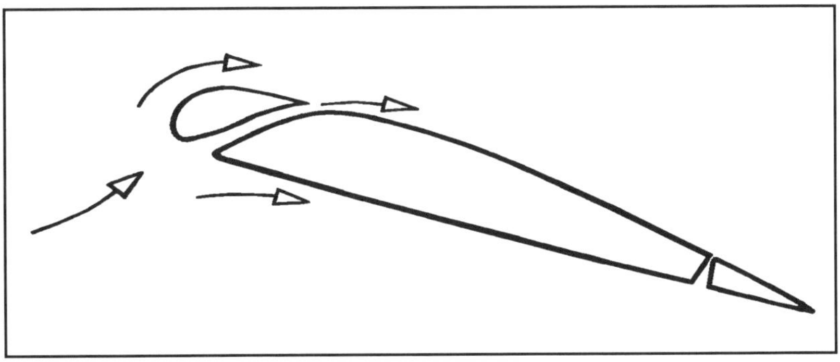

Figure 5.8 **Slot**
Slots are included across entire leading edges of some planes intended primarily for slow and short field operation. On others, slots are included only ahead of ailerons for aileron effectiveness when operating at high lift coefficient.

As pictured in Figure 5.9, the passive slats of an F-86 fighter plane droop open under force of gravity when the plane is not moving. They serve to delay stall of outer regions of the tapered wing and preserve aileron function at low airspeeds.

Figure 5.9 **F86D Open Slat (WPAFB Museum)**
F-86D slats are forced forward by downward movement of the stagnation line as lift coefficient increases, leaving a *slot* between slat and wing through which rising air is directed rearward to reinforce upper surface flow. Here the slat is held open by gravity.

Figure 5.10 shows a plane with full span passive slats in open position. This plane is used for missionary work which requires operation from very short fields. It can fly slowly in such small circles that items can be lowered and retrieved on a very long rope between plane and persons on the ground.

Helio-Courier with full span slats in open position

The need for boundary layer flow reinforcement is at the reduced pressure surface, which in the case of a downward-lifting horizontal stabilizer, as pictured in Figure 5.11 on a Cessna Cardinal, would be the lower surface. Thus a horizontal stabilizer may have slot inlet openings in the upper leading edge with the outlets directed rearward below for boundary layer reinforcement.

Figure 5.11 **Stabilator**
This Cessna Cardinal stabilator has slot openings at the upper leading edge. The outlets below are directed rearward for lower surface boundary layer flow reinforcement.

Instead of having a hinged elevator attached to a fixed horizontal stabilizer, the Cardinal has combined the stabilizer and elevator functions into a pivotable *stabilator*. Because the stabilator is of rectangular form, it would tend to stall first in the midsection. This tendency is made greater by having the midsection operate in disturbed flow near the fuselage. Thus the slots are located in the midsection where the need is greatest.

Blown Wings

Slots which passively serve to energize upper wing surface flow are dependent on pressure differences generated by the wing. Alternatively, greater effectiveness can be gained if air is forced through rearward facing slots on a wing upper surface, as indicated in Figure 5.12. This arrangement, known as "blown wing," is not often used because of system complexity.

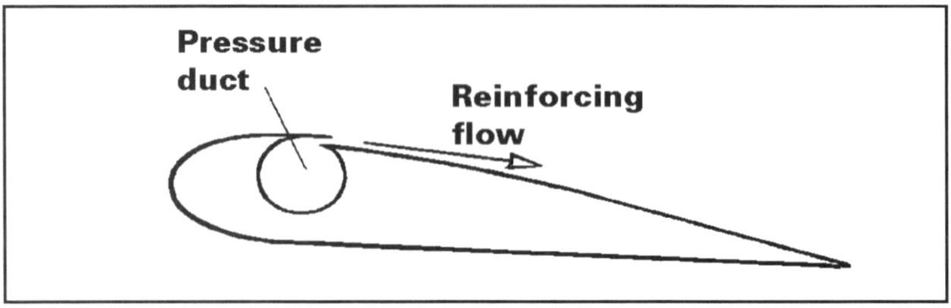

Figure 5.12 **Blown Wing**
High pressure air from a central blower is discharged rearward over the upper wing surface to reinforce boundary layer flow.

Introduction to Aerodynamics

VI. Stability and Control

Stability and Controllability

An airplane must be controllable and stable around its three axes of roll, pitch and yaw. As described earlicr in connection with round cylinder aerodynamics, stability is enforced by negative feedback, otherwise known as degenerative feedback, in which an action produces opposition to that action. Oppositely, positive feedback, otherwise known as regenerative feedback in which action tends to cause further action, promotes instability.

Ground Stability

Even before flight, regenerative and degenerative feedback effects influence airplane behavior in ground operation. Steering in taxi to the runway is mainly through tire traction, although rudder can be involved when air flow is sufficient to make it effective. Landing gear configuration is an important factor. In the first half of the 20th century, most airplanes had *conventional* landing gear, with the airplane CG aft of the main wheels, as depicted in Figure 6.1. Steering was controlled by a small wheel at the tail, usually connected to the rudder. This gear configuration became known as "conventional" when then unconventional *tricycle* geared planes, with a nose wheel and CG forward of the main wheels, were introduced. Tricycle gear steering is controlled by the nose wheel, which is usually coupled to the rudder system.

The conventional gear airplane, now often referred to as a "taildragger," was simpler than a tricycle geared airplane to build, and the tail wheel had less weight and less drag than a nose wheel mounted on a sturdy strut extension. Another taildragger advantage is a nose high attitude on the ground, providing good ground clearance for the propeller during taxi to minimize chances of propeller strike and damage. Disadvantages of conventional gear are that ground steering is unstable due to positive feedback, and over-the-nose visibility may be impaired by low pilot position in relation to the forward fuselage or engine cowling.

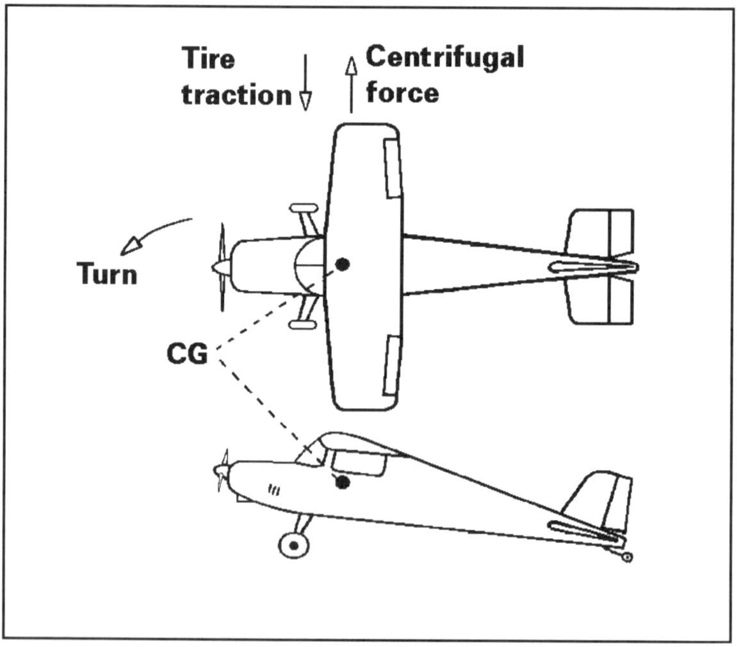

Figure 6.1 **Conventional Gear Turning Forces**
In a turn, outward centrifugal force of the center of mass and countering inward force of tire traction constitute a moment couple for increasing the rate of turn.

In a turn, the couple between centrifugal force and the force of tire traction tends to turn the airplane into increasingly tighter turn and possible *groundloop*, in which turning tips the plane sufficiently for a wing to strike the ground. Wingtip drag may then initiate an excessive turn rate in the opposite direction so that both wings are damaged. A ground loop seldom results in occupant injury, but airplane repair can be costly.

Improper control procedure in crosswind landing can also initiate a groundloop. If the aircraft is made to track down the middle of the runway by angling it into the crosswind, then it will touch down with a sideward component of travel which immediately produces a turning couple and possible groundloop. This aiming of the plane off of runway heading in crosswind landing is known as "crabbing," in reference to the sideways walking of a crab.

A good landing can be made by the "crab and kick" method, wherein crabbing is maintained until a second or two before touchdown, when quick rudder application is used to orient the plane into runway alignment. This requires good timing judgement, especially if the crosswind is strong.

A better method is to maintain runway alignment with the plane banked into the wind so that sideslip maintains runway tracking. This requires opposite rudder application to prevent turning in the banked direction. If the crosswind is strong enough to threaten to lift the upwind wing after touchdown, the upwind bank can be maintained with one wheel in surface contact until aileron effectiveness is lost as speed diminishes.

Steering instability of a taildragger makes it fun to learn. A pilot learns to anticipate adverse turning tendencies before they become readily apparent, and to apply corrective action. After a few hours of experience, the necessary pilot actions become automatic without conscious thought.

The "tricycle gear" airplane of Figure 6.2 has its CG ahead of the main gear. In landing touchdown with a positive pitch attitude, only the main wheels make initial contact. The couple between mass inertia and wheel friction then tends to automatically align the plane in the runway direction. This negative, or degenerative, feedback tendency aids stability in steering, while a level airframe attitude provides good forward visibility. In recognition of these factors, insurance companies apply lower rates to nosewheel airplanes.

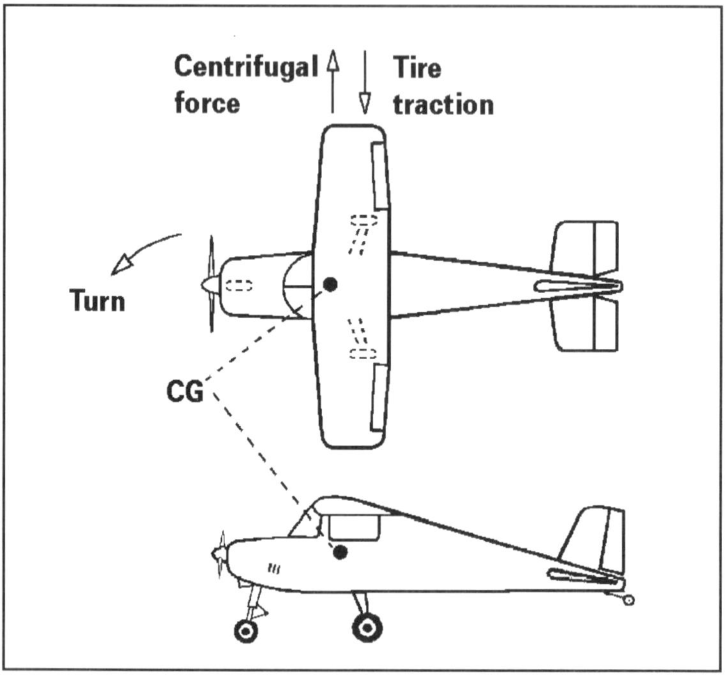

Figure 6.2 **Negative Feedback in Ground Steering**
In a turn, outward centrifugal force of the center of mass, and countering inward force of tire traction constitute a moment couple for decreasing the rate of turn.

Although the nosewheel plane is more stable in ground handling, it is still possible to groundloop. In crosswind landings, banking for sideslip into the wind for runway alignment at touchdown is still good practice.

Flight Stability

Two classes of stability in flight are *static stability*, related to slow effects, and *dynamic stability*, related to rapid transient effects such as gusts. Both types of stability are achieved with negative feedback in the lift process. With no feedback, an airplane has *neutral stability*, requiring constant attention of the pilot in adjusting for minor disturbances in holding flight attitude. With negative feedback an airplane can have *positive stability*, tending to resist changes in flight attitude so that less reaction is required of the pilot. Dynamic stability for gusts involves automatically aiming the plane into relative passing flow, much like tail feathers hold arrows aimed into relative flow. Like arrows, most airplanes gain this effect by having the aerodynamic surfaces of wings and tail centered aft of the center of mass, or the center of gravity, CG, as it is more often called in aerodynamics.

Although aerodynamic surfaces on conventional airplanes are centered aft of the CG, the center of lift must coincide with the CG for stable flight. This is accomplished by having forces of an upward lifting wing shortly aft of the CG, and a downward lifting tail farther aft, as represented in Figure 6.3. The trimmed moments, of upward wing lift and lesser downward tail lift farther back, when summed about the CG produce zero net pitching moment in stable level flight.

In event of an upward gust, the net force change is upward on both wing and tail, causing a negative pitching moment about the CG, tending to re-pitch the plane attitude downward toward alignment with relative passing flow, thereby reducing stress on airplane and passengers. In the case of a downward gust, the tendency is for positive pitching, again reducing the gust effect.

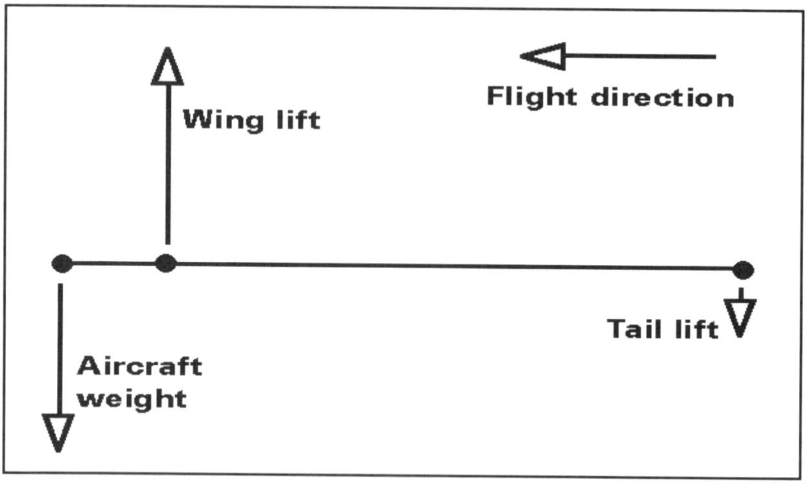

Figure 6.3 **Stable Configuration**
The negative moment about the CG due to upward wing lift is balanced against positive moment due to downward smaller tail lift farther aft. Net lift for supporting the airplane weight is equal to wing lift minus downward force of the tail.

The downward-lifting tail configuration which produces transient dynamic stability also produces long term static stability. If a plane trimmed for level flight begins a slight descent, gravitational work energy causes airspeed increase. Speed increase causes increase of both upward wing lift and downward tail lift, resulting in an increase couple tending to pitch the plane upward to regain level flight. Oppositely, climb reduces airspeed and causes pitch downward, again toward level flight.

Inherent pitch stability is not without cost. Downward tail force adds to wing load, and thus affects wing drag and fuel economy. The effect can be minimized by loading with the CG farther aft, so that less down force is required at the tail. Limits of CG location are established by airplane manufacturers to assure adequate stability. Commercial cargo planes are loaded for CG location near the rear limit for best fuel economy achievable with stability.

A competition sailplane operates at the edge of aft CG stability in order to minimize drag, but this requires constant attention of the pilot in maintaining flight attitude.

Pitch trim of an airplane is usually accomplished by having a pilot-adjustable "trim tab" located on an elevator trailing edge, as in Figure 6.4. Lift force of the trim tab, leveraged by aft location, moves the elevator in direction opposite to tab movement. For example, a downturned tab causes the elevator to move up, pitching the airplane attitude upward. Rudders and ailerons may also have trim systems. The airplane in the picture has a bungee system for pilot controlled rudder trim, and has no aileron trim system accessible to the pilot. If the airplane is reported to have an undesired roll tendency it is corrected by a mechanic with slight change of wing angle. For that purpose this plane has adjustable eccentric bushings on the wing rear spar attach bolts.

Figure 6.4 **Elevator Trim Tab**
The small section at the rear of the elevator is a "trim tab," adjustable by the pilot. Leveraged lift at the tab moves the elevator in opposite direction.

Positive stability decreases pilot work load. A condition of neutral stability would require constant pilot attention and reaction to attitude changes. Negative stability, in which pitch change would cause further pitch change could make a plane uncontrollable by direct human action. For example, a tailless airplane based on a simple cambered wing would have positive feedback instability, in which angle of attack change would produce center of lift movement promoting further angle of attack change. This is overcome by the downward-lifting aft tail arrangement. However, a plane with an upward-lifting tail can be successful if equipped with fast reacting servos operated automatically by computer control. Such systems are sometimes referred to as "fly by wire." Fighter planes with computer control can sometimes be seen in movies of carrier landings in which very rapid stabilizer movements occur as touchdown is approached.

A tailless, or "flying wing," airplane wing with camber develops stabilizing rearward down force behind the center of upward lift by having the wing upturned in the trailing edge region, as in Figure 6.5. This upturning of the trailing edge, known as *reflex*, provides stability in the manner of an aft-tail arrangement.

Figure 6.5 **Reflex** With forward CG, the lift-load couple for pitch down can be countered with upturn of the trailing edge.

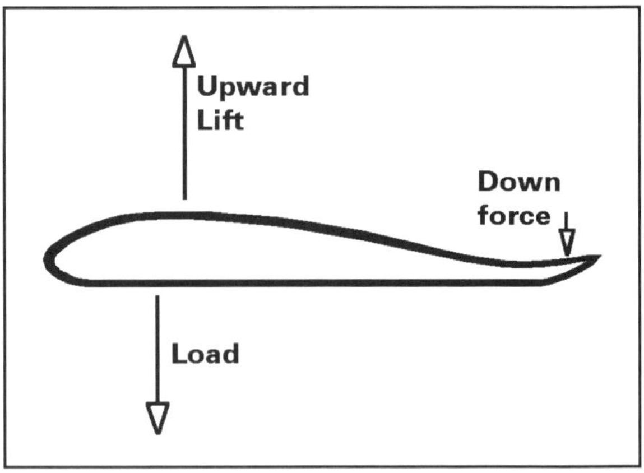

In adding reflex to a wing trailing edge, downward force at the trailing edge must be compensated by increased lift of the forward portion of the wing. The drag penalty of this is greater than with lesser down force required of a farther aft tail. In the case of a rearward-swept wing, the drag penalty can be less if the reflex is limited to outer wing regions farther aft.

Tail Designs

A *spin*, sometimes called a *tailspin*, is a condition in which an airplane descends vertically in a steep down pitch attitude while rotating around an axis nearly parallel to the airplane longitudinal axis. A spin begins when one wing becomes stalled while the other is not. The plane then drops off toward the stalled wing and descends in nose-down rotation. The direction of rotation adds to the high angle of attack on the stalled wing , while reducing the angle of attack on the other wing. To recover, the stalled wing must be unstalled. This usually is done by application of rudder against the direction of rotation, but may also require nose down elevator application to steepen the dive for increased airspeed.

Longitudinal mass distribution may produce centrifugal forces tending to level the airplane attitude even though the downward spin continues. The level attitude spin, known as a *flat spin,* is more difficult or perhaps even impossible for recovery. In testing of a plane design having potential for flat spins, a parachute may be packed in the tail. If the plane enters a nonrecoverable flat spin, the parachute can be deployed to oppose the spin and pitch the plane downward for recovery. After achieving recoverable attitude the parachute is released.

Because rudder is the primary control for spin recovery, some airplanes have the horizontal stabilizer situated high on the vertical stabilizer, as in Figure 6.6. The intent is to alleviate horizontal stabilizer interference with flow past the vertical stabilizer and rudder while in a spin. This configuration has not become popular on lightplanes. Being above the propeller backwash makes the elevator less effective in takeoff. However, on multi-engine jets the horizontal stabilizer may be mounted high to clear engine discharge hot gases.

Figure 6.6 **High Mounted Horizontal Stabilizer**

A “V” tail design, as in Figure 6.7, found on older Beech Bonanzas, has “ruddervators” combining functions of elevator and rudder through a mechanical system in which elevator operation moves the ruddervators in unison, while rudder operation moves them differentially.

Figure 6.7 **Bonanza “V”” Tail**

In producing down force for normal pitch control, the two stabilizers also produce useless opposing lateral forces. Thus, while it might be imagined that doing the job of a vertical stabilizer and two horizontal stabilizer elements with only two "V" configured elements would be more efficient, the results aren't so simple.

The useless lateral thrusts unavoidably have undesirable associated vortices loss drag. Also the laterally opposed elements compete for shared circulation. In straight and level flight, forward circulation paths between the two elements are shared in the space between them. The regenerative nature of lift causes the element with greatest lift to attract a greater portion of the shared circulation to further increase its lift. This can cause tail yaw toward the side of greater lift until the yaw angle produces straightening action, causing shift of circulation to the other element which again can overcorrect to produce opposite yaw. In smooth air the system may have sufficient natural damping for smooth flight, but in turbulence may develop yaw oscillation sometimes described as "tail-wagging."

Loading

To achieve stabilized and controllable flight, an airplane must be loaded for appropriate balance of lift and weight forces. Airplane operating handbooks include information to enable a pilot to determine proper loading and balance. A plane may be loaded well below its rated load carrying capacity, but still not acceptably balanced for safe flight. Permissible CG location varies with total weight load, and is specified in relation to a reference position within the aircraft. For single engine lightplanes the reference position is usually the *firewall*, the fuselage closure behind the engine. The moment arm to be balanced with aerodynamic forces of wing and tail is calculated as the total of products of all weights times distances aft of the reference position. For weight of engine oil, which is usually included in weight and balance, the position relative to the firewall is negative.

Figure 6.8 represents a typical loading graph for a two-place lightplane. The range of possible weight total can vary from 1100 pounds for an empty plane, without fuel or oil, to 1600 pounds maximum weight limit. The moment arm referenced from the firewall might include two passengers of 320 pounds total, centered at 38 inches aft, 135 pounds of fuel, centered at 42 inches aft, 11 pounds of oil centered at minus nine inches forward of the firewall, and 1120 pounds airplane empty weight centered at 32.5 inches aft.

Adding:	Weight-pounds	Location-inches	Moment arm-inch pounds
Passengers	320	38	12,160
Fuel	135	42	5,670
Oil	11	-9	-99
Plane E.W.	1120	32.5	36,400
Totals	1586		54,131

To simplify calculation of moments, the operating manual includes graphs of moments as functions of weight, or perhaps in the case of fuel and oil, a function of volume. The total load of 1586 pounds, located at the total moment position of 54,131 inch pounds on the loading graph, as indicated by the asterisk, indicates loading within acceptable limits.

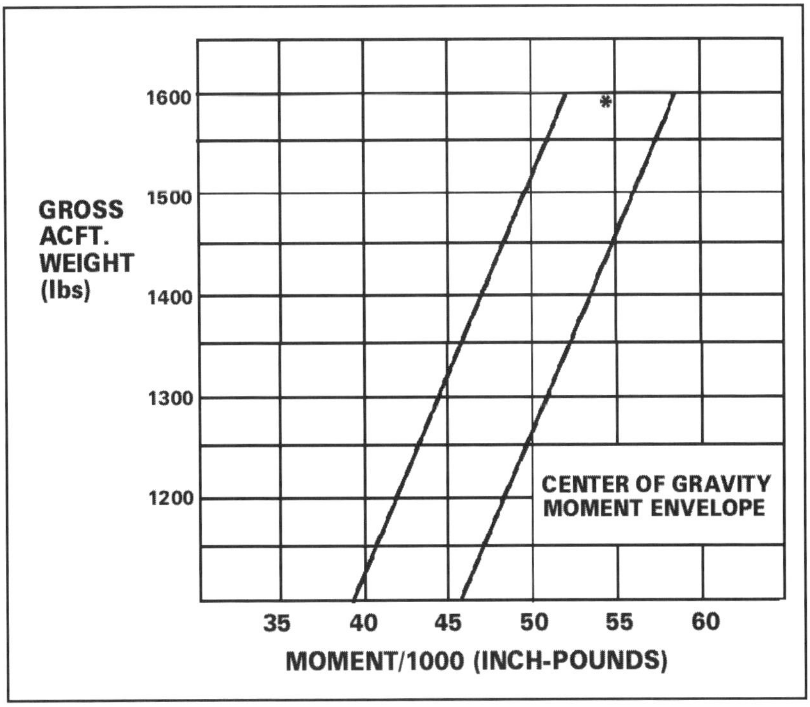

Figure 6.8 **Lightplane Loading Graph**
For safe flight the load and its moment must fall within boundaries of the parallelogram to assure stability and controllability.

Beech T-34 military trainer based on the Beech Bonanza.

VII. Planform and Configuration

In addition to airfoil variations incorporated into wing design, the shape of wings as viewed from above, called "planform," is quite important to operation. Also, placement of wing and stabilizer assemblies is important to performance. Popular single engine Cessna airplanes have "high wings" located above the passenger cabins, while also popular single engine Piper airplanes have wings located below passenger cabins. Other planes have both high and low wings, and a few have "mid-wings" located at cabin level.

Engine placement is another variation. Most propeller driven planes have engines in front, but some have engines at the rear. As shown in Figure 7.1, some airplanes have both front and rear engines. *Voyager*, the airplane which flew around the world unrefueled, had both front and rear engines, but shut down the front engine when fuel load had decreased sufficiently for efficient performance on the rear engine alone.

Figure 7.1 **Cessna 337 Skymaster**

Canard Airplanes

A canard airplane, pictured in Figure 7.2, derives upward lift from both the horizontal stabilizer, located near the front, and from a larger wing located farther aft. The CG is located between the two sources of lift, as indicated in Figure 7.3, with the greater portion of load carried by the main wing.

Figure 7.2 **A *Variviggen*, Canard Airplane**

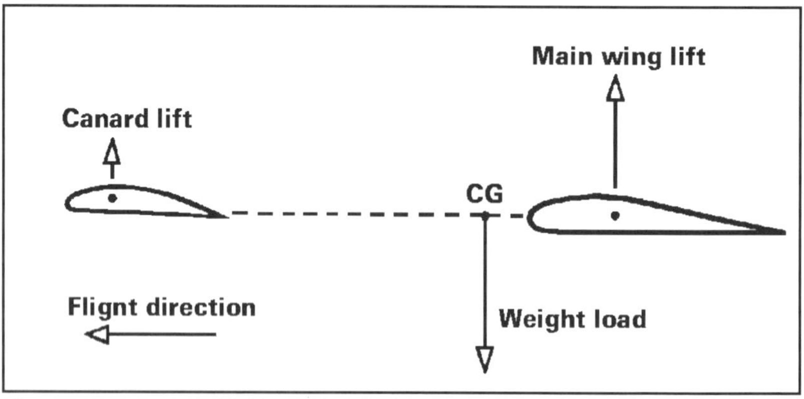

Figure 7.3 **Canard Lift and CG Centers**
The center of lift of a canard airplane lies between the centers of lift of canard wing and main wing, closer to the main wing, which carries the greater load. Stability is derived from differing slopes of lift vs angle of attack curves of the two wings.

The Wright brothers' first successful powered airplane was of canard configuration. The designation, "canard," is from the French word for "duck," apparently because the outstretched forward horizontal stabilizer support made the plane resemble a duck in flight with outstretched neck. Why the Wrights chose a canard configuration is not clear. Many successful gliders built by earlier experimenters had aft tails. While a plane having an aft stabilizer is easily made stable, a canard configuration requires more careful choice of airfoil characteristics.

For stability the canard horizontal stabilizer operates in the curved portion of its lift slope with shallow lift versus angle of attack, as indicated in Figure 7.4. Because of slope difference, the main wing lift is more responsive than the canard to angle of attack change resulting from an upward or downward

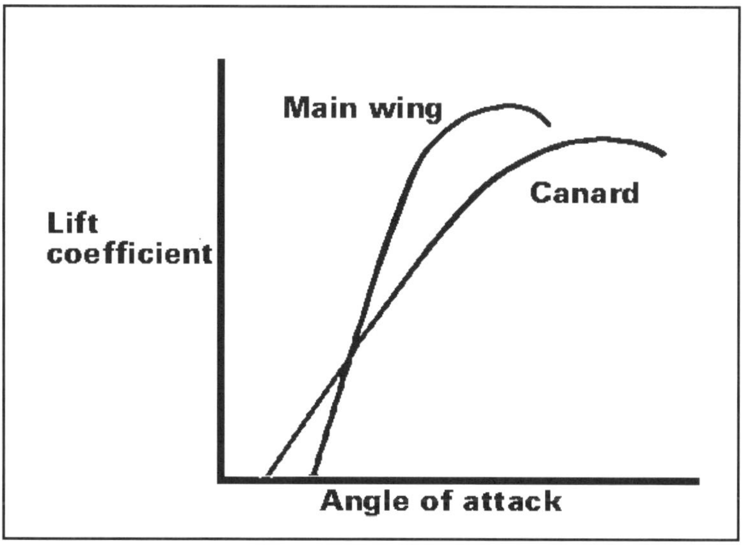

Figure 7.4 **Different Slopes vs Angle of Attack**
Lift versus angle of attack curves for wings of different camber and aspect ratio have different slopes. Dynamic stability derives from shallower slope of the canard wing curve. Static stability derives from Reynolds number sensitivity in curvature of the canard slope.

gust. In an upward gust angles of attack of both wings are increased, but the canard increase of lift is proportionally less than lift increase of the main wing. The result is a negative pitching tendency which moderates the gust effect. In a downward gust, lift decrease of the canard is again proportionately less than that of the main wing, tending toward pitch up and moderation of the downward gust effect, thus promoting dynamic stability with negative feedback effects.

An airplane also needs static stability. This is achieved through canard sensitivity to Reynolds number due to operating at high lift coefficient in a more curved region of lift vs angle of attack plot. If the plane develops a small downward pitch, so that speed increases, both main wing and canard will be operating with some degree of increase of Reynolds number, but the canard, operating in a less linear region with greater area loading and sensitivity to Reynolds number change, will produce a greater proportionate lift increase, tending to re-pitch the aircraft upward. Conversely, speed decrease will tend to re-pitch the airplane downward.

The relatively heavily loaded canard stabilizer can be sensitive to effects of rain. Its aft pressure gradient tends to promote not only accumulation of boundary layer flow, but accumulation of rain water as well. This has created problems for some canard airplanes. *Voyager*, the first airplane to fly around the world unrefueled was nearly lost in testing when it flew into rain and developed uncontrollable pitch downward. Fortunately it came out of the rain before it was lost.

Turbulators, or small vertical vanes, added to the canard upper surface solved the problem with Voyager. These created turbulence for mixing surface water into passing flow. This may not be an appropriate modification for canards in general, for stability depends on curvature of lift slope, which turbulators can change.

Unlike a more conventional plane having a downward-lifting horizontal stabilizer which increases wing load, a canard stabilizer shares the upward load. Thus a canard airplane could be expected to be more efficient. There are, however, other factors to consider. The canard horizontal stabilizer is loaded heavily and usually has low aspect ratio, which would cause it to have lower lift-to-drag ratio than the main wing. Also the main wing operates in the wake of the canard stabilizer, and loses some degree of efficiency as a result.

Another consideration is that a properly designed conventional aft stabilizer produces forward thrust. The stabilizer intercepts some of the wing downwash and redirects it in a more rearward direction, as in Figure 7.5. In this redirection, downward lift is produced, and rearward air momentum is increased. Forward thrust of the stabilizer occurs in reaction to rearward increase of air mass momentum, in accordance with Newtonian principles.

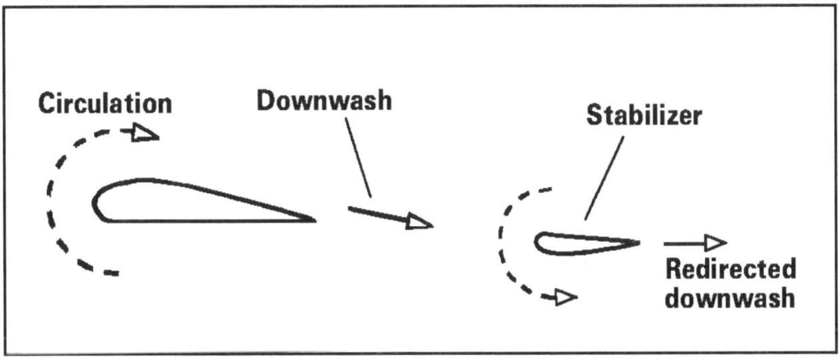

Figure 7.5 **Stabilizer Redirection of Downwash**
In producing downward lift the horizontal stabilizer intercepts downward momentum of wing downwash and redirects it to more rearward direction.

Stabilizer forward thrust can alternatively be described as lift normal to passing flow. Slope of wing downwash approaching the downward-lifting stabilizer causes lift to be in a somewhat forward direction so that forward thrust is produced. For the thrust to be optimized, the stabilizer should be designed as a proper airfoil with a rounded leading edge. Stabilizers on some older light airplanes are thin and flat, so that they have flat plate drag, with lost energy dissipated in production of turbulence.

Circulation Convergence

Circulation convergence is an important factor in wing operation. An example of convergence occurs in flow through a propeller. During engine run-up in rain, flow of mist shed from a propeller may be seen to converge aft, and water drawn up from the surface into a propeller continues inward movement behind the propeller, leaving visible curving flow of mist. Propeller flow convergence is illustrated in Figure 7.6

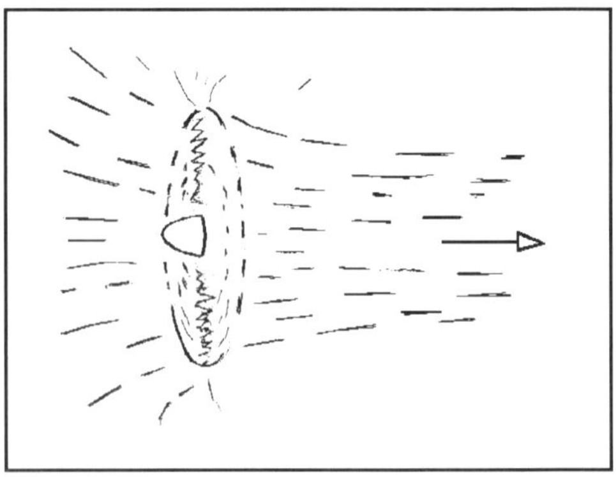

Figure 7.6 **Convergence** Air drawn into a propeller maintains inward momentum which causes convergence of flow aft of the propeller. In flow convergence, rearward air velocity is greater behind the propeller than in air passing through it.

Air flow accelerating into reduced pressure above a wing converges toward mid-wing, as illustrated in Figure 7.7. Because of inward momentum, convergence in downwash continues aft of mid-wing. Since air movements are circulatory, the region of downwash convergence is also the region of re-divergence of below-wing circulation into forward circulation around the wing and laterally into trailing vortices.

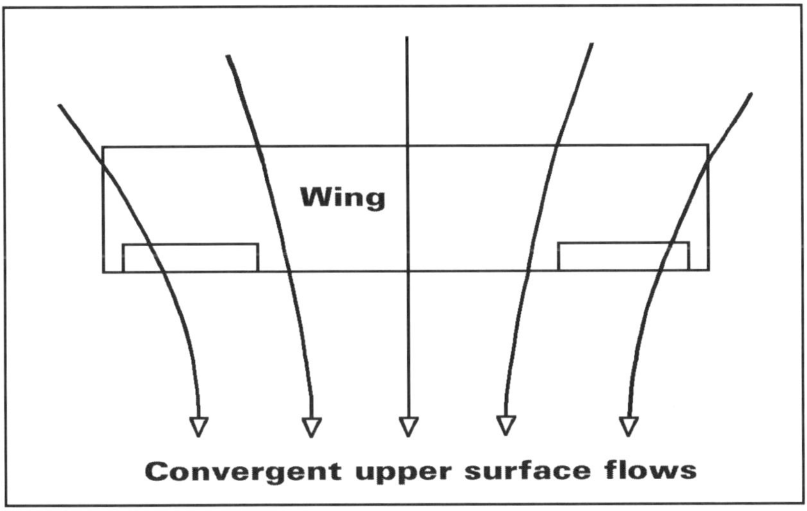

Figure 7.7 **Upper Surface Flow Convergence**
In perspective of passing flow over a stationary wing, above-wing flows, reacting to pressure reduction, converge aft of the wing to concentrate downwash momentum behind the wing midsection. From this convergence, below-wing flows re-diverge laterally and forward into circulation around the wing and into trailing vortices along the lines of passage of the wing ends. Convergence angularity is related to lift coefficient.

A plain rectangular wing produces greatest concentration of lift in its midsection because lateral losses are greater in the end regions, and because forward recirculation into upwash is more concentrated in the midsection, which is closest to the center of circulation convergence and re-divergence.

Because a rectangular wing has greatest lift concentration in its midsection, it tends to stall there first as lift coefficient is increased. This is advantageous for purposes of safety, for it allows maintaining effective roll control when wings are partially stalled, by virtue of ailerons being located in unstalled outer regions.

Wing taper is advantageous in reducing recirculation energy loss at the tips. It is also advantageous structurally, for it concentrates lift in the larger and presumably stronger mid-wing region and reduces cantilevered load. However, if the wing has sharp taper, upwash lateral divergence from the center region, as illustrated in Figure 7.8, where more lift and circulation are concentrated, can result in greater outboard region lift concentration, in terms of lift per unit area, resulting in earlier stall there. In stalling, a sharply tapered wing, losing outer region lift and aileron control, may abruptly enter a spin followed by difficult recovery. Sharp taper is therefore to be avoided, but the stall problem potential can be reduced with inclusion of preventive features.

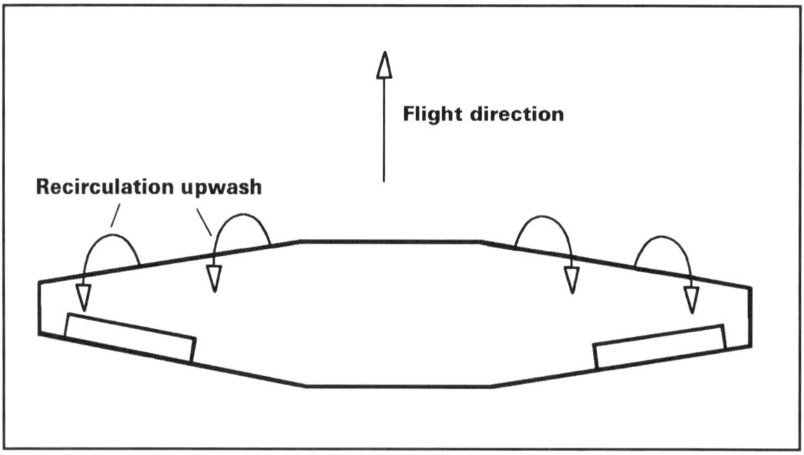

Figure 7.8 **Tapered Wing Loading**
Recirculation from the longer chord mid-span region diverges laterally to increase the concentration of upwash interception and loading in the shorter chord outer span regions. The outer span regions, being more heavily lift-loaded, are more likely to stall first, resulting in loss of aileron effectiveness.

Washout

The tendency for early outboard region stall can be reduced with *washout*, otherwise known as *twist*, as illustrated in Figure 7.9. Wings can be constructed with varying angle of attack along the span, such that outer regions lift is not so concentrated as to produce early stall. The inner regions with greater angle of attack then have more tendency to stall first, leaving ailerons effective for roll control. Washout, however, can producc drag at higher Reynolds numbers when lift coefficients are low, and can even produce negative lift in wing outer regions. Thus there is a need for variable washout, which is accomplished with flaps. Extended flaps increase angle of attack of flapped inboard sections of a wing when needed for operating with high lift coefficient, in effect producing washout when needed.

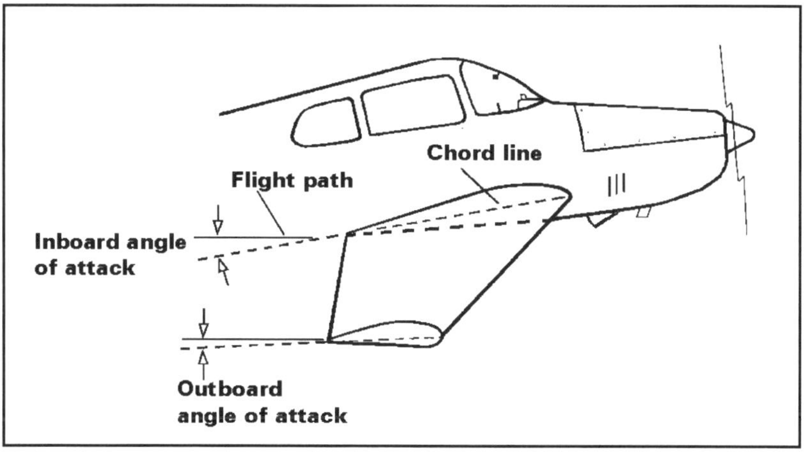

Figure 7.9 **Washout**
To reduce the tendency of tapered wings to stall first in the outboard regions, they may be designed with *twist* or *washout* which makes the outer region angle of attack less than that farther inboard.

Sweep

Most jet airplanes have wings angled rearward in relation to the roots in what is known as rearward sweep. This usually is for reduction of drag at speeds approaching that of sound. However, sweep also influences aerodynamic performance at low speeds. For example, it is known that rearward sweep contributes to earlier stall of outer wing regions, and conversely, forward sweep can inhibit outer region stall.

Differing effects of fore and aft sweep can be explained in relation to how recirculation upwash concentration differs along the wing span in relation to angle of sweep.

As illustrated in Figure 7.10, rearward sweep moves the wing midsection relatively farther from recirculation sourcing. Thus upwash concentration and loading are lessened in the midsection as compared to the outer sections, and with reduced mid-wing loading, stall there is inhibited so that outer span regions may stall earlier with loss of roll control, especially with taper. The tendency can be corrected with washout and perhaps other features such as slots or other boundary layer energization methods.

Forward sweep could be used to offset, at least partially, the effect of taper. A deterrent of concern with forward sweep is the effect of wing flexing on angle of attack. As forward-swept wings flex upward under loading, the twisting effectively increases angle of attack in the outer regions. With lift suddenly increased by an upward gust, resultant twist may cause more angle of attack increase, making the gust-induced lift transient worse. Structural stresses and ride discomfort may be exacerbated if the wing is not sufficiently rigid and resistant to distortion. With rearward sweep the effect of wing flexing is to twist the wing end to lesser angle of attack, thus reducing transient lift stress.

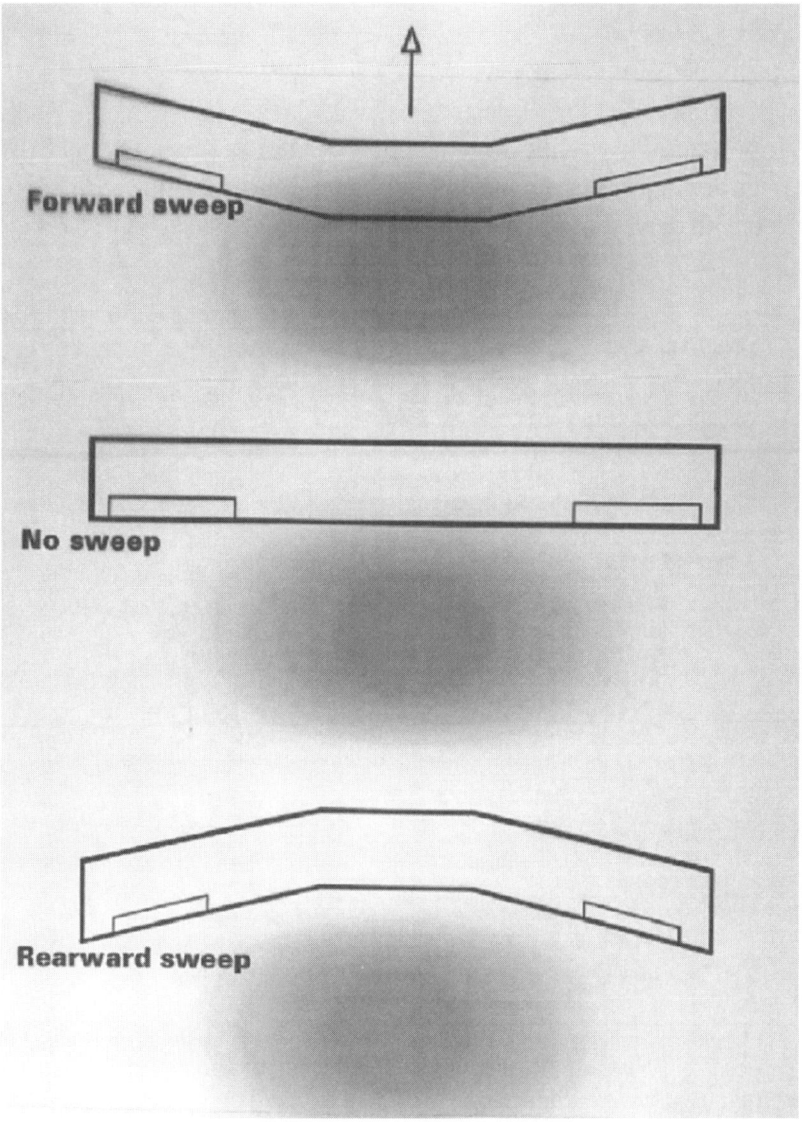

Figure 7.10 **Circulation Convergence vs Sweep**
As compared to a rectangular wing in the middle illustration, forward sweep moves circulation convergence and re-divergence relatively closer to the wing midsection for earlier stall there. Oppositely, rearward sweep decreases the tendency for mid-region stall, so that outer regions may stall first.

Spanwise Flow

Another concern with sweep is that of lateral, or spanwise, flow of boundary layer air along the upper wing surface. Pressure forces involved in spanwise flow are illustrated in Figure 7.11.

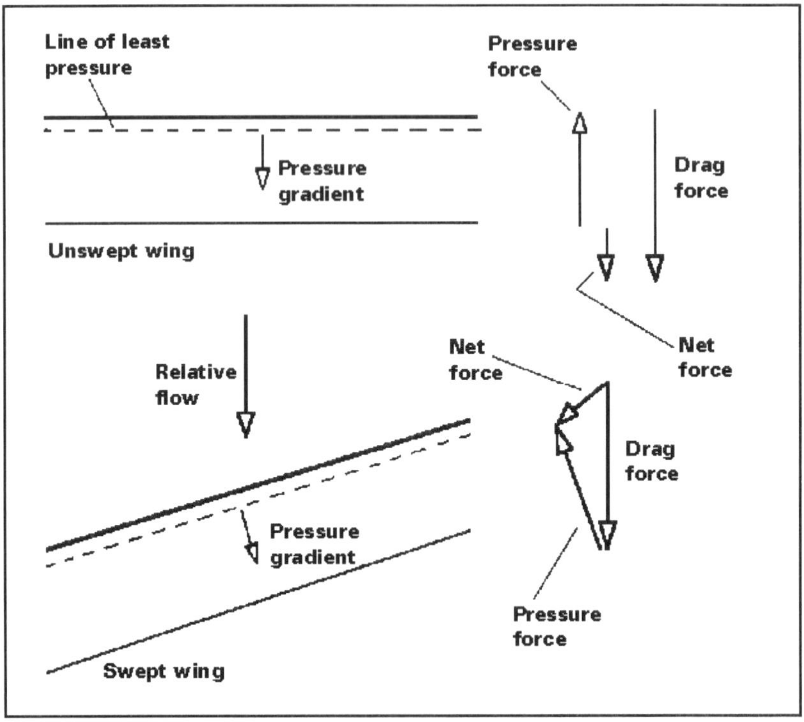

Figure 7.11 **Forces Affecting Boundary Layer Flow**
In the straight wing case, rearward force exerted on the boundary layer by drag of passing flow is in direct opposition to forward force exerted on the boundary layer by the rearward pressure gradient. In the lower case with sweep, the combination of pressure gradient and viscous drag from passing flow add to produce net force in the downspan direction, as indicated in the force diagram at the right. Resulting spanwise movement of the boundary layer builds in depth, and contributes to greater flow separation problems as compared to a straight wing.

The position of least pressure on an airfoil upper surface is along a line essentially parallel to, and shortly behind, the leading edge. From this line rearward exists a gradient of pressure rise, which exerts opposing force on boundary layer rearward movement. On an unswept airfoil, forward force of this pressure gradient is directly opposed by rearward drag of passing flow. Thus for a straight wing, pressure gradient and drag forces acting on the boundary layer are substantially in direct opposition, except in end areas where lateral flows are of significance. In the case of a swept wing, the rearward pressure gradient is not parallel to passing flow, and hence the pressure gradient effect on boundary layer is not directly opposed by rearward drag. The combined forces of pressure gradient and rearward drag produce a net lateral component which promotes downspan flow with consequently greater buildup of the boundary layer. This problem may be dealt with by adding vortex generators to break up and disperse spanwise flows, or short sheet metal *fences*, oriented along the desired flow paths to block downspan flow, as in Figure 7.12.

Figure 7.12 **MIG Fences (Kalamazoo Air Zoo Display)**
Fences on this MIG 17 wing block downspan (outward) flow of the boundary layer. Outer regions, where precessing flow is better oriented in relation to aft adverse pressure gradient, need no fences.

Precession

Recirculation diverging outward from downwash convergence produces forward, upward and rearward circulation directly ahead of mid-wing, but as indicated in Figure 7.13, recirculation along the wing end paths is angled outward below and inward above. This angular movement precesses toward becoming lateral as the wing passes, but does not become effectively lateral until some distance aft. In the case of rearward sweep, the wing end pressure gradient is better oriented in relation to precessing recirculation, so that downspan flow may not be a problem there, as indicated by absence of outer region fences on the MIG of Figure 7.12.

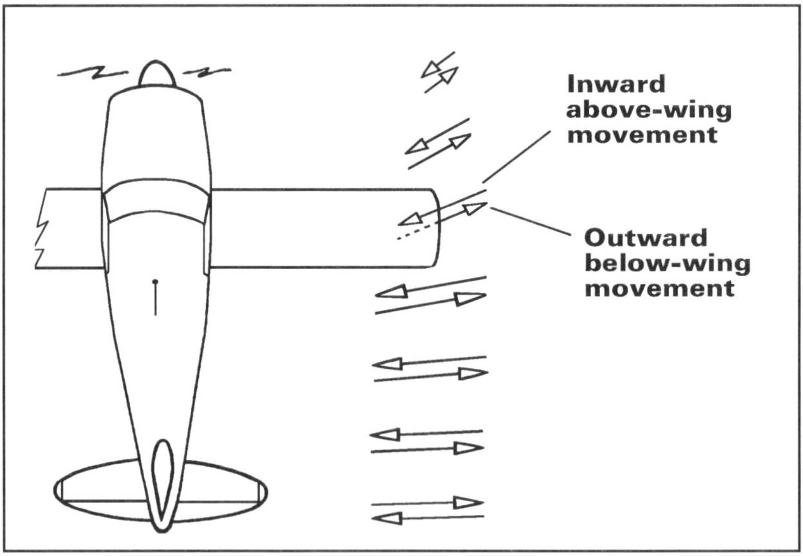

Figure 7.13 **Recirculation Precession**
Ahead of the wing, air is accelerated outward from the region of increased pressure below in circulatory movement, and inward toward the above-wing region of decreased pressure. Ahead of mid-wing position, recirculation is purely longitudinal, but ahead of wing outer regions, recirculation develops angularly with both lateral and longitudinal components. As the wing passes, pressure forces cause recirculation paths to *precess* toward becoming more nearly lateral.

With forward sweep, the spanwise flow problem is compounded in the wing outer regions where precessing recirculation more closely corresponds with downspan direction, as indicated in Figure 7.14. Angles of flow in the wing end regions vary with lift coefficient, and hence vary with airspeed, loading and air density. Where lift coefficients are less, recirculation convergence angles would be less. Thus sweep of wings is of greatest influence on flow separation during takeoff and landing, in "high G" maneuvers, and in high altitude operation.

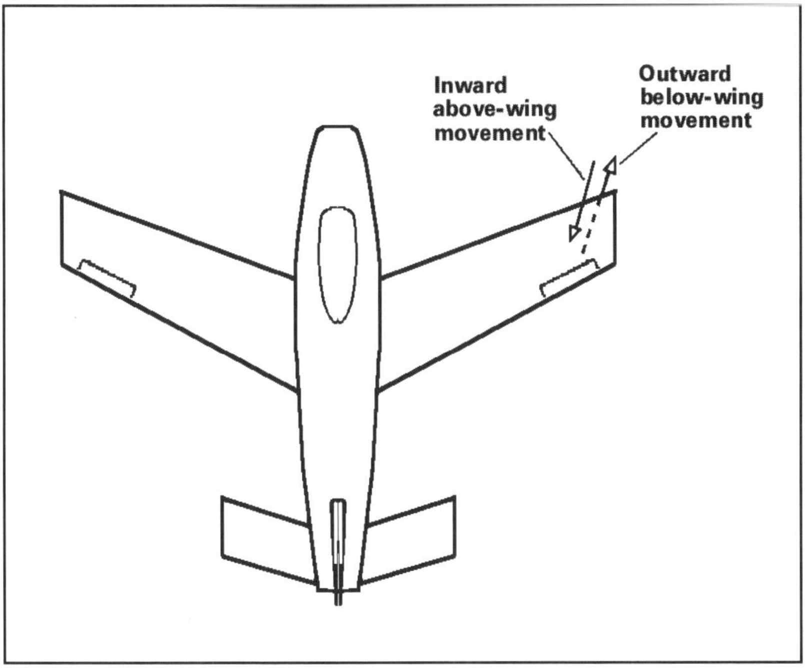

Figure 7.14 **Forward Sweep**
With forward sweep, precessing circulation at upper surfaces of wing outer regions would be toward the downspan direction, and hence would increase downspan boundary layer flow. Because of this, and the problem of wing flexing adding to instability, forward sweep of such degree as indicated here may not be practical.

Downspan boundary layer flow is cumulative, and hence becomes greatest in the aft wing regions so that with forward sweep, both downspan flow and greater midsection loading would contribute to early stall of a wing midsection. This would tend to allow aileron function during partial stall, but in stall forward shift of the center of lift would tend to cause pitch-up and greater stall. Thus a plane with a high degree of forward sweep might require stall preventive methods such as computer control. Rearward sweep with relatively greater outer span lift loading and downspan flow would also tend to promote aft region stall and consequent pitch-up into greater stall. In addition, taper would further contribute to outer span stall. Thus tapered and rearward-swept wings may require stall preventive features. Figure 7.15 shows the wing end region of a retired US Navy fighter plane in the Blue Angels exhibit at Pensacola. It has both sweep and taper. To inhibit stall and extend performance in wing end regions it has vortex generators for energizing the boundary layer.

Figure 7.15 **Vortex Generators**
This retired Blue Angels plane is exhibited in the Naval Air Museum at Pensacola.

Vortex generators are short angled vertical vanes which produce vortices at the wing surface. The vortices mix higher velocity flow into the boundary layer so that the layer is carried rearward and stall is thereby inhibited. Vortex generators are effective and necessary on some wings, but add drag at lower lift coefficients when they are not needed.

Combined Fore and Aft Sweep

Wings with a degree of forward sweep may be more efficient by virtue of concentrating lift in the midsection where the longer lateral recirculation loss path has greater mass resistance. However, the potential disadvantage of greater downspan boundary layer flow and accumulation makes forward sweep less attractive. A design compromise, with forward sweep in the midsection for efficiency might be combined with rearward sweep in the outer regions for orientation with respect to precessing flow. If evolution has found that combination advantageous for birds, as illustrated in Figure 7.16, then perhaps it could also be advantageous for some airplanes. A disadvantage would be complexity of construction.

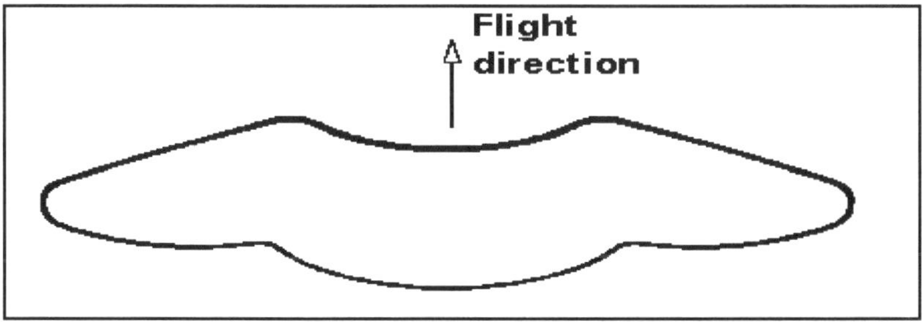

Figure 7.16 **Combined Forward and Rearward Sweep**
Many bird species have wing planforms similar to the outline above. Presumably this combination of forward sweep in the mid-region and rearward sweep in the outer regions is advantageous for survival.

Dihedral

Airplane wings, especially those of low wing craft, are usually designed with upward slope in the outward direction, so that the middle is lower than the ends, as in Figure 7.17. This characteristic is known as *dihedral.* There are different reasons for dihedral. On low wing planes, dihedral provides more ground clearance for avoiding wingtip strike during landing and takeoff. On high wing planes, dihedral allows better in-flight lateral visibility for sighting other aircraft. With both high wing and low wing planes however, dihedral affects flight characteristics.

When a plane is turned with rudder alone, it tends to slip, or skid laterally. Wings with dihedral develop differential lift in a slip or a crosswind, tending to roll in opposite direction to the direction of slip. A high degree of dihedral is sometimes built into model airplanes which have no ailerons. With sufficient dihedral, steering can be accomplished quite well with rudder alone. When rudder produces a skid, dihedral effect produces roll into the turn.

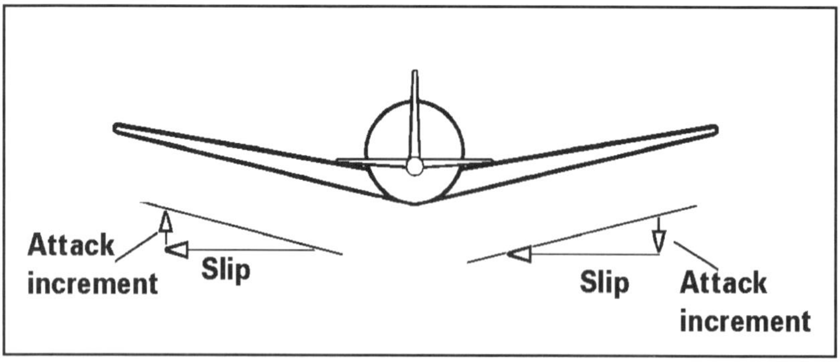

Figure 7.17 **Dihedral**
In slipping or skidding to the left, the inclined left wing has an added positive increment of angle of attack, while the wing opposite has an angle of attack decrement. Resulting greater lift on the left wing rolls the craft away from the slip, or toward the right as seen above.

Excessive dihedral may produce *Dutch roll.* If dihedral causes a plane to roll past the point of stopping a slip, it can develop a new slip in the opposite direction, followed again by excess roll. The process may continue with yawing, rolling oscillation. In occupied airplanes the tendency can be countered by the pilot, but the action increases pilot work load and can be nauseating.

Dihedral Effect of Sweep

An effect similar to that of dihedral is produced by rearward sweep, as diagrammed in Figure 7.18.

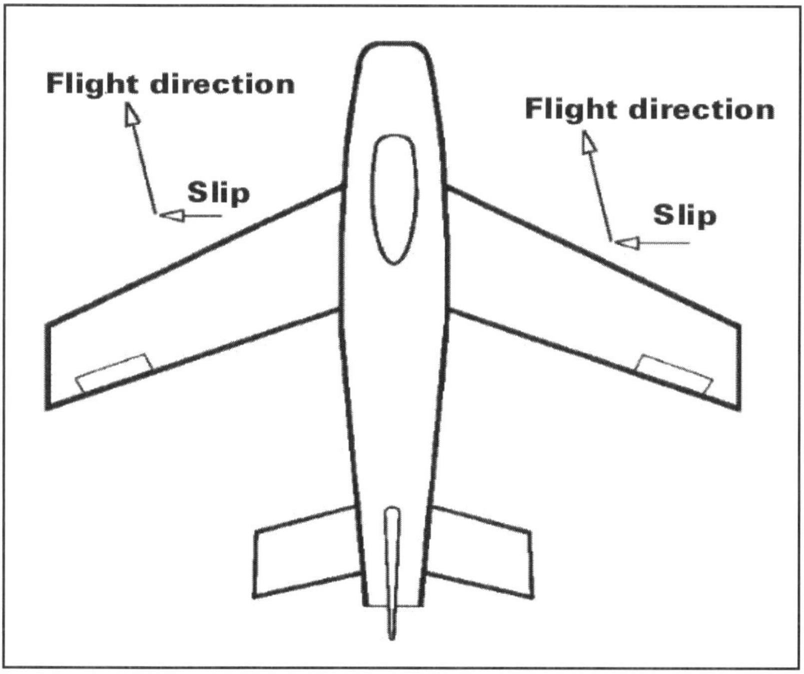

Figure 7.18 **Dihedral effect of Sweep**
With rearward sweep, the wing on the side toward which a plane is slipping is better oriented for lift in relation to forward motion as compared to the other wing. As a result the plane will roll in direction opposite to the slip.

As sideslip of a rearward-swept wing develops, the wing on the side of the direction of skid is oriented more nearly normal to the direction of travel through the air, so that lift is increased. At the other wing there is an opposite effect. Resultant lift differential then rolls the plane in direction opposite to the skid.

Even without dihedral or sweep a straight wing has a tendency to roll in direction opposite to a slip. This occurs as the wing end slipping into lateral upwash, as indicated in Figure 7.19, generates more lift, while the opposite wing, slipping away from upwash generates less. This can be related to the Kutta-Joukowski theorem, according to which relative movement between a vortex and air around it produces lift normal to the direction of relative travel. Thus crosswind movement of a wing with its end vortices produces differential lift and a dihedral-like effect.

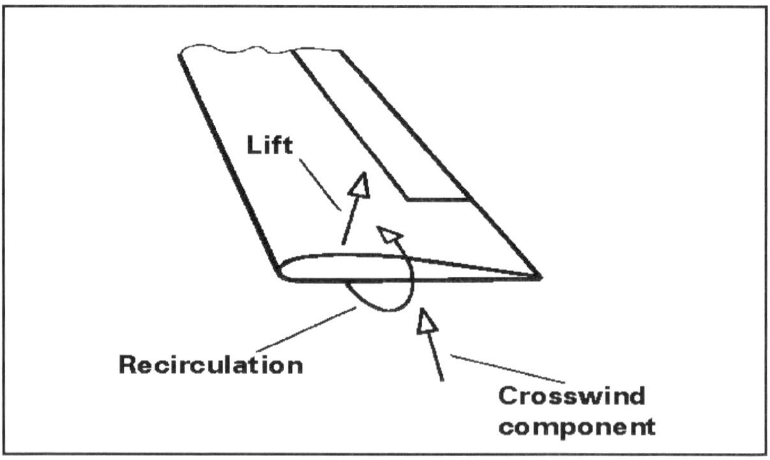

Figure 7.19 **Wingtip Kutta-Joukowski Lift**
Wingtip recirculation, in combination with air flow across the tip, produces lift in accordance with the Kutta-Joukowski theorem. In the illustration above, wingtip recirculation from below to above the wing is subjected to lateral flow, which may be from a crosswind or from lateral slipping. Resultant lift adds to that of wing operation at the upwind end, and detracts from lift at the downwind end.

Because of dihedral or dihedral-like effects, airplanes can be steered by rudder alone. When rudder orients the airplane longitudinal axis in the desired flight direction, dihedral causes the plane to roll toward the desired turn until sideslipping stops. Thus a pilot may turn a plane with rudder alone while hands are occupied with other matters, such as holding maps. This also makes possible a simplified autopilot which can control plane heading by rudder alone. A plane may also hold heading by aileron alone. When banked for turning, sideslip will produce vertical stabilizer force to reorient the longitudinal axis in the turn direction.

Wing Location

A popular airplane type, in terms of numbers, is the single engine propeller driven monoplane having its wing located above the passenger cabin, and having an aft tail for stability and control of pitch and yaw. The center of lift is above the airplane's center of gravity, which helps airplane stability; the wing shades the passenger compartment from sunlight, which is desirable in hot weather; unrestricted downward visibility through the side windows aids in navigation and sight-seeing, and fuel can be fed from wing tanks by gravity, which by eliminating the need for a fuel pump, adds to reliability. The principal disadvantage of a high wing is that when the plane is banked for turning, the lowered wing can block visibility in the direction of turn.

A low wing single engine light plane, having unrestricted upward and lateral visibility, feels less confining as compared to a high wing. The wing below the passenger compartment restricts downward visibility at the sides, but in a turn the wing on the side of the turn is banked downward so that visibility for seeing other planes or runway approach in the turn direction is unrestricted. Having fuel tanks in the wings below engine level requires a fuel pump, and requires that the pilot remember to switch fuel feed between tanks as necessary so that maximum fuel is available without interruption caused by pumping air from an empty tank.

A low wing may have engine and propeller above the wing, as in Figure 7.20, where rearward propwash above the wing aids circulation and thus has a beneficial effect on lift during takeoff, while in power-off glide, a windmilling above-wing propeller opposes circulation and lift. These effects can be beneficial in takeoff and in descent to landing. For high wing airplanes, with engine below wing level, the effects are opposite to the need for lift in takeoff, and for reduction of lift in landing, but are not really very significant.

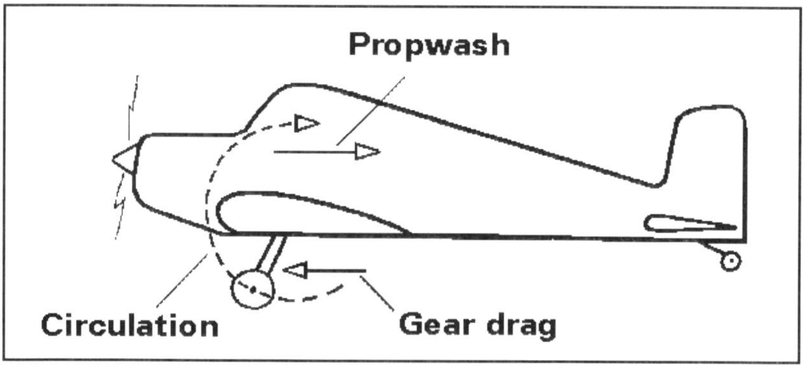

Figure 7.20 **Propeller Wake and Gear Drag In Circulation** Circulation around an aircraft wing is affected by engine location. As indicated here, the propeller wake of a low wing plane assists circulation, as does landing gear drag.

Most small high-wing planes have fixed landing gear, the drag of which adds to forward air movement below the wing, and thus to circulation. With floats for water operation, below-wing drag is even greater, adding more to circulation so as to slightly reduce minimum flying speed in operations of takeoff and landing.

Because circulation upwash energy is recovered in lift and forward thrust, energy lost to below-wing drag is partially recovered from upwash at the wing leading edge. Without this recovery it is likely that more planes would have retractable gear or other modifications to minimize gear drag.

Rear Engine

By locating the airplane propeller at the rear of a plane, as on the canard airplane of Figure 7.2, the fuselage is not subjected to propeller wash. Thus airflow passing the fuselage is slower and less turbulent as compared to a forward propeller airplane, so that fuselage drag is reduced, and noise in the cabin is reduced as well. There is a possibility of greater thrust efficiency of a rear propeller operating in air having forward movement due to having been dragged forward by the fuselage. However, this advantage may be compromised by operating the propeller in turbulent flow. The principle might also apply to a jet engine having annular air intake around a fuselage.

Most of the increasingly popular canard designs have rear engine-propeller combinations which seem to work quite well. It is likely that part of the success is due to smooth, well streamlined composite fuselages creating minimal turbulence in which propellers operate.

Biplanes

A *biplane,* pictured in Figure 7.21, has both high and low wings. Most successful early airplanes, including those of the Wright brothers, were biplanes. The chief advantage of a biplane is that its inter-wing bracing system provides for great

Figure 7.21 **Hatz Biplane**
A biplane has both high and low wings. Favored by some for nostalgic reasons, but not in great demand, they are built by individuals, usually from purchased plans.

strength in a light structure, but aerodynamically, as a class, biplanes are not very efficient. Circulation paths of the two wings mutually interfere, as indicated in Figure 7.22, and the inter-wing bracing system has high drag. Biplanes may have docile stall characteristics, with the upper wing stalling before the lower, but this is dependent upon relative properties of the two wings and the way they are rigged. For aerobatics, the biplane is often chosen for the structural strength of inter-wing bracing, and because the high drag permits diving maneuvers which otherwise might produce dangerously high airspeeds. Docile stall is not desired in an aerobatic show-plane which may be required to perform abrupt snap rolls. A snap roll is a sort of spin with one wing stalled, usually horizontal in motion, entered abruptly with high angle of attack and applied full rudder.

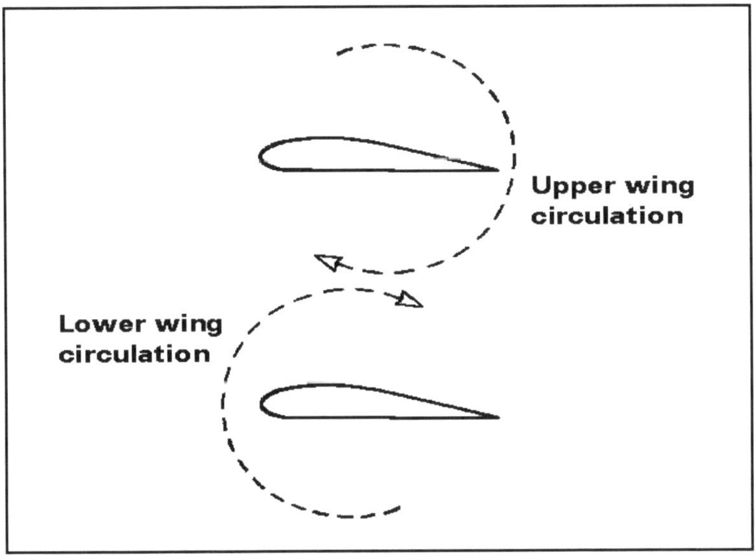

Figure 7.22 **Biplane Circulation Interference**
Between biplane wings circulation moves in opposite directions. Mutual interference between the circulation patterns has a negative effect on performance.

Cessna 185 on Amphibious Floats

Introduction to Aerodynamics

VIII. Rotary Wings and Surface Effect Vehicles

Airplanes, having wings fixed in position with no relative movement, except for those of control appendages, are classed as "fixed wing" aircraft. Other aircraft having wings which must move in flight include autogyros, helicopters and ornithopters. Autogyros and helicopters, having wings which rotate around axes substantially perpendicular to flight direction, are referred to as "rotary-wing" aircraft.

Juan de la Cierva's Autogyros

First successful man-carrying rotary wing flight was that of an autogyro engineered by Spanish Juan de le Cierva. An autogyro, also known as autogiro, gyroplane and gyrocopter, receives forward thrust for flight from an engine-driven propeller, while lift is obtained from an overhead rotor powered by the flow of air passing through it. In January 1923 Cierva succeeded in takeoff, turns and return to landing. His autogyro was a modified forward engine airplane from which he had removed the wings and replaced them with an overhead rotor. Other early autogyros which followed had forward engines, but in later years rear engine propulsion became popular, as indicated in Figure 8.1.

As in the case of sailplane or glider wings, rotary wings of autogyros gain lift and thrust from passage of air. This requires rearward tilt of the rotational axis, which also tilts rotor lift rearward. The rearward component of rotor lift appears as drag, which must be overcome by forward thrust of the engine-driven propeller. Thus rotor drive power is indirectly derived

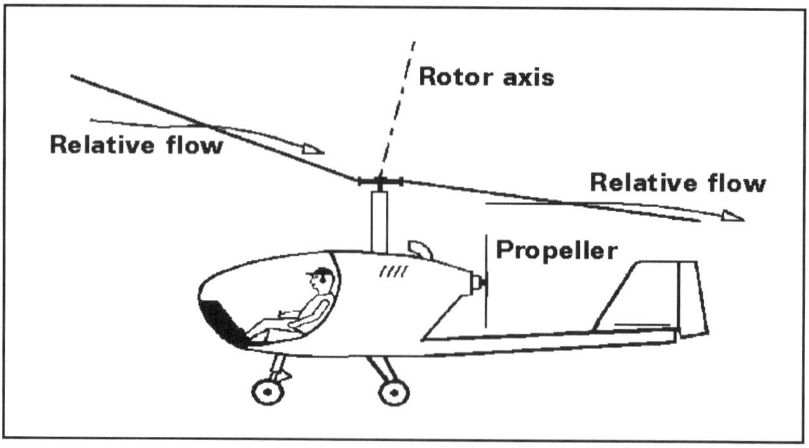

Figure 8.1 **Autogyro**
The autogyro has a freely turning rotor with axis tilted rearward. Relative flow through the rotor blades, from bottom side to top side, provides rotor drive.

from the engine through aerodynamic processes. Without engine power, the rotor continues to be powered by air passage in gliding. It should be noted that although air passes through the rotor from bottom side to top side, the air is moved downward in rotor passage, as Figure 8.1 indicates. Thus autogyros, like airplanes, produce downwash, as Newtonian principles demand.

Roll Control and Coning

A problem recognized before Cierva's success was that of greater lift of a forward-moving rotor blade as compared to a rearward moving blade. Velocity of a forward-moving blade is equal to rotational velocity plus forward velocity of the craft, while rearward velocity is equal to rotational velocity minus velocity of the craft, as indicated in Figure 8.2. Lift, proportional to relative speed squared, would be laterally unbalanced, tending to cause the craft to roll.

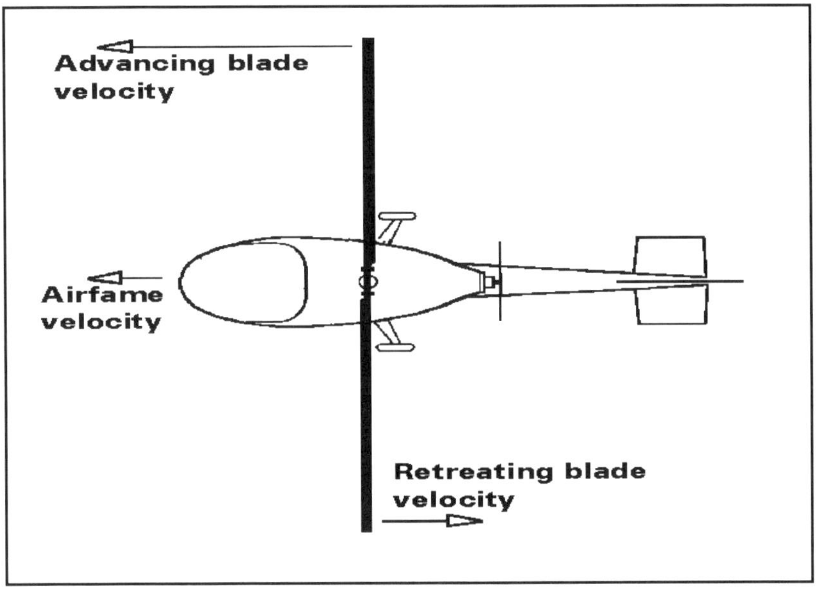

Figure 8.2 **Blade Velocities**
Forward velocity of the autogyro adds to velocity of the advancing blade and subtracts from velocity of the retreating blade. Without some means of lift equalization the craft would roll to the left. Blade tip speeds differ by twice the airframe velocity.

Cierva solved the roll problem in his three blade rotors by allowing blades to flap independently. Each blade would automatically rise or fall as lift force for flap angle increase would balance against outward centrifugal force for angle decrease, as illustrated in Figure 8.3. With this freedom, an "advancing," or forward-moving blade rises, reaching peak angle at foremost position, and descends while moving rearward, or "retreating," reaching lowest position at the rear. In this process the path of blade rotation becomes a cone. The upward slope of blade rotational paths is referred to as "coning." Because of mass inertia, position change lags behind accelerative forces, so that maximum upward and downward blade positions occur with ninety degrees rotational offset from maximum and minimum lift positions. In gyroscopes this offset between applied force and resultant attitude is known as *precession*.

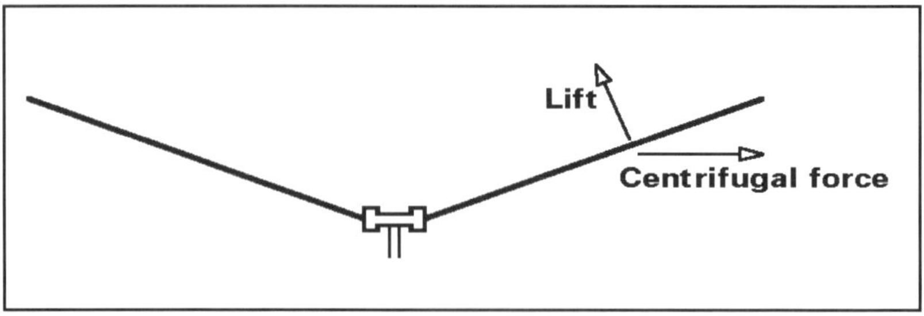

Figure 8.3 **Coning**
Blades, subjected to upward lift force for angle increase and outward force for angle decrease, automatically adjust angles to balance the opposing forces. The resulting blade passage path forms a conical shape, known as "coning."

Upward movement of a blade reduces its angle of attack and decreases downward redirection of flow, thereby reducing lift, while downward blade movement on the opposite side increases angle of attack and downward redirection of flow to increase lift. Thus allowing blades to flap alleviates difference in lift between blades on opposite sides of the rotor.

Freedom to flap also greatly reduces blade vertical bending moments, allowing blades to be of lighter construction. However, as can be deduced from Figure 8.3, blade vertical movement also involves radial movement. This results in conflict involving conservation of angular momentum. In flapping, inward radial movement tends to increase rotation rate, while outward movement tends to decrease rotation rate, just as an ice skater increases or decreases rotation rate by moving arms inward or outward. If blades differ in radial movement the forces for speed change also differ, creating a torque between blades which the coning rotor design must accommodate or relieve.

Lead-Lag

Cierva solved the problem of flapping blade torque stress in 1927 by adding "lead-lag" hinges to allow freedom for forward and rearward accelerations of blades as they rotate. As a forward-moving blade flaps upward, moving closer to the rotation axis, its lead-lag hinge allows the blade to maintain angular momentum by moving ahead, or leading, and as it flaps down in moving rearward it is allowed to lag. Thus blade rotational speeds can vary individually to maintain angular momentum without torque stress conflict between them.

To achieve roll control, Cierva's first autogyro, and others which followed, included side ailerons as in the manner of airplanes. For pitch and yaw control, the airplane which Cierva modified into an autogyro retained its original airplane tail assembly.

In 1933 Cierva added phased blade pitch variation, known as *cyclic pitch*, which has become common on larger autogyros and helicopters. Cyclic pitch is controlled by the pilot's stick, usually through a swash plate system, functionally illustrated in Figure 8.4. Angling of the swash plate controls blade pitch cycling for rotor tilt control, while the pilot's *collective pitch* lever may raise or lower the plate to increase or decrease all blade angles of attack equally. By cyclic pitch control the pilot can tilt the rotor plane, or cone, as desired for maneuvering.

Two blade rotors are often, but not always, "semirigid," with opposing blade spans aligned in a plane. A central gimbal or hinge system allows blades to operate as a "seesaw" where one blade rises as the other descends, so that tilt angles are equal and opposite. Equal and opposite tilt angles produce equal tendencies for increase and decrease of speed in conserving

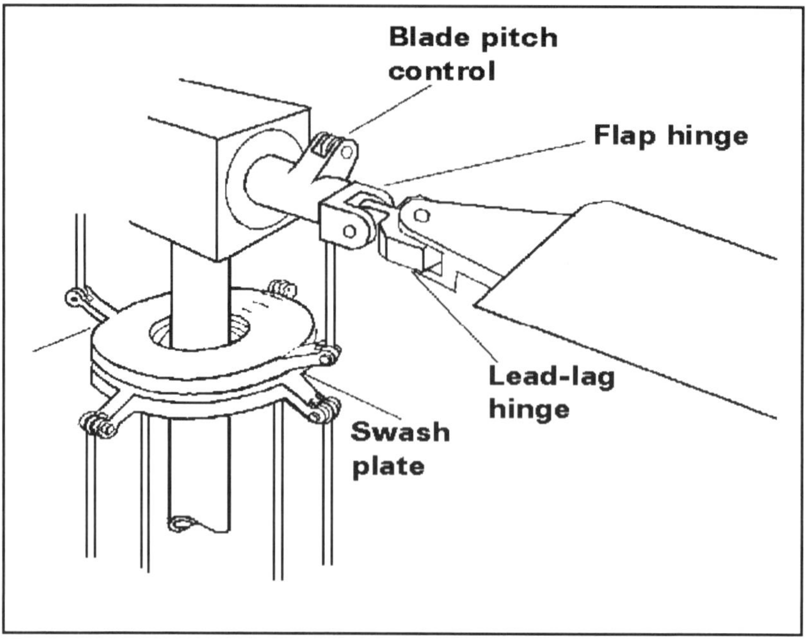

Figure 8.4 **Rotor Control**
Pilot command for rotor disk lift and tilt control couples through a swash plate bearing to a follower which rotates with the rotor. Tilt direction of the swash plate controls angle of attack phasing, while degree of tilt controls magnitude of angle of attack cycling. This illustration is of function only. Actual swash plates are usually enclosed to protect them from contamination.

angular momentum. Thus there is no torque conflict between blades, and no need for lead-lag hinging. Blade lifts are also equalized across the tilt axis without requiring separate tilt hinges. As the forward-moving blade rises, its angle of attack is decreased, while the other blade in descent has angle of attack increase to maintain equal lift. Simplicity of the two blade semirigid rotor makes it attractive for smaller autogyros and helicopters.

More blades are desirable for a smooth ride. Two-blade helicopter pilots are often recognizable in radio transmissions as they sometimes sound like truckers on bumpy roads.

Cierva pioneered "jump takeoff" in 1933. Jump takeoff requires coupling of engine power to the rotor for pre-rotating it to an overspeed condition. Kinetic energy stored in the rotor can then be converted into lift for vertical ascension. Cierva used a hinging system in which applied torque during run-up caused the blades to pitch into non-lifting angles of attack. When driving torque ceased, centrifugal force caused the blades to assume a lifting angle for vertical takeoff. One might imagine that in this system a loss of engine power could cause premature lift off followed by a drop as rotor speed decreases.

In more modern autogyro jump takeoff, brakes are locked, cyclic pitch is neutralized and collective pitch is adjusted to minimize rotor lift and drag. The rotor is then driven to desired speed, whereupon rotor drive is disconnected, and collective pitch is applied for takeoff. As the autogyro rises, the propeller thrusts it forward, and as forward speed is gained, the rotor is tilted rearward to gain drive from passing flow. The jump takeoff mechanism is not required to carry the high level of power required for helicopter drive. Thus its requirements of ruggedness and reliability are much less than those of a helicopter.

Without jump takeoff an autogyro rotor still needs some pre-rotation, perhaps by hand before takeoff run, because a non-rotating blade would be stalled and unable to develop required rotation. On an early autogyro, Cierva initiated rotation by pulling a rope wound around a drum. In many small autogyros pre-rotation power is coupled from the engine through a flexible sheathed drive cable. A small gear at the end of the flexible drive is mated to a larger gear on the rotor hub. The flexible drive allows for rotor hub tilting.

In landing, autogyro flight does not normally become vertical. Landing of an autogyro involves a slow and steep power-off glide, phasing into a pronounced nose-up landing "flare." During nose-up flaring the rotor intercepts a greater volume of air, providing deceleration of forward speed and cushioning of touchdown into a soft landing and short ground roll. In the deceleration, conversion of kinetic energy of aircraft motion into increased rotor speed enhances control. Because autogyros normally have forward motion on the ground they usually have wheels, whereas a light helicopter may have only skids.

No large autogyros have been produced in the United States, perhaps because they are not perceived as sound by the public. However, reliability and safety has been demonstrated for several years by the sturdy Air & Space model 18A, pictured in Figure 8.5.

Figure 8.5 **Air & Space Model 18A**
Air and Space Model 18A is a proven certificated two-place autogyro, and an economical rotorcraft rating trainer for pilots who may then upgrade to helicopter rating in the more expensive machines.

Autogyro training and sales have been available in past years, as has been helicopter training, at the Air and Space facility located near Paducah, Kentucky.

Small autogyros without jump takeoff capability may have semirigid two-blade rotors without flap hinging, lead-lag hinging or collective pitch, but still be capable of cyclic pitch variation for rotor tilt control. As illustrated in Figure 8.6, as one blade increases pitch the other decreases. This still requires a swash plate system, as pictured in figure 8.7, or a mechanism of similar function.

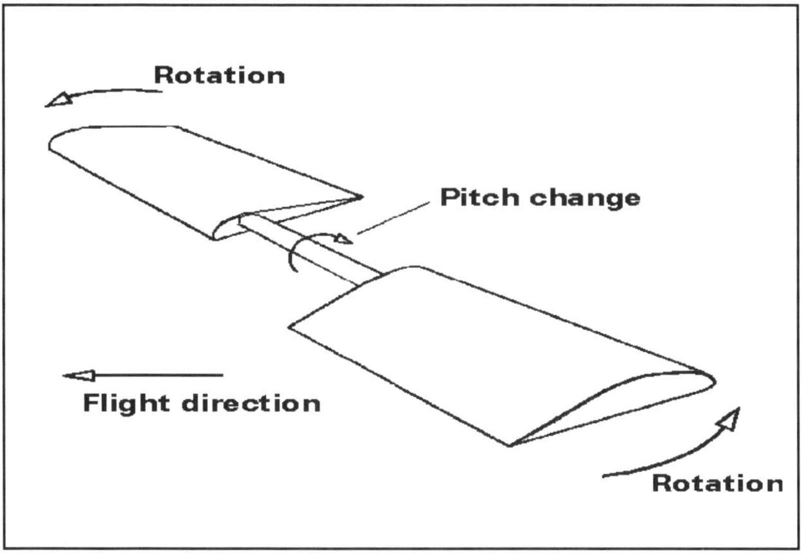

Figure 8.6 **Pitch Changes**
On a rigid two-blade rotor, pitch changes of the two blades occur in opposite direction. With clockwise pitch rotation as indicated above, angle of attack of the advancing blade is increased while that of the retreating blade is equally decreased.

Figure 8.7 **Semirigid Rotor Swash Plate** (Courtesy PRA Museum, Mentone Indiana) The swash plate has two control rods connected to the pilots stick for tilting the rotor in two axes.

In a simpler system requiring a bit more pilot effort, rotor tilt control in two axes may be accomplished by directly tilting the rotor hub with manual force coupled from the pilot's stick. The limit of rotor simplicity, that of a "glider," is pictured in Figure 8.8. Rotor tilt of this machine is controlled directly by effort of the pilot, which, because of precession, is applied in direction ninety degrees offset from desired tilt change. A glider may be towed by rope from a surface vehicle to altitude and released to glide back to landing. One model of glider was used in wartime by the Germans. Towed behind a German U-boat, or submarine, the pilot was able to see farther over the horizon than could a conning tower observer. We can assume the pilot was expendable if sighting of a fast-approaching destroyer made a quick dive of the submarine necessary.

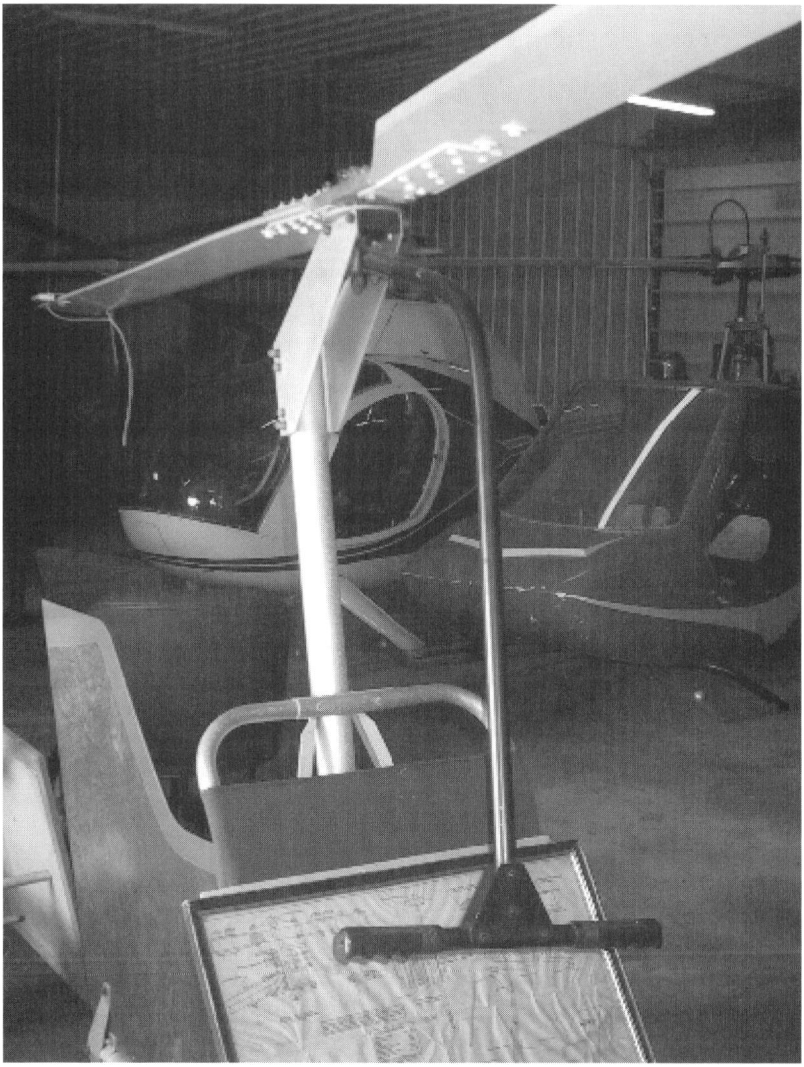

Figure 8.8 **Autogyro Glider**
(Courtesy PRA Museum, Mentone Indiana)

Helicopters

Cierva's pioneering of autogyros led the way to development of helicopters, which are in present times more often seen by the public. Rotors of more complex autogyros, and rotors of helicopters are similar, differing mainly in that autogyro rotors are driven by passing flow, while helicopter rotors are driven directly from the aircraft engine. The autogyro thus must move forward at some minimum speed to maintain rotor drive and lift, while the helicopter can hover in place.

Although advantageous in being able to hover and land in very little space, low and slow helicopter flight is to be avoided for safety reasons. In event of engine failure in flight, a helicopter can convert to autorotation and glide as an autogyro, but with a faster rate of descent because helicopter disc loading (weight to rotor disc area) is usually greater. Conversion to autorotation involves some delay as failure is recognized and the rotor is tilted rearward to obtain drive from passing flow. If helicopter flight is slow at low altitude, it may not be possible to gain autorotation lift needed before impacting the ground. In autorotating, the descent rate requires good timing of collective pitch increase just before touchdown to bring smooth landing.

In event of engine failure an autogyro has a safety advantage of always being in autorotation mode, and can make a normal landing if the terrain permits. For close-up observation of ground activities, an autogyro cannot hover, but can fly in slow circles to maintain forward speed. If ground observation involves continuous travel, as in pipeline patrol or monitoring highway traffic, the autogyro can be effective with lower costs of purchase and maintenance as compared to a helicopter. Because of lower cost, one may save by training initially in autogyros to obtain a rotorcraft license, and then upgrade to helicopter license in the more expensive machines.

Tail Rotors

Torque of driving a helicopter rotor is equaled by opposite torque on the airframe, in accordance with Newton's third law. To counter this torque and prevent airframe rotation, helicopters usually have a horizontal axis tail rotor at the rear which exerts sideward thrust. Like the main rotor, a tail rotor can be subject to greater lift on an advancing blade moving in the forward direction, than on a retreating blade moving rearward. Thus one may find tail rotors with flapping hinges, and perhaps even lead-lag hinges.

Another torque correcting means, known as "NOTAR," meaning no tail rotor, uses air jets directed downward tangent to the surface of a cylindrical tail boom. In combination with rotor downwash, the tangential jets produce circulation around the tail boom and sideward lift to offset rotor drive torque. Another means of eliminating a tail rotor is to have rotor drive produced by jets at the rotor blade ends so that there is no torque between rotor and airframe.

Torque compensation by means of sideward thrust at the tail tends to cause sideward helicopter drift. This is corrected by main rotor tilt to produce opposite direction equal thrust. Because a tail rotor is located below a main rotor for convenience in keeping tail length reasonably short, the opposing forces cause slight fuselage leaning. The effect can be seen in first ground contact of a left wheel in landing, except in the UK where rotor rotation is of opposite direction.

Future Rotary Wing Aircraft

It would seem that safety, economics and simplicity would favor more popularity of autogyros as compared to helicopters.

Although the helicopter is advantageous in requiring no runway, autogyros runway requirements can be minimal. In the 1930s autogyros operating from post office roofs carried the U.S. Mail. It has been speculated that greater popularity of helicopters as compared to autogyros is more due to public perception than to practical utility. Average citizen may regard engine-powered rotors as understandable, but consider autorotation a bit mysterious.

A shortcoming of both autogyros and helicopters is speed limitation. As described previously, rotational speed of a blade adds algebraically to aircraft airspeed. Thus the blades must operate between advancing blade sonic speed limitation and retreating blade stall, where blade speed relative to the air becomes too low for effective lift. This limits rotorcraft speeds to usually less than 200 miles per hour.

Hybrid Vertical Takeoff Machines

Through the years there have been several efforts to produce winged aircraft which would rise vertically in the manner of helicopters, and then transition to airplane-like flight as forward airspeed is gained. Some have been autogyros with fixed wings attached, while others have been essentially airplanes having propellers re-orientable for vertical lift or forward propulsion, or combination thereof.

The Navy *Osprey* has two 6000 horsepower turbine engines and propellers mounted on opposite wing ends. An interconnecting drive shaft between propeller drives can couple power between engines for maintaining drive of both propellers in event of one engine failure. The engines with attached propellers can rotate to vertical thrust position for takeoff and landing, as in Figure 8.9, or can be positioned for forward flight, as in Figure 8.10. The Osprey can take off vertically with up to 8,300 pounds of cargo, or can carry 24 equipped soldiers.

Figure 8.9 **Osprey in Vertical Lift Mode**
(Courtesy U.S. Navy)

Figure 8.10 **Osprey in Horizontal Flight Mode**
(Courtesy U.S. Navy)

A much smaller, but promising project is the *CarterCopter*, pictured in Figure 8.11. This machine can take off like an autogyro, but as forward speed is gained, so that lift is achieved from the fixed wings, the rotor is slowed to keep tips well below sonic speed. The intent is to achieve fuel economy and speed of fixed wing airplanes while retaining the autogyro advantage of short or vertical takeoff. Future speeds up to 500 miles per hour are contemplated.

Figure 8.11 **CarterCopter** (Courtesy CarterCopters)
Because lift at low speeds is obtained from the rotor, wing area requirement is lessened. As a result, low parasitic wing drag at high speed should permit good fuel economy.

Wing In Ground Effect Planes

As previously described, the process of wing lift becomes more efficient near the Earth surface. A student airplane pilot finds this in landing. In takeoff, an airplane with insufficient power to climb may rise off the surface, only to impact an obstruction beyond the runway if flight is not aborted in a timely manner.

On one occasion, in the middle of the 20th Century, a transoceanic airliner experienced power loss and could not maintain altitude. As the plane descended, preparations were made for ditching in the ocean, but as the plane neared the surface it gained surface effect lift, and was able to continue near the surface until reaching land and safe landing.

Airplanes have been built specifically for near-surface operation. These have become known as "wing-in-ground-effect" (WIG) vehicles. Usually used in transit over large lakes or other bodies of water, they can float on the water, but as speed is gained they rise to become airborne. In flight just above the water, drag is low and high speeds can be achieved with good fuel economy. WIG vehicle efforts are increasing in countries of Japan, Korea, Australia and Germany, and at the time of this writing, Boeing Aircraft in the U.S. has proposed a WIG plane of 500 feet wingspan, capable of carrying 750 tons. Though intended for flight as low as 20 feet above the surface, it could fly as high as 20,000 feet with reduced efficiency.

The greatest progress in operational WIG planes has been in Russia. When our CIA learned of Russia's very large WIG plane, they dubbed it the "Caspian Sea Monster." Since the end of the cold war, that name has also been adopted by the Russians. It was capable of flying ten feet above the water with gross weight of 540 tons, and could travel over 300 miles per hour.

In other countries WIG planes are mostly in experimental stages, usually of sizes carrying from one to eight people. These craft usually have wings of negative dihedral, with floats at the ends, as indicated in Figure 8.12. In flight, float proximity near the surface restricts recirculation loss around the wing ends. Regarded by authorities as marine craft, rather than airplanes, they do not require pilot's license.

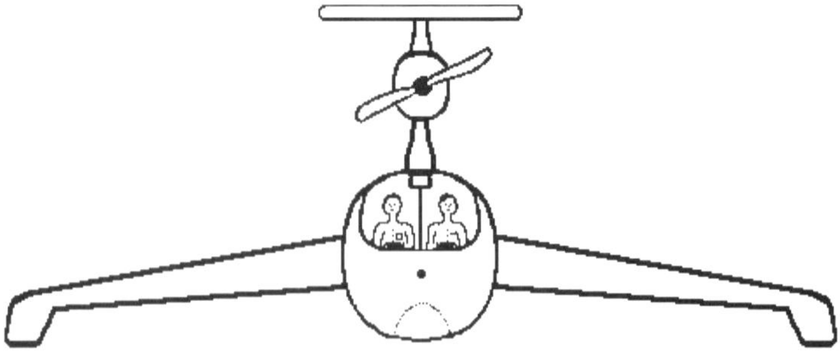

Figure 8.12 **WIG (Wing in Ground Effect) Plane.**
Intended for flight near water surfaces, a WIG plane floats on the water when at rest, but rises a few feet above the surface in ground effect as speed is gained.

Hovercraft

Another type of vehicle held up by a cushion of air is the Hovercraft, which can operate on water or land. These have been built in wide range of sizes, some small ones by individuals who have purchased plans. The craft is held above the surface by an air cushion contained within a flexible skirt. Thus it might be classed as an aerostatic vehicle rather than an aerodynamic one. The craft rises to a height limited by escape of air from beneath the skirt. Skirt flexibility allows the craft to move between water and a shore which may be irregular.

Probably the most important commercial hovercraft was the SRN4, which carried passengers and vehicles between Dover England and Calais France before the tunnel between countries was completed. Service began with four craft in 1969. Air cushion depth was ten feet. Propulsion was by four 3,800 horsepower gas turbines driving four propellers, providing speed greater than fifty knots. Capacity was 254 passengers and thirty cars.

For loading and unloading on land, the SRN4 sat on its hull with the air cushion not pressurized. Then, before moving to the water, the skirt was inflated to raise the hull clear of the surface for transit down the ramp to the water.

A smaller, but still impressive hovercraft is used by our armed forces for amphibious landings. Such a craft is pictured in Figure 8.13. The picture shows water spray issuing from beneath the skirt as it is blown outward by escaping air.

Figure 8.13 **Amphibious Landing Craft**
(Courtesy U.S. Navy)
Designated LCAC, this craft, powered by turbine engines, can carry a load of 75 tons of military equipment at high speed, including tanks and troops.

Two Place Tandem Autogyro
Training for autogyro flying is essential to safety, even if one has extensive fixed wing experience. This craft, pictured at the Oshkosh EAA convention, is intended for that purpose.

Introduction to Aerodynamics

IX. Flapping Wings

Man's earliest recorded thoughts of flying involved flapping wings, presumably inspired by the flight of birds. Flapping wing flight is still being attempted in machines referred to as *ornithopters*, from the Greek *ornis*, meaning bird. The myth of Icarus is perhaps the oldest recorded story of man's flapping wing flight. Icarus' life is said to have tragically ended when he flew too close to the Sun, causing the wax which held wing feathers in place to melt. In real life, occupied ornithopters, as of the time of this writing, have not been flown despite sporadic efforts extending far back in time. Success in flight of flapping wing machines has thus far been limited to imitation birds, such as the rubber band powered toy pictured in Figure 9.1, and a much larger engine-powered Pterodactyl by Paul MacCready in 1984.

In past sections we have considered lift associated with flow following the mean camber line of an airfoil, and have considered forward and rearward thrusts associated with fore and aft circulation. These phenomena are also the sources of lift and thrust in flapping wing flight

Figure 9.1 **Toy Bird** Powered by a windup rubber band, this plastic bird can fly more than fifty feet.

For a beginning concept, let us suppose we might wish to have flapping wings propel a blimp in forward flight, perhaps for reasons of attention-getting advertising, or for night stealth. Once launched into forward flight, a buoyant blimp, not needing aerodynamic lift, could gain forward thrust from flapping symmetric airfoil, non-articulated wings. As indicated in Figure 9.2, a symmetric airfoil with chord parallel to flight direction could produce upward lift and forward thrust in the downstroke, and equal downward lift and forward thrust in the upstroke. The lift forces would average zero while forward thrust would be positive. Such a machine might have no practical value, but the idea serves here to illustrate simplest possible flapping wing propulsion.

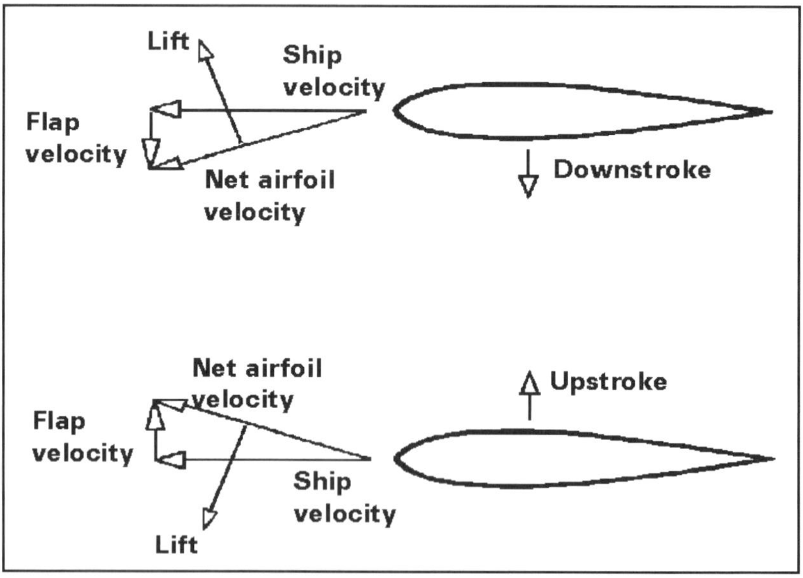

Figure 9.2 **Flapping Wing Airship Propulsion**
Lift of the airfoil is perpendicular to its velocity relative to the air. In both downstroke and upstroke, the lift has a forward thrust component useful for propulsion.

Ornithopter flight would require upward lift as well as forward propulsion. Thus net downward air acceleration of air is required. Net downward air acceleration might be achieved by operating with angle of attack variation which would produce downwash with lift and propulsion on downstroke but no flow deflection on upstroke. This could suffice for net lift and propulsion, but the oscillations of intermittent lift could be quite uncomfortable and nauseating for a pilot.

Ideally, for maximum comfort, lift should be constant through the flap cycle, with no difference between magnitudes of downstroke lift and upstroke lift. At first consideration this might seem an unreasonable goal. However, in movies of some birds in flight, such as geese for example, the bodies seem to move rather smoothly ahead with little vertical oscillation associated with wing flapping.

One possibility would be to have a non-flapping wing center section for lift control combined with flapping outer sections. The outer flapping sections would produce lift and propulsion on downstroke while the center section could have angle of attack articulated as necessary for maintaining lift during upstroke.

The center section might even move vertically in direction opposite to the flapping outer sections. At the time of this writing, an ornithopter project of this type is under way at the University of Toronto. Schematically illustrated in Figure 9.3, this ornithopter has a mid-wing section engine-driven in vertical oscillation. The vertical movement is linked to outer sections anchored to pivot points so as to invert the vertical movement of the center section into opposite vertical flapping of the outer sections.

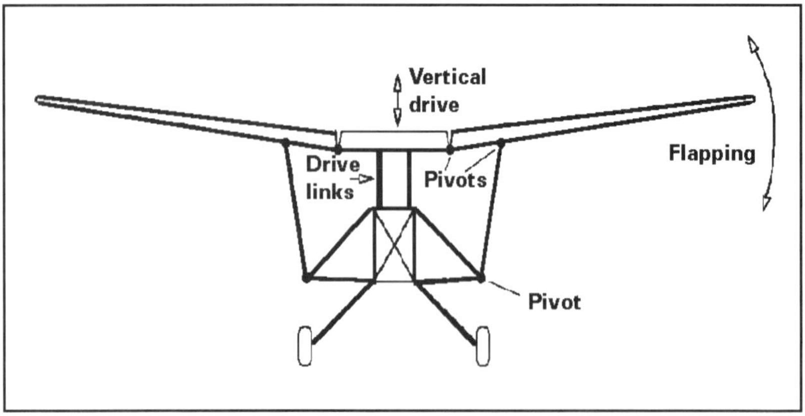

Figure 9.3 **Ornithopter Wing Drive Schematic**
Drive links move the wing center section in vertical oscillation. The motion is coupled through hinged pivots to flapping section roots so as to produce outer region wing flapping opposite in direction to that of the center section.

The University of Toronto project may have motions which are rather unbirdlike, but it has the birdlike feature of concentrating lift more inboard during flapping upstroke and more outboard on downstroke. For birds this is somewhat automatic. During upward flapping upward movement tends to reduce angle of attack, while during downward flapping angle of attack is increased. Thus forward thrust and greater lift would tend to be produced during downstroke. For a smooth ride the tendency of lift variation must be minimized.

Birds can vary wing camber, twist and angle of attack during the stroke cycle so as to reduce lift variation. Partly this is automatic. In the downstroke, feathers flex to reduce angle of attack and moderate lift increase, while inner wing regions can be articulated to also reduce lift. Then on the upstroke, midsection lift can be increased by increased angle of attack as outer section lift is reduced by upward movement. In concentrating lift more in outer wing regions on the downstroke and more on inner regions on the upstroke, more energy is expended in the work of downward movement than is recovered in upward movement. Thus net energy expenditure in each flap cycle provides for losses involved in lift and forward propulsion.

Achieving smooth ride with flapping wings which provide both lift and forward propulsion would require variation of spanwise lift distribution throughout the flap cycle. This would be the aerodynamic challenge. There would be other challenges of a mechanical nature, such as gaining initial speed to unstall the wings. Birds can gain some initial speed through foot propulsion and can articulate wing angles of attack over a wide range for takeoff. An ornithopter with sufficient initial forward speed through wheel propulsion might avoid the necessity of wide range variation in wing angles of attack.

Another concern would be the recycling of flapping energy. During upstroke, wings absorb energy from aerodynamic lift. This energy needs to be stored and returned on the downstroke. For the purpose, birds store energy in elastic cartilage. An ornithopter might use springs for storage. Another possibility is to store energy in a flywheel which would be incorporated into the flap drive system.

The problems to be solved may not be easy, and flapping wing flight may have no practical utilitarian value, but it undoubtedly will be accomplished someday, for perhaps no useful purpose other than satisfaction of accomplishment.

WW1 Sopwith Camel Replica at Lakeland, Florida Sun-N-Fun

Introduction to Aerodynamics

X. Classical Analogy

Analogy and Approximation

It is reported that long ago, Lord Kelvin stated: "When you can measure what you are talking about and express it in numbers, you know something about it. But when you cannot measure it and express it in numbers, your knowledge is of a meagre and unsatisfactory kind."

Mathematical expression of physical phenomena is indeed quite important to advancement of science and engineering. However, perfect mathematical expression of physical processes is often not possible. In such cases, approximation to reality is usually sufficient for engineering purposes, and is helpful in development of academic concepts. For useful approximation, compressibility of air was dismissed herein as a factor of minor consequence. In incompressible air, pressure waves would travel with infinite speed, instantly conveying aerodynamic forces through the atmosphere and to the surface. We know this is not true, but for lower speeds of flight, resulting calculations are close to real-world measured effects.

In some instances, physical situations can be represented quite nicely by analogy. In radio transmitting antenna design, for example, antennas are regarded as reflecting complementary image antennas beneath the ground. Again this is quite useful as mathematical analogy which produces good calculation results, even though it is assumed that in reality radio waves hardly penetrate the surface. Similarly in aerodynamics, analogy has been of great value in advancing aircraft engineering and flight. Classical aerodynamic theory invokes reflected in-ground images to explain ground effect. However mathematically useful, analogy confused with reality can impede real learning and understanding of the basic principles of flight, which might better be explained in Newtonian terms.

Instead of explaining wing aerodynamics in conventional Newtonian physics context, classical explanation of lift begins with viscous drag producing a starting vortex which is left behind a wing trailing edge. Since Helmholtz' vortex laws require a simultaneous second vortex, circulation around a wing is described as a reaction to the starting vortex. The mechanism of this connection is not made clear. Starting vortex and circulation are real, but the connection can be simply explained in Newtonian terms of action and reaction.

Another vortex rule is that vortices must either terminate at surfaces, as in the manner of a tornado, or must be continuous in a closed loop. A wing vortex loop is said to include starting vortex, a wing vortex and trailing vortices between. This is

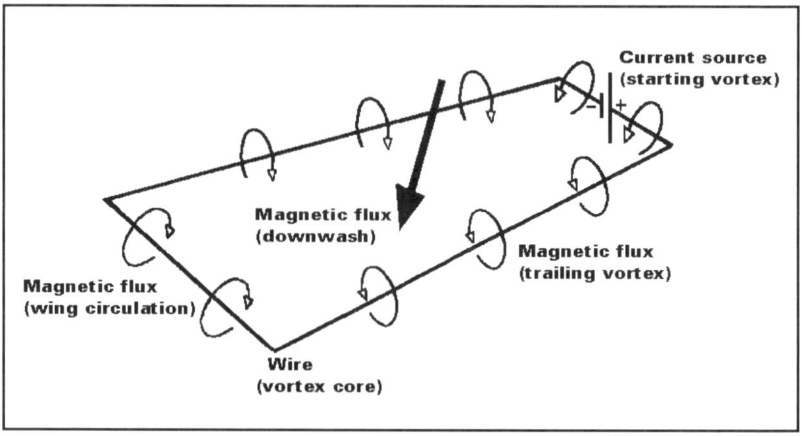

Figure 10.1 **Electromagnetic/Air Vortex Loop**
The concept of an aerodynamic vortex loop, consisting of starting vortex, trailing vortices and wing circulation, is analogous to an electrical current loop in a wire around which magnetic vortices are induced.

analogous to a closed loop of electrical current, as in Figure 10.1. Classical theory regards wing circulation as being "induced" from a central core of vorticity, in a manner comparable to magnetic vortices bring induced from a central electrical current. Some texts depict an imaginary line passing spanwise through a wing as representing the vortex core from which circulation is induced.

Classical theory recognizes upward air movement around a wing tip as due to pressure difference, but only as a small localized core of vorticity which "induces" greater air mass rotation encompassing the wing, as indicated in Figure 10.2. One prominent textbook author states the "...vortices tend to drag the surrounding air around with them..." To one who has witnessed the magnitude and speed of mass rotation in vortices made visible by smoke at airshows, it would seem the viscous drag in surrounding air could never be great enough for this to be true, and accompanying torque concentration at the wing tip would require an extremely strong structure.

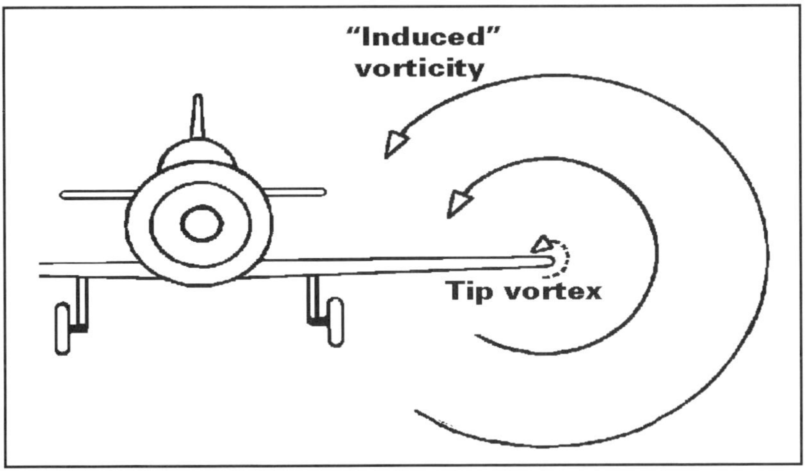

Figure 10.2 **"Induced" Vorticity**
Classical theory describes the large mass rotation about a wing end as "induced" by a small vortex at the tip.

We may justifiably question the logic in effectively denying air acceleration away from the below-wing overall region of increased pressure and toward above-wing region of reduced pressure in response to pressure difference, as Newtonian principles would demand, while acknowledging localized acceleration of air immediately around the wing tip in response to the pressure difference.

The "induced" downward air movement encompassing the wing is said to add a downward velocity component to passing flow, as in Figure 10.3, so that the angle of attack is changed. This is referred to as "induced angle of attack." Since lift is perpendicular to the direction of relative flow, the lift is said to tilt rearward. The rearward component of lift then is said to cause "induced drag" opposing forward flight. Thus the analogy for mathematical representation of induced drag is that a tip vortex causes "induced downwash" which produces an "induced angle of attack" which causes "induced drag." It is worth noting that the basic premise can be presented without mathematical complexity which would preclude description to lay persons.

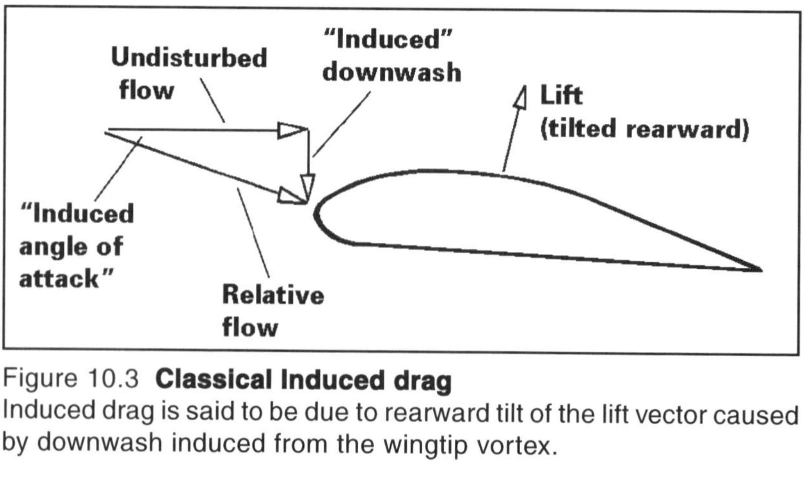

Figure 10.3 **Classical Induced drag**
Induced drag is said to be due to rearward tilt of the lift vector caused by downwash induced from the wingtip vortex.

In a sense, there actually is downward tilt imparted to passing flow, due to lateral circulation loss, which causes downwash movement behind a wing to be greater than that of upwash ahead. Thus the concept of "induced" downward tilt of passing flow, though confusing, serves its mathematical purpose.

Classical Ground Effect

In classical analogy of ground effect, the apparent reduction of drag and increase of lift as an airplane nears the surface in landing, is explained by invoking image vortices below the surface. It is said that surface blockage of circulation is equivalent to having an image reflected in the surface having inverted downwash which opposes downwash of the real wing, as indicated in Figure 10.4.

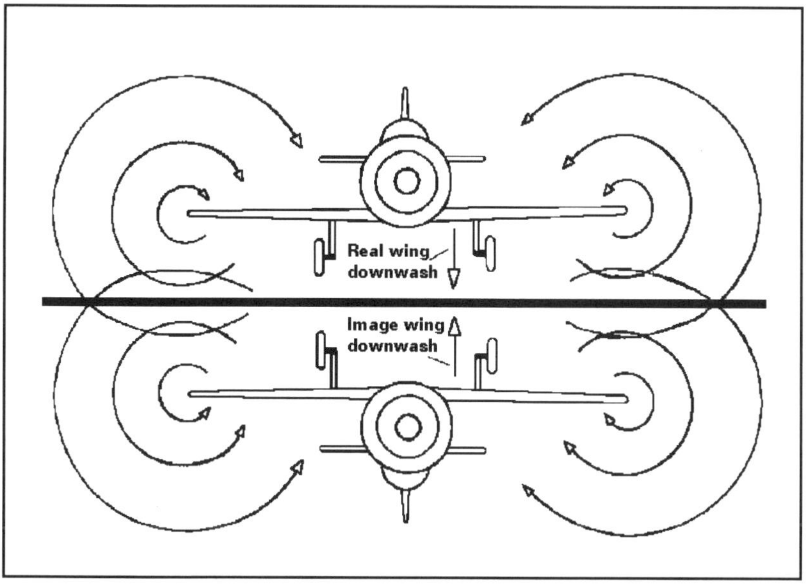

Figure 10.4 Classical Ground Effect
Image vortices regarded as reflected in the surface are said to produce upward downwash which partially cancels downwash of the real airplane. With induced downwash reduced, lift is said to be increased while drag is reduced.

Partial cancellation of wing downwash is said to reorient the lift vector more perpendicular to the line of wing travel so that upward lift is increased and drag is decreased.

If it had been found that wing performance, instead of being enhanced, were degraded in ground effect, the image analogy could just as well be adapted to explain that. We could regard the image vortices as assisting outward flow from beneath the wing so that below-wing pressure would be reduced.

It is perhaps unfortunate that the induction principles of electromagnetism provided mathematics which fit so well into aerodynamics. Normally, scientific methods involve formulating a theory to explain phenomena, developing mathematics to represent the theory, and then experimenting to verify the mathematical representations. In the case of aerodynamics, it seems the mathematics of electromagnetic theory came first, and was adopted into aerodynamics, complete with the induction hypothesis.

Though the applicability of induction to aerodynamics is only imaginary, the process has been useful in producing numerical results. Classical theory based on induction principles has been of great value in flight development, but however useful in application, induction analogies are of little or even negative value in teaching how wings really work.

The induction analogy, if stripped of mathematical obfuscation for pilots and lower level students, can be interpreted as legerdemain, while the concept of equal upper and lower wing surface transit times, even though untrue, apparently seems intuitively more reasonable. Little wonder then that basic induction theory minus the mathematics has not been adapted to teaching at the lower levels. This leaves the lower levels to continue the falsehood of equal transit time, as has been the case for about a century.

Education is normally a progressive process in which learning at each level is based on learning at a preceding level. Learning begun in high school physics could continue into college aerodynamics if the subject at that level were also based on credible Newtonian principles. Unfortunately, continuity of reasoning is impeded when false hump theory in high school is followed by abstract analogy in college.

For higher level learning, mathematically rigorous aerodynamic texts based directly on Newtonian principles may not currently exist. Although this may present a problem for continuity of Newtonian-based learning, solution of the problem offers incentive and opportunity for renewed academic vitality.

Introduction to Aerodynamics

INDEX